TURNER PUBLISHING COMPANY

Paducah, Kentucky

TURNER PUBLISHING COMPANY
Publishers of Military History
412 Broadway, P. O. Box 3101
Paducah, Kentucky 42002-3101

Editor: Bart Hagerman

U.S.A. Airborne Fiftieth Anniversary
Foundation Inc. Executive Committee:
    Co-Chairman: Charles Pugh
    Co-Chairman: William Weber
    Secretary: Joseph Quade
    Treasurer: Donald Parks

Copyright © 1990 Turner Publishing Company
This book or any part thereof may not be reproduced without the written consent of the publisher.

The materials for this publication were compiled using available information: the publisher regrets it cannot assume liability for errors or omissions.

Library of Congress Catalog Card No. 90-070635
ISBN: 0-938021-90-7
Printed in the U. S. A.

First Printing 1990

Rakkasans landing after a jump at Fort Campbell.

Brig. Gen. Anthony C. McAuliffe, artillery commander, 101st Abn., gives his glider pilots last minute instructions before the take off on D plus 1, 18 September 1944. (U.S. Army photo, courtesy of 101st Airborne Division Association)

TABLE OF CONTENTS

U. S. A. Airborne Insignia 6

President Bush's Message 14

Dedication 16

In Memoriam 17

Preface 18

Introduction 19

Acknowledgements 20

U. S. A. Airborne Fiftieth Anniversary Foundation, Inc. 21

Airborne Through the Ages 23

Airborne Operations 71

Airborne Unit Histories 169
    Parachute Test Platoon 171
    11th Abn. Div. 176
    13th Abn. Div. 184
    17th Abn. Div. 186
    82nd Abn. Div. 193
    101st. Abn. Div. 211
    1st Air Cav. Div. 226
    173rd Abn. Inf. Bde. 239
    187th Abn. Inf. 242
    88th Glider 249
    501st P.I. B. 251
    503rd PRCT 253
    505th P.I.R. 260
    507th P.I.R. 264
    508th Parachute Inf. 278
    509th P.I.B. 291
    517th PRCT 294
    541st P.I.R. 300
    542nd P.I.R. 301
    550th Abn. Inf. Bn. 304
    551st. P.I.B. 305
    75th Inf. Ranger Regt. 312
    Ranger Inf. Co. (Abn.) Korea . . . 316
    First Special Service Force 318
    Special Forces 324
    Office of Strategic Services 350
    Vietnamese Abn. Advisory Det. . . 355
    Marine Parachute Bn. 361
    Abn. Troop Carrier Command . . . 365
    Airborne Pathfinders 371
    Air Assault 373

Airborne All the Way 377

Chronology–U.S. Army Airborne Forces . . . 401

Campaign Streamers awarded to Airborne Units . 474

Appendix 494

Airborne Units Affiliated with the U.S.A. Airborne Fiftieth Anniversay Foundation . . . 497

Airborne Fiftieth Anniversary Celebration . . . 498

Publisher's Message 506

Index 507

U. S. A. Airborne Insignia

Insignia are not in order of precedence.

Parachute Badge, Basic

Glider Badge

Air Assault Badge

Senior Parachute Badge

Master Parachute Badge

Pilot Wings

Glider Pilot Wings

Pathfinder Badge

Parachute Rigger Badge

HALO Badge
(High Altitude-Low Opening)

Paratroop Cap Patch
Blue=Infantry
Red=Artillery

Glider/Paratroops (WW II) Cap Patch
Blue=Infantry
Red=Artillery
Airborne Troops (Current)

Glider Troops Cap Patch
Blue=Infantry
Red=Artillery

U. S. A. Airborne Insignia

(Shoulder Sleeve)
Insignia are not in order of precedence.

1st Allied ABN Army   *1st US ABN Army*   *US Army ABN Command*   *Airborne Troop Carrier Command*   *1st ABN Troop Carrier Command*

IX ABN Troop Carrier Command   *US Army Special Operations Command*   *1st Special Operations Command (Airborne)*   *XVIII ABN Corps*

6th ABN Div.   *9th ABN Div.*   *11th ABN Div*   *11th Air Assault Div.*   *13th ABN Div.*

U. S. A. Airborne Insignia

(Shoulder Sleeve)
Insignia are not in order of precedence.

15th ABN Div. 17th ABN Div. 18th ABN Div. 21st ABN Div.

80th ABN Div. 82nd ABN Div 84th ABN Div. 100th ABN Div.

101st ABN Div. 108th ABN Div. 135th ABN Div. 2nd ABN Bde.

173rd ABN Bde 71st ABN Bde./36th Inf. Div. (TXNG) 187th ABN Regimental Combat Team 508th ABN Regimental Combat Team

U. S. A. Airborne Insignia

(Shoulder Sleeve)
Insignia are not in order of precedence.

503rd PRCT   *513th P.I.R.*   *517th P.I.R.*   *551st P.I.R.*

555th P.I.B.   *First Special Service Force*   *Ranger Infantry Company (ABN)*

75th Ranger Regiment (ABN)   *1st Cavalry Div. (Air Mobile)*   *MACV (ABN)*   *Special Forces Groups (Airborne)*

U. S. A. Airborne Insignia

(Distinctive Unit Insignia)
Insignia are not in order of precedence.

XVIII ABN Corps · 82nd ABN Div. · 101st ABN Div. · 108th ABN Div. · 173rd ABN Brigade

Special Forces Groups · 75th ABN Ranger Regiment · 88th Glider Infantry Regiment · 187th ABN Infantry Regiment · 188th ABN Infantry Regiment

317th Parachute Infantry Regiment · 318th Parachute Infantry Regiment · 319th Parachute Infantry Regiment · 325th ABN Infantry Regiment · 326th Glider Infantry Regiment

327th Glider Infantry Regiment (1942-45)
516th ABN Infantry Regiment (1948-53)
327th ABN Infantry Regiment (Current) · 333rd Glider Infantry Regiment · 334th Parachute Infantry Regiment · 335th Parachute Infantry Regiment · 397th ABN Infantry Regiment

398th ABN Infantry Regiment · 399th ABN Infantry Regiment · 401st Glider Infantry Regiment · 501st ABN Infantry Regiment · 502nd ABN Infantry Regiment

503rd ABN Infantry Regiment · 504th ABN Infantry Regiment · 505th ABN Infantry Regiment · 506th ABN Infantry Regiment · 507th ABN Infantry Regiment

U. S. A. Airborne Insignia

(Distinctive Unit Insignia)
Insignia are not in order of precedence.

508th ABN Infantry Regiment   509th ABN Infantry Regiment   511th ABN Infantry Regiment   518th Parachute Infantry Regiment   519th Parachute Infantry Regiment

550th Glider Infantry Bn.   81st ABN Field Artillery Bn.   88th ABN Antiaircraft Artillery Bn.   89th ABN Field Artillery Bn.   98th ABN Field Artillery Bn.

319th ABN Field Artillery Bn.   320th ABN Field Artillery Bn.   321st ABN Field Artillery Bn.   376th ABN Field Artillery Bn.

377th ABN Field Artillery Bn.
(515th ABN Field Artillery Bn. 1948-56)   456th ABN Field Artillery Bn.   457th ABN Field Artillery Bn.

462nd Para. Field Artillery Bn.   463rd ABN Field Artillery Bn.
(516th ABN Field Artillery Bn. 1948-56)   544th ABN Field Artillery Bn.

674th ABN Field Artillery Bn.   675th ABN Field Artillery Bn.
(675th Glider Field Artillery Bn. 1942-49)   907th ABN Field Artillery Bn.

U. S. A. Airborne Insignia

*Streamers Authorized Airborne Units for Service in
WW II, Korea, Vietnam, Grenada and Panama.*

*Presidential Unit
Citation (Army)*

*Presidential Unit
Citation (Navy-Marine Corps)
(Star indicates 2nd Award)*

*Meritorious Unit
Commendation (Army)*

*French Croix de Guerre
with Palm, World War II*

*Belgian Croix de
Guerre 1940*

*Military Order of William
(Degree of a Knight of
the Fourth Class)*

*Philippine Presidential
Unit Citation*

*Republic of Korea Presidential Unit
Citation*

*Republic of Vietnam Cross
of Gallantry with Palm*

*Republic of Vietnam Civil Action Honor Medal,
First Class Campaign Streamer*

U. S. A. Airborne Insignia

*Streamers Authorized Airborne Units for Service in
WW II, Korea, Vietnam, Grenada and Panama.*

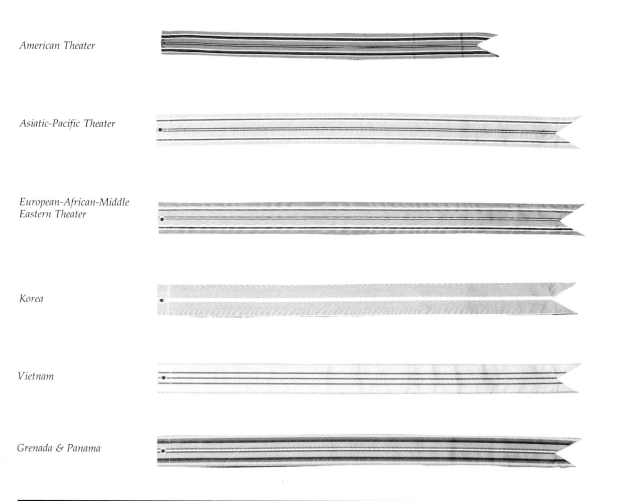

American Theater

Asiatic-Pacific Theater

European-African-Middle Eastern Theater

Korea

Vietnam

Grenada & Panama

The following is a list of airborne units for which no distinctive unit insignia was authorized.

INFANTRY REGIMENTS

- 189th Glider Infantry
- 190th Glider Infantry
- 191st Glider Infantry
- 192nd Glider Infantry
- 193rd Glider Infantry
- 194th Glider Infantry
- 485th Glider Infantry
- 513th Airborne Infantry
- 514th Airborne Infantry
- 515th Parachute Infantry
- 516th Parachute Infantry
- 517th Parachute Infantry
- 541st Parachute Infantry
- 542nd Parachute Infantry
- 543rd Parachute Infantry
- 544th Parachute Infantry
- 545th Parachute Infantry

ARTILLERY BATTALIONS

- 458th Parachute Field Artillery
- 459th Parachute Field Arritllery
- 460th Parachute Field Artillery
- 464th Parachute Field Artillery
- 466th Airborne Field Artillery
- 472nd Glider Field Artillery
- 549th Airborne Field Artillery
- 550th Airborne Field Artillery
- 551st Airborne Field Artillery
- 676th Glider Field Artillery
- 677th Glider Field Artillery
- 678th Glider Field Artillery
- 679th Glider Field Artillery
- 680th Glider Field Artillery
- 681st Glider Field Artillery

President of the United States, George Bush

THE WHITE HOUSE

WASHINGTON

May 1, 1990

I am delighted to send my warmest greetings to all those gathered in our Nation's Capital to celebrate the 50th Anniversary of the Airborne.

I have always had the greatest respect for the unique talent, courage, and magnificent camaraderie of our American Airborne forces. From Normandy to Panama, whether by glider, parachute, air assault, air landing, or special operations, Airborne has always risen to the occasion. Your special fraternity has set -- and continues to set -- the standard for elite forces of the world. Time and again, the gallant daring of the Airborne has played an invaluable role in helping to bring freedom to oppressed peoples the world over. Each of you can be proud of the significant role our Airborne forces have played in service to our Nation, and all Americans are in your debt.

As I join you in celebrating this milestone in Airborne history, I also join you in remembering and paying tribute to your brothers-in-arms who made the ultimate sacrifice in the defense of liberty. Let us resolve to ever honor their memory by our promise to preserve the freedom they purchase for our Republic at so great a price.

God bless you.

George Bush

Dedication

This commemorative volume is dedicated to all who have earned and wear the "Silver Wings of Courage" that mark the American soldiery of the airborne family. For fifty years those who wear the wings of "Airborne" have brought to our nation's military history page after page of valor, dedication and selfless service.

It matters not whether the wings were those of the glider trooper, paratrooper, air assault trooper, airborne troop carrier pilots and crewmen, riggers, Special Forces, Rangers or Army, Navy, Air Force or Marine Corps. It was the earning of the wings that distinguished those who placed them on their chest. From the moment of pinning on "Blood Wings" the wearer stood taller, walked prouder, fought harder and was committed to whatever it took to uphold the honor and traditions of his unit, his commander, his service and "Airborne"!

As we gather on our fiftieth year of service to country and people, it is proper that we take pride in our accomplishments. No nation can survive unless there are those willing to go the extra step and give the extra measure. That has been the hallmark of airborne and the standard by which others are judged.

We cannot predict with certainty what the future has in store for airborne as we have known it and as it is evolving today. But, we can predict without question that whenever and wherever there is a need for a soldier's soldier, there you will find the kind of American that bears the proud title of "Airborne" today!

For those of our comrades who are unseen, but nonetheless next to we who are present during our fiftieth, let them know that we are aware of their presence and mindful of the sacrifices they made so that we might make this formation. And if you listen closely, you will hear their voices echo ours as we call out the motto of our fraternity — Airborne, all the way, always, in all ways!

William E. Weber
Col., USA (Ret)

187th Airborne RCT Munsan-ni Combat Jump, Korea, March 23, 1951.

In Memoriam

"No man lives unto himself, and no man dies alone."

Those immortal words are reminiscent of the great John Donne, dean of St. Paul's Cathedral, London, in the early 17th century. Ernest Hemingway popularized the words with the title of his book *For Whom the Bell Tolls*. But listen to John Donne say it: "Any man's death diminishes me, because I am involved in mankind, and therefore never send to know for whom the bell tolls, it tolls for thee."

There is no fraternity of men to equal the airborne fellowship with its particular mystique of brotherhood; whether it be Africa or Europe, Korea or Vietnam, we don't have to look for heroes. They are here today. We are in the midst of them. When we qualified as airborne, we were indelibly stamped with the mark of hero. We didn't have to work at it; we were that.

And you can't separate us of today from those of yesterday. *No man dies alone*. We identify with them. We remember them. We glory in their bravery, in their heroics, in their daring to fight for freedom anywhere on the face of this earth. The airborne trooper doesn't go into combat to survive. His purpose is to win, because he is involved in mankind. There is an intricate web of cohesion, which binds all mankind into a unity, and the airborne trooper knows he cannot live to himself. If need be, he dies, but he does not die alone. His death diminishes me, and it diminishes you. He does not die alone.

Julius Fucik, a Czeck journalist who was executed in Berlin, Sept. 8, 1943, for his participation in resistance activities, has something to say to us even now. Hear his haunting appeal spoken to all who survive the horror of Nazism:

"I ask for one thing. You who will survive this era, do not forget. Forget neither the good man nor the evil. Gather together patiently the testimonies about those who have fallen. One of these days the present will be the past, and people will speak of 'the great epoch' and of the nameless heroes who shaped history. I should like it to be known that there were no nameless heroes, that these were men-who had names, faces, desires and hopes. May they always remain close to you like acquaintances, like kinsmen, like yourselves."

George B. Wood
Chaplin

The truth is that our lives are built on the foundation of previous deaths. There is a continuity in the life and death sequence on this earth. We give thanks and praise for the good examples of those who fought and died in the preservation of freedom. But those who lived through combat to die in living a life of freedom were also good examples for whom we give thanks and praise. They are all the seed that dies and comes to life and bears much fruit. We are here to memorialize our dead, the dead who fought in combat and the dead who fought in life.

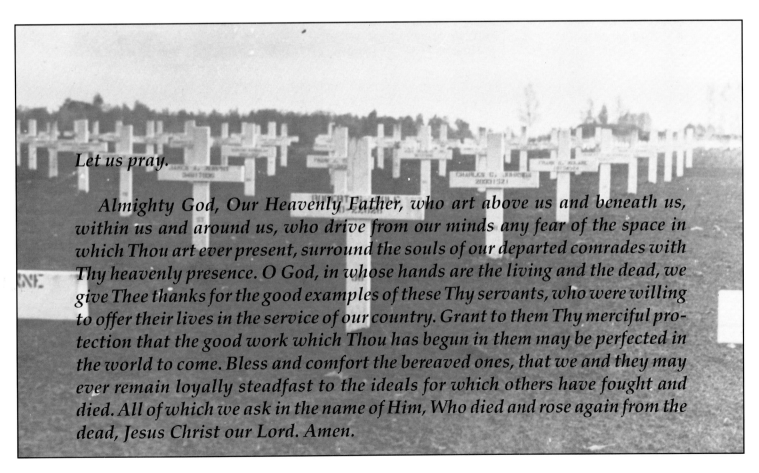

Let us pray.

Almighty God, Our Heavenly Father, who art above us and beneath us, within us and around us, who drive from our minds any fear of the space in which Thou art ever present, surround the souls of our departed comrades with Thy heavenly presence. O God, in whose hands are the living and the dead, we give Thee thanks for the good examples of these Thy servants, who were willing to offer their lives in the service of our country. Grant to them Thy merciful protection that the good work which Thou has begun in them may be perfected in the world to come. Bless and comfort the bereaved ones, that we and they may ever remain loyally steadfast to the ideals for which others have fought and died. All of which we ask in the name of Him, Who died and rose again from the dead, Jesus Christ our Lord. Amen.

Preface

As we mark the 50th birthday of our Army's Airborne Forces, starting with the original Test Platoon in 1940, I would like to point out our comparatively late acceptance of the concept and reflect on my airborne service. Though Germany and Russia had created such forces as early as the 1930s, I can recall our simulated use of paratroops for the seizure of a critical bridge for an advancing armored force in August of 1936. This occurred in Michigan and was under U. S. Second Army control in our largest two-sided troop maneuvers in years. I was then Assistant Chief of Staff, G-3, Second Army, and wrote the scenario. General Marshall commanded one side.

To no individuals do our Army Airborne Forces, both parachute and glider, owe more than to General of the Army, George C. Marshall, and General of the Air Force, Henry H. Arnold. It was their early and sustained support that brought our airborne forces into being for the large scale combat operations in Europe and Asia in World War II, and subsequently in Korea, Vietnam, Grenada and Panama.

Having just acceded to command of the 82d Infantry Division, I vividly recall first having been told the 82d would become a mechanized division and then suddenly being informed in secrecy that it would be an Airborne division.

Matthew B. Ridgway
Lt. Gen., USA, (Ret.)

Asked if I would like to have it, my reply was "Fine! I would like to have it, though I have never heard of one." I must have forgotten Crete and Germany's airborne success.

This was the beginning of my privileged combat service, first with the 82d, then with the XVIII Airborne Corps. I remember having to split off half of the 82d to form the 101st under General William C. Lee, and later upon his illness and death, General Maxwell B. Taylor taking command. I remember our warm friendships with the people of the Leicester, England, and the Ste Mere Eglise area in Normandy; of our delightful association with the British Airborne, briefly with the Polish Parachute Brigade, and with some of the finest of the British and U. S. Infantry and Armored Divisions, along with U. S. Rangers and British Commandos. I remember especially my close, warm personal friendship and delightful working relations with Major General Hal C. Clark, Commander, 52d Troop Carrier Wing, associate in training and combat of my 82d Airborne Division.

No military man could have dreamed of a finer or more satisfying period of combat service, nor with more gallant men. My appreciation to each of you, and to those of you who have since joined our Airborne ranks.

The 82nd Airborne fills the sky as they jump from the sides of C-46 airplanes.

Introduction

The development of airborne military operations is remarkable for the magnitude of its impact, over a relatively short span of time, on the overall history of warfare. With the advent of practical aviation, the concept of an airborne force loomed as an exciting, achievable capability of dominating proportions. Early demonstrations of combat effectiveness soon worked dramatic changes in strategic and tactical doctrines.

Here was a magnificent multiplier of the field of commander's prime assets of choice — speed and surprise. While airborne could be perceived as elegant in its simplicity, its complexity in combat applications—particularly as to the necessity for precise total—force management—became evident, at heavy cost, in World War II operations.

It is fitting that we reflect on a half century of airborne development and use by the United States Army. It was in 1940, at Fort Benning, Georgia, that there began the first serious official effort to produce airborne troops for this nation's defense. Under Lee recruited 48 volunteers from the 29th Infantry Regiment. (Three years later, he would find himself a major general commanding the 101st Airborne Division—but denied by a heart attack the opportunity to lead it at Normandy on D-Day)

Within two months, after training on a tower of a type originally designed for wire-controlled drops at the 1939 New York World's Fair, the test platoon was making jumps from a Douglas B-18. A mass demonstration jump intensified official interest in airborne capabilities.

The time was late. Germany had begun training paratroops in 1936; ran trials of cargo gliders the next year and, by the end of 1938 had in being, albeit short-trained, an airborne division of two parachute and seven air-landed divisions. The Soviet Union which. like Italy, had taken up airborne troop training in the 1920's, dropped a force of 1,500 in a 1936 demonstration.

In retrospect, the catch-up challenge we faced in 1940 (as did our allies-to-be in the European conflict) seems catastrophically daunting. The story of how that challenge was met is an inspiring one which deserves to be retold and reread at this time of anniversary. It is an epic of high-risk, heavy-loss undertakings which forged a powerful instrument of victory.

Perhaps the vigor, professional competence and audacity of U.S.A. Airborne brought comfort at last to the shade of General Billy Mitchell, who had proposed, in 1918, parachuting troops behind the German lines from the early bombing planes of World War I.

In 1940, with the paratroop program at Benning just underway, the Germans already were committing airborne troops effectively in the invasions of Denmark, Norway and Holland, and in the masterful reduction of Belgium's Eben Emael fortress by 78 men.

In 1941, in a major airborne assault, the Germans took Crete, despite heavy losses.

The next year, though, U.S.A. Airborne began to write its

John O. Marsh, Jr.
Secretary of the Army, 1981-1989

own combat records. The 509th Parachute Infantry Regiment jumped in Algeria and Tunisia.

The list of U.S. Army combat airborne operations now stretches over more than 40 years. Among the World War II place names, the most prominent, of course, is Normandy—D-Day, 6 June 1944. Others to conjure up Airborne veterans' recollection include Sicily, Salerno, Holland, Germany and, on the other side of the globe, Markham Valley, Noemfoor Island, Luzon, Corregidor and Los Banos.

There were to be combat operations again in the Korean War and in Viet Nam, and most recently, on 25 October 1983 in Grenada and in Panama on 20 December 1989.

The Airborne tradition, grounded in rigorous training and tempered under fire, now is firmly established and proudly guarded within the Army. It testifies to capabilities honed by cruel confrontations with battleground reality. It is difficult now to envision any threat of significant magnitude in which Airborne would not be parted of the military response.

When we recognize Airborne as in the front line of the national defense, we need to keep in mind that it does not stand alone. Its maximum effectiveness depends on coordinated planning among all Army elements assigned to the military problem being addressed, and with elements of the other armed services participating.

In the evolution of the Army, the Airborne represents a major expansion of strategic and intensify, by use of the air, the prized military advantages of speed and surprise. Even so, the ground remains paramount. The Army core objective—take ground and hold it—is a constant.

We see, therefore, that what is of intrinsic importance in the Airborne concept is the evolutionary advance of the individual soldier. It confers on the fighting man accelerated access to the battlefield, the ground to be taken and held. The concept provides reasonable assurance, as well, that supporting weaponry, ammunition and maintenance supplies will accompany him by air delivery.

It is reasonable to assume, then, that the soldier with the Airborne patch—intensively conditioned physically; trained diligently for proficiency and survival n a wide range of battle zone environments, and inculcated with a clear sense of mission—is a soldier buoyed by confidence and pride.

This spirit pervades the entire Army, providing each soldier, in whichever branch serving, with a supportive base of confidence and understanding of purpose. A high order of morale—force-wide—is critical to self-reliance, endurance, team effort and eventual victory.

During the more then eight years I was privileged to serve as Secretary of the Army, I had frequent opportunity to observe the enhanced proficiency and reliable commitment of the American soldier. The Airborne story is one of many testimonials to the historic determined and valorous response of the American soldier in time of national danger.

Acknowledgements

It would be extremely difficult to thank everyone who helped make this 50th Anniversary of Airborne Commemorative Book possible. Many of the unit histories carry the name of the author and certainly they are to be thanked, but photographs, statistics and accounts of personal experiences and bravery under fire have come in from dozens of troopers. All have contributed immeasurably to the finished product. We gratefully acknowledge all of these great people who obviously still have the "Old Airborne Spirit."

To Clark Archer, who conceived and orchestrated the original plan for the book and helped in the collection of material, we give a special thanks. Clark got us involved at the beginning and we will always be glad we "hooked up."

Many of the great airborne commanders of the World War II era are now senior in age and were unable to participate in this book to the extent that they had first planned. However, in several cases they called on some of today's leaders who had served under them and these people responded in true Airborne fashion. Our leaders of yesteryear were with us all the way. That spirit of cooperation was an inspiration and a guiding force in preparing this book.

Some of the histories and operations accounts are noticeably longer and in more detail than others. This should not be taken to mean they are more important than the shorter ones. The authors simply sent us more material than we had requested and we couldn't bring ourselves to edit out what is a part of our nation's history.

Our only regret is that representatives of a few of the airborne units could not be located, or if located, they failed to provide us with any material. Consequently, some units are not included in the book. Numerous inquiries and requests were sent out to association officials, various leaders and recognized representatives, but still some failed to reply. Maybe the years have taken their toll and slowed their reaction time for this type of mission. However, in the years past when the call went out for men to jump or glide into battle, these same men were ready, willing and able!

And, that's what "Airborne" is all about.

Bart Hagerman, Editor
LTC, AUS (Ret)

Bart Hagerman lives in Bowling Green, Kentucky, he is a former glider trooper and master parachutist. His Airborne service includes the 17th Airborne Division, The Parachute School at Ft. Benning, Georgia and the 19th Special Forces Group (Airborne). He retired as a Lieutenant Colonel after 31 years of active, National Guard and Reserve duty.

11th Airborne troopers land during a practice jump.

U. S. A. Airborne
Fiftieth Anniversary Foundation, Inc.

Executive Committee

Charles Pugh	(517th PRCT)	Co-Chairman
William Weber	(187th ARCT)	Co-Chairman
Joseph Quade	(17th Abn. Div.)	Secretary
Donald Parks	(187th ARCT)	Treasurer
Clifford Allen	(555th PIB)	
Hal Bailey	(Dept. of Army)	
Dan Campbell	(82nd Abn. Div.)	
Gerald Hasbargen	(503rd PRCT)	
George Hurchalla	(11th Abn. Div.)	
Robert Jones	(101st Abn. Div.)	
John Marr	(507th PIR)	
Charles Rambo	(503rd PRCT)	
Robert Sommers	(13th Abn. Div.)	
Frederick Thompson	(11th Abn. Div.)	

Advisory Board

John Costello	(509th PIB)
Paul DeVries	(Society VN Abn.)
Doug Dillard	(551st PIR)
John Grady	(542nd PIR and Abn. Cmd.)
Vito Pedone	(Abn. Troop Carrier)
Gerrell Plummer	(1/8 1st Cav. Div.)
Ed Segassie	(542nd PIR and Abn. Cmd)
Clyde Sincere	(Special Forces Assn.)
Bill Story	(First Special Service Force)
Joseph C. Watts	(Assn. Rngr. Inf. Cos-Korea)
Gene Weeks	(Society of 173rd Abn. Bde.)

50th Anniversary Emblem Designed by:
Robert M. Baldwin 507/504

Special Committee Chairmen

John Costello	(509th PIB)	Parade
Bart Hagerman	(17th Abn. Div.)	Commemorative Book
Steve Rosenstock		Smithsonian Exhibit
Bob Sommers	(13th Abn. Div.)	RFK Stadium Events
Bill Story	(FSSF)	Publicity
Ivan Worrell	(101st Abn. Div.)	Fund Raising

Part of a crew of a glider, 101st Airborne, who took part in the landing behind German lines in Holland, pose beside their craft before the take off at an airdrome in England, September 18, 1944. (Courtesy of 101st Airborne Division Association)

Airborne
Through The Ages

Communications Training Signal Corps radio man on field maneuvers with 17th Airborne Division at Camp Mackall, North Carolina. (Courtesy of Engineer Reproduction Section, Airborne Headquarters)

Airborne Through the Ages

by Scott Garrett, Ph.D.

Beginnings: Antiquity—World War I

Man has long dreamed of flying. In the mythologies of many cultures over the sweep of time and place, warriors from the heavens are often described. Either winged themselves or mounted on winged steeds of one sort or another, these sky troopers were the first practitioners of vertical envelopment. But the notion of airborne fighters was not confined only to ancient myth.

In the years prior to World War II, popular science fiction had a host of supermen, hawkmen and others whose powers included flight. Comic strips like Flash Gordon and Buck Rogers as well as Saturday matinees featuring these fantasy heroes helped to give boys coming of age during this time the notion of warriors coming from the sky. When this generation went to war, it is little wonder that many of them volunteered for airborne, turning fantasy into reality.

In addition to myth makers, some sought to develop means that would actually enable man to fly. One of Leonardo da Vinci's greatest disappointments was that his splendid mind could not devise a means for him to do so. The origins of this dream are lost in the far reaches of the past, but in Greek mythology mortal man could fly. In this legend Daedalus, the Athenian, and his son Icarus were imprisoned in the Labyrinth by Minos, King of Crete. To fly like birds was their only chance for escape. Daedalus crafted wings of bird feathers, held in place with wax, for himself and his son. Fitted with these, both flew from the Labyrinth. Icarus, however, ignored his father's injunction not to fly too high. As a consequence, the heat of the sun melted the wax that secured the feathers. Icarus fell into the sea and was lost. Daedalus continued his flight, landing safely at Cumae. There he built a temple to Apollo as a thanksgiving for his deliverance.

Although da Vinci had not mastered flight by the time of his death in 1519, he had designed a workable parachute. In one of his notebooks he wrote:

> If a man have a tent of linen of which the apertures have all been stopped up, and it be 12 braccia [about 36 feet] across and 12 in depth, he will be able to throw himself down from any great height without sustaining any injury.

Accompanying his text da Vinci sketched a "tent of linen," four-sided and rigid, with a man suspended beneath.

The dream of flight became reality in 1783 when the Montgolfier brothers of France rode into the sky aboard a hot-air balloon. In the following year the notion of attacking from the sky—thus adding a third dimension to the geometry of war—was expounded upon by Benjamin Franklin. In 1784 he wrote:

> Where is the prince who can afford so to cover his country with troops for its defense, as that ten thousand men descending from the clouds might not, in many places, do an infinite deal of mischief before a force could be brought together to repel them?

The first to parachute, in this case from a balloon, was an

Daedalus and his son Icarus. From a Roman relief.

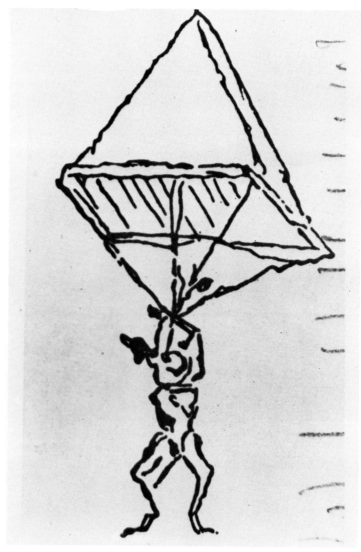
Leonardo da Vinci's design for a parachute that he sketched in his notebook.

animal. In 1785, Jean Pierre Blanchard, a Frenchman, dropped a small dog from his aircraft. Equipped with a parachute patterned after da Vinci's design, the dog made a safe landing and ran swiftly away. Neither dog nor parachute was ever seen again. Some authors claim Blanchard parachuted from a balloon shortly after his successful experiment with the dog. Yet, it is not definitively known who the first human was to make a parachute jump. Nevertheless, within a few years of Blanchard's test, stunt parachutists were making thousand-foot drops from tethered balloons. These early skydivers used rigid, pyramidal parachutes suspended beneath the balloon and whose design was essentially that proposed by da Vinci. The major drawback at the time was that the rigid parachute rocked violently from side to side during descent.

In 1835, Robert Cocking designed a rigid parachute that looked very much like an inverted umbrella. This Englishman believed that his design would provide a gentle ride to earth. He secured the necessary backing and was ready to test it by the summer of 1837. So it transpired that Cocking has the unfortunate distinction of being the first recorded unsuccessful parachutist.

Suspended beneath the huge balloon "Nassau," Cocking and his parachute began their ascent late on Monday afternoon, 24 July 1837. A vast crowd waited below in London's Vauxhall Gardens to witness the event. At 4,000 feet Cocking was released and, after a successful beginning, the parachute's struts proved unequal to the strain. Amid a tangle of broken spars and torn linen, the sixty-year-old inventor and showman fell to his death.

Nearly fifty years later two American high-wire circus performers, the brothers Thomas and Samuel Baldwin, developed the concept of an all-cloth, flexible parachute that could be carried in a small container. Using a table napkin, string and the cork from a wine bottle they built a "prototype" and dropped it from the fifth floor of their New Orleans hotel. After that test proved successful, they set about building a parachute large enough for a man. The result of their efforts was the first modern parachute.

Tom Baldwin, at 29 the younger brother, was the first to test their new design. On 30 January 1887, a sellout crowd gathered at Golden Gate Park in San Francisco to watch what many thought would be Tom's death. From a jump altitude of 5,000 feet Baldwin leaped from a tethered balloon. Everything worked as planned. As Tom floated toward earth, the crowd cheered and cheered.

By 1912, great strides in aircraft development had occurred since the first tentative flight at Kitty Hawk, North Carolina in 1903. Another aviation first happened on 28 February 1912, when Albert Berry made a parachute jump from an airplane. At about 1500 hours, he jumped off the plane's landing gear axel. The aircraft was somewhat above 1,000 feet altitude and flying at 50 m.p.h. Berry had selected the parade ground at Jefferson Barracks, Missouri, as his landing zone. A large number of soldiers watched the spectacle. Most had never seen an airplane, much less a parachutist. Berry overshot the parade ground but landed safely behind the messhall.

Although parachute development continued up through World War I, most interest was fixed on the aircraft, not ways to get out of them. Even so, there were those who gave thought to the combat potential of the parachute. Winston Churchill advocated air dropping fast-moving units behind German lines to disrupt communications, destroy bridges and factories. Nothing, however, came of his suggestion.

Another early advocate of airborne operations was Colonel William "Billy" Mitchell, chief of all air units in the American Expeditionary Force (A.E.F.). On 17 October 1918, Mitchell had a meeting with General John J. "Black Jack" Pershing, commander-in-chief of the A.E.F., in which he outlined his concept of an airborne assault to reduce the key German positions, like the fortress-city of Metz. As Mitchell recounted this meeting after the war:

> I . . . proposed to him [Pershing] that in the spring of 1919, when I would have a great force of bombardment airplanes, he should assign one of the infantry divisions permanently to the Air Service, preferably the 1st Division; that we should arm the men with a great number of machine guns and train them to go over the front in our large airplanes, which would carry ten or fifteen of these soldiers. We could equip each man with a parachute, so that when we desired to make a rear attack on the enemy, we could carry these men over the lines and drop them off in parachutes behind the German position. They could assemble at a prearranged strong point, fortify it, and we could supply them by aircraft with food and ammunition. Our low flying attack aviation would then cover every road in their vicinity, both day and night, so as to prevent the Germans falling on them before they could throughly organize the position. Then we could attack the Germans from the rear, aided by an attack from our army on the front, and support the whole maneuver with our great air force.

The armistice of 11 November 1918, ended fighting on the Western Front before any action was taken on Billy Mitchell's revolutionary proposal. Nonetheless, he is still given credit for originating the concept of vertical envelopment. That is, dropping paratroopers behind enemy lines and effecting a linkup with them by means of a coordinated ground attack.

Between World Wars, 1919-1938

For the first decade after the Great War, as it was called until one greater was fought, advocates of the vertical envelopment concept were virtually voices crying in the wilderness. To be sure, not all airborne experiments ceased: on 6 November 1927, nine fully-equipped Italian military parachutists dropped on Cinisello Airfield in Italy; during the 1928 maneuvers three American soldiers with a disassembled machine gun jumped at Brooks Airfield, Texas; the Soviet Red Army staged a drop of twelve men with their combat equipment on 2 August 1930.

During this time there were also experiments in air trans-

Parachute jumping from balloons in the 19th century.

The first parachute jump from a plane in flight was made by Albert Berry on 12 March 1912, at Jefferson Barracks, Missouri. (Smithsonian)

Mass jump of Soviet paratroopers from TB-3 transport planes. Mid-1930s.

porting combat troops and equipment. In 1931, the men and guns of an American field artillery battery were flown from one end of the Panama Canal Zone to the other. The following year the British flew an infantry battalion and all its equipment from Egypt to Iraq. Although the military potential of air transport was thus demonstrated, neither country had plans to create airborne units of any type.

It was in the Soviet Union during the early 1930s that the first serious efforts were made to develop airborne formations.

Volunteers from the 11th Rifle Division were selected to form the Test Airborne Landing Detachment, which was set up in the Leningrad Military District during March 1931. Soviet military planners were pleased with the success of the Test Detachment and alert to the unrealized potential of vertical envelopment. Consequently, in January 1932, they formed four airborne motorized detachments, each of battalion strength. The military districts of Belorussia, Ukraine, Moscow and Leningrad were each assigned one airborne unit. In the

Leningrad Military District the Test Airborne Landing Detachment formed the cadre of the new unit. So it was that the Soviet Union created the first airborne formations in the world.

Soviet airborne doctrinal development and training continued. In one training exercise during 1934, forty-six men jumped and a light tank was dropped. Foreign military attaches were invited to observe the 1935 autumn maneuvers in the Ukrainian Military District. At a training area near Kiev these foreign experts were astonished when two battalions of "locust warriors"—as the Soviets nicknamed their paratroopers—made a mass jump. The Red Army continued experimenting with drops of larger and larger groups. For instance, in the fall 1936, around 5,500 men jumped as part of one training exercise.

Soviet developments sparked major interest in many countries but only a few followed through and established airborne units. The Germans made an immediate commitment to develop a significant airborne capability. During1936, the Luftwaffe (German Air Force) established a paratroop training school at Stendal, west of Berlin, by the River Elbe.

Even though no parachute units were formed by the United States during the decade of the 1930s, American military thinkers were not oblivious to their development in other countries. An article entitled, "The Employment of and Defense Against Parachute Troops", appeared in the September 1937 issue of the U. S. Army Command and General Staff School Quarterly. It caused much discussion in army circles. One officer went so far as to unequivocally state that someday the parachute would be the army's "sword of silk."

By the outbreak of war in 1939, only the Soviet Union, Germany, Italy, France and Poland had constituted airborne units. But only the first two of these countries had significant airborne capacity.

World War II, 1939-1945

In the Polish Campaign of 1939, none of the combatants employed their airborne forces because it was over so quickly. After completing their part in the conquest of Poland, the USSR turned its attention to Finland. Leningrad, cradle of the Bolshevik Revolution, was within artillery range of the Finnish border. To secure the northwestern flank of their country, the Soviets wanted the southern part of the Karelian Isthmus as well as other pieces of Finnish territory. In the face of increasing pressure, the Finns mobilized against the USSR along their common frontier. Without a formal declaration of war, the Soviets attacked Finland on 30 November 1939. On the opening day of the "Winter War," small detachments of Red Army parachute troops jumped near Petsamo, in the far north of Finland. Although they failed to achieve any real tactical success, this was the first time in history that airborne troops were used in combat.

Despite the fact that Soviet Russia established the first airborne units, including military gliders, and first used them in war, Soviet airborne operations played only a relatively minor role from the Russo-Finnish War of 1939-1940 through World War II.

During the war the Soviets primarily used paratroopers for rather small-scale sabotage and partisan-support missions. Even so, over a two-day period in October 1941, some 6,000 Red Army troops were airlanded near Tula, about one hundred miles south of Moscow to help halt the German drive on the Soviet capital.

Soviet combat jump in early World War II.

As events turned out, the Red Army staged only two large-scale airborne operations in all of World War II. The first was as a part of the Soviet counter offensive launched on 6 December 1941, from Finland to the Crimea. A segment of this grand offensive involved operations in the Vyazma area in which some 7,000 troops were air landed and about 3,500 paratroopers jumped in over the period from 3 January 1942, through mid-March. Although Soviet airborne elements caused the Germans considerable difficulty, they were ultimately destroyed.

The second major airborne attack began in late September 1943, to support the Soviet assault crossing of the Dnieper River. A total Red Army force of around 10,000 parachute and glider infantry landed in the great river bend southeast of Kiev. Misfortune beset the operation from the beginning and, in the end, the Germans eliminated the Soviet airhead. Despite their destruction, these airborne troops successfully distracted and tied down large numbers of German troops and helped enable the Red Army to cross the Dnieper elsewhere in force.

There are two significant reasons that probably account for the failure of the Red Army, despite their pioneering efforts, to use airborne forces on a larger scale. The first is the desperate straits in which the Soviets found themselves during the first part of the war. That necessitated committing their airborne troops as regular infantry in an attempt to stop the German onslaught. Secondly, throughout the war, the Red Army was chronically short of the transport aircraft necessary to launch and sustain airborne operations of great magnitude.

In the first part of the Second World War the Germans held the dominant hand in all phases of airborne operations. With dramatic swiftness, German glider and parachute troops made themselves famous with battlefield victories during the 1940

World War II German paratrooper stands in the door of a JU-52 transport plane.

campaigns in the West. Even though they played key roles in the invasions of Denmark, Norway and Holland, it was in Belgium that they achieved their most spectacular conquest.

Fort Eban Emael was the powerful northern anchor of the main Belgian frontier fortifications facing Germany. The German General Staff estimated that it would take at least two weeks, if not longer, to reduce this key fort by conventional military means. That was unacceptable if the tempo of blitzkrieg were to be maintained. Obviously, an unconventional approach was required.

In the pre-dawn darkness of 10 May 1940, Adolf Hitler opened his great offensive in the West and Eban Emael was a vital German target. A company of paratroopers from the 1st Parachute Regiment and a paratroop engineer platoon silently landed on the open area inside the fort in DFS-230 gliders. This was the first glider assault in history.

Before the Belgian fortress troops could react, the German assault force had them hard pressed. The engineer platoon quickly set about destroying gun turrets and casements with shaped charges while the rest of the paratroopers put suppressive fire on the defenders. Shortly after daylight, 300 paratroopers jumped in to support the German attack. About noon the next day, the supposedly invincible Fort Eben Emael surrendered, with its garrison of over a thousand. Thus, German tank and transport columns were able to cross the Albert Canal and drive west, unimpeded by the fort that was supposed to prevent just such an occurance.

However, it was not until after the fall of France that Hitler permitted the story of Eben Emael to be told. Hitler did not initially reveal the role of gliders lest he want to use them for other operations.

Hardly had France surrendered in late June 1940, before Prime Minister Winston Churchill ordered creation of a parachute corps at least 5,000 strong. The German airborne victories had vindicated what Billy Mitchell and Churchill had advocated during the latter part of World War I.

Meanwhile, on the other side of the Atlantic, a still-neutral United States moved from talk to deeds regarding airborne. On 26 June 1940, three days after Churchill's order, action was taken to form a Parachute Test Platoon at Fort Benning. Volunteers for this unit came from the 29th Infantry Regiment and were put on special-duty status and attached to the Infantry Board. Because there was risk of serious injury or death, only unmarried men could join the Test Platoon. The Test Platoon was formed immediately thereafter, on 1 July 1940, and has the distinction of being the first such organization in the US Armed Forces.

Although the men of the 29th were surprised, the decision to begin developing an airborne capability was not made precipitously. In the late 1930s, U.S. Army attaches reported the growing military interest in and development of parachute

One of more than two dozen large bunkers that were a part of the Eben Emael Fortress complex.

Above and below: Eben Emael – armored observation dome and artillery piece destroyed by German paratroopers using shaped charges.

and air-landing infantry units in various European countries. Army Chief of Staff George C. Marshall was impressed and grasped the potential of such troops. In the spring of 1939, he sent a memorandum to his Chief of Infantry, Major General George A. Lynch, to report on the status of "air infantry" developments world wide.

General Marshall received this study very promptly. In addition to discussing foreign airborne experimentation and units, it reviewed earlier U.S. exercises in air-transporting troops and equipment. Even though Marshall favored going further, the Army Air Corps lacked the transport planes and funding to immediatly follow up with active research and development programs for airborne. It was not until January 1940, that the situation had improved to the point that things could move forward.

General Lynch named Major William C. "Bill" Lee project officer to develop an airborne capacity for the U.S. Lee's first priority was to develop a parachute suitable for troop use. That was done by early June 1940. The next phase was to develop a coherent airborne doctrine and appropriate training program, which was initially based on the obviously successful German model. In the beginning, the U.S., among others, was simply striving to catch up with the Germans. Obviously, a major part of this necessitated conducting tests with live parachutists to develop suitable equipment and delivery techniques; hence, the Parachute Test Platoon. Given Major Lee's important role, he is considered to be the father of American airborne.

On the basis of competative evaluations, Lieutenant William T. Ryder was selected to command the Test Platoon. The platoon was ready to make its first jump from an airplane after seven weeks of intensive training. As platoon leader, Lt. Ryder was the first to jump, thus making history. The honor of being the first enlisted man to jump went to Private William "Red" King.

A few days later, when the Test Platoon made its first mass drop, the battle cry "Geronimo!" originated. Private Aubrey Eberhardt yelled it so loudly when he jumped that observers on the ground heard him. Eberhardt shouted it to prove that he was not struck dumb with terror when he jumped. He said that he got the idea from a western movie he had seen at the post theater the night before the jump.

Things went so well with the Parachute Test Platoon that the War Department authorized creation of a regular airborne unit. The 501st Parachute Infantry Battalion was activated at Fort Benning on 1 October 1940. The 501st P.I.B. was the first

Taken in the Spring of 1941, this photo shows what the well-dressed paratroopers of that early era wore. The sateen-finish jump coveralls and the ankle-strap jump boots were shortly replaced by a more practical suit and the present paratrooper boot. (Courtesy of 101st Airborne Division Association)

tactical airborne unit established in the U.S. Army. Even though there was considerable official enthusiasm, the War Department was slow to set up additional units. Funding, always a consideration, as well as a lack of suitable transport aircraft slowed expansion. For example, a year after activating the 501st, there were only sixty-six C-47s in service, and only twelve of those were available for paratroop use. However, in keeping with the need to continue to develop practical experience in airborne techniques, the 88th Airborne Infantry Battalion was soon formed to develop air landing techniques. Early on, "airborne" meant all forms of vertical envelopment, whether by parachute, glider or air landing.

Still, by selection and training as well as by virtue of the exceptional physical and mental toughness required, paratroopers and glidermen were a breed apart. The Germans, later the British, consciously instilled a sense of eliteness and esprit de corps in their airborne troops. American paratroopers came to be set apart from regular infantry by their high-topped jump boots and insignia, in addition to distictive combat uniforms. Early in 1941, the Department of Heraldry proposed a special qualification badge for those who finished jump training. The design proposed was completely uninspiring.

Lieutenant William P. Yarborough developed the design for jump wings that is still used. With Major Lee's help, he hand-carried the design through the appropriate places in the War Department. Next, Yarborough took the approved specifications to Bailey, Banks and Biddle in Philidelphia. Not only was this a well-known jewelry manufacturing firm, but they also had made military decorations and insignia since the Civil War. Within a few days, Lt. Yarborough was on his way back to Ft. Benning with 350 sterling silver jump wings.

The year 1941 proved to be one of pivotal significance for airborne development world wide. On 6 April 1941, Hitler simultaneously invaded Yugoslavia and Greece. Yugoslavia fell in only ten days and the Greek army was compelled to surrender on 23 April. British and Commonwealth forces, in addition to Greek troops who refused to surrender, continued to withdraw southward. German parachute and glider troops were used on 26 April to seize the Corinth Canal and block their enemies' retreat. Nevertheless, the Allied forces managed to fight their way through, and some 43,000 troops were evacuated by sea from the Greek mainland. Most of those who escaped the Germans landed at Crete.

To the Germans, Crete posed both threat and opportunity. Allied air and naval forces could use bases there to interdict Axis shipping over a large part of the Mediterranean Sea. Hitler also viewed with grave concern the possibility that Allied bombers, operating from airfields on Crete, could bomb the Rumanian oil fields upon which the German war machine depended. Conversely, German conquest of Crete would enable the Germans to secure a large part of the Mediterranean as well as launch their own air attacks against the huge Royal Navy base at Alexandria, Egypt.

Hitler decided Crete was too important a prize to ignore, so invasion planning went forward. "Operation Mercury," as the invasion was code named, was a combined-arms assault that involved all three branches of the Wehrmacht (Armed Forces) but in which Luftwaffe airborne units had the major role.

We were just learning how when this picture was made in November 1941. These troopers of the 502nd Parachute Battalion were captured after a drop during the First Army maneuvers at Fort Bragg. The B-18 bomber in the background was often used as a troop transport. (Courtesy of 101st Airborne Division Association)

During the activation ceremony for the first American tactical parachute organization - the 501st Parachute Battalion - in early September 1940, General Omar N. Bradley (left) and Maj. William M. Miley render a smart salute. In the back row are General Bradley's aide and three staff officers of the fledgling 501st. (Courtesy of 101st Airborne Division Association)

Col. Robert Hugh Williams, Senior U.S. Marine paratroop commander during World War II. (Courtesy of U.S. Marine Corps)

The invasion of Crete began on 20 May 1941, when elements of the 7th Parachute Division began landing near their principal objectives, the airfields of Maleme, Retimo (Rhethymnon) and Herakleion. The defenders were alert to just such an attack and inflicted heavy casualties on the paratroopers. The rest of the 7th jumped in the next day. Continuing to suffer heavy casualties, they secured part of the airfield at Maleme. Elements of the 5th Mountain Division were then flown in from Greece. Most of the troop transports were shot down, but enough men survived the crashes to take a significant part in the fighting.

Twice, 21-22 May, the Germans tried to send the rest of the 5th Mountain Division to Crete by sea. Both times they were repulsed with heavy losses by the Royal Navy. Eventually, with high casualties on both sides, the Germans began to take the airfields, and increasing numbers of reinforcements were flown in. By 28 May, the issue was no longer in doubt. On 31 May, those British and Commonwealth troops unable to be evacuated from Sfakia surrendered to the Germans.

The invasion of Crete was the first major airborne assault in history. Although the Royal Navy completely dominated the waters around the island, the Germans had air superiority. That enabled them to initiate, support, and ultimately win the battle for Crete. However, their combat losses were very high. German sources show that a third of the airborne forces were killed or wounded and 220 planes were lost.

Despite the high casualties, General Kurt Student, commander of airborne forces, was confident that the worth of vertical envelopment had been conclusively established. After all, his men had assaulted and captured a large, well-defended island of great strategic value without naval support or the usual supporting ground forces such as tanks and heavy artillery.

Some two months after the battle for Crete, Hitler summoned Student and some other paratroop commanders to his headquarters. It was a very cordial affair. Hitler awarded Knight's Crosses of the Iron Cross to these officers and was most complementary of their achievement.

Therefore, Student was surprised when, over coffee after dinner, Hitler quietly said to him:

> Of course, you know, General, that we shall never do another airborne operation. Crete proved that the days of parachute troops are over. The parachute arm is one that relies entirely on surprise.... The surprise factor has exhausted itself.

Thereafter, despite General Student's diligent efforts to the contrary, his forces were never again used for a major airborne operation. Later, the worsening war situation for Germany precluded such operations anyway. Paratroopers continued to be trained, but usually were committed to battle as regular infantry.

The Allies were profoundly impressed, as was Hitler, by the successful airborne offensive against Crete. Nevertheless, they drew conclusions very different from those of the Fuehrer. Crete provided further impetus to large-scale U.S. and British airborne expansion.

In Germany, regular airborne troops were a part of the air force. But in England, as well as in the United States, such units were usually part of the army. Nonetheless, other branches of service determined that they had need for some airborne capability. During the spring of 1940, the U.S. Marine Corps observed with interest beginning developments in the Army airborne program as well as what the Europeans had already accomplished in this area.

The Commandant of the Marine Corps, in May 1940, ordered a staff paper prepared that examined the potential use of Marine parachute troops. Marine interest was spurred on later that summer by two primary factors. One was the positive report made by several Marine officers concerning their inspection of the Army parachute training program and facilities. The other was the well-publicized combat use of German airborne units that spring.

Within a relatively short time, Major General Holcomb, the Commandant, decided that the Marines could use some airborne units as an adjunct to their usual assault role. Consequently, in October 1940, he authorized one battalion in each infantry regiment to be trained as "air troops." One company in this air battalion would be parachutists, while the others would be air landing. Given the small size of the U.S.M.C. in the fall of 1940, the total number of paratroopers required by the entire Corps was estimated at 750.

As with the Army, there was no lack of volunteers even though application standards were high and the wash-out rate was also high. In June 1941, jump pay was authorized for all qualified paratroopers on jump status, Army and Marine Corps alike. Officers received an additional $100 a month and enlisted men $50.

By mid-summer 1941, enough "paramarines," as they were often called, had been trained to start filling out airborne units. The 1st Parachute Battalion was constituted on 15 August 1941, commanded by Captain Marcellus J. Howard. The 1st Parachute Battalion shipped out from New River less than a year later, June 1942.

On 7 August 1942, American troops invaded the Japanese-held Solomon Islands. This was the first major U.S. island offensive in the Pacific war. While landings were occuring on Guadalcanal and Tulagi, the 1st Parachute Battalion went ashore at Gavutu. These were the first American parachute troops committed to action in World War II. Yet, their attack was launched from seaborne landing craft, not airplanes. The Solomon Islands were simply beyond the operational range of the transport planes. By nightfall of the first day, the paramarines had captured Gavutu.

The 1st Parachute Battalion was withdrawn from Guadalcanal on 18 September and shipped to New Caledonia for rest and refitting. They named their temporary base, outside Noumea, Camp Kiser. This was to honor Second Lieutenant Walter W. Kiser who had died in the assault on Gavutu. Kiser was the first U.S. paratroop officer killed in action during the Second World War.

Meanwhile, in October 1942, the Marine 2d Parachute Battalion was sent to New Zealand from San Diego. A few weeks earlier the 3d Parachute Battalion was formed. The Marine airborne program continued to grow. On 1 April 1943, the 1st Parachute Regiment, composed of the 1st, 2d and 3d Parachute Battalions, was created. The next day, the 4th Parachute Battalion was activated but, as events transpired, was never sent overseas.

The 1st Parachute Regiment continued to carry out a variety of combat missions until they were relieved in early January 1944. Shortly before that, on 30 December 1943, the Commandant of the Marine Corps had ordered disbandment of all Marine parachute forces. Such a course of action had been under consideration for several months, in view of operational experience gained in island fighting and manpower needs elsewhere.

That there were no combat jumps by Marines in World War II was due to the nature of the Pacific Theater of Operations. The lack of sufficient air transport and the long distances involved from staging areas to drop zones as well as the objectives themselves made paramarines a "luxury" the Corps believed it could ill afford. After all, jumping into heavily defended, relatively small areas with broken and overgrown terrain made, as the German experience on Crete had shown, for very high casualties. Moreover, the three thousand Marines in airborne units could, the Commandant thought, be better utilized elsewhere. They subsequently made up the cadre of the then-forming 5th Marine Division.

Even though they had not jumped into combat, Marine paratroopers had made noteworthy military contributions where ever they were engaged. Their high morale, professionalism and training were evident when, as members of the 5th Marine Division, they fought on Iwo Jima the following year. For example, a former platoon leader in the 3d Parachute Battalion, Robert H. Dunlap, received the Medal of Honor on Iwo. Another paramarine was Corporal Harlon H. Block. He was one of the six men who raised the second American flag on Mt. Suribachi, 23 February 1945. Of the six, three were killed in action on Iwo Jima. Corporal Block, a few days later, was killed northward of Suribachi while attacking Japanese positions along Nishi Ridge.

Before the end of 1941, and not long after creation of the 501st P.I.B., the War Department activated five more airborne battalions; three parachute and two air landing. The German victory on Crete, using just such units, provided impetus for the American expansion. Airborne appeared so promising that Bill Lee began advocating establishment of whole divisions even before the U.S. became a combatant. After the

Bougainville Island, 30 November 1943, U.S. Marine paratroop raiding party takes a short rest break during a lull in the fighting. (U.S. Marine Corps photo)

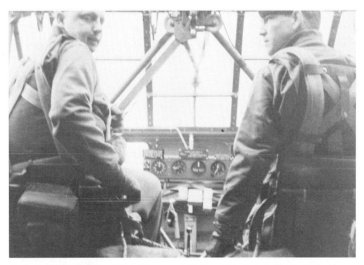

Cockpit of CG4A Glider. Pilots Henry and McDowell in their "office". Cottesmore, England in August 1944. (Courtesy of Douglas W. Wilmer)

Japanese attack on Pearl Harbor on 7 December 1941, and the declarations of war by Germany and Italy four days later, the U.S. began swiftly shifting to a war footing.

Airborne forces became an increasingly important factor in U.S. military calculations. Within two years, for example, glider forces had over 10,000 pilots trained by the Army Air Corps and more than 13,000 "Waco" (CG-4A) gliders had been contracted. The Waco carried thirteen fully-equipped men, in addition to a pilot and co-pilot.

For most of the war, glider infantry did not have any sort of distinctive uniform or insignia nor did they receive extra pay even though their airborne role was more dangerous than that of regular infantry. This state of affairs often led to animosity between glider and parachute troops. At one American camp a humorously bitter poster reflected this. The home-made sign had photographs of wrecked gliders along with the caption: "Join the Glider Troops! No Flight Pay, No Jump Pay, but Never a Dull Moment." It was not until shortly before the 1944 Normany Invasion that glider troops got extra pay comparable to that of the paratroops, and got their own distinctive qualification badge.

Glider troops had to aquire skills unique to their service. Specifically, they had to master loading and lashing equipment in gliders to that the combat load was at the center of gravity. Moreover, their physical conditioning was just as tough as that for paratroopers. For once on the ground in the airhead, their mission and need to sustain themselves in combat was the same as for the parachutist.

The evolution of the glider, like that of the parachute, is a complex story. Although there are mythical and semi-mythical stories about man's attempts to soar, it was not until the late nineteenth century that there was major work in glider development. Otto Lilienthal, a Prussian mechanical engineer, carefully observed birds in flight and attempted to translate the lessons he learned into a workable, man-sized glider. In 1891, he began glider flights near Berlin. Over the next five years Lilienthal's experiments enabled him to improve both design and gliding technique. He was banking his glider during a flight on 9 August 1896, when he stalled at an altitude of about fifty feet and was unable to recover. Severely injured in the resulting crash, he died the next day in a Berlin hospital. Lilienthal's last words were both stoic and prophetic: "Sacrifices must be made."

Advances continued to be made in gliding until Wilbur and Orville Wright achieved powered flight in late 1903. Thereafter, relatively little interest was paid to gliders. It was in the aftermath of World War I that a defeated Germany focused on gliders, because the Treaty of Versailles prohibited Germany from having military aircraft. In 1919, military aircraft meant only ones that were powered. The Germans actively sought ways to circumvent all the military restrictions imposed by that treaty. In the course of the 1920s, glider clubs sprang up all over Germany, some of which were secretly supported by military funds. Qualified glider pilots could readily make the transition to powered aircraft when Germany could again have military aviation.

It was not until Germany began openly to rearm after Hitler came to power in 1933, that consideration was given to the military potential of gliders. The assault and transport gliders developed by the Germans were considered combat aircraft and armed with one or more machine guns. The value of these gliders was graphically proven in the early days of World War II at such places as Eben Emael and Crete.

The British, in conjunction with their paratroop program, also began developing a glider arm. Their Horsa Glider carried thirty-two fully-equipped troops. The U.S. Marine Corps, just like the Army, began developing a glider capability in 1941. Marine glider efforts were terminated in June 1943. The nature of the island campaigns in the Pacific war had shown that the Marines had no operational use for gliders. The few training gliders that they had were turned over to the Army, as were their purchase contracts.

When the United States went to war in December 1941, it was involved in a conflict that was literally global in scope. Vast as its resources were, the U.S. still could not prosecute the war with equal vigor simultaneously throughout the world. Consequently, the decision was made in Washington, D. C. about how the military effort would be apportioned. True it was that Imperial Japan had attacked Pearl Harbor and was on the offensive throughout the Western Pacific. Yet, Japan was a poor country with limited resources and thus limited ability to sustain its huge offensive and hold what it conquered. Much blood and treasure would have to be expended over several years, but there was no serious doubt at top command levels that the United States would eventually defeat Japan.

Nazi Germany, however, was an entirely different case. Hitler's conquests between 1939 and the end of 1941 had added vast industrial and agricultural areas in addition to sources of raw materials and manpower to the Third Reich. Moreover, the Reich was not dependent on overseas colonies or suppliers for its needs because it occupied a continental heartland. A blockade was, therefore, largely irrelevant. These circumstances, coupled with varying types of alliances and economic arrangements with most of the rest of Europe, gave Germany the potential ability to wage war indefinitely.

Consequently, the decision was made in Washington, D.C. to give the war against Nazi Germany top priority in men and materiel. In the spring of 1942, even before the U.S. had undertaken major offensives in any theater of operations, the War Department began considering an invasion of German-occupied Europe in 1943. Lieutenant General Lesley J. McNair, Chief of Army Ground Forces, decided that an airborne infantry division would be useful in such an invasion. However, as first conceived, such a unit would simply be a regular infantry division with air transport elements attached.

McNair sent Brigadier General Bill Lee overseas to learn what developments were occuring in British airborne planning that would be beneficial to the U.S. Lee brought back word that their English allies were establishing specific airborne divisions. Impressed, General McNair authorized

In double tow - At Camp Mackall airfield, two CG-4A cargo gliders take off in double tow behind one C-47 airplane. Troop Carrier pilots did a great job for Airborne Troopers. (U.S. Army photo)

German glider club in 1937. Even after Germany had openly begun rearming, gliders remained an important part of military planning. (Bundesarchiv)

creation of the first two American airborne divisions on 15 August 1942. On that date, two regular ground units, the 82nd and 101st Infantry Divisions were designated airborne. At that time they were stationed at Camp Claiborne, Louisiana. But, within a few weeks, were transferred to Fort Bragg, North Carolina.

In the beginning, American airborne divisions were much smaller than the 14,253 men in a normal infantry division. The table of organization and equipment (TO&E) originally drawn up provided for a strength of only 8,505 men, but was essentially a regular infantry division in miniature, less most of its ground transport and heavy weapons. For example, the standard 105mm howitzers were replaced by light 75mm pack howitzers. However, it did have an anti-aircraft battalion. Airborne troops also had a lot more automatic weapons than regular infantry. Each airborne division had one parachute infantry regiment and two glider infantry regiments. Although Brigadier General Lee strongly opposed this ratio, the view of Lieutenant General McNair prevailed.

During the course of World War II, five airborne divisions were activated: the 11th, 13th, 17th, 82nd and 101st. Another ten divisions were designated. The 6th, 9th, 18th, 21st and 135th Airborne Divisions were designated but not constituted or activated in the active army. The 15th Airborne Division was designated and constituted but never activated in the active army. The 80th, 84th, 100th and 108th Airborne Divisions were eventually designated, constituted and activated in the army reserve shortly following World War II to ensure a viable airborne capability if needed in the event of mobilization.

The activated airborne divisions were considerably reinforced for combat operations with the attachment of various other airborne elements having special skills. Consequently, their operational strength was always considerably greater than the TO&E called for. Combat experience in North Africa and Europe showed the need for reconfiguring the airborne divisions. The regimental ratio was reversed to one glider and two parachute regiments and organic manpower increased to 13,906. Introduced on a test basis on 16 December 1944—the opening day of the Battle of the Bulge—all airborne divisions in Europe were permanently converted to the new TO&E by 1 March 1945. The divisions effected were the 13th, 17th, 82nd and 101st. The 11th Airborne Division was in the Pacific Theater of Operations and retained the old table of organization for the duration of the war. However, even in the 11th Airborne Division, unique conditions prevailed. The Division Commander, MG Joseph Swing, required all combat personnel to be dual qualified (parachute and glider) to the greatest extent possible. This requirement was met in most instances. By common usage, the glider regiments of the 11th Airborne Division (187th and 188th) came to be known as paraglider infantry.

Changing the ratio of glider and parachute infantry regiments within the airborne division had more to do with logistics than tactical concepts. There was a shortage of shipping space that made it difficult to get gliders overseas. In addition, a lack of both gliders and Troop Carrier Command aircraft to tow them for training exercises hampered deployment of glider units.

In the first part of 1942, the Americans and British had a major difference of opinion on the appropriate strategy for the global war in which they were allied against common enemies. The Anglo-American allies were agreed that they should make every effort to hold their vital interests in the Pacific and Far East. Nevertheless, both viewed Nazi Germany as their paramount enemy and thus their primary military effort would be against Hitler. Toward that end it was also essential that the Soviet Union not be knocked out of the war. Were the Soviets to utterly collapse, then Hitler would add vast new resources to his war machine as well as have huge numbers of

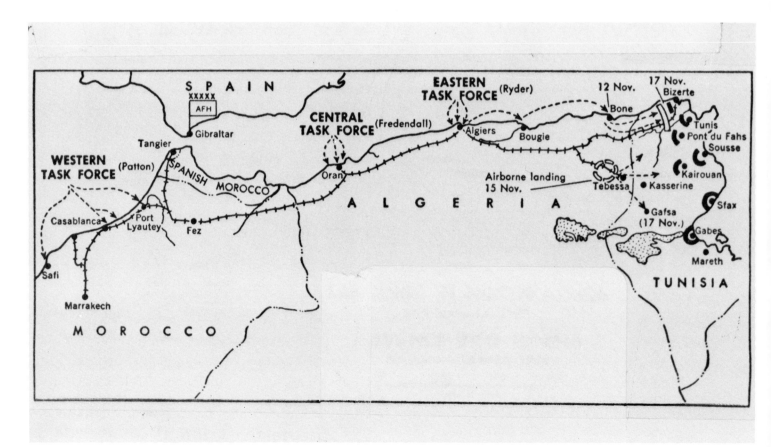

Map of North Africa showing the landings and initial operations. November-December 1943.

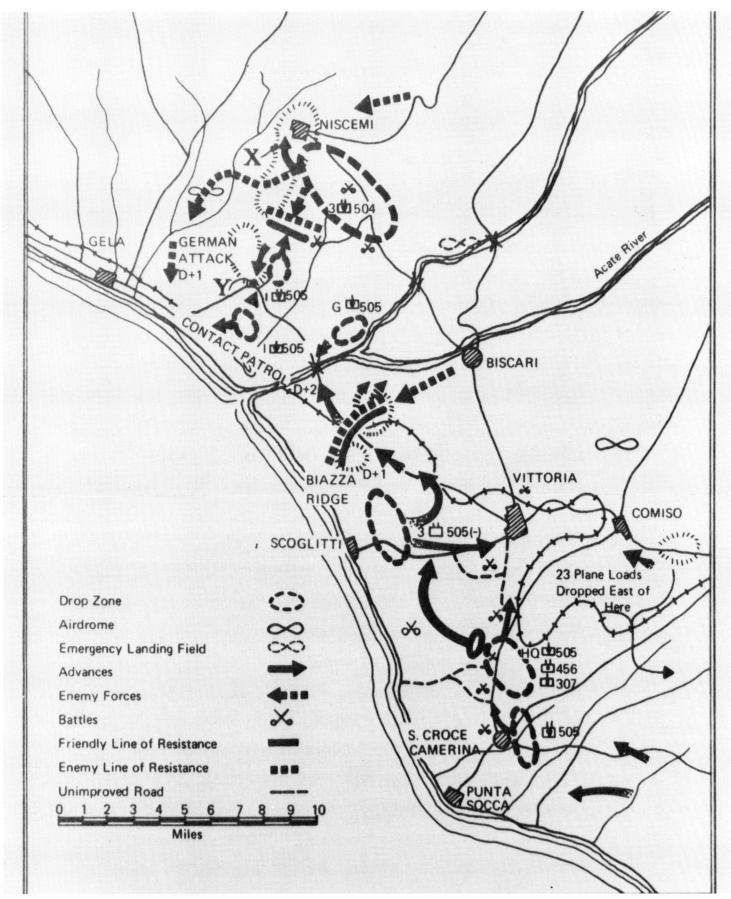

Invasion of Sicily. Map shows the main areas where Airborne troops landed. (Courtesy of Gen. James M. Gavin)

troops available for deployment against the Western Allies. Where Britain and the United States disagreed was on how and where to best strike against Germany.

Joseph Stalin, communist dictator of the USSR, wanted a second European front in 1942, in order to take some of the pressure off his country. The resulting Anglo-American decision was a compromise that especially displeased the hard-pressed Soviets. The US Chiefs of Staff had wanted the earliest possible invasion of Western Europe. Their British counterparts had opposed such a move, fearing that a premature invasion of the continent would be defeated. British fears were confirmed on 19 August when an Anglo-Canadian force of some 5,000 men supported by tanks conducted a major amphibious raid on the French coastal city of Dieppe. The attackers lost around 3,500 killed and wounded, numerous landing craft and twenty-eight tanks.

It was politically as well as militarily necessary to launch a major invasion of Axis territory before the end of 1942. President Franklin D. Roosevelt opted in favor of the British desire to strike, in Churchill's phrase, at the "soft underbelly," of Europe. Thus, from August through November 1942, planning was conducted for "Operation Torch"—the Allied invasion of North Africa. Such a landing would serve the dual purpose of helping to defeat Field Marshal Erwin Rommel and the German Africa Corps and to secure bases from which to operate against Axis air, sea and land forces in the Mediterranean basin. An ancillary goal was to destroy or capture the bulk of the Italian military forces that were heavily committed to the African campaign and, by so doing, encourage the downfall of Mussolini and force Italy from the Axis alliance.

Allied planning for "Torch" had not originally considered using airborne units as a part of the assault forces. Barely a month prior to the scheduled invasion date of 8 November, with the attitude of the Vichy French still uncertain, it was considered essential to secure the two French airfields, La Senia and Tafaroui, in western Algeria near Oran. That mission was assigned to the 509th Parachute Infantry Battalion, commanded by Lieutenant Colonel Edson D. Raff.

Raff's battalion had been sent to England in April 1942, to finish its airborne training with the British. Not only was there a traditional animosity between Great Britain and France, there had been occasions of armed hostility between them since the fall of France in June 1940. Consequently, Allied planners believed that Vichy French forces in North Africa would be less likely to resist Americans than British. Nevertheless, the American paratroopers would be prepared to fight.

On the evening of 7 November 1942, the 509th Battalion took off from England in thirty-nine C-47s for the fifteen hundred mile flight to their drop zone in Algeria. This turned out to be the longest single flight by American paratroopers in the Second World War. At dawn on the 8th, thirty-two of the thirty-seven planes were still in some sort of formation. This in itself was remarkable. Most of the pilots had never flown in a night formation and some of the navigators had joined the crews only a few days before.

As the C-47s flew across the Straights of Gibraltar they were supposed to receive a coded message that definitely told them what sort of reception they should expect from the French troops in Algeria. The radio signal was not received. Their first indication of what the French attitude might be came as they approached the La Senia airfield and found Allied planes bombing it.

The situation was confused. The paratroopers in twelve planes jumped between the two airfields, which were several miles apart. Critically low on fuel, the C-47s, with the rest of the battalion, landed on a dry lake bed several miles beyond La Senia. However, an American motorized column from the landing beach secured both airfields before the paratroops, now on foot, arrived. Despite this inauspicious beginning, Raff's men had shown dash and initiative. The star of airborne was in the ascendant.

Within in a few months the German Africa Corps was obviously doomed, caught between two major Allied armies, and planning progressed on the next invasion target—Sicily. Airborne troops had been employed for "Operation Torch" as a hasty afterthought. Since then, the number of airborne units had increased and their potential was better appreciated.

Capturing Sicily would at least improve the Allied naval position in the Mediterranean Sea and provide a springboard for operations against Italy. There was also the possibility that Italy, a very unhappy member of the Axis, would be knocked out of the war.

For "Operation Husky" the Allied commanders had decided to use their airborne forces against tactical objectives in order to facilitate the success of the seaborne assaults. George S. Patton, Jr., the US commander, decided to drop the 505th Parachute Infantry Regiment, 82nd Airborne Division, on the high ground behind Gela to shield the landing beaches from Axis counterattacks. In the early morning darkness of 10 July 1943, the paratroopers were en route to Sicily from North Africa.

Gale-force winds scattered the transport planes causing them to miss check points and lose both formation and orientation. They approached Sicily from all directions and, once crossing the coastline, received anti-aircraft fire. This futher scattered the transport planes and added to the confusion. Darkness and ground haze combined with smoke and dust from the preliminary naval bombardment of Sicily obscured the landmarks. Under those circumstances simply dropping the troops over land rather than the sea became a critical issue.

At about 0230 hours the paratroopers began jumping, just a scant fifteen minutes before the amphibious troops began coming ashore. High winds and the already dispersed transport planes caused the 505th to be scattered all over southeastern Sicily, far from their designated drop zones. Only one battalion landed relatively intact but it was about twenty-five miles from its target.

The men found themselves in a strange land that bore little or no resemblance to their maps. Colonel James Gavin, commander of the 505th, was at first not even sure they had landed in Sicily. In such a confused situation, it was individual initiative and daring, two of the hallmarks of airborne, that manifested themselves. The widely dispersed drop meant normal unit cohesion was impossible. The paratroopers coalesced into small groups and began moving toward the sound of the guns.

Moving toward the invasion beaches, the airborne troopers cut communication lines and engaged German and Italian troops where ever they encountered them. Now the Axis commanders were confused. Reports of airborne attacks were coming in from all over Sicily. Where was the focus of the main Allied invasion?

While most of the paratroopers headed toward the sounds of battle, some two hundred of their regiment had actually landed on the high ground on the landward side behind Gela. Possession of this high ground could quite literally spell success or failure for the seaborne assault in that sector. This small but determined group held off an Italian infantry attack until a battalion from the invasion beach could reinforce them. Together, they fought off a strong German assault later in the day.

Although the US 15th Army Group had established a beachhead on Sicily, they were still in danger of being driven back into the sea. The next day, 11 July, the German "Hermann Goering" Division and Italian "Livorno" Division launched a strong counterattack against the landing beach. Impressed by the stubborn American defense on the high ground on the first day, these two Axis divisions bypassed that area when they attacked. Driving to within about two thousand yards of the shore, the Italo-German offensive was stopped by naval gun fire and that from newly-landed tanks.

The Axis thrust had been stopped but the American hold

Map showing the Allied invasion of Italy in September 1943.

on Sicily was far from secure. To reinforce his position, Patton ordered the rest of the 82nd Airborne Division flown over that night from Tunis by an air convoy composed of 144 C-47s. There ensued a major tragedy of war.

The Allied invasion fleet had been informed that a large formation of C-47s would be passing overhead on their way to Sicily. However, two days of combat and a German air raid just forty-five minutes before the arrival of the friendly aircraft had made the naval anti-aircraft gunners, who were still at battle stations, very apprehensive. The first wave of C-47s flew over without incident. As the next flight approached, a single anti-aircraft gunner began firing. Within moments the whole fleet opened up.

Six C-47s were shot down before the paratroopers aboard could jump. Seventeen more were shot down as they flew back over the fleet after dropping their men. Thirty-seven aircraft were damaged by flak. In addition to materiel losses, eighty-one men were killed, sixteen were listed as missing and presumed dead, and 132 wounded.

Despite these losses to friendly fire, most of the paratroopers made it that tragic night. The following day Axis forces began a systematic withdrawal with the intention of evacuating Sicily after causing the Allies as much delay as possible. Before the end of the Sicilian Campaign, one of the Anglo-American goals was realized. Mussolini had been deposed, 26 July, and continued Italian membership in the Axis was now in question. By 17 August 1943, Sicily was entirely in Allied hands and the Mediterranean again open. German resistance had been both tenacious and professional while the Italians had shown little desire to fight.

Overall, Allied operations in Sicily had been skillfully conducted. The role of airborne was significant enough that even Hitler was impressed. The Fuehrer mused that perhaps he had been hasty in reducing his own airborne after Crete. The premier German paratroop general, Kurt Student, stated that Allied airborne troops played a decisive role in the battle for Sicily.

The American view was quite different. The Sicilian Campaign caused some top commanders to question the usefulness of keeping airborne divisions. General Dwight D. Eisenhower, Mediterranean Theater commander, wrote the US Army Chief of Staff, George C. Marshall that, "I do not believe in the airborne division."

Eisenhower went on to state that dropping an entire division would result in such troop dispersal that the divisional commander would never gain control. General McNair, Chief of Army Ground Forces, was even more negative than Eisenhower. McNair viewed the wide dispersal, confusion and high losses among the Allied airborne forces during the Sicilian operation as validating his earlier stand against division-sized airborne units. Both generals thought that smaller airborne formations, perhaps none larger than regiments, would be more useful.

Lieutenant General F. A. M. Browning, commanding officer of British airborne forces, strongly disagreed with the Americans. He saw the problem as one of coordination and control between the air transport and the airborne units. Browning argued that there was nothing inherently flawed with vertical envelopment as a tactic that made command and control impossible. Browning advocated creating an airborne army with its own troop transport planes. The US War Department established a special board to examine the issue. Its conclusion was not as far reaching as General Browning's proposal, but did advocate much more joint training between airborne and troop carrier units as a possible solution to the problems. However, the question of whether or not to have airborne divisions was still open.

On the other side of the world from Sicily, there was a combat operation that influenced the decision on the fate of airborne divisions. New Guinea was the scene of this action. Australian forces launched a three-pronged drive to capture the Japanese-held port of Lae. On 5 September 1943, the 503rd Parachute Infantry Regiment, supported by an Australian parachute artillery battalion, jumped in to seize the airstrip at Nadzab in the Markham River Valley. This was some twenty miles inland, northwest of Lae. They were to block the Japanese escape route.

The scale of opposition was not comparable to Sicily. Even so, the excellent coordination between airborne troops and airmen in this operation was most impressive. Fast attack bombers laid smoke to screen the drop zones while medium bombers pounded the landing area. Specially fitted heavy bombers dropped weapons and equipment to the paratroopers on the ground. Above all, swarms of friendly fighter planes flew cover. The Nadzab operation was a quick, complete success, and demonstrated what proponents of airborne had long maintained. Among those impressed was the Secretary of War, Henry L. Stimson. He instructed Army leaders to carefully study and evaluate the organizational arrangements of the airborne phase in the New Guinea operation.

Meanwhile, back in the Mediterranean, "Operation Avalanche" began on 9 September 1943. This was the US Fifth Army's amphibious assault on the Italian mainland in the Gulf of Salerno. The German defense was skillful and determined. By nightfall on the first day, the Allies held only four, narrow, unconnected beachheads.

Moreover, German reinforcements were quickly moving from other areas and concentrated nearly six divisions in the vicinity of the invasion beaches.

Santa Maria Capua Vetere, Italy. First Special Service Force Base from December 1943 - August 1944. This was a former Italian barracks that had been used by the elite German "Hermann Goering" Division. (Courtesy of Bill Story)

Since the Germans held the high ground, they were able to direct their artillery with devastating effect. Three days after the seaborne assault, the beachheads were still not securely consolidated. Then, on 12 September, the Germans launched a violent counterattack all along the lines, but it fell heavest on the American VI Corps. The Allied situation became desperate. They urgently needed reinforcements, but it would take several days to bring them in by sea. In some areas, the Americans were literally conducting a last-ditch defense with their backs to the sea. Eisenhower ordered the 82nd Airborne Division to drop paratroopers on the beachhead to strengthen the lines.

In order to prevent the tragedy of friendly fire like that at Sicily, anti-aircraft gunners on ship and shore were ordered not to fire at any aircraft, Allied or Axis, near the time of the drop. The first paratroopers jumped near midnight on the 12th. These were special teams of "pathfinders," whose mission was to mark the drop zones. They had portable radar sets to guide the C-47 pilots to the target. The pathfinders also formed a giant "T" of flaming oil drums to guide in the paratroopers. Before daylight on the 13th, thirteen hundred paratroopers jumped with great accuracy on the drop zone, while more came in after dark that same day. The German attack was halted after much bitter fighting.

Twenty miles beyond the beachhead was the town of Avellino. The 509th Parachute Infantry Battalion jumped near there on 14 September, to block roads that the Germans were using to support their operations against the Salerno beaches. Unfortunately, at night in mountainous country, one tiny village looked much like another. Also, pathfinders were not used because of the terrain. In the resulting confusion, the 509th was dropped over a hundred square miles. Its battalion commander landed in the middle of a German tank park and was swiftly made prisoner.

Always able to cope with apparent disaster, the American paratroopers came together in small groups and set about causing the Germans as much trouble as they possibly could. They attacked German troops where ever encountered, cut communication wires, destroyed bridges and any other militarily valuable targets that they could find. Of the 640 men who jumped on the Avellino mission, all but about a hundred eventually made their way back to Allied lines. Considering the circumstances, casualties were relatively light.

Back in the United States, the 11th and the 17th Airborne Divisions had been activated and plans were being made to create the 13th Airborne Division, even though there were still high-ranking skeptics who saw no need for it. One who especially continued to doubt the viability of airborne divisions was General McNair. He had begun considering that no airborne unit larger than a battalion should be established. Given his post as Army Chief of Ground Forces, McNair was in a key position to influence future airborne developments.

In December 1943, General McNair and Secretary of War Stimson were observers at a maneuver conducted by the 11th Airborne Division. Within the space of thirty-nine hours over twelve thousand men were delivered by parachute and glider from four widely separated airfields to thirteen different landing zones. By utilizing pathfinder techniques, only about fifteen percent of the airborne troops came down outside their designated drop zones. Although there were some problems, the observers, including General McNair, called the exercise an unqualified success. A few weeks later the 17th Airborne Division successfully staged a similar maneuver for high-ranking observers. McNair was finally convinced that airborne divisions were worthwhile.

Also in December 1944, the first black parachute test platoon was activated at Fort Benning, Georgia with eighteen volunteers from the 92nd Infantry Division. This unit was later expanded to become the 555th Parachute Infantry Battalion ("Triple Nickel"). Although not deployed into combat, the 555th helped the war effort. It was used by the US Forestry Service to fight fires in the virgin forests of the American northwest that resulted from Japanese incendiary balloons. Immediately after World War II, the 555th was integrated into the 82nd Airborne Division as the 3d Battalion, 505th Airborne Infantry Regiment.

Especially under the impetus of General Browning, the British airborne commander, Anglo-American efforts to improve coordination between airborne and troop carrier units

Anzio, 1944, two First Special Service Force troopers with a M1941 Johnson Light Machine Gun. Only the Force and Marine Raiders had the "Johnny Gun." (Courtesy of Bill Story)

and to correct deficiencies noted in the Sicilian Campaign went forward. Improved communication was a priority. Airborne troops were issued VHF radios so that they could maintain contact with the aircraft even after reaching the ground. One goal was to provide all transport aircraft and airborne units with both navigational and pathfinder radar. Most of the pathfinder techniques were joint efforts developed by the Americans and the British.

The procedure finally worked out was for pathfinder teams of ten men each to jump on the designated drop zone half an hour before the main airborne assault force arrived. That gave the pathfinders time to get oriented and set up their equipment but, hopefully, not for the enemy to have time to react to their presence.

Standard pathfinder markings were also established that were both simple and effective. For a parachute drop zone the marker was a "T" formed by five lights with a "Eureka" radar set at the top of the "T." A "Rebecca" beacon aboard each aircraft transmitted on the "Eureka's" frequency. The "Eureka" automatically responded to the signal, giving the aircraft a fix on its position relative to the "Eureka." The standard marker for a glider landing zone was a straight line of seven lights, with alternating colors, set out along the main axis of the landing area. A "Eureka" was positioned on the down-wind side of the lights. Colored smoke, rather than lights, could be used for daylight operations.

Besides developing and improving pathfinder techniques, closer coordination between airborne and air transport units was worked out. The US troop carrier wings in Great Britain were brought together in a single formation, the IX Troop Carrier Command. This was a component of the Ninth Air Force. Also in 1943, Lieutenant General Lewis H. Brereton assumed command of the Ninth Air Force. He appreciated the need for joint training between the airborne units and the aircraft squadrons assigned to deliver them on target. After all, Brereton, as a junior officer in 1918, had been ordered by Billy Mitchell to do the tactical planning for the proposed airborne assault on Metz. In November 1943, James Gavin was sent to England. Newly promoted to brigadier general, he was to deal directly with the Troop Carrier Command and the airborne units in order to facilitate joint training, command and control.

As 1943 ended, the Axis was still a formidable enemy but far less powerful than it had been at the beginning of the year. Italy had switched sides and German defeats in North Africa and Russia increasingly forced Hitler on the defensive. In the Far East, overextended Imperial Japanese forces suffered increasing losses that they could not made good. Ever-increasing numbers of American men and quantities of materiel poured into all theaters of operation. However, even though the tide of war was turning, Allied victory was not at hand. Much more blood and treasure had to be expended before Nazi Germany and Imperial Japan were finally vanquished.

As 1944 opened, the strategic situation in western Europe was obvious to both Allied and Axis leaders. Considering the build-up of forces and supplies in the British Isles, an Allied invasion of Fortress Europe, as the Fuehrer called it, was a foregone conclusion. When, where, and how this massive attack would come were questions about which Hitler could only speculate. The Allies, in order to keep Hitler uncertain, carried out an elaborate plan of deception.

On 12 February 1944, in one of the most succinct operations orders ever given, General Eisenhower was instructed: "You will enter the continent of Europe and, in conjunction

Prime Minister Winston Churchill inspecting the 101st Airborne Division in the early spring of 1944. He is escorted by General Maxwell D. Taylor, who had been named division commander in March 1944, after Bill Lee suffered a heart attack. (U.S. Army photo)

with the other United Nations, undertake operations aimed at the heart of Germany and the destruction of her armed forces." Set for the spring of 1944, planning for "Operation Overlord," as this massive assault was codenamed, went swiftly forward.

The ground work for an amphibious assault on the European continent had begun in 1940, even as British forces were being driven out. Prime Minister Winston Churchill was considering offensive operations even as Great Britain was fighting desperately to stave off defeat. From the beginning, serious consideration was given to using airborne troops as a component of the invasion force. In the late summer of 1942, the Soviet Union was under intense German pressure. The Western Allies had a contingency plan for an emergency invasion of France to draw German troops westward, should the Red Army start exhibiting symptoms of imminent collapse. Stalin was very displeased and complained ever-more strongly to his Anglo-American Allies when 1942 and 1943 slipped away without a cross-channel invasion. Although the grand invasion did not happen in either of those years, planning continued.

Even though the use of airborne forces in the invasion of France had always been envisioned, initial planning anticipated that their role would be modest. By the end of 1943, however, the decision had been made that a significant portion of the assault forces would be airborne.

Even though much expanded over earlier plans, General George C. Marshall thought the airborne segment of the invasion plans too conservative. Many, including Eisenhower, thought it extremely risky. Some predicted that airborne losses could run as high as eighty percent. Air Marshal Sir Trafford Leigh-Mallory, commander of the Allied Expedi-

These gliders and C-47 aircraft are lined up in preparation for the invasion of France. (U.S. Air Force photo)

Paratroopers of a 75mm howitzer section Battery C, 377th Fld. Art., 101st Abn. put parachutes on para-cases containing their 75mm howitzer. They are loaded into the bomb racks of the C-47 cargo plane. (U.S. Army photo)

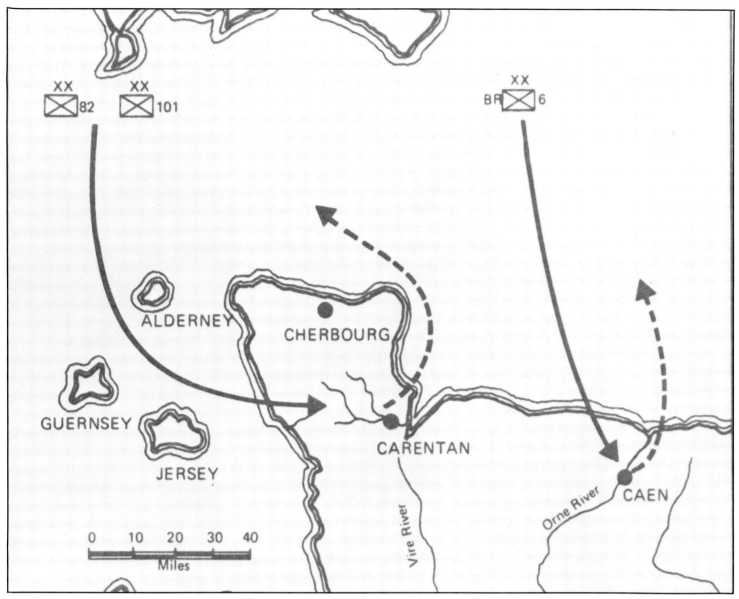

Map showing air routes used to deliver the three Allied airborne divisions on D-Day. (Courtesy of Gen. James M. Gavin)

tionary Air Force, was increasingly pessimistic. Shortly before D-Day he told Eisenhower that it would be imprudent to risk his troop carriers because their loss rate might be greater than fifty percent. Despite his apprehension, Eisenhower believed that the airborne role was indispensible and was thus prepared to accept extremely high losses.

As the fates would have it, neither of the founders of the Allied airborne were destined to take part in the Normandy Invasion. In October 1942, the organizer of the British airborne forces, John Rock, had been killed in a glider crash. Bill Lee was in England to command the American 101st Airborne Division. However, in February 1944, he suffered a heart attack and had to return to the United States. General Maxwell Taylor, former artillery commander of the 82nd Airborne Division, replaced the stricken Lee. Out of respect for Lee as the founder of US airborne, Taylor urged his paratroopers to shout "Bill Lee" as they jumped into combat.

The pilots of the IX Troop Carrier Command were more experienced than those of previous airborne operations. They had taken part in various airborne maneuvers during the spring of 1944, but demands for air transporting freight were still heavy and this limited the number of planes and pilots available for training. Constant shifting from airborne training with day and night formation flying to air freight missions was hard on the pilots. Nonetheless, despite difficulties, it became apparent that coordination between airborne and carrier units was improving. The "sword of silk" was aquiring a keener edge.

All invasions have a D-Day. But after 6 June 1944, D-Day used by itself meant the Allied landings in German-occupied France. Preceeded by three airborne divisions, this was the greatest amphibious assault of all times. The IX Airborne Troop Carrier Command Pathfider Group dropped 82nd and 101st Airborne Divisions pathfinder units on designated DZs at 0006 hours on 6 June, about thirty minutes in advance of the Troop Carrier armada that followed. Some 176,000 troops and their equipment was transported in about 4,000 ships and landing craft. Six hundred warships protected the invasion armada at sea level while the skies above were patrolled by around 7,000 fighter-bombers and fighters. The Luftwaffe was swept from the skies and the German navy from the path of the Allied fleet. By nightfall, five Allied divisions were ashore and a tenable hold established on the invasion beaches. The liberation of the continent had begun, but many airborne troopers paid in blood for that beginning.

The Allies knew that when their three airborne divisions landed in France the Germans would realize that the long-awaited invasion was at hand. Consequently, it was important to deceive the enemy as to the true landing beaches. Using dummy paratroopers rigged with noisemakers to mimic small-

Pictured on the ground where they came to rest near a road in Normandy, these gliders of the IX Troop Carrier Command were the first to land glider infantry on French soil. (U.S. Air Force photo)

arms fire, the Allies simulated airborne attacks of up to brigade strength. The Pas de Calais area was the narrowest part of the English Channel, thus an obvious invasion target. In order to draw German attention north, away from the genuine targets, one such fake-paratroop drop was made near Le Havre, some twenty miles inland. To further enhance the illusion of Allied intentions against the Pas de Calais area, Anglo-American sea and air forces operated aggressively in the same vicinity. Directly across the channel, meanwhile, heavy radio traffic simulated the presence of another Allied combat command large enough to execute an invasion of France. General Patton was the commander of this phantom army that included two phantom airborne divisions, the 6th and the 9th. Hitler and Rommel, his commander charged with defending the invasion front, both anticipated the main invasion effort would be at the Pas de Calais.

Allied pathfinders jumped into the night sky over Normandy shortly after midnight, 6 June 1944. They had only an hour to mark the drop zones before the main paratroop forces arrived overhead. The course followed by the Troop Carrier Command planes en route from England required several turns, thus increasing the potential for pilot error. Once they crossed the coast, navigation errors were made as a result of enemy anti-aircraft fire and low cloud cover. Many pathfinders landed nowhere near their assigned targets. The same problems that beset the pathfinder drops also effected the main body. Dispersion was so great that some units were as much as thirty-five miles off target.

Between the deliberate Allied deceptions and the unintentional dispersion of the paratroopers, the German commanders were very uncertain. Indeed, early reports, many conflicting, of paratroop drops began to pour into the various enemy headquarters in such volume that it was initially impossible for the Germans to determine any sort of pattern to the landings. Unable to form any sort of clear picture of what was going on, the Germans sent out patrols and waited. In some cases paratroopers were captured. One pathfider team for the British 6th Airborne Division landed right in front of the headquarters building of the German 711th Infantry Division and were immediately taken prisoner. The paratroopers did not reveal their mission and the German divisional commander could only speculate.

In general, the mission of all three Allied airborne divisons was to shield the invasion beaches by preventing the Germans from bringing up reinforcements. Specifically, the 82nd was to establish an airhead and hold it against all counterattacks as well as to seize bridgeheads over the Merderet River to facilitate an eventual attack westward. The 101st was to capture the area behind Utah Beach and to silence a heavy coastal battery at Varreville. This division was also to seize causeways over flooded areas so that the 4th Infantry Division, landing by sea, could advance. East of the Americans, the British 6th Airborne Division was to capture two vital bridges over the Orne River intact and to destroy all other bridges in their area.

By about 0330 hours, all the paratroopers were on the ground. Despite all the confusion and wide dispersal, they began forming up and organizing their positions. Anticipating the difficulty of telling friend from foe in the darkness, each

This is the glider which carried BG Don F. Pratt to his death at 0400 hours on D-Day morning. Co-pilot also lost his life. Pilot was LTC Mike Murphy. Another passenger was aide to Gen. Pratt, LT Lee Mays. Note that the glider was named "Fighting Falcon" and was purchased for Armed Forces by War Stamp and War Bond contributions of the school of Greenville, Michigan. (U.S. Army photo)

D-Day: Crashed glider near Ste. Mere Eglise. Dead are from 82nd Airborne Division. (U.S. Army photo)

paratrooper was issued a toy metal "cricket." Metallic clicks helped the troops sort things out.

The paratroopers had seized an airhead and awaited gliderborne reinforcements. At roughly 0400 hours the gliders began landing. The hedgerows of Normandy cut up much of the region into a patch-work of small fields, and caused many gliders to crash upon landing. One of those killed in a glider crash was Brigadier General Don Pratt, assistant division commander of the 101st. Fewer than half the gliders came down in their assigned landing zones. The Germans had flooded much of the area and rigged a variety of obstacles in the larger open fields as a defense against airborne troops. Many men died upon landing, killed in crashes or drowned.

Nevertheless, in spite of the great confusion and difficulties, the airborne attackers were able to carry out many of their assigned missions. The 82nd captured St. Mere Eglise, the first town liberated, and raised above its smoking ruins the same American flag that they had flown over Naples. By dawn they had cut the St. Mere Eglise-Cherbourg main highway and secured crossings over the Douve and Merderet Rivers, in addition to securing the area inland from Utah Beach.

While the 82nd prepared to hold against expected German attacks from the north; fifty miles eastward, the British 6th Airborne Division had fulfilled their initial combat assignments and occupied vital defensive positions on the eastern flank of the beachhead codenamed "Sword." By 0530 hours, the amphibious invasion of the continent was underway.

All three Allied airborne divisions were reinforced by more glider troops near dusk. The twenty-four hours that was D-Day witnessed, among other things, the largest airborne assault the Allies had ever staged. Over 17,000 airborne troops of all types were delivered to France on that day by 2,435 powered aircraft and 867 gliders. On D-Day airborne casualties were fifteen percent, which was far less than the possible eighty percent projected by some.

As night approached, neither commanding general of the 82nd nor the 101st Airborne had real control of their respective divisions. The disperal and overall confusion precluded truly effective command and control from the top. It was under these circumstances, however, that the high state of individual training, initiative and esprit de corps of the airborne paid off. More often than not, individual paratroopers banded together when they could not find their units, and carried the war to the enemy. They realized that it was crucial to disrupt German communications and, if not halt, at least delay enemy troop movements. This led to incidents of a few paratroopers attacking a whole convoy. If the enemy halted and deployed for battle, it did not matter that there were only a handful of Americans that were quickly killed or driven off. What did matter was that the convoy had lost time, time that could not be reclaimed.

There were hundreds of individual acts of heroism in countless small, fierce engagements. The disadvantages of the airborne were counterbalanced by German confusion as to the true situation. The aggressiveness of the airborne troops compensated in some ways for their lack of numbers and added to the enemy's uncertainty.

There is no point in belaboring everything that went wrong with the airborne phase of the Normandy Invasion. Rather, what should be stressed was that these Allied airborne troops proved again and again that they were worthy of their elite status. They went into battle knowing the tasks to be done.

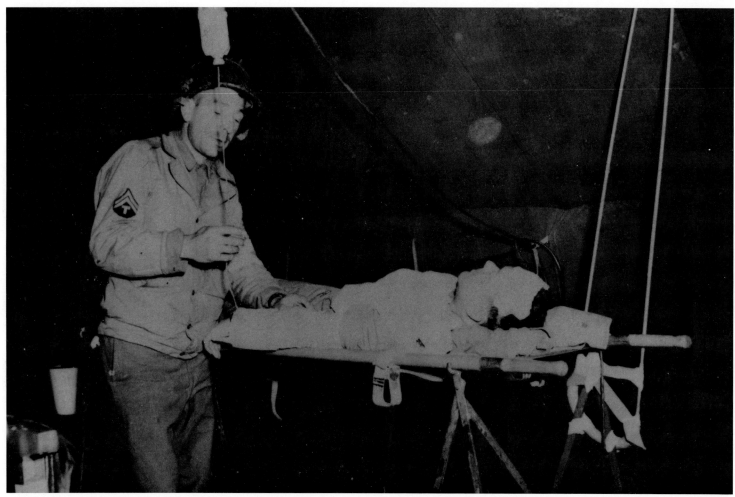

An aidman from the 326th Airborne Medical Company administers blood plasma to a wounded paratrooper shortly after the invasion of Normandy. (U.S. Army photo)

LST type landing craft bring the 426th Airborne Quartermaster Company and other supporting elements of the Division into Omaha Beach on the morning of 8 June 1944. (U.S. Army photo)

D-Day: British Horsa Gliders on Landing Zone "N" (Imperial War Museum photo)

American glider pilots speed back to England aboard a troopship. Among the first to strike during the successful landings in France, 6 June 1944, they fought their way back through German-held territory to the beachhead to catch a boat. They are going back to fly more gliders into action. L to R: Lieutenant Charles Dellington, Flight Officer Joe Gilreath, Flight Officer Kenneth Ensor. (Courtesy of 101st Airborne Division Association)

Where they did not have command and control on the ground, they showed consumate initiative and skill at arms. In short, the airborne assault on D-Day played a significant role in bringing about the overall success of "Operation Overlord."

Once commited to action, they were not withdrawn after they had served the purpose for which they were trained and equipped. The airborne troops fought on as regular infantry. The 82nd was pulled out of the line thirty-two days after D-Day and the 101st thirty-four. For eighty-three days, the British 6th Airborne Division stayed in the line.

This highlighted a problem that was never satisfactorily resolved during the Second World War. American and British doctrine called for airborne troops to be withdrawn from the battleline as soon as possible after link up with ground forces. However, the doctrine was not generally followed. Once in contact with the enemy, hard-pressed ground commanders were understandably reluctant to release the airborne troops even after their specific missions had been accomplished.

The majority of casualties in these units were not sustained in the airborne-specific stages of a combat assault, but during their long commitment afterwards as regular infantry. This problem was also encountered by other elite units such as the American Rangers and British Commandos. Yet, in emergency situations elite troops should certainly be used as line infantry. The fact that they were light formations and very mobile made them especially suited for quick deployment at critical spots. In such situations it was perfectly justified to use them. However, there were too many times that airborne troops were kept in the line because they were available, not because their special military expertise was needed.

The stark reality was that airborne troops were above average in mental and physical abilities in addition to which they received a lot of highly specialized training and equipment. This meant that it was more difficult to replace casualties in airborne than in line units. So the question was, why spend a lot more time and expense to select and train an elite soldier if he were to be used mostly in a role that did not require specialized training and equipment. This question was never satisfactorily addressed.

Another issue was what to do with the glider pilots after they had landed. British glider pilots had been given basic infantry training and, after landing, assembled themselves into units to support the rest of the troops. On the other hand, American glider pilots were given no particular mission after landing. They could voluntarily attach themselves to the units they brought in or simply stay low until such time as contact was established with ground troops and they could be evacuated.

Although the invasion was a great success, fighting in the hedgerow country of Normandy was costly and slow until the breakout from the beachhead. Preceeded by massive carpet bombing, Allied forces launched a fierce offensive against German forces in the vicinity of St. Lo. "Operation Cobra," as it was called, broke through the enemy lines. George Patton's US Third Army then exploited the breach. Logistical support of the widening and ever longer Allied front became increasingly difficult. This was anticipated and another Allied invasion of France had been under discussion for many months.

"Operation Dragoon," formally called "Anvil," was an invasion of the French Mediterranean coast. There were two main objectives. First, capture Marseilles in order to provide another major port through which to supply the Allied armies in France, and, secondly, to secure their increasingly longer southern flank. A provisional airborne division was assembled to participate in "Dragoon."

Designated the First Airborne Task Force, this was an Anglo-American unit. In the pre-dawn darkness of 15 August 1944, the approximately 10,000 men of the provisional airborne division took off from ten different airfields in Italy. They were transported to the invasion area in 465 gliders and 535 troop-carrier aircraft. Some of the American elements of this composite formation, such as signal, anti-tank and engineer, were regular ground units. They had been given a quick course in glider operations at an ad hoc glider training school set up near Rome. Their flight across the Mediterranean was uneventful.

The landing zone of the 1st Airborne Task Force was in the vicinity of Le Muy, about fifteen miles inland. Their mission was to seize the communication centers in the area and to prevent the Germans from moving reinforcements to the invasion beaches, which extended from St. Tropez almost to Cannes. The German coastal defenses were similar to those in Normandy, but there were fewer defenders to man them.

Just as the Allies had tried to deceive the enemy as to their true intentions at Normandy, they also employed deception for "Dragoon." They conducted naval demonstrations between Cannes and Nice as well as between Marseilles and Toulon. West of Toulon five hundred dummy paratroopers with noise-makers to simulate small-arms fire, just as were used on 6 June 1944, were dropped.

When the airborne assault force crossed the coastline, they discovered thick ground fog. Only three of the ten pathfinder teams that preceeded the main drop had come down on the correct drop zones. The main drop was dispersed, but only fifteen percent of the airborne troops failed to land on or near their assigned zones. Fortunately, German forces in southern France were far weaker than Allied intelligence had estimated. As it turned out, "Operation Dragoon" came at a time when the enemy was already considering withdrawal from the region.

After the paratroopers landed, they began clearing anti-landing obstacles from the fields for the gliders that followed. Many, however, were still in place when the gliders came in. A large number of gliders wrecked upon landing, accounting for most of the 240 airborne casualties that day. Some landed around a German corps headquarters, further hindering coordination of the enemy's defense efforts. Luckily, no aircraft were lost to German fire that day.

Not only was the airborne assault very successful, the main amphibious landings also went very well. Three days after the invasion, Hitler ordered a major withdrawal. By that time, ground forces from the invasion beaches had already passed beyond the airhead some fifteen miles inland.

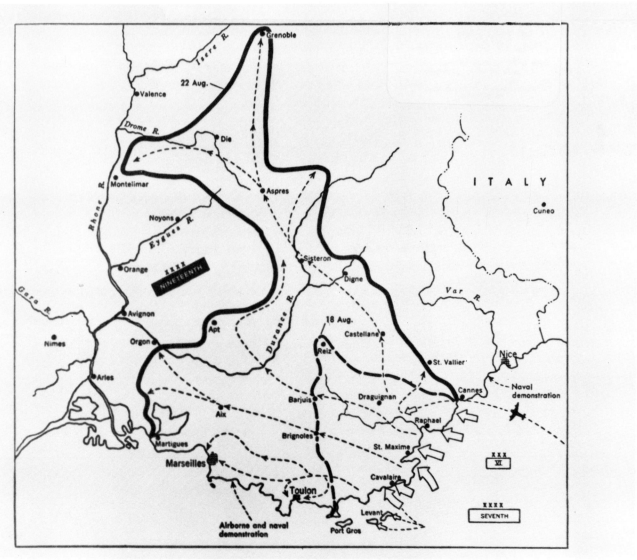

Map of "Operation Dragoon" and the campaign in southern France.

On the evening of 15 August 1944, the 551st Parachute Infantry Battalion making a combat jump behind the French Riviera during "Operation Dragoon." (U.S. Army photo)

15 August 1944. Elements of the 550th Glider Infantry Battalion shortly after landing in southern France. (U.S. Army photo)

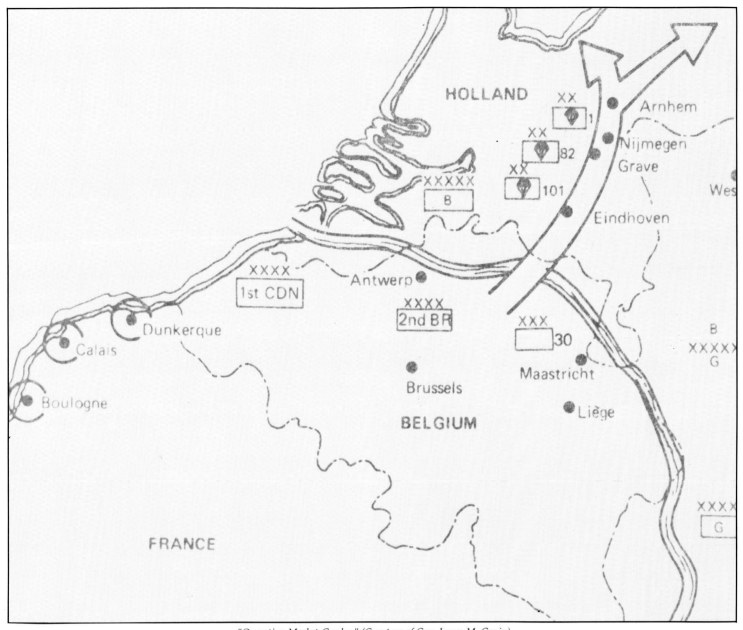
"Operation Market-Garden" (Courtesy of Gen. James M. Gavin)

"Dragoon" exceeded all command expectations and proved to be an important victory. The invasion force linked up with Eisenhower's right flank two months ahead of schedule. All German military presence in the south of France had been eliminated and several ports in addition to the big one at Marseilles were available to support the drive against the German fatherland. This airborne operation, so effectively carried out, was a significant element in the overall success of the invasion.

In "Overlord" Allied airborne troops had certainly demonstrated their value. Although there had been some serious problems, such as wide dipsersal, many leaders were confident that there was nothing additional training and coordination could not remedy. Eisenhower, the Supreme Allied Commander in Europe, was obviously pleased and impressed. In July, he ordered that an army of airborne troops and the aircraft necesary to carry them into battle be created. He designated Lieutenant General Lewis H. Brereton its commander. Brereton had long supported the airborne concept and had successfully lead the 9th Air Force for several months prior to his new assignment. A short time later in August, the 1st Allied Airborne Army was established.

After the Normandy breakout, however, the pace of Allied advances was such that an additional airborne attack was not necessary. Even so, contingency planning for as many as ten potential airborne operations was underway. For example, at the beginning of September 1944, Eisenhower asked Bradley if he wanted to launch an airborne operation in the Maastricht-Aachen area to help breach the Siegfried Line. Bradley replied that he did not want airborne troops but wanted to keep their transport aircraft carrying the fuel to the fast-moving US 3rd Army.

Bradley's answer highlighted a major problem confronting the new 1st Allied Airborne Army. Their aircraft were increasingly taken from them to transport supplies to sustain the fast-moving front. This frustrated Brereton's plans for intensified training in air navigation as well as joint training for air and ground personnel. In addition, the feeling that a valuable weapon, the airborne, was not being properly utilized was shared by airborne troops and generals alike.

Airborne forces got back into the war on 10 September 1944, when British Field Marshal Bernard Montgomery met with his boss, General Eisenhower, and proposed a daring operation. Mongomery proposed dropping three airborne divisions behind German lines in order to seize bridges across the Rivers Maas, Waal (Rhine) and Lek (lower Rhine). Led by

17 September 1944. Part of the Allied air armada en route to Holland. (Courtesy of the 101st Airborne Division Association)

A glider unloading a jeep near Grave, Holland, 23 September 1944. Note broken right wing. (Courtesy of Douglas W. Wilmer)

Late in the morning of 17 September, 1,545 troop carrier aircraft and 478 gliders took off from twenty-two airfields in England on their way to landing zones in Holland. Over twelve hundred fighters escorted the air armada. The evening before, about one thousand bombers had attacked German anti-aircraft emplacements along the flight paths and LZs. The Allies achieved complete tactical surprise, and at about two in the afternoon, troops began landing. Within a few minutes, some 3,500 glider and 16,500 parchute troops were on the ground. Most of them came down in the designated places.

The 101st, in the most southerly drop zone, landed near Eindhoven and took four of the assigned five bridges intact. The middle stepping stone, the 82nd, hit the ground near Nijmegen, taking the high ground and the bridge at Grave. At Arnhem, in the most northerly drop zone, the British 1st had been able to capture only one end of the road bridge.

Although the Germans had been surprised, they were better organized and in much greater strength than Allied military intelligence had estimated. What is more, thanks to poor security on the Allied side, the Germans captured a complete copy of the operational plans and so knew exactly what Allied intentions were.

At 1425 hours the ground phase of the attack opened. XXX Corps artillery opened a rolling barrage a mile wide and five miles deep. It was oriented on the only major road on their route of advance. This hard-surface road, flanked by swampy terrain that hindered movement, was soon christened "Hell's Highway." It proved an apt description.

The next day, 18 September, bad weather prevented reinforcements and supplies from being flown into the three airheads. For example, it was not until Thursday evening, 21 September, that the Polish Airborne Brigade was able to jump in an attempt to reinforce the British. Such delays were critical setbacks to the Allies. Despite gains by the 82nd and 101st, the

XXX Corps, the British 2nd Army would destroy or push aside all enemy resistance and advance the length of this "airborne carpet." Once across the Rhine, Allied forces could turn the northern flank of the Siegfried Line. Then they could drive south and southeast to the industrial centers of the Ruhr as well as head for Berlin across the north German plain.

Montgomery's uncharacteristically bold plan had much to commend it. In addition to the primary mission of seizing a bridgehead across the Rhine, this operation would free the Scheldt estuary and enable the Allies to use Antwerp, the third largest port in the world, to support their push into Germany. Moreover, Allied troops would capture the German V2 missile sites and thus end their attacks on England. If successful, Montgomery's thrust would shorten the war considerably. Eisenhower approved "Operation Market Garden."

The airborne part of this grand operation would initially employ the American 82nd, 101st and British 1st Airborne Divisions to seize and hold key bridges over rivers and canals, then await link up from the attacking Allied ground forces. These divisions would be stepping stones on the path to the other side of the Rhine.

The Germans expected an Allied offensive to clear the approaches to Antwerp. They had considered but dismissed the possibility of a thrust toward Arnhem. General Student, father of German airborne forces, made an off-hand comment that the Allies might try a vertical envelopment in that area. Field Marshal Model replied that "Montgomery is a very cautious general, not inclined to plunge into mad adventures."

Kamiri Airstrip, Noemfoor Island. Sergeant Alton W. Davis, 503rd Parachute Infantry Regiment secures his parachute after landing on 3 July 1944. (U.S. Army photo)

Near Grave, Holland, 23 September 1944, just unloaded from a glider, this jeep is on its way to rescue a glider that had "nosed" into a nearby dyke and had trapped the personnel inside. The jeep shoved the glider's tail around, freeing the nose for unloading - then the jeep hooked up to the trailer aboard the vehicle, with troops clinging to it, joined a departing convoy. About that time the area came under German 88mm artillery fire. (Courtesy of Douglas W. Wilmer)

situation for the British 1st grew increasingly precarious. "Operation Market Garden" anticipated that the 1st Airborne Division would have to hold for only two days before elements of XXX Corps reached them. Events did not coincide with Allied plans.

After ten days of bitter fighting against a tenacious and numerous foe, the 1st Airborne was ordered to withdraw. About 2,200 crossed the Lek in assault boats on the night of 25-26 September. They left behind some 7,000 of their men dead or prisoners. Many of the prisoners were wounded.

The rest of "Operation Market Garden" had gone rather well. The drive up Hell's Highway had come within ten miles of Arnhem. The 82nd and 101st had fulfilled their missions. The stand of the British 1st Airborne Division was gallant but, since the necessary support could not be brought in, doomed.

Although initially taken off guard, the German reaction to "Market Garden" had been swift, skillful and effective. They had lost much in men and materiel but had prevented the Allies from crossing the Rhine and turning their flank. It would be seven more months before Montgomery's troops were again in Arnhem. The degree of success or failure of "Operation Market Garden" was and still is controversial.

A world away, airborne units in the Asiatic-Pacific Theater of Operations were used more there during 1944 than previously. On 2 July 1944, American forces invaded Noemfoor Island, which is near New Guinea. The next day the 503rd Parachute Regimental Combat Team jumped in to seize the island's airstrip. At Noemfoor, jump casualties were unusually high. The 503rd discovered that the airstrip had just been overrun by US seaborne forces. A fast-moving front was often a deterrent to airborne operations in Europe as well. Too often the objective was overrun before before an airborne operation could be mounted.

Also that summer, the 11th Airborne Division arrived in the western Pacific. It took part in the fighting on Leyte, in the Philippines. This invasion had begun on 20 October. However, it was not until some weeks later that they went into action. They were to relieve the 7th Infantry Division on Bito Beach. On 29 November, three Japanese planes crashed on or near the San Pablo airstrip. The bodies of enemy paratroopers were found among the wreckage. Apparently, they had intended to raid the airfield. However, those who were not killed in the

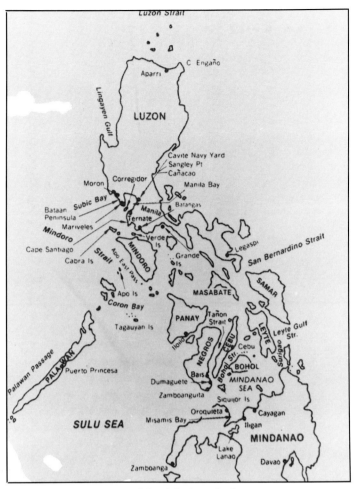

Map of the Philippines Islands.

crashes, died fighting or disappeared into the jungle without doing any particular damage.

Jumping without pathfinders, a much larger Japanese attack came at dusk on 6 December. Some three hundred Japanese paratroopers jumped in an attempt to capture San Pablo and another airstrip. They destroyed some aircraft on the ground and did some other damage to the facilities. A seaborne Japanese invasion was supposed to link up with the paratroopers. However, it never materialized. The next day, the 187th Glider Infantry Regiment of the 11th Aiborne Division counterattacked and drove the enemy from the one airfield they had managed to capture. When the fighting ended, the Japanese paratroopers had died to a man. Thereafter, the 11th was part of the American force that slowly pushed its way westward through the central mountains of Leyte.

United States forces were now firmly lodged in the Philippine Islands, but much hard fighting remained before they were liberated. MacArthur's next target was northward, the main island of Luzon. On 9 January 1945, four US army divisions landed at Lingayen, north of Manila. At the end of the month, airborne units went into action during the final stages of the drive on Manila.

In conjunction with amphibious landings north and south of Manila, the two glider infantry regiments of the 11th Airborne Division landed at Nasugbu, southwest of the Philippine capital, on 31 January. They swept aside Japanese opposition. Three days later, the 511th Parachute Infantry Regimental Combat Team jumped onto Tagaytay Ridge, on the northern edge of Lake Taal. The 511th overcame determined Japanese resistance, linked up with the 187th and 188th Glider Infantry Regiments, and advanced rapidly north toward Manila along Highway 17. That same evening, American forces entered the city. Not long afterwards, the glidermen and paratroopers of the 11th were caught up in vicious street fighting. It was not until 3 March that organized enemy resistance was overcome in Manila, now a city of rubble.

There were still strong Japanese forces in northern Luzon. MacArthur decided to clear Manila Bay so that he could use the port to support further operations. To do this required capturing Bataan and Corregidor. Bataan was assaulted by conventional means. However, the terrain of Corregidor did not lend itself to a regular amphibious attack.

The Gibraltar of the East or "the Rock," as Corregidor was variously called, was a tadpole-shaped rocky island in Manila Bay. Three-and-a-half miles long by one and a half at its widest point, much of its shoreline was cliffs. There was an abandoned airfield on the "tail" of the tadpole which was the best drop zone on the island. Even so, it was not selected because it was a lower elevation than the rest of Corregidor. Troops landing there would be exposed to enemy fire.

The only other drop zone was the former parade ground and golf course on the "head" of the tadpole. However, it was about a half-mile long by a quarter-mile wide and surrounded by bombed-out buildings. It was also close to the cliffs. Without even considering enemy resistance, simply jumping on to Corregidor was very hazardous.

The 503rd Parachute Infantry Regimental Combat Team was assigned the mission of assaulting the Rock. Staff planners estimated that the paratroopers would suffer a jump casualty rate of twenty percent. Colonel George M. Jones, the regimental commander, thought it would be closer to fifty percent. A factor that influenced the Americans to accept this level of risk was the intelligence estimate that the enemy garrison on Cor-

Leyte, Philippine Islands, January 1945. Five paratroopers of Company E, 511th Parachute Infantry Regiment, pose with Japanese flag captured during their trek through the mountains of Leyte. Seated (L to R): Neil Rutherford, Jack Lafone. Standing (L to R): Johnson Cail, Eugene Barrett, Marivin Samples.(11th Airborne Division Association photo)

Private James Harper, a member of Company G, 187th Regiment, rests beside a dud para-bomb during a lull in the fighting in Luzon, 9 April 1945. (11th Airborne Division Association photo)

A stick of paratroopers, led by Sergeant Albert Baldwin, prepares to bail out over Corregidor. (U.S. Army photo)

American paratroopers landing on Corregidor's Topside. Other troopers have already discarded their chutes. (U.S. Army photo)

Near Appari, Luzon, 23 June 1945. Six of seven gliders used by Gypsy Task Force and discarded parachutes on Camalaniugan Airstrip. (U.S. Air Force photo)

regidor was only about 850. As it turned out, there were more than 5,000 Imperial Japanese troops on the Rock.

On 16 February 1945, paratroopers began dropping on Corregidor. Some two hours later, elements of the 38th Infantry Division made an amphibious assault on the south shore of the Rock. Complete tactical surpise was achieved over the Japanese. Drop losses were twenty-five percent plus combat casualties. Nevertheless, the assault was completely successful. Still, it took twelve days of bitter fighting to secure the island.

The day after the 503rd jumped on Corregidor, another element of the 11th Airborne Division participated in a daring combined amphibious-ground-parachute operation. They swiftly moved to take the infamous Los Banos internment camp near Manila. The Americans feared that the Japanese might kill the approximately 2,300 Allied prisoners rather than permit them to be liberated. This lightning strike was a total success.

The last operation of the 11th Airborne Division was carried out on 23 June 1945, in northern Luzon. To assist the 37th Infantry Division, the 511th Parachute Infantry Regiment jumped on Camalaniugan Airfield south of Aparri. The jump was successful and, three days later, the 511th P.I.R. made contact with elements of the 37th I.D.

In Europe, meantime, there was no lack of proposed airborne operations. Yet, a number of factors precluded staging another major airborne offensive after "Operation Market Garden." The Battle of the Bulge, which began early on 16 December 1944, involved the 17th, 82nd and 101st Airborne Divisions in heavy fighting. However, they were committed to battle as regular infantry.

The only American airborne operation during this period took place on 23 December when twenty pathfinders jumped

Shortly after jumping onto The Rock, paratroop Sergeant Frank Arrige and Private First Class Clyde Bates unfurl Old Glory. (U.S. Army photo)

Northern Luzon, 23 June 1945. Parachute and glider troops assemble on the edge of Camalaniugan Airfield south of Aparri. (11th Airborne Division Association photo)

Members of the 101st Airborne Division pathfinder team set up radio equipment on a brick pile just outside Bastogne, Belgium to help guide in the aerial resupply planes. (U.S. Army photo)

C-47s over Wesel, Germany during "Operation Varsity." (U.S. Army photo)

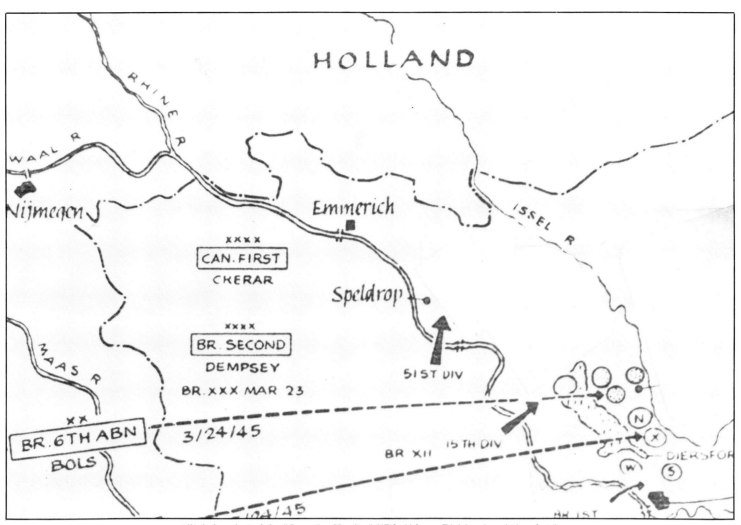
Allied plan of attack for "Operation Varsity." (17th Airborne Division Association photo)

Glider troops of the 17th Airborne Division, First Allied Airborne Army soon after landing. (U.S. Army photo)

24 March 1945, Winston Churchill observing the Rhine crossings and airborne operations. On his right Field Marshall Sir Alan Brooke and, on his left, Commander Thompson. (Imperial War Museum)

into encircled and beseiged Bastogne. They marked drop zones so that the 101st could be resupplied by air.

The last Allied airborne operation in Europe was "Operation Varsity," a drop east of the Rhine. This was to support Montgomery's 21st Army Group in its thrust across the North German Plain. However, the unexpected capture of the Ludendorff Railroad Bridge at Remagen on 7 March 1945, enabled the Americans to cross the rhine in force and develop a rapidly-expanding bridgehead on the east bank, south of the 21st Army Group.

Nevertheless, Eisenhower ordered Montgomery to proceed with "Varsity." The 1st Allied Airborne Army, under General Brereton, was responsible for pre-assault planning and training for the role it would play. Once on the ground, however, command would pass to General Ridgway, XVIII Airborne Corps. For this operation the corps was made up of the US 17th and British 6th Airborne Divisions commanded by Major Generals William M. Miley and Eric Bols, respectively. With attachments, this was over twenty-one thousand men.

Montgomery intended to force the Rhine on a broad front. The mission of Ridgway's corps was to seize the high ground northwest of Wesel in order to deny the enemy observation of the main river-crossing sites. Moreover, they were to act as a blocking force north and northwest of Wesel, thereby preventing movement of German reinforcements to the Rhine.

On the morning of 24 March 1945, the the two airborne divisions flew to their drop and landing zones. The air armada took more than two and one-half hours to pass overhead. The river assault had begun the night before so that the artillery preparation and support would not endanger the airborne troops.

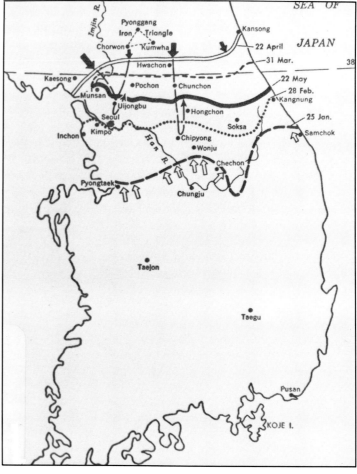

Combat operations in Korea: January - May 1951.

The first men to jump were from the 507th Parachute Infantry Regiment of the 17th Airborne Division. Ground haze and drifting smoke lowered visibility but did not seriously interfere with the airborne assault. These first troops received little anti-aircraft fire, but the men of the 513th Parachute Infantry Regiment and 194th Glider Infantry Regiment that followed took heavy fire. Moreover, German resistance on the ground was often fierce.

Even though some of the airborne drops and landings were not directly on target, the units came down in compact groups. Thus, they were able to quickly assemble and move decisively toward their targets. By nightfall, all objectives had been taken and contact established with the ground troops advancing eastward from the Rhine.

Without a doubt, "Operation Varsity" was a complete success. General Eisenhower and Prime Minister Churchill had watched the airborne assault from the west bank of the Rhine. At that time Churchill commented to Eisenhower, "My dear General, the German is whipped. We've got him. He is all through."

Just before the unconditional German surrender, elements of the 13th Airborne Division were slated to participate, as part of the 1st Allied Airborne Army, in an operation to capture Copenhagen, Denmark from its German occupiers. But World War II in Europe ended without additional airborne operations. Contingency planning had gone on, but Allied ground forces were so swiftly advancing on all fronts that there proved to be no need for their futher deployment. It was envisioned that the 11th and 13th Airborne Divisions would be used in "Operation Olympia," the Allied invasion of the Japanese Home Islands. There, as in Europe, the war ended before futher airborne offensives could be launched.

When World War II ended in 1945, millions of American service personnel were demobilized, and many units deactivated. Hot war was soon replaced by cold war and much of the world remained uncertain and dangerous. Although the military was not maintained at wartime levels, airborne forces had proved their worth. Research and development of airborne equipment and tactics continued.

Interlude: Between Wars

At the end of World War II, the United States possessed the largest airborne force in the world. Five active divisions, four separate regimental combat teams, and five separate regiments, plus an airborne troop carrier command encompassing the greatest armada of troop transport planes and gliders ever assembled.

Less than two years later, this force had been decimated by the presumption that peace would prevail and that the need for elite forces was no longer viable. Accordingly, units whose battle honors included many of the major airborne campaigns soon found their colors furled and cased and their status that of inactive units.

However, those leaders who had cut their teeth on the battle to develop major unit airborne configurations during the war, now made their influence felt in the halls of the pentagon and developed a concept of creating, within the reserve structure of the army, an airborne capability. This, coupled with the retention of two division forces (albeit at peacetime staffing), kept a semblance of airborne in the nation's forces.

The 11th Airborne Division was continued on active status and stationed in the Far East (Japan) to provide an airborne capability in the Pacific Basin area. The 82nd Airborne Division was designated as the primary strategic reserve division and was oriented towards possible operations in the Atlantic Basin. To provide a mobilization capability, the 80th, 84th, 100th and 108th Airborne Divisions were activated as reserve units and developed and maintained trained airborne cadre at peacetime troop strength. A similar procedure was adopted by the airborne troop carrier elements of the U.S. Air Force except that the bulk of squadrons were assigned to the Air National Guard.

To support the foregoing, the army instituted a practice of maximizing the training of new entry soldiers as qualified parachutists even though their anticipated assignments were to non-airborne organizations. This ensured a mobilization filler capability to round out the reserve and active airborne divisions in the event of mobilization.

More importantly however, was the significant effort given to development and refinement of the lessons learned from the employment of airborne forces during the war. From this effort the demise of the glider and the advent of the helicopter resulted. Morever, greatly improved aircraft, heavy drop techniques and improved parachutes evolved in the late 1940s.

Korean War, 1950-1953

Less than five years after the end of the Second World War, the US was again involved in a major war. On 25 June 1950, Communist North Korean forces crossed the 38th Parallel, invading South Korea. Greatly outnumbered and under intense pressure, US and South Korean troops were pushed relentlessly down the peninsula until they held a relatively small perimeter around the port city of Pusan on the southeastern coast.

It was during this grim time that General MacArthur initiated one of the most daring, brilliant, amphibious assaults in military history. At dawn on 15 September 1950, the US X Corps landed on the treacherous beaches at Inchon, and cut its way inland. The 187th Airborne Regimental Combat Team, formed from the 11th Airborne Division at Fort Campbell, Kentucky, arrived and took part in this operation as ground troops. This was on the western coast of Korea and more than one hundred and fifty miles behind enemy lines.

Supplies cut off and in deadly peril from widening fronts at both Pusan and Inchon, the North Korean Army began to lose cohesion. Losing much materiel in addition to tens of thousands of men as casualties and prisoners, the Communist

Commanding general of the 187th Airborne Regimental Combat Team, Brigadier General Frank S. Bowen, Jr., with his staff in discussion with his battalion commanders. Munsan-ni, Korea on 23 March 1951. (U.S. Army photo)

Helicopter assault on 22 January 1971, in Military Region 1, Vietnam. Troopers of the 101st Airborne Division (Air Mobile) rapel into dense jungle in search of the enemy. (U.S. Army photo)

forces scattered into the rugged, usually roadless, countryside.

Seoul was liberated on 26 September when American forces from Inchon and Pusan linked up. President Harry S. Truman directed MacArthur to carry the war into North Korea. By 16 October it was evident that Pyongyang, capital of the Communist North, was within the grasp of advancing tank and infantry units.

At this juncture, MacArthur decided to commit the 187th Airborne Regimental Combat Team. Their mission was to drop some thirty miles north of Pyongyang to block the North Korean forces withdrawing that way. Even though they had fought against North Korean troops defending Kimpo Airport in Seoul, this would be their first combat jump.

On 20 October the North Korean capital was overrun. That same day the 187th jumped at Sukchon and Sunchon in mid-afternoon. The drop zones were in mountainous country, about twelve miles apart. At both DZs, the paratroopers achieved complete tactical surprise.

Skirmishing countinued throughout the day. By 1700 hours, however, all objectives had been achieved. The rail lines and roads running north through Sukchon and Sunchon had been cut. The paratroopers dug in and prepared to halt a fleeing enemy.

The next day a reinforced battalion of the North Korean 239th Regiment attempted to fight its way through. This initial engagement developed into a running battle that lasted for two days. In this fight, the 239th was destroyed. On 23 October, ground forces made contact with the airborne troops. Mission accomplished, the 187th was withdrawn from the line and sent to Pyongyang to rest and refit. In addition, the 187th blocked raids by the 10th North Korean Corps on the main supply route.

By this time it was obvious that the North Korean Army was being cut to pieces and, barring some unforeseen circumstance, would never regain the initative. Unwilling to see their ally destroyed, the Chinese Communists sent some thirty divisions south of the Yalu River. On 25 November 1950, they fell upon United Nations troops. A stunning surprise, Chinese intervention in such numbers forced the United Nations forces, which were mostly American, to lose much of what they had conquered. Nevertheless, by mid-January 1951, the front stabilized roughly fifty miles south of the 38th Parallel. Communist forces launched a counterattack 11-18 February. At Wonju the 187th ARCT acted as a firebreak to protect withdrawing elements of the US 8th Army.

Shortly thereafter, UN troops launched a series of limited counteroffensives. "Operation Ripper" began a few weeks later. Intended primarily to inflict heavy casualties on the enemy and, secondarily, to eliminate a major Chi-Com base of supply and to relieve Seoul, it was carried out 7-31 March.

As communist forces withdrew northward, Seoul was liberated again on 14 March. Just north of the South Korean capital, enemy resistance stiffened. It was decided to employ the 187th Airborne Regimental Combat Team to cut off and hold the Chi-Com forces in that area so that UN ground units could smash them.

The objective of this operation was the town of Munsan-ni, over twenty miles northwest of Seoul, at the mouth of the Imjin River. On 23 March, Good Friday, the 187th along with the 2nd and 4th Airborne Ranger Companies, jumped at Munsan-ni. Initial resistance was light. However, the North Korean 36th Infantry Regiment offered strong resistance when the paratroopers moved to secure Hills 228 and 229.

The paratroopers were prepared to hold out alone for two days before the advancing columns reached them. Yet, it was after dark on just the first day when tanks from I Corps linked up with them. The paratroopers continued in the line as regular infantry for some time to come. They could be proud that their drop at Munsan-ni was an instrumental factor in forcing a general communist withdrawal across the Imjin and on northward. This was the last airborne operation of the Korean War.

Even so, it was not the last use of the 187th. On 7 May 1952, there was a carefully organized revolt of Communist prisoners at Koje-do. There were more than 80,000 Communist POWs held on this island just down the coast from Pusan. After some initial delay in deciding what course to take, the 187th was brought in to quell the riots and restore order. That is exactly what was done. The 187th had repeatedly been used as the "fire brigade" for the 8th Army during the first eighteen months of the war and had been thrown into the lines no less than six times.

Korea was an important watershed in American geopolitical and military policy. Various international alliances were formed and the US Armed Forces was rebuilt. During Korea and its aftermath, the 11th Airborne Division was deployed to Europe and the 101st was reactivated. Furthermore, the Special Forces were established. The Green Berets came about as the result of recommendations from General Lawton Collins, Army Chief of Staff, who had started training airborne Ranger infantry companies at Fort Benning in September 1950.

Four volunteer Ranger companies (1st, 2nd, 3rd, 4th) were the first cycle trained. Three companies were sent to Korea after eight weeks of training, arriving there in December 1950. Each company was attached to an infantry division. The 1st Ranger Company was attached to the 2nd Infantry Division, the 2nd (the first and only black airborne ranger company in the history of the Army) to the 7th Infantry Division, and the 4th to the 1st Cavalry Division. The 3rd Ranger Company was retained at Fort Benning to help train the second cycle of Ranger companies.

The 5th, 6th, 7th and 8th Ranger Infantry Companies (Airborne), also made up of volunteers from the 82nd Airborne Division, began training on Thanksgiving Day, 1950. After eight weeks of training at Fort Benning, the 3rd, 5th, and 8th were sent to Fort Carson, Colorado for mountain and cold weather training. Completing their training there, these Ranger companies arrived in Korea on 24 March 1951. The 3rd Company was attached to the 3rd Infantry Division, the 5th to the 25th Infantry Division, and the 8th to the 24th Infantry Division. The 7th Company stayed at Fort Benning to train following cycles of Rangers. The 9th through 15th Ranger Companies were also trained but not committed. The Ranger Companies in Korea were disbanded on 1 August 1951. The 6th Company was deployed to Germany and attached to the 1st Infantry Division, but the remaining companies were disbanded during the fall of 1951.

Vietnam, 1965-1973

Communist insurgency in Indochina after World War II ultimately forced out the French. Concerned about the threat of communism in the region, the United States became involved in Vietnam. The original US role of military support became, in 1965, one of combat operations. In what proved to be America's longest war, airborne units were involved, but most often as regular infantry. The nature of the war usually precluded conventional airborne operations.

Nevertheless, a battalion of the 173rd Airborne Brigade carried out a jump at the beginning of "Operation Junction City Alternate," on 22 February 1967. The purpose of this combat operation was to capture or destroy enemy bases north of Tay Ninh City. The decision to use paratroopers was based on the need to get as many troops on the ground as quickly as possible with the rest of the 173rd being committed to action by other means. The jump was followed closely by heliborne infantry as well as ground units sweeping through the zone of operations. Four days after the jump, phase one of "Junction City Alternate" was complete. The only airborne unit to fight in World War II, Korea and Vietnam was the 3rd Battalion of the 187th.

Although parachute-type operations were exceedingly rare in Vietnam, the helicopter was well suited to operational requirements there. Thus, airborne assault by helicopter came to be common. A brigade of the 101st Airborne Division arrived in South Vietnam in July 1965. By December 1967, the entire division had been committed. There were, nonetheless, many Americans that participated in airborne operations. The US Airborne Advisory Group to the Republic of Vietnam Airborne Division took part in more than thirty combat jumps with the Vietnamese between 1965-1972.

The 101st did not make any combat jumps during the Vietnam War. For rapid insertion onto the battlefield they were heliborne. In order to recognize their new role, they were redesignated on 1 July 1968, as the 101st Air Cavalry Division. Later, on 29 August of that year, they were again redesignated to more closely reflect the complete transition from parachutes to helicopters: 101st Airborne Division (Air Mobile).

On 1 February 1974, a new airborne award was authorized, the Air Mobile Badge. In early October of the same year, the unit designation changed for the last time. It was now the 101st Airborne Division (Air Assault) and the Air Mobile Badge became the Air Assault Badge.

Air Assault technique was refined and the 101st Airborne Division (Air Assault) was continuously revamped. So it is that in this one unit, speed, mobility and firepower combine to make it one of the most powerful divisions in the world.

It was in the post-Vietnam period that many important airborne developments took place. Long-range troop carrier aircraft such as the C-5A and refinement of air assault techniques proved very important to later airborne operations. During this period the XVIII Airborne Corps was formed as a rapid-deployment force of strategic significance. Of further significance was the integration of women into airborne units. Although prohibited by law from combat arms, female airborne personnel perform vital support missions that would take them onto the battlefield. Indeed, many female airborne troops are now deployed in Saudi Arabia as part of "Operation Desert Shield."

Grenada, 1983

Long an island colony of Great Britain, Grenada was granted its independence in early 1974. A combination of factors contributed to political unrest which produced a nearly bloodless coup in 1979 by a political group calling itself the New Jewel

82nd airborne Trooper with freed American Medical Students in Grenada. (U.S. Army photo)

Movement. The new regime did not keep its promises and moved away from English-style democracy toward the Cuban model of revolutionary democracy.

From 1980 onward, Grenada was increasingly involved with both Cuba and the Soviet Union. Domestically, revolutionary democracy meant press censorship, political repression, and a developing military-police state.

By October 1983, a power struggle split the Grenadian cabinet over the pace of economic socialization. This internal problem became acute when the deputy prime minister assassinated the prime minister and established a Revolutionary Military Council. The situation within Grenada deteriorated rapidly.

The United States had been concerned for some months about the relatively large number of Cuban military advisors and workers in Grenada as well as about construction projects with military applications. The most notable example of the latter was the Point Salines airport. Additionally, large quantities of military weapons and materiel were being imported and stockpiled.

Added to this was concern about the safety for upwards of a thousand Americans in an unstable, potentially violent situation. The Grenada airport and radio station were closed, and a shoot-on-sight curfew imposed. There was every indication that matters would get progressively worse.

The Governor-General of Grenada appealed to the Organization of Eastern Caribbean States (OECS) to intervene before Grenada descended into bloody anarchy. The OECS sought additional support from Jamaica, Barbados and the United States.

"Operation Urgent Fury" was implemented. The mission was to militarily invade Grenada in order to neutralize Cuban and Revolutionary Military Council personnel, protect foreign nationals, and to maintain order until such time as a new government could be set up.

The American airborne elements in this operation were the 82nd Airborne Division and the 1st and 2nd Battalions of the 75th Rangers. The Rangers jumped onto the Port Salines Airfield from an altitude of five hundred feet. There were antiaircraft guns guarding the field but they could not bring their fire to bear on targets lower than six hundred feet. As a result, the US planes flew below an ineffectual curtain of enemy fire.

The Rangers landed on target and set about consolidating their position. As anticipated, the airfield had been blocked to prevent aircraft from landing. Two members of the 20th Engineer Brigade, XVIII Airborne Corps, jumped in with the Rangers to deal with just such a contingency. Using captured bulldozers, they quickly cleared the runway.

Shortly thereafter, transport aircraft carrying elements of the 82nd Airborne Division landed at Port Salines. Ultimately, the entire 82nd took part in "Operation Urgent Fury." The paratroopers carried out combat operations to secure the island and eliminate all resistance. During the course of this mission, the American civilians, most of whom were medical students, were found unharmed. The 82nd protected them until they could be evacuated from Grenada.

Ther are alway lessons to be learned from any military operation and Grenada was no exception. Despite various shortcomings, the value of airborne was again demonstrated. After all, it is vital to have highly mobile forces that can undertake combat missions with little or no prior notice in order to respond to fluid situations.

Panama, 1989

The developing situation in Panama that led to US military intervention was non-conventional and, in some ways, quite bizarre. General Manuel Noriega, the President of Panama, had been indicted in the United States on thirteen counts of drug trafficking and racketeering in February 1988. Thereafter, the already tense US-Panamanian relations grew progressively worse.

Incidents between US personnel and those of the Panamanian Defense Force (PDF) and the so-called Dignity Battalions increased in frequency and violence. In October 1989, a coup against Noriega failed and the situation continued to deteriorate. Noriega proclaimed himself "President for Life" and declared that a state of war existed with the United States. The next day, 16 December, a Marine officer was fatally wounded by the PDF. The security of American lives and property grew even more precarious.

US contingency plans for various options in Panama had long been developed and were continuously updated. All other attempts to resolve the increasingly violent situation in Panama had failed. President George Bush then chose to exercise the military option. US objectives were four-fold: protect American lives; safeguard the Panama Canal in accordance with the treaties signed by both countries; restore order and democracy to Panama; bring Manuel Noriega to trial.

Massive and quick, "Operation Just Cause" began shortly after midnight on 20 December 1989. Some 7,000 troops were airlifted into Panama to join the over 12,000 already there. During the first hour after midnight, American units initiated airborne assaults on several key objectives.

A battalion of the 75th Rangers parachuted onto the runways of the Torrijos-Tocumen International Airport on the eastern outskirts of Panama City. Far to the west of the capital, a second Ranger battalion dropped on Rio Hato where some of Noriega's most loyal units were based. Moreover, there was an armory there. A fierce firefight ensued before surviving PDF troops surrendered to the Rangers.

Part of the airborne phase of "Operation Just Cause" called for a brigade of the 82nd Airborne Division to jump in and reinforce the Rangers at Torrijos-Tocumen Airport. Half the brigade arrived over the drop zone on schedule but the other half was delayed by bad weather back in the states, but did eventually arrive.

Other elements of the 82nd Airborne were part of the force that captured Gamboa Prison and released forty-eight political prisoners. These paratroopers also helped seize a key Panama Canal facility, the electrical power center at Sierra Tigre by the Madden Dam.

Although sporadic fighting lasted several days, "Just Cause" was swiftly and effectively carried out. Without doubt, a major factor in the success of this operation was the airborne elements. Panama proved again what so many already knew: airborne forces are most excellently suited for this sort of mission. Therefore, the US must always maintain an airborne capability.

AH-1 Cobra attack helicopter at a U.S. base camp in Central America. (Courtesy of SSG Joe Owens)

The Next Fifty Years, 1990-2040

It has often been observed that the past is prologue. If that holds true in the case of the United States airborne forces, the next fifty years will see some remarkable innovations in techniques of vertical envelopment. As the means to insert airborne forces onto the battlefield have become more sophisticated, so too have the means to counter the airborne threat.

However, increased defense capability does not mean that airborne no longer has a role in modern battle. It simply means that new methods must be developed to meet changing needs. If Benjamin Franklin could envision airborne operations more than two centuries ago, surely we can conceive of and implement new ways to use this concept.

The moden commander needs a variety of different force capabilities from which he can choose in order to fight and win. To eliminate airborne from his list of choices would limit his options and, very possibly, his chances of victory. What the human mind can imagine, may ultimately come to pass. In whatever form, future airborne developments should be spectacular.

Epilogue

On 2 August 1990, Iraq invaded Kuwait. Soon afterwards, in order to forestall further aggression in the region, the United States initiated "Operation Desert Shield." The 82nd Airborne Divison was immediately airlifted into the Arabian desert. Now, the 101st Airborne Division (Air Assault) is deployed in Saudi Arabia along the Kuwaiti border. Moreover, Special Forces and Rangers will most certainly play a role there.

U. S. troop levels continue to be enhanced, thanks in large part to the capability of the Military Airlift Command. Whether or not there is war in this region remains to be seen. Nevertheless, if peace prevails, it will be because of the ability of the Air Force to deploy major airborne and other forces in such a brief time.

With the deployment of airborne troops to Saudi Arabia, the next fifty years has already begun. Airborne history is not over. It continues.

On the ground in Saudi Arabia. The small mounds are camouflaged positions occupied by U. S. airborne forces. In the distance two CH-47 helicopters transport supplies.

Stinger Missile team from the 2nd Battalion, 44th Air Defense Regiment, 101st Airborne Division (Air Assault) in Saudi Arabia.

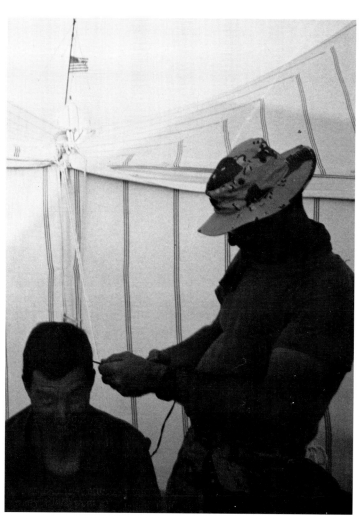

Life in the desert–Captain Kent C. Curtsinger, 101st Airborne Division, originally of Fancy Farm, Kentucky gets a haircut somewhere in Saudi Arabia.

Operation Desert Shield: Vulcan anti-anticraft gun, 2-44 ADA, emplacement in the Arabian desert.

Paratroopers of a 75mm howitzer section, Battery C, 377th Fld. Art., 101st Abn., put on their parachutes before loading into C-47 cargo plane, Newburg, Berks, England, March 1944. (101st Airborne Division Association)

Airborne Operations
1942 – 1990

U. S. MILITARY COMBAT JUMPS

By Donald A. Parks, Ph.D.
88th ABN AAA BN, 187th ARCT

Whenever two or more paratroopers are gathered together, inevitably, the subject of combat jumps becomes a part of the conversation. Without thorough research and a statistical analysis of the data, conversation between paratroopers is often inaccurate, prejudiced by unit pride and charged with emotion. This chapter attempts to analyze the history of U.S. combat jumps, consolidate existing data with new data, provide current information for incorporation in this publication and place the data into proper perspective.

Previous authors, as well as myself, have had difficulty in securing exact data. Either the information is difficult to obtain because of the time that has elapsed, misplacement of data by authorities, inaccurate data or information which has been recorded but is not complete. For instance, the identification of units that made combat jumps is not always precise, nor the number of troops that made the jumps and the exact dates. Some data is very precise and well documented, while other data may be subject to scrutiny. Nevertheless, sufficient research has brougtht the data into more current focus and provides an interesting insight into the overall discussion of who did what, when and where!

Table I tabulates 93 U.S. combat jumps. Many paratroopers have always thought Korea, Vietnam, Grenada, and most recently, Panama. Many discussions only considered the major mass unit drops of either division, regiment, battalion, company or platoon size drops. However, in addition to the mass tactical unit combat jumps, there were many jumps made in World War II by OSS teams, as well as small United Nations Partisan Infantry Korea (UNPIK) groups that jumped in Korea, and small groups of American advisors, principally Special Forces, who accompanied Vietnamese paratroopers. The operation, or mission column, is brief but self-explanatory and when available includes the operational code name in parentheses. An analysis of the missions generally corresponds with the stated missions of airborne forces.

The question of what constitutes a combat jump is also often a part of conversation between paratroopers. Apparently, there is no clear definition in any official publication. However, the very nature of airborne forces and their missions defines what might qualify as a combat jump. A reasonable definition would be any jump made into an unsecured area where there is a liklihood of enemy presence, and certainly an element of risk. There is also no specific designation as to how many paratroopers are involved to qualify the event as a combat jump. Therefore, it would appear that even one person jumping under the conditions described above could be a combat jump.

There is also the consideration that a combat jump is more commonly considered a drop made by a tactical unit with an existing TO&E. Covert operations are also made into unsecured areas with an enemy presence and an element of risk. The principal difference between a jump made by a tactical unit and a covert operation would seem to be that covert operations, such as OSS drops, are primarily intended to avoid contact with the enemy, whereas tactical units are dropped with the specific intent to contact the enemy and engage in combat. There are undoubtedly many ways to interpret the statistical data presented in this article and the reader is left to draw some of his own conclusions.

More paratroopers could have been dropped in different operations, but there were limitations on airlift capability which had a distinct effect on operational planning. Almost every major mass drop has experienced airlift limitations.

The analysis shows, as tabulated in Table I, that the 93 U.S. combat jumps involved 69,257 paratroopers, during five wars, over forty-seven years, in 25 countries. A major military achievement.

Table II identifies the 25 countries in which U.S. paratroopers were dropped. Twenty percent, or 19, of the 93 jumps were made in Korea, whereas 27%, or 18,757 paratroopers, were dropped into France.

Table III consolidates combat jumps by unit. The greatest number of jumps, 32, or 34%, were made by the OSS, followed by 15 jumps, or 16%, made by UNPIK forces, followed by 7 jumps, or 7% made by the 82nd Airborne Division. While this analysis shows that more jumps were made by small units, it is not surprising that the majority of paratroopers were dropped by the larger tactical units, such as 36% of all paratroopers dropped, or 24,959, by the 82nd Airborne Division, followed by 19%, or 13,407, from the 101st Airborne Division, and 10%, or 6,730 paratroopers, dropped by the 187th Airborne Regimental Combat Team.

Table IV is also not a surprise, but simply quantifies what most readers already knew. The majority of jumps, 31, or 33%, were made in 1944, and dropped 50% of all paratroopers ever dropped, or 34,469 paratroopers, in that year.

The U.S. military has relied on U.S. airborne operations and the deployment of troops by vertical envelopment in every major U.S. engagement since World War II. How long parachute infantry and artillery will be used by the military is for the future to decide. However, the first fifty years have clearly demonstrated the practicality of airborne warfare. There are probably other small airborne combat operations which have not been recorded, at least several which have been spoken of but have not been documented and remain unpublished, perhaps for security reasons.

It is interesting to contemplate how U.S. airborne operations have significantly helped to change and, therefore, make history. It is also interesting to contemplate what additional changes could have been made with the extended use of existing or proposed airborne commands and how other countries, particularly our enemies at the time, might have more effectively used their own airborne forces. It is even more interesting to consider the relatively small percent of trained U.S. airborne forces that have been committed to combat by parachute. While the number of U.S. troops that have been trained as paratroopers is elusive, it is considered that the 69,257 trained parachutists that jumped into combat represent perhaps 3% or 4% of all those who have been trained as pararoopers. These 93 combat jumps represent a miniscule percentage of all training and practice jumps that have occurred in our fifty year history, which must number in the thousands.

This chapter deals only with parachute operations. Glider, air assault by helicopter and air landings are also important airborne operations, but are not a part of this particular analysis. Perhaps future studies can include these other operations which would surely enhance the overall perspective and understanding of the total airborne concept.

Existing publications, which to some extent have considered the same subject, as well as original research by this author, were used to assemble the data. Two authors are credited with excellent references on the subject, namely R.J. Bragg and Roy Turner. Their publications are, "Parachute

Badges and Insignia of the World" and "Parachute Wings." Both publications contain an interesting tabulation of jumps made by every country through 1978 in the first reference, and through 1983 in the second reference.

The history of U.S. paratroopers is well documented in this publication and others. Most nations of the world are well aware of the impact that U.S. paratroopers make in battle. There are 25 countries of the world that have experienced the impact of the U.S. paratrooper. The future will determine the roll that other U.S. paratroopers will play in helping to decide conflicts and dramatize evermore the spirit of the U.S. paratrooper.

Like a bud about to burst into bloom, this 101st Airborne Division paratrooper's parachute is filling with air following his leap from the aircraft. Four seconds after jumping he will look up to check the proper deployment of his chute and then prepare to land during this training jump at Fort Campbell, Kentucky, Circa 1958. (Photo by SFC Joe M. Gonzales)

TABLE I   UNITED STATES AIRBORNE COMBAT JUMPS

No.	Year	Date	Country	Drop Zone	Unit	Troopers	%	Operation
1	1942	8 Nov	Morocco	Tafaraoui/La Senia	509th PIB	556	0.80	Seize airfield
2		15 Nov	Algeria	Youks les Bains	509th PIB	350	0.51	Seize airfield
3		24 Dec	Tunisia	El Djem	509th PIB	32	0.05	Blow bridge
4	1943	26/27 Jan	Burma	Myitkyina	OSS, Det. 101	30	0.04	Establish base (Piccadilly)
5		7/8 Feb	Burma	North	OSS, Det. 101	12	0.02	Establish base
6		? Mar	Burma	Lawsawk	OSS, Det. 101	12	0.02	Establish base
7		9 Jul	Italy	Gela, Sicily	505th PIR	3,406	4.92	Invasion of Sicily
8		10 Jul	Italy	Gela, Sicily	504th PIR	2,304	3.33	Invasion of Sicily
9		? Aug	Burma	Naga Hills	USAAF Med Evac	5	0.01	Aid aircrash
10		5 Sep	New Guinea	Nadzab	503rd PIR	1,700	2.45	Seize Markham Valley
11		12 Sep	Italy	Cagliari, Sardinia	OSS	4	0.01	Arrange surrender
12		13 Sep	Italy	Salerno	504th PIR	1,300	1.88	Reinforcement
13		13 Sep	Italy, Sardinia	Decimomannu airfield	OSS Group	30	0.04	Negotiation
14		14 Sep	Italy	Salerno	505th PIR	2,105	3.04	Reinforcement
15		14 Sep	Italy	Avellino	509th PIB	640	0.92	Raid
16		2 Oct	Italy	Gransasso	OSS	66	0.10	Free POW's
17	1944	6 Jan	France	Rhone Valley	Inter Allied	40	0.06	Aid resistance
18		15 Jan	Burma	Taro	OSS, Det. 101	6	0.01	Establish base (Tramp)
19		7 Feb	Yugoslavia	Ticeuo	USAAF Met Ops	8	0.01	Establish station (Bunghole)
20		14 Feb	Italy	Udine	OSS	3	0.00	Liaison
21		16 Mar	Italy	Northern	OSS Team	4	0.01	Liaison (Apricot)
22		16 Mar	Italy	Alps	OSS Team	2	0.00	Liaison (Orange)
23		17 Mar	Hungary	Unknown	OSS	6	0.01	Information
24		? May	Yugoslavia	Kupresko	OSS Team	4	0.01	Prepare Balkan airstrip
25		6 Jun	France	Normandy	82nd Abn. Div.	6,418	9.27	Invasion (Overlord)
26		6 Jun	France	Normandy	101st Abn. Div.	6,638	9.58	Invasion (Overlord)
27		20 Jun	France	Cantal	OSS OG Section	20	0.03	Destroy Bridge (Emry)
28		23 Jun	Yugoslavia	Unknown	OSS Team	4	0.01	Liaison (Spike)
29		28 Jun	France	Vercors	OSS Ops. Gp.	30	0.04	Aid resistance
30		3 Jul	New Guinea	Noemfoor	503rd, 1st Bn.	739	1.07	Assist landing (Table Tennis)
31		4 Jul	New Guinea	Noemfoor	503rd, 3rd Bn.	685	0.99	Reinforcement (Table Tennis)
32		17 Jul	France	Finisters	Jedburgh Team	2	0.00	Liaison (Horace)
33		2 Aug	Yugoslavia	Pranjane	OSS	5	0.01	Rescue aircrew (Halyard)
34		3 Aug	Italy	Northern	OSS Mission	3	0.00	Try block Brenner Pass (Eagle)
35		? Aug	Bulgaria	Greek border	OSS Team	2	0.00	Contact resistance
36		11 Aug	Italy	Val D'Ossoga	OSS Mission	5	0.01	Aid Partisans (Chrysler)
37		14 Aug	France	Unknown	Jedburgh Team	2	0.00	Liaison (Bruce)
38		15 Aug	France	Cote d'Azur, Riviera	517, 509, 551	5,607	8.10	Invasion, S. Fr. (Dragoon)
39		17 Sep	Holland	Eindhoven	101st Abn. Div.	6,769	9.77	Seize corridor (Market Garden)
40		17 Sep	Holland	Nijmegen	82nd Abn. Div.	7,250	10.47	Seize corridor (Market Garden)
41		25 Sep	Czechoslovakia	Tatra Mountains	OSS Group	30	0.04	Contact Partisans
42		26 Sep	Italy	Milan, outskirts	OSS Group	30	0.04	Aid Partisans
43		13 Oct	Italy	Venice, northern	OSS Group	15	0.02	Aid Partisans (Aztec)
44		? Nov	Malaya	Highlands	OSS Group	4	0.01	Liaison (Cairngorn)
45		4 Dec	Philippines	Leyte, Manarawat	457th FA, A Btry	120	0.17	Establish gun position
46		27 Dec	Italy	Venice, northern	OSS Group	15	0.02	Aid Partisans (Azetc)
47		23 Dec	Belgium	Bastogne	Pathfinder teams	30	0.04	Aid resupply
48	1945	? Dec	Burma	Mongmit	OSS, Det. 101	12	0.02	Intelligence
49		? Dec	Burma	Mandalay area	OSS, Det. 101	12	0.02	Inteligance
50		3 Feb	Philippines	Tagaytay, Luzon	511th PIR	1,455	2.10	Seize bridge (Shoestring)
51		16 Feb	Philippines	Corregidor Island	503rd PIR	2,019	2.92	Assault, destroy guns
52		23 Feb	Philippines	Los Banos, Luzon	511th PIR, B Co.	130	0.19	Rescue prisoners
53		? Mar	China	East Coast	OSS Team	8	0.01	Survey coastline (Akron)
54		24 Mar	Norway	Jaevsjo	OSS Ops. Gp.	20	0.03	Aid resistance (Rype)

UNITED STATES AIRBORNE COMBAT JUMPS — TABLE I

No.	Year	Date	Country	Drop Zone	Unit	Troopers	%	Operation
55		24 Mar	Germany	Wesel	17th Abn. Div.	4,964	7.17	Rhine crossing (Varsity)
56		25 Apr	Germany	Altengrabon	Sp. Recon. Force	6	0.01	POW protection (Violet)
57		23 Jun	Philippines	Aparri, Luzon	511th PIR	1,030	1.49	Cut off enemy (Gypsy)
58		16 Jul	Indochina	Thai-Nguyen	OSS Team	15	0.02	Contact Ho Chi Minh
59		? Aug	China	Mukden	OSS Team	15	0.02	POW rescue
60		22 Aug	Indochina	Gialam	OSS Team	15	0.02	Clear airfield
61	1950	8 Sep	Laos	Luang-Prabang	OSS Team	15	0.02	Liaison
62		20 Oct	Korea	Sunchon	187th ARCT, 2/Bn	1,203	1.74	Cut off enemy (DZ Easy)
63		20 Oct	Korea	Sukchon	187th, 1,3/Bn	1,470	2.12	Cut off enemy (DZ William)
64	1951	21 Oct	Korea	Sukchon	187th ARCT	671	0.97	Reinforcement
65		17 Mar	Korea	Na-Osan-Ni	UNPIK	4	0.01	Blow tunnel
66		23 Mar	Korea	Munsan-Ni	187th ARCT	3,486	5.03	Cut off enemy
67	1952	17 Jun	Korea	Koksan	UNPIK	15	0.02	Establish base
68		22 Jan	Korea	North	UNPIK	11	0.02	Attack railroads (Mustang III)
69		16 Mar	Korea	North	UNPIK	8	0.01	Attack railroads (Mustang IV)
70		14 May	Korea	North	UNPIK	25	0.04	Attack railroads (Mustang V,VI)
71		31 Oct	Korea	North	UNPIK	6	0.01	Attack railroads (Mustang VII, VIII)
72		28 Dec	Korea	North	UNPIK	12	0.02	Attack railroads (Jesse James II, III)
73	1953	30 Dec	Korea	North	UNPIK	6	0.01	Attack road traffic (Jesse James I)
74		25 Jan	Korea	North	UNPIK	45	0.06	Establish base (Green Dragon)
75		7 Feb	Korea	North	UNPIK	14	0.02	Attack rail traffic (Boxer I, II)
76		9 Feb	Korea	North	UNPIK	7	0.01	Attack rail traffic (Boxer III)
77		11 Feb	Korea	North	UNPIK	7	0.01	Attack rail traffic (Boxer IV)
78		31 Mar	Korea	North	UNPIK	3	0.00	Establish base (Hurricane)
79		1 Apr	Korea	North	UNPIK	15	0.02	Attack rail traffic (Rabbit I)
80	1960	6 Apr	Korea	North	UNPIK	3	0.00	Attack rail traffic (Rabbit II)
81	1966	? Jan	Indonesia	Sumatra	CIA	2	0.00	Aid rebels
82		14 Jun	Vietnam	Chu-Lai	USMC 1st Recon	13	0.02	Establish base
83	1967	27 Dec	Vietnam	Chuong-Thien	Special Forces	33	0.05	Raid (w/1,200 ARVN para Bns)
84		22 Feb	Vietnam	Katum	503rd PIR, 2/Bn	750	1.08	Aid ground troops (Junction City Alt.)
85		2 Apr	Vietnam	Camp Bunard	Special Forces	25	0.04	Establish camp (w/330 ARVN)
86		13 May	Vietnam	Nui Giai	Special Forces	21	0.03	Raid (w/373 ARVN paras)
87		5 Sep	Vietnam	South	USMC 1st Recon	10	0.01	Reconnaissance
88	1969	5 Oct	Vietnam	Bu Praug	5th SF Gp., Mike	35	0.05	Raid (w/Vietnamese paras)
89	1983	17 Nov	Vietnam	Nui Tran	USMC, Team 51	6	0.01	Reconnaissance (Night Cover)
90		25 Oct	Grenada	Point Salines	75th Rgrs 1, 2/Bn.	500	0.72	Seize airport (Urgent Fury)
91	1989	25 Oct	Grenada	Governors Residence	SEAL Detachment	11	0.02	Rescue Governor
92		20 Dec	Panama	City environs	82, 1, 2/504, 4/325	2,176	3.14	Seize city (Just Cause)
93		20 Dec	Panama	City environs	75th Ranger Rgt.	1,900	2.74	Seize airport (Just Cause)
						69,257	100.00	

TABLE II — COMBAT JUMPS BY COUNTRY

	COUNTRY	JUMPS NO.	%	TROOPS NO.	%
1	Algeria	1	1.08	350	0.51
2	Belgium	1	1.08	30	0.04
3	Bulgaria	1	1.08	2	0.00
4	Burma	7	7.53	89	0.13
5	China	2	2.15	23	0.03
6	Czechoslovkia	1	1.08	30	0.04
7	France	8	8.60	18,757	27.08
8	Germany	2	2.15	4,970	7.18
9	Grenada	2	2.15	511	0.74
10	Holland	2	2.15	14,019	20.24
11	Hungary	1	1.08	6	0.01
12	Indochina	2	2.15	30	0.04
13	Indonesia	1	1.08	2	0.00
14	Italy	16	17.20	9,932	14.34
15	Korea	19	20.43	7,011	10.12
16	Laos	1	1.08	15	0.02
17	Malaya	1	1.08	4	0.01
18	Morocco	1	1.08	556	0.80
19	New Guinea	3	3.23	3,124	4.51
20	Norway	1	1.08	20	0.03
21	Panama	2	2.15	4,076	5.89
22	Philippines	5	5.38	4,754	6.86
23	Tunisia	1	1.08	32	0.05
24	Vietnam	8	8.60	893	1.29
25	Yugoslavia	4	4.30	21	0.03
		93	100.00	69,257	100.00

TABLE III — COMBAT JUMPS BY UNIT

	UNIT	JUMPS NO.	%	TROOPS NO.	%
1	OSS Missions	32	33.68	454	0.66
2	UNPIK	15	15.79	181	0.26
3	82nd Abn. Div.	7	7.37	24,959	36.04
4	503rd PIR	5	5.26	5,893	8.51
5	509th PIB	5	5.26	2,328	3.36
6	187th ARCT	4	4.21	6,830	9.86
7	511th PIR	3	3.16	2,615	3.78
8	Special Forces	3	3.16	79	0.11
9	101st Abn. Div.	2	2.11	13,407	19.36
10	Jedburgh Team	2	2.11	4	0.01
11	USMC 1st Recon	2	2.11	23	0.03
12	17th Abn. Div.	1	1.05	4,964	7.17
13	457th FA Bn.	1	1.05	120	0.17
14	517th PIR	1	1.05	3,800	5.49
15	551st PIB	1	1.05	1,057	1.53
16	5th SF Gp., Mike	1	1.05	35	0.05
17	75th Ranger Rgt.	1	1.05	1,900	2.74
18	75th Rangers	1	1.05	500	0.72
19	CIA	1	1.05	2	0.00
20	Inter Allied	1	1.05	40	0.06
21	Pathfinder Teams	1	1.05	30	0.04
22	SEAL Detachment	1	1.05	11	0.02
23	Special ReconForce	1	1.05	6	0.01
24	USAAF Med. Evac.	1	1.05	5	0.01
25	USAAF Med. Opns.	1	1.05	8	0.01
26	USMC, Team 51	1	1.05	6	0.01
		95	100.00	69,257	100.00

TABLE IV — COMBAT JUMPS BY YEAR

YEAR	JUMPS NO.	JUMPS %	TROOPS NO.	TROOPS %
1 1942	3	3.23	938	1.35
2 1943	13	13.98	11,614	16.77
3 1944	31	33.33	34,496	49.81
4 1945	14	15.05	9,716	14.03
5 1950	3	3.23	3,344	4.83
6 1951	3	3.23	3,505	5.06
7 1952	6	6.45	68	0.10
8 1953	7	7.53	94	0.14
9 1960	1	1.08	2	0.00
10 1966	2	2.15	46	0.07
11 1967	5	5.38	841	1.21
12 1969	1	1.08	6	0.01
13 1983	2	2.15	511	0.74
14 1989	2	2.15	4,076	5.89
	93	100.00	69,257	100.00

The 187th Airborne RCT over Korea.

OPERATIONS IN THE CHINA-BURMA-INDIA (CBI) THEATER

By Bart Hagerman, LTC, AUS (Ret)

In this history of airborne operations in the United States military, something needs to be said about the China-Burma-India Theater of Operations. The U.S. involvement here was minimal, but the results obtained and the lessons learned proved invaluable to the development of airborne techniques. It was particularly important to the proponents of glider-transported troops.

In February 1943, British General Orde Charles Wingate led an assortment of British troops and "Chindits" on a long range penetration into Japanese-controlled territory in Burma. A force of five C-47s, two old bombers and some light aircraft flew 178 sorties over the dense jungle terrain and dropped some 300 tons of supplies to sustain the brigade-size force.

When the operation proved to be a tremendous success—at least from the Allies' point of view—Wingate became anxious to mount a large operation which would really strike a blow at the Japanese and hopefully chase them from Burma.

Wingate's enthusiasm apparently sold Winston Churchill on more extensive air-support operations as a means of breaking the stalemate in Burma. Soon Churchill enlisted the support of President Roosevelt and the plan was put into motion. At this time, both Britain and the United States were hungry for a victory in this theater and this offered the best opportunity.

Named to organize the air arm of this unique new force were two daring young fighter pilots, Colonels Phillip G. Cochran and John R. Alison. Cochran was later to become the model for the adventure comic strip by Milton Caniff, "Terry and the Pirates." The pair was daring and ingenious and the new unit rapidly began to take shape.

Cochran, with Alison as his deputy, selected 75 glider pilots, a like number of light plane pilots and a number of enlisted men. By October 1, a force of 523 officers and men—designated as the 1st Air Commando Force—was on the way to Burma. Included in the unit's equipment were 150 CG-4A gliders, 75 TG-5 training gliders, a liaison force of L-1s and L-5s, a transport force of C-47s and C-64s and even some P-47s and B-25s.

As Cochran began training his new unit for the coming missions, Wingate initiated a grueling training schedule for his troops. The plan he had in mind was basically simple: Suitable jungle clearings would be selected within striking distance of Japanese strong points; pathfinders would be flown into clear and secure; engineers and bulldozing equipment would be flown in to prepare landing strips for the transports; and then he would fly in his main body of troops and go on the attack.

The whole operation would take two or three days, but Cochran was confident he could resupply the troops from the air. The trick was not to land too close to the enemy so that the buildup could take place without being under fire.

Wingate's mission was to cut Japanese communication capabilities in Central Burma and force their withdrawal from all areas north of the 24th parallel. It was part of a broad strategy to bring into coordination General Joseph Stilwell's forces and Colonel Merrill's Marauders, and to cut Jap supply lines once and for all. Wingate planned to land troops by air in open areas of the jungle between Chowringhee and Myirkyina, code-named Broadway and Piccadilly.

The 77th Indian Infantry Brigade moved to airfields and

Near Chindwin River at a drop zone British troops man the anti-aircraft batteries.

Troops of the 5307th Composite Unit of Merrill's Marauders, move to the front via the Ledo Road, February 1944.

began preparations for a glider assault. The 11th Indian Infantry Brigade also moved to airfields and prepared to be either moved by glider or to be air-landed at another clearing named Aberdeen.

A month before the air assault, the British 16th Infantry Brigade began marching south to meet the airborne force. Finally, on March 5, 1944, the 80 gliders rolled down their runways, all grossly overloaded and struggled into the air.

The assault was on.

The story from here is fraught with problems. Many of the overloaded gliders were forced to cut loose. Others went down with broken tow lines leaving the occupants to their fate. Probably faring the best were the mules that had been carefully loaded into padded stalls improvised in the gliders.

Landings took their toll. Many of the fields, although looking fine from air reconnaissance photos, were laced with logs and ruts lurking below the waves of grass covering the open areas. Gliders crashed and heavy equipment, including the valuable bulldozers, tore through the thin fabric and were damaged beyond repair. Casualties were not as high as expected, but many good men were lost in just the landing phase.

Fortunately for all, the landings went virtually unknown to the Japanese and the troops were not brought under fire during these critical first few days. As a result, the troops quickly secured and then cleared the fields so that the transport planes could air-land the reinforcements to secure the area. Transport aircraft, using the glider "snatch" technique, were removing the injured even before the landing strip was readied.

From the advantage points secured by the Commando Force, the airport at Myitkyina was seized by Merrill's Marauders and air transport operations into Myitkyina was initiated and continued from May 17 until the end of the war.

In all, the Air Commandos flew 74 glider sorties for General Wingate's forces and a total of 96 by the end of the Myitkyina operation. Although this number hardly compares with later airborne assaults in Europe, it is highly significant if for no other reason than for the experience obtained and the lessons learned.

The glider had proved itself as a transport of men and materials, and the feasibility of supplying ground troops by air transport was no longer a theory. Airborne troops were seen as a viable way to spearhead an invasion, to reinforce already engaged ground troops, to change the strategy of the battlefield and to gain the initiative.

As a proving ground the Burma operation probably set the scene for the coming airborne invasion of Normandy. General Eisenhower immediately called for some of the Wingate-Cochran team to proceed directly to Europe to aid in the airborne preparations for D-Day.

THE INVASION OF NORTH AFRICA

By Bart Hagerman, LTC, AUS (Ret)

When the U.S. Army issued Field Manual 31-30, *Tactics and Techniques of Airborne Troops*, in May 1942, it stated that parachute drops were considered "the spearhead of a vertical envelopment of the advance guard element of air landing troops or other forces." Although the statement envisioned the seizure of airfields there was still some doubt about how paratroops would be deployed. At that early date, the United States had never committed airborne troops to combat and most of our airborne doctrine was spawned by the use our enemy Germany had made of this new tool of warfare.

Needless to say, our commanders were anxious to see what parachute troops could do in combat, but not any more anxious than those early parachute troops longed to prove their capabilities. This then was the situation when planning began for the invasion of North Africa. It would be a dress rehearsal for the eventual invasion of the European mainland, and the future of parachute troops in the American scheme of things would be at stake.

Not until early October 1942 did General Eisenhower decide to commit a parachute unit to the North African invasion. The 2nd Battalion, 503rd Parachute Infantry Regiment (later redesignated the 509th Parachute Infantry Battalion) had arrived in England in June 1942 and had been attached to a British Airborne Division for training. This battalion was selected to make the first American combat parachute assault.

From the beginning things went wrong. The battalion, led by Lieutenant Colonel Edson D. Raff, was assigned a troop carrier group of young, inexperienced personnel who had had no training in dropping paratroops. In addition, the mission required that the aircraft had to be flown some 1,500 miles from England to a point near Oran in Algeria, a tactical flight that would be a tough assignment for a veteran carrier group.

Raff was a tough, Gung-Ho type of commander, and the men were anxious to prove themselves, so they went about their difficult assignment with a spirit of enthusiasm. They were determined to make a difference in the invasion and also to vindicate the theory of parachute troops.

Adding to the confusion was the question as to whether the French forces would fight or permit the Americans to land unopposed. Two plans were drawn-up by Major William P. Yarborough, airborne advisor to General Mark Clark. One was called "Plan Peace" which would be put into effect if the French chose not to oppose the Allied invasion. The other was designated "Plan War" and as suggested, if the French opposed the landing, this plan meant a real baptism of fire for the troopers.

Still uncertain about the French, the paratroops took off at 2130 hours on 7 November 1942, from two airdromes in southern England in 39 C-47 transport planes of the 60th Troop Carrier Group. Major Yarborough had received permission from General Clark to accompany the troops.

The mission was to seize the Tafaraoui and La Senia airports near Oran. Whether the troopers jumped in or were airlanded depended on the decision of the French as to whether they would fight or welcome the Americans. The troopers were ready, however, and primed for a fight if necessary.

It was the first night formation flight for many of the pilots and many of the navigators had just joined the group before the operation. The planes formed well and maintained good order until they neared Spain. Then fog and the difficulties of night flying began to break up the formations. By dawn there were no formations of more than six planes.

Although just before takeoff the word had been passed that "Plan Peace" was to be in effect, the French and Vichy soured at the last moment and decided to resist. Hurried messages to the planes, now in flight, were sent but never received. Someone had given the operator the wrong frequency and the calls went unanswered.

Homing signals from a war vessel some 25 miles off the coast were also ineffective as the operator there was sending on the wrong frequency. Everything that could go wrong, seemed to. Even the operator at Tafaraoui, thinking the mission has been cancelled, turned off his radar set at 0100. The planes were left flying into a hostile area, unarmed, unescorted and thinking their landing would be unopposed.

Major Yarborough's and another plane, flying low over the coast trying to ascertain their location, saw one plane of the group with a train of parachutes on the ground nearby. The jumpers had all been taken prisoner by the Spanish authorities. Another plane was still aloft. The three-plane group continued up the coast, realizing that they were several hundred miles west of Oran.

Meanwhile, another group of 21 planes which had joined together over the Mediterranean Sea headed for the airport at LaSenia expecting to peacefully land at their objective. As the lead ship arrived, the French opened up with anti-aircraft fire and made it quite evident that "Plan War" was in effect. Turning away from the field, the group flew a short distance away and landed on the Sebkra D'Oran, a large, dry salt lake bed west of Oran. Just about all of the planes were low on gas and there was a need to talk things over to decide the proper course of action.

While they conferred, a couple of hostile riflemen in the hills began to fire on the planes. One round hit a land mine and the aircraft went up in a cloud of smoke. The explosion wounded the pilot, but the troopers fortunately had unloaded and were not harmed.

Two squads were dispatched to wipe out the snipers as six more planes appeared in the sky. This was the group led by Colonel Raff. One of the pilots on the ground radioed the approaching planes that they were under fire from the north. Raff looked down from his plane and saw a column of tanks approaching the location from the north. He mistakenly assumed that this was the source of fire being directed toward the troops on the ground.

Making a split-second decision, Raff gave the word for the troopers in the six planes to jump to help protect the landed element from attack. Although the ground looked smooth, it was covered with rocks and the troopers all had a rough landing. Raff crashed into a sharp rock and cracked two of his ribs.

As the tank column approached, the troopers saw the large white stars and realized that the column was an American unit and not enemy after all. The column was part of American Combat Command B, an armored force that had landed on the beaches west of Oran earlier that morning.

The planes that had carried Raff and his troopers landed. Shortly after, Major Yarborough and his three-plane group approached and also landed. The two officers met and tried to get things sorted out. One of the missing planes had landed at Gibraltar after becoming lost over Spain during the night.

The 509th PIR marched in review in England for Eleanor Roosevelt. Three days later they were parachuting into North Africa. (Photo from <u>Operation Torch</u> by William Breuer)

Two others, running low on fuel, had landed and had been taken prisoner by the French.

Two other planes had landed in French Morocco at Fez Airdrome and their occupants had been interned. The remaining four were forced to land in Spanish Morocco and they were rounded up and taken prisoner. They were finally released three months later, but they were lost to this operation. Only about 300 of the original 556 paratroopers were present and ready for duty.

Since most of the planes were nearly out of gas, the troopers, weighed down with their heavy equipment and ammunition, started out on a forced march some 35 miles to Tafaraoui, their original objective. Colonel Raff, spitting blood from his jump injury, was forced to ride in one of the armored column's jeeps. Meanwhile, Major Yarborough led the troopers on the gruelling march.

About two hours later, the marching troopers caught up with Raff and the armored column. They had received word by radio that Tafaraoui airfield was under the control of American tank units and over 500 prisoners had been taken. The armored force commander was requesting that some of the paratroops be flown in to take over control of the airfield so that they could pursue the retreating enemy.

As the troopers were still some 25 miles from the airfield, Yarborough suggested that they get three of the planes—that had enough gas for a short hop—back at the dry lake bed, to pick up a company of troopers and fly to the airfield. He volunteered to lead the force. The remainder of the troops would continue the long march overland.

A half hour later the planes picked up the troopers and headed for the airfield. Before reaching the field, the planes were attacked by two French Dewoitine fighter aircraft. Machine gun bullets cracked through the crowded transport planes killing and wounding the troopers.

Evasive action helped little and the three transports finally were forced to crash-land on the rough terrain. The troopers scrambled for what cover was available and the fighters strafed their ranks, wounding and killing more.

One of the troopers wounded here later died of his wounds.

He was 21-year-old Private John "Tommy" Mackall of Wellsville, OH. The Airborne Center in North Carolina would later be named Camp Mackall for this young trooper.

Still some 15 miles from Tafaraoui, the troopers, led by Yarborough, set out for their objective. After an all-night march, the exhausted force reached Tafaraoui early the next morning. They took over the guard duties and the armored force left immediately to pursue the retreating Germans.

The first combat assault by American paratroops had been a confused and disappointing action. The men felt robbed of their chance to show their capabilities. They were not to have long to debate the action, however, as another mission was not long in coming.

In early November the Germans and Italians were busy reinforcing their lines in Tunisia, and Eisenhower's troops were reading their drive toward Tunis. On November 10, Colonel Raff was alerted to conduct a second parachute assault somewhere to the east within five days. The battalion was to be placed under the operational control of General Anderson's British 1st Army.

Raff reported to Anderson and learned that the unit's next mission was to seize the north coastal road network leading into Tunis and thus cut off the flow of supplies to Rommel's Afrika Korps. It was a critical mission and there was precious little time to prepare.

The troopers were moved to Maison Blanche airfield near Algiers and in record time were readied for the operation. Final plans called for the troopers to jump to secure the airfield at Tebessa. It was known to be controlled by the French and although it was believed the landing would be unopposed, they were to be ready in the event resistance was met.

Some 24 hours before the time the plans were to take off, Raff learned of another airfield some 15 kilometers from the one at Tebessa. This airfield, named Youks les Bains, was larger than the one at Tebessa. After a hurried conference at General Anderson's headquarters, the mission was alerted to designate the Youks les Bains airfield as the primary objective. From Youks les Bains a company would be sent overland to secure the field at Tebessa.

At 0730 hours on November 15, 22 C-47 transport planes carrying two companies of troopers lifted off the runway at Maison Blanche and headed for Youks les Bains. A fighter escort insured an uneventful flight over the Mediterranean and then toward the African coast.

Dropping to an altitude of 300 feet, the planes came in over the airfield. Things were quiet below as the 350 men of the battalion leaped from the planes and landed without a shot being fired. Twenty minutes later, they were off the drop zone and assembled, awaiting orders.

As one company began digging in around the airfield, Raff sent the other company off toward Tebessa. Then Raff met with the French commander who welcomed the Americans and readily agreed to join them in defense of the field against the Germans. It was then that Raff saw that the French had the field zeroed in with interlocking machine gun fire and artillery. Had the French elected to oppose the battalion's landing they would have been cut to ribbons. Again the trooper's luck had held.

For the next week, Raff and his troopers guarded the airfields. The only contact with their German enemy came when a German pilot, thinking the field at Youks les Bains was in German hands, landed his JU-88 in the field. He was greeted with a hail of small arms fire and immediately took off only to crash in the nearby hills.

From Youks les Bains, Raff led his forces on short forays across the border into Tunisa to hit the Germans and Italians defending around Gafsa. These forays became more and more frequent and deeper into enemy-held territory as the paratroopers found themselves involved in straight infantry missions.

As the German buildup gained headway, the Allied offensive slowly ground to a halt. The flow of supplies to Rommel's troops became a primary concern and an all-out bombing effort was mounted. Despite repeated missions, the materials seemed to continue to get through and finally, the paratroopers were called upon to assist in the effort.

A bridge near the city of El Djem seemed to be a weak link in the north-south railroad line supplying the Germans, but try as they did, the bombers had been unable to score a hit on the structure. It was decided that a special raiding party was to be dropped by parachute to place charges and blow the bridge.

Raff's paratroopers, now officially designated as the 509th Parachute Battalion, was given the task. A special team was formed consisting of an officer in charge, five demolition experts, 25 riflemen and two French Army paratroopers. The two Frenchmen knew the area and spoke fluent Arabic. The team was also to carry 400 pounds of TNT which would be dropped in two equipment bundles.

The jump went smoothly although some valuable time was lost in hunting for one of the equipment bundles. Only after several hours of marching along the railroad track, did they finally realize that the plane had dropped them south of the bridge instead of north. They had been marching away from the bridge instead of toward it!

As it began to get light, the party realized they had no chance of reaching the bridge, so they decided to set their charges on the railroad track, scatter and by pairs make their way back to American lines some 110 miles away. The decision was a good one as just as the charges had been completed, enemy patrols were sighted.

The explosives ripped up some 100 yards of track, destroyed switching controls and effectively cut the supply line for weeks. The troopers were not so lucky, however. Most of them were picked up by German or Italian troops and confined as prisoners of war. Out of the 33-man raiding party, only eight made their way back safely. Later, as the 509th fought on, 16 others escaped their captors or were liberated by Allied forces. The remaining nine were presumed killed by the enemy.

This was the last airborne mission for the 509th in Africa. The unit fought in several engagements as straight infantry, but the airborne phase of the African campaign was over. Only the European continent lay ahead and, with it, a new chapter in airborne warfare.

Major General Mark W. Clark planned and commanded Operation Torch, the invasion of North Africa. Major William P. Yarborough (R) planned and participated in the airborne phase. (Photo from <u>Operation Torch</u> by William Breuer)

SICILY OPERATIONS

By James M. Gavin, Lt. Gen., USA (Ret)
Edited from his book, <u>On To Berlin.</u>

Editor's note: Lieutenant General (Ret) James M. Gavin passed away Thursday, 22 February 1990. Stricken with Parkenson's Disease, which was complicated by a stroke and pneumonia, the General displayed great courage and determination. He was laid to rest at his beloved West Point on February 27th, with private services for the family. A memorial service, attended by the members of the government and many of his former associates, was held at the New Chapel, at Fort Meyer, Virginia, on March 6, 1990.

The beloved commander of the 82nd Airborne Division passed away in his 82nd year, during the 50th Anniversary of the Army Airborne forces for which he had worked and had served. Rest easy, General, your job is done. A grateful nation salutes you for your lifelong commitment and dedication. Airborne!!

The plan was simple. Taking off from Tunisia in a long column of aircraft, we were to fly via the island of Linosa to Malta. There we were to dogleg to the left, coming in on Sicily's southwestern shore. This was an important point—the island was to come into sight on the right side of the approaching aircraft. The orders were that every man would jump even though there might be some uncertainty in his mind as to his whereabouts. No one but the pilots and crews were to return to North Africa.

Due to the high winds, the entire air armada was blown far east of its intended landing zones. Some pilots made landfall along the eastern coast of Sicily and, having done so, turned back to find their way around to the southwest coast. Several planeloads actually jumped in front of the British Army. The troopers were from the 3rd Battalion and Regimental Headquarters.

The 2nd Battalion, commanded by Major Mark Alexander, was the next farthest to the east to land. It landed about 15 miles east of Gela, near the town of S. Croce Camerina, a town that figures prominently in the Thucydides' account of the Peloponnesian Wars.

Although they landed amid a number of huge pillboxes and areas organized for defense, they were quite successful in reorganizing the battalion. The Italian pillboxes were formidable affairs, several stories high, with apertures here and there, and so sited as to overlook other pillboxes. The troopers quickly learned that the way to reduce them was to keep firing at the slits until a trooper could get close enough to throw a grenade into them. The battalion fought most of the night and by daylight had assembled a majority of its men. It then moved toward the coast near a village overlooking the town of Marina di Ragusa, and they organized an all-around defense for the night.

At daylight, July 11, his battalion turned north and moved in the direction of S. Croce Camerina, using donkeys, donkey carts and wheelbarrows to help carry the weapons and ammunition. At noon his battalion captured S. Croce Camerina and later that afternoon captured Vittoria. They rejoined the 82nd Airborne Division on July 12.

My own flight with the Regimental Headquarters group was uneventful until Linosa was due. It was not to be seen. Malta, which was to be well-lighted to assist our navigators, could not be seen either. Suddenly, ships by the score became visible in the ocean below, all knifing their way toward Sicily. Obviously, we were off course, since our plan called for us to fly between the American fleet on the left and the British on the right. In fact, the Americans told us that we would probably be shot down if we flew over them. We continued on, finally doglegging to the left on the basis of time calculation. Soon the flash of gunfire could be seen through the dust and haze caused by the pre-invasion bombing, and finally the coast itself could be seen off to the right. Unfortunately, many of the planes overflew the Malta dog-leg, and the island first became visible on the left, thus causing confusion and widespread dispersion of the troopers.

We turned inland; the small-arms fire increased; the green light over the jump door went on, and out we went. The reception was mixed. Some of us met heavy fighting at once, others were unopposed for some time, but all were shaken up by the heavy landings on trees, buildings and rocky hillsides.

I continued to move cross-country in a direction that would take me around the area where we had had the fire fight. We could hear intense firing from time to time, and we were never sure when we would walk into another fire fight or how we would get into the fight since we couldn't tell friend from foe. Then there was the problem of enemy armor. I decided to look for a place where tanks would be unlikely to travel and where we could get good cover to hold up until dark. I wanted to survive until dark and then strike across country again to the combat team objective.

It had been a long day. We waited and waited for the setting sun. Soon the Sicilian sun was low in the sky and quickly disappeared like a ball of fire into a cauldron. We went into the Sicilian night, heading for what we hoped was Gela, somewhere to the west.

About 0230 hours we were challenged by a machine-gun post of the 45th Division, and at last, we had re-entered our own lines. We learned that we were about five miles southwest of Vittoria. In about another mile we came to the main paved road from the beach to Vittoria, passing by a number of foxholes and dead Italian soldiers. By then I had about eight troopers with me.

We then went on to the edge of Vittoria, where I was able to borrow a jeep. I had heard rumors that there were more paratroopers a few miles away in the direction of Gela. I continued on toward Gela and to my surprise came across the 3rd Battalion of the 505th, in foxholes in a tomato field and just awakening. The battalion commander Lieutenant Colonel Edward Krause, whose nickname was "Cannonball," was sitting on the edge of a foxhole dangling his feet. I asked him what his battalion was doing. He said that he had been reorganizing the battalion and that he had about 250 troopers present. He had landed nearby and had rounded everybody up. I asked him about his objective, several miles to the west near Gela, and he said that he had not done anything about it. I said we would move at once toward Gela and told him to get the battalion on its feet and going. In the meantime I took a platoon of the 307th Engineers, commanded by Lieutenant Ben L. Wechsler. Colonel Krause said that there were supposed to be Germans between where he was and Gela and that the 45th Division had been having a difficult time.

Using the platoon of Engineers as infantry, we moved at once on the road toward Gela.

By then it was broad daylight, about 0830. In less than a mile, we reached a point where a small railroad crossed the road. On the right was a house where the gatekeeper lived. There was a striped pole that could be lowered to signal the automotive and donkey-cart traffic when a train approached. Just ahead was a ridge, about half a mile away and perhaps a hundred feet high. The slope to the top was gradual. On both sides of the road were olive trees and beneath them tall brown

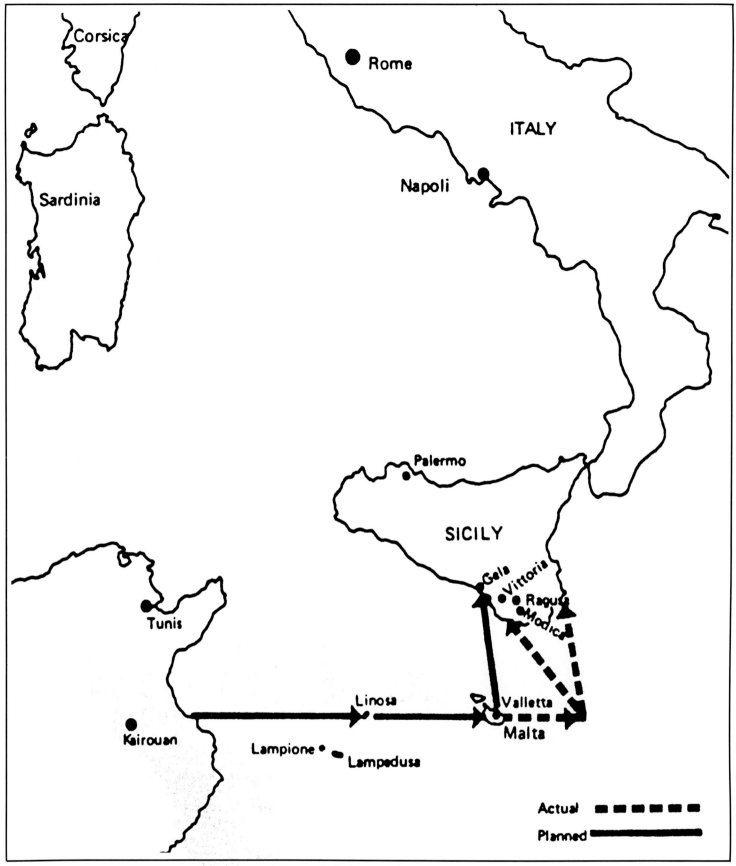

Air Route to Sicily. (Courtesy of James M. Gavin)

and yellow grass, burnt by the hot Sicilian summer sun. The firing from the ridge increased. I told Lieutenant Wechsler to deploy his platoon on the right and to move on to seize the ridge. In the meantime I sent word to Cannonball to bring his battalion up as promptly as he could.

We moved forward. I was with Wechsler, and in a few hundred yards the fire became intense. As we neared the top of the ridge, there was a rain of leaves and branches as bullets tore through the trees, and there was a buzzing like the sound of swarms of bees. A few moments later Wechsler was hit and fell. Some troopers were hit; others continued to crawl forward. Soon we were pinned down by heavy small-arms fire, but so far, nothing else.

I made my way back to the railroad crossing, and in about 20 minutes Major William Hagen joined me. He was the battalion executive for the 3rd Battalion. He said the battalion was

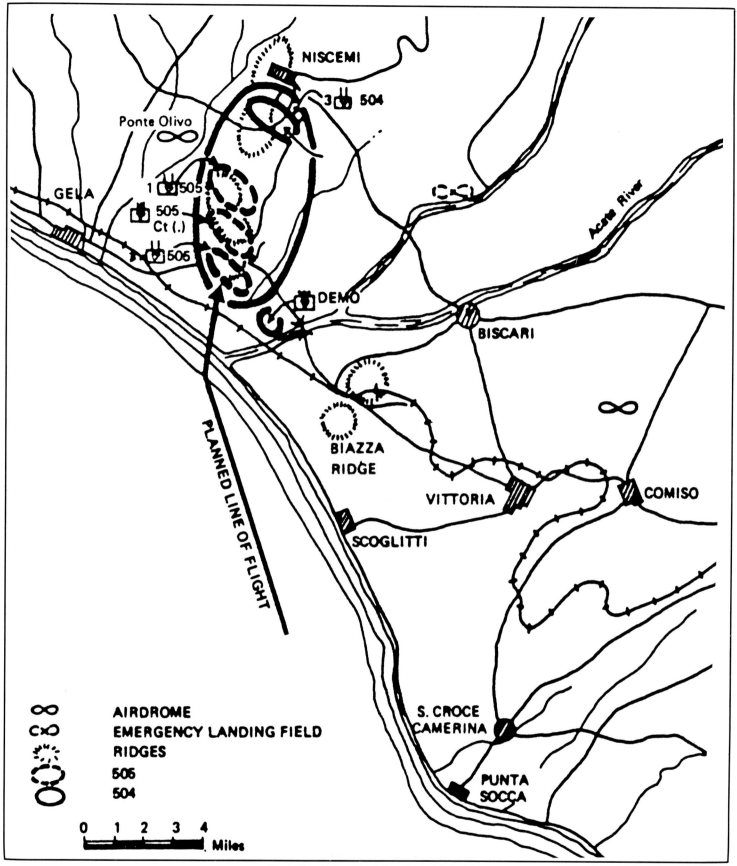

Attack Plan, Sicily. (Courtesy of James M. Gavin)

coming. I ordered Hagen to have the troops drop their packs and get ready to attack the Germans on the ridge as soon as they came. By that time we had picked up a platoon of the 45th Division that happened to be there, part of a company from the 180th Infantry. There were also a sailor or two who had come ashore in the amphibious landings. We grabbed them also.

The attack went off as planned, and the infantry reached the top of the ridge and continued to attack down the far side. As they went over the top of the ridge, the fire became intense. We were going to have a very serious situation on our hands. This was not a patrol or a platoon action. Mortar and artillery fire began to fall on the ridge, and there was considerable machine-gun fire. I was worried about being enveloped on the right; some of the 45th Infantry Division should have been down on the left toward the beaches, but the right was wide

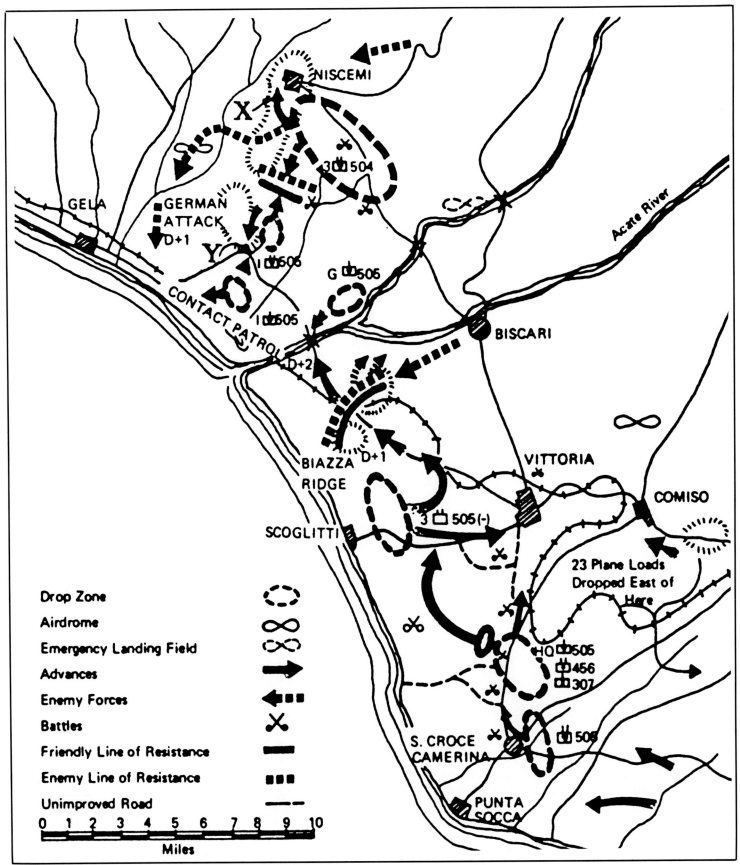

Actual Landings, Sicily. (Courtesy of James M. Gavin)

open, and so far, I had no one I could send out to protect that flank. If the German column was coming from Biscari, the tactical logic would have suggested that they bypass me on the right and attack me from the rear. At that time I had a few engineers I kept in reserve and two 81mm mortars. They were commanded by a young officer, Lieutenant Robert May, who had been my first sergeant almost a year earlier when I had commanded C Company of the 503rd Parachute Infantry. He sent two or three troopers off to the right as a security patrol. Later a Mountain Pack 75 mm. artillery piece from the 456th Parachute Artillery joined me, then another. Occasionally troopers, having heard where we were, would come in from the direction of Vittoria.

The first German prisoners also came back. They said they were from the Hermann Goering Parachute Panzer Division. I remember one of them asking if we had fought the Japanese

82nd Airborne troopers on the attack along the ridgeline the morning after the jump. (Photo taken from <u>Out of the Blue</u> by James A. Huston)

in the Pacific; he said he asked because the paratroopers had fought so hard.

At the height of the fighting the first German Messerschmitts appeared overhead. To my surprise, they ignored us and attacked the small railroad gatekeeper's house repeatedly.

About four o'clock a young ensign, who had parachuted with me the first night, came up with a radio and said he could call for naval gunfire. I was a bit nervous about it because we didn't know precisely where we were and to have the Navy shooting at us would only add to the danger and excitement of what was turning out to be quite a day. We tried to fix our position in terms of the railroad crossing over the road, and he called for a trial round. He then called for a concentration, and from then on the battle seemed to change.

In about an hour I heard that more troopers were coming, and at six o'clock I heard that Lieutenant Harold H. Swingler and quite a few troopers from Regimental Headquarters Company were on the road. He arrived about seven o'clock. In his wake appeared half a dozen of our own Sherman tanks. All the troopers cheered loud and long; it was a very dramatic moment. The Germans must have heard the cheering, although they did not know then what it was about. They soon found out.

I decided it was time to counterattack. I wanted to destroy the German force in front of us and to recover our dead and wounded. I felt that if I could do this and at the same time secure the ridge, I would be in good shape for whatever came next—probably a German attack against our defenses at daylight with us having the advantage of holding the ridge. Our attack jumped off on schedule; regimental clerks, cooks, truck drivers, everyone who could carry a rifle or a carbine was in the attack. The Germans reacted, and their fire increased in intensity.

Soon, we overran German machine guns, a couple of trucks, and finally, we captured twelve 120mm Russian mortars, all in position with their ammunition nearby and aiming stakes out. Apparently the troopers had either killed, captured or driven off the German crews. The attack continued, and all German resistance disappeared, the Germans having fled from the battlefield.

In their postwar interrogation report, the Germans claimed to have made a vigorous attack only to be met head-on by American parachutists. It was later in the day that the Americans counterattacked and destroyed the German command. A number of Germans and some vehicles fled across the Ponte Dirillo. As reported by the Office of the Chief of Military History, Washington, D.C., in *Axis Tactical Operations in Sicily*, the remainder "became panicky and fled in disorder. Some officers succeeded in bringing them to a stop just short of Biscari, where they took up defensive positions." The regimental commander went to the headquarters of the Hermann Goering division, under which he commanded a battle group, to explain his actions. The postwar interrogation report comments cryptically, "He was relieved and subsequently court-martialed." Finally, the disaster that befell his eastern column caused General Conrath to change his plans entirely and to pull back his column near Gela, which on July 11 had been meeting with significant success. As reported in the official history of the war, "The paratrooper stand on Biazza Ridge prompted Conrath to change his plans. Learning of the heavy losses being sustained by his infantry-heavy force, he decided, apparently on his own initiative, to break off contact with the Americans near Gela. So the troopers at Biazza Ridge did made a contribution to the accomplishment of the regimental mission, which was to block the movement of the German forces near Gela. In a roundabout way perhaps, and in a way that none of them realized at the time, it is now clear that they accomplished just that.

The 1st Battalion of the 505th, commanded by Lieutenant Colonel Art ("Hardnose") Gorham, had the mission of landing on the high ground, north of the Y, so as to block the movement of any Axis forces southward, to control the Ponte Olivo airfield by fire and to assist in reducing the Y if necessary.

Hardnose Gorham had about a hundred troopers under his control. He was in the valley astride the road from Niscemi down to the Y, and he decided to consolidate his position there for the time being. About 0700 a German armored column was seen about 4,000 yards away, coming from Niscemi. It was the Hermann Goering western column. It was preceded by a small advance guard of two motorcycles and a Volkswagon. The paratroopers kept concealed and let the advance guard get into their positions, where they killed or captured them. At that point the armored column stopped, having heard the firing on the advance guard. It was then about 3000 yards away. The Germans deployed their infantry, which looked to the troopers to be about two companies, approximately 200 men. The troopers kept concealed and let the Germans approach to within 100 yards. They then pinned them down on open terrain. Most of the Germans were killed or captured. The German tanks came on, however, but two of them were knocked out by bazookas, and two were damaged. The tanks then withdrew. Thus ended the first day's advance of the Hermann Goering Division's western column against the beaches.

After defeating the German force, Colonel Gorham reassessed the situation and decided that he should move on to his objective, Piano Lupo—which was the high ground off to his left—since it was terrain that controlled both the airfield and the road. Using about 50 prisoners, the troopers carried their wounded with them, moved on to the high ground and organized it for defense. Having accomplished that and remembering his mission, Colonel Gorham sent Captain Sayre with a small detachment (about a squad) to attack the Y.

As Sayre approached the fortification around the Y, heavy naval gunfire began to fall near the road about 100 yards north of the pillboxes. The pillboxes appeared to be

82nd troopers, using any transportation available, move through Vittoria, Sicily on 13 July. (Photo from Out of the Blue by James A. Huston)

sheltered in defilade. Sayre then directed one of the Italian prisoners to go to the pillboxes and demand their surrender. He ordered the prisoner to tell the occupants that if they did not surrender, the colonel would bring the naval gunfire right down on top of them. (Actually, he didn't have any communications with the Navy, but the men in the pillboxes didn't know that.) They surrendered, and the paratroopers occupied the pillboxes at about 1045.

Colonel Gorham and his small group of troopers and the lieutenants from the 3rd Battalion, 504th, accomplished all the missions assigned to the entire regimental combat team. It was a remarkable performance, and I know of nothing like it that occurred at any time later in the war. Sadly, in the fighting the following morning, Colonel Gorham engaged a tank with a bazooka, and he himself was killed. His death was a great loss to the division.

That same night, at Biazza Ridge, learning that the Germans had completely withdrawn from the action, I moved my command post from the top of the ridge back about a half mile under the olive trees. I deployed the troopers for the night, expecting an attack from the direction of Biscari to come into our right flank, probably at daylight.

It must have been about 2200 hours when all hell broke loose in the direction of the beaches. Anti-aircraft fire was exploding like fireworks on the Fourth of July, tracers were whipping through the sky, and as we were observing the phenomena, the low, steady drone of airplanes could be heard. They seemed to be flying through the flak and coming in our direction. Everyone began to grasp their weapons to be ready to shoot at them. A few of us cautioned the troopers to take it easy until we understood what was going on. Suddenly, at about 600 feet the silhouettes of American C-47s appeared against the sky—our own parachute transports! Some seemed to be burning; and they continued directly overhead in the direction of Gela. From the damaged planes some troopers jumped or fell. At daylight, we found some of them dead in front of our positions.

Later we learned that it was the 504th Parachute Infantry that was being flown to a drop zone near Gela to reinforce the 1st Infantry Division. General Ridgway had been there to meet them. Unfortunately, the Germans had sent in parachute reinforcements on the British front to the east the same night. In addition, there had been German air attacks on our Navy, so when the parachute transports showed up, our ships fired at them, 23 were shot down. Many were damaged.

That morning of July 13 found Patton well-established in Sicily. It had not been easy. The presence of the Hermann Goering Division had been a complete surprise to us; none of the pre-battle intelligence summaries of the higher headquarters reported the Goering Division as even being in Sicily. Not only had it been there, but it had been placed in exactly the right location to maximize its chances for throwing the amphibious forces back into the sea. Having defeated it, Patton at once began to drive his divisions forward. On July 16 Agrigento was seized, and on the following day, the 82nd Airborne Division relieved the U.S. 3rd Infantry Division. Patton began to see the possibilities of an end run to Palermo. He flew back to La Marsa, Tunisia, to place before General Sir Harold Alexander his plan for such an attack on July 17. It began to appear to the staffs of Patton's army that unless his mission was changed, he would soon be relegated to the dull role of guarding the British 8th Army's rear.

He gave Patton the green light, and on the following day, July 18, Patton organized a Provisional Corps consisting of the 2nd Armored Division, new to battle; the U.S. 82nd Airborne Division; and the 39th Regimental Combat Team from the U.S. 9th Infantry Division.

At 0500 on July 19 the attack jumped off. According to the official report of the war, it was "little more than a road march," but it was a strange affair, unlike anything else I encountered during the war. True, it did seem like a road march, but suddenly, a machine gun or antitank weapon would open up, then the white flags would appear. A shot had been fired for "honor," but it was just as likely to cause casualties as a shot fired in anger. The 2nd Armored Division moved rapidly. It was a spectacular affair. Clouds of dust billowed into the sky for miles and at times obscured the sun. By late evening of July 22, Palermo was occupied.

By July 22 the U. S. 1st Infantry Division was well-established ashore. Fortunately, it was a veteran division; otherwise, it surely would have broken under the weight and magnitude of the German attack.

The 505th Parachute Infantry, in an all-night march, arrived in the town of San Margherita late on July 22. The following morning, moving by truck, the 505th Parachute Combat Team swung west to the town of Trapani. As we reached the outskirts of Trapani, we encountered a roadblock and heavy artillery fire. The regiment deployed, swept into the town with no casualties and accepted the surrender of the admiral in command. Thus ended the Sicilian campaign for the 82nd Airborne Division.

With the fall of Messina, the battle for Sicily was over. The 82nd Airborne Division moved back to Africa to ready itself for the next campaign: Italy. Who would our opponents be— Would the Italians fight once we landed on the mainland? We were especially concerned with German Panzers; there were rumors that many of them had escaped across the Strait of Messina. As we later learned, they had. And we were to fight them once again.

So the battle of Sicily was not only a military success; the military lessons learned were to prove invaluable to the Allies in the battles that were to come.

Most important were the lessons learned in how to organize and deliver our airborne troops. We at once set up a professional training unit at Biscari Airfield in Sicily. Its purpose was to develop specially-trained pathfinder units, consisting of experienced pilots and reliable paratroopers who would land about 20 minutes before the main assault. A pathfinder team consisted of one officer and nine enlisted men, reinforced by enough protective troopers to insure the success of their mission. The team would be equipped with electronic gear on which the following troop carrier pilots would home. They would also be equipped with lights both to mark the drop zone and to help the landing paratroopers to reorganize.

In charge of the training at Biscari were Lieutenant Colonel Joel Crouch of the Air Corps and Captain John Norton of the 82nd Airborne Division. The program was very successful, and as the war went on, more and more pathfinder teams were trained to land ahead of resupply missions whenever they became necessary. We also learned to move on our objectives immediately on landing; we observed that the first minutes in the parachute operation, when the paratroopers have the initiative, are important to both their survival and the capturing of their objective. The Airborne experience in Sicily proved valuable to us in our later battles and in helping train the green units and individuals coming from the United States.

PACIFIC OPERATIONS
By Richard N. Loughrin

The Leyte, Philippine Islands, Campaign

On 18 November 1944, after a seven-day convoy on nine ships from Oro Bay, New Guinea, the 11th Airborne Division, commanded by Maj. Gen. Joseph M. Swing, landed amphibiously without opposition on the east coast of Leyte, Philippine Islands, between Abuyog and Terragon, 44 miles south of Tacloban.

The division was not originally scheduled for employment during the Leyte campaign and was to have staged for subsequent operations. On 22 November the division was ordered to relieve 7th Infantry Division units along the line Burauen-La Paz-Bugho. The general mission was to seize and secure all eastern exits from the mountains into Leyte Valley in its area. It was then to advance through the central mountain range, secure the western exits from the mountains and assist the 7th Infantry Division in its drive north toward Ormoc.

Intelligence indicated that the enemy was planning a coordinated ground and airborne attack to seize U.S. airfields in the vicinity of Burauen and deny their use to U.S. forces in a Luzon invasion. It was also reported that the Japanese had been constructing a supply trail from Albuera, on the west coast nine miles south of Ormoc, across the mountains to Burauen. The 11th was also to secure that trail, the Japanese combat lifeline.

From the east coast at Dulag, 10 miles west to Burauen, the sandy beaches turned into dense swamps and then just west of Burauen the central mountain range rose abruptly from Leyte Valley to peaks over 4,000 feet in height. The mountains were covered with lush tropical rain forest or jungle which limited visibility to a few feet except in rare clearings. Many of the deep gorges were impassable even for foot soldiers. No roads went through the mountains but there were narrow foot and game trails from one locality to another which led over rocky, swiftly running streams and deep gorges. A slip meant a drop of 30 to 40 feet and the trails were often so steep that foot holes had to be cut into the hillsides, and soldiers were forced to pull themselves up with vines and branches. The dense foliage made flank security virtually impossible on the trails.

On 25 November, after relieving the 17th Infantry, the 511th Parachute Infantry Regiment, commanded by Col. Orin D. Haugen, marched west from Burauen for Mahonag, 10 miles away. They moved on two routes with 1st Battalion, 511th, commanded by Lt. Col. Ernest H. LaFlamme on the right and 3rd Battalion, 511th, commanded by Lt. Col. Edward H. Lahti, on the left. The almost impassable terrain, incessant monsoon rain and numerous Japanese ambushes made passage extremely difficult. It was not possible to move as units. In small parties, sometimes even less than a squad, the 511th, later followed by 2nd Battalion, 187th Glider Infantry Regiment, commanded by Lt. Col. Arthur H. Wilson, moved west.

The journey defied description. Sucking clay mud, jungle vines and vertical inclines exhausted men before they had marched an hour. Long before they reached the west coast, the

The 511th attacked Nichols Field over the Paranaque River and the Paranaque River Bridge became a hotly contested structure.

11th Airborne Division, Mike VI Operation, Map C: Defenses of Southern Manila. (Courtesy of Richard N. Loughrin)

constant rain and marching in and out of streambeds would rot the clothing and boots from the soldiers' bodies and feet. Contact with the enemy was limited to small engagements with scattered patrols until 27 November. C Company, 511th, was moving west on the northern route accompanied by Col. Haugen and a small regimental command group when they were ambushed in a deep gorge by a strong Japanese force. The company was pinned down except for small groups that were scattered and eventually made their way back to 1st Battalion, 511th.

On 28 November the lead elements of 3rd Battalion, 511th, closed at Manarawat, a small deserted village atop a cliff above the Daugitan River and 1st Battalion, 511th closed in the vicinity of Lubi, across the river. The next seven days were spent locating and relieving C Company, 511th, patrolling forward to Mahonag and clearing a small drop zone/light airstrip at Manarawat. It was soon clear that overland resupply and medical evacuation were impossible. Efforts to use native carabao for transport proved fruitless as even those hardy beasts were unable to negotiate the rugged trails and streambeds. From that point on, all resupply was provided by parachute drop or pushed from aircraft. During the period the 511th was reinforced by A Battery, the 457th Parachute Field Artillery Battalion which was parachuted in on the drop zone.

Some walking wounded were escorted over the rugged trails to Burauen, a few were airlifted out in light observation aircraft, but the majority of those unable to walk remained with the regiment until the campaign ended and were carried out on improvised litters to the west coast. On 4 December a division forward command post and tent hospital were established at Manarawat.

On 5 December the 1st and 3rd Battalions, 511th moved west on parallel routes to Mahonag. 3rd Battalion, 511th, closed in Mahonag on 6 December, followed closely by 1st Battalion, 511th, and Regimental Headquarters. 2nd Battalion, 511th, commanded by Lt. Col. Norman M. Shipley, departed Burauen on 29 November and located the Japanese Supply Trail near Anonang on 3 December. On 6 December, 2nd Battalion, 511th, moved west against heavy resistance toward Mahonag on a third route on the right. They were followed by 2nd Battalion, 187th Infantry.

At dawn on 6 December, the predicted attack on U.S. airfields in the Burauen area began when an estimated 200 Japanese attacked Buri strip. At dusk on the same day, 350 to 400 Japanese parachutists jumped on San Pablo and Buri airfields. The 127th Airborne Engineers, division supply personnel, elements of Division Headquarters Company, 511th Airborne Signal Company, 408th Airborne Quartermaster Company, Headquarters Division Artillery and the entire 674th Para-Glider Field Artillery Battalion, acting as infantry, destroyed the Japanese force in five days of heavy fighting.

On 7 December, the 511th continued the attack west from Mahonag. I Company, 511th, led and met heavy resistance as it cut the Japanese main supply trail on a small hill, later called

Maloney Hill. After a brief fire fight, 3rd Battalion, 511th followed the trail south and seized and occupied a large hill, later called Rock or Haugen Hill, which dominated the trail. H Company occupied a perimeter on Maloney Hill. That night, and for the next three days, the 3rd Battalion was subjected to numerous Banzai attacks as they moved south to continue the attack toward the coast, and the Japanese frantically fought to regain control of the trail.

On 8 December, after numerous scattered fire fights, 2nd Battalion, 511th, closed in the vicinity of Mahonag. On that day Private Elmer E. Fryar, E Company, 511th, was posthumously awarded the Medal of Honor for extraordinary gallantry above and beyond the call of duty. 2nd Battalion, 511th, patrolled to the north and west and conducted ambushes in the area. A second hospital and resupply point was established at Mahonag.

A strong enemy force was encountered in defensive positions on the next large ridgeline south of Rock Hill. 3rd Battalion, 511th, conducted unsuccessful assaults against the position beginning 8 December and on 11 December, G Company was ordered to bypass to the west of the hill and move across country to envelop the position and make contact with 7th Division units holding in defensive positions at the western base of the mountains. Due to a lack of radio batteries, G Company laid wire; however, shortly after leaving the perimeter, the wire was cut. There had been no resupply to the Battalion since 5 December and G Company would be out of contact with other U.S. forces and completely without rations or ammunition resupply until it entered friendly lines on 17 December.

As the 511th attempted to continue the attack, Maloney Hill was temporarily abandoned and was immediately reoccupied by a strong enemy force that constructed heavily fortified positions. This isolated regimental units on Rock Hill from 511th at Mahonag. 2nd Battalion, 187th, closed in the Mahonag area on 15 December and established ambushes and patrolled to the north. During the period 12 through 18 December, numerous attacks were conducted by units of all three battalions of the 511th and Maloney Hill was finally retaken on 18 December by B Company, 511th, with heavy supporting artillery fire. As 2nd Battalion moved forward they received heavy artillery fire and sustained numerous casualties. Lt. Col. Shipley was seriously wounded.

Throughout this period aerial resupply was hampered by rain and fog and the destruction of L-4 and L-5 aircraft in the Japanese attacks on the airfields. Severe shortages of rations, medical supplies and some types of ammunition existed. All units endured days without food and some received no rations for up to a week or more. The 511th Medical Detachment, commanded by Regimental Surgeon Maj. Wallace L. Chambers, performed surgery on the bare ground and accomplished medical miracles in keeping the swelling numbers of wounded in action alive.

On 20 December, following heavy artillery and mortar preparatory fires, H and I Companies, 511th, attacked up the almost vertical, vine-entangled side of the ridge, later known as McGinnis Hill, and seized the position. I Company, 511th, continued the attack south along the ridge and advanced 250 yards against strong opposition.

On 21 December, 2nd Battalion, 511th, now commanded by Maj. Frank S. Holcombe, passed through the 3rd Battalion perimeter and forward positions to continue the attack along the Japanese supply trail down the ridge, later known as Hacksaw Ridge. F Company, 511th, followed by D Company encountered Japanese machine guns about 100 yards from I Company's position. They drove the enemy from four successive knolls and advanced 900 yards southwest.

On 22 December 2nd Battalion, 511th, continued the attack with D Company followed by F Company assaulting at dawn, with what later was called the "Rats Ass Attack". They pushed the attack vigorously and the Japanese were unable to reorganize. About 500 yards from its night position D Company assaulted a Japanese bivouac area, killing some 200 Japanese and advancing 2,500 more yards. 2nd Battalion, 511th, killed 750 Japanese that day and the Burauen-Albuera trail was cleared of all organized hostile forces. As 2nd Battalion, 511th, approached within sight of 32nd Infantry Regiment units, they were held up by division order; and 2nd Battalion, 187th, passed through 2nd Battalion, 511th, made contact with the 7th Infantry Division and moved on to the beach.

On 23 and 24 December, the days were spent carrying litter patients down to the beach and mopping up Japanese stragglers. On Christmas day the 511th moved out of the mountains to the beach on Ormoc Bay. With the mission completed, units were motor marched back to the east coast and closed in a bivouac area near Abuyog on 27 December.

While the 511th and 2nd Battalion, 187th, were fighting their way through the mountains, 2nd Battalion, 188th, and 1st Battalion, 187th Glider Infantry Regiments had contained a large enemy position in the hills northwest of Anonang. They attacked on 22 December supported by the fires of three field artillery battalions. The position was neutralized on 27 December.

The total enemy dead counted by the division during the campaign reached 5,760; prisoners of war totaled 12; and the ratio of enemy killed in action to division killed in action was 45:1.

The Luzon, Philippine Islands, Campaign

After the Leyte Campaign the 11th Airborne Division was made part of the new 8th Army, commanded by Lt. Gen. Robert L. Eichelberger, which was assigned to execute a landing on Luzon as a subsidiary to the main invasion effort at Lingayen Gulf. Earlier in the month, the 6th Army had landed at Lingayen Gulf far to the north and was now approaching Manila. The division, less the 511th Parachute Infantry, staged on the shores of Leyte Gulf and departed for Luzon 27 January 1945 in APDs, transport ships, as part of a convoy.

On 31 January the division (minus), with the 188th Regimental Combat Team in the assault waves, began landing at Nasugbu along Luzon's west coast, 55 miles south and west of Manila, against light opposition. The successful landing was turned into a full-scale invasion by Gen. Eichelberger who ordered the division commander to land the remainder of the division and to advance inland along Highway 17 to Tagaytay Ridge with all possible speed.

The 188th preceded rapidly inland, surprised the Japanese and captured the important Palico River bridge intact. The 188th, with the 1st Battalion of 187th Infantry attached, continued the march through the night. On 1 February increasingly heavy resistance was met from the Japanese line extending across the highway from Mt. Carilao to Mt. Batulao. Company A of the 188th broke through the enemy line and seized Mt. Aiming, the terrain feature dominating the center of the line. The suddenness of the penetration and power of the regimental attack scattered the Japanese to the wooded slopes of the mountains where they never again offered organized resistance.

By 1800 on 2 February, the 1st Battalion, 187th, leading the attack along Highway 17 was two miles short of the west end of Tagaytay Ridge. On the morning of the 3rd the troops were hit by a heavy concentration of fire from enemy positions on Shorty Ridge, just west of Tagaytay Ridge. After intense fighting the battalion and the 188th, supported by accurate A-20 airstrikes, succeeded in overrunning and capturing the deeply caved enemy position on Shorty Ridge.

On the morning of 3 February, about 1,750 troops of the 511th Parachute Infantry began dropping along Tagaytay Ridge, unopposed by the enemy. The last echelon, 457th Parachute Field Artillery Battalion, jumped on the 4th. Contact by the 511th with the 188th Infantry was made abut 1300 on 3 February by patrols westward. The 511th secured the eastern end of Tagaytay Ridge where Highway 17 turned sharply north and downhill toward Manila.

Generals Eichelberger and Swing decided to have the reinforced 188th Infantry hold Tagaytay Ridge and reduce the enemy pocket on Shorty Ridge while the 511th Infantry pushed north toward Manila with all possible speed, the 188th to follow when ready. This plan was a change in mission for the division, originally to contain enemy forces in southern Luzon and to patrol in its area of responsibility.

On 4 February, the 511th alternately marched and was moved by motor transportation in battalion-size shuttles on Highway 17 toward Manila. The 2nd Battalion cleared Imus and Las Pinas and the 1st Battalion passed through and was stopped at 2130 in Paranaque on the outskirts of Manila at the bridge over the Paranaque River by heavy enemy mortar and artillery fire from the Genko Line on the north side of the river.

The Genko Line was designed to defend Manila from the originally expected American landing in southern Luzon. It consisted of a line of concrete mutually supporting pillboxes, commencing at Paranaque and extending in depth 6,000 yards to include the Manila Polo Club at Libertad Avenue. The line extended east across Nichols Field and was anchored on the high ground along Laguna de Bay, south of Mabato Point. The rear of the line was based on the high ground of Ft. McKinley. Five-and six-inch naval guns were emplaced in concrete structures, facing south, and 20, 40, and 90mm. antiaircraft cannon were strategically placed to aid the ground defense.

On the morning of 5 February, the 511th crossed the bridge over mined roads and under heavy fire from Japanese artillery which started north along Highway 1 over a quarter-mile wide strip of land lying between the river on the east and Manila Bay on the west. During the next several days the regiment fought its way 6,000 yards northward house by house and pillbox by pillbox. Supported only by light artillery the 511th depended heavily upon direct assaults with flame throwers, demolitions and 60 and 81mm. mortars in its advance.

The 2nd Battalion, 187th Infantry arrived on 8 February to support the attack of the 511th. The 188th, with the 1st Battalion of the 187th attached, closed in the vicinity of Las Pinas on 6 February. Under cover of darkness the regiment executed a rapid wide envelopment of the southern defenses on Nichols Field and attacked toward the center of the enemy line on 7 February.

The division sent the 188th Infantry against Nichols Field from the south and southeast while one battalion of the 511th attacked from the west across the Paranaque River. For four days they attacked in the face of concentrated artillery, mortar and machine gun fire from the enemy defenses. Support fires of A-20s and the division light artillery had not destroyed enough Japanese weapons to permit the infantry to advance without taking unduly heavy casualties. The volume of fire from Japanese naval guns was still so great that one infantry company commander requested: "Tell Halsey to stop looking for the Jap fleet. It's dug in on Nichols Field." On 10 February the division passed to 6th Army control and quickly coordinated heavier artillery fire plans.

On 10 February, the 188th Infantry succeeded in breaking through the southern defenses of Nichols Field.

On 11 February, the 511th attacked north along the bay front in its sector to Libertad Avenue, losing its commander Col. Orin D. Haugen who was mortally wounded in the chest by a 40mm. shell. Lt. Col. Edward H. Lahti took command of the regiment.

The division advance north was halted on Corps order on 11 February when front lines extended from Libertad Avenue east to Taft Avenue Extension. Patrols made contact with elements of the 1st Cavalry Division the same day.

On 12 February the division mounted a concerted attack against Nichols Field with artillery and mortar concentration and an air strike. The 1st and 2nd Battalions, 187th Infantry, and the 188th Infantry attacked and cleared most of the field by dusk. The airfield formed the center of the Genko Line. The positions were mainly a network of concrete pillboxes and gun emplacements connected by underground tunnels. Barbed wire ringed the area. Naval guns of various caliber supported the defenses.

From 13 through 19 February, the 11th Airborne's three regiments and elements of the 1st Cavalry Division attacked and cleared all the approaches to Fort McKinley. During the attack PFC Manuel Perez, Jr. of Company A, 511th Parachute Infantry, was awarded the Medal of Honor for heroic action in reducing Japanese pillboxes that held up the advance of his company. By 19 February, the bulk of the Japanese had fled the fort.

From 18 through 23 February, a division task force of three infantry battalions closely supported by artillery, tank destroyers and Marine aircraft besieged the enemy Abe Battalion on high ground at Mabato Point on the northwest shore of Laguna de Bay. Enemy losses were about 750 killed; the division had 60 casualties.

On the night of 21-22 February, a special task force composed of the 1st Battalion, 511th Infantry; 1st Battalion, 188th Infantry; 675th and 472nd Field Artillery Battalions; Battery D, 457th Field Artillery; and 672nd Amtrack (amphibious tractors) Battalion and Provisional Division Reconnaissance Platoon, moved to concealed positions in the vicinity of Mantinlupa on the west coast of Laguna de Bay, a large freshwater lake southeast of Manila, for an attack on the Los Banos Prison Camp early on the 23rd. The purpose was to rescue 2,147 American and Allied civilian internees who were being held prisoner by the Japanese.

The prison camp was 25 miles behind enemy lines at Los Banos on the southern shore of Laguna de Bay. An estimated 8,000 enemy soldiers were in the immediate area. A move in force against the camp would have given ample warning to the Japanese and probably have resulted in a mass slaughter of the internees.

At 0700, shortly after dawn on 23 February, 89 paratroopers of B Company, 511th Infantry, jumped from C-47s on a small field 1,000 yards from the camp. At the same moment the 1st Battalion, 511th, hit the beach in amphibious tractors and the Division Reconnaissance Platoon, set at key points the night before, annihilated the Japanese garrison of more than 250 soldiers who were doing their morning exercises, with their rifles stacked as the first trooper jumped from the plane.

The 1st Battalion, 188th Infantry, attacked enemy forces in the Lecheria Hills area on the north flank prior to establishing a road block to protect the evacuation. The 675th and 472nd Field Artillery supported the ground operations, and Battery D, 457th, supported the amphibious attack.

By 1700 the same day all the internees had been evacuated by Amtrack to Muntinlupa and the last soldier left the Los Banos beach a few minutes before Japanese reinforcements arrived. Casualties were two internees and a few soldiers, slightly wounded. The speed of the daring, carefully-timed attack and rescue came as a complete surprise to the Japanese.

On 24 February, the 511th and 187th Infantry deployed in the area between Lake Taal and Laguna de Bay to develop the enemy main line of resistance. The 158th Regimental Combat Team, which was attached to the division, on 3 March attacked to the south and east along the route Tuy-Balayan-Lemery-Batangas. The combined effort reduced the strong ring of enemy outposts of Mt. Bijiang, southwest of Los Banos; Hill 580, northeast of Santo Tomas; Mt. Sungay; Hill 660, west of Tanauan; and Cuenca and Batangas. This forced the troops of Lt. Gen. Fujishige back to the line: Los Banos, Santo Tomas, Tanauan, Malvar and Mt. Macolod.

Meanwhile, in order to secure the southern shores of Manila Bay, the 1st Battalion, 188th Infantry and 457th Field Artillery Battalion attacked a Japanese force in the Ternate area southwest of Cavite. The American force captured the town of Ternate and 40 Japanese Q boats on 2 March and pushed southwest to reduce the enemy cave defenses in the jungle covered mountains of Mt. Pico de Loro. Fighting in this sector consisted of the systematic reduction of a series of well-concealed underground defenses, under adverse conditions and in difficult terrain. Aerial resupply for this operation was conducted by L-4 planes. Mopping up operations were completed on 1 April.

On 23 and 24 March, the division moved from the area between Lake Taal and Los Banos to west of the lake in the Cuenca-Batangas area. The 158th Regimental Combat Team was released from division control and the 511th Infantry, less one battalion, was held in 6th Army reserve. The 1st Battalion, 188th Infantry, continued operations in the Ternate area.

The 187th Infantry assaulted the formidable defenses in the Macolod-Dita-Bukel Hill triangle while the 188th, less the 1st Battalion and with the 3rd Battalion, 511th attached, protected the division right flank by a series of lightning-like moves which destroyed the enemy in the Lipa Hill-Rosario-Tiaong area. Lateral contact was established with the 1st Cavalry Division in Lipa on 30 March.

Successive task forces (Division Reconnaissance Platoon, Company B, 188th Infantry, Battery B, 457th Field Artillery Battalion and 90 guerrillas) were dispersed by air and landing craft at Lucena, 22 miles behind enemy lines. These units organized guerrilla forces in that sector and moved west to contact elements of the 188th which were heading eastward along Highway 1 toward Candelaria. By 8 April, the division had secured southern Luzon to the east coast and had sealed off the southeast Bicol Peninsula of Luzon as far south as Calauag.

The general defensive plan of Lt. Gen. Fujishige in Southern Luzon was a gradual retirement of Japanese forces in the well-prepared positions in and around Mt. Malepunyo. On April 6. the 3rd Battalion, 511th Infantry and 2nd Battalion, 187th Infantry, attacked northeast from Lipa in the face of enemy artillery fire to seize Japanese positions in the Malepunyo foothills. On 12 April, the 511th Infantry, less one battalion, returned to division control and concentrated in the Lipa area relieving the 2nd Battalion, 187th and developing the southern flank of the enemy main line of resistance.

The 187th Infantry's two battalions launched a coordinated attack on the enemy stronghold at Mt. Macolod and annihilated that garrison on 20 April.

On 22 April, the corps commander assigned the division to take over the mission of the 1st Cavalry Division in the Malepunyo hill mass, and the next day the 188th Infantry was assigned the mission of the 8th Cavalry.

On 27th April, the division with the 8th Cavalry and F Troop, 7th Cavalry attached, launched a coordinated attack from four directions against the Fuji Heidan position in the Malepunyo Hill mass. The 511th Infantry closed in from the south, the 188th Infantry from the north and the 8th Cavalry Regimental Combat Team held on in the northwest. A composite battalion of C Company, 511th Infantry, D Battery, 457th Field Artillery Battalion and F Troop, 7th Cavalry, advanced from the east. Those attacks were supported by air strikes, seven battalions of artillery and 4.2 mortar fire. The forces, supplied by air and native carrier, moved across the difficult terrain to seize intermediate objectives and converge on Mt. Malepunyo on 1 May. This junction ended all organized resistance in Southern Luzon and "mopping up" operations cleared the areas of hostile remnants.

The final operations of the division in the Luzon campaign were conducted on 23 June to seize Aparri, the last Japanese port in Northern Luzon, and to block the enemy retreat before the northward advance of the 37th Division. The reinforced battalion combat team, Gypsy Task Force, was selected for the mission. It included the 1st Battalion and Companies G and I of the 511th Infantry and Battery C, 457th Parachute Field Artillery. The parachute elements jumped on Camalaniugan Airfield and seven gliders, used for the first time in the Southwest Pacific area, brought in artillery and equipment. Guerrillas had already seized the port of Aparri, so the entire force turned south and by forced marches contacted the 37th Division near the Paret River on 26 June.

Operations of the division during the Luzon campaign were greatly aided by the use of 5,000 organized Filipino guerrillas. Various units were attached to the 11th Airborne Division during the campaign. The light airborne division successfully completed an extended ground action of 105 days of continuous contact with the enemy.

In the operation from 31 January 1945 to 30 June 1945, the division freed over 7,200 square miles of southern Luzon from Japanese control. Enemy casualties were 9,458 killed and 128 captured. The division's casualties were 353 killed and 1,326 wounded in action.

Selected Sources:
Records of the 11th Airborne Division and of the 511th Parachute Infantry include the following:

World War II Records Division, National Archives and Records Service (NARS), General Service Administration (GSA)

Flanagan, Maj. Edward M., Jr. The Angles: A History of the 11th Airborne Division, 1943 - 1946. Washington: Infantry Journal Press, 1947.

Robert Ross Smith. The United States Army In World War II, The War in the Pacific, Triumph in the Philippines. Office of the Chief of Military History, Department of the Army, Washington, D.C., 1963.

Report After Action with the Enemy, Operation Mike VI, Luzon Campaign, from 31 January, 1945 to 30 June, 1945. Headquarters, 11th Airborne Division.

M. Hamlin Cannon. THE UNITED STATES ARMY IN WORLD WAR II, The War in the Pacific, Leyte: The Return to the Philippines. Office of the Chief of Military History, Department of the Army, Washington, D.C. 1954.

Pacific Operations of the 503rd PRCT
By Robert Flynn, Unit Historian

Early in September 1943, General Douglas MacArthur and his chief of air operations, Lt. General George C. Kenney, were in Port Moresby, New Guinea planning strategy which would provide the necessary airstrips from which fighter planes and bombers could pound Rabaul. Kenney proposed to use the 503rd Parachute Infantry Regiment in a drop on the airstrip at Nadzab, north of Lae, New Guinea. General MacArthur insisted on going along with Kenney as an observer.

On 5 September 1943 seventy-nine C-47s emptied the entire regiment from 700 feet precisely over the drop zone in 4 1/2 minutes. The first mass parachute combat jump in the Pacific Theater during World War II in MacArthur's words, "was a honey."

The estimated 10,000 Japanese troops in the area around Lae were being pinched between Australian troops and the American paratroopers. During the fighting that ensued the enemy was dispersed and disapperared into the jungle apparently seeking an escape route. The first mass combat jump in the Pacific had been an overwhelming success and Nadzab was built into one of the largest U. S. airfields in New Guinea.

On Biak Island which borders the northeastern shores of New Guinea, an estimated 10,000 Japanese troops had constructed an elaborate network of caves, tunnels and gun emplacements. In May 1944, the 41st Infantry division ran into a stiff Japanese defense and intelligence indicated that Biak was being reinforced with troops and supplies from Noemfoor Island 75 miles to the west.

The 503rd Parachute Infantry Regiment, patrolling the jungle around Hollandia, was alerted to prepare for a parachute drop on Kamiri Airfield. Navy scouts reported 3,000 Japanese troops on the island and other estimates placed the number of enemy at twice the amount. Since there were not enough transport planes available to drop the entire regiment at one time, one battalion would be dropped on three successive days.

On 3 July Colonel George M. Jones was in the lead plane approaching Noemfoor Island. The jump was to be made at 500 feet but a faulty altimeter caused the lst Battalion to jump at less than 400 feet. Of the 739 troopers who had jumped, 72 were injured, some seriously. When the 3rd Battalion jumped the following morning from 500 feet, 56 men were injured on the rock-hard surface. Colonel Jones then requested that the 2nd Battalion be brought in by landing craft.

As patrols fanned out from Kamiri airstrip, violent and brutal warfare erupted in the snake-infested swamps and jungles all over Noemfoor. In an attack on a Japanese gun position Sergeant Ray Eubanks of Company D distinguished himself by single-handedly wiping out a gun position while protecting the lives of the men in his platoon. Sergeant Eubanks was awarded the Medal of Honor posthumously.

After Noemfoor was secured, the 503rd Parachute Infantry Regiment was joined by the 462nd Parachute Field Artillery and Company C, 161st Airborne Engineer Battalion after which the unit would be known as the 503rd Parachute Regimental Combat Team.

Fighting on Leyte had already commenced when the 503rd Regimental Combat Tean arrived in November, 1944. Colonel Jones and some other officers had preceded the unit to prepare for military operations proposed for the invasions of Mindoro.

The large island lay 200 miles south of Manila and 300 miles north of Leyte Gulf. The Pentagon had advised MacArthur that the mission was too risky because it was within easy reach of powerful enemy airfields. Admiral Kincaid also objected because in order to get to Mindoro, the task force would have to pass through the dangerous Suirigao Strait and the Sulu Sea where the vessels would be easy marks for Japanese land-based planes.

Nearly 18,000 assault troops and 9,500 Air Force personnel were assigned to the task force and were being transported in a convoy consisting of a cruiser, destroyers, destroyer transports, landing craft of every size, minesweepers and some small craft. On 13 December, near the Sulu Sea a Kamikaze pilot plunged his plane into the flagship cruiser Nashville killing 135 and wounding 200 others.

The landing on Mindoro was unopposed. The airstrip was captured and a perimeter was established. Allied P-38s battled Kamikaze planes overhead while naval vessels tried to ward off the suicide bombers that were plunging toward them. General George Marshall was to report later that the Mindoro operation had forced the Japanese to reckon with the possibility of an invasion from the south.

Corregidor fell to the combined airborne assault of the 503rd Parachute Regimental Combat Team and the amphibious landing of the 3rd Battalion 34th Infantry.

At 0830 hours on 16 February 1945, the Combat Team jumped onto two tiny drop zones on the island's 600 foot plateau. For a week the fighting was so fierce that one officer described it as, "a massacre in an insane asylum." Demolition teams systematically sealed off caves only to have the enemy dig themselves out and continue to fight in vain. Finally on 26 February a number of the enemy estimated at about 200, holed up in a large cave and a lucky round fired into the cave entrance ignited a huge mass of explosives. The blast was so great that it showered debris upon the offshore fleet and tossed an M-4 tank almost 50 yards. Utter carnage was everywhere, American and Japanese. Total Japanese losses counted on Corregidor were 4,509. The 503rd suffered 163 killed and 620 wounded or injured.

The final campaign was to seize the central and southern islands of the Philippines. MacArthur had pledged to the people that the entire archipelago would be liberated. The 503rd Regimental Combat Team made an amphibious landing on Negros Island at Bacolod City on 7 April 1945. The Japanese commander elected to withdraw all forces into the mountains of north-central Negros and prepare for a long siege. Token forces were left in the coastal plains to harass approaching American troops and to demolish bridges and supplies. The 503rd paratroopers doggedly pursued the enemy attacking each day as the enemy withdrew further into the mountains. By mid-May no further organized enemy resistance could be found. The fighting wound down and following the capitulation of the Japanese government on 15 August 1945, General Kono surrendered more than 6,000 of his troops to Lt. Col. Joe Lawrie of the 503rd Parachute Regimental Combat Team.

Selected Bibliography:

Belote, James H. and William M., Corregidor - The Saga of a Fortress: Harper & Row, 1967.

Breuer, William, B., Geronimo American Paratroopers in World War II. New York: St. Martin's Press, 1989.

Devlin, Maj. Gerard M., Paratrooper. New York St. Martin's Press, 1986.

Flanagan, Edward M., Jr., Corregidor. California: Presidio Press, 1988.

Guthrie, Bennett, M., Three Winds of Death. Chicago: Adams Press, 1985.

Manchester, William, American Caesar. Boston: Little, Brown and Company, 1978.

Spector, Ronald, H., Eagle Against the Sun. New York: MacMillan, 1985.

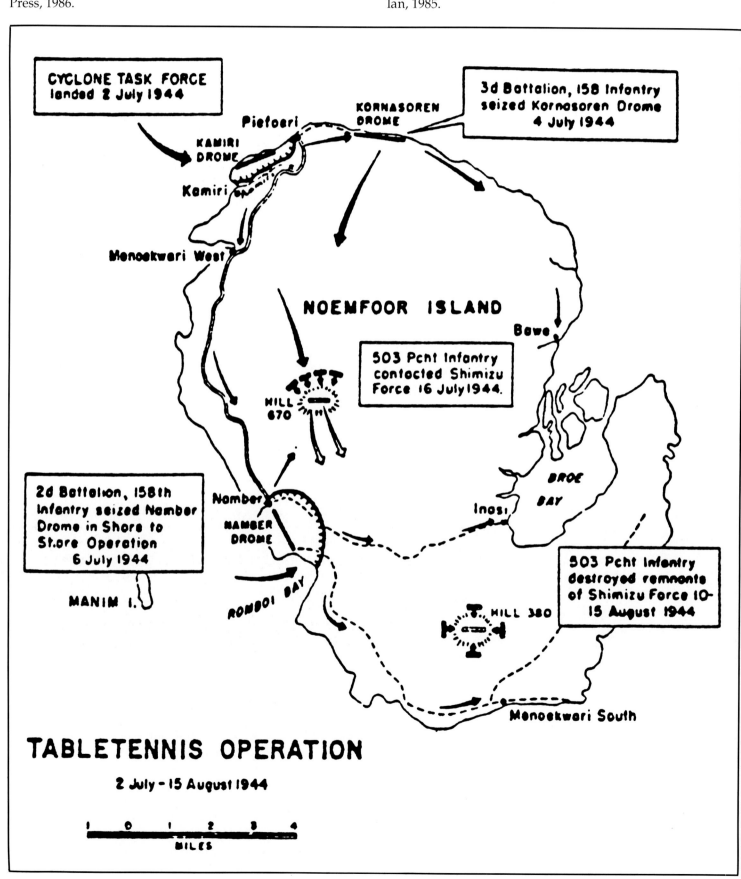

OPERATIONS IN ITALY

By Lt. Gen. James M. Gavin, edited from his book, On To Berlin

Immediately after the battle of Sicily the 82nd Airborne returned to North Africa, arriving there on 20 August 1943. It received reinforcements and equipped and prepared itself for the coming invasion of Italy. For the division commander Major General Matthew Ridgway, and for the staffs of the division, this was an extremely busy period. Plans were made, and then there were changes and more plans and changes. Yet all that was typical of what usually happens in the planning stages of an airborne operation.

The final plan for the amphibious invasion of Italy contained the following major elements: (1) the main assault by the U.S. 5th Army with the mission of landing at Salerno and moving northwestward and capturing Naples; (2) two attacks by the British 8th Army, one against Calabria and another against Taranto; (3) a naval diversionary attack in the Gulf of Gaeta. This was to be known as operation AVALANCHE: named, one can suppose, for the avalanche of combat troops soon to swarm onto the war-weary Italian Peninsula.

To carry out its mission, the 5th Army planned to invade Italy at the Gulf of Salerno, fight northwestward, capture Naples and continue the fight northward until all Italy was overrun. The 82nd Airborne Division was made available to the commander of the 5th Army, General Mark Clark, for this operation. How best to make use of the division now became a problem of much speculation and concern.

The first mission presented to the 82nd called for the seizure by an airborne task force of the towns of Nocera and Sarno at the exits to the passes leading northwest from Salerno. The purpose was to cover the landing and debouchment of the 5th Army from the Salerno area. All available transport planes and gliders, 318 of each, were to be used on the operation.

The staffs made hurried studies to select drop zones for the parachutists and landing zones for the gliders. Our airborne assault was to be supported by an amphibious assault by troops from the infantry division landing in the Amalfi-Maiori area on the Sorrento Peninsula. After considerable study in conjunction with the Air Force of the terrain, ack-ack and possible flight routes, it was decided that the paratroopers of the 82nd would have to be dropped in Nocera Pass on the Sorrento Ridge at altitudes from 4,500 to 6,000 feet, in moonlight.

It was decided, however, on 12 August to throw this whole plan out and use the 82nd Airborne farther inland.

On 18 August General Ridgway was told that the new

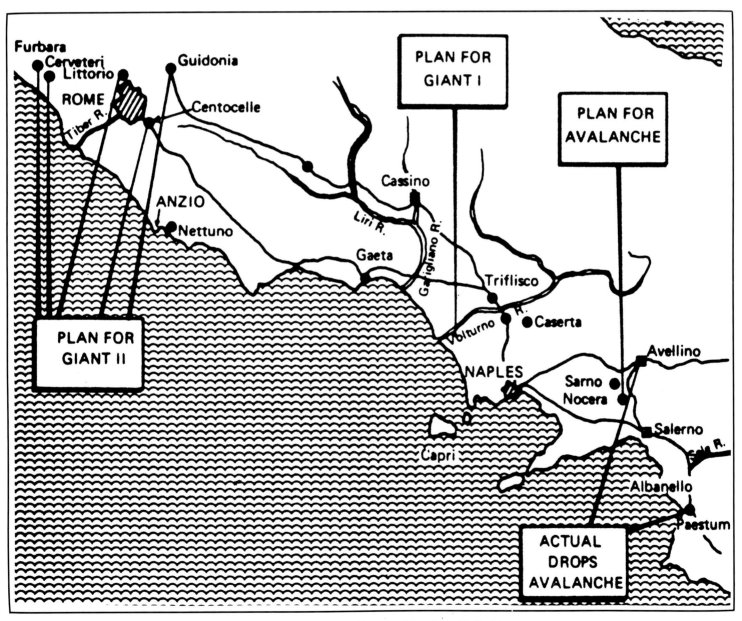

Planned Landings in Italy. (Courtesy of Gen. James M. Gavin)

Troopers of 505 enter Naples, Italy, Sept. 20, 1943. (Courtesy of Steve Mrozek)

decision had been reached to conduct an airborne operation on the Volturno River, northwest of Naples and some 40 miles from the nearest beach landings in Salerno. Initially, this new plan gave the 82nd Airborne Division the sea and of a hold on the Volturno itself against all enemy attempts to move south.

I was designated as the airborne task force commander. Staff planning for the new operation, known as GIANT I, was undertaken without delay at all staff levels concerned.

The planning for the airborne operation was carried through down to the smallest detail. General Ridgway emphatically presented the resupply problem to the 5th Army staff. Approximately three groups of troop carrier transports, about 145 airplanes in all, would be needed to bring in our daily resupply by parachute. Because of the great likelihood of interception by hostile fighters, the supplies would have to be delivered at night on a time schedule that was changed daily.

Concurrently with all this, the 82nd Airborne Division was directed to be prepared to drop a separate parachute battalion on either Battipaglia, Avellino, Nocera, or Sarno, for the purpose of blocking enemy movements to these localities. The 2nd Battalion, 509th Parachute Infantry, was attached to the division. This was the battalion that had jumped in North Africa after a 1,500-mile flight from England in November 1942. Under the able leadership of Colonel Edson Raff, it had fought with great courage and effectiveness.

A final conference was held on 31 August in the headquarters of the 5th Army at Mostaganem. All the senior officers taking part in AVALANCHE were present: General Eisenhower, Admiral Hall, Air Chief Marshal Tedder, General Clark, all British and American corps commanders, and a number of division commanders. General Ridgway was allowed three minutes to talk about the Volturno plan. It immediately caused considerable discussion, after which it was decided by General Eisenhower that the Volturno plan should be cancelled. For the time being, however, the part having to do with the seizure of Capua and the bridges was left in, and it was anticipated that they could be held with five days' of supplies. This was a two-battalion mission, so General Ridgway decided to give it to the 504th. The 505th was kept on an alert status, and I was told that dock landing at Naples or air landing at Rome was possible.

In preparation for AVALANCHE, an extensive bombing program was already under way. The big bombers of the North African Air Corps were working over the boot of Italy from top to toe.

Air Chief Marshal Tedder and General Clark decided in the end that the possible gain from our Volturno mission would not be worth the probable losses. The mission was called off, and thus ended GIANT I.

On the evening of 2 September the division commander and several staff officers were called to the Headquarters, 15th Army Group at Siracusa, Sicily, to receive the first information of a new mission. It was to be the seizure of Rome, one of the most interesting airborne plans of the war, a plan that was extensively discussed and argued about many years after the war. This operation, known as GIANT II, called for placing the strongest airborne task force that the available aircraft could carry, on and near three airfields immediately east and northeast of Rome. Because of the distance of Rome from the proposed take-off airfields in Sicily, there was no possibility of fighter support. The date for GIANT II was the night of September 8-9. The mission was to secure Rome by operating in conjunction with the Italian forces in the Rome area. The airborne lift was to be repeated the next night and as directed thereafter until the mission of the 15th Army Group was accomplished. The airborne part of the operation was to be supported by a landing at the mouth of the Tiber River, also staged by troops of the 82nd Division.

The planning moved at a rapid pace, and the afternoon of 8 September found the troops of the 82nd busily checking orders and plans, loading the para-containers, and making a last check on rations, ammo and weapons.

But I think the fact is that the Italians at Siracusa had simply promised everything—much more than they could possibly have delivered. And it became fully evident that there were no guarantees of the needed support at the airfields where the regiments of the 82nd were to land, so the plans were changed almost at the last minute.

The latest plan provided for the leading assault regiment, the 504th Parachute Infantry, to land on the Furbara and Cerveteri airfields near the seacoast. From there they were to push inland toward Rome. The regiment that was to jump on the second night was the 505th Parachute Infantry. It was to land on Guidonia, Littorio, and the Centocelle airfields, all of which are considerably nearer the center of Rome.

Brigadier General Maxwell D. Taylor, chief of artillery of the 82nd Airborne Division, had been sent through to Rome 24 hours earlier to confer with the leading Italian authorities. He was to inform the Allied commander by code whether or not in his opinion the operation should be attempted. General Taylor was taken by a British PT boat to the island of Ustica, where he was transferred to an Italian corvette that landed him at Gaeta, the scene of much bitter fighting later in the war. He was quickly taken from Gaeta to Rome where he conferred with General G. Carboni, commanding the Italian troops in the Rome area, and the aging Marshal Pietro Badoglio. The marshal and General Carboni agreed that in recent days the German forces in Italy had been greatly increased in strength. They said further that the Italian troops had little or no gasoline and only enough ammunition for a few hours of fighting, that they could not guarantee that all the airfields would be in Italian hands, and that, anyway, our airborne landings would cause the Germans to take drastic steps against the Italians. Therefore, the whole plan as proposed would be nothing less than disastrous.

General Taylor decided they were right. He radioed the prearranged code message, and the operation was called off.

On 13 September about 1330 hours, a tired, begrimed pilot from the Salerno beachhead landed a fighter plane at Licata Field in Sicily. He had an urgent message for the division commander and refused to give it to anyone else. I talked to him on the field, but finally had the chief of staff radio General Ridgway, who had taken off for Termini. General Ridgway came back immediately.

The message was a personal letter from General Mark Clark, 5th Army Commander. It contained an appeal for immediate help. Specifically, he wanted one regimental combat team dropped inside the beachhead south of the Sele River that night, another drop on the mountain village of Avellino, far behind the German lines, on the night of September 14.

The airborne and troop carrier staffs went into a hasty huddle. They reallocated the departure fields, reshuffled troops as necessary and prepared flight plans. An immediate check was made to insure that our own ground troops and our Navy received clear warning of our routes and times with descriptions of our flights. The messenger with General Clark's letter also delivered a plan for marking the drop zone prepared by a 5th Army Airborne staff officer. The troops already in the area would use cans of sand soaked with gasoline, laid out in the form of a large letter "T." They would light them upon the approach of the first flight of transports over the drop zone and douse them out with dirt when the transports had gone.

In addition, special pathfinding homing equipment was to be released on the Sele River beachhead drop zone with the stick from the first airplane. This would then be used to assist the following airplanes to home accurately on the drop zone. All pilots and jumpmasters were carefully briefed on the plan. Such pathfinder refinements could not be used, of course, for

Men of 376 Para. Field Artillery Bn. "enjoy" Thanksgiving dinner in Italy, Nov. 1943. (Courtesy of Steve Mrozek)

the Avellino drop zone, which was well behind the German lines.

All planes were complete eight hours after the request for reinforcements came from General Clark, and the troops were loaded with their complete equipment, rations and ammunition, and the C-47s were rolling down the runways on their way. Shortly after midnight, the 504th Parachute Infantry, with Company C of the 307th Airborne Engineers attached, had made its landing and assembled near Paestum. By daylight it was in the front lines, fresh, eager and looking for a fight.

That same night the 509th Parachute Infantry landed at Avellino. The following night the 505th Parachute Infantry landed at Paestum.

I flew with the 505th. After the Sicilian experience we were all quite apprehensive. However, we were in such a rush to get our proper orders out and assemble the necessary arms, equipment and ammunition that we had little time to think about what was going to happen to us. We took off on schedule. It was a beautiful, clear night with considerable moonlight. Soon after we left the northwest corner of Sicily, the Italian mainland came into view off to the east. We crossed a peninsula jutting out into the Tyrrhenian Sea. In the plane the red warning light came on to tell us that we were approximately four minutes out from the drop zone. We seemed to have been flying over the peninsula forever when a white beach and a river mouth appeared. The scene looked exactly like that in the photos of the correct drop zone. The green light flashed on. There was no burning "T" on the ground as we had been told there would be, but the area appeared to be correct in every way, so out we went.

The first parachutes had barely opened when the great "T" did light up directly under us. To the Germans who occupied the hills, the operation must have appeared bizarre. Units began to reorganize; they assembled without any interference. A combat team was in action by daylight.

The regiment that had jumped the first night, the 504th, had had little sustained combat in Sicily; it was commanded by Colonel Reuben Tucker, a tough, superb combat leader, probably the best regimental combat commander of the war.

Colonel Tucker deployed his two battalions and moved out against the German positions during the darkness. He completely overran the Germans, leaving a number of them in his rear. They reacted vigorously at dawn. Four successive counterattacks, well supported by artillery and tank fire, were made against Tucker's 504th, but it proved to be just as tough as its sister regiment, the 505th, at Biazza Ridge. It destroyed all the Germans and held on to its position.

The other unit that parachuted the same night as the 504th was the 509th Parachute Infantry. The 2nd Battalion of the 509th had been attached to the 82nd Airborne Division. It had participated in the landings in North Africa in the fall of '42. Since then, under the able leadership of Colonel Edson Raff, it had given a very good account of itself. It was now to drop on the small Italian town of Avellino.

Avellino is a typical Italian town nestled in a deep mountain pass about 20 miles from the Salerno beaches. It lies at the junction of several important roads, to Salerno and Battipaglia to the north, and toward which German reserves were likely to come from farther south where hard-pressed German divisions were withdrawing under the pressure of the British 8th Army. As a road center, Avellino was a stopping point for numerous transient units. There were, however, no suitable drop zones in the area; what few flat cleared areas the photographs showed were too small, and the mountains were so high that it was impossible to jump at proper low altitudes.

When the 2nd Battalion, 509th, flown by the 64th Troop Carrier Group, rolled down the runways at Licata Field, Sicily, early on the evening of 13 September and headed for Italy, it was undaunted and confident. But the small town of Avellino looked like many other small mountain towns and the lower air was full of battle haze. From the high altitude at which the mission had to be flown, Avellino proved too hard to find. Few of the transport ships reached their proper drop zones.

The troopers were scattered over an area of more than a hundred square miles. Despite this first great handicap, individual troopers and small units gave a good account of themselves. They mined roads, blew up bridges and destroyed German communications. They ambushed small enemy columns and shot isolated German messengers. The 509th caused a considerable number of German troops to be committed to antiparachute and search work. Of the 640 troopers who jumped, approximately 510 eventually filtered back to the Allied lines.

The battalion had accomplished what General Mark Clark had had in mind. It disrupted German communications and partly blocked the German's supplies and reserves. It also caused the Germans to keep units on antiparachute missions that otherwise could have been used at the point of their main effort at Salerno. In fact, the Germans used many more troops for corrective and preventive purposes against the airborne troops that were committed by the Allied high command.

Salerno had been a touch-and-go affair. It came very close to being a disaster. Higher headquarters believed the fact that the 45th Infantry Division had done so well in Sicily indicated that the inexperienced 36th should be able to match its performance at Salerno. There were two big differences between these operations. First, in Sicily the paratroopers had landed in front of the 45th Division and intercepted enemy troops, knocked out pillboxes and cleared mines until the 45th was well established ashore. In addition, the paratroopers had protected the left flank of the 45th Division from the attack of the Hermann Goering Division. Second, in Italy, German intelligence had discerned the Allied plans well, and German panzer troops were poised and waiting for the amphibious landing of the 36th Division.

General Clark personally took charge on the beachhead, and through his endeavors the situation was stabilized. With the arrival of the 82nd Airborne Division and then the linkup with the British 8th Army, which was working its way up the boot, the battle for the landing had been won.

The 82nd Airborne Division was moved to Amalfi, and the 505th Parachute Infantry moved up the mountain road from Amalfi to the top of the Sorrento Peninsula. There, near the small town of Agerola, where the Amalfi-Castellammare road goes through the tunnel at the top of the mountain, I established my command post. We were at one end of the tunnel and the Germans at the other. On 28 September I climbed to the top of the Sorrento Peninsula and there had my first view of Naples. Billowing black clouds of smoke covered the waterfront, and buildings burned throughout the city. A pall of smoke and dust, so characteristic of a battlefield, hung over the area. I was told to attack and seize the town of Gragnano at the foot of the mountain the following morning.

Fortunately, at daylight on 29 September, I learned that the Germans had withdrawn. I moved at once down the road into the town and then on to Castellammare. The following day we occupied Pompeii and Torre Annunziata. A battalion of the 36th Division had been attempting to make its way up the coastal road, but with little success. It had been working closely with the British 23rd Mechanized Brigade commanded by Brigadier R.H.E. Arkwright. During the night of 30 September the 505th Parachute Infantry was attached to the British 23rd Mechanized Brigade and plans were made for an early jump-off in the direction of Naples.

The day began, as often happened during the early period of the battle for Italy, with the Germans completely out of sight, and our first attacks were blows into the air. When we reached the city limits, we were ordered to halt while a unit of the 23rd Brigade made a reconnaissance into the city. The 505th had been attacking in two columns, the main column on the primary road and the right column, the 1st Battalion of the 505th, on a poor road up along the slopes of Mount Vesuvius. I was standing with the advance guard, discussing the situation, when the regimental S-3 Major John Norton approached. It was almost noon.

"Colonel," he said, "we are to wait until a triumphant entry is organized."

"A triumphant entry!" I exclaimed. "How in the world can we organize such a thing? It takes participation of the natives."

Word came down that General Clark was going to come to the head of the column and that he would lead the triumphant march into the city.

I finally decided that I would enter in a lead jeep, since I had to find my way through the streets, followed by a half-track in which General Clark would ride standing, with General Ridgway beside him, and that the 3rd Battalion of the 505th Parachute Infantry would be immediately behind them in trucks. The plan was to move directly into Garibaldi Square, and as General Clark's vehicle entered the square, the vehicles of the 3rd Battalion of the 505th would make a complete circle around the square and thus seal it off. The troopers would then jump from the trucks and clear all the people out of the square.

It was midafternoon before we were fully organized and Generals Clark and Ridgway took their place in the column. Finding my way in was not as difficult as I had anticipated, and the streets were ominously empty.

Later I learned that thousands of people had massed at the Plaza Plebiscito about a mile away in another part of the city. It was there that the conquerors traditionally had been received, and the people had assumed that that was where the Allied generals would make their triumphant appearance.

The 82nd Airborne Division was at once committed to the task of cleaning up the city and restoring law and order. It was not easy, for there was a great deal of private fighting going on. The next morning the situation was well under control, and we began to clear up the debris, clear the port, get the utilities back in operation and provide food.

The 505th Parachute Infantry continued to attack northward to the Volturno River. The fighting was not too costly, and the Germans were obviously withdrawing. They would usually make a stand by late morning, and after we drove them back and prepared for a heavy attack the following morning, we invariably found that they had withdrawn during the night. Just before we reached the Volturno, General Ridgway called me back to division headquarters and informed me that I was to be the assistant division commander. I hated to leave the 505th, since I had been through so much combat with it, but it would still be in the division with me. On Sunday, 10 October, I was promoted to assistant division commander and made a Brigadier General. General Ridgway arranged for a brief star-pinning ceremony in front of the Questura, the city police station, which we had been using as a headquarters.

As Mark Clark's 5th Army made its way from the Volturno northward and Montgomery's 8th Army attacked northward on the other side of the boot, a gap slowly developed in the center. To fill this gap, Clark asked General Ridgway for a parachute regiment. Ridgway chose the 504th. I made several visits to the 504th when it was in an unbelievable situation in the mountains. The mountains were very high, totally rocky and generally devoid of trees and cover. One hill they fought over was 1,205 meters high. It was very cold at night, frequently rainy and soon the first snow appeared. Unfortunately, most of the troopers were still wearing their summer jumpsuits. The fighting was extremely difficult, with frequent personal encounters and surprises for the unwary combatant. All supplies had to be brought up by mule. There was a chronic shortage of water, food and ammunition, and, of course, the wounded had to be taken out by mule.

The 504th continued to fight in the mountains until well after Christmas. By the time it was taken out, the casualties had been heavy and the weather and terrain the worst in which it was ever to fight. From there, the 504th was brought back to the Naples area and readied for the Anzio landing.

The Anzio landing was made on 22 January 1944, by the U.S. VI Corps commanded by Major General John Porter Lucas. Colonel Tucker originally expected his regiment to remain in floating reserve and possibly not be committed to the battle at all. Very soon, however, his men were ordered over the side and went into the line on the right flank of the beachhead under the command of the U.S. 3rd Infantry Division. They were generally along the Mussolini Canal. The day of the landing was cold and wintry. Anzio itself was a small port from which the land stretched, flat and exposed, to a line of hills some distance away. The hills were never taken; they remained in the hands of the Germans and were used as observation posts. The troopers often talked to me about their Anzio fighting after the regiment joined the division in England. The mud, the cold, the constant harassment by artillery fire and the heavy nighttime infantry attacks lingered in their memories.

Anzio remains one of the most controversial campaigns of World War II. Its launching was a consequence of a number of events that made it almost inevitable. By late November 1943 the German Winter Line in the mountains was well established and was proving to be very effective. By mid-November General Clark was of the opinion that continuing frontal attacks would exhaust his divisions to a dangerous degree. In order to break the stalemate, on 8 November General Alexander had issued instructions to the 5th Army to plan an amphibious landing to seize Anzio, 35 miles from Rome. General Clark, remembering the Salerno experience, was initially quite skeptical, believing the inadequate troops would be put ashore and that there would not be enough shipping available to support them. Nevertheless, he tried to carry out the wishes of General Alexander and began to plan for the operation.

The landing on 22 January was virtually unopposed. The veteran divisions moved out to their initial battle lines, supplies and reinforcements were brought ashore, and the landings could only be described as having been successful.

In mid-November General Ridgway talked to me about the forthcoming landings in Normandy. They were to take place in the spring of 1944, and they would have a large parachute and glider contingent. In response to a request from General Eisenhower he decided to send me to London to participate in the planning for OVERLOAD, the code name for the Normandy landings.

While we were in Italy, we continued to conduct experiments in both parachute landing and glider landing in Sicily. In

Troopers of the 509th Parachute Infantry Battalion practice amphibious assault landing techniques in preparation for their landing at Anzio. (U. S. Army photo)

order to help our glider pilots we trained parachute pathfinders to jump with lights that were then used to mark a landing strip. The lights were placed in pairs about 50 yards apart, and about four pairs were laced the length of the landing strip. We found by experiment that they gave the glider pilots a sense of perspective, and we were quite confident, in view of our training experience, that we could successfully conduct large-scale glider landings at night. When we thought we had achieved a good state of efficiency, we invited a number of the senior officers to come to the troop carrier headquarters in Agrigento, Sicily, to see a night demonstration. Unknown to us, the troop carrier unit that was to provide the glider pilots had been changed and the pilots called upon to make the night landing were entirely inexperienced. The demonstration was a shambles, with gliders landing all over the place. Fortunately, only one pilot was hurt, and not too seriously. We were still convinced, however, that night landing of gliders was entirely feasible.

We changed our minds concerning one aspect of entry into combat. Before Sicily, individuals and small groups were taught to attack the enemy at once, whenever encountered and regardless of the size of the force. With the Sicilian experience in mind, we decided that in some circumstances it would be better for individuals or small groups not to attack overwhelming numbers of enemy; that would result only in their death or capture. Instead, we now began to train troopers to avoid engagements with units they could not cope with and instead to make their way carefully to their objective area and once there gain contact, if possible, with friendlier troops. This proved both realistic and helpful in the Normandy operation.

OPERATION OVERLORD NORMANDY, D-Day 6 June 1944

By Geoffrey T. Barker, LTC, USA (Ret)

Prior to the greatest airborne/amphibious armada of all time, many key integral activities were falling into place. It was apparent to any sound planner, that an allied invasion from the British Isles, had limited possibilities for successful execution. The key to this entire operation was to establish a strong foothold in Europe by breaching the German defenses, and thus allowing entry of the allied forces. Documents have established that Adolph Hitler fully anticipated such an action. The two elusive details that evaded him were - where and when. From these two pieces of information he could ascertan how the allies would attempt to accomplish their mission, (by sea, air or overland), and prepare his forces to repel them. He knew who the enemy was and had a fair estimate of their forces, and was fully aware that their primary objective was to pour into Europe and overthrow the axis forces and the Third Reich.

Less than a week prior to the Normandy invasion, the German High Command had pinpointed the possible invasion site as either Normandy or Brittany, both in Northern France, with the type beaches favorable for such an activity, easily accessible by sea from England. The most favored beach for the allied invasion, according to German planners, was the Pas-de-Calais, a mere twenty-two miles across the English Channel from Dover. Allied planners capitalized upon intelligence reports indicating these beliefs of the Germans.

Appointed as the Supreme Allied Commander on 7 December 1943, General Dwight D. Eisenhower pondered carefully over the courses of actions for the invasion plans. Air Chief Marshall Sir Arthur Tedder was the Deputy Supreme Commander, and the Chief of Staff was Major General Walter Bedell-Smith. Air Chief Marshall Sir Trafford Leigh-Mallory was appointed as commander of Allied air forces and Admiral Sir Bertram Ramsey the commander of Allied naval forces for the invasion. Initially, General Sir Bernard Paget was designated as the coordinator of British and US land forces. He was the acknowledged authority concerning assault landings into Continental Europe. General Paget was not available from his assignment activities in North Africa, and thus could not fill this position. General Bernard Montgomery was named to command the assault forces, and the two primary ground commanders once the beachhead was established, were Lieutenant Generals Omar Bradley (and his US First Army), and Sir Miles Dempsey (with his British Second Army).

Five beaches comprised the target areas of the amphibi-

General Eisenhower visits paratroopers of the 502nd PIR, 101st Airborne Division in the marshalling area prior to D-Day's Normandy jump. 1st Lt. Wallace C. Strobel braces for the General.

82nd Airborne Pathfinders pose for photograph prior to boarding their C-47, 5 June 1944 - Normandy. (Courtesy of Steve Mrozek)

ous forces in the Normandy area of Northern France, along a sixty mile stretch of the beach west of Le Havre. The beaches were code named UTAH to the West, approximately ten miles from the city of Ste-Mere-Eglise; SWORD on the East flank is some fifteen miles from the inland city of Caen. Between UTAH and SWORD, from West to East are OMAHA, GOLD and JUNO beach landing sites. Large ferro-concrete caissons, known as Mulberries, supplemented by block-ships and floating piers, were manufactured in England. These were designed to be towed across the Channel, to form artificial harbors, designed to be connected to the shore line with 'off-ramp' metal bridge works.

A major factor while considering the decision to mount this offensive, was the weather. The Channel crossing would involve some 3,000 vessels to cross the English Channel, and acceptable weather was essential for the airborne phases of the invasion. Because the weather was bad, and predicted to remain so, the Germans were confident that an invasion was not imminent. Field Marshall von Runstedt had corresponded his confidence to the German High Command that the invasion could not be mounted during the early part of June. The weather prior to June, mild and sunny, had changed drastically, and rainstorms lashed England, with visibility too poor to even consider either amphibious or airborne operations. The date fixed for the invasion had been set for 5 June 1944, however on the afternoon of 4 June, despite a forecast of possibly clearer weather, General Eisenhower reluctantly put the operation on a twenty-four hour delay.

Field Marshall Gerd von Rundstedt was recalled from retirement to become the Commander in Chief, West. His responsibilities included the coastal defenses between Continental Europe and the British Isles. Field Marshall Erwin Rommel, commanding Army Group B, with the responsibilities of securing and safeguarding the Northern coastline of France, departed for Germany. His wife Lucie-Maria's birthday was on 6 June, and he too, like Field Marshall von Runstedt, was confident that any invasions were not forthcoming in the immediate future. Coincidentally, again due to the dismal weather, a number of the senior staff members of General Friedrich Dollman's Seventh Army were on leave.

British bombers had been bombing the railway marshalling yards at Trappes, providing serious damage to the rail system, and indicating that invasion plans were 'favoring' the Pas-de-Calais. Operation FORTITUDE was the largest and most successful deception plan executed during the Second World War. This plan consisted of two parts - Operation SKYE, the northern (British) element; and Operation QUICKSILVER, the southern (US) element. A nonexistent 300,000 man strong allied force was reported to be readying to invade into Norway in the north and the Pas-de-Calais in the South. Five of the twenty fictitious US Divisions were reported to be airborne divisions - the 6th, 9th, 18th, 21st and 135th Airborne Divisions. Although totally nonexistent, great detailed plans were formulated and 'passed' to German agents and spies, as part of the diversion from Normandy. Although never activated, orders and even shoulder sleeve insignia were manufactured in support of this plan. A number of German units were moved from the Normandy area to the Pas-de-Calais area as a result of this detailed misinformation. The fictitious Fourth Army held imaginary training drills in Scotland. These non-existent activities were carefully leaked through calculated indiscretions, to diplomatic sources and German agents.

The 5th of June provided little respite, resulting in air reconnaissance and bombing sorties on both sides being restricted to an absolute minimum, however that evening more than 1,000 bomber missions were conducted by Allied aircraft, to include key anti-aircraft batteries along the Normandy coast. Simultaneously, almost 1,000 of approximately 1,100 railroad lines were cut by the underground and partisans supporting the Allies.

At five minutes prior to midnight on 5 June 1944, pathfinders and gliders of the British 6th Airborne Division landed approximately five miles north of Caen. The Normandy invasion - code named Operation OVERLORD, had commenced. As the first invaders landed, approximately 3,500 assorted ships from seven countries, were ferrying the main invasion across the English Channel - Operation NEPTUNE was committed. General George C. Marshall, Chief of Staff of the US Army, had positioned close to 1,000,000 US soldiers and their necessary provisions in Great Britain. More than 156,000 would be ashore in France by night fall of the first day of the invasion. The first US pathfinders commanded by Captain Frank L. Lillyman of the 101st Airborne Division, landed in Normandy at 0015 hours, on D-Day, the 6th of June.

In concert with the Operation FORTITUDE deception plan, bombers dropped aluminum strips in the Pas-de-Calais area, and naval vessels towed reflecting balloons which provided radar. Dummy parachutists were dropped along the French coast, away from the actual invasion sites, contributing more confusion to the German defenders. The 5th Parachute Brigade, commanded by Brigadier J. H. N. Poett, similarly scattered, re-assembled, and secured the area to the east of the bridges. General Richard N. Gale arrived with the gliders of the airlanding brigade of the 6th Airborne Division that evening. The British parachute battalions assembled between the Dives and Orne Rivers, establishing themselves and the bridgehead across the Orne River. This became the semi-permanent home of the 6th Airborne Division for the next two months. During the three months following D-Day, the 6th Airborne Division would lose more than two thirds of its originally assigned 6,000 man force. More than eight hundred would die, almost three thousand would be seriously injured, and almost one thousand would be reported as missing in action.

The massive armada of Operation NEPTUNE, assembled in mid-channel, between Portsmouth, England and Le Havre, France. the First (US) Army, commanded by Lieutenant General Omar Bradley would assault the western sector, with the 1st Infantry Division, (under V Corps), landing at Omaha Beach, and the 4th Infantry Division, (under VII Corps), at Utah Beach. The 2d Ranger Battalion would land midway between Omaha and Utah beaches at Point-du-Hoe. The Brit-

After D-Day eight glider troopers, their faces covered with parachute canopies, lie dead in France with the wreckage of their Horsa Glider in the background. (Courtesy of 101st Airborne Division Association)

ish 2d Army, commanded by General Sir Miles C. Dempsey, would attack to the East. The 50th Infantry Division, (under XXX Corps), in the center of the invasion area at Gold beach; the Canadian 2d Infantry Division and the 48th Royal Marine Commandos would assault Juno beach; and the 3d Infantry Division, the British 1st Special Service Brigade, the 41st Royal Marine Commandos, and the French 4th Commando, would assault the eastern flank at Sword Beach. Elements attacking both Juno and Sword beaches were under the command of I Corps, and the British 4th Special Service Brigade would attack at both Sword and Juno.

As the complicated maneuvering was underway crossing the English Channel, and the amphibious assaults were being aligned, intricate coordination was under way in the air. At approximately 2230 hours, 5 June, air transport aircraft began departing from England, carrying the intrepid paratroopers of the 6th, 82d and 101st Airborne Divisions towards their bloody destinations.

The British pathfinders, with a party of infantry and engineers successfully executed a *coup-de-main* against two vital bridges which crossed the Orne River and the Canal de Caen. The lead glider landed with pin-point accuracy, within fifty meters of the bridge crossing the canal. Following a quick but devastating fire fight, the British possessed both bridges. The bulk of the 6th Airborne Division was badly scattered, with battalions taking several hours to re-group, assemble and close in to hold Pegasus Bridge. It was shortly following noon that the first wails of a bagpipe were heard by the British paratroopers, and two hours later, a contingent of Lord Lovatt's Royal Marine Commandos marched across the bridge following their Regimental Piper.

The 3d (British) Parachute Brigade, led by Brigadier James Hill, accomplished their mission of securing the German coastal artillery at Merville. This battery was a significant target, whose destruction was vital to the overall invasion plan. Lieutenant Colonel Terrence B. H. Otway performed with gallantry, leading his men of the 9th Parachute Battalion. He charged the concrete encased battery, breaching the barbed wire perimeter, and engaged the German defenders in hand-to-hand combat. Informal communications between the British elements was accomplished through the use of hunting horns. Destruction of the bridges spanning the Dives River completed the secondary mission of the 3d Parachute Brigade.

Inland from Utah beach, the US 82d and 101st Airborne Divisions, commanded by Major Generals Mathew B. Ridgeway and Maxwell D. Taylor, were preparing airborne assaults onto St. Mere Eglise and Carenton respectively. General Richard N. Gale, who arrived with the 6th Air-Landing Brigade, commanded by Brigadier the Honourable Hugh Kindersley, took charge of the 6th Airborne Division. He led the 6th Airborne Division to their destination, east of Sword beach and Caen, to the Pegasus Bridge. Phase I of Operation OVERLORD was for his airborne forces, to disrupt, confuse and overpower the coastal defenses as the amphibious landings occurred.

It was planned for the US Airborne Divisions to seal and

D-Day wounded are moved by the transport Queen Express to hospitals in England. A number of these men are from the 327th GIR.

This stick of 506th PIR, 101st Airborne Division troopers receive final instructions from their jumpmaster, Lt. Alex Bobuck, before boarding for the D-Day drop.

block the Cotentin Peninsular. The 101st Airborne Division was to secure the four main water routes, enabling the seaborne forces to continue forward and attack the city of Cherbourg from the water. Some elements of the 101st Airborne Division were dropped up to 30-35 miles from their target areas. Due to high, gusty winds, some heavily laden members of the 101st drowned landing in the marsh lands around the Douves River. In addition to the marsh lands, the Germans had flooded likely airborne drop zones.

The 82d Airborne Division was to drop inland, into a fortified area and overcome the German defences, holding out until reinforced by elements of the invasion force from Utah beach. The 82d Airborne Division quickly consolidated and had the Village of St. Mere Eglise captured by 0430 hours. Until captured, Private John Steele had dangled from the steeple of the church, where his parachute had entangled.

Major maneuver elements parachuting in with the 82d Airborne Division included the 505th Parachute Infantry, commanded by Colonel William E. Eckman; the 507th Parachute Infantry led by Colonel George V. Millett Jr; and Colonel Roy Lindquist with his 508th Parachute Infantry Regiment. The airborne elements assaulting with the 101st Airborne Division included the 501st Parachute Infantry commanded by Colonel Howard R. Johnson; the 502d Parachute Infantry, commanded by Colonel George Van Horne Mosely; and Colonel Robert F. Sink with the 506th Parachute Infantry. Major F. C. Kellem and Lieutenant Colonels Benjamin Vandervoort and Edward Krause commanded the 1st, 2d and 3d Battalions, 505th Parachute Infantry. Lieutenant Colonels William L. Turner, Robert L. Stayer and Robert M. Wolverton commanded the 1st, 2d and 3d Battalions, 506th Parachute Infantry.

Three Drop Zones (Able, Charlie and Dog), and one Landing Zone (Easy) were identified for the 101st Airborne Division. The 502d Parachute Infantry and the 377th Parachute Field Artillery would jump on DZ Able. The 506th Parachute Infantry Regiment was assigned to DZ Charlie, and the 501st Parachute Infantry, with Company C, 326th Airborne Engineer Battalion, would parachute into DZ Dog. Two glider assaults were scheduled into LZ Easy, consisting of more artillery, antitank and communications elements. The 327th Glider Infantry and the 1st Battalion, 501st Glider Infantry were linking up with the Division after landing in the amphibious phase at Utah beach. It had been determined that the many hedgerows in the Normandy area were not conducive to gliders landing in the quantities assigned to the airborne divisions.

Like the 101st Airborne Division, the 82d Airborne Division was assigned three Drop Zones (Nan, Oboe and Tom), and one Landing Zone (Whiskey). The 508th Parachute Infantry was targeted to DZ Nancy; the 505th Parachute Infantry was to infiltrate at DZ Oboe; and the 507th Parachute Infantry would assault into DZ Tom. The 325th and the 2d Battalion, 501st Glider Infantry were slated to land on LZ Whiskey.

Colonel Edson Raff, former commander of the 509th Parachute Infantry Battalion, headed up a task force of almost a hundred glidermen of the 325th Glider Infantry, a battery of the 319th Glider Field Artillery, and a company from the 746th Tank Battalion, which he would lead from Utah beach to link up and defend the 82d Airborne Division headquarters at St. Mere Eglise.

Of the pathfinder teams dropped into the six Drop Zone and two Landing Zone areas, only two Drop Zones received their first visitors with any accuracy. This resulted in DZ Charlie (the 506th Parachute Infantry), and DZ Oboe, (the veteran combat tested 505th Parachute Infantry, led by Colonel William E. Ekman), having reasonable dispersion rates around their drop zone areas. Even so, Lieutenant Colonel Edward Krause and the 3d Battalion, 505th Parachute Infantry were quite a distance from DZ Oboe. All other units were also dropped wide of their targets. The 3d Battalion of the 505th rejoined the 82d Airborne Division at St. Mere Eglise where they hoisted an American flag in the center of the town, a flag they had previously flown when the 505th had captured the city of Naples in Italy eight months earlier. The 1st and 2d Battalions, 505th Parachute Infantry fought their way to their objectives, when Major F. C. Kellem, commanding the 1st Battalion, was killed taking the bridge crossing the Merderet River. The 2d Battalion, under Lieutenant Colonel Vandervoort, received many injuries during their landings, (to include Vandervoort who had broken a bone in his foot). They gallantly fought their way, at the cost of many lives, to reach and defend the town of St. Mere Eglise. The 2d Battalion linked up alongside the wounded Lieutenant Colonel Krause.

The 507th Parachute Infantry, destined to be dropped West of the Merderet River, landed mostly in the river, where many brave, heavily ladened paratroopers drowned, to the chagrin of the Regimental Commander, Colonel Millett. The 507th Parachute Infantry, through no fault of their own, were largely ineffective. Colonel Millett was himself captured two days following the jump, still gathering his lost troopers. One exception was a small force of the 2d Battalion, commanded by Lieutenant Colonel Charles J. Timmes, who established a perimeter in the area of Cauquigny, on the western bank of the Merderet River. Lieutenant Colonel Timmes was to retire two decades later, as a Major General, after serving as the Chief, Military Assistance Advisory Group, Vietnam.

Colonel Roy Lindquist was concerned as his 508th Parachute Infantry Regiment prepared to exit over Drop Zone Nancy. Colonel Lindquist prayed for a smooth and uneventful landing. His prayers went unanswered as full aircraft loads landed up to fifteen miles off target, some landing in the area of the 101st Airborne Division. Collecting his scattered troopers, he began trekking towards La Fiere, where he encountered Major General Matthew B. Ridgeway, the Commanding General of the 82d Airborne Division, gathering his forces as he moved along. Shortly after landing and organizing his 3d Battalion, Lieutenant Colonel Malcolm Brennan and his men opened fire upon a German staff car. This vehicle later proved to be carrying the Commanding General of the German 91st Infantry Division, General Wilhelm Falley, the first General officer to lose his life during the invasion of Normandy.

Brigadier General James M. Gavin, heading a three regiment parachute infantry and artillery task force, was scheduled to land approximately twenty miles inland from Utah beach at Sauveur-le-Vicomte. General Gavin did not land where he had anticipated - he had been dropped more than three miles away from his target area, in the marshlands along the Merderet River. He began assembling an ad-hoc force, moving to an area between the Chef-du-Pont and La Fiere, and capturing both bridges. He was astonished to meet Major General Ridgeway, who was similarly somewhat off-course. Directing Colonel Lindquist, who had also arrived on the scene, to take appropriate actions, General Ridgeway caused the 508th Parachute Infantry to attack the bridge at La Fiere. By this time however, the bridge was heavily guarded by German armor.

The 2d Battalion, 508th Parachute Infantry, also dropped way off target, had managed to consolidate two companies at their objective, capturing the bridge crossing the Douve River in the area of Pont l'Abbe.

As the 82d Airborne Division continued to organize their positions and defenses, the 101st Airborne Division was experiencing similar problems. The 101st had been dropped scattered from their intended drop zones and landing zones, although not so overall widely spaced as the 82 Airborne Division. Units of the 101st Airborne Division experienced from the most accurate to the most inaccurate landing patterns of the entire D-Day operation. Major General Maxwell D. Taylor found himself alone when disengaging from his parachute after landing. He gradually pieced a force together, including Brigadier General Anthony MacAuliffe, his Division Artillery Commander, and Colonel Gerald Higgins, his Chief of Staff. General Taylor established his Division Command Post in the vicinity of Drop Zone Easy, where Lieutenant Colonel Julian Ewell and the 3d Battalion, 501st Parachute Infantry were consolidating. Brigadier General Don Pratt was the second General officer to be killed as his glider smashed into Landing Zone Whiskey.

The 501st Parachute Infantry Regiment was close to being on target, with a minimum dispersion pattern.

The 502d Parachute Infantry was the most widely dispersed. Colonel George Van Horn Mosely Jr, broke his leg upon landing, and was unable to command his regiment. Elements of the 502d were up to six miles away from their target areas, misoriented, trying desperately to reorganize. The primary objectives of the 502d Parachute Infantry and the 377th Parachute Field Artillery, were the two northern causeways leading inland from Utah beach, and also to silence the German coastal artillery in the vicinity of Saint Martin-de-Varreville. The mission to capture the two causeways had been assigned to the 3d Battalion, 502d Parachute Infantry. Lieutenant Colonel Cole finally arrived at his objective area with the first lights of dawn, following one small fire fight on the way. His small force numbered approximately 100 men, which he had pieced together as he travelled from his landing site. It was a mixture of troopers from different elements of both the 82d and 101st Airborne Divisions, that had been separated during their respective insertions.

The two southern causeways from Utah beach had been assigned to Colonel Robert F. Sink and the 506th Parachute Infantry Regiment. Like many other units that night, the troopers of the 506th were well scattered to the winds. As the 1st and 2d Battalions began to slowly consolidate, only sixty personnel were initially accounted for in the 1st Battalion. The 2d Battalion realized that they had landed in Drop Zone Able, three miles north of their intended Drop Zone Charlie. The 3d Battalion was decimated as it landed in one of the few areas designated by the Germans as a likely airborne landing area. Pre-positioned machine gun emplacements placed withering fire, killing most of the battalion, including the battalion commander, Lieutenant Colonel Wolverton.

Colonel Sink landed uninjured, but unaware of the predicament of his regiment. He linked up with Lieutenant Colonel Turner and the 1st Battalion, who he had immediately directed towards their mission to secure the two causeways. He tried desperately, and was unsuccessful in establishing radio contact with either the Division Headquarters or his 2d and 3d Battalions. In the meantime, General Taylor, likewise unable to communicate with Colonel Sink, committed the 3d Battalion, 501st Parachute Infantry, from Drop Zone Easy to secure the causeway objectives originally assigned to the 506th Parachute Infantry.

The attention of the German defenders of Utah beach was diverted to the 82d and 101st Airborne Divisions, allowing the 4th Infantry Division to wade ashore with liitle or no opposition. It was a different proposition on Omaha, Gold, Juno and Sword beaches. German artillery from the vicinity of Vierville, on the western edge of Omaha, was scoring direct hits on the advancing landing craft. Company A, 116th Infantry of the 29th Infantry Division was severely hit. Only seven out of 32 men from one boat of Company F, 116th Infantry, made it to the shore line. Some 2,500 Allied military died at Omaha beach, paying the supreme sacrifice to allow 34,000 of their comrades to pass through their fallen ranks, to route the enemy.

At Omaha beach, the German 352d Division had been alerted, and had moved to the cliffs overlooking the beach at Omaha. They arrived and were emplaced before the battle hardened 1st Infantry Division, and the inexperienced (but well trained) attached 116th Infantry, 29th Infantry Division, headed for the beaches.

Elsewhere, at the western end of Omaha beach, 225 Rangers had reached the base of the hundred foot cliffs below Pointe-du-Hoe. Fifteen of their number had been hit between the landing craft and the cliffs. Only ninety of the Rangers remained after toiling hand over hand up the cliffs under withering fire from the Germans above. Commanded by Lieutenant Colonel James E. Rudder, the Rangers clearly demonstrated their toughness, preparedness and determination to beat the enemy. It was discovered that the guns had yet to be emplaced inside the concrete bunkers, and the sterling performance of the 2d Ranger Battalion had been for nothing. Moving inland, the Rangers located the German cannons in an apple orchard, and totally demolished them

Another German gun emplacement was discovered on the cliff tops, closer to Vierville. First Lieutenant William D. Moody, assigned to Company C, 2d Ranger Battalion, was engaged by enemy machine gun fire as he and his men hit the beach. Many of his unit were killed as they scrambled ashore. Taking two volunteers, Lieutenant Moody found a crevice and scaled the cliffs, passing information down to their company commander, Captain Ralph Goranson, who led the rest of the company to the cliff top. Company C, 2d Ranger Battalion, was the first Allied element to assault the high ground of Normandy.

Gold Beach faired no better than Omaha. The 1st Hampshire and the 5th East Yorkshire Regiments of the British 50th Division, landing close to the towns of Le Hamel and La Riviere, were stopped by hails of machine gun bullets. The HMS Bulolo was engaged in an naval/artillery gun duel, until the HMS Ajax came upon the scene and demolished the artillery battery. It was believed that Field Marshall Rommel had located the seasoned German 352d Division inland at St. Lo, to be positioned to counterattack any invasion forces. It was also anticipated that the 352d Division, could not be moved to counterattack, any earlier than before the evening of D-Day. It has been stated that plans would be great if both sides followed the script. This does not happen as both sides normally wish to be the winner.

On beach Juno, the 3d Canadian Division came ashore facing four German strongpoints, comprising 50, 75 and 80 millimeter guns and heavy artillery. Despite the battlefield preparations through multiple bombing runs, almost eighty percent of the German defenses were intact and functioning. A primary mission of the 3d Canadian Division was to head inland to relieve the British 6th Airborne Division.

That evening, the 325th Glider Infantry Regiment, commanded by Colonel Harry Lewis, was scheduled to arrive at Landing Zone Whiskey at 2100 hours. Less than one hour before their scheduled arrival, Colonel Raff and his armored/gliderman task force, were attempting their third unsuccessful attempt to dislodge the German occupants from the landing zone. Between Colonel Raff's force and the hastily organized elements of the 325th Glider Infantry, the Germans were defeated.

The following day, the combined forces of Colonels Raff, Lewis and the 4th Infantry Division, forced their way through to join the 82d Airborne Division. Colonel Raff was appointed to command the 507th Parachute Infantry in place of the captured Colonel Millett.

The Allies had their toe in the doorway of France, leading into Europe. A hole had been punched through the Atlantic Wall, and the end of World War II had begun. The strategic 5 German Divisions of reserve forces, and the 19 Divisions of the Fifteenth Army, were positioned to repel the Allied invasion at the Pas-de-Calais.

On the second night following the initial assaults, glidermen of the 325th Glider Infantry Regiment moved across the Merderet River to engage the Germans that had dug in on the west bank. During the ensuing fire fight, Company C became cut off from the rest of the Regiment. The Germans maneuvering to pincer the unfortunate Company C, did not reckon on Private Charles N. DeGlopper, who single handedly halted the German attack. Keeping the Germans pinned down with his machine gun, he allowed the remaining members of his unit to escape under his covering fire. Private DeGlopper was posthumously awarded the Congressional Medal of Honor for his gallantry. The following morning, the 325th and 501st Glider Infantry retook the bridge at La Fiere.

By the end of the third day, paratroopers and glidermen alike were exhausted from their toil. The 6th Infantry Division had lost two thirds of its 6,000 man force; the 82nd and 101st Airborne Divisions could each only account for approximately 2,500 personnel. The price of valor, paid by the magnificent airborne forces, enabled the doorway into Europe to be thrown wide open. The march towards the end of the war was on its way. With the Normandy Landings, the Third Reich was now opposed from three fronts.

The airborne forces initially remained in the area of Northern France, with the 82d Airborne Division securing the area to the west. They cut off the Cotentin Peninsular, continuing on to seize Saint Sauveur-le-Vicomte and La Haye du Puits. The 101st Airborne Division was ordered to the south, where they attacked and took Carentan on 12 June. Both Airborne Divisions returned to England in mid-July, with Cherbourg and Normandy securely controlled by the Allies.

Through the valorous performance conducted on 6 June 1944, and the ensuing battles in which those first invaders fought, the Allied foothold was secure. By 25 July, more than 812,000 Americans, and more than 640,000 British and Canadians, advanced into Europe towards victory.

OPERATION DRAGOON: THE INVASION OF SOUTHERN FRANCE

By Thomas R. Cross, Col., AUS (Ret)

Our generation, the first to employ parachute and glider-borne troops in armed combat against an enemy force, should remember certain actions in certain ways. We should refuse to accept any changes in the description or in the account of the employment of these forces and their subsequent operations in any manner or form that will modify, interpret or change what we actually saw and what we did.

If we are not vigilant in the true preservation of our historical actions, then others might try to do it by attempting to correct our memories. How then will we and our descendents know who we were, what we did and when and the true significance of our actions. With this in mind this rendering of Operation DRAGOON will, for the most part, deal with actual official after-action reports written by airborne units and their higher echelons of command directly involved in the planning and in the conduct of operations for Operation DRAGOON.

This account will primarily present the big picture aspects of Operation DRAGOON. Most airborne and troop carrier units that participated in this operation have already published their own individual unit histories that include their specific roles and actions. What is presented here compliments these unit histories. When the unit historical reports are used they are intended to fill out and to amplify the broader aspects of Operation DRAGOON.

The initial concept for and the planning of Operation DRAGOON were shrouded in controversy beginning as far back as 1943. It was during the QUADRANT Conference, held in Quebec during 14-24 August 1943, that both President Roosevelt and Prime Minister Churchill, their Special Advisors and the Combined Chiefs of Staff, decided on a target date for OVERLORD; the invasion of Normandy. The date for OVERLORD was initially set for 1 May 1944. Also, the invasion of Southern France was decided and the code name of ANVIL was assigned to this operation.

Prime Minister Churchill, even at this very early date, voiced his anguish over the selection of the Southern France approach and strongly championed a military inroad into the Balkans. Churchill resisted ANVIL—the code name of which was later changed to DRAGOON—right up until 9 August 1944, only six days before the actual invasion of Southern France. The change in the code name from ANVIL to DRAGOON mischievously reflects Prime Minister Churchill's views that he was pressed by the Americans into doing something that he did not favor. The Webster Dictionary definition for "dragoon" was used by Churchill to verbally accentuate his opposition to the Southern France operation.

The EUREKA Conference in Teheran from 28 November to 1 December 1943, between Marshall Stalin, Prime Minister Churchill and President Roosevelt once again confirmed that there would be a landing in Southern France. The exact details of the plan were left in limbo due to the still-present feeling of opposition by Churchill. Roosevelt did insist, however, that both OVERLORD and ANVIL, now DRAGOON, would receive top priority from here on out.

The introduction of additional combat and support units to the troop list for OVERLORD as well as the current state of operations in Italy, indicated that DRAGOON would have to be executed at a date later than that set for OVERLORD. The operation for Anzio caused 68 LSTs to be diverted from the OVERLORD resources. Thus the shortage of LSTs for the seaborne lift and the requirement for maximum troop carrier support for the airborne phase of OVERLORD dictated a change in the invasion sequence as developed at these earlier strategic conferences.

General Eisenhower was a strong supporter for DRAGOON even if the operation had to be conducted at a later date, for his main concern was the achievement of a military victory in Europe in the shortest possible period of time. DRAGOON offered General Eisenhower the opening of the valuable port of Marseilles, the second largest in France, plus a chance to open a second front in Europe which would detain a large number of German forces in the Southern France area and prevent them from reinforcing their troops in Normandy. Last but not least was the important factor of giving the Free French Forces a meaningful and important rule in the liberation of their homeland.

Thus we see the political vs. military controversy surrounding perhaps one of the best planned and executed invasion operations in World War II. The success of the airborne phase of DRAGOON, as we will note from here on, came about through the military professionalism and leadership displayed at all levels of staff and command from the highest headquarters and command level right down to the airborne squad level and, yes, even to the individual trooper.

In spite of a last minute plea from Prime Minister Churchill, the U.S. Joint Chiefs of Staff did, on 8 August 1944, cable General Eisenhower that Operation DRAGOON would proceed as planned. To us in the 1st Airborne Task Force it meant "green light and go" and that we did over the drop zones and landing zones of Southern France.

Initial planning for the airborne operation in DRAGOON was begun by the planning staff of the U.S. 7th Army in February 1944. The status of airborne units in the Mediterranean Theater at that time materially influenced planning at this stage. At this time nine of the units, airborne or troop carrier, were actually prepared for airborne operations.

The 51st Troop Carrier Wing, composed of three groups, had remained in the theater after the inactivation of the XII Troop Carrier Command. However, only a portion of the 51st Wing was available for airborne training because of the demand for Troop Carrier aircraft for special operations, air evacuation and general transport requirements. A few of these aircraft were intermittently attached to the Airborne Training Center located in Trapini, Sicily and later at the new location of the Airborne Training Center at Lido de Roma for a program of limited airborne training. At the Center, the 1st French Parachute Regiment, two pathfinder platoons and the American replacements received limited airborne training.

The British 2nd Independent Parachute Brigade Group commanded by Brigadier C.H.V. Pritchard, batteries A and B of the 463rd Parachute Artillery Battalion commanded by Lt. Col. John Cooper and the 509th Parachute Infantry Battalion commanded by Lt. Col. William P. Yarbrough were withdrawn from the front in Italy and given intensive training with a full troop carrier group of the 51st Wing.

The War Department was requested to provide an airborne division for employment in DRAGOON, but in lieu of this, a number of separate airborne units were shipped to the theater. These were the 551st Parachute Infantry Battalion commanded by Lt. Col. Wood G. Joerg, the 517th Parachute Regimental Combat Team commanded by Colonel Rupert D.

Graves and the 550th Glider Infantry Battalion commanded by Lt. Col. Edward I. Sachs.

The 517th consisted of three parachute infantry battalions: the 1st Battalion commanded by Major William J. Boyle, the 2nd Battalion commanded by Lt. Col. Richard J. Seitz and the 3rd Battalion commanded by Lt. Col. Melvin Zais. The 460th Parachute Artillery Battalion of the 517th Parachute Regimental Combat Team was commanded by Lt. Col. Raymond L. Cato and the remaining contingent of the 517th Parachute Regimental Combat Team which was the 596th Parachute Combat Engineer Company, was commanded by Captain Robert W. Dalrymple.

The 517th Parachute Regimental Combat Team arrived from the States after just completing its final large scale airborne training exercise and participation in the Tennessee Maneuvers. The 517th Parachute Regimental Combat Team was attached to the U.S. 5th Army for 10 days of battle experience in the lines and the 550th and 551st battalions were attached to the Airborne Training Center, then located in Sicily, for further training.

Thus, by the middle of June, there were considerable airborne forces in the theater which could be considered for airborne operations. To secure the utmost cohesion and to obtain the optimum results, it was decided to move the Airborne Training Center with its attached units, as well as the troop carrier aircraft (now increased to two full groups of the 51st Troop Carrier Wing) to the Rome area. Here was established a compact forward base for all airborne forces.

ORGANIZATION

Airborne Elements

Toward the first of July 1944, the plans for operation DRAGOON were made firm, including the use of a provisional airborne division made up of available units in the theater. Major General Robert T. Frederick, formerly commander of the 1st Special Service Force and later the commander of the 36th Infantry Division, assumed command of the composite force. Conferences were held immediately to secure the additional supporting units needed to organize such a balanced airborne force. Certain units on the DRAGOON troop list were earmarked for this purpose. Authority was requested from the War Department to activate those units which were not authorized on the theater troop list. By 7 July initial instructions relative to the organization of a provisional airborne division were issued to General Frederick. However, on 21 July 1944 this provisional airborne division known officially, for the record, as the 7th Army Airborne Division (Provisional) was disbanded and the 1st Airborne Task force was constituted by a secret adjutant general order of 18 July 1944. Assignment was to the North African Theater of Operations, and activation was accomplished by the reassignment of personnel from the 7th Army Airborne Division (Prov). The action was concurrent.

The following units, which just a short time before had been assigned to the provisional airborne division, were then assigned to the 1st Airborne Task force: 2nd British Independ-

Paratroop drop between Nice and Marseilles. (U.S. Army photo)

Glider troops of the First Airborne Task Force land near La Motte in Southern France on D-Day, 15 August. (Photo from Out of the Blue by James A. Huston)

ent Parachute Brigade Group; 517th Parachute Infantry Regiment; 509th Parachute Infantry Battalion; 550th Airborne Infantry Battalion; 1st battalion, 551st Parachute Infantry Regiment (Reinf); 460th Parachute Field Artillery Battalion; 463rd Parachute Artillery Battalion; 602nd Pack Field Artillery Battalion; 596th Parachute Combat Engineer Company; 887th Airborne Aviation Engineer Company; 512th Airborne Signal Company; Anti-tank Company, 442nd Infantry Regiment; Company A, 2nd Chemical Battalion (Motorized); 172nd DID British Heavy Aerial Resupply Company; 334th Quartermaster Depot Company, Aerial Resupply; Detachment, 3rd Ordnance Company (MM); 676th Medical Collecting Company (Designated 29 July 1944); five unit pathfinder platoons. (These were unauthorized but were formed from 1st Airborne Task Force resources.)

The 1st Airborne Task Force was then given a five percent overstrength by the assignment of parachute filler replacements from the Airborne Training Center. Meanwhile, the activation of the Task Force Headquarters and Headquarters Company, two additional batteries of the 463rd Parachute Artillery Battalion, the 512th Airborne Signal Company, and an anti-tank unit to be designated the 552nd Anti-tank Company all proceeded on schedule.

It was then decided not to prepare the 552nd Anti-tank Company for operation because of the short time remaining before D-Day. The 442nd Anti-tank Company was therefore substituted and added to the troop list. The 442nd Anti-tank Company was well-trained prior to arrival in the theater and when it was necessary to re-equip them with the British six-pounder, since the 442nd's 57mm anti-tank gun would not fit in the CG-4A glider, this well-trained unit was able to make the transition in record time.

Due to the shortage of qualified airborne officers in the theater, it was necessary to ask the War Department to make available a divisional staff for General Frederick. Thirty-six qualified staff officers arrived in the theater by air toward the middle of July. Most of these came from the 13th Airborne Division and a few from the Airborne Center at Camp Mackall, North Carolina.

Certain other organizations were made available to the 1st Airborne Task Force on the basis of their being employed in the preparatory stage but not in the operation itself. These units included detachments from a signal company, a quartermaster truck company and some 400 replacements from the Airborne Training Center.

Troop Carrier Elements

As of the middle of July there were available in the theater for airborne operations, two groups of the 51st Troop Carrier Wing. The third group was occupied with special operations. To provide sufficient lift for DRAGOON, additional troop carrier groups were called for by Allied Force Headquarters. The total minimum of aircraft required for the operation was 450. On 10 July 1944 orders were issued placing the 50th and 53rd Troop Carrier Wings of the IV Troop Carrier Command (then located in the United Kingdom) on temporary duty with the theater. Each wing contained four groups of three squadrons each, reinforced by self-sustaining administrative and maintenance echelons and by the IX Troop Carrier Command Pathfinder Unit, a total of 413 aircraft.

In addition to the personnel and equipment moved in organic aircraft, the Air Transport Command augmented the movement by transporting the 819th Medical Air Evacuation Squadron, various signal detachments, assorted parapack equipment and 375 organic glider pilots.

The move, made in eight echelons via GIBRALTAR and MARRAKECH required but two days. Two aircraft were lost enroute. Brigadier General Paul L. Williams, in command of two wings from the United Kingdom, arrived on 16 July 1944 and activated the Provisional Troop Carrier Division. By 20 July the entire Provisional Troop Carrier Division had arrived in the theater and was stationed at the designated airfields, prepared to carry out its mission.

Since there were approximately 130 operational CG-4As and 50 Horsa gliders on hand, hurried steps were taken to secure the additional number required for the operation. Fortunately, a previous requisition for 350 CG-4A gliders from the United States had been made. It was necessary only to expedite this requisition in order to provide the glider lift. The British airborne forces had sufficient Horsa gliders on hand in theater to provide for the needs of the 2nd Independent Parachute Brigade Group. The shipment from the United States arrived as scheduled, and the gliders were assembled in record time. They were ready for operational use 10 days before D-Day. Unfortunately, because of the limited time constraints and the fact that only 40 percent of the new reinforced Griswold glider nose modification kits were available, it was decided not to attempt any modifications prior to D-Day.

After considerable discussions, a request was made to the United Kingdom for approximately 350 additional glider pilots. Previous arrangements made to secure these pilots on three days notice were carried out and all the glider pilots arrived as requested.

Resupply Equipment

Hurried preparations were required to assemble the necessary cargo parachutes and aerial delivery equipment needed to organize and prepare for the contemplated aerial resupply effort. As late as the 10 July, the acting staff for General Frederick submitted an overall requisition for this equipment to higher headquarters. By air and by special water transport, some 600,000 pounds of these supplies arrived in the theater in time for the operation. The last freight shipment was delivered to the 334th Quartermaster Depot on D-4. Every item that was requested arrived in time and the preparations for the operations were carried out as scheduled.

Concentration of Units

The Airborne Training Center and the 51st Troop Carrier Wing had been ordered to the Rome area and established a compact airborne base at Ciampino and Lido de Roma Airfields. By the 3 July, an advance echelon of the Airborne Training Center was established at Ciampino Airfield, ready to operate.

By the 10 July 1944, the center with its attached units, the 551st Parachute Infantry Battalion, and the 550th Glider Infantry Battalion were completely located at the airborne base. The divisional staff ordered from the United States for the 1st Airborne Task Force Headquarters would not arrive until approximately the 15 July. Therefore, all other American units in the theater were attached to the Airborne Training Center so that its staff could be used to assist in expediting the concentration of airborne troops.

The 517th Parachute Regimental Combat Team was ordered out of the line from U.S. 5th Army control and arrived in the Rome area by 5 July 1944. The 509th Parachute Infantry Battalion, already located at Lido de Roma was similarly attached to the Airborne Training Center for instruction. The various supporting arms and services which had been placed at the disposal of the 7th Army Airborne Division (Provisional) were likewise attached.

By 17 July 1944, General Frederick had moved his Headquarters to Lido de Roma and was ready to proceed with the final organization and training of the airborne force and the detailed planning of the operation. On the 21 July, General Frederick requested that the name of the provisional organization be changed to the 1st Airborne Task Force, since the use of the term "division" was considered a misnomer.

PLANNING

Although continuous planning for DRAGOON had been under consideration since February 1944, no final detailed planning was possible until the 1st Airborne Task Force and the Provisional Troop Carrier Air Division were organized and prepared to function. Consequently, the final planning could not commence until almost 20 July. On his arrival, General Williams, the commanding general, Provisional Troop Carrier Air Division, approved the suggested plan of utilizing the previously selected Rome area as the training site. He also concurred in the choice of previously selected take-off airfields, at Ciampino, Galera, Marcigliano, Fabrisi, Viterbo, Tarquinia, Voltone, Montalto, Canino, Orbetello, Ombrone, Grosseto, Fallonica, and Piombino. Subsequently, the Provisional Troop Carrier Air Division undertook primarily the planning aspects of the operation involving high level coordination, timing, routes, corridors, rendezvous and traffic patterns. In general, planning for the details of the selection of drop zones (DZs), landing zones (LZs), and the composition of lifts were left to the airborne and troop carrier units involved.

It was at first decided that a pre-dusk airborne assault on D-Day should not be made, as this might jeopardize the success of the entire operation. Second, it was decided that it would be neither necessary nor advisable to launch the initial vertical attack after the amphibious assault had begun. The latter decision was reached in view of the wide experience of our troop carrier crews in night take-off operations and because of marked improvement that had been made in the pathfinder technique. Consequently, the basic plan called for the pre-dawn assault.

One proposed plan contemplating an immediate staging in Corsica was rejected because of lack of available Corsica airfields, and also because those fields available were located on the eastern side of the island, and their use would have necessitated a flight over 9,000-foot mountain peaks. Such a flight would be difficult even for unencumbered transport aircraft, and for C-47's towing gliders, it was considered excessively dangerous. A further consideration was the fact that such an intermediate staging would have required the establishment of the airborne corridor south of the main naval channel and would have necessitated the adoption of a dog-leg course for the flight.

After several conferences had been held at the 7th Army Headquarters, with all Army, Navy, Air and Airborne commanders concerned, the rough plan was drawn up and approved about 25 July 1944. This plan envisioned the use of an airborne division prior to H-Hour with the dropping of the airborne pathfinder crews beginning at 0323 hours on D-Day. The main parachute lift of 396 plane loads was to follow, starting at 0412 hours and ending at 0509 hours. The first glider landing was to take place at 0814 hours and continue on through until 0822 hours. Later in the same day a total of 42 paratroop plane loads was to be dropped, followed by 335 CG-4A gliders starting at 1810 hours and ending at 1859 hours.

The automatic air resupply planned for late afternoon D-Day, was postponed at a late stage of the planning because insufficient troop carrier aircraft were available and because Troop Carrier Command would not drop supplies from aircraft towing gliders in the afternoon glider lift. The final plan provided that 112 plane loads were to be brought in automatically on D+1. The remainder of the supplies were to be packed and held available for emergency use by either the 1st Airborne Task Force or by any U.S. 7th Army unit which might become isolated.

The troop carrier routes selected were carefully chosen after due consideration of the shortest distance, prominent terrain features, traffic control for the 10 troop carrier groups, naval convoy routes, position of assault beaches, primary aerial targets, enemy radar avoidance, excessive dog legs, prominent landfalls and position charts of enemy flak installations. This route logically followed the Italian coast generally from the Rome area to the island of Elba, which was used as the first over-water check point, followed by the tip of the Island of Corsica and proceeded on an azimuth course over naval check points to the landfall just north of Frejus at Agay.

Complete plans were made with the Navy on the position of this airborne corridor, and detailed information concerning it was widely disseminated among the naval forces.

Because of high terrain features in the target area, it was decided that it would be necessary to drop the paratroopers and release the gliders at exceptionally high altitudes varying from 1,500 to 2,000 feet. Glider speeds for towing were set at 120 mph and dropping speeds at 110 mph The formation adopted for the parachute columns was the universal "V of Vs" in nine ships in serials of an average of 45 aircraft, with five-minute intervals head to head between serials. The glider columns adopted a "pair of pairs" formation echeloned to the right rear. Serials made up of 48 aircraft towing gliders in trial were used with eight-minute intervals between serial lead aircraft. Parachute aircraft employed a maximum payload of 5,430 lbs.; Horsa gliders, 6,900 lbs.; and the CG-4A gliders, 3,750 lbs.

Difficulty in the procurement of maps and models proved to be serious inconveniences in the planning and preparations for the operation. Map shipments in many instances were late in arriving or were improperly made up. Terrain models on a scale of 1:100,000 were available, but the most useful terrain model, a photo-model in scale of 1:25,000, was available only in one copy which was wholly inadequate to serve both the Provisional Troop Carrier Division and the 1st Airborne Task Force. The blown-up large scale photographs of the DZ-LZ areas in particular were excellent, but these arrived too late for general use. The original coastal obliques were not of much assistance to the Provisional Troop Carrier Air Division since the run-in from the IP (First Landfall) was not adequately covered. These late photographic studies uncovered the previously unknown element of anti-glider poles installed in the

Glider troopers of the 550th Airborne Infantry Battalion land in Southern France and move off to their assembly area. (Photo from Out of the Blue by James A. Huston)

LZs. All earlier photo studies had failed to reveal this pertinent information. An excellent terrain model was turned out by the 2nd British Independent Parachute Brigade Group and it was of great assistance to that unit for the operation.

PRE-OPERATIONAL TRAINING

By the middle of July, nearly all the airborne units to be employed on operation DRAGOON had been assembled in the Rome area. An intensive final training program had been begun by the 1st Airborne Task Force in conjunction with the Airborne Training Center. Of the airborne units to be used in the operation, only the 509th Parachute Infantry Battalion and the 2nd British Independent Parachute Brigade Group had received any recent combined airborne training with the troop carrier units. The 517th Parachute Regimental Combat Team had just come out of the line with 5th Army as had the 463rd Parachute Artillery Battalion. Other units, such as the 551st Parachute Infantry Battalion and the 550th Glider Infantry Battalion had but recently arrived overseas and they were given a course in ground and airborne refresher training at the Airborne Training Center.

Particularly urgent was the task of training the newly chosen glider-borne troops. A combined glider school was set up and instruction commenced in loading and lashing for the units involved. The units involved in this difficult, last-minute procedure were the 602nd Pack Field Artillery Battalion, the 442nd Infantry Anti-tank Company, the 887th Airborne Aviation Engineer Company, Company A, 2nd Chemical Battalion, Company D, 83rd Chemical Battalion, and various other units such as the Division Ordnance Detachment and the Medical Collecting Company. Once these troops had finished the course in loading and lashing they were given orientation flights and finally one reduced-size practice operational landing on a simulated LZ.

The Provisional Troop Carrier Air Division Pathfinder Unit went to work with the three airborne pathfinder platoons and thoroughly tested the radar and radio aids to be used in the operation. This training was divided into three phases, the first being concerned with technical training with "Eureka" sets, M/F Beacons, lights and panels. Tests were made to locate any deficiencies in either the training or the equipment to be used. The second phase was devoted to practice by the crews in using the equipment as a team. All teams practiced in setting up and operating the equipment under all possible conditions. The third phase emphasized actual drops with full equipment in which every attempt was made to secure the utmost realism in the preparatory exercises. Small groups of parachute troops were dropped on the prepared DZs to test the accuracy of the pathfinder aids.

Due to the lack of time and difficulty of re-packing the parachutes for the operation, it was impossible for the 1st Airborne Task Force to stage any large scale realistic final exercise. The various individual units participated in practice drops to the fullest possible extent by generally using a skeleton drop of two or three men representing a full "stick" of paratroopers, while the remainder of the unit was already on the DZ so that the assembly procedures could be tested and experienced. A combined training exercise with the Navy was scheduled. All naval craft carrying water-borne navigational aids were placed in the same relative positions as in the actual operation. A token force of three aircraft per serial were flown on the exact timing schedules, routes and altitudes as were to be used in the operation. Two serials of 36 aircraft each were flown over this same route in daylight in order that the naval forces would become acquainted with troop carrier formations. Further practice runs were made by the troop carrier units in conjunction with the 31st and 325th fighter groups so as to work out the details of the fighter cover plan and the air-sea rescue plan.

In view of the fact that the Task Force was composed of units that had not previously worked together, training of combat teams, as organized for the operation, was emphasized to enhance successful operations after landing. Training of each newly-organized combat team was conducted on terrain carefully selected to duplicate, as nearly as possible, the combat team's sector in the target area.

The problem of securing and organizing qualified personnel and then training those units that had to be activated on short notice proved to be difficult. Such highly specialized personnel as are required for an airborne signal company or for an airborne divisional headquarters were extremely difficult to find in an overseas theater. Consequently, these personnel had to be located in area replacement depots or at the Airborne Training Center and then trained for the specific positions they were to fill. The highest praise is due General Frederick and his staff and the Airborne Training Center for the manner in which this task was accomplished.

Fortunately, the larger elements of the task force, particularly the combat teams, were already well-trained. Some of them were battle-seasoned organizations, and nearly all were accustomed to providing for themselves, since each was basically designed to be a separate regiment, battalion or company. This latter fact allowed the units not only to look after their own requirements, but also permitted them to aid the 1st Airborne Task Force as a whole in many ways during this period of training.

THE OPERATION
Airborne Phase

The night of D-1 was clear and cool in the take-off areas used by the airborne forces in DRAGOON. The troop carrier units were at their stations at 10 airfields extending from Ciampino near Rome to Fallonica, north to Grosseto. Due to the serious lack of ground transportation, it was necessary for the bulk of the Task Force, except for the British 2nd Independent Parachute Brigade Group, to commence the movement to the dispersal airfields by D-5. By D-2 the airborne forces had been shuttled from their training and concentration areas near Rome to their designated airfields. The C-47 aircraft had all been deadlined and checked thoroughly and were considered in excellent condition for the invasion flight. Also, all of the preliminary checks had been completed at the glider marshalling airfields and they were ready to roll. A feeling of assurance as to the outcome of the operation prevailed among all elements.

As to be expected in any airborne operation, prevailing weather was to be important and could influence the parachute drops. Once the target date had been set it could not be changed for the benefit of the airborne forces, even though it necessitated a drop without the assistance of moonlight. It had been hoped that the drop could be made on a clear night so that the troop carrier aircraft could identify large hill masses and coastal features as possible check-points. However, on 14-15 August 1944, all of western Europe was covered by a large, flat, high-pressure area centered over the North Sea. A portion of this "high" had broken off and had settled over the main target area. This did preclude the probability of any sizeable storm or heavy winds, and the only threat was one of accumulating fog or stratus. Consequently, the forecast for the operation was clear weather to Elba, followed by decreasing visibility until the DZs were reached, at which time the visibility was expected to be from two to three miles. Actually, the haze was heavier than anticipated and the visibility was less than a half mile on the DZs.

The valley fog which completely blanketed the early parachute operation later dispersed by 0800 hours in time for the morning glider mission. Considerable navigation difficulties were to arise from the fact that the wind forecast was almost 90 degrees off the direction initially indicated. Consequently, the navigators could make necessary corrections only by use of checkpoints over the water route. Fortunately, the wind did not reach high velocity and was less than six mph on the DZs.

The operation was prefaced by a successful airborne diversion designed to serve two purposes in the cover plan. First, it was to create the illusion of a southern airborne corridor; second, it was to simulate a false airborne DZ by dropping rubber parachute dummies in selected areas. The six aircraft used on this mission dropped "window" enroute to give enemy radar the effect of a mass flight and at 0205 hours on D-Day they dropped 600 parachute dummies as planed on false DZs located north and west of Toulon. German radio reports indicated the complete success of this ruse. The rifle simulators and other battle noise effects used in the diversion functioned well and added realism to this feint.

The airborne operation began shortly after midnight on 14-15 Aug. 1944. Aircraft were loaded, engines were warmed up and the marshalling of aircraft for takeoff was underway at 0030 hours. At the same hour the first troop carrier aircraft took off with their load of three pathfinder units. During the aircraft marshalling phase several aircraft received minor damage and one was demolished. One aircraft from the 439th Troop Carrier Group crashed and burned on take-off, two aircraft struck trucks and received minor damage; and two other aircraft from the same group had a collision while rolling out on the taxi ramp. One other aircraft suffered damage to a wing when one of its wheels hit a poorly filled hole on the airfield, and another suffered damage due to the premature release of a parapack. Considering that the take-off was from prepared landing strips without any moon to aid the poor visibility caused by excessive dust, the success of this phase of the operation is unquestioned. An estimated eight gliders had difficulty on take-off and had to be unloaded so that substitute gliders could be employed to lift the loads.

The pathfinder mission is of interest in that, of the reported complete success of the mission made early in the operation, later facts revealed that it was less than 50 percent effective. Nine pathfinder aircraft divided into three serials were employed in such a manner that three teams would be dropped on each DZ at 0323 hours. The lead serial, supposed to drop pathfinders, from the 509th and 551st Parachute Infantry Battalions and the 550th Glider Infantry Battalion, lost its way and circled back to sea to make a second run. After circling for about half an hour one aircraft dropped its pathfinder team and went home. The other two aircraft separated after that and one dropped its team at about 0400 hours. The last team jumped at 0415 hours on the sixth run on the target. Lost in the woods and far from their objective, none of the three teams were able to reach the Le Muy area in time to act as pathfinders.

The Second Serial was to drop its pathfinders from the 517th Parachute Regimental Combat Team at 0330 hours. The actual drop was at 0328 hours. The 517th Pathfinders landed in the woods 3 1/2 miles east of DZ A and just east of Le Muy itself. A two-minute jump delay would have put them right on target. To make matters worse they were attacked and spent time beating off the assault. They arrived at DZ A at approximately 1630 hours and set up a Eureka Beacon, a MF Beacon, and a T panel which was of assistance to the afternoon missions for this area.

The third serial carrying the pathfinders from the British 2nd Independent Brigade Group dropped on DZ O at 0334 hours, exactly on schedule. By 0430 hours they had two Eureka Beacons in operation about 300 yards apart along the axis of the approach. They also set up two lights for assembly purposes. This pathfinder drop was the most accurate of the three.

Approximately one hour after the pathfinder aircraft took off, the main parachute lift, composed of 396 aircraft in nine serials averaging 45 aircraft each, took off and proceeded on its courses. The flight toward the designated DZs was unmarked by incident. Amber downward recognition lights were employed until the final water checkpoint had been crossed. Wing formation lights were similarly employed and no instance of friendly naval fire on the troop carrier aircraft was reported. Likewise, no enemy aircraft were encountered during the flight.

Of particular note is the fact that over 400 troop carrier aircraft had flown in relatively tight formation under operational strain for some 500 miles without accident. The many hours of time devoted to training in night formation flying had produced excellent results.

The radio, radar and other marker installations undoubtedly helped to save the day for the success of the mission. The Eurekas which had been installed at each Troop Carrier Wing Departure Point, the Command Departure Point, the Northeast tip of Elba, Giroglia Island (North Corsica) on three marker beacon boats spaced 30 miles apart on course from Corsica to Agay, France, the first landfall checkpoint, all worked exceedingly well with an average reception of 25 miles. Holophane lights similarly had been placed at these positions and aided the navigators in their work with the contrary wind currents. Their reception was an average of eight miles until the DZs were reached at which time they became invisible because of the haze and ground fog. MF Beacons (Radio Compass Homing Devices) were installed at Elba, North Corsica and on the central boat marker beacon and dropped on the DZs along with the Eurekas and Holophane Lights. Many pilots reported that they picked up these signals up to 50 miles and often kept the aircraft on beam when they occasionally lost the Rebecca signal on their Eurekas which in many cases exhibited a tendency to drift off the frequency despite constant operational checking.

It should be emphasized again that the entire parachute drop was made "blind" by the troop carrier aircraft who had to depend on these MF Beacons and Eureka sets for their signal to drop the paratroopers. Brigadier C.H.V. Prichard, commanding officer, 2nd British Independent Parachute Brigade Group, felt that this single possible deficiency could have jeopardized the complete operation.

Initial reports stated that the parachute drop was eminently successful and it was given a score card rating of 85 percent accuracy. This was later downgraded as the incoming reports became more accurate and detailed in content. Despite the accuracy of the reporting system, sufficient paratroopers landed on the DZs or in the immediate vicinity thereof, in areas which for all practical purposes can be considered as contiguous to the DZs, and from which terrain the parachute forces were in position to allow them to carry out their assigned missions. This was accomplished despite the handicaps of no moon, general haze and heavy ground fog. An estimated 45 aircraft completely missed their designated DZs. Some of these dropped their paratroops as far as 20 miles from the selected areas.

Among the aircraft which missed the DZs were 20 in Serial Number 8, which released their troops prematurely on the red light signal. The only plausible explanation that can be offered is that a faulty light mechanism in one of the leading aircraft must have gone on green prematurely and the troops in the lead aircraft jumped according to this signal. The paratroopers in the following aircraft, on seeing the leading aircraft's paratroopers jump, probably did likewise, and jumped even though the red signal still showed in their own aircraft.

This group principally comprised elements of the 509th Parachute Infantry Battalion and about half of the 463rd Parachute Field Artillery Battalion. Two "sticks" of paratroopers landed in the sea off St. Tropez, near Cannes. The remainder made ground landings in the vicinity of these two towns. Although far from the designated DZ, these elements organized themselves, made contact with the French Resistance Forces and proceeded to seize and hold St. Tropez.

Approximately 25 aircraft from another Troop Carrier Group mistakenly dropped their paratroopers some 15 miles north of Le Muy near Fayance. The troops in this instance comprised part of the 5th (Scots) Para Battalion of the British 2nd Independent Parachute Brigade Group, and elements of the 3rd Battalion, 517th Parachute Regimental Combat Team. Although some 20 miles from their DZ, these troops either undertook individual missions or fought their way back to their own units in the proper objective area. By evening of D-Day, most of this group were reassembled on DZs A and O. The Task Force Chief of Staff, along with the Task Force Surgeon and other key staff officers were among this group

DZ A, generally west of Le Muy had a tendency, during the drop, to become merged with DZ O slightly northwest of this key town on the Vargennes Valley, which caused considerable confusion later in the day. This inadvertent merging of the two DZs also produced some confusion and difficulty during the period of the equipment bundle recovery and this was increased because the British 2nd Independent Parachute Brigade Group was using different equipment from that of the 517th Parachute Regimental Combat Team on DZ A.

The terrain of the DZs on which the paratroopers landed was, in general, excellent for such an operation. DZs A and O covered an area of small cultivated farms consisting mainly of vineyards and orchards. There were very few large buildings, telephone wires, tall trees, or other formidable obstacles. The anti-airborne poles established in the parachute drop areas were not sharp or placed in sufficient density to obstruct the parachute landings to any material degree. A total of 175 paratroopers, scarcely more than two percent, suffered jump casualties. DZ C, on which the 509th Parachute Infantry Battalion Combat Team jumped, was a hill mass more rugged than the ground of the other DZs, but even this rougher terrain did not interfere with the success of the jump.

Serial Number 14 (the first of the glider serials), made up of supporting artillery and anti-tank weapons for the British 2nd Independent Parachute Brigade Group, departed as scheduled for its 0800 hours glider landing, but was recalled because of heavy overcast. The flight circled for one hour and landed at 0900 hours. One glider and tug aircraft had to turn back. One glider ditched off shore and another disintegrated in midair over the water (cause was laid to structural failure).

The stakes driven into the ground all over the LZs did not prove to be difficult obstacles, even though the poles did cause considerable damage to the gliders and in some instances to their loads. The anti-glider poles served in many instances as additional braking power for the gliders, since the poles were small, planted at shallow depth, and were too widely dispersed to perform their intended mission. Evidently, the French farmers who were forced to plant the anti-airborne (glider) stakes had done the minimum work they could in this forced construction. These poles were on the average of 12 feet high and 6 inches in diameter. They had been driven in the ground less than two feet and were generally more than 30 to 40 feet apart.

Serial Number 16 was a parachute load made up of the 551st Parachute Infantry Battalion. It dropped accurately on DZ A at 1800 hours as planned. This drop was followed rapidly by continuous glider serials numbered 17 through 23. Nine gliders were reported to have been released prematurely, four of which made water landings. A large percentage of their crews and personnel were saved by prompt action on the part of the Navy.

The landing skill of our glider pilots was outstanding. Although the 1,000-foot interval adopted for towing caused considerable jamming over the LZs, these pilots effected excellent landings. Several pilots ground-looped their gliders to avoid obstacles and still brought in personnel and cargoes safely. Another reason for the crowded conditions over the landing zones was a notable tendency on the part of successive flights to seek additional altitude as a result of the "accordion movement" of the flights en route. In turn this progressively created a greater mass of aircraft and gliders being over the LZs at any one time than had been contemplated. Further difficulty arose because the pilots of the early glider lifts landed on the best and most obvious sections of the landing zones instead of in their own designated sectors. On their arrival, the later lifts consequently found that their assigned landing areas were almost entirely occupied with gliders which forced them to seek alternate and less desirable areas. This was further compounded by the situation wherein two glider serials were released so close together that they both were in the air at the same time over the same already crowded LZ. Quick reaction was the order of the day for one did not make another landing run in a glider. The pilots simply had to "dig in" on their landings because of limited space.

Although these abrupt, heavy landings did cause excessive damage to the gliders because of the lack of the "Griswold Nose" modification, the glider pilots by presence of mind, prompt action and skillful maneuver, saved many lives and much valuable airborne equipment. It was established by D+6 that not more than 125 glider-borne personnel were injured in these landings.

Although not encountered in the immediate objective area, as a matter of interest it is worthy to note that in the Frejus area outside of the drop zones, there was a second type of anti-glider obstacle which consisted of small but sturdy sharpened stakes some 18-inches high, firmly imbedded in the ground and connected by wire which could play havoc on the belly of any glider landing on such obstructed terrain.

In general, the problem of air resupply did not become the urgent problem as had been expected. Absence of serious enemy opposition caused ammunition expenditure to fall below the anticipated amount. The initial plan for bringing in the first supplies by air on D-Day was consequently changed so that it was not until 1100 hours on D+1 that two troop carrier groups brought in 116 aircraft loaded with supplies. The aircraft arrived over the DZs on schedule but at an altitude well over 2,000 feet which made accurate dropping extremely difficult. A rather stiff breeze, coupled with the high altitude and the merging of DZs A and O, caused much of the dropped equipment and supplies to get into the hands of the wrong units on the ground. Well over 95 percent of the 1,700-odd bundles dropped by parachute landed safely, but much of the specialized equipment failed to reach the units which had requested it. Subsequently, resupply missions carrying emergency signal and medical supplies were flown again on the night of D+1.

Although these drops had to be made at night by pathfinder aids the success of these missions was above the average, except that again the high altitude caused excessive scattering of equipment. The 334th Quartermaster Depot Company, Aerial Resupply, aided by the Parachute Maintenance Section of the 517th Parachute Regimental Combat Team, packed over 14,000 parachutes and 1,000 tons of equipment for the operation and deserve commendation for the outstanding work accomplished. The British Allied Air Supply Base deserves the very same credit and a well done for their contribution.

Ground Phase

While in the grand scheme of the whole operation, enemy strength and action could be considered slight, it was a far different matter from the individual trooper's standpoint. He and his comrades, in individual actions and small unit combat situations, struggled to get back to their designated DZs and assembly areas, due to over a 50 percent dropping error during the final run into the objective area. To these "mission oriented" troopers this was in every sense of the word a "big war." How the airborne forces accomplished their missions under adverse conditions can best be illustrated by taking the largest airborne unit involved, in this phase, the 517th Parachute Regimental Combat Team (517th PRCT), and highlighting its D to D+1 combat actions.

The 3rd Battalion of the 517th PRCT dropped approximately 15 to 20 miles northeast of DZ A and scattered over eight miles of rough terrain from Seillans east through Fayence to Tourettes and Callian, all a good day's march from DZ A. As this unit began to assemble and look for its equipment, three major sub-sections of this unit emerged; the first sub-section consisted of the first 10 aircraft loads dropped near Seillans which included most of Company I, the battalion commander, plus a Battalion Headquarters contingent all totaling about 160 troopers; the second increment consisted of 60 troopers from Battalion Headquarters Company and G and H companies that dropped in the vicinity of Tourettes; the third group was composed of over 200 troopers from units within the battalion plus members from the 596th Parachute Combat Engineer Company and members of Regimental Headquarters and Service Companies who had been attached to the 3rd Battalion for the drop.

All of this last group had been dropped in the vicinity of Callian. The combined strength of these subsections totaled about 480 troopers. About 35 troopers were injured during the drop and had to be left behind, with an escort, due to the nature of their injuries. Seventy-five troopers caught up later and about 50 others were too far away to join up and so they resorted to independent actions of opportunity. By 0800 hours the battalion commander Lieutenant Colonel Melvin Zais, with almost all of his battalion now intact, commenced a forced march to get back to his objective area some distance away.

The 1st Battalion, 517th PRCT was also dropped in a scattered pattern over an area 30 to 40 square miles west-northwest, and southwest of Trans-en-Provence. At daylight Captain Charles La Chaussee of Company C and Lt. Erle Ehly of Battalion Headquarters Company had managed to get together about 150 troopers in the battalion assembly area and had made the decision to move out to the battalion objective area without further delay. At this point the battalion executive officer Major Herbert Bowlby, joined them and took command.

The battalion commander Major William Boyle, landed about 4 or 5 miles from Trans and having lost about two-thirds of his "stick" in the dark gathered the remaining five troopers and took off for his objective area. He entered the outskirts of Les Arcs in the afternoon and found an additional 20 troopers that included the battalion S-3 and the assistant battalion surgeon. Gathering up this group, Major Boyle again started out for the objective area. On the way he and his party got in a firefight with about 300 of the enemy and the situation became "touch and go." Additional 1st Battalion troopers coming in from the northwest joined Major Boyle until his force numbered about 50 or more troopers. He was forced to set up a perimeter type defense on the southeast edge of Les Arcs and "dig in."

A third group from the same Battalion managed to meet up with Captain Donald Fraser whose Company A had been designated as the 517th PRCT Reserve Force. Captain Fraser managed to get together about 200 troopers to take over the area that was previously assigned to the 3rd Battalion, 517th PRCT on the west of the Chateau Ste Rosaline. This broke the 1st Battalion 517th PRCT into three separate groups: Major Bowlby, the Battalion executive officer, with about 200 in the objective area; Captain Fraser with over 200 troopers in the area west of the Chateau Ste Rosaline; and Major Boyle encircled in the vicinity of Les Arcs.

The 2nd Battalion, 517th PRCT, commanded by Lt. Col. Richard Seitz, at about 50 percent strength, was on its way to its objective area by daylight having landed about a mile from its drop zone which was DZ A. By noon of D-Day the 2nd Battalion minus Company F was on its objective. Company F which had dropped with part of the Regimental Headquarters and part of the 596th Parachute Combat Engineer Company in the vicinity of Le Muy rejoined the 2nd Battalion, 517th PRCT on the battalion objective later in the afternoon.

The versatility of the airborne troopers was demonstrated even more when the 2nd Battalion, 517th PRCT was ordered on D+1 to relieve the Regimental Reserve commanded by Captain Fraser and to extricate the 1st Battalion commander and his group in the vicinity of Les Arcs.

Meantime, the 3rd Battalion, 517th PRCT arrived at the 517th PRCT Command Post at Chateau Ste Rosaline at about 1600 hours on D+1. Despite exhaustion from the forced march the regimental commander Colonel Rupert Graves ordered his 3rd Battalion to take Les Arcs by nightfall. This action was essential as General Lucien Truscott, commanding general, VI Army Corps, wanted Les Arcs cleared of the enemy by the following day. At daylight the 3rd Battalion was 800 yards short of Les Arcs. The advance was resumed and Les Arcs was taken.

The town of Le Muy which had been assigned to the British 2nd Independent Parachute Brigade Group as a D-Day objective proved rather difficult to take even though the British troopers had captured the main bridge leading into town well ahead of schedule. General Frederick, commanding general of the 1st Airborne Task Force, reassigned this mission to the newly arrived 550th Glider Infantry Battalion which captured Le Muy on D+1.

The 551st Parachute Infantry Battalion with the 602nd Pack Artillery Battalion (Glider) attached, was ordered by 1st ABTF at 1100 hours on D+1 to attack Draguigan and seize the town. Throughout that afternoon and into the night the 551st fought its way into the town. Under the cover of darkness one company managed to get into town by coming in from the east, and by morning, the 551st was in control of the town. The commanding general of the German Army LXII Corps, General Neuling, and his staff along with several hundred troops that included a special officer cadet class, surrendered. This was done in sufficient time to permit a special mobile force from the U.S. Army VI Corps to pass through the town on D+3 on their way to the Rhone and beyond.

The 509th Parachute Infantry Battalion was of great assistance to the amphibious forces landing by making early contact with these forces and subsequently easing their movement inland. Also of note is the fact that 11 out of 12 pack howitzers of the 463rd Parachute Artillery Battalion were operative in less than an hour after landing. Likewise the 4.2 Mortar Companies and the 602nd Pack Artillery Battalion (Glider) which came in by glider were ready for action shortly after landing.

The rigorous parachute training for the artillery units paid off and was exemplified by the 460th Parachute Artillery Battalion in the way they man-handled their howitzers in order to keep up with the ever-changing fluid situation that characterized this operation. Although insufficient targets materialized to require much use of this firepower the artillery units nevertheless were prepared to furnish artillery support when and as needed. It was comforting for the infantry troopers to know that they were there and ready.

By D+3 the 1st ABTF had commenced to reorganize in the vicinity of Le Muy. Following the reorganization it proceeded to advance along the Riviera towards Cannes, Nice and the Italian Border. The British 2nd Independent Parachute Brigade Group was taken out of action and preparations were made to return it to its base in the Rome area for further operational use. The 1st Special Service Force replaced the British Paras, and the 1st ABTF then continued to advance along the coast, meeting determined enemy rear guard opposition.

These operations of the 1st ABTF toward the Franco-Italian border were not restricted to the coast, but extended to a point some 65 miles inland. As has always been the case when airborne troops are retained in the line in an offensive role, they experienced back-breaking difficulty in transporting their heavy supporting arms and ammunition. The fluid, rapid advance of the U.S. 7th Army as a whole made it difficult for 7th Army to provide necessary vehicles for the 1st ABTF. As a result, the paratroopers in many cases hauled their 75mm pack howitzers for some 60 or 70 miles over the rugged Riviera coastline. Fortunately, a number of captured enemy vehicles, together with the unit's organic transport brought in by gliders, did make the movement feasible.

While the 517th PRCT was used to portray the versatility and aggressiveness of airborne units the very same qualities were found and ably demonstrated time after time by the other 1st ABTF units in DRAGOON. The 1st ABTF, at the close of the airborne operations phase of DRAGOON, had all the necessary credits to recommend its retention as a separate airborne force, to be retained at theater level, for future airborne operations.

For all practical purposes the airborne phase of DRAGOON was over by D+2 due to the earlier-than-expected-link-up with the amphibious forces late on D+1. Combat actions continued in the airborne objective area due to the scattered nature of the combat actions but the enemy was in complete disarray and had difficulty in conducting any coordinated actions of significance. From this point on the 1st ABTF was to play the role of a light special type divisional-size force as it protected the Franco-Italian border from enemy inroads.

It was at this point in time that the lightness of airborne units that made them so well-suited for airborne assaults now went against them due to their lack of ground mobility which was the key to staying power in extended ground combat operations. It was this overpowering lack of mobility that caused General Devers, commanding general of 6th Army Group, to relegate the 1st ABTF to a security-type role.

STATISTICS

During Operation DRAGOON, the Provisional Troop Carrier Air Division flew 987 sorties and carried 9,000 airborne troops, 221 jeeps and 213 artillery pieces. The sorties flown also included 407 towed gliders and over 2 million pounds of equipment carried into the battle area for the 1st ABTF. One aircraft was lost as a result of the operation itself and the losses in aircraft from the period of movement from the United Kingdom to the conclusion of the operation totalled only nine. No troop carrier personnel other than glider pilots were known to have been killed; four were listed as missing and 16 were hospitalized. The balance of the 746 glider pilots dispatched on the operation had returned to their organizations.

On the airborne side of the picture, 873 American airborne personnel were listed as killed, captured or missing in action with 327 hospitalized on D+2. By 20 August (D+5), this figure had fallen to 434 still listed as killed, captured or missing in action while many of the hospital cases had returned to their units for further action. The British 2nd Independent Parachute Brigade Group had 181 troopers listed as missing in action and 130 men hospitalized. Later reports indicated that 51 British paratroopers had been killed and that 500 replacements had been requested by the American units and 126 replacements by the British unit. By D+2 over 1,000 prisoners had been taken by the American Forces and nearly 350 by the British Brigade Group. By D+8 this figure was well over 2,000. The total jump and glider crash accidents amounted to 283 or approximately three percent of the operational personnel involved.

The recovery of parachutes both of the personnel and cargo types was very low. As of 1 September not more than 1,000 parachutes had been sent to the Rome Base for salvage and repair. Similarly the gliders that could be used again was very small in number. A survey of the LZs indicated that fewer than 50 of the gliders could be salvaged without excessive cost.

MARKET–GARDEN: THE OPERATION IN HOLLAND

By James M. Gavin, Lt. Gen. (Edited from his book, <u>On To Berlin</u>)

By the third week of July the 82nd Airborne Division was back in its old billets in mid-England. Division headquarters and three infantry regiments were near Leicester, and other divisional troops were in Nottingham and Market Harborough. Immediately following their return from Normandy, all the troopers were given short furloughs. On their return we settled down to dealing urgently with the problems of taking in new volunteers and giving them parachute training, welcoming back our wounded, re-equipping—and, finally, intensive training. Our casualties had been heavy; some infantry companies lost more than 50 percent—killed, wounded, or missing. Offsetting those losses was the extensive battle experience gained by the survivors. Now the problem was to transmit this experience to the new young troopers who were joining us.

In early August, arrangements were being made for General Ridgway to assume command of the XVIII Airborne Corps, and I was pleased to learn that I was to be given command of the 82nd Airborne Division. On 16 August he formally relinquished command, and I was designated division commander. General Ridgway had given the division attributes that would serve it well for the remainder of the war. His great courage, integrity and aggressiveness in combat all made a lasting impression on everyone in the division and on all commands.

In late August, we began to receive a series of alerts to ready ourselves for an airborne landing. The British 2nd Army was making rapid progress, and the Wehrmacht was in a precipitate retreat. On 1 September we were given an outline plan for landings near Tournai, Belgium. The landings were canceled and the objective was changed to Liége. Troopers were dispersed to take-off airfields. The mission was canceled at the last moment.

On Sunday afternoon, 10 September I was in London visiting some friends when I received a phone call from General Brereton's headquarters in Sunningdale about an hour's drive west of London. I was told to be at his headquarters for a meeting as promptly as I could get there. There was to be another airborne operation, and they considered it to be imminent. It was 1800 hours when I arrived at the meeting; it had already been under way for a few minutes. General Browning had just flown over from the Continent, from Montgomery's headquarters, and he was holding forth. In addition to General Brereton there were present General Maxwell D. Taylor of the U.S. 101st Airborne, General Robert F. Urquhart of the British 1st Airborne, and all the senior troop carrier commanders, British and American. General Browning continued to outline the plan for the proposed operation. It envisioned seizing bridges over five major waterways, as well as a number of other tactical objectives extending from the present front of the British 2nd Army along the Albert Canal, 64 miles into Holland, to the farthest bridge, over the lower Rhine in the town of Arnhem.

After a brief discussion about who would take what mission, it became apparent that the present locations of the divisions in the U.K. would determine what objectives would be assigned to each division. As they were located, the British 1st Airborne Division was best positioned for the Arnhem drop, the 82nd Airborne Division for the operation between Nijmegen and Grave, and the U.S. 101st Airborne Division for all the southern bridges. The next decision had to do with the allocation of airlift and the bringing together of the appropriate parachute-glider troops with the proper troop-carrying units.

The mission assigned to the 82nd Airborne Division was to seize the long bridge over the Maas River at Grave, to seize and hold the high ground in the vicinity of Groesbeek, to seize at least one of the four bridges over the Maas-Waal Canal, and, finally, to seize the big bridge over the Waal at the city of Nijmegen.

The U.S. 101st Airborne Division had the mission of seizing the bridges of several canals and rivers south of Grave. Finally, the British 1st Airborne Division was to seize the bridge over the lower Rhine at the town of Arnhem.

On Friday, 15 September all troops were moved to take-off airdromes. They were sealed in by darkness Friday night; thus, they had one full day to ready themselves finally for the battle. In the meantime I assembled all the battalion commanders in one room and went over once again the detailed plans of each battalion.

Sunday, 17 September was a beautiful day. The sky was clear, and it was sunny and moderately warm. Everyone was up before daylight, busily trucking bundles of weapons and supplies to airplanes and taking care of final personal preparations. Because of our experiences in Normandy, the troopers loaded themselves with all the ammunition and antitank mines they could carry.

Finally all the gliders, parachute planes and towplanes were marshaled, and from all over England, they began to take off and converge on the cross-Channel flying routes.

About a half hour after we crossed the coastline, down beneath us a flight of C-47s came into view, flying across our path. Suddenly, parachutes began to blossom from them. I knew of no planned flight of this character, and I was disturbed about it. I decided that it must be the 101st that was to jump farther south. As it turned out, it was the 101st. As the ground rose, it seemed to be very close to us, and everything that I had memorized was coming into sight. The triangular patch of woods near where I was to jump appeared under us just as the jump light went on. Although we seemed quite close to the ground, we went out without a second's delay, and we seemed to hit the ground almost at once.

Early indications were that the drop had been unusually successful. Unit after unit reported in on schedule, and with few exceptions, all were in their pre-planned locations.

One of the first units to land was E Company of the 504th Parachute Infantry. It had the mission of landing south of the bridge over the Maas River, not far from Grave. The platoon leader who captured the bridge was Lieutenant John S. Thompson.

Thompson had his men break up into two teams, working their way around the end of the bridge, cutting all wires leading to the bridge and crossing it. In a matter of minutes he was contacted by elements from his battalion which had landed on the other side of the bridge, and with the cooperation of Thompson's platoon they captured the flak towers at the northern end of the bridge. They, too, cut all wires, and the first bridge was captured intact. To us this was the most important bridge of all, since it assured us of a linkup with the British 30th Corps, provided, of course, that General Taylor captured his bridges. More than 1,100 feet long, spanning the wide-flowing Maas (Meuse) River, its capture was essential to the division's survival.

The 504th Parachute Infantry Regiment, commanded by

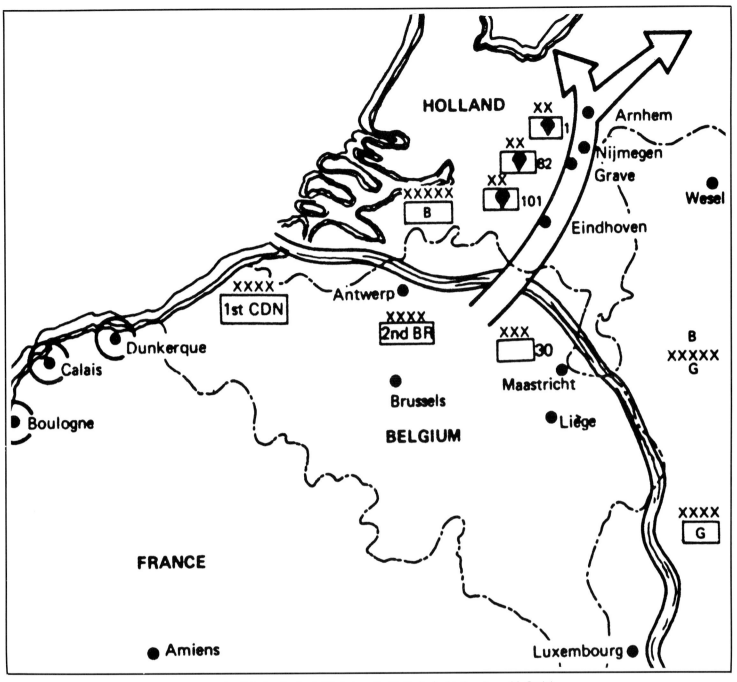

Operation "Market-Garden". (Photo from On To Berlin by Lt. Gen. James M. Gavin)

Colonel Reuben Tucker, landed exactly as planned between the Grave Bridge and the Maas-Waal Canal. Colonel Tucker immediately sent patrols to the main Grave-Nijmegen road. The patrols had no contact with the Germans. Colonel Tucker also sent strong patrols to seize the bridges over the Maas-Waal Canal.

They cut all the wires they could find. They kept the Germans pinned down and did not permit them to move about. In the midst of this the 505th sent a patrol which attacked the bridge from the other side, and together both forces overcame the German garrison at 1800 hours, thus capturing the bridge intact. Now we had our second bridge, making it possible for armor to get to Groesbeek and Nijmegen when the 30th Corps arrived.

Reports coming to the Division Command Post late in the afternoon of 17 September continued to be favorable. Units seemed to have landed as planned with minor exceptions. Both glider troops and parachute troopers landing behind the Germans were, in most cases, able to fight their way back with little difficulty.

I had a jeep by the late afternoon, and I drove toward Groesbeek and beyond. The first unit I encountered was the 376th Field Artillery. The battalion commander, Lieutenant Colonel Wilbur Griffith, had broken his ankle and was being moved about in a wheelbarrow. As he saw me approach, he laughed, saluted and said, "General, the 376th Field Artillery is in position with all guns ready to fire." Bringing in the 376th Field Artillery had been an experiment. We had reasoned that the first units the Germans would commit against us would be make-up formations of soldiers on furlough and local home guards, and that good, accurate field-artillery fire would keep them deployed and far from our infantry for some time. It turned out to be a correct assessment of what was to follow. The 376th fired 315 rounds in the first 24 hours and was most effective in keeping large German formations at some distance. In addition to fighting as artillerymen, it captured 400 German enlisted men and eight German officers.

Reports continued to come in from the widely dispersed parachute battalions. All were getting on with the tasks assigned to them.

During the night all three parachute regiments reported that they had the situation well in hand. Colonel Tucker with the 504th had captured the big Grave bridge and helped capture bridge No. 7 at Molenhoek. He was patrolling aggressively toward the West, expecting a major German reaction from that direction. Colonel William E. Ekman in the 505th, with two battalions, was organizing a defensive position from Mook through Horst, swinging back toward the town of Kamp, approximately a mile out of Groesbeek. Patrols had been sent to the Reichswald. The 2nd Battalion, 505th, under Colonel Ben Vandervoort, was in division reserve on high ground about a half mile from the Division Command Post. Colonel Lindquist's 508th organized a defense from Kamp to Wyler and established several roadblocks along the south of the high ground at Berg-en Dal.

As we had seized all our other bridges in the division area the key to the success of the battle was now the Nijmegen Bridge. It absolutely had to be seized and its destruction prevented as well.

During the first night, 17 September the staff, with some concern, watched Lindquist make significant troop commitments against the big bridge. There were high hopes that he would get the bridge quickly.

At daylight on 18 September when I went to the 508th Command Post, the report was grim. My heart sank. They had failed to get the bridge.

I had to get troops back to clear the glider landing zone without delay. I therefore instructed Colonel Lindquist to disengage the 3rd Battalion and move it back to clear the glider landing zone of Germans. It was a big order, for already Germans were in the woods between Berg-en-Dal and Wyler, attacking. The 3rd Battalion had been moving and fighting most of the night, and now they had to march six or seven miles back to Wyler, attack and destroy Germans in the woods and travel on to clear the drop zone beyond.

Once I was satisfied that I had done all I could with the 508th, I hurried back to Groesbeek to check that front. The 505th took me to an observation post in the upper story of a building on the outskirts of Groesbeek.

If the Germans had been aware of how thin the dispositions were, they could have moved quite readily through the huge gaps in the 505th front, but they did not probe deeply and the 505th was able to hold its positions. It reminded me once again of Biazza Ridge in Sicily.

The 505th at Groesbeek was having a busy time of it but seemed to have the situation under control. Heavy firing increasing in volume could be heard in the direction of Kamp-Wyler. I hurried back to the Division Command Post to check on the over-all situation. The glider reinforcements were due to land in early afternoon. I therefore took the immediately available reserves of two engineer companies and moved them through Groesbeek and toward Kamp. The battalion commander Captain Johnson was in command of the two companies, and I accompanied him for several miles through broken country. I ordered him to clear the landing zones of Germans and to watch out for the 3rd Battalion of the 508th, which was due to come in on his left. It was imperative that we get the Germans off the glider landing zone before the gliders arrived; otherwise the casualties among the glidermen would be devastating.

Considerable firing was taking place all along the front of the 505th and the 508th. As I watched, I wished there were some

Holland 1944, Pathfinder Crew L to R: John Zamanakos, Tom Walton, Glen Braddock, John Kleinfelder, Raymond Smith, Sgt. O'Shaughnessy, Capt. Frank Brown, Buford Williams, Delbert Jones, Cpl. Murphy, Fred Wilhelm, Wm. B. Mench, John Hopke, Wm. Casey. (Courtesy of 101st Airborne Division Association)

Sniper hunting after the Holland jump of 17 September 1944. (Courtesy of 101st Airborne Division Association)

way I could get word to the gliders so that they would not be surprised by the German fire immediately upon landing. Four hundred and fifty gliders were expected to land, and there was no way to communicate with them, since they were already en route.

Shortly before 1400 hours the great air armada could be seen approaching from England, 900 aircraft in all: 450 gliders and 450 C-47 tugs. The drone of the engines reached a roar as they came directly over the landing zones. I experienced a terrible feeling of helplessness. I wanted to tell them that they were landing right on the German infantry. Soon they were overhead, and the gliders began to cut loose and start their encircling descent. As they landed, they raised tremendous clouds of dust, and the weapons fire increased over the area. Some spun on one wing, others ended up on their noses or tipped over as they dug the glider nose into the earth in their desire to bring them to a quick stop. Glidermen could be seen running from the gliders and engaging the Germans. Others were attempting to extricate their artillery and jeeps.

It seemed almost a miracle when the battle was over and a count was taken of the men and equipment. The 319th Glider Field Artillery Battalion recovered 12 of its 12 howitzers and 26 of its 34 jeeps; the 320th Glider Field Artillery Battalion recovered eight of its 12 howitzers and 29 of its 39 jeeps. The 456th Parachute Field Artillery Battalion recovered 10 of its 12 howitzers and 23 of its 33 jeeps. Some of the glider units landed far behind the drop zones in Germany, but most of them fought their way back. In addition, the medical battalion brought in 26 jeeps. The 307th Engineers Battalion had 100 percent recovery, five out of five jeeps; the Signal Company eight out of 10 jeeps. And finally, most important of all, eight of eight 57 mm. antitank guns and eight out of nine jeeps were recovered by Battery D of the 80th Airborne Antiaircraft Battalion. A highly creditable performance and one that I never would have thought possible as I watched them approach the landing zone. The landings had begun at 1400 hours, and the last glider had landed by 1430 hours.

We were still very short of infantry, but we expected the 325th Glider Infantry to land the following day, 20 September. It took an hour or so to get everything arranged; then I moved once again to the 508th sector in which combat was the most intense in the division area.

So far, the 508th had been unable to do anything with the southern end of the Nijmegen bridge, and it was fully preoccupied with the Beek-Wyler front. I had great confidence in Vandervoort's battalion, probably one of the best in the division, and the Grenadier Guards, but it was evident that they would not be able to launch a full-scale coordinated attack until early the following morning, 20 September.

I decided that I somehow had to get across the river with our infantry and attack the northern end of the bridge and cut off the Germans at the southern end. The question was how.

General Horrocks, commanding British 30th Corps; General Browning, commanding the Airborne Corps; and General Allan Adair, commanding the Guards Armored Division; and I had a meeting near the sidewalk in front of the Malden Schoolhouse late that afternoon.

We did not know at the time, but the German general opposing us, General Student, had in his hands a complete copy of our attack order within an hour of the landings on D-day. It had been taken from a wrecked glider. It told him exactly what roads we were to use and what troops were to be given what specific missions. He at once organized counterattacks that cut the road at several places.

Glider troopers of the 907th GFA Bn. front load a 75mm pack howitzer in the CG-4A glider prior to Operation Market-Garden.

In the American Army, the corps acting in an independent role, such as 30th Corps in the Holland situation, would have an engineer battalion or regiment attached to it, and that would include a company of boats. As we stood talking, I asked General Horrocks about it, and he said he thought they had some boats well down the road in the train somewhere.

I told General Horrocks that if he could get the boats to me, I would move the 504th Parachute Infantry to the riverbank and make a crossing as rapidly as possible, and thus we would be in a position to attack the far end of the bridge. He accepted the plan, and I immediately began preparations for the river crossing.

Thus 20 September turned out to be a day unprecedented in the division's combat history. Each of the three regiments fought a critical battle in its own area and won over heavy odds, but the most brilliant and spectacular battle of all was that of the 504th to get across the Waal River.

The assault crossing was to be made by the 3rd Battalion under the command of a 27-year-old West Pointer, Major Julian Cook.

Finally the boats arrived. To the paratroopers they were flimsy-looking affairs—folding canvas with the sides held erect by wooden slats. Each craft weighed about 200 pounds, and it was expected that each would carry 13 paratroopers and a crew of three engineers to row them across. That may have been someone's idea of how the crossing was to be made, but in effect, every paratrooper tried to row, using the butt of his rifle for a paddle when necessary. The width of the river startled, if not shocked, those who saw it for the first time.

Twenty-six boats were assembled, someone yelled, "Go," and there was a rush for the water's edge. The troopers had a hard time getting the boats into deep water while they climbed over the sides with their weapons. To add to their difficulties, German small-arms fire began to intercept the fragile flotilla. Never having rowed together, the troopers sometimes worked against each other, and boats were spinning in the river. The German firing steadily increased, heavy artillery fire joining the machine-gun and mortar fire.

Nevertheless, as we expected, the 504th kept battling its way across the bloody river. There were many individual acts of courage, many casualties, and later, the troopers told me of stuffing handkerchiefs in the bullet holes to keep the water from pouring into the boats, and using their helmets to bail out water, while all around them men were being killed and wounded.

To those watching the crossing, it seemed forever before the first boats touched down on the northern bank. Men struggled out of the boats, waded and made their way through the mud, and ran forward. Some of them said later that they were so glad to be alive that they had only one thought: to kill the Germans along the embankment who had been making the crossing so difficult. As reported by Cornelius Ryan in *A Bridge Too Far*, Lieutenant Colonel Giles Vandeleur, who was watching the landings, later said, "'I saw one or two boats hit the

The British 2nd Army moves across the bridge into Nijmegan, Holland 19 September 1944 to link up with the 506th PIR.

Troopers of the 101st Airborne Division crash to the ground in Holland during Operation Market-Garden.

Major General Maxwell D. Taylor, Commanding General of the 101st Airborne Division, enplanes for Holland during Operation Market-Garden.

Members of the Dutch underground inspect the glider TARFU II on the 82nd LZ. (Courtesy of Douglas W. Wilmer)

An airborne catapillar, from the 326th Airborne Engineer Battalion, works on a muddy road in front of the Division command post area at Slijk Ewijk, Holland, on 6 October 1944.

MG Maxwell D. Taylor in his jeep between Zon and St. Oedenrode in Holland, 1944.

A patrol of the 506th PIR moves through Nijmegen, Holland on 28 September 1944.

beach, followed immediately by three or four others. Nobody paused. Men got out and began running toward the embankment. My God, what a courageous sight it was! They just moved steadily across that open ground. I never saw a single man lie down until he was hit.' Then, to Vandeleur's amazement, 'the boats turned around and started back for the second wave.' Turning to Horrocks, General Browning said, 'I have never seen a more gallant action.'"

By 1700 hours, two hours after the start of the assault, they had gotten control of the northern end of the bridge, and the Germans were fleeing across it. The German losses on the bridge were heavy; many of them were killed in the girders, and others jumped into the river below.

By 1800 hours Vandervoort and the Grenadier Guards were closing in on the last of the German foxholes. The fighting was heavy, but the 505th troopers moved forward from rooftop to rooftop, supported by the Grenadier Guards' tanks. In the final all-out assault they overran all the German positions. Of the 500 Germans Captain Euling had south of the river as a bridgehead, only 60 survived.

Although we did not know it, on Wednesday evening, 20 September, one of the most courageous actions of the battle was being fought by a handful of British troopers at the northern end of the Arnhem Bridge. When General Urquhart's British Airborne Division was landed eight miles to the west of the Arnhem Bridge, three battalions were dispatched at once by three separate routes with the mission of converging on the bridge. Two of them quickly ran into heavy German resistance and were stopped. The third, the 2nd Battalion of the 1st Parachute Brigade, under the command of Lieutenant Colonel John Frost, moved by a secondary road along the norther bank of the Rhine River. He occasionally encountered German resistance, and it was quickly overcome. Nevertheless, it took him seven hours to reach the northern end of the bridge. Already the Germans had crossed the bridge and held the southern end. He at once deployed his battalion around the northern end, and occupied buildings in a small bridgehead. He arrived in that position shortly after 2000 hours. His battle began that night and continued relentlessly through Wednesday evening. By that time, he was badly wounded and had more than 200 casualties that he was trying to save in the cellars of buildings. Finally the buildings were set afire, and when burning timbers began to fall and it was obvious that the troopers would be roasted alive, he sent out a Red Cross flag and asked for a truce. It was a tragedy that help was so near but did not reach him.

In the meantime the Polish Parachute Brigade had been south of the lower Rhine River, about five miles to the west of Arnhem. The remains of Urquhart's command had been compressed into a small pocket around the town of Oosterbeek, 2 1/2 miles west of the Arnhem Bridge. Finally on the night of 25-26 September more than 2,000 survivors of Urquhart's command of 9,500 men were ferried across the river. We brought them to Nijmegen, where we provided them with blankets, shelter and food. They had fought an extraordinarily gallant action and a desperate battle for survival in their small bridgehead. We in the 82nd had a lasting regret that we had not reached them. They were brave men and they had done all that human flesh and human spirit could accomplish. Thus, the great gamble to end the war in the fall of '44 came to an end.

When the 504th was relieved north of the bridge, it was withdrawn and committed to the front between the 508th and the 505th. The 82nd then had its left flank resting on the Rhine River, extending directly across the flatland to Wyler. Here the 504th took up the front and held it until it was near Groesbeek, where it was taken up by the 505th. The right flank of the 505th was on a small hill mass close to the Reichswald at Reithorst.

The 325th Glider Infantry Regiment finally landed in the old 504th drop zone area on Saturday, 23 September, six days after our first landing—five days late. I had been asked by General Horrocks to clear a bridgehead over the Maas River in the vicinity of Mook. The British intent was to bring another corps into that sector on the right of Horrocks's 30th Corps. To establish the bridgehead, the 325th attacked on the right of the 505th. With good support from the Sherwood Ranger Yeomanry Tank Battalion, it finally occupied a front from Reithorst to the Maas River. This put the division in a defensive position extending from the Rhine River on its left approximately 12 miles to the Maas River on its right.

I was directed to fly to Rheims, France, to look at billets that were proposed for the 82nd Airborne Division. They were old French army cantonments that had been used by the Germans. They were at Sissonne and Suippes, both about 20 miles from Rheims. They did not offer much in the way of comfort, but they were far better than foxholes in the fall and winter in Holland. On 13 November I turned over our sector facing the Reichswald to the 3rd Canadian Infantry Division, and the 82nd moved back to France to its new billets.

After the war, and as the memories of the battle of Nijmegen lingered in our minds, there was much talk about it among the veterans. There has been a lasting discontent over the historical record of the Americans in MARKET-GARDEN. From the viewpoint of the 82nd Airborne Division, it was its most difficult battle in extremely difficult circumstances, a battle in which the division was frequently outnumbered and yet won every tactical engagement. People who know their history comment on the fact that the 82nd captured the Grave bridge at Nijmegen. And while capturing these objectives, the 82nd had to be capable of beating off any attacks the Germans might launch against it.

As the battle developed, thanks particularly to the skill of the individual troopers, the NCOs and junior officers, we were able to beat major German counterattacks by patrolling actively the huge gaps around our perimeter and building up before each German counteroffensive began to take shape. And after each German attack we counterattacked and destroyed them.

After the war I received a letter from one of our opponents, General Von der Heydte, who enclosed a letter from a German general officer who had opposed us at Molenhoek. He commented that "in forest fighting and night fighting the members of the 82nd Airborne Division proved superior to the Germans." Finally, in a desperate effort to reach our comrades at Arnhem, the 504th Parachute Infantry made a gallant crossing of the Rhine River, and the huge Nijmegen bridge was ours. Although our losses were heavy, they were about the same as those sustained by the 505th Parachute Infantry in the defensive actions in Sicily against the Hermann Goering Division. We never could have fought and won the battle of Nijmegen without all the combat experience from the battles of Sicily, Italy and Normandy.

Paratroopers of the 101st Airborne Division move up "Hell's Highway" in Holland to attempt to break a German roadblock between Veghel and Uhen late in September 1944.

OPERATION VARSITY: THE AIRBORNE ASSAULT OVER THE RHINE

By Bart Hagerman, LTC, AUS (Ret)

A conference held near St. Germain, France, on 6 February 1945, briefed the general plan for Operation VARSITY. It was announced that the 17th Airborne Division would be ordered out of the battlefront commencing 10 February and moved by rail and motor to the vicinity of Chalons sur Marne to prepare for this operation, which would force a crossing of the Rhine on or about 1 April.

At this time, the division was engaged in combat along the Our River where it was holding a small bridgehead just south of Clerveaux, Luxembourg.

The camp at Chalons was not completed. The spring thaws had made the ground a sea of mud, and roads were practically impassable in the new camp area. The division was faced with the problems of housing in a casern and tent area where housekeeping facilities were either poor or lacking altogether.

The area was cleaned up; tents pitched; walks and roads fixed; latrine facilities built or improved; stoves, cots and other necessary equipment drawn and issued; and hard-standings for motor parks built. The division was given four days to settle in its new area and complete all arrangements for housekeeping. At the end of this allotted time, the camp was in splendid condition and the division ready to start training.

Weather hindered the movement of the base echelon and supplies from England, however, 500 planes finally arrived. The Parachute Maintenance Company cleaned a warehouse and by 10 March they were operating in the Chalons area.

At this time, the stupendousness of the task ahead in preparing for an airborne operation on such short notice was apparent. The division had just completed the most rigorous campaign the U.S. Army had ever fought: The Battle of the Bulge. Casualties had been so heavy that some rifle companies had less than 40 men of their original strength, and some were without officers. The division was about 4,000 officers and men understrength. The loss of key personnel and specialists had been particularly heavy. Equipment had been lost or turned in, and the remainder was in need of complete overhaul.

As reinforcements arrived and men returned from the hospital, an intensive training program was undertaken. Individual training, particularly of specialists and key personnel, was immediately started. This was followed as soon as possible by small unit training. All individuals were given familiarization firing; crews of crew-served weapons were picked and trained. Parachute exercises for each parachute regiment were drawn and conducted. Two complete division CPXs were held. Experimental work in jumping from the new C-46 was necessary when it was announced that one regiment would have to use these airplanes. Gliders were obtained and the new replacements were given glider training.

During this period, equipment was drawn to replace losses. A complete showdown inspection was held for every piece of equipment. All weapons were given a thorough inspection and overhaul. Motors of the entire division were inspected and overhauled in the unbelievable short time of 15 days. Combat loads of ammunition weighing more than 400 tons were hauled and distributed.

Orders and directives were given from time to time. As Allied forces moved rapidly toward the Rhine River, the necessity for an early airborne operation became evident. D-Day was advanced to 24 March.

The division was given new orders to reorganize. The major change was the disbandment of the 193rd Glider Infantry Regiment, with the subsequent reorganization of the 194th Glider Infantry Regiment with three battalions, the reorganization of the Glider Field Artillery Battalions with three firing batteries, the conversion of the 680th Glider Field Artillery Battalion to a 105mm howitzer battalion, the addition of another squad to the rifle platoon of parachute infantry regiments, and the addition of another parachute field artillery battalion, the 464th Parachute Field Artillery Battalion.

Liaison officers were sent to First Allied Airborne Army and IX Troop Carrier Command, 6th British Airborne Division, 1st British Commando Brigade and 15th Scottish Division. The last three liaison officers went into the operation with the units to which they were attached and continued to maintain liaison during the operation.

The 411th Airborne Quartermaster Unit coordinated a collecting point for the dead. It was designated on one side of the resupply drop area. Here the Quartermaster personnel could operate with the greatest degree of protection, and supply transportation could be utilized for hauling bodies. The location was adjacent to, and readily accessible to, a proposed bridge at Wesel. Ample space was also available for a cemetery should temporary burial be necessary.

The prisoner of war collecting point was designated near the burial collecting point, so that prisoner labor could be used, if necessary, in preparing and maintaining the temporary cemetery.

The Division chaplain was to come in by glider with the Division Medical Clearing Station. His clerk, with driver and transportation, were to come overland and join him with the other land elements.

The Red Cross was to travel overland and join the airborne group on the afternoon of D-Day and immediately begin dispensing needed articles to the men.

Mail deliveries were planned to cease upon the movement of a unit from the division area. Mail was to be held at the base at Chalons until delivery was possible after the operation. Preparation of mail for delivery was made at the Chalons base. Each unit left a mail clerk at the base post office who assisted in sorting the mail and tying it in company or battery bundles. Then each unit's mail was put in separate sacks to ease the distribution at the ration distribution points. For delivery after the operation, the division C-47 was scheduled to make a daily flight with Class I mail to a forward field near Venlo. There it would be picked up by a truck from the Division Quartermaster Company and taken to the ration dump where it would be distributed with the rations. It was estimated that one plane per day would carry all the Class I mail.

The general plan for supply and evacuation called for all categories of supplies sufficient for one day to be carried by air as basic unit loads.

Immediately following the airlifts, a resupply mission consisting of 270 tons of all classes of supplies for one day was to be flown by B-24 bombers based in the United Kingdom. An automatic resupply mission of 540 tons of supplies was also scheduled. Further items were to be requested by division through XVIII Airborne Corps to First Allied Airborne Army which would be responsible for procuring, preparing the aerial delivery and dropping it into the drop zones. All subsequent aerial resupply was to be upon radio request by the

Elements of the 17th Airborne Division ready to jump.

Division. The materials to sustain the division for nine days of operation were to be prepared at the three continent resupply airstrips.

Additional supplies and replacement items of critical equipment, particularly radios and crew-served weapons, were carried in the overland echelon. Ground resupply arrangements were made with the 9th Army to establish a dump west of the Rhine River for this purpose. The dump would consist of all classes of supplies and was to be maintained at a three-days' operating level until normal bridge traffic across the river was established.

Planning for the signal communications of VARSITY began on 23 February when the division commander notified the signal officer of the pending operation. The planning was divided into three sections: ground communications after the drop, supply and resupply and communications in the marshalling areas.

During the period 1-17 March, all elements of the division signal system ran continuous rehearsals simulating as nearly as possible conditions expected at VARSITY. To care for immediate needs upon landing, each unit carried a three-day supply of wire, batteries and tubes.

No other rehearsals were held except for a communication exercise involving all wire and radio communication of the division artillery, including the special forward observer and translator on 14 March. Since all personnel involved in this rehearsal had not been briefed, it was staged as a type of communication exercise.

Movement to the marshalling areas began on the evening of 19 March. Twelve camps were scattered throughout four base sections. The capacity of each camp was 1,200 with the exception of the camp where the C-46s were to depart. This camp had a capacity of 2,600. The camps were completed by 15 March and were in operating condition when troops began arriving 21 March. The construction of these areas at the airfields, the movement of troops, and the billeting of the troops were worthy of note. The service forces charged with these missions did extremely fine and efficient jobs in making those parts of the operation a success.

Upon arrival at the fields, the men had numerous facilities available at Oise Base fields. Movies were shown as often as the commanding officer desired in a movie tent that seated 500. Athletic equipment, soccer, volleyball, softball, horseshoes and footballs were available in quantities that any man could play if he so desired. Two dayrooms were set up with radios, victrola, records and reading material. The *Stars and Stripes* was delivered and gratuitous exchange rations issued 24 hours before D-Day. Mail was censored by unit officers and turned over to the Camp CIC. Loud speakers were set up in all mess halls, and music was played and news broadcasts heard during meal hours.

One father shared the life and last letter of his eldest son, Pvt. Jerzy S. Spitzer. His early life did not reflect that of most American soldiers, but the thoughts expressed in the letter could have been written by almost any man in the Airborne.

After the Germans marched into Poland, the Spitzer family of four fled. After harrowing experiences, they found their way to the U.S.A. in April 1941. Jerzy, the oldest son, was admitted to the University of California at Berkeley and graduated with honors. The day after graduation he enlisted in the U.S. Army. After basic training he was attached to the anti-aircraft artillery. He volunteered for the paratroopers, quali-

17th Airborne Troopers hear a brief message from their chaplain on the morning of 24 March before Operation Varsity.

fied as a jumper and completed demolition training with the highest classification. In early February, 1945 he was in France, assigned to the 17th Airborne.

Five days before the air offensive over the Rhine River, he wrote a letter, in Polish, to his parents and younger brother, in which he declared the ideals for which he fought and his willingness to die if necessary.

"I am not afraid of anything, and I do not worry about myself. I am fighting and working for my ideals and believe that God will help me. My ideal is that all people are individuals and should be treated as such, regardless of their origin, understanding of God, and appearance. That is my faith, ideal and religion. The goal of my life is to do whatever I can to spread that faith among people, so that finally everybody may recognize that truth and unite as citizens of the world and live together in friendship, so that the ideals of peace on earth, goodwill toward everyone and brotherhood among men should be realism in the life, thinking and deeds of men."

On 30 March he was severely wounded in Germany and died of wounds on 6 April 1945. He was only 22 years old.

At the marshalling area, a large operation tent was maintained. It contained the Division War Room for staff use and another section for distribution of maps, photos, documents, etc. During the three-day period in the marshalling areas, a large volume of material was distributed to the units, and two separate intelligence reports were published. Because of the rapidly changing enemy situation and difficulty of positive identification, it was necessary to carefully evaluate enemy defenses and continually revise previous estimates.

Extreme efforts were made to prevent local security violations, despite the difficulties caused by the close association with French civilian personnel. Every effort was made to conceal the nature of the unit in its movements and to prevent leakage of information as to specific date and place of the operation. Use of 12 separate airfields presented a security problem, but, in general, security was good and no evidence of overt acts of espionage was documented.

At each airfield briefings were conducted. Joint briefings between pilots and crewmembers of the planes, and between glider pilots and the members of their glider loads were held. In addition, each plane and glider crew was briefed individually with the latest air photos available. This briefing of glider pilots as well as members of the glider loads proved extremely valuable. Individuals knew exactly what was expected of them and went rapidly about their business upon landing. The briefings created perfect teamwork between glider pilots and the airborne troops.

D-Day arrived for the comprehensive campaign that was code named VARSITY. The 17th Airborne Division prepared to take off for its third campaign—and its first airborne mission—from a dozen airfields in France on 24 March 1945.

The 507th and 513th Parachute Infantry Regimental Combat Teams were the first elements of the 17th to take off. They were followed by the 194th Glider Infantry Regiment. They were dropped over the Rhine River by glider and parachuted into the enemy's position in Westphalia, in the vicinity of Wesel, Germany. This was the first airborne mission conducted by the Allies into Germany.

Orders described the mission of the 17th as being "to seize, clear and secure" the division area with priority to the high ground just east of Dieresfordt and the bridges over the Issel River; to protect the right (south) flank of Corps; to

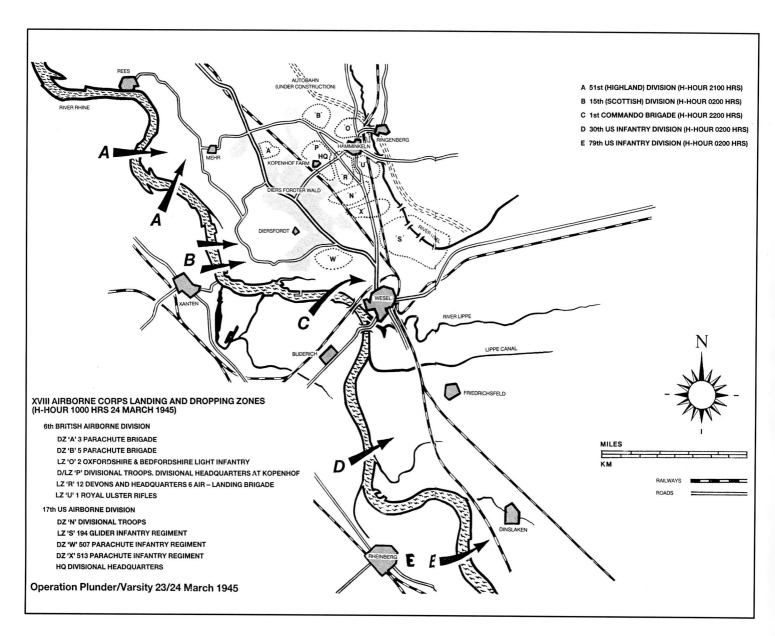

LZs and DZs for the British 6th Airborne and U.S. 17th Airborne Division's drop across the Rhine into Germany.

establish contact with 1st Commando Brigade (British) northeast of Wesel and with the 15th British Division and the 6th British Airborne Division.

The day started early, and never really ended. At 0700 hours everyone was aboard the various aircraft, and at 0730 hours the first plane was airborne. The entire column was two hours and 18 minutes in length and consisted of 226 C-47s and 72 C-46s carrying parachute troops, while 906 gliders were towed by 610 C-47s. To the left of these formations were 42 C-54s, 752 C-47s, and 420 gliders of the British 6th Airborne Division. And above and around this huge assembly of aircraft were 676 U.S. fighter planes and 213 RAF fighter planes flying escort for the "Airborne Carpet." They were banking and flitting around the long train of aircraft, keeping a constant watch for enemy planes.

It was a sight never before seen. Stretched across the sky as far as one's eye could see was the largest sky armada ever assembled, then or now. Several records were set: first time C-47s double-towed gliders in combat, first and only time troopers jumped the double door C-46 in combat, and first time glider combat troops landed on fields not previously secured by paratroops.

The panorama from the planes showed a beautiful European landscape, colorfully roofed hamlets, large industrial centers and cozy little farms. War seemed so far away. But then came the 20 minute warning.

For the next several minutes, each man preparing to jump did the same thing: He checked his gear—helmet, flak vest, parachute harness, weapons, and equipment. The clang, jangle, and ring of men checking their gear filled the planes. Then the men stood up and hooked up. Many paused to say a brief prayer, cross themselves and wish a buddy best of luck. It would not be long now.

Suddenly, as if a curtain had been lifted, the view changed. The countryside was beaten and gray, barren, a graveyard look. The smoke of battle wafted drearily upward. Shots could be heard. As the number one men were standing in the doors of their planes ready to jump, the ack-ack and anti-aircraft fire rattled and shook the planes as it cracked nearby, frequently sending flak through the wings and sides, and occasionally hitting troopers waiting to jump.

Then the green light flashed, signaling to jump. There was a rush of air, a jerk, a jolt, a look above, and the men of the 507th were beginning to hit the ground. They were the first paratroopers to land on German soil. The 17th Airborne Division was committed to the Central European Campaign as it began to take up its ground positions in the vicinity of Wesel.

The departure of the 17th from the marshalling areas in France ran generally as scheduled, with some last minute changes due to wind velocity so as to maintain correct times over the objectives. Weather and visibility were good, although, during the first 90 minutes of flight time, the air was rough. As the planes approached the drop and landing zones, visibility was reduced by smoke and haze in the vicinity of the Maas River to the Rhine River laid in order to mask the landings. Smoke and heat from enemy artillery also hampered good visibility.

The weather itself did not adversely affect the landings,

After the drop over the Rhine, these 17th Airborne troopers move to their assembly area. Note equipment and personnel chutes draped over wires.

Trooper Bill Gamble, 513th PIR, complete with equipment bag, ready to go for Operation Varsity.

Sgt. Franzek was one of the 17th Airborne troopers who got a Mohican haircut prior to Operation Varsity.

but the paratroopers and glidermen faced very heavy enemy anti-aircraft fire from small arms to 88mm anti-aircraft artillery. The enemy was well-prepared and waiting. It was later reported that the Germans knew eight days before the attack that the Allied paratroopers and glidermen were preparing to drop and where they were going to land. The Allied veil of secrecy had been pierced.

The initial drop was made by the 507th Parachute Infantry Regiment Combat Team, at 0948 hours, and the glider landings of the last units of the 194th Glider Regiment Combat Team were completed by 1210 hours. The gliders, too, were harassed by flak and small arms fire, which belted through the wood and canvas crates with alarming ease. The men would sit, strapped in their seats, anxiously watching the pilot to see when they were cut loose. They, too, wished each other well, and most of them said a prayer.

Except for the 1st Battalion of the 507th, most of the landings were made in the assigned drop zones and landing zones. This battalion landed one and a half to two miles off course, in an area near Dieresfordt, northwest of the planned drop zone. But reorganization upon the ground was immediate, and within four hours, all of the 17th Airborne Division's designed objectives had been seized. Contact was also made with the 6th British Assault Brigade to the south and southeast.

The 17th Airborne Division's plan called for the 513th Regimental Combat Team to land in the northern part of the division's zone and to seize and clear the territory in that zone and then to attack west and seize the wooded high ground north of Dieresfordt. The 3rd Battalion of the 513th on a separate mission was to move east after landing near the Issel Canal.

The 194th Glider Combat Team was to land by glider in the southeastern part of the division's zone, to clear that portion of the division's area and to seize and defend the crossings over the Issel River and the Issel Canal. One battalion was to be held in reserve and to be committed only upon division order.

The 507th Parachute Regimental Combat Team was to land in the southern part of the division's zone and to clear that zone and attack northeast, clearing the woods in the area.

Both the 507th and 513th Combat Teams were to assist the advance of the ground troops after seizing their objectives.

The 17th Airborne Division artillery had three battalions, and their mission throughout the day included direct fire on self-propelled guns and buildings as well as normal artillery missions. By 1800 hours, 38 pieces of artillery of the 51 pieces brought in were in action against the enemy. Of the 38 pieces in action, one had been pieced together from salvageable parts from three damaged artillery pieces. Communications were established and a fire direction control center was in operation by 1500 hours.

At the close of the day (24 March), all the units of the 17th Airborne Division had taken a total of 2,873 prisoners (by 0800 hours on 25 March, that number increased to 3,100 prisoners), many of whom were high-ranking German officers. They had destroyed at least 13 enemy tanks, two self-propelled guns, and several batteries of artillery. They also captured considerable other equipment, including several personnel carriers and other smaller vehicles.

The 17th, however, also suffered heavy casualties on the first day: killed in action, 159; wounded in action, 522; and missing in action, 840, but 600 of those reported as missing in action were located and rejoined their units.

Operation VARSITY had been accomplished as pre-

After the airborne operation, 17th Airborne troopers loaded aboard the tanks of the British Coldstream Guards to keep the Germans on the run.

The 17th Airborne troopers grabbed any transportation they could find after the drop. Here Sgt. Lupoli, a medic, utilizes a captured German motorcycle.

Part of the POW bag taken by the 17th Airborne after the airborne operation.

After the drop, 24 March 1944, the 17th Airborne carried on into Germany. Here they pass through Essen.

This 17th Airborne trooper moves ammunition on his captured weapons carrier.

scribed without a single amendment or change to the original scheme of maneuver. The coordination require precise liaison with the 6th British Airborne Division's base. The efficiency demonstrated by all the staff and operational echelons made this the most successful airborne operation to be conducted in the European Theater of Operations.

The drop across the Rhine into Germany was a feat which will be remembered by historians as one of the truly great military maneuvers of all time. But the untold story is the support given the paratroopers and glidermen by the Division Land Echelon. Without the support of the Land Echelon the drop could not have taken place. It was imperative that the paratroopers and glidermen travel light, and that all of their heavy supplies and equipment, medical support and food provisions get to the battlefront at the same time. The men of the Division Land Echelon are among the unsung heroes of the crossing of the Rhine.

The Division Land Echelon, under the command of Brig. Gen. John L. Whitelaw, assistant division commander, left base on the morning of D-3 and closed in its appointed assembly area at 0400 hours of D-2 with no major difficulties encountered.

The organic vehicles of the division moved to the bivouac area in Bree, Belgium, 21 March, to be dispatched to the Issum-Kapellen area when called. The attached units reported directly to the area approximately 24 hours prior to the contemplated crossing time.

Road clearances and routes were established for the Overland Echelon, which was to concentrate west of the Rhine River in the vicinity of Issum and Kapellen, Germany. It was necessary to provide 30, 2 1/2-ton trucks from sources outside the division. The 2,005 vehicle convoy involved in the movement overland used two routes and covered the 246 miles without any overnight halt. Using only blackout lights they drove over roads controlled by the British. At the designated time they were to cross the Rhine either by ferry or by bridge.

In order to avoid congestion near the crossing sites on the west bank of the Rhine a system of staging areas was organized. An assembly area between Issum and Kapellen was assigned. From there vehicles were moved by stages on a strictly controlled road to the river. The number of vehicles in the assembly at any one time was kept at a minimum in order to insure a steady flow of traffic.

Priorities were established for vehicles to cross the river. The general plan was for the tactical units, transportation and organic ambulances, plus a few control vehicles for supply officers, to cross first. They were to be followed by heavy transportation of the division service organizations. An attached tank battalion and attached tank destroyer battalion were also given high priority, but their crossing depended upon heavy river craft and heavy capacity bridges being available.

The British provided wire communications for the staging area, including the division assembly area and liaison officer for the assembly area. The 17th Airborne sent liaison officers to be with the British bank control officer and at all secondary areas where vehicles were to stop on their way to the river. Two officers were sent to the far bank to supervise the movement of the vehicles from the river to a division contact point at Muhlenfeld. At this place unit guides took control of their own vehicles.

A Quartermaster Amphibious Truck Company (DUKWS) was augmented by 27 additional DUKWS. This provided movement of supplies across the river to dumps on the far shore until bridges were established. Supplies via DUKWS began crossing the Rhine by 1700 hours D-Day. Tactical reasons caused the cancellation of air resupply. This required a shift in the loading plan and a change in the traffic priority. It also required the establishment of supply dumps nearer the east bank of the river.

Ferrying the bridge construction was started by the British on D-Day north of Xanten and continued without interruption in spite of strafing and bombing by hostile aircraft. In accordance with the predetermined schedule, ambulances crossed on D-Day followed by most of the self-propelled AT guns and tanks. By the third day the division had its organic transportation and was joined by all attached units.

Ground resupply arrangements were made with the 9th Army which established a dump on the west bank of the Rhine River in the vicinity of Issum. The dump consisted of all classes of supplies, which were to be maintained at three days' operating level until normal bridge traffic was established.

Edward Dorrity, a member of the Overland Echelon, wrote: Disguised as 'Plain old Infantrymen' they moved swiftly and secretly through France into a bivouac area in Belgium.

Early morning, 24 March, the forward group of jeeps moved out to be ferried across the Rhine at the first opportunity. The heavy vehicles followed.

Eyes were constantly to the sky—looking, hoping, waiting. Moving through Holland and into Germany, they spotted the first planes of the sky train, the huge, graceful C-46s. The gliders came behind. The men in the vehicles nodded to each other knowingly. They knew what those innocent unarmed planes had in their cargo bellies. In a few minutes the Nazis would be handed another surprise from the mailed fist of American ingenuity.

The motor echelon moved on into a different land from what they had been in before. Here there was not the two-fingered sign of victory, the cheering groups of little children, or the smiling, waving young girls and gazing old folks. These people were Germans. They did not smile. They did not wave. The few that were on the road stared grimly or turned their backs. These people were not happy to see Americans.

The commander of the ground unit of the 194th Glider Infantry called the men together and told them they were now in Germany. They went into the woods to sleep and to await the call that a bridge had been thrown across the Rhine. The call came at 0300 hours, after an exciting night of watching an aerial circus at the expense of the ever-vanishing Luftwaffe.

The land movement was well-ordered. Trucks lined up and groaned over the dirt roads toward the Rhine. Xanten was a snarl of traffic with each unit eager to get to the bridge. Glider pilots were dribbling back, each telling his own story of hours spent behind enemy lines. The convoy finally reached the Rhine and the bridge. The vehicles were bumper to bumper; the bridge was never empty. Amphibious jeeps and trucks scuttled back and forth across the river.

Here was the evidence of the airborne landings. Parachutes draped from trees; parapacks, used as resupply, dotted the fields. Farther on were the fields crowded with empty gliders.

Across the Rhine the convoy split and each vehicle went to its own unit knowing that tomorrow morning they would have hotcakes, eggs, and hot coffee for breakfast. This would be a pleasant change from the K-rations and D-bars. The ground unit had accomplished its primary mission. It had successfully reached the airborne troops. Now was left the job of winning the war.

Operation VARSITY having been a success, the 17th moved deeper into Germany.

KOREA OPERATIONS

By Edward M. (Fly) Flanagan, Jr., Lt. Gen., USA (Ret.)

The five-year old fragile peace in the "Land of the Morning Calm" was brutally shattered by the eruption of concentrated North Korean artillery and mortar shells along the rain-soaked 38th Parallel in the pre-dawn darkness of Sunday, 25 June 1950.

The mobile and tank-heavy North Korean strike force that rumbled south after the massive artillery preparation was a formidable one. Between 15 and 24 June, the North Koreans had, without detection by the South Koreans, infiltrated an additional 80,000 men and their equipment close behind the arbitrary line dividing North and South Korea. Then General Chai Ung Jun, the North Korean commanding general, carefully and methodically deployed his total battle force along routes to the south which he and his staff painstakingly selected during the preceding weeks. General Chai's main battle force included seven infantry divisions, one armored brigade, one separate infantry regiment, one motorcycle brigade and one border constabulary brigade, an army totaling 90,000 men supported by 135 T-34 tanks.

The main North Korean effort, along the Uijongbu Corridor, an ancient invasion route, was aimed directly at Seoul, the South Korean capital. From Kaesong in the west to Chorwon in the east, an arc of 40 miles, General Chai concentrated more than half his divisions and artillery and most of his tanks for the attack on Seoul. The remainder of the attacking force, in a move well coordinated with the attack in the west, moved south in a wide, frontal attack along main roads down the center of the peninsula and along the east coast.

On that rainy Sunday morning, the South Koreans had four infantry divisions and one separate regiment deployed along the Parallel. But many of the officers and some of the men as well as many of the U.S. KMAG advisors were in Seoul or other towns on weekend passes (the North Koreans were obviously well aware of the weekend laxity along the South Korean side of the border). Only one third of the South Korean force along the Parallel was in its defensive positions. The rest was in reserve 10 to 30 miles below the Parallel. Once launched, the North Korean onslaught overran the thinly manned South Korean defenses with surprise, ease and speed.

Within three days, a North Korean tank-infantry force leading the main thrust rolled into Seoul, 35 miles below the Parallel. The drives down the center and the east of the peninsula kept pace with the attack in the west. All along the crumbling front, the South Korean Army retreated in haste and disorder, abandoning most of its arms and equipment on the way.

On 30 June, President Truman authorized General MacArthur to use all U.S. forces available to him to stem the advance of the North Koreans. His main units in Japan were the 7th, 24th, 25th Infantry Divisions and the 1st Cavalry Division, none of which was anywhere near a state of combat readiness in June 1950. Nonetheless, in early June, MacArthur sent the 24th Division to Korea piecemeal; it took up blocking positions north of Taejon. On 14 July, he added the 25th Division which moved to extend the 24th's lines to the east. On the 18th of July, he deployed the 1st Cavalry Division which joined the other U.S. forces so hastily deployed to Korea. These units failed to stem the North Korean onslaught and retreated slowly south into the Pusan Perimeter.

On 13 July, Lt. General Walton H. Walker established his 8th Army advanced command post at Taegu and assumed command of all U.S. forces, and, with the consent of the South Korean President, Syngman Rhee, all South Korean Forces. By mid-August, 8th Army had been forced back into the 140-mile Pusan Perimeter where it dug in and defended stubbornly for a month and a half against General Chai's repeated heavy attacks, with a force now swelled to 13 infantry and one armored divisions. In the interim, the U.N. Force within the perimeter built up its strength with the addition of the U.S. 2nd Division, the 1st Marine Brigade, four battalions of medium tanks and the 5th Regimental Combat Team from Hawaii. Great Britain added the 27th Brigade from Hong Kong. With some effort, the American advisors restored some semblance of order among five South Korean divisions. Additional forces would continue to build up the United Nations Command in the coming weeks. Thus, as the North Koreans extended their lines, the U.N. Command strengthened itself into an offensive posture.

Even though his command had been ignominiously beaten back into a restricted corner of the Korean Peninsula, General MacArthur was not pessimistic. He knew that the North Korean lines were over-extended down the length of Korea and that, at Pusan for the time being, he had the advantage of interior lines. And based on his vast World War II experience and success in landing his forces amphibiously in the rear of the enemy, it was only natural that he would use a proven plan to extricate his Pusan-bound forces involving a waterborne sweep around the flank and an inland attack on the enemy's trailing and extended lines of communications.

In fulfillment of this tactic, on the 15th and 16th of September, General Edward M. Almond floated ashore his X Corps—the 7th Division and the 1st Marine Division—at Inchon, more than 200 miles, as the crow flies, from Pusan. By the 29th of September, X Corps had recaptured Seoul.

On the 16th of September in conjunction with X Corp's amphibious landing, General Walker launched his attack north out of the Pusan Perimeter with two U.S. Corps and two ROK Corps abreast. 8th Army initially ran into heavy opposition, but when the North Koreans recognized that their units could be squeezed and totally smashed between 8th Army and X Corps, they fled north in disarray. At 0826 on the 27th of September, Sgt. Edward C. Mancil of L Company of the 7th Cavalry met elements of H Company, 31st Infantry, 7th Division on a small bridge north of Osan. X Corps and 8th Army had linked up.

By the 1st of October, most of the surviving North Koreans had retreated above the 38th Parallel or had escaped into the mountains of South Korea. "For all practical purposes," wrote Roy E. Appleman in *U.S. Army in the Korean War*, "the North Korean People's Army had been destroyed. That was the real measure of the success of the Inchon landing and 8th Army's correlated attack—General MacArthur's strategy for winning the war."

The stage was thus set for the pursuit of the remnants of the North Koreans above the Parallel and for the possible use of an airborne force to parachute behind the retreating enemy forces, to cut off their escape, to capture high ranking military and civilian leaders fleeing north out of P'yongyang, and to rescue U.N. POWs presumably being evacuated to the north.

On the 1st of August 1950, in Theatre No. Three at Fort Campbell, KY, halfway around the world, Colonel Frank S. Bowen, Jr., who had assumed command on the 21st of June, announced to his troopers of the 187th Airborne Infantry Regiment that the 187th had been alerted for overseas movement. By the 27th, the 187th officially became the 187th Airborne Regimental Combat Team with the addition of the 674th Airborne Field Artillery Battalion, Company A of the 127th Airborne Engineer Battalion, Battery A of the 88th Airborne Anti-aircraft Battalion, and detachments of MPs, Parachute Maintenance, and Medics. By rail to Ft. Lawton, WA and by airlift across the Pacific to Japan and Korea, the 187th moved to Korea. On 20 September, five days after the Inchon landings, C-119s carrying the advanced elements of the 187th began to land on Kimpo Airfield, about 10 miles due west of Seoul.

J.H. Alexander, a rifleman assigned to A Company of the 1st of the 187th, described the arrival of the battalion at Kimpo this way:

This move was really a hurry-up operation. We first heard that we were going to make a combat jump. It looked like the real thing when we were issued parachutes. The 1st Battalion was fully loaded for combat when we went aboard our C-119s, including monorail bundles rigged for paradrop. Flying into some anti-aircraft fire on approaching the coast of Korea, we turned out to sea and our serial came in from another direction. About mid-morning, we landed in Kimpo. U.S. Marines had partially cleared the strip of North Korean soldiers and guerrillas, but we landed under small-arms fire from the edges of the field. All the buildings were smashed and burning. Destroyed aircraft littered the runways.

When the plane came to a halt, the last man in the left stick opened the door and was promptly killed by a sniper bullet between the eyes...this was war.

By the 26th of September, all combat elements of the 187th had arrived in Korea. For the next two weeks, the 187th fought in the Inchon Peninsula Campaign and helped clear the Kimpo Peninsula. In early October, General MacArthur pulled the 187th Regimental Combat Team back into Kimpo Airfield as General Headquarters Reserve. In the 187th, the rumors flew that the Regimental Combat Team was going to make a combat jump—somewhere. The rumors had some justification: General MacArthur was a believer in the airborne concept, and he knew how to use paratroopers. In World War II, he had employed airborne forces successfully in a number of areas, amoung them Nadzab, Tagaytay Ridge on Luzon, and Corregidor. Therefore, he was ready to insert the 187th in a classic airborne role: Drop behind the enemy, attack him from an unexpected direction, block his escape routes and crush him between the paratroopers and an advancing friendly force.

General MacArthur's plan was simple: Parachute the 187th Regimental Combat Team onto two drop zones astride two main highways and railroads running north from the North Korean capital to block the main routes of escape of the North Koreans. One DZ was at Sukch'on, about 25 miles north of P'yongyang and 17 miles to the east of Sukch'on. Another was at Sunch'on, a secondary mission of the paratroopers, and one vital to General MacArthur, was to intercept American POWs moving with the retreating enemy columns. (When General MacArthur returned to the Philippines in 1944, one of his major priorities was the immediate liberation of interned civilians and soldiers.) General MacArthur had to assume that, with the imminent fall of P'yongyang, the North Koreans would move the American POWs northward. In fact, General MacArthur's Intelligence Division had informed Colonel Bowen that a trainload of American POWs, travelling only at night and then very slowly, was on its way north from P'yongyang. That information heightened the sense of mission for the 187th. Considering the rate of advance of his forces and the retreat of the North Koreans, General MacArthur set the date for the drop on the morning of 20 October.

At Kimpo, the 187th went through the usual pre-combat jump preparations: well-guarded, top secret briefings on hastily constructed sand tables and maps; packing of equipment

187th Rakkasans, September 1950, clearing Kimpo Peninsula. Just took position, rooting out North Korean from bunker complex.

bundles; drawing and fitting of personal chutes; oiling and cleaning weapons; and finally, striking camps. At 1900 on the 18th, pilots and jumpmasters went through a final, detailed briefing at Kimpo. A heavy drizzle dampened the spirits of the keyed-up paratroopers. Even though weather reports were unfavorable for the 20th, plans proceeded as if the sun were shining.

At 0230 on the 20th, in a heavy rain and darkness the 187th troopers fell out for reveille, ate the traditional pre-combat jump breakfast, and then trucked by echelons to their planes on the parking strips at Kimpo. At 0400, the troops drew parachutes and began to adjust them. The rains still came; the jump was postponed for three hours.

Shortly before noon the sky began to clear and, at 1030, the troopers were ordered to "chute up." The Regimental Combat Team loaded into 73 C-119s of the 314th Troop Carrier Wing and 40 C-47s of the 21st Troop Carrier Wing, both based in Japan. Once the troops had shuffled aboard, the planes were crowded. A typical C-119 load was 46 men in two sticks, 15 monorail bundles and four door bundles, two for each of the jump doors. Besides his main chute and reserve, each trooper carried a .45-caliber pistol and a carbine or M1 rifle, water and rations, ammunition, and some carried an extra Griswold container filled with small arms or light mortar ammunition. Before jumping, a combat-loaded paratrooper was relatively immobile.

At noon, the first aircraft carrying Colonel Bowen, 13 pathfinders, riflemen, unit guides and part of the Regimental Combat Team staff, was airborne. It was followed in rapid takeoffs by planes headed for DZ William, southeast of Sukch'on. The planes rendezvoused in nine plane Vee of Vee's over the Han River and then flew north along the west coast in waves of 18 and 36 planes spaced about 15 minutes apart. The monorail doors opened 20 minutes out from the DZ. At the four minute warning light, the planes were still over the ocean but headed for the beach. "The colonel stood up and gave the jump commands just like a training operation," remembers SFC Ignatz, one of the Pathfinders in the first plane. "I felt pretty good about jumping with Colonel Bowen. If it was good enough for the colonel, it was good enough for me."

As the troop carriers approached DZ William, southeast of Sukch'on, fighter planes rocketed and strafed the ground. At 1400, the lead pilot flipped on the green light. Colonel Bowen, standing in the door with the slip stream wrinkling his face and tugging at his strapped-down helmet, leaned back from the jump door and yelled, "Go!" He leaped out the door and was gone in the "peculiar sucking swish of sound that accompanies a paratrooper into the prop wash." The two sticks in the plane shuffled rapidly to the rear with the characteristic side-slip motion of combat-loaded paratroopers and jumped out the two doors. In seconds, the sky was filled with strings of the opened chutes of the rest of the serial. There was no enemy anti-aircraft fire and only occasional sniper fire from the ground.

The second serial of 17 C-119s also dropped on DZ William. It was under the command of Lt. Colonel Gerhart, and was composed of the 1st Battalion, Regimental Headquarters, Support Company, Company A of the 127th Engineers, Medical Company and Service Company. The first two serials dropped 1470 men and 74 tons of equipment; one man was killed in his parachute by enemy fire, and 25 men were injured. One group landed a mile and a half east of the DZ.

After the troop drop on DZ William came the heavy equipment—105mm howitzers, jeeps, 90mm anti-tank guns, a mobile radio transmitter equivalent in weight to a 2 1/2 ton

Supply drop onto rice paddies in Korea for U.S. troopers. (U.S. Air Force photo)

truck. The 674th Airborne FA Battalion (minus B Battery) dropped seven 105mm howitzers and 1,125 rounds of ammunition. This marked the first time that heavy equipment and 105mm howitzers had been dropped in combat and that the C-119 had been used in a combat parachute operation.

In the meantime, the 3rd Battalion had also jumped into DZ William, assembled, headed south, took up defensive positions on low hills two miles south of the town and set up road blocks across the highway and railroad in its area. It secured its objectives by 1700 and killed five of the enemy and captured 42 others with no losses. The 1st Battalion, during the same period, against light resistance, seized Hill 97 east of Sukch'on and Hill 104 north of the town, cleared the town of Sukch'on, and set up a roadblock to the north. General Bowen (he was notified of his promotion to brigade general just after the jump) established his command post on Hill 97. By the end of the day, the 187th had dropped about 2,800 men onto DZ William.

DZ Easy was two miles southwest of Sunch'on. At 1420, the Second Battalion, reinforced by the Second Platoon of A Company, 127th Engineers, the 4.2 Mortar Platoon of Support Company, one section of 90mm anti-tank guns, B Battery of the 674th, a Pathfinder Team, and one Forward Air Control Party, began its drop on DZ Easy. By nightfall, the battalion had secured its objectives against very light resistance. "As the Rakkasans marched into Sunchon in a column of twos", one trooper, PFC Kirksey of F Company, remembers, "the Koreans tossed rifles and other weapons out onto the street. The din was terrific."

By 21 October, the 1st Battalion had occupied the dominant terrain north of Sukch'on, thereby blocking the main highway running north. The enemy, in some strength, held the next line of hills to the north. By the afternoon of the 21st, the 1st Battalion had made contact with the 2nd Battalion at Sunch'on.

A total of approximately 4,000 paratroopers and more than 600 tons of equipment and supplies were dropped at Sukch'on and Sunch'on including 12 105mm howitzers, 30 jeeps, 38 1/4-ton trailers, 4 90mm anti-aircraft guns, 4 3/4-ton trucks, and 584 tons of ammunition, gasoline, water, rations and other supplies. The 187th Regimental Combat Team had made airborne history with the first heavy drop of vehicles and 105mm howitzers in combat.

The airborne operation had, of course, not gone unnoticed by General MacArthur. In fact, he and generals Stratemeyer, Wright and Whitney had flown in from Japan to watch the airdrop from the air. After he saw the success of the landings, he flew into P'yongyang where he told reporters that the airborne operation had apparently taken the enemy completely by surprise. He suggested that some 30,000 enemy troops, about half those remaining in the north, were trapped between the 187th and the 1st Cavalry and the ROK 1st Division at P'yongyang to the south, and that he had every expectation that they would be trapped and captured or wiped out. He said that the airborne operation was an "expert performance" and that it "closes the trap on the enemy." The next day, back in Tokyo, General MacArthur, flushed with the success he had witnessed over the 187th DZs, predicted too optimistically: "The war is very definitely coming to an end." Unfortunately, the Chinese would deem otherwise.

On 24 October, the 187th moved into P'yongyang; the Sukch'on-Sunch'on airborne operation was over. The 187th's new mission was the control of P'yongyang, Chinnampo, the P'yongyang airfield and the main supply route.

By the end of October, brief clashes with Chinese troops in both the 8th Army and X Corps, that had landed amphibiously at Wonsan on the east coast of North Korea, posed a new, ominous but as yet unappreciated threat to the U.N. Forces. Nonetheless, between 8 and 23 November, 8th Army and X Corps advanced slowly against moderate resistance as far north as the Ch'ongch'on River in the west and the Yalu in the east. On 25 November, however, the Chinese stopped 8th Army in its tracks with a furious, massive attack. On the 27th, two Chinese armies hit X Corps in the east, attacking along both sides of the Changjin Reservoir. By 5 December, 8th Army had withdrawn to positions along the Imjin River near the 38th Parallel. By 24 December, X Corps had evacuated from Hungnam by ship under heavy fire and had unloaded at Pusan. Meanwhile, by the end of December, 8th Army had organized a 140-mile coast-to-coast defensive line just below the Parallel.

The 187th evacuated P'yongyang in December and moved into 8th Army reserve with the mission of providing security across the Han River, conducting operations in the Hoengsong, Wonju, Chech'on and Chungju areas and protecting and assisting the evacuation of the Kimpo airfield and Inch'on.

On 1 January 1951, Chinese troops in overwhelming numbers attacked along 8th Army's entire front. The major effort was directed again at Seoul. General Matthew B. Ridgway, who had assumed command of 8th Army on the death of General Walker in a motor vehicle accident on 23 December, pulled back from the defensive line established in December, put a bridgehead around Seoul and set up a new defensive line on the south bank of the Han River below Seoul. The enemy continued its relentless attack with massed armies and forced Ridgway to evacuate Seoul and to pull back again. This time he set up a new line across the peninsula about 40 miles to the south of the capital.

Beginning in late January, Ridgway ordered a series of cautious attacks to probe slowly and methodically through a series of phase lines. Using this tactic, 8th Army moved steadily north and, by 1 March, had regained the lower banks of the Han River, just below Seoul.

The 187th, meanwhile, had fought south through Kyongan-ni, Chonsen, Yonju and Wonju. On 28 February, the Regimental Combat Team passed to the operational control of X Corps, closed into its rear assembly area at Taegu and, throughout the first half of March, remained in administrative bivouac at K-2 Airstrip near Taegu.

By 19 March, Ridgway's forces had succeeded in recapturing Seoul and had fought forward to a line across Korea just below the Parallel. 8th Army was back to a line approximately where the war had started some nine months before. But above that line, the enemy had assembled troops and equipment for the resumption of their offensive. Ridgway reasoned that his best defense was the offense. On 22 March, Ridgway informed General MacArthur that he had prepared a plan, Operation COURAGEOUS, to advance to a line that, except for a short stretch in the west, lay just above the Parallel, generally between the confluence of the Han and Yesong rivers on the west coast and the town of Yangyang on the Sea of Japan. On 22 of March, the 8th Army moved out and advanced steadily forward all along the front. According to General Ridgway, "The spirit of the Army was at its peak."

At Taegu, the 187th prepared for another combat jump. Initially, Ridgway planed to drop the Regimental Combat Team at Ch'unch'on because 8th Army Intelligence had located the Chinese III Field Army south of the Parallel and had established concentrations around the Uijongbu area. The 187th was making final preparations for this drop when, on the 19th, 8th Army cancelled the Ch'unch'on jump and directed General Bowen to drop the 187th on its alternate DZs near Munsan-ni. On that same day, the same aircraft that had dropped the combat team on Sukch'on-Sunch'on arrived at the strips near Taegu.

The mission of the 187th was to drop behind the Chinese and North Koreans, who were apparently withdrawing across the Injin River and were establishing in-depth defensive positions near Munsan-ni, to destroy the enemy at the restricted withdrawal route at the Injin River crossing. In coordination with the drop of the 187th, an armored column would attack north and link-up with the paratroopers within 48 hours. D-Day for the drop was 23 March 1951.

The assault elements of the Regimental Combat Team dropped onto two drop zones just to the northeast of Munsan-ni. The drop area was on flat terrain between three hill masses. The 1st Battalion, with the exception of its commander, Lt. Colonel Harry Wilson and his command group, dropped on the north DZ by error. The 1st had been scheduled to land on the south DZ but, during the flight from Taegu, Colonel Wilson's plane dropped out of the formation with engine trouble, and the flight erroneously continued on to the northern DZ. Later, B Company escorted Colonel Wilson and his staff through the lines on foot. The 1st Battalion's immediate mission was the capture of hills 228 and 229, just to the north of the drop area.

M/Sgt. Kenneth E. Ryals, who led a 75mm recoilless gun section in Support Company and who jumped in with the 1st

187th Rakkasans, March 1951, loading up for combat jump at Munsani-ni, Korea. Note the combat packs slung below reserve parachute.

187th Rakkasans en route to drop zone Munsan-ni, Korea, 23 March 1951 - Good Friday.

Battalion, remembered: "My serial was airborne at 1000 hours. One plane had engine trouble and crashed into the sea during the flight. Another was lost from the preceding serial.

"We flew out to sea for our rendezvous, then flew north in volume. Crossing the coast, I could see Chinese forces dug in in trenches surrounding the DZ. The Air Force, prior to the jump, had reported the enemy, in groups of one thousand men, moving in the Munsan-ni Valley. USAF pilots called Munsan-ni 'Happy Valley' because of the large number of targets. I remember that it was a clear, sunny day. Below my plane small bunches of our people were moving out to the assembly points with scattered rounds coming in around them. The village of Munsan-ni was burning in the near distance. Farm houses ringed the drop zone.

"In some amazement, I saw that the green light had flashed on.

"Flipping out my 75mm door bundle, I followed after. As the engine noise subsided, I could hear a considerable amount of small arms fire below. Landing in soft ground, I cleared my parachute harness and headed for my assembly area on the southwestern section of the drop zone. A few minutes after we had secured the high ground in our sector, the heavy drop arrived and with it the attached Medical Battalion from India."

After landing and reassembling, the battalions moved out to their objectives: The 1st took the high ground to the north; the 2nd moved to a hill on the east of the DZ; the 3rd took the northern side of a hill to the west. "The 1st Battalion marched all night," according to Sgt. Alexander, a rifleman with A Company, "to reach the 2nd Battalion which was heavily engaged. We immediately went into the attack and took the critical terrain to establish blocking positions to cut off the retreating Chinese and North Korean forces.

"The enemy was well dug in around Munsan-ni. It was later learned that the CCF were withdrawing in that sector in order to draw the U.N. Forces north, so that an envelopment by enemy forces could be accomplished and thereby cut off friendly units. Later reports revealed the Chinese Communist forces had been in the area for two or three days before the drop."

By 1450 of the first day, Task Force Crowder, the armor unit ordered to link up with the 187th, was some 15 miles to the south of Munsan-ni; by 1855, patrols from the 187th had contacted the lead elements of TF Crowder; and by the end of the day, the regiment was dug in on its objectives.

SFC Maria was with B Battery of the 674th. His battery had dropped pack 75s instead of the heavier 105s to facilitate more rapid movement in support of the infantry. Maria recalled that "There were many Chinese dead on the drop zone where we landed."

Tanks from Task Force Crowder entered the Regimental Combat Team Headquarters area at 0400 on the 24th. Shortly thereafter, General Bowen mounted his paratroopers on the tanks and moved out on a new mission: block the retreating Chinese in the vicinity of Uijongbu, 26 miles to the southeast. The tanks shuttled the regiment along roads which had become streams of mud in a cold driving rain.

On 26 March, the 3rd Battalion jumped off in its attack to the east. The battalion ran into very heavy resistance late in the afternoon as it assaulted Hill 228. By daylight of the 27th, the 3rd Battalion had secured its objective. The 2nd Battalion forced the enemy to withdraw from Hill 507, a dominating piece of terrain whose capture collapsed the organized enemy defenses in the area. On the 28th, the Regimental Combat Team linked up with elements of the U.S. 3rd Infantry Division which cleared the road from Uijongbu to Chapmon. On the 29th, the 3rd Division occupied the positions of the 187th, and the 187th moved back to its base at Taegu.

Generally speaking, the airborne operations in Korea were smaller and more accurate than those of World War II in Europe. In Europe, most airborne operations were massive drops of division sized units, entering combat by both parachute and glider over wide-spread areas. Many of the drops were at night and many of the flight formations were scattered by enemy ground anti-aircraft fire, inaccurate navigation and, as in Sicily, by fire from our own Navy and shore-based anti-aircraft guns. The airborne units in Europe accomplished their missions against great odds, scattered as they were and fighting, with the light weapons of the paratrooper/gliderman, formidable, heavily equipped and armed enemy concentrations.

The airborne operations in Korea, conducted during the daylight hours, and unhampered by any concentrated enemy aircraft and anti-aircraft opposition, were more precisely executed. The airborne effort in Korea had some other advantages. General MacArthur was a staunch supporter of the airborne capability, and, or course, General Ridgway had been the XVIIIth Airborne Corps Commander in Europe. The use of the C-119 permitted the drop of heavy equipment. Speaking of the Munsan-ni operation, General Ridgway wrote: "It was a good drop. We had improved our techniques some since World War II. We were dropping jeeps now, under big cargo canopies, and 105 howitzers, a heavier gun than we'd been able to take in on the drops in Europe."

He summed up the situation this way: "The American flag never flew over a prouder, tougher, more spirited and more competent fighting force than was 8th Army as it drove north beyond the Parallel. It was a magnificent fighting organization, supremely confident that it could take any objective assigned to it." The 187th Airborne Regimental Combat Team was a proud and gallant member of that Eighth Army.

Airborne Rangers In Korea

0400 hrs., 25 June 1950, the North Korean Army attacks south across the 38th parallel. Within days it is evident to the world they will drive the Army of the Republic of Korea into the sea. On 1 July elements of the United Nations forces intervene, delay, and finally stop the NK at what becomes the Pusan perimeter.

24 August 1950, the UN is on the defensive in Pusan as 8th Army welcomes General J. Lawton Collins, Chief of Staff, U. S. Army. He is there to discuss General MacArthur's requirement for an airborne regimental combat team to be in Japan by 10 September so as to participate in the Inchon operation. General Collins advises him the Department of Army can not field an airborne RCT until after that date. During this trip Gen. Collins visits the 27th Infantry Regiment near Teague.[1] There he discusses the North Korean guerrrrila infiltrations and their success in disrupting United Nations' lines of communications. The solution to this problem is the 8th Army Ranger Company then being recruited.[2]

29 August 1950, General Collins now back in CONUS, is thinking airborne troops, guerrillas, and economy of force. Confronted with expanding personnel requirements versus diminishing available, trained personnel, he orders the reestablishment of Ranger units in the Army; one Company to each combat Division. In a memorandum this date he outlines the mission of these Ranger Companies, in essence, to: infil-

187th Airborne Rakkasans chuting up - Korea 1951.

trate enemy lines, attack command posts, key communications centers or facilities, artillery and tank parks.

September 1950, recruiting begins for airborne personnel to join Ranger units. Concurrently the Ranger Training Center commences operations at Harmony Church, Fort Benning, Georgia. On 27 September, the first airborne Ranger Company Morning Report entries are recorded.[3]

2 October 1950, training begins for the first three companies. The following Monday the fourth Company, the first all black Ranger Unit in the Army and later to become the 2nd Ranger Infantry Company, Airborne, begins training.[3]

25 October 1950, the U.S. Army assigns the Ranger Infantry Companies to the lineage of the lst Special Forces and the WW II Ranger Battalions.[4] In the ensuing weeks and months a total of fifteen Airborne Ranger Companies are recruited and trained. Nine of these companies deploy overseas, six to Korea.

25 December 1950, three of the companies are operational in Korea.

23 March 1951, six companies are operational in country. The 2nd and 4th Ranger Companies parachute into Munsonni with the 187th Regimental Combat Team.

As the UN police action wears on, the ill-use of these six spirited, elite, 115 man companies by their respective divisions threatens to dissipate their effectiveness and becomes a point of controversy. Pamphlets on the proper feeding, care and employment of The Ranger Infantry Company, Airborne, are prepared by the Department of Army and the Ranger Training Center. Distribution is made to 8th Army Headquarters and the parent Divisions. Mission tasking improves as each division realizes the value of their company for special, short-fuse, long-range, task force operations commensurate with their size and support requirements. Still, none of the companies and few of the personnel are performing the missions prescribed by General Collins through the Ranger Training Center.

1 August 1951, as the forward edge of the battle position is stablized by a United Nations edict and the rear-area guerrrilla activity is under control, the Ranger Infantry Companies, Airborne, in Korea, are inactivated.

There are Companies that remain active beyond that date due to contingencies of the service. However, by 6 November 1951 the U.S. Army Ranger Infantry Companies, Airborne, follow the same trail as Rogers' Rangers, Darby's Rangers, and the 8th Army Rangers —— history.

The Ranger School established at Fort Benning in 1950, has continued to turn out a polished product. Proof of the Army's need for Airborne Ranger skills is evident through their enployment as teams and individuals in Vietnam, Grenada and Panama. Today our Army prepares for contingencies worldwide. It must be able to deploy on short notice, operate inland, and be self-sustaining. To meet these requirements it needs light contingency forces capble of being inserted by air anywhere in the world in a relatively short time —— the Ranger School is meeting this challenge.

---

1. J. Lawton Collins, War in Peactime (Boston: Houghton Mifflin Co., 1969).

2. John Hugh McGee, Rice and Salt (San Antonio: the Naylor Co., 1963). In addition: call to McGee and Ralph Puckett, 21 Apr. 90.

3. Telephone conversation with Bob Channon, 21 Apr. 90.

4. Army Lineage Series, INFANTRY, Part I, Regular Army, Office of the Chief of Military History, U.S. A., WDC. 1972.

VIETNAM OPERATIONS

By Edward M. (Fly) Flanagan, Jr., Lt. Gen., USA (Ret.)

Vietnam was the proving ground for a new use of Army rotary wing aircraft—the air mobility concept. The introduction of the helicopter in a combat role—carrying men into battle and supporting them with medical evacuation with helicopters, helicopter gunships, helicopter command and control ships—opened the way for the rapid insertion of air mobile troops behind the enemy lines or in battle areas where other troops already in a fight required reinforcement. The nature of the Vietnam battles, the combat tactics and the political limitations on the objectives of the U.S. Forces, almost entirely ruled out the traditional use of massed "airborne" forces. There were some examples, however.

In the early stages of the war, a number of small covert South Vietnamese teams had been dropped into North Vietnam by parachute to foment guerilla war against the Communist North. Unfortunately, most of these teams were immediately wiped out; a few were even "doubled" back against the South.

Standard Vietnamese parachute units were also used a number of times in combat in the South. In January 1963, for example, the Vietnamese dropped a battalion of their elite airborne division at Ap Bac. Unfortunately, the battalion landed in an area from which it could have little effect on the outcome of the battle.

One U.S. parachute assault occurred on the 22 of February 1967 when Brigadier General John R. Deane, Jr., Commander of the 173rd Airborne Brigade, led one of his battalions, the 2nd Battalion of the 503rd, in a jump which signalled the beginning of Operation JUNCTION CITY ALTERNATE. The overall operation employed a large contingent of forces: The 1st and 25th Divisions, the 11th Armored Cavalry Regiment, the 196th Light Infantry Brigade, elements of the 4th and 9th Infantry Divisions, South Vietnamese units and the 173rd Airborne Brigade. The target of this large force was enemy bases north of Tay Ninh City in War Zone C. The decision to use paratroopers was based on the need to insert a large number of men on the ground as soon as possible and still have enough helicopter assets to make a large heliborne, immediate follow-up. The 2nd of the 503rd used 13 C-130s for the personnel drop and 8 C-130s for the heavy drop. Jump altitude was 1,000 feet.

The 173rd was under the operational control of the commander of the 1st Infantry Division. Before the operation, a tightly controlled deception plan concealed the actual drop zone until 1900 the night before the assault. The battalion dropped on schedule at 0900, and, by 0920, of the 780 men who made the drop, only 11 had minor injuries. The heavy equipment drop began at 0925 and continued throughout the day. The 1st Battalion of the 503rd, part of the 173rd, began landing by helicopter at 1035 on the 3rd Battalion's DZ. These units had no direct contact with the enemy in the early hours of D-Day. Another of the 173rd's battalions, the 4th of the 503rd, made a helicopter assault into two other nearby landing zones at 1420; this operation essentially completed Phase One of JUNCTION CITY ALTERNATE.

During the operation, the 173rd was supported by the 11th, 145th and 1st Aviation Battalions who flew over 9,700 sorties, lifting 9,518 troops and a daily average of 50 tons of cargo. Some operational problems resulted from mixing parachute and heliborne assaults on the same terrain. One accident and several near accidents occurred when helicopters tried to land on DZs littered with personnel chutes.

Phu Bai, Vietnam–BATTLE PLAN. With communication contact to 101st Airborne Division paratroopers and members of the 1st ARVN Division, Capt. James G. Shepard, Biloxi, Miss., coordinates a planned reconnaissance-in-force mission in northern I Corps during Operation Caentan II. Spec. 4 Geoffrey G. Grant, San Bernadino, California, radio-telephone operator, stands ready for instructions. (USA photo by Spec. 4 Ben Croxton)

In summing up the role of "airborne" versus heliborne assault forces, General John J. Tolson, commander of the 1st Cavalry Division in Vietnam and later commander of XVIIIth Airborne Corps, wrote: "The employment of the airborne parachute force is historically visualized as a theatre-controlled operation aimed at achieving tactical surprise. Although parachute delivery of troops and equipment is a relatively inefficient means of introduction into combat, the very existence of this capability complicates the enemy's planning and offers the friendly commander one more option of surprise."

The fact that airborne techniques were not used more often in Vietnam can be attributed to many factors. The most obvious restraint to an airborne operation in Vietnam was the time lag inherent in airborne operations in responding to intelligence on the elusive enemy. The relatively unsuccessful French parachute operations already had pointed this out to us. A much more important restraint was the nature of the war itself and the limitations imposed on U.S. forces. From a strategic point of view, the U.S. posture in Vietnam was defensive. U.S. tactical offensive operations were limited to the confines of South Vietnam. Had the rules been changed, the parachute potential could have been profitably employed by planning an airborne assault into enemy territory at a distance within the ferry range of the Huey. This would have allowed the parachute force to secure a landing zone and construct a hasty strip. Fixed-wing aircraft would have air-dropped or air-landed essential fuel and supplies. Then the helicopters could have married up with this force, refueled and immediately given them tactical mobility out of the airhead. These circumstances never came about.

Every man with jump wings was eager to prove his particular mettle in Vietnam. However, this special talent was not suited for that enemy, that terrain and that situation. Nevertheless, I firmly believe that there is a continuing requirement for an airborne capability in the U.S. Army structure."

In summary, airborne operations in Korea were a follow-up to World War II-style airborne operations but on a much lesser scale, with more accuracy, with more concentration on the drop zones because of the larger personnel aircraft and with much longer sticks, with heavier equipment drops, and under more favorable conditions of daylight and generally lesser enemy antiaircraft and fighter opposition.

The airborne operations in Vietnam were a totally different story than either World War II or Korea because of the tactical situations, the limited objectives, the dispersion of U.S. Forces throughout the country, the size of the country, and the proliferation of mobile, ready, rapidly reacting, trained, heliborne assault forces able to concentrate on and flood a landing zone.

SP4 Leland W. Brooks A/2-502 IN, gives a wounded trooper a drink of water as they await a medavac helicopter during Operation Nevada Eagle.

This member of a Ranger patrol fades into the background as he evades the VC in the jungles of Vietman. (Courtesy of 101st Airborne Division Association)

UH-1D helicopters from the 101st Assault Helicopter Bn., 101st Abn. Div. (Airmobile), transport troops of the 502nd Inf. to a fire support base southwest of Hue., 1970. (Courtesy of Bob Jones, Col., USA (Ret)

OPERATION URGENT FURY: GRENADA

By Geoffrey T. Barker, LTC, USA (Ret)

The Island of Grenada was formed as a colony of Great Britain, receiving self-government status in 1967. Grenada attained independence in early 1974, which resulted in both internal and external controversy, based upon the politics of Sir Eric Gairy, a flamboyant strong-arm politician. The opposition party - The New Jewel Movement - formed in 1973, accused Gairy of ballot-rigging the national elections in 1976. Heading the New Jewel Movement was Maurice Bishop, a strong protestor, supporting civil rights, and decrying the poor social and meager economic conditions throughout the country. Bishop continually sought to exploit the bully tactics of the Gairy regime. The New Jewel Movement ousted the Gairy government, and Bishop was appointed as the Prime Minister.

Bishop was initially welcomed by the Grenadian populace, who eagerly looked forward to his promises of early elections, stressing the freedom of voting and appointing a democratically elected leadership. The elections did not transpire, and Bishop leaned more heavily away from the original British Law - the basis of Grenada Law - and embraced the Fidel Castro methodology by establishing a 'revolutionary government'. Gradually the relationships between Grenada and the communist strongholds became stronger and stronger. Civil rights, an original platform of Bishop, became almost non-existent.

At the turn of the decade into 1980, almost 100 Cuban advisors were resident in Grenada, and almost 100 political prisoners were languishing in Grenadian custody. Airfield construction commenced in 1980, as part of the Grenadian national upgrade, to promote tourism and commerce. Intelligence authorities quickly ascertained that the majority of the 'commercial' airport facilities resembled military structures. A closer look revealed that the construction employees were former or active military personnel, and the Cuban advisors were deeply involved in this venture.

Maurice Bishop continued his involvement with his Marxist masters, but lost favor due to his inability to meet construction time schedules. During his tenure, Bishop had entered into negotiations and treaties with Russia, Cuba and North Korea. Meanwhile in the civilian sector, he curried immense disfavor due to his non-compliance with requests for democratic elections. He met with Fidel Castro for three days on 6 October, after a whirlwind tour of European political headquarters. Four days after his return from Cuba, he was involved in a political altercation with Deputy Bernard Coard, who strongly advocated Bishop's resignation.

Maurice Bishop was placed under house arrest on 13 October, and a purge of supporters of Bishop commenced. Radio Free Grenada narrated the activities, with continual updates of arrests from within the Bishop Administration. Radio Free Grenada, (owned by Phyllis Coard), announced that Bernard Coard had replaced Bishop as Prime Minister. Crowds began to gather and protest.

Messages from the Antigua Caribbean Liberation Movement expressed their concern about the safety of Bishop and his party. Cabinet Ministers, formerly serving Bishop, resigned on 18 October, after the self-proclaimed leader - Bernard Coard, appeared determined to follow a course of violence to retain his position. Maurice Bishop and a number of his Cabinet Ministers were murdered on 19 October 1983. Bernard Coard resigned from office, following allegations of his support for the assassination of Bishop. That day, Grenada became incommunicado with the outside world, and all movement to and from the island ceased. Intelligence reports described total chaos, with public utilities, government offices, business and schools closed. More former supporters of Bishop were detained and killed.

A Revolutionary Military Council was established under Army General Hudson Austin, with a strict curfew imposed. The island remained incommunicado, accepting no outside visitors. Foreign news reporters were arrested and detained.

Outside speculations of the outcome of this take-over were grim. The government of seven adjacent Caribbean countries joined London, England, in their expressions of horror, condemning the Revolutionary Government of Bernard Coard. Tensions continue to escalate, as mobilization increased in Grenada. Reports indicated that the military had split views regarding the current situation, and that another coup may be imminent.

The Organization of Eastern Caribbean States (OECS), expressed their concern that the situation in Grenada posed a threat to the security of other countries in the Caribbean. The OECS officially requested assistance from the United States, to resolve the situation in Grenada. The US Government was already concerned for the safety of the many US citizen students attending medical training in Grenada. The Governor-General of Grenada had appealed directly to the OECS for assistance in restoring normalcy to the country.

By 24 October, no resolution had been attained. The students in Grenada were not allowed to leave, and continued reports of violence and brutality prevailed. The decision was reached to employ US military to secure the island, ensure the safety of the US citizens on the island, and assist the legal government of Grenada, with the cooperation of the OECS, to restore order to the people of Grenada.

On 25 October 1983, US military forces deployed to the island of Grenada. Operation URGENT FURY was underway. Leading the way, in true Ranger fashion, were the 1st and 2d Battalions, 75th Infantry (Ranger). The option to assault by parachute was made as the units were preparing to embark. Lieutenant Colonel Wesley Taylor, and his Ranger battalion were heading towards their objective at Point Salinas airfield when navigational problems occurred in the leading USAF C-130 Hercules aircraft. Rather than return to home station, the lead aircraft dropped to the rear of the flight formation, relying on visual observations to place the paratroopers correctly onto the Drop Zone. Rangers had been mobilized from Fort Lewis, Washington and Fort Stewart, Georgia, and were assigned to the recently activated US Army 1st Special Operations Command (1st SOCOM). The 1st SOCOM had been organized to provide a single command and control for US Army Special Operations Forces, and was commanded at Fort Bragg, North Carolina, by Brigadier General Joseph C. Lutz, a veteran Special Forces officer.

As the Rangers headed towards Grenada, the US Army's elite 82d Airborne Division had also been alerted at Fort Bragg, and their alert Brigade was in close pursuit.

Closing in towards their objective area, it was determined that the jump altitude would be at 500 feet, thus eliminating time in the air - the most vulnerable time for a paratrooper - and minimize ground dispersion, thus supporting rapid assembly on the drop zone. The lead aircraft came in to make the final approach to the drop zone, Point Salinas Airfield. Intelligence reports had already indicated that barrels and poles had been placed across the runways to prevent aircraft from landing.

The inbound approach at 500 feet became a significant tactical decision. Cuban anti-aircraft guns, which had been emplaced on the hill tops surrounding the airfield, could not depress their barrels to less that 600 feet above ground level. Ineffective anti-aircraft fire was placed one hundred feet above the Rangers as they exited their C-130s. Lieutenant Colonel Taylor and his staff quickly consolidated on the ground, but even more quickly discovered that his lead elements were nowhere in sight. The original first aircraft, which had experienced navigational problems, was now in trail behind the flight pattern. It did not take long to adapt to the changed situation.

With the Rangers, were two engineers from the 20th Engineer Brigade, XVIII Airborne Corps. They were immediately put to work with commandeered bulldozers, and cleared the runways sufficiently to allow the follow-on aircraft carrying the 82d Airborne Division to land.

Although under the control and authority of the US Atlantic Command (USLANTCOM), the actual command and control of ground forces was through Lieutenant General Jack V. Mackmull, Commanding General, XVIII Airborne Corps. The XVIII Airborne Corps senior officer in Grenada was Brigadier General Jack B. Farris. The 82d Airborne Division, commanded by Major General Edward L. Trobaugh, quickly consolidated, and commenced patrolling the south western area of the island, through to the northern points of the island. Troopers of the 82d Airborne Division were rapidly rotated to Grenada, allowing almost the entire 82d Airborne Division to participate in Operation URGENT FURY.

The civilian commercial radio station was destroyed during the invasion. Broadcasts were provided to calm the civilian populace by transmitters of the 4th Psychological Operations Group, from Fort Bragg, and from airborne transmitters provided by the USAF Reserve.

The American students were liberated from the medical school where they remained during the pre-invasion curfew and hostilities, and during the actual invasion phase. Elements of the 82d Airborne Division provided protection until evacuation flights could be scheduled.

Civil Affairs volunteers from the US Army Reserve were flown to the island, and elements of the USAR 358th Civil Affairs Brigade, were rushed in from Norristown, Pennsylvania.

More than 10,000 rifles and machine guns, and almost 40 crew served weapons were captured during this operation. Also seized were 5.5 million rounds of rifle ammunition; 9,000 mortar rounds; almost 120,000 rounds of machine gun ammunition; more than 2,250 grenades; and 2 armored fighting vehicles.

As the invasion forces departed, mobile training teams from the 7th Special Forces Group, were dispatched to assist training the Grenada peacekeeping militia. US Army psychological operations units from the 4th Psychological Operations Group, and nation re-builders from the 96th Civil Affairs Battalion, commenced assisting in returning the island to a peaceful environment.

An interesting situation occurred after the dust had settled, and representatives of the participating units were preparing to attend a Presidential Awards Ceremony at the White House in Washington, DC. It was realized, that although the 1st and 2d Battalions, 75th Infantry (Ranger), had been wearing the Ranger-scroll shoulder sleeve insignia since activation in 1974, the Ranger-scroll had never been authorized by the Department of Heraldry. The black, white and red Ranger-scrolls had been worn in place of the diamond shaped authorized shoulder sleeve insignia, since before the end of World War II. A rush priority finally authorized the famed shoulder sleeve insignia of the Rangers, that is so well known today. The second calamity, was that when General George C. Marshall had been Chief of Staff of the US Army during World War II, he had disapproved the authorization to wear small stars on the parachutist wings, to indicate combat jumps. This 'tradition' had been in violation of General Marshall's decision for so long, that it had become legend. Combat paratroopers from privates to General Officers had worn combat jump stars on their wings. The dilemma was identical to the authorized shoulder patch. How could unauthorized stars be worn by the Ranger representatives, during an award ceremony to be conducted by the Commander-in-Chief of the US Military - President Ronald W. Reagan? Again the impossible was accomplished. Not only was the wearing of combat stars approved, it was made retro-active to World War II.

Grenada: U.S. paratroopers advance past two abandoned Soviet-made BTR-60-PB armored personnel carriers. (U.S. Army photo)

OPERATION JUST CAUSE: PANAMA

By Geoffrey T. Barker

Like most major military engagements, the United States invasion of Panama had been building up for some time prior to the actual invasion date.

Almost a year earlier, in February 1989, Panamanian President, General Manuel Antonio Noriega, had been indicted in the courts in the State of Florida. He was charged with thirteen counts of racketeering and drug trafficking. One indictment, unsealed on 5 February, in the Miami, Florida courts, alleges that General Noriega and Colonel Luis del Cid, accepted a $4.6 million bribe from the Medellin Drug Cartel of Columbia. General Noriega provided security to protect shipments of cocaine to the United States, and laundered drug money, supplied drug laboratories, and shielded drug traffickers from the law. In Tampa, Florida, similar (and separate) charges were filed, to include a $5.4 million bribe from the Medellin Drug Cartel. The United States could not oust General Noriega; he could not be extradited to the United States as extradition agreements were not in effect between the US and Panama.

During the next several months, relationships between the US and Panama deteriorated, becoming extremely strained. Severe economic sanctions were imposed by the United States, and in retaliation, the Panama Defense Forces (PDF), commanded by General Noriega, commenced an increasing pattern of harassment against the 35,000 US military and civilian personnel in Panama. The tension increased, culminating in a series of events.

More than a thousand incidents occurred between US personnel and the Panamanians. Largely this was petty harassment. Periodically severe beatings, sexual harassment, and threats with weapons, were made upon off-duty US military personnel and their dependents. A crucially tense period occurred in November 1988, when Mike Nieves, a Navy Damage Controlman First, was attacked and severely beaten while on duty. He was ambushed by PDF police at the Panama International Airport, and after being severely beaten, was forced to his knees at gunpoint, to beg for his life. The following February, a US Military employee was beaten while handcuffed, by the PDF, suffering lacerations and a broken eardrum. Assisting the PDF were the Dignity Battalions, an armed but non-uniformed organization created by General Noriega.

This state continued, interspersed with unprovoked harassment by traffic police, bomb threats against US facilities, and minor skirmishes in off-duty areas. Retention of diplomatic relations with General Noriega appeared to be the order of the day, at least until after the forthcoming elections on May 7th. Election Day arrived, the voters visited the polls, and Guillermo Endara, leader of the opposition, won by a margin of 3-1. The election was then annulled by General Noriega, and Guillermo Endara and his running mates, Ricardo Arias Calderon and Guillermo Ford, were beaten by PDF thugs on May 9th, during a protest march through Panama City.

The PDF expanded their harassment role, to impede official US government and military vehicles, in violation of the Panama Canal Treaty covering freedom of movement. US combat troops stationed in Panama included the 5th Battalion, 87th Infantry and the 1st Battalion, 508th Infantry (Airborne), of the 193d Infantry Brigade. As the Viet Cong in Vietnam became known as 'Victor Charlie' or just plain 'Charlie', the Dignity Battalions of General Noriega, were re-christened as the 'Dingbats'. The Dingbat cadre had been trained in Cuba by Cuban Special Forces, under the tutelage of Jose Antonio Arbesu Fraga, then the Chief of the Cuban Interests Section of the Czechoslovakian Embassy in Washington DC during the invasion.

As tensions continued to escalate, a Brigade sized force was deployed in May from the United States to beef-up the garrison in Panama. The 4th Battalion, 17th Infantry and a battery of field artillery of the 2d Brigade, 7th Infantry Division (Light), from Fort Ord, California, were dispatched. The 4th Battalion, 6th Infantry, 5th Infantry Division (Mechanized), was deployed from Fort Polk, Louisiana, also a light armored marine company was included with the 2d Marine Expeditionary Force from Camp Lejuene, North Carolina. A Brigade Headquarters element from the 7th Infantry Division (Light), was deployed as the command headquarters for these almost 2,000 additional forces. This show of force resulted in a considerable curbing of PFD harassment, dropping by approximately 75% during the first month.

General Frederick Woerner was relieved as commander of the US Southern Command, being replaced by General Maxwell Thurman, who assumed command on 1 October.

On 3 October, an unsuccessful coup attempt was led by several PDF officials. Assisting the rebels, at their request, US Forces had assisted in blocking two access roads. US aircraft were in the air, prepared to provide support if ordered. Unfortunately, a third access road, which was unguarded, allowed the Noriega supporters to escape. General Noriega was actually held in custody for several hours before the coup failed.

Admiral William J. Crowe, former Chairman of the Joint Chiefs of Staff under the Reagan Administration had publicly denounced military action against General Noriega. Army General Colin I. Powell, who had recently succeeded Admiral Crowe as the senior military leader, appeared more politically attuned, having served as the National Security Advisor to President Reagan. Even as he commenced his duties as Chairman of the Joint Chiefs, he initiated discussions with General Maxwell Thurman, who was to shortly replace General Woerner as the Commander in Chief, US Southern Command. General Powell quietly and methodically prepared the battleground, determined to bring order to this troubled area. Tanks and helicopters were quietly positioned in Panama, and contingency rehearsals began.

The failed coup resulted in many defamatory remarks being issued against President George Bush, for not allowing the US to play a more active role.

On 15 December, General Noriega declared war against the United States, and proclaimed himself 'President for Life.'

The following day, 16 December, four unarmed US officers, in civilian clothes, drove up to a PDF road block. The PDF attempted to forcibly eject the officers from the car. The officers drove away, however First Lieutenant Robert Paz, a marine, was fatally wounded, and pronounced dead upon arrival at the US Gorgas Military Hospital. A naval officer and his wife witnessed the shooting. He was captured and beaten, his wife was sexually harassed. In another incident a Panamanian soldier pointed his loaded rifle at a patrolling American Military Policeman.

On 18 December, a US Army lieutenant shot and wounded a PDF soldier in the vicinity of a US installation. The officer was illegally armed, but claimed that the PDF soldier was reaching for his weapon.

President Bush outlined four US strategy objectives regarding the situation in Panama. These were: the protection of

Aboard a C-141 Starlifter in route to Panama. View of troops on right side of the aircraft. (Courtesy of SSG Joe Owens)

American lives; safeguarding the integrity of the Panama Canal Treaties; restore democracy in Panama; and bring General Noriega to Justice. In a statement published several days later, President Bush declared his intention to wage an all-out military assault because General Noriega had become so belligerent, that the danger to American lives in Panama was unacceptable. In the days to follow, through his decisions, President Bush earned his spurs as the Commander in Chief of the US Armed Forces.

The US Government recognized the deposed, but democratically elected, Guillermo Endara, who supported the US intervention. The Endara Government had been sworn in on the evening of 19 December 1989. The first official act of President Endara was to impose a curfew. At 0155 hours, 20 December, lead elements of the 82d Airborne Division departed Pope Air Force Base, aboard C-141 Starlifter aircraft.

On the morning of 20 December 1989, US Forces invaded Panama. President Bush demonstrated superb leadership and statesmanship by adequately preparing for this decisive action. This was a coordinated attack, using both conventional and special operations forces, of all services. A contingency Operational Plan, code named BLUE SPOON, was taken from the shelf, quickly revised, and changed to Operation JUST CAUSE. Some 14,000 US military personnel invaded Panama, to link up with approximately 7,000 (of the 13,000 US military stationed in Panama) other participants. All air insertions were conducted with precision, with more than 4,500 paratroopers being dropped on target. Some paratroopers had flown more than 3,500 miles to their drop zones. A bitterly cold drop in the temperature in North Carolina delayed take-offs from Pope Air Force Base, this resulted in a protraction of the drop times, with the first sticks leaving their aircraft at 0100 hours (local time), and the final drops exiting their aircraft at approximately 0500 hours. Less than 40 personnel received injuries upon landing, the major injuries being twisted ankles or wrenched knees. This injury rate was less than 2%, as compared with approximately 1% during training exercises.

Concurrent with the arrival of US paratroopers, the PDF surrounded the US Embassy. At 0122 hours, the PDF fired six rocket-propelled grenades at the US Embassy in Panama City, where some 30 employees were present. The Embassy personnel expressed outrage, as their only defense during the invasion activity, was comprised of a US Marine Corps guard and a security officer. No one was killed or wounded prior to the arrival of an Army security element. The guard force was later recommended for awards and promotions for their defense of the Embassy.

US Army Special Forces from the 3d Battalion, 7th Special Forces Group, Task Force Black, stationed at Howard Air Force Base, Panama, conducted reconnaissance missions of parachute drop zone (DZ) sites, blocking a bridge over the Pacora River, leading to the Ranger DZ, east of Panama City. The Special Forces soldiers engaged the PDF relief forces in a fierce fire-fight, assisted by US Air Force AC-130 (Specter) gunships. Ranger elements moved across and secured the Pacora Bridge. Task Force Black conducted reconnaissance and surveillance of the majority of the key targets prior to H-Hour, forming a nucleus for follow-on operations, and provided medical, engineering and communications support as required in the outlying regions.

The United States order of battle included a command and control headquarters from the XVIII Airborne Corps, commanded by Lieutenant General Carl W. Stiner. Also on the ground was Major General Ed Scholes, Chief of Staff, XVIII Airborne Corps. The 1st Brigade, 82 Airborne Division; 1st Corps Support Command; 16th Military Police Brigade; and

C-5A Galaxy lifting off from an airfield in Central America. (Courtesy of SSG Joe Owens)

525th Military Intelligence Brigade deployed from Fort Bragg, North Carolina. Elements of the 5th and 7th Infantry Divisions from Forts Polk and Ord, were already in place. The 75th Ranger Regimental Hq and the 3rd Ranger Battalion deployed from Fort Benning, Georgia. The 1st and 2d Ranger Battalions parachuted in from Hunter Army Airfield, Georgia, and Fort Lewis, Washington. Other deployed special operations organizations included elements of the 4th Psychological Operations Group (Airborne), and the 96th Civil Affairs Battalion (Airborne), both from Fort Bragg. Eleven Special Operations personnel lost their lives during the operation, and 129 were wounded in action.

The Combat arms elements of the 82d Airborne Division included the 1st, 2d and 3d Battalions, 504th Parachute Infantry; 4th Battalion, 325th Airborne Infantry; Company A, 3d Battalion, 505th Parachute Infantry; 3d Battalion, 73d Armor; the 3d Battalion, 319th Field Artillery; and 3d Battalion, 4th Air Defense Artillery. Eight light M-551 Sheridan Tanks were parachuted into Panama on 20 December. The 3d Battalion, 73d Armor, commanded by Lieutenant Colonel James J. Grazioplene, is the sole airborne armor unit remaining in the US Army inventory. The Sheridans, like the Specter gunships, were combat veterans of South Vietnam. Combat support units included the 82d Division Support Command, also the 307th Engineer Battalion, Company B, 82d Signal Battalion, and the 313th Military Intelligence Battalion. Company B, 307th Medical Battalion provided medical coverage. The 82d Aviation Brigade provided personnel participating in the airborne assault and flying support for the Division.

Although the estimated strength of the PDF numbered in the low thousands, the US order of battle was not excessive to accomplish the planned coup-de-main. As opposed to a phased invasion, a coup-de-main, (normally a Special Operations type activity), is designed to produce immediate results. US planners carefully selected the correct mix of ground troops, exploiting the unique capabilities of the airborne, light and mechanized infantry, Rangers, Special Forces and marines. The true objectives of a coup-de-main, is to defeat or paralyze the enemy in one major thrust. In Grenada, in 1983, Rangers were parachuted into one end of the island, and marines landed at the other. The 82d Airborne Division was air landed in the center. In all, some 6,000 US military deployed to the island of Grenada. In the Dominican Republic, in 1965, approximately 15,000 personnel were deployed over several weeks. The invasion of Grenada and the Dominican Republic, were classic examples of a phased defeat of the enemy through attrition.

The US Marine Corps contingent included Company D, 2d Light Armored Infantry Division; Company K, 3d Battalion, 6th Marines; Marine Corps Security Force Company, Panama; Detachment G, Bridge Service Support Group Six; and the First Fleet Anti-Terrorist Security Team.

The US Air Force employed Military Transport Command (MAC) aircraft from 21 different wings; the 919th Special Operations Group (AF Reserve), from Eglin Air Force Base; and 26 refuelling squadrons from the Strategic Air Command. Tasking requirements from MAC were accomplished by the 21st Air Force, McGuire AFB, New Jersey, and the 22d Air Force, Travis AFB, California. The 23d Air Force of Hurlbert Field, Florida, the Air Force Component of the US Special Operations Command, executed special operations requirements. AC-130 Gunships supported operations at La Commandencia; Rio Hato; Torrijos International Airport; the Pacora River Bridge; Fort Amador; and Colon. Commencing 19 December, it required 111 aircraft to insert the initial forces into Panama, with 78 follow-on sorties during the next three days. Airdrop missions totalled 84 sorties. More than a thousand flying hours were accomplished by the 1st Special Operations Wing. MAC ferried more than 3,000 tons of equipment to Panama, including three tons of medical supplies.

Other firsts hit the book of records - US military performed tactical infiltrations from jet aircraft - C-141 Starlifters. The Panama Canal was closed for the first time since opening 75 years ago, to be reopened, and for the first time, under Panamanian control. Significantly, the 7th Infantry Division (Light), commanded by Major General Carmen J. Cavezza, deployed the remaining 4,000 Light Fighters and their equipment, from Fort Ord to Howard AFB and the Torrijos-Tocumen Airport in less than two days. This is the first deployment of an entire US Army Division since the Vietnam War.

Task Force Atlantic, a brigade sized Task Force, including the 4th Battalion, 17th Infantry, of the 7th Infantry Division (Light), demonstrated their proficiency in urban warfare techniques in Panama City, and Colon. Fierce fighting and opposition was encountered by the Light Fighters before they successfully contained the enemy. The 3d Battalion, 504th Parachute Infantry, 3d Brigade, 82d Airborne Division, was targeted against Gamboa Prison, and its vicinity. The PDF 8th Infantry Company was no match for theLight Fighters and paratroopers. Heavier fighting ensued shortly after between the American forces and a naval infantry company, again the Americans were the victors. The 3d Battalion, 504th Parachute Infantry had been conducting training at the Jungle Operations Training Center when the invasion was mounted. An American housing area was located by the Gamboa Prison, and political prisoners from the unsuccessful coup attempt of 3 October, were inside the prison. In a lightning operation, Special Operations Forces entered the Gamboa prison, rescued an American agent from custody, and released 48 prisoners who had been captive since the attempted coup in October.

In addition to the Ranger element, Task Force Pacific included, the 4th Battalion, 325th Infantry, and the 1st and 2d Battalions of the 504th Parachute Infantry Regiment, all assigned to the 82d Airborne Division. Task Force Pacific lead elements parachuted into the Tocumen-Torrijos Airport at 0210 hours. Moving from their assembly areas by helicopter, the 1st Battalion, 504th Infantry overpowered the PDF 1st Infantry Company and captured the Las Tigres Military Installations, at Tinajitas. The 2d Battalion took the barracks at the Panama Viejo, defeating a Cavalry Squadron. The 4th Battalion, 325th Airborne Infantry Regiment, seized Fort Cimarron, from where many members of the PDF elite Battalion 2000 had donned civilian clothes and deserted.

These were precautionary measures to restrict PDF reinforcements being flown in, and to deter General Noriega from using the airport, where his personal airplane was based.

Task Force Red comprised the three Ranger Battalions, conducting airborne assaults, supported by the 4th Psychological Operations Group (Airborne), and the 96th Civil Affairs Battalion (Airborne).

The 1st Ranger Battalion, and one company of the 3d Ranger Battalion, 75th Ranger regiment, parachuted into the Torrijos International Airport at 0100 hours, 20 December 1989, supported by elements of the 4th Psychological Operations Group (Airborne), and the 96th Civil Affairs Battalion (Airborne). Ground opposition included the PDF 2d Infantry Company, and other Panamanian support forces. Reconnaissance Teams from the 3d Battalion, 7th Special Forces Group had maintained surveillance of the airport area. A late arriving flight from Brazil, carrying 347 civilian passengers, on a routine commercial flight, were taken into custody by the PDF. The passengers were released following hostage negotiations between the US Special Operations Forces ground elements and the PDF. A significant role was played by the 4th Psychological Operations Group during the hostage exchange. The PDF 2d Infantry Company and ground crews were defeated by the Rangers, with more than fifty prisoners captured.

The airport fell to the Rangers, and was secured in preparation for follow-on incoming units. Elements of the 82d Airborne Division followed less than an hour later, the first US soldiers to parachute into combat via UIS Air Force C-141 (jet) Starlifter aircraft.

Simultaneously, the 2d Ranger Battalion, commanded by Lieutenant Colonel Alan Maestas, and two companies from the 3d Ranger Battalion, leaped from their C-130 Hercules air transports, over Rio Hato, Panama. As their parachutes deployed, rifle fire came up to meet them from the ground, unfortunately for the PDF, they did not know what unit they were engaging, and what support they had on hand. As the Rangers of Task Force Red approached their Drop Zone, a supersecret US Air Force F-117A stealth fighter was on hand to maximize their safety. The 1st Special Operations Wing had rehearsed their role for the scenario now being played out. The F-117A stealth fighter, assigned to the 37th Tactical Fighter Wing, Tonopah Test Range Airfield, Nevada, dropped two 2,000 lb bombs in the vicinity of Rio Hato. The stealth fighters are designed to penetrate radar and air defenses, and perform single aircraft attacks on high priority targets, deep behind enemy lines. From Rio Hato, the Rangers seized eleven ZPU-4s (anti-aircraft guns), 16 armored cars, eight 81 mm mortars, 55 machine guns, 48 rocket-propelled grenade launchers, and literally tons of ammunition.

On 18 December, the Ranger Regiment had recalled its 2d Ranger Battalion to Fort Lewis in an alert status. They boarded C-141 Starlifter aircraft to an isolation area on the east coast, where they were given their mission, and prepared for their insertion on the morning of the 20th. Companies A and B, 2d Battalion, were to seize three military compounds of Rio Hato, housing the PDF Training Camp and the 6th and 7th PDF Infantry Companies. Company C was to secure General Noriega's beach house, and be the reserve force, prepared to assist Companies A and B as required. The two companies from the 3d Ranger Battalion were tasked to secure the Pan-American Highway , crossing the Rio Hato runway, and an ammunition storage area to the northwest of the primary objective area. The Rio Hato military airfield is located approximately 75 miles west of Panama City, comprising a one and a half mile runway and a military compound.

Colonel William F. "Buck" Kernan, Commander of the 75th Ranger Regiment, would command and control his assets on the ground, and coordinate the overall plan with the US Special Operations Command forward headquarters, provided by the Joint Special Operations (JSOC), from Fort Bragg, North Carolina. Major General Wayne A. Downing, commanding JSOC, was the first Regimental Commander when the 75th Ranger Regiment was activated in 1986. Alas, Colonel Kerrigan smashed the heel of his foot upon landing

Following their seven hour flight from the isolation area, personnel of the 2d Battalion were ready to hit the silk. As they flew towards the drop zone, the jumpmasters could see the burning and smoke from the two 2,000 pound bombs dropped by the F-117A stealth aircraft. The Rangers exited their aircraft at 500 feet.

Quickly reforming on the ground, the Rangers organized into pre-assigned fire teams, assaulted their targets, attacked vehicles attempting to escape, captured the airfield control tower, and re-grouped in their assembly areas. Company A, guided and assisted US Air Force AC-130 (Specter) gunships and routed the 6th and 7th PDF Infantry Companies. Approximately 300 prisoners were captured from the two PDF companies, part of the PDF Battalion 2000.

The Specter Gunships are armed with 105mm howitzers and automatic cannons, targeted through highly sophisticated and complex laser and infra-red scopes and range finders. The Specters have been in the Special Operations inventory since the latter part of the Vietnam conflict, and like the B-52 bombers, are getting old. The US Special Operations Command, and its USAF component, the 23d Air Force, rely upon Specter as an integral support for Special Operations. Despite their antiquity, and excessive requirements for in-flight refuelling due to non-pressurization, they cannot afford to discard these warrior birds, as they proved their value to literally be worth their weight in gold.

Task Force Bayonet had two major objectives: capture of La Commandancia, (headquarters of the Noriega Regime and the PDF), and securing the central canal zone of the city. Task Force Bayonet comprised the 5th Battalion, 87th Infantry and the 1st Battalion, 508th Infantry (Airborne), of the 193d Infantry Brigade (Light); the 4th Battalion, 6th Infantry, deployed from the 5th Infantry Division (Mechanized), Fort Polk, Louisiana; the 519th Military Police Battalion; and a platoon of the 3d Battalion, 73d Armor, 82d Airborne Division. Fighting was intense throughout the night, resulting in many of the surviving PDF fleeing by daybreak. The 1st Battalion, 508th Infantry (Airborne), commanded by Lieutenant Colonel B. R. Fitzgerald, sized Fort Amador on 20 December, and were credited with the capture of 141 prisoners, more than 2,000 weapons, a V-300 Armored Vehicle and a ZPU-4 Anti-aircraft Gun. This gallant battalion sustained two soldiers killed in action and six wounded while seizing this objective. The Panamanian 5th Infantry was housed at Fort Amador. The 1st Battalion, 508th Infantry (Airborne), rapidly restrained all opposition.

The 4th Battalion, 6th Infantry Regiment, 5th Infantry Division (Mechanized), commanded by Lieutenant Colonel James Reed, was quickly positioned, and assisted by Company C, 1st Battalion, 508th Infantry (Airborne). They breached and captured La Commandancia. M551A1 Sheridan Tanks of the 3d Battalion, 73d Armor (82d Airborne Division), and LAV-25s of the 2d Light Armored Infantry Division (2d Marine Division) were employed in a direct fire role, devastating the western end of La Commandancia. The Sheridans were an ideally suited armored vehicle for this type of operation, which required the firepower and mobility in confined urban areas. Accepting that this light armored vehicle, is more vulnerable then the heavily armored modern family of tanks, the demonstrated tactical deployments of the 3d Battalion, 73d Armor, provided many accolades for their obviously sound training. The sophisticated M1 Abrams Tank could not have performed as well in these restricted areas without totally reducing the civilian homes to rubble. AC-130 Gunships competed the jointly coordinated attack against La Commandancia.

US Navy SEALs of Naval Special Warfare Groups 2 and 8, of Task Force White, destroyed a PDF Patrol Boat in Balboa Harbor. Three SEAL platoons with their Zodiac assault boats parachuted into the ocean shortly after midnight. They conducted riverine security along the approaches to Howard Air Force Base, scuttling a fleet of small boats. Beaching their assault boats, the SEALS advanced onto Punta Paitilla Airport. Two of the SEAL platoons established security along the airfield, with the third platoon, commanded by Lieutenant (jg) John Connors, moved stealthily across to the terminal and parked aircraft, farther inland. Their objective was to disable the aircraft belonging to General Noriega and other aircraft in the vicinity, and then capture the airfield control tower.

Following precisely the orders that they had been given, the SEALS challenged two approaching PDF soldiers, who immediately ran into the cover of a nearby hanger. The SEALs continued their mission, moving towards their target aircraft. As they passed the hanger, establishing a hasty perimeter, bursts of fire from the hanger killed Lieutenant Connors and three of his men. Eight more SEALs were wounded. As Panamanian reinforcements hurried to the scene, a major firefight erupted, allowing the SEALs to demonstrate their superior combat abilities. The other SEAL platoons moved to provide assistance, and the fight was over. The control tower was captured and secured. Paitilla Airport was where the 7th Infantry Company of the PDF had flown into, in response to the unsuccessful coup on 3 October. The SEALs were quickly reinforced by Captain Kurt Runge and his Company D, 4th Battalion, 325th Airborne Infantry, who had parachuted into Torrijos International Airport.

US Special Forces and Navy SEALs, Task Forces Green and Blue, raided other prime targets, suspected to be used by General Noriega and his mistress in Panama City, but again, the self-styled 'Maximum Leader', was not to be found. Similar searches were concurrently conducted elsewhere in Panama City and in the neighboring Colon. Specific intelligence pinpointing the exact whereabouts of General Noriega was not available. It was known that routinely, prior to the invasion, General Noriega changed his overnight locations as many as five times in a single evening. The single greatest flaw in the US intelligence, prior to and during the early days following the invasion, was the lack of knowledge regarding General Noriega's whereabouts. Numerous leads and reported sightings were followed by personnel assigned to Task Forces Green and Blue. Another SEAL preventive strike against a seaside villa used by General Noriega and his mistress, was apparently late by mere minutes. Smoking cigarettes were still smouldering in the ashtrays.

Task Force Semper Fidelis included two US Marine Corps elements. A rifle company from the 3d Battalion, 6th Marine Regiment, Camp Lejeune, North Carolina, and an Armored Infantry Company from the 2d Light Armored Infantry Battalion formed Task Force Semper Fi.

Rapidly deploying, the US Marine Corps quickly secured the Bridge of the Americas, crossing the Panama Canal, and Howard Air Force Base.

As the tempo of activity increased, the US leadership became more perplexed. General Noriega was nowhere to be found. It was certain that he had not fled the country, but that he was still within the confines of Panama City. By the end of the first day, 15 Americans had lost their lives (to include three Navy SEALs, one marine and seven soldiers). Casualties were 59 US service members were wounded and one was reported missing in action. Three US helicopters had been shot down, and a fourth hit a wire and crashed.

Over a thousand refugees streamed into makeshift shelters at a local high school on 21 December, and over 800 PDF soldiers were taken prisoner. The streets of Panama City were overrun with looters. Storefronts were demolished, and the residents of the neighboring shanty-towns and slums, ravaged everything in sight. Between $500 Million and one billion dollars worth of looting was carried out. US Forces were neither blamed for looting activities or for not preventing these crimes. As General Thurman pointed out, the invasion task force had higher priority military targets (142 sites were designated and secured by the US military). US targets included Panama City power, light, water, sanitation, telephones and other public utility services, in addition to military targets. The US troop strength was increased to approximately 24,000 personnel. 'Light Fighters' of the 7th Infantry Division were well trained and prepared to conduct urban operations within Panama City against the PDF and the armed, non-uniformed 'Dignity Battalions' of the Noriega regime. Some 3,000 members of the 16th Military Police Brigade, including the 503d Military Police Battalion, of the XVIII Airborne Corps were deployed and well prepared to assist in the return of law and order. The military police of the 16th Military Police Brigade assumed responsibilities for police duties throughout Panama City, assisting the Panama based 92d Military Police Battalion. The 92d Military Police Battalion had been assisted by the 534th and 549th Military Police Companies of the 193d Infantry Brigade, and the 324th Support Group.

The 519th Military Police Battalion had also arrived a few days earlier from Fort Meade, Maryland. Accompanying the 519th Military Police Battalion were its assigned five companies: the 209th MP Company also from Fort Meade; the 401st MP Company from Fort Hood, Texas; the 511th MP Company from Fort Drum, New York; the 555th MP Company from Fort Lee, Virginia; and the 988th MP Company from Fort Benning, Georgia. The 988th Military Police Company, from Fort Benning, Georgia, had arrived on 17 December, commanded by Captain Linda Bray. Captain Bray and 12 of the 15 females assigned to her company, are reported to have participated in actual combat operations. Captain Bray has discussed her actions, when she learned that soldiers of her command were meeting heavier resistance than originally estimated.

On 22 December, enemy losses were released as 122 dead, 45 wounded and approximately 1,400 prisoners. US casualties stood at 21 service members killed in action and 221 wounded in action. Two US civilians had been killed, including Gertrude Helin, a Department of Defense contracted school teacher, who was caught in cross-fire on her way home. Military dependents and US civilians had been advised to remain indoors and under cover. The remainder of the 96th Civil Affairs Battalion (Airborne) was moved to Panama from Fort Bragg with the 528th Support Battalion, both are assigned to the 1st Special Operations Command.

On 24 December, Colonel Eduardo Herrera Hassan, formerly of the Panamanian Army, who had been exiled in Florida for several years, returned to Panama at the request of the new Government. Colonel Hassan was appointed as the Deputy Commander of the newly structured Panamanian military. The Commander of the Fuerza Publica Panamena, (as the PDF had been redesignated) was Colonel Roberto Armijo, the former commander. Colonel Armijo, who had been loyal to General Noriega through the unsuccessful coup attempts, would remain as head of the 15,000 member strong Panamanian Military.

US military evacuated more than 100 people from the Marriott Hotel in Panama City, which had been placed under siege by pro-Noriega forces. The day following their airborne assault, Task Force Pacific secured the Marriott Hotel freeing 14 Americans barricaded inside.

Fierce fighting was reported in the vicinity of US Southern Command Headquarters at Quarry Heights. Soldiers of Task

CH-47D Chinook loaded with troops flying above the jungle. (Courtesy of SSG Joe Owens)

Force Bayonet repelled a joint PDF/Dingbat attack targeted against the recently organized Transit Police. Company A, 1st Battalion, 504th Parachute Infantry established a defensive perimeter for the power station, patrolling the surrounding suburbs and villages.

On 25 December, General Noriega was located. He had taken refuge at the Papal Nuncio, the resident of the Vatican's representative, Monsignor Jose Sebastian Laboa, arriving at 1530 hours on the afternoon of Christmas Day. General Noriega was accompanied by Captain Eliezer Gaytan, his chief of security and special forces. US troops and tanks quickly sealed off the area, located in the upper-class suburb of Paitilla. The 21st Military Police Company, 16th Military Police Brigade, established a guard force. The debates to extradite General Noriega commenced, and discussions soon started between the Papal Nuncio's representatives and General Maxwell Thurman, Commander in Chief, US Southern Command. General Noriega had sought advice and counsel from Monsignor Laboa previously. Initial indications were that General Noriega was attempting to gain political asylum in Spain or Cuba.

Preventive measures were taken against potential Panamanian message traffic between General Noriega and the PDF. In addition to the psychological responses elicited against General Noriega from observers, the 4th Psychological Operations Group played continuous loud and blaring music outside of the Papal Nuncio. Regretfully this commenced to play upon the nerves of the papal Nuncio staff in addition to General Noriega.

The Spanish Government responded that asylum would not be granted to General Noriega in their country. Deputy Foreign Minister Ricardo Alarcon of Cuba, confirmed in Havana, Cuba, that the Cuban Government would grant asylum to General Noriega if safe passage could be coordinated. The Cuban Embassy representative in Panama, Luis Delfin Peres, charge d'affaires, reported that no direct request had been made by General Noriega for asylum in Cuba.

The United States continued to press for extradition, however this was repeatedly denied. It was learned however from Vatican sources, that the Vatican did not grant asylum outside of its territory, but does grant 'temporary refuge' to individuals claiming political or religious persecution.

On 26 December, eighteen more US casualties were flown to the Wilford Hall and Brooke Medical Centers in San Antonio, Texas. This brought the total up to 229 US wounded treated at these military hospital centers. Of the total 101 soldiers treated at Brooke Medical Center, Fort Sam Houston to date, thirty-eight had been discharged. At Wilford Hall Medical Center, Lackland Air Force Base, 42 of the 129 service members admitted, had been released.

Colonel Luis de Cid, the Panama Defense Force commander for Chirriqui Province, surrendered to the US forces. Colonel Cid had been indicted with General Noriega on drug trafficking charges. He was turned over to the Drug Enforce-

AH-1 Cobra gunship at a Central American airfield being readied for another mission. (SSG Joe Owens)

ment Agency, and was flown to Miami, Florida for trial arraignment. US Military Police commenced joint patrolling with the newly reconstituted Panamanian police. Few American troops patrolled in residential areas, but heavily armed road blocks were established on the Trans-Isthmus Highway and the Via Espana. In many areas PDF renegades and Dingbats were stopping and harassing automobiles, breaking into buildings, starting fires, and shooting at people in their homes.

On 27 December, US forces surrounded the Cuban Embassy in addition to the Papal Nuncio, in order to apprehend General Noriega should he attempt to gain political asylum at that Embassy. Lieutenant Kimberley Thompson, a female platoon leader of the 988th Military Police Company was in charge of security for this operation. Panamanian and foreign officials had announced that negotiations for asylum with the Cubans had been discussed in Panama and with other foreign officials. President Endara maintained that he preferred General Noriega to be retained in US custody, rather than be tried under Panamanian laws. Dissent against General Noriega was maintained at fever pitch by the 4th Psychological Operations Group through broadcasts and the news media. This Fort Bragg based special operations unit has clearly demonstrated the power of the pen. General Maxwell Thurman continued his discussions with the Reverent Sebastian Laboa in front of the Papal Nunciature.

Pentagon sources reported that the number of refugees ranged from 5,000 to 13,000 as Panamanians moved in and out of refugee camps. More than 34,000 enemy weapons had been captured or turned in to US military forces. PDF losses were reported as 297 killed and 129 wounded.

On 29 December, a third US civilian was confirmed to have been killed by the Dignity Battalions. The body of 47 year old Raymond Dragseth, a computer-science teacher, employed at the Panama Canal College, was recovered from a grave containing other unidentified victims. Mr Dragseth is reported to be related by marriage to the Press Secretary of President Endara. The grave held other Panamanian victims slain by the Dingbats.

As the new decade commenced on 1 January 1990, Panamanian losses were hard to confirm. Sources reported 314 dead, 124 wounded, and 5,313 prisoners. As of 30 December, more than 72,000 weapons had been captured, turned in, or confiscated, and 33 armored vehicles captured. The US cost of JUST CAUSE, although not so high, resulted in the deaths of some brave servicemen. Of the US Military involved in this Operation, 23 had been killed in action, and 323 wounded in action. Additionally, three American civilians had died. Fourteen of the Military Airlift Command transports sustained damage, with only two still undergoing repairs. Three of the damaged aircraft were C-141 Starlifters, one of which received some damage to a tail cone during an airborne insertion; eleven were C-130 Hercules, which had been subjected to small arms fire.

Two artillery batteries from the 7th Infantry Division (Light), totalling 141 soldiers, commenced their homeward trip to Fort Ord, California.

Ambassador Deane R. Hinton was named by President Bush to replace Arthur H. Davis as the Ambassador to Panama on 3 January 1990. Ambassador Hinton was transferred to Panama from Costa Rica. On this day, the Vatican announced three principal rules that must be fulfilled prior to General Noriega leaving its Embassy in Panama. These principles included:

* General Noriega must be assured that his life is not in danger, and that he will be guaranteed a fair trial.
* Panama must file charges against General Noriega.
* General Noriega must leave the embassy and surrender himself of his own free will.

Investigations by the recently installed Panamanian Government, revealed that almost all of the 900 officers of the 15,000 person strong PDF, had committed some crimes. Typically, their crimes included corruption and abuse of power.

At approximately 1600 hours, an estimated 20,000 Panamanians marched to the Vatican Embassy, demanding that General Noriega be turned over to the authorities. Two hours later, the Papal Nuncio passed a message to the US National Security Advisor, Brent Scowcroft, that General Noriega had agreed to turn himself over to US authorities rather that face a Panamanian trial.

On 3 January 1990, at approximately 1850 hours, General Manuel Antonio Noriega freely turned himself over to US authorities. He did so with the full knowledge of the Panamanian Government, sworn in at the outset of the invasion. General Noriega surrendered under four conditions:

* An officer of equal rank be present when he surrenders.
* He be allowed to wear his military uniform.
* There be no coverage by the news media at the time of his surrender
* He be allowed to communicate telephonically with his wife and mistress.

The terms were agreed upon, and General Noriega surrendered to officials of the US Drug Enforcement Agency (DEA). General Maxwell Thurman was present for the surrender, but no conversation was held between the two generals. Shortly after his surrender, General Noriega was transported by helicopter to Howard Air Force Base, where he was formerly arrested by DEA officials, and transferred to a USAF C-130 aircraft to fly him to Homestead Air Force Base, Florida.

As General Noriega surrendered, Colonel Roberto Armijo resigned as head of the Panamanian Forces. The deputy commander, formerly exiled Colonel Eduardo Herrera Hassan, assumed control of the Fuerza Publica Panamenia.

Approximately 500 members of the 1st Ranger Battalion, 75th Ranger Regiment, returned to Fort Stewart, Georgia.

On 9 January, Peruvian officials confirmed that a group of twelve Panama citizens requesting asylum, had taken refuge in the Peruvian Ambassador's residence. Included were Major Gonzalo Gonzalez, a former aide to General Noriega; Major Heraclides Sucre, with his wife and three children; Major Edgardo Lopez; and Captain Marcela Tason, a secretary of General Noriega, and her husband Captain Narcisa Castillero. There was currently no Peruvian Ambassador in Panama, after the former ambassador, Mario Castro, was recalled to Peru following the debacle after the Panamanian elections in May.

The Pentagon announced on 10 January that 220 Panamanian civilians, and 814 members of the PDF, had been killed during the Panama invasion.

On 12 January, major portions of the invasion force returned to the United States. As a demonstration of their preparedness, approximately 2,000 members of the 82d Airborne Division arrived back at Fort Bragg by parachute. As the invasion force gradually returned home, the US Army Reserve Special Operations Command, guided by Brigadier General Joseph C. Hurteau, at Fort Bragg, swung into action. The USAR Special Operations Command had coordinated for more than 130 USAR Civil Affairs volunteers to assist in the aftermath. The initial 113 person contingent of Civil Affairs specialists had departed for Panama on the 3d and 4th of January, as General

Targets marked with smoke (W. P.) rockets by a Cobra. (SSG Joe Owens)

Noriega was turning himself over to US authorities. The follow-on activities to assist in the safe and orderly return of Panama to the control of the elected government, was aptly named: Operation PROMOTE LIBERTY.

General Noriega was arraigned for trial, and redesignated to Prisoner of War (POW) status, and US military missions in support of the invasion of Panama were complete. As a POW, General Noriega is authorized fair treatment by an impartial humanitarian group; he is authorized to be incarcerated in a military prison instead of a civil prison; he is authorized the treatment and respect worthy of a military general; he has a right to wear his uniform, military badges and decorations; and he is authorized two hours of outdoor exercise daily.

The nation-building process, performed by the Civil Affairs specialists, will be continuing as the 50th Airborne Anniversary celebrations are underway in Washington, DC. The United States has demonstrated to the world, that it has successfully mastered joint operations, combining conventional and unconventional forces of the Army, Navy, Air Force and Marine Corps into a finely honed and coordinated fighting machine.

A primary ingredient resulting in this successful operation, was the limitation to one major ground force. In the past, both in Grenada and Desert One (the Iran rescue attempt), multiple services were involved, to ensure they all received part of the 'glory'. Command, control and communications were certainly not enhanced under these conditions. The Panama invasion was basically an Army ground operation, with two US Marine Corps companies, and a Navy SEAL unit. Unlike Grenada, however, which was planned and executed in less than a week, the Panama invasion was planned well in advance, units and equipment were pre-positioned, and units were able to rehearse many of their missions. The biggest airborne operation since World War II was well coordinated, dispite minor mishaps, (especially with weather conditions), and was a prime example of unity of command. The US Army will not always experience the luxuries of battlefield preparation that were available prior to the execution of Operation JUST CAUSE.

The US Army Personnel Center announced on 19 January 1990, that those Army personnel who had participated in Operation JUST CAUSE between the dates 20 December 1989 and 13 January 1990, were authorized to wear their organizational shoulder sleeve insignia on the right shoulder of their Army uniform, as a 'combat patch', and receive the Armed Forces Expeditionary Medal. Approved Army SSI as combat patches for Operation JUST CAUSE included: US Special Operations Command; US Army South; XVIII Airborne Corps; 5th Infantry Division (Mechanized); 7th Infantry Division (Light); 82d Airborne Division; 193d Infantry Brigade; 1st Corps Support Command; 16th Military Police Brigade; 18th Aviation Brigade; 35th Signal Brigade; 44th Medical Brigade; 470th Military Intelligence Brigade; 525th Military Intelligence Brigade; 1109th Signal Brigade; 75th Ranger Regiment; and the 7th Special Forces Group. Soldiers assigned to units not listed above, will wear the SSI of the unit to which attached, or the unit that had operational control over them. Soldiers not attached or under the operational control of any of the units listed, are authorized to wear the SSI of US Army South.

AIRBORNE CASUALTIES
OF WORLD WAR TWO
By William C. Mitchell

"There's no future in this paratrooper business."
Robert Capa, Life Magazine after participating in the Rhine Jump, March 24, 1945.

"Not in vain may be the pride of those who have survived–the epitaph of those who fell."
Winston S. Churchill

S.L.A. Marshall's *Men Against Fire* is still among the leading works in military organization and tactics. Despite the deserved importance of the book its distinguished author never considered the role of casualties in battlefield decisions, i.e., as an element in the choice of tactics or as an outcome of battle. When a leading thinker is remiss what is to be expected from lesser minds? Regardless, casualties must be included in any serious assessment of wars past as well as future commitments.

Although knowledge about and a better understanding of the costs of war is imperative my task is an unenviable one. While I have no desire to become an accountant of the gruesome and no intention to invite invidious comparisons a truthful assessment and explanation of varied casualties among military units and wars is necessarily fraught with powerful emotions. When life is precious it is painful to contemplate let alone compare human losses. Given the purposes of this commemorative volume I have avoided any systematic effort at explaining diverse casualty rates. Such explanations are not only scientifically difficult but, again beset with emotion. Sometimes men die as a result of courageous choice; at other times, inexperience and stupidity, whether of command decision or the individual soldier. In the scientific study of casualties all this is reduced to "force-ratios," such as whether a unit is on the offensive or not, terrain, prevailing weather, weapons, etc. Such a treatment is not systematically pursued in these pages.

Compounding the difficulty of objective treatment is the fact that the historical records on battle loses are not entirely error free; indeed, they and the secondary works, especially, are replete with error, confusion, contradictions, and incompleteness. It is not likely that we shall ever obtain a genuinely accurate and complete accounting for the casualties of World War II or other wars. Readers of this chapter must, therefore, be appreciative of the fiendish and frustrating difficulties faced by researchers and tolerant of their errors. Fortunately, for most purposes, perfection is neither required nor worth the additional cost of attainment. Finally, inadequate attention is given to more specialized Airborne units including the Special Forces, etc. Data is not readily available.

Although my overall concern is with Airborne casualties during its entire fifty years of existence I pay most attention to the Second World War. But even a concentration on the Airborne during one war benefits considerably when placed in some comparative framework. We begin, therefore, by noting that the United States Army during the Second World War incurred a total of 234,874 deaths on or resulting from the battlefields, contrasted to 27,704 in Korea and another 30,889 in the Vietnam fighting. Most of the casualties incurred by the Army ground forces during World War II were sustained in the European Theater (133,673 KIA and DOW), followed by the Mediterranean Theater (39,433) and the Pacific (38,900). The grand total for the Second World War including all branches of service and types of casualties numbered over a million (1,078,162), with the Army, including the Air Force having 884,135, the Navy 100,392 and the Marines, 91,718. The Coast Guard sustained losses of 1,917.

During the Second World War, the Army ground forces incurred the most casualties and except for certain Air Force combat crews had the highest rates. In terms of total losses, the Infantry led all other ground forces. The Marine Corps which had total casualties numbering about 91,000 sustained about nine percent of the total casualties while having but five percent of the troops. During the Korean venture the Marines provided seven percent of the manpower while sustaining approximately 19 percent of the casualties. In Vietnam the Marines fielded 794,000 men and women (about 125,000 more than during WW-II!) and incurred casualties totalling 66,213. That meant the Marines contributed about 9 percent of the troops and 31 percent of the casualties. The Army made up about half of the total personnel while sustaining 64 percent of the casualties.

Although the Second World War was America's "biggest" war we would be both in error and remiss if mention were not made of our Civil War. During that great national tragedy more soldiers were killed and died as a result of wounds and disease than in all our other wars, together. In a nation of only 32 million people (4 million were slaves) nearly 605,000 men lost their lives with about 60 percent from the Northern states. Nearly 141,000 of the Northern dead were killed in action. Southern casualty rates tended to be higher with the troops of General Lee sustaining the highest averages—20 percent—while those of General Grant took 16 percent. Both of these figures were well above those of World War II. Despite vast increases in firepower since 1865, battle death rates in the Civil War dwarfed those of World War II with some 40 men killed per 1,000 engaged in combat, whereas, in World War II the figure was 11.6 per thousand. In Korea the ratio dropped still more to 5.5 men per thousand in combat. In Vietnam it was still lower.

Every wartime generation tends to believe that "its" war was the most demanding and costly and that is understandable. Those who die in light as contrasted to heavy and prolonged combat die and their families grieve as much if not more. So, the point is not invidious comparison but to more fully and humbly understand the tragic costs of war. In doing so we must shed false perceptions and honor facts and truth. We must recognize that costly as was the Second World War for us it was much more so for many other nations including the USSR whose human losses totalled nearly 20 million people. Germany lost about 6 million individuals with more than half being civilians, while in China 9 million perished. Close to 55,000,000 people died in World War II.

"A LEGEND WAS BORN"

A "Legend Was Born" on the drop and landing zones of the Second World War, a legend founded on the courageous actions of but four small, lightly-equipped divisions (a total of 45,000 men and officers) who routinely met materially superior forces under adverse conditions and invariably defeated them. But more often than not they paid a fearful price in the loss of nearly 7,500 killed-in-action and another 26,000 wounded. Another roughly estimated 3,550 were listed as Missing in Action of whom approximately 2,000 were captured and interned. In rough terms, the Airborne constituted less than one percent of the Armed Forces but sustained 5 percent of the casualties. About 29 percent of those who served in Airborne combat ended up on the casualty lists. That fraction stands in rather sharp contrast with the overall casualty rates suffered by other ground force units and, on many occasions, the Marines. For a listing of casualties among the top thirty divisions see Table 1.

Table 1
TOTAL DIVISIONAL LOSSES: TOP 30
(KIA, DOW, WIA)

RANK	DIVISION	TOTAL CASUALTIES
1	3rd Infantry	25,977
2	9th Infantry	23,277
3	4th Infantry	22,660
4	45th Infantry	20,993
5	1st Infantry	20,659
6	29th Infantry	20,620
7	36th Infantry	19,466
8	90th Infantry	19,200
9	30th Infantry	18,446
10	1st Marine	17,908
11	80th Infantry	17,087
12	2nd Infantry	16,795
13	28th Infantry	16,762
14	34th Infantry	16,401
15	83rd Infantry	15,910
16	82nd Airborne	15,832
17	35th Infantry	15,822
18	4th Marine	15,390
19	79th Infantry	15,203
20	8th Infantry	13,986
21	88th Infantry	13,111
22	5th Infantry	12,818
23	2nd Marine	12,770
24	26th Infantry	10,701
25	3rd Marine	10,416
26	5th Marine	9,573
27	6th Marine	9,330
28	101st Airborne	9,328
29	3rd Armored	9,243
30	7th Infantry (Pacific)	9,212
TOTAL		474,896

Sources:
Army Battle Casualties and Non-Battle Deaths in World War II: Final Report (Washington, D.C.: Statistical and Actuary Branch of the Adjutant General, 1953); *MSMC Combat Casualties by Division* (Washington, D.C.: Department of Navy, Headquarters United States Marines Corps, 1950).

Although fourteen Army Infantry Divisions and one of six Marine Divisions (1st Marine) incurred larger total losses than the 82nd Airborne none exceeded the rates of the three Airborne Divisions that served in the Mediterranean and European Theaters of Operations. Here it is important to know that Army Infantry Divisions had normal complements of about 14-15,000 men while Marine divisions in 1942-43 typically numbered 19 to 20 thousand and in 1944-45 as many as 24,000 men and on one occasion 26,000. During the four battles in which the 4th Marine Division was engaged, nearly 82,000 men saw combat. On average, a WW-II Airborne division had 10 to 12,000 men of whom 9,000 constituted the units in combat.

Because of the replacement policy most of the top thirty Army divisions serving in the ETO spent much more continuous time in battle than did the Airborne or Marines. For example, the 29th Infantry Division landed on D-Day in Normandy and spent 337 days of the remaining eleven months of the war in combat. The 34th Infantry Division fought for 500 days in the Mediterranean and European Theaters.

The six Marine divisions, together, fought for approximately 600 days with highs of 286 days for the First and about 68 days for the Second. The Third put in about 50 days of combat while the Fourth had 66. The Fifth fought for 40 days while the Sixth had a total of 82 days. Contrast the 82nd Airborne's 316 days of frontline duty or the 101st Airborne with about 160 days. The 17th Airborne had 65 days. Oddly enough, both the 504 and 505 PIR of the 82nd had more combat time than the parent Division. In Europe, the independent Airborne units (509 and 551 Parachute Battalions and the 517th PIR) fought for a total of 400 days. In the Pacific Theater, the 11th Airborne and the independent 503 PIR fought 112 and 138 days, respectively, for a grand Airborne total of more than 1,191 days. As with the Army, Marine casualties were reflected but, roughly, in the number of days fighting with the First Marine Division incurring the largest number (17,908) followed by the Fourth Marine Division with 15,390 casualties for a rate of 19 percent. In third place, was the Second Marine Division with about 12,770 while the Third Division with 10,416 was in fourth place. The Fifth with 9,573 and the youngest Marine Division—the 6th—with 9,330 all incurring during some 82 days at Okinawa.

DIVISIONAL LOSSES

There seems to be a consensus that the average overall casualty rate for army infantry divisions in the European fighting was rarely more than ten percent and in the Pacific considerably less. In one regard, these rates were artificially low because the actual proportion of front-line companies in a division was rarely more than one-half of the division personnel with riflemen constituting about a third. But, the front-line units typically incurred 90 percent of divisional casualties. This meant that losses at the levels of regiment, battalion, and, especially, at the company, platoon, and squad levels could and frequently did exceed one hundred percent. In fact, most infantry and Airborne companies and platoons were replaced several times. While armored divisions were more likely to incur about 5-6 percent casualties their infantry battalions losses were far greater. The renowned 2nd and 4th Armored Divisions which fought for 238 and 150 days, respectively, across France during 1944-45 had about the same numbers killed in action as the 17th Airborne (1,382) in about 65 days. By my calculations, the 82nd Airborne's overall average rate of casualties was nearly 27 percent; that of the 101st—26 percent, and of the 17th, 31 percent. The 11th Airborne in the Pacific theater incurred a casualty rate of 18 percent which was far higher than infantry divisions in the Pacific. The Marines, overall, would seem to have incurred casualties at a rate of about 26 percent. Obviously, these rates are approximations and varied considerably among units of each division and from one campaign to another. All this we examine in greater detail, below.

Table 2 sets forth selective basic data for casualties among the four Airborne divisions. The Table does not include such conventional entries as "Missing in Action" and "Captured" or "Prisoner of War"; it also ignores injuries, illnesses and psychiatric disabilities suffered during combat all of which actually exceed those killed and wounded. These figures and categories are either too ambiguous to be relied upon and are, therefore, left out, or are as the injuries, often included among the wounded; needless to say, all constitute serious reductions in effective fighting strength. We might also note that various diseases, mostly tropical, and mental breakdowns, apparently affected more soldiers in the Pacific than was the case in the Mediterranean and European theaters. The First Marine Division lost over 5,000 men to malaria and only 621 killed in five months on Guadalcanal. At Iwo Jima, the Third, Fourth, and Fifth Marine Divisions lost 1,986 men to "fatigue" while the Fifth evacuated 1,275 men because of sickness. More generally, combat fatigue plagued our forces not only in the Second World War, but in Korea and Vietnam. One authority has contended, however, that our elite forces incurred at least one-third fewer such losses.

wounded in action generally ran around 3 to 3.5 WIA for each KIA. These ratios varied a good deal more when one considers not divisions but regiments and smaller units and particular engagements. And, the ratio increased significantly in Vietnam. This development will be taken up, shortly. At the divisional as well as lesser unit levels, the ratio appears to be chiefly affected by the types, quantities and quality of weapons at the disposal of the enemy and whether a unit is on the offense or defense. When the enemy depends upon small arms firepower the ratio decreases, i.e., more of the casualties will be killed and fewer wounded, whereas, when artillery, mines, mortars, and tanks are used the ratio of wounded increases dramatically. These weapons cast their fragments over a considerable area but not often lethally. Students of such matters are also of the opinion that German artillery as well as auto-

Table 2

DIVISIONAL CASUALTIES

DIVISION	KIA/DOW	WIA/IIA	TOTAL	DAYS IN COMBAT	DAILY RATE
11th	620	1,806	2,426	112	24
17th	1,226	4,904	6,130	65	96
101st	2,155	6,388	8,543	160	53
82nd	3,228	12,604	15,832	316	50
TOTALS	7,229	25,702	32,931		

Sources:

Army Battle Casualties and Non-Battle Deaths in World War II: Final Report (Washington, D.C.: Statistical and Actuary Branch, Office of the Adjutant Genera, 1953), p. 80. More complete figures for the 82nd Airborne Division are set forth in W. Forrest Dawson, Ed., *Saga of the All American* (Chicago: 82nd Airborne Association, 1947), no pagination. It is important to note that since the 11th Airborne was considerably smaller (8,400 men) than the other Airborne divisions its casualty rates must be adjusted. So adjusted, the totals would be the equal of about 900 KIA and 3,000 wounded in a division of 12,000 men, and the daily rate would be equal to 33 casualties.

Airborne divisional casualty numbers should occasion little surprise; again, they reflect, however, roughly, the respective lengths of time spent in combat. The 82nd began its combat (July, 1943) a full year before the 101st and the latter about six months before the 17th Airborne. The relatively low numbers for the 11th airborne reflect the fact that it had but 8,400 men. Nevertheless, nearly all combat units in the Pacific tended to have much longer periods between battles than was the case in the ETO and, therefore, fewer combat days. In addition, the 11th did not make any divisional jumps so its operations were on a reduced scale. Since the 503 PIR fought in the Pacific as an independent unit, its losses are not reported in Table 2. Although independent Regiments or Combat teams, the 507 and 517th were frequently teamed up with Airborne or other divisions for their Mediterranean and ETO operations. Their casualties are included in the overall figures for the three divisions with which they fought: for example, the 507th served with the 82nd in Normandy and, thereafter, with the 17th in the Ardennes, Rhineland, and Central Europe.

It should be noted that the ratio of killed-in-action to

matic weapons were vastly superior in numbers and quality to that of the Japanese and, therefore, far more lethal. In stressing weaponry we must not ignore the human element—the courage and skill of the opponents.

Divisional losses also varied greatly from one campaign to another, although, perhaps, not as much as at the lesser unit levels. In any event, Table 3 displays the Airborne losses (both KIA and WIA/IIA) for most of their major operations, including pure Airborne and conventional infantry confrontations.

Table 3
DIVISIONAL LOSSES BY CAMPAIGN

DIVISION	OPERATION	KIA/DOW	WIA/IIA	TOTAL
1st Airborne (British)	Arnhem	1,350	3,250	4,600
82nd	Normandy	1,182	5,133	6,351
101st	Normandy	868	2,303	3,171
6th Airborne (British)	Normandy	820	2,709	3,529
101st	Holland	752	2,151	2,903
6th Airborne (British)	Rhine-Wesel	700	1,500	2,200
82nd	Ardennes/Rhineland	670	4,082	4,752
82nd	Holland	658	3,140	3,798
17th	Ardennes	691	3,000	3,691
17th	Rhine-Wesel	623	1,900	2,523
11th	Luzon	517	1,440	1,824
101st	Ardennes	482	2,449	2,931
1st Airborne (British)	Sicily	454	240	694
82nd	Italy	327	1,939	2,266
82nd	Sicily	206	810	1,016
11th	Leyte	97		
TOTALS		10,397	36,046	46,249

Sources:
Data for the 82nd Airborne are contained in W. Forrest Dawson, ED., *Saga of the All American* (Chicago: 82nd Airborne Association, Inc., 1946) and reprinted by Battery Press, Nashville, Tenn.) while those for the 101st Airborne are found in Leonard Rapport and Arthur Northwood, Jr., *Rendezvous with Destiny: A History of the 101st Airborne Division* (Fort Campbell, Ky.: 101st Airborne Association, 1948), pp. 820-21. For the 17th Airborne Division, see *Casualty Lists, 17th Airborne Division* (Washington, D.C.: Military Archives Division, 1947). The figures for the 11th Airborne Division are found in Major General Edward M. Flanagan, Jr., *The Angels: A History of the 11th Airborne Division* (Nashville: Battery Press, 1988), pp. 62; 134. G.G. Norton presents figures for the British Airborne in *The Red Devils* (New York: Hippocrene Books, 1984), p. 299. Because figures on missing and captures are less reliable, I have not included them.

Here, we see in fairly dramatic terms, the costs absorbed by the 82nd and 101st Divisions. Losses for the former at Normandy nearly reached the 42 percent mark while for the 101st they were considerably less—27 percent. Only the tragic events at Arnhem left greater losses; the British 1st Airborne lost more than 7,000 of 10,000 men. Approximately 1,350 were killed in action—the largest single Allied Airborne loss of the War. More than 3,000 were wounded. And, all this took place in nine days! These tragic losses were, however, exceeded by those of the German Airborne during thirteen days in Crete (May, 1941), when their one division had 3,764 fatalities, including many senior officers, and lost more than half of its air transport planes.

As Table 3 makes abundantly clear, 33 days in Normandy exacted the highest price in total Airborne casualties. Interestingly, the then inexperienced 101st and British 6th Airborne incurred fewer casualties than the more battle-tested 82nd. Although the British 6th Airborne was to remain on the line for another month after the American airborne returned to England its losses were similar to those of 101st with 820 killed and 2,709 wounded. It seems fair to say that the 82nd suffered around 50 percent casualties, the highest during any single American Airborne action of the War. While the combined casualties of the two U.S. Airborne Divisions in the Cotentin Peninsula constituted 41 percent of VII Corps' losses the Airborne made up but one-fourth of its personnel. Although considerable, the Holland losses (1,410 KIA) were not only much lower for both divisions but incurred over a period about twice as long (17 September-11 November 1944, for the 82nd and 17 September to 25 November 1944 for the 101st). While the 82nd's rate was about 26 percent, the 101st lost 22 percent of its men. Much of the action after the first ten days was essentially static, defensive small unit patrolling conducted under Montgomery's overall command.

If Normandy inflicted the most killed in action (2,505), the Ardennes was in second place with a total of 1,843 KIA among the three Airborne divisions. The Bulge also inflicted many more wounded and injured (9,531) than Holland (5,291) and because of the horrendous weather also caused substantial losses due to injuries and frost-bite, etc. But, the divisional rates were quite different from those of Holland with all three divisions losing a little less than a third of their men. It should also be pointed out that all four Airborne divisions lost varying but substantial numbers of men due to jump and glider injuries suffered in both training and combat. In some operations including Corregidor, Normandy, and Sicily and several smaller Airborne actions in the Pacific, numerous accidents, etc., killed or injured hundreds of men with rates in the Pacific in excess of 10 percent. Not the least was the shooting down by our own forces of 24 C-47's in the Sicily jump staged by the 504 PIR. In Normandy as in Varsity many if not most of the gliders simply crash landed injuring and killing scores of men. In each major operation considerable numbers of planes and glider flights were aborted on take-off or in flight to the drop and landing zones. And, of course, training accidents were commonplace as the Airborne learned its job.

REGIMENTAL LOSSES

In general, as shown in Table 4, regimental losses reflected those of the divisions of which they were a part, but at higher rates. By one account, losses of the 506th PIR led all and by a substantial number (714 KIA/DOW). By another, it was the 504th and 506th with the most KIA. The 505th PIR of the 82nd had the next largest number of casualties and was followed by the 504th PIR. Perhaps, the single most interesting finding is that one of the less experienced regiments (513th) had not only a high percentage of casualties but measured on a per day of combat rate may have led all parachute regiments with about 7 KIA per day. Other leading regiments experienced KIA losses of around half or less that figure or 3 to 4 per day. The 504th with 371 days in combat averaged 1.53 KIA per day, and the 508th with 154 days had 3.97 killed per day and 17.7 total casualties each day of its action. Of course, days in "combat" often includes being in various states of "reserve" so some casualty rates appear low, perhaps, misleadingly low. And, averages often conceal more than they reveal for they do not trace highly variable day-to-day losses of smaller units. And, of course, casualties include far more than killed-in-action.

Clark L. Archer, Ed., *Paratrooper's Odyssey: A History of the 517th Parachute Team* (Hudson, FL: 517th Parachute Team Association, 1985), p. 218. Figures in brackets indicate other estimates, usually a Regimental history.

One fact not found in Table 4 is made clearer in Table 5, i.e., varying regimental losses in both the same and different battles. Here we learn, perhaps, for the first time, that the single most costly engagement for any American Airborne Regiment was in Normandy and for the Regiments (507 and 508) in their first and for the 507 only affiliation with the 82nd Airborne Division. Some 303 officers and men of the 507 lost their lives, a loss in the same engagement exceeding that of the highly experienced 505 PIR by 117 men! This loss as well as that of the 508 (total casualties equalled 55%) was far greater than those suffered by any of the three Regiments of the inexperienced 101st in Normandy.

Table 4 REGIMENTAL CASUALTIES: KIA/DOW		
REGIMENT	KIA/DOW	WIA
501st (101st)	608	1,639 (Includes IIA)
502nd (101st)	465	1,590 (Includes IIA)
503th (Ind Regt.)	210	
504th (82nd)	626 (714)	
505th (82nd)	627 (544)	2,841
506th (101st)	641 (714)	1,735 (Includes IIA)
507th (82nd: 3rd03) (17th: 220th)	523	1,099
508th (82nd)	463 (614)	1,788 (Includes IIA, MIA)
509th (Ind Bat)	204	
511th (11th)	216	
513th (17th)	453	1,297
517th (Ind Regt.)	219	1,178
193rd GIR (17th)	106	1,213 (Includes IIA)
194th GIR (17th)	220	864 (Includes IIA)
TOTALS	5,581 (2,586)	15,244

Sources:

The National Military Archives Division of the National Archives supplies casualty lists for the following Regiments: 501; 502; 503; 504; 505; 507; 508; 511; 513; 193; 194. Unofficial sources include Leonard Rapport and Arthur Northwood, Jr., *Rendezvous with Destiny: A History of the 101st Airborne Division* (Fort Campbell, Ky.: 101st Airborne Division Association, 1948), pp. 820-21; Lou Hauptfleisch, "Devils in Baggy Pants," *The Static Line*, May, 1988, p. 35; Allen Langton, *Ready: The History of the 505th Parachute Infantry Regiment* (Indianapolis: Western Newspaper Publishing Co., 1986), pp. 133-37; *3rd Anniversary: 508 Parachute Infantry, 1945*, pp. 11-12, 35, 54, 58; Charles H. Doyle and Terrill Stewart, *Stand in the Door: The Wartime History of the Elite 509th Parachute Infantry Battalion* (Williamstown, PA: Phillips Publications, 1988), pp. 386-91;

This assault gun (75mm on a Mark IV chasis), shown here after the Wesel operation, took its share of American airborne casualties.

Table 5
REGIMENTAL LOSSES (KIA/DOW) IN INDIVIDUAL BATTLES

PIR	KIA/DOW	BATTLE
507	303	Normandy
5Pl Brig (British 6th)	268	Normandy
Glider BNS (British 6th)	267	Wesel
508	258 (334)	Normandy
513	247	Bulge
506	231	Normandy
504	217	Italy (Includes Anzio)
501	213	Normandy
3Pl Brig (British 6th)	207	Normandy
502	200	Normandy
513	190	Wesel
505	186	Normandy
506	176	Holland
501	172	Holland
502	162	Holland
501	158	Holland
327 Glider (101st)	158	Holland
Glider BNS (British 6th)	155	Normandy
508	146	Holland
503	143	Corregidor
505	144	Holland
194 Glider (17th)	140	Wesel
505	139	Bulge
508	139 (134)	Bulge
501	132	Bulge
517	124	Bulge
507	122	Wesel
504	119	Sicily
504	108	Holland
193 Glider (17th)	106	Bulge
502	103	Bulge
327 Glider (101st)	103	Normandy
506	102	Bulge
504	100	Bulge
507	98	Bulge
504	82	Sicily
517	81	France
194 Glider (17th)	80	Bulge
505	64	Sicily
504	43	Central Europe
505	35	Central Europe
508	28	Rhineland
TOTAL	6,249	

Sources: Figures for this Table were compiled from various unit histories and/or official Reports of the Military Archives Division, cited in previous tables. Data for the British units was provided by their Airborne museum and the Public Record Office as well as General R.N. Gale's, *With the 6th Airborne Division in Normandy* (London: Sampson Low, Marston, 1947), p. 160. The bracketed figures indicate different counts as found in Regimental histories. Some of these accounts list the names of the deceased and in the case of the 507, the battle as well and date of death. The 505 lists incidences of death and battles.

Another surprise in Table 5 is the fact that the third highest single battle losses were sustained by the 513th in its initial combat operations during the Bulge, a total of 247 KIA, plus another 17, or 263 (total casualty rate was 62 percent) if the Rhineland or closing phase of the battle of the Bulge is included. Once more, inexperience showed its ugly head. This is also suggested by the fact that the regiments of the 101st suffered their highest casualties in their first combat engagements at Normandy.

Inexperience is not, however, the sole cause of higher casualties. The 505 PIR of the 82nd Airborne had considerable experience in Sicily and Italy prior to the Normandy jump but sustained far more casualties in June of 1944, than on previous operations. The Germans were in a far better position to put up a tremendous and defensively skilled fight and did. One other illustration: after having had their baptism of fire in Normandy and losing a total of 800 men the 507 PIR lost but 98 KIA and a total of 303 WIA in the Bulge, but then lost 122 KIA in Varsity.

In short, varying casualties depend upon a variety of factors of which experience, especially, among the officers is paramount. Or, to illustrate another anomaly involving an artillery battalion, the 466 PFA of the 17th Airborne: The unit lost five killed-in-action during the 45 days in the Bulge, but during the first two hours on the drop-zone at Wesel lost 49 men killed in action, 52 wounded and 6 injured or a total of 107 from among its 400 men. The Battalion had the misfortune of being dropped in the midst of strong German positions without the drop-zone having been cleared as intended by the 513th which had been misdropped by some two miles. Had the 513th been dropped in its designated area it would have sustained more casualties but, perhaps, not as many because they were better prepared, trained, and equipped to defend themselves. Such were the misfortunes of war. Airborne operations during World War II were replete with such disasters and, many, were far worse, including the shooting down by friendly forces of 24 C-47's carrying two battalions of the 504 and an artillery battalion to Sicily on 11 July 1943. The 504, numbering 1,900 men, lost 229 of whom 82 were killed and 132 wounded. Ninety air crew became casualties.

No matter how casualties are sustained each reduces the fighting effectiveness of a unit. In this connection it should, again, be pointed out that Airborne divisions were particularly prone to incur injuries and no more so than with the five glider regiments. With the possible exception of the initial Holland operations, gliders, their pilots, and the glider infantry sustained substantial injuries and damages. Merely landing a glider safely in unknown territory takes some doing and no inconsiderable amount of luck. In Varsity, for example, less than two percent of the 1,305 gliders used were deemed salvageable and recovered for future use. Even before gliders land they and their occupants must face danger in transit to the land-zones. In each of the major operations, glider flights were aborted because of malfunctioning in towing, equipment failures, bad weather, and human error. In Normandy, for example, while only eight glider flights on D-Day were aborted in the short flight from England to the landing zones, in Holland—a much longer distance—a total of 24 glider flights were aborted during the first two days and another 139 were lost during the resupply missions through the next two weeks. During Varsity, the 17th Airborne's glider regiment (194th) lost 18 gliders in transit and another 23 to enemy fire while the British 6th Airborne Division had 35 glider flights aborted, ten shot down and 284 riddled by machine guns, killing 38 of their pilots and wounding another 38. Only 80 or 20 percent of the gliders landed untouched. And, even if a glider escaped enemy fire it had to land on small, rough, irregular fields, often at night as was true during the operations of 1943 and 1944. Confronting small drop zones, hedgerows, woods, farm buildings, power lines, etc., was far more productive of excitement than a sense of security. Since the glider was not very maneuverable and the landing zones of the Pacific Islands pitifully small they were employed only once (Camalamiugan on 23 June 1945) when seven gliders carried equipment.

Regimental losses, as to be expected, constituted the bulk of Airborne divisional casualties; casualty rates among the other components of a division were considerably smaller in almost all operations. Thus, a typical parachute regiment numbered a little over 2,000 men or about one-fifth to one-sixth of the strength of the division, but each of those regiments sustained a far larger percentage of the total casualties. It could not be otherwise; the regiments formed the fighting core of the division. The 513th, for example, constituted about 17 percent of the 17th Airborne Division during the Ardennes fighting but sustained over a third of the Division's killed-in-action and nearly half of its wounded. Its casualties for the Varsity operation were smaller but the proportions remained roughly similar.

These observations may be made more concrete by an examination of Tables 6 and 7 which list casualties within the 17th and 101st—divisions. Each of these divisions experienced similar unit distributions with the parachute regiments incurring the highest casualties, then glider regiments, and various support and administrative units incurring the fewest.

Table 6
101ST AIRBORNE DIVISION: KIA

UNIT	NORMANDY	HOLLAND	BULGE	TOTAL
501st PIR	213	172	132	517
502nd PIR	200	162	103	465
506th PIR	231	176	102	509
327th GIR	103	158	84	345
377th PFA Bn	59	12	5	76
463rd PFA Bn			6	6
321st GFA Bn	0	13	2	15
907th GFA Bn	3	12	3	18
81st AA Bn	13	16	8	37
326th Eng Bn	16	21	27	64
326th Med Co	13	3	1	17
426th QM Co	1	3	0	4
801st Ord Co	0	1	0	1
101st Sig Co	4	1	3	8
Div MP Pl	4	0	1	5
Div Hdq & Hdq Co	7	1	5	13
TOTALS	868	752	482	2,100

Source: Leonard Rapport and Arthur Northwood, Jr., *Rendezvous with Destiny: A History of the 101st Airborne Division* (Fort Campbell, Ky: 101st Airborne Division Association, 1948), pp. 820-21. The figures for the Bulge are on the low side since they were counted only thru January 14, 1945. The Division fought on in the Bulge and later in Alsace well into February, 1945.

further polished in the highlands and jungles of South Vietnam during the late 1960's and early 70's. But things were never quite the same; no full divisions were committed in either war and the jumps were not spectacular ones.

In Korea the 187th Regimental Combat Team consisting of nearly 4,000 men made two jumps, one at Sunchon on 20 October 1950, and the other, at Munsani-ni on 17 March 1950.

Table 7
17TH AIRBORNE DIVISION: KIA

UNIT	ARDENNES/RHINELAND	WESEL	TOTAL
513 PIR	263	190	453
507 PIR	98	122	220
193 GIR	121	(Absorbed into 194 GIR)	121
194 GIR	80	140	220
550 Abn Inf. Bn	46	(Absorbed into 194 GIR)	46
464 PFA	0	9	9
466 PFA	6	51	57
680 GFA	3	20	23
681 GFA		23	23
155 AA Bn	31	15	46
139 Engr	12	13	25
224 Med Co	0	15	15
411 QM Co	0	3	3
17th Para M Co	0	1	1
517th Abn Sig Co	0	0	0
17th Div MP Co	0	0	0
Div Hdq Art	1	7	8
17th Div Hdq	3	5	8
TOTALS	664	614	1,278

Source: This table is based on figures contained in various casualty lists for the Division and its units and supplied by John Manning, Assistant Chief, Mortuary Affairs, Department of the Army and by National Military Archives. At the time of writing this report, I was unable to account for 107 deaths.

As always, there were exceptions due to exceptional circumstances. For example, while most Airborne artillery units (generally, battalions of about 400 men), had low casualty rates some did incur vastly greater ones as was true of the 377th (101st) during Normandy where substantial numbers were killed and wounded and as noted, above, with the 466 (17th) during the Varsity operation. It should also be pointed out that the medical companies and engineer battalions not infrequently sustained substantial losses. Such losses were more often incurred on airborne than purely infantry missions. Such units might have been misdropped or scattered and, therefore, more susceptible to heavier losses. In general, it would appear that the internal distributions of casualties among Airborne divisions was not at variance with what is generally known about the distributions within infantry, armour, and Marine divisions.

KOREA AND SOUTH VIETNAM

The mystique of the Airborne was confirmed but five years after World War II on the rugged hills of Korea and its luster

Neither operation achieved its goal of entrapping a large, retreating enemy force but both were conducted with considerable efficiency. In the three years that the 187th was in Korea three of its men won the Congressional Medal of Honor and it suffered 2,197 casualties, of whom 492 were killed-in-action and 1,705 wounded.

Because South Vietnam was not well suited for airborne warfare but one jump was conducted on 22 February 1967, by the 173rd Airborne Brigade, a brigade that was to be the first American Army unit committed to battle. The 173rd (5,000 men and officers) arrived in May, 1965 and did not leave until August, 1971. During those nearly six and one half years twelve of the 35,000 paratroopers who served earned Congressional Medals of Honor (10 posthumous), a partial reward for 1,533 killed-in-action and another 6,000 plus wounded for an overall casualty rate of 22 percent. At Dak To in 1966, the Brigade had 833 casualties of whom 191 were KIA. Put another way, the 173rd made up less than one percent of the Army but had 6 percent of its casualties. In addition to the 173rd, a former Regiment in the 82nd Airborne Division—the 505—was committed as the 505 Airborne Infantry Brigade from February, 1968 to December, 1969. Its casualties included 184 killed-in-action and 1,009 wounded in action. In total, the Airborne constituted one percent (40,000) of the Army yet sustained 8,726 casualties or about 6 percent its casualties.

On average, the casualty rates in Korea and South Vietnam

were lower than during World War II and in the case of South Vietnam were caused less by the artillery, aircraft, tanks and mortars of World War II and more often by small arms fire. In fact, the Viet Cong and the North Vietnamese had little artillery, aircraft, or tanks. Casualties in Korea appear to have been inflicted by means similar to the Second World War. Then, too, dramatic improvements in the speed (helicopters) and quality of medical aid in both the Korean and Vietnam fighting served to decrease the number who died of wounds. A wounded man was never far from first-class medical facilities.

IN MEMORIAM

It is now fifty years since the Airborne was established in the United States Army. Since that time, paratroopers and glidermen have fought in three major wars, one minor one, and served the United Nations as peacekeeping forces in the Middle East. One of our divisions, the 82nd, has continued the proud tradition it established during this century's greatest and, surely, most noble conflict. American airborne forces participated, along with those of our allies as well as the German in establishing a combat record of unparalleled skill and devotion of duty. The maroon beret of the world's airborne is an assured symbol of prowess and courage on the fields of battle. No military force relishes the thought of going up against the "paras," the "Parachutists," "fallschirmjagers," "Rakkasans," "those devils in baggy pants," or the "Red Devils."

It was not long after World War II, that career military men and officers decided that without earning the wings and boots one had not arrived professionally and was less likely to win promotions and esteem. The paratrooper had come to replace the cavalryman as the gallant and romantic swashbuckling soldier of a bygone era. Many airborne commanders of World War II went on to distinguished careers at the head of American and other forces. And, some became notable civil advisers and even heads of leading private industrial firms. Those troopers who returned to more mundane civilian activities did not forget their past and in their quiet but proud ways preserved the Airborne mystique. The veterans of no other Army units are as well organized or active as the Airborne. That same long-lived pride is also true of Airborne veterans in Great Britain, France, Germany, and wherever the Airborne exists.

Of all the countless individual medals and unit awards bestowed upon the Airborne, perhaps, the single greatest tribute was extended by that able and most demanding of battlefield commanders, Field Marshal Montgomery of El Alamein:

"What manner of men who wear the maroon beret?"

They are firstly all volunteers and are then toughened by hard physical training. As a result they have that infectious optimism and that offensive eagerness which comes from physical well-being.

They have jumped from the air and by doing so have conquered fear.

Their duty lies in the van of the battle; they are proud of this honor and have never failed in any task. They have the highest standards in all things, whether it be skill in battle or smartness in the execution of all peace time duties. They have shown themselves to be tenacious and determined in defense as they are courageous in attack.

They are, in fact, men apart—every man an Emperor.

I have great affection for these men, who were my comrades-in-arms on many battlefields in the Second World War. And those occasions where I myself wear the maroon beret, I regard it as an outward sign of respect to grand fighters and good comrades.

Of all the factors which make for success in battle, the spirit of the Warrior is the most decisive. The spirit will be found in full measure in the men who wear the Maroon Beret."

Montgomery's esteem for the Airborne was shared not only by those who witnessed their deeds, but some who were not convinced of the need for such troops. No less than Generals Eisenhower and Bradley, themselves, skeptics, came to appreciate troops on whom they could depend to "beat the German." General Marshall, a friend and mentor of General Ridgway, was from the beginning a vigorous proponent of Airborne forces and dramatic new ways in which to employ them. This greatest of American military men was our very best friend.

His deep and lasting regard was shared by those against whom we fought. When at last at Arnhem, without ammunition the beleaguered British First Airborne had to lay down its arms, the German captors doffed their helmets and saluted those undaunted fighting men. We recall with poignant pride the tenacious defense of Bastogne, the valiant stand of the 82nd to the north, the valor at St. Mere Eglis, Carentan, the River Waal, Corregidor, Hamburger Hill, and Dak To. Or, General Middleton's (VII Corps) tribute to the 17th when fighting under the worst possible conditions during the Bulge and against the best armored troops the Germans could muster said the paratroopers were "aggressive to the point of recklessness." At the very same time, some other troops deserted their posts.

But such courage exacts its toll. Still, the losses and hurt were not without value. Terrible tyrannies were ended and two new nations-states (West Germany and Japan) emerged from them and the War to become vigorous democracies and trusted allies. Foreign citizens now care for the final resting places of our fallen comrades. The "Legend that was Born" in the hedgerows of Normandy was renewed and sustained on the frigid mountains of Korea and in the rice paddies and jungles of South Vietnam. And, it goes on, today, as before at Benning, Bragg, and Campbell.

Casualties were moved to the Battalion Aid Stations via a stretcher-rigged jeep.

187th Airborne RCT - Korea

Airborne Unit Histories

Members of the Parachute Test Platoon at Ft. Benning, 1940. (Courtesy of 101st Airborne Division Association)

The first enlisted paratrooper, Sgt. William "Red" King, stands at the far right as an early trainee gets a workout on the slide landing trainer. (Courtesy of 101st Airborne Division Association)

THE PARACHUTE TEST PLATOON

By William T. Ryder
BG, USA (Ret)

On 1 July 1940, at Ft. Benning, GA, a volunteer group of one officer and 48 enlisted men were assigned to the Infantry Board from the 29th Inf. Regt. for special duty as a Parachute Test Platoon and Air Infantry. This small group was destined to become an embryonic nucleus for today's airborne forces. I was its commander and what follows is an account of the Test Platoon's creation, development and disposition.

It all began on 2 January 1940, when the War Dept. directed the chief of Infantry to undertake a project in collaboration with the chief of Air Corps (then part of the Army) "to determine the feasibility, practicability, organization, command, control and development of equipment of parachute troops and air infantry."

Between January and June 1940, the Chief of Infantry planned for a Parachute Test Platoon at Benning under Infantry Board supervision, activated Air Corps development and procurement of suitable parachute troops and had the Infantry Board investigate parachute jumping techniques of the U.S. Forest Service in Lola National Park, MT.

On 25 June, the 29th Inf. Regt. at Benning was directed to designate one first or second lieutenant, six sergeants and 42 privates first class or privates, for special duty with the Infantry Board to work on organization, equipment and tactics for a parachute platoon. The directive specified that all designated would be "volunteers."

The 29th Inf. notified commanders that the War Dept. was contemplating organization of parachute troops and Air Infantry "to operate in rear of hostile lines on special missions" and that the regiment was directed to designate personnel for a Test Platoon. Personnel designated would receive "flying pay" (50 percent increase in base pay), but must be volunteers and have the following qualifications:

At least one year's service (preferably one enlistment), not over 30 years of age (preferably under 25), good physical condition (athletically inclined), desire to be transferred to parachute organizations formed, willing to ride in and jump from airplanes after instruction, unmarried desirable (and be personally recommended by their commanding officers).

I submitted my application, as a first lieutenant, from Co. D of the 29th Inf. Although my commander favorably endorsed the application and it noted, "I have had experience with Air Corp units, done research on Parachute Troops and am highly interested in their development," I was not optimistic about my chances for selection.

This was because over 30 eligible officers had applied; some had received Air Corps training at Kelly Field, and most were unmarried. I was not only married but the father of a baby daughter! But, by a combination of luck and initiative, I was selected and my name appeared on Special Orders No. 127. Hq. 29th Inf. Regt. dated 1 July 1940, detailing me and 48 enlisted men on special duty with the Infantry Board.

The luck derived from an earlier chance meeting with a fellow officer ordered to take a rifle squad to Lawson Field and practice recovery of weapons and equipment from canvas bundles being dropped by parachute from transport aircraft. I and a classmate, Bill Yarborough, hurried to Lawson Field to observe those drops. We quickly deduced that the Infantry Board had a project underway involving parachute troops. Guarded but helpful responses to questions we asked various Infantry Board members confirmed our suspicions.

Such a prospect excited me because it would provide an opportunity for Air Corps affiliation. I'd applied for the Air Corps at West Point but didn't qualify because of poor depth perception. However, I never lost an affinity for flying. During three previous years in the Philippines, I logged quite a few "observer" hours by soliciting flights with Air Corps classmates at Nichols Field. Thus, I could validly claim "experience with Air Corps units" in my parachute-troop application.

The chance meeting leading to my early observation of Infantry Board test drops triggered initiatives that later stood in my favor. Being convinced that some form of parachute activity was in the offing, I began voluntarily to submit memoranda on the subject to the Infantry Board. These included recommendations for training and equipping a parachute platoon, mission potentials for such forces, commentary on German airborne operations, and other parachute-related subjects. It was the presence of those memoranda in Infantry Board files, I learned later, that prompted my selection from among the extensive list of officer volunteers.

On 2 July, I reported to the Infantry Board and, with Colonels Taylor and Baldwin, began work on a TO&E (Table of Organization and Equipment) for a parachute platoon and company.

On 5 July, I assembled the platoon for the first time, for lectures on the type of training they might expect, interviews to verify personal interests and desires for the project and written examinations to establish intelligence levels. All men evidenced high enthusiasm. Six of them had requested voluntary reductions from corporal to private, to qualify as volunteers, because no corporal positions were authorized for the Test Platoon.

On 8 July, the platoon was given Air Corps physical examinations by a flight surgeon. Three men were lost due to physical disqualifications, but replacements were made immediately available from the 29th Inf. For the next two days, the platoon prepared a tent campsite at Lawson Field and prepared to move there from 29th Inf. quarters.

Concurrently, another officer was assigned to the platoon I'd requested, and the Board approved an assistant platoon leader to assure command continuity in the event of training injuries. I asked for Bill Yarborough, who had previously agreed to seek the other's assignment should either of us be selected for the Test Platoon. However, by then Bill had received orders to Camp Jackson, and 2nd

Lt. James Bassett, another volunteer, was assigned as second-in-command. Jim Bassett was a very serious, industrious and competent young officer. He was later to take a contingent to Chanute Field and organize a training infra-structure for early parachute packing and rigging instruction at Benning. Fortunately, Bill Yarborough was not lost to the parachute troops. He joined a few months later, when the 501st Parachute Bn. was activated. He was later to design our parachute "wings", the two-piece jump suit replacing the original one-piece jump suit, and command the first test section of the Provisional Parachute Group.

On 9 July, Warrant Officer Harry "Tug" Wilson and four riggers from Chanute Field arrived to instruct the platoon in parachute packing and jumping. An old hangar at Lawson Field was converted into a packing shed and training building. Four 43-foot packing tables and two suspended harness rigs were installed and work began on a "door mock-up" of a C-33 transport aircraft. Additionally, eight condemned T-3 parachute assemblies were provided to get us started on parachute panel folding and linestowing practice. With Tug Wilson's help, I prepared training schedules to accommodate physical training, parachute packing, jump and tactical training, along with orientation aerial flights.

By 11 July, the platoon was encamped at Lawson Field and ready to start its specialized training, and for the following weeks, worked intensely thereon. Here are brief descriptions of basic equipment, methods and training aids involved in the training that followed.

The T-4 main chute issued for our use was a modification of the Air Corps' 28-foot diameter, back-pack, rip-cord-actuated T-3 training parachute. It had 28 silk panels, an elastic banded "pucker vent" at its apex and was "static line" actuated. This was managed by stowing suspension lines and canopy on a wire-ridged, rectangular canvas frame that attached to the parachute's rear harness. A canvas pack-cover was placed over the stowed canopy and lines, its interior tied to the suspension lines by a break-cord and its edges sewed to those of the back-pack's canvas frame to completely enclose the canopy and lines. A static line sewn to the pack-cover exterior ended in a strong metal snap-fastener that could be secured to a shoulder-high cable, strung along the aircraft's interior between rear door and front-compartment bulkhead. Thus, when the jumper had left the plane and reached the end of the static line, it remained attached to the aircraft.

The jumper's forward motion (about 100 mph) and propeller blast fully deployed the canopy within two seconds of exit. The shock of that opening was quite severe and produced shoulder bruises when the jumper exited in a poor body position. Although faster descending and more susceptible to oscillations than larger parachutes later adopted, the old T-4 was a dependable workhorse that had the confidence of all who jumped with it. (Ed. Note: Due to the guaranteed bruises for anyone who jumped with it, as would its "shocking" replacement, the T-5. Newer chutes don't do this).

Unlike the British and German parachutists, we had a reserve chute that gave us the assurance, and insurance, of another chute in the unlikely event of main-chute failure or malfunction. The reserve was a separate 22-foot, front-pack, rip-cord-actuated chute in a rectangular canvas pack, with D rings allowing its attachment to the jumper's front harness. Later, when jumping with British parachutists, I missed the reserve's comforting presence.

Belief that personal packing enhanced jumper respect for, and confidence in, the parachute led us to require each man to pack his own chute for jumps. This entailed over 40 hours' instruction on panel folding, line stowing, twist removals, pack sewing and closing, tacking, harness adjustment, etc. Such practice was later abandoned and supplanted by "rigger" packing. But it did serve its purpose of instilling great confidence in the parachute as a reliable means of transport among our early parachute units.

Early jump training was simple, functional and suited to exit and landing attitudes then deemed most effective. A door mock-up of a C-33 transport aircraft was used for exit training, and suspended harness rigs for landing training.

For exit, jumpers were taught to crouch in the aircraft's door, with right foot slightly rear of left, and hands placed shoulder-high outside the door's surface. Pushing off the right foot and hand, the jumper was to leap out vigorously, turn to face the tail, tuck his head down, and bring his feet together. He was to hold the reserve firmly to his body with both arms and count aloud, "ONE THOUSAND - TWO THOUSAND - THREE THOUSAND - FOUR THOUSAND." If he had not felt the opening shock by the end of that four-second interval, he was to pull the reserve's ripcord and manually assist the canopy deployment if required. Vigorous exits were stressed, because weak exits risked poor body positions or even possible foulings on parts of the door or plane's fuselage. This could cause premature canopy exposure which could result in its entanglement on the planes' tail surface. With a proper exit, such entanglement was impossible.

A correct landing position called for the jumper to "prepare for landing" about 100 feet off the ground, by assuring that he faced his drift direction. If confronted by a backward drift or a back-swing oscillation, he could turn his body immediately prior to impact, he was to position his feet, shoulder-width apart, pull down on his risers to lessen landing impact, and try to land on the balls of both feet simultaneously. Landing with knees flexed, he was to tumble in the direction of drift, regain his feet and collapse his canopy. This was no problem in still air; in a light wind, air could be spilled from the canopy by gathering the lower risers towards the body; in high wind, the jumper could run around the canopy to face its surface towards the wind, thus collapsing it. With canopy collapsed, the jumper removed his harness, recovered his equipment from a separately-dropped container and proceeded on his mission. [Ed. Note: Certain elements of the above procedure have been changed over the years.]

Parachute manipulation, body manipulation, body turns, landing attitudes, etc. were taught on a suspended harness rig. This was a harness with seven-foot risers attached to a four-foot diameter iron ring. The ring, suspended from roof or rafter by rope and pulley, could be hauled aloft until the harnessed jumper was about three feet off the ground. In that suspended position he responded to manipulation commands and, on being released, dropped to the ground to perform correct landing procedures.

Our jump uniform was the Air Corps one-piece twill coverall, a canvas flying helmet with chin strap and jump boots with reinforced ankle sections akin to those used by the Forest Service smoke-jumpers.

To supplement training with these basic aids, the platoon went through special physical exercises, jumps from six- and-ten-feet platforms, forced runs and marches to strengthen legs and ankles and recurrent push-ups to build arm and shoulder strength. Initial training focused almost exclusively on packing, jump training and physical exercises.

In late July, the platoon went to

Hightstown, NJ, for a week's training on Commander Strong's two parachute towers. One was a "controlled" tower, from which an opened descending canopy was guided down along cables to a controlled landing. The other was a "free" tower, which enabled the jumper to be hauled aloft with an open canopy secured to a large wooden ring. At the tower's top an automatic mechanism released the parachute, the canopy pulled loose from the ring and the jumper made a free descent and landing. At Hightstown the men had 15 tower jumps each and continued tumbling, landing drills and exercises in preparation for live jumps on our return to Benning.

By Monday, 5 August, the platoon was back at Benning, packing parachutes for the initial jump. By then we'd received a full quota of T-4 parachute assemblies and issued one to each platoon member. The Air Corps had provided two C-33 cargo aircraft. This military-configured DC-3 had removable metal-frame seats so it could be used for passenger or cargo hauls. During our first two weeks we'd witnessed numerous T-4 demonstration drops with wind dummies, and one of the Air Corps riggers (Cpl. Wallace) had made a live T-4 jump for the platoon's benefit. So physically, technically and psychologically, the platoon was ready to go!

I'd organized our novice jumps by squads in numerical sequence, with me jumping first and Bassett jumping last. Between 5-8 August, we carefully packed our respective chutes and were set to begin jumps the next day. But in the late afternoon of 8 August, a telephone call from the Materiel Command at Wright Field directed that I cancel planned jumps and return all chutes to Wright Field for reinforced harness stitching!

That afternoon at Wright Field, during some T-4 test drops with wind dummies, a dummy had fallen out of its harness and plunged to the ground without a parachute. The dummy was a weighted, hemp-filled canvas bag, configured in human-torso shape to take a parachute harness. It was used for test-dropping new parachutes and as a wind indicator. In this case, some stitching in the harness had given way, allowing the dummy to pull free on opening shock. Taking no chances that one of our chutes might be defective, the Materiel Command decided to re-inspect all chutes and therefore, some supplemental stitching reinforcement was ordered as a safeguard. Since such work would not be done at Benning or nearby Maxwell Field, the chutes had to be sent to Wright the next day. The Materiel Command expressed regrets but promised to put emergency crews on the job, work through the weekend and return the chutes as soon as possible.

This unexpected turn was a great disappointment as well as a psychological blow for me. But, next day I explained matters as diplomatically as I could to the platoon, assuring them we'd soon have our chutes back in perfect condition. Since a weekend was coming up, I gave the men a weekend pass to ease the situation.

The Materiel Command was as good as its word. On Sunday, 11 August, about half our chutes were returned with the remainder to follow the next day. But alas, the chutes were not returned in the squad order by which they'd been issued to the men. Accordingly, I revised the planned squad jumps and decided to let the men draw lots for a jumping order, with me still making the first jump and Bassett the last. Numbered slips were placed in a box (with the number 13 removed), and the men drew for a jump position. Available chutes were re-issued accordingly, and harnesses were adjusted as required.

On Tuesday, 13 August, the first jump unit, myself and 10 men, had completed packing our chutes and were ready for our novice jump that afternoon, Again, I was to encounter another disappointment, a drastic but short-lived psychological blow, when the man who had drawn the No. 1 jump position refused to jump! This is what happened.

Tug Wilson was jump master for all of our novice jumps, which were individual jumps from 1,500 feet, the altitude prescribed for Air Corps training jumps. Tug would drop a wind dummy as our aircraft drew abreast of two red ground panels on its first pass over the turf-surfaced Lawson Field. Ground crews would move the panels to adjust for the dummy's drift, and the pilot would adjust his next flight path to spot the jumper in the desired landing area.

Although Tug was jump master, I wanted to be at the door with him as each man made his first jump. This would not only give me a chance to observe closely each man's reactions and performance, but somehow it seemed a proper and fitting thing to do. Accordingly, it was arranged that after I'd jumped, the plane would land and pick me up so that I could be with each jumper as planned.

We had a beautiful, quiet summer afternoon for that first jump. We emplaned, seated ourselves somewhat awkwardly on the metal-framed kapok-padded seats and had a smooth takeoff for our eventful flight. Our C-33 gained altitude, made a wide circling turn to the left, passed over nearby Columbus, GA, and leveled off for its wind-dummy run over Lawson Field. Tug pushed "Oscar" (the wind dummy) out the door, right on target and such was his smooth descent that no panel adjustment was necessary.

On the next run in, about five miles from the field, Tug called me to the door. I stood up, hooked up, checked equipment and on Tug's order took the prescribed position in the door. I felt the prop blast rushing past, tugging at my coveralls and flattening my helmet against my cheek. Looking ahead I saw the red ground panels and felt the adrenalin rise, as I awaited Tug's shout of, "Go!" and the whack on the behind that were the exit commands.

When they came I leapt out, felt the prop blast full at my rear and began my count. Before I completed "TWO THOUSAND," I felt that violent but welcome opening shock, glanced up and saw my opened chute.

It's difficult to do justice to the elation one feels the first time he looks up into a fully-deployed canopy and down toward the silent and seemingly unmoving earth beneath one's feet. There is a joyous and almost dreamlike thrill of personal flight as one hangs and swings gently in a sunlit silence. I heard the soft, sibilant sound of air swishing around the skirt of my canopy. I felt an overwhelming desire to hover there indefinitely. But all too soon it was over. I noted the earth beginning to move with increasing rapidity beneath my feet as I approached the ground. A slight pull on the riser enabled me to align my landing direction. About 50 feet off the ground, I reached up, grasped the riser, pulled down and landed lightly on my novice jump.

Almost immediately thereafter the plane landed, and I re-embarked to join the first jump unit. As we climbed and circled for the next jumper, I conveyed to the men, as best I could, the elation, satisfaction and confidence I'd experienced from my jump. As we headed towards Lawson, I joined Tug by the door and, with eager anticipation, awaited the historic occasion of the first enlisted man's qualifying "parachutist" jump.

All went smoothly. Tug called the man to the door; he stood up and moved quietly to join us. Properly he hooked up, checked equipment and calmly crouched in the door, awaiting Tug's command. But when it came, he remained immobile. Tug again shouted, "Go!" and slapped his backside, thinking that perhaps the first command had not been heard. But the man remained fixed and made no move. Quietly we removed him from the door and placed him in a rear seat. It was vital that every parachutist jump willingly and with determination, on his own volition. A hesitant or balky jumper in combat could be a disaster.

At that moment my elation was deflated and I was engulfed by the depressing thought that there might be other refusals within the platoon. But those fears proved groundless. "Red" King, one of the men who had taken a "bust" from rank of corporal to be an eligible volunteer, had drawn the No. 2 position. Red was a flamboyant, devil-may-care type whose enthusiasm and vitality had boosted morale throughout our training. He moved smartly to the door when called and on Tug's command, leaped immediately and vigorously from the door, thus earning the distinction of being the first U.S. Army soldier (enlisted man) to jump as a parachutist.

Units 2, 3 and 4 jumped thereafter on successive days with no evidence of doubt on the part of any man. Accordingly, by 16 August, the Parachute Test Platoon had completed its novice jumps.

During the next six weeks we managed a jump per week, except for a week in early September, when we underwent a concentrated demolitions course. These included both mass and individual jumps made from lower altitudes. We were relieved from the 1,500-foot altitude constraint prescribed for Air Corps training jumps, and allowed to jump at 750 feet to obtain tighter landing patterns in our mass jumps. Accordingly, by late September 1940, Test Platoon members had completed the five individual/mass jumps that were to become the qualification standard for a parachutist's rating.

The platoon made its first "demonstration" jump at Lawson Field on 29 August. It involved a mock attack on two old "enemy" hangars at the far end of the field, from the Base Operations Office. Our planes flew a good tight formation; jumpers exited exactly on time; and equipment bundles dropped perfectly from the equipment plane. However, a freak thermal and shifting wind landed one man (Pvt. Leo Brown) right on top one of the hangars! Brown managed a stand-up landing and was unhurt. The remainder of the platoon gathered their rifles and light air-cooled machine guns and proceeded to take the objective in short order, ignoring Brown and his awkward position. Later, I was pleasantly surprised to hear that some visiting South American officers who observed the exercise went away doubly impressed by the accuracy of U.S. parachutists who could land men on top of buildings. They assumed this was part of the attack plan.

Also during August and September, the platoon experimented with jumps from B-18 aircraft, worked on improving container drops and test-jumped new "Pioneer" T-4 chutes. On 21 September, Lt. Jim Bassett took 25 men to Chanute Field for a month's course in parachute repair and maintenance. I remained at Benning with the remainder of the platoon and began preparations for us to serve as

Suspended harness training. (Courtesy of 101st Airborne Division Association)

The parachute landing fall replaced the tumble demonstrated here by the parachute pioneers. (Courtesy of the 101st Airborne Division Association)

jump-training instructors for the 501st Parachute Bn. soon to be activated.

On 1 October, Major "Bud" Miley, formerly Master of the Sword at West Point (and later as major general, the Commander of the 17th Airborne Division,) was appointed to command the 501st Parachute Bn. On 7 October the commanding officer, Ft. Benning, GA, announced its constitution and activation at that post. The Test Platoon was assigned, in its entirety, as a nucleus group for this first parachute battalion.

This then, was the end and disposition of the U.S. Army's first Parachute Test Platoon. During its brief three-month life span, it had developed methods and techniques for training men to jump and pack parachutes, fixed standards and requirements for parachutist qualifications and begun the long, extensive train of events that were to foster the development to today's Airborne forces.

Secretary of the Army John Marsh commended the platoon's efforts. The citation read, in part:

"The Test Platoon, Parachute Troops and Air Infantry, United States Army, is commended for meritorious conduct in the performance of hazardous service from July to September 1940...It pioneered experimentation to determine the feasibility of employing paratroopers in modern warfare...Long and laborious application to dangerous assignments was necessary to carry out the tactical experiments...Although several men were injured, the steadfastness and loyalty of purpose of every rank never faltered. The intricate problems of parachute technique were solved; special parachute equipment was designed, test-jumped and refined; and the organization of minor tactics of parachute infantry established. Highly successful airborne operations in all theatres of operations throughout the war attest to the achievements of the Test Platoon."

Since 1963, surviving members of the Test Platoon have gathered annually at Ft. Benning for reunions. Here they are warmly welcomed and received. Here, their Airborne Spirit is renewed and revitalized by the esprit, enthusiasm and dedications encountered in today's young paratroopers.

And here, in familiar surroundings, they can vividly recall their shared pride and satisfaction of a half-century ago, when they were designated as volunteer members of the U.S. Army's first Test Platoon Parachute Troops and Air Infantry.

11TH AIRBORNE DIVISION

By Edward M. Flanagan Jr.
Lt. Gen. AUS (Ret)

SYNOPTIC HISTORY

ACTIVATION
25 February 1943 at Camp Mackall, N.C.

CAMPAIGN CREDITS
New Guinea, Leyte and Luzon; Arrowhead on Pacific Campaign Ribbon for amphibious and parachute landings on Luzon.

UNIT CITATIONS, AWARDS AND DECORATIONS
The following units of the 11th Airborne Division were awarded Distinguished Unit Citations for "heroism and outstanding performance of duty during the period 31 January to 5 February 1945" (the Invasion of Southern Luzon):
Headquarters & Headquarters Company
1st Battalion, 187th Glider Infantry Regiment
Headquarters & Headquarters Company, 188th Glider Infantry Regiment
1st Battalion, 188th Glider Infantry Regiment
2d Battalion, 188th Glider Infantry Regiment
Headquarters & Headquarters Company, 511th Parachute Infantry Regiment
1st Battalion, 511th Parachute Infantry Regiment
2d Battalion, 511th Parachute Infantry Regiment
3d Battalion, 511th Parachute Infantry Regiment
Air Section, 457th Parachute Field Artillery Battalion
Battery D, 457th Parachute Field Artillery Battalion
674th Glider Field Artillery Battalion
675th Glider Field Artillery Battalion
511th Airborne Signal Company

The following units were awarded Distinguished Unit Citations and were cited for outstanding performance of duty in the Los Banos Raid on Luzon on 23 February 1945:
Provisional Reconnaissance Platoon
Company B, 511th Parachute Infantry Regiment

MEDAL OF HONOR RECIPIENTS:
1. Private Elmer Fryar, E Company, 511th Parachute Infantry Regiment was awarded the Medal of Honor for conspicuous gallantry on Leyte on 8 December 1944. Pvt. Fryar exposed himself to the direct fire of a machine gun on an early morning Japanese Banzai attack; then, under heavy enemy fire recovered a wounded sergeant outside the perimeter; and later in the day purposely stepped in front of his platoon leader and took a fatal burst of automatic fire in his chest from a sniper whom he spotted at the last moment.
2. Private First Class Manuel Perez, Jr., A Company, 511th Parachute Infantry Regiment was awarded the Medal of Honor for "Conspicuous gallantry and intrepidity at the risk of his life and beyond the call of duty near Fort McKinley, Luzon on 17 February 1945. Pfc Perez single-handedly attacked a heavily defended Japanese pillbox and killed 18 enemy soldiers with his own rifle, grenades, and a Japanese rifle and bayonet which an enemy soldier had javelined at him, knocking his own rifle to the ground. Perez neutralized the position which had been holding up his entire company. Unfortunately, he was killed a week later near Fort McKinley.

DISTINGUISHED SERVICE CROSS RECIPIENTS:
1. Sgt. Pat Berardi*, 511th PIR
2. 1st Ltg. Henry G. Hynds, 188th GIR
3. Major Charles P. Loeper*, 188th GIR
4. 2d Lt. Mills T. Lowe, 511th PIR
5. S/Sgt. Edward B. Reed, 511th PIR
6. Colonel Irving R. Schimmelpfennig*, Division Headquarters
7. Major General Joseph M. Swing, Division Headquarters
8. Pfc Joe R. Siedenburg*, 187th GIR
10. T/Sgt. Robert C. Steele*, 511th PIR
(* indicates posthumous awards)

COMMANDING OFFICERS
11th Airborne Division: Maj. General Joseph M. Swing (from activation until 2 February 1948, a near record for length of command of the same division; retired as a Lt. General)
Assistant Division Commander: BG Albert Pierson; retired as a Maj. General
Division Artillery Commander: BG Francis W. Farrell; retired Lt. General
511th PIR: Colonel Orin D. Haugen (KIA 11 February 1945, near Manila); Colonel Edward H. Lahti (until August 1947; retired Colonel)
1st Battalion, 511th: Lt. Colonel Ernest LaFlamme, retired Colonel; Major Henry A. Burgess
2d Battalion, 511th: Lt. Colonel Norman M. Shipley (WIA Leyte); Lt. Col. Frank S. Holcombe
3d Battalion, 511th: Lt. Colonel Edward H. Lahti; Lt. Colonel John Strong
187th GIR: Colonel Harry Hildebrand; Colonel George Pearson, retired Maj. General
1st Battalion, 187th: Lt. Colonel George Pearson; Lt. Colonel Harry Wilson, retired Colonel
2d Battalion, 187th: Lt. Colonel Harry Wilson; Lt. Colonel Norman E. Tipton, retired Colonel
188th GIR: Colonel Robert H. Soule, retired Maj. General; Colonel Norman E. Tipton
1st Battalion, 188th: Lt. Colonel Ernest LaFlamme
2d Battalion, 188th: Lt. Colonel Thomas L. Mann
457th PFA Bn: Lt. Colonel Douglass P. Quandt, retired Maj. General; Lt. Colonel Nicholas G. Stadtherr
674th GFA Bn: Lt. Colonel Lukas E. Hoska, Jr., retired Colonel
675th GFA Bn: Lt. Colonel Ernest L. Massad
152d Abn AA Bn: Lt. Colonel James Farren
127th Abn Engr Bn: Lt. Colonel Douglas C. Davis

CASUALTY SUMMARY

Leyte: Approximately 130 men, KIA; approximately 421, WIA

Luzon: KIA, 445; DOW, 72; WIA, 1440; MIA, 3

Enemy: KIA, Leyte: 5,760; captured, 12,

KIA, Luzon: 9,458; captured, 128

TRAINING

The 11th Airborne Division trained at Camp Mackall, NC (and other locations—Fort Benning, Camp Davis, Laurenburg-Maxton Air Base, Fort Bragg—for parachute, glider, artillery and AA training) from 25 February 1943 until 1 January 1944; at Camp Polk, LA from 1 January 1944 until 20 April 1944; in New Guinea from June 1944 until October 1944.

CREST

Dark blue shield with white, double-winged center circle enclosed in a red center with a white "11." Matching blue tab with white "AIRBORNE" lettering.

INDIVIDUAL DECORATIONS

CMH, 2
DSC, 10
SS, 483
LM, 10
SM, 63
AM, 108
PH, 2,510
(No data on Bronze Stars)

CHRONOLOGY

General Orders No. 1, Headquarters 11th Airborne Division, signed by Major General Joseph M. Swing, dated 25 February 1943, contained the following information:

Authority: War Department Letter dated 24 November 1942;

Location: Camp Mackall, North Carolina;

Date at time of Activation: 0001, 25 February 1943;

Commanding General: Major General Joseph M. Swing;

Assistant Division Commander: Brigadier General Albert Pierson;

Division Artillery Commander; Brigadier General Wyburn D. Brown;

Chief of Staff: Colonel Francis W. Farrell.

Organic Units: Headquarters, Headquarters Company, Military Police Platoon, 408th Airborne Quartermaster Company, 511th Airborne Signal Company, 711th Airborne Ordnance Maintenance Company, 221st Airborne Medical Company, 127th Airborne Engineer Battalion, 152d Airborne Anti-aircraft Battalion, Headquarters and Headquarters Battery, 11th Airborne Division Artillery Band, 457th Parachute Field Artillery Battalion, 674th Glider Field Artillery Battalion, 675th Glider Field Artillery Battalion, 187th Glider Infantry Regiment, 188th Glider Infantry Regiment, 511th Parachute Infantry Regiment Band.

Initial Regimental Commanders:

511th Parachute Infantry Regiment—Colonel Orin D. Haugen

187th Glider Infantry Regiment—Colonel Harry Hildebrand

188th Glider Infantry Regiment—Colonel Robert H. Soule

Significant Dates:

5 January 1943: Activation of 511th PIR;

25 February 1943: Activation of 11th Airborne Division;

February-December 1943: Jump schools at Benning for parachute units; basic, advanced and unit training for all units at Mackall; glider training for glider units; major unit maneuvers;

6 December 1943: Knollwood Maneuver during which the 11th Airborne proves the feasibility of the Airborne Division Concept by landing the entire division by parachute and glider on various DZs and LZs after a four-hour flight out over the Atlantic and back and maintaining itself in the field by aerial resupply for three days; the War Department thereafter retained airborne divisions rather than reverting to battalion-sized parachute units as favored initially by General Eisenhower;

1 January 1944: 11th Airborne Division Headquarters at Mackall burns to the ground;

2-10 January 1944: 11th Airborne Division moves in 22 trains to Camp Polk, LA;

January-April 1944: 11th Airborne trains and maneuvers in the Calcasieu Swamps; passes War Department final tactical tests and administrative and inspector general inspections; opens its own jump school at DeRitter Army Air Corps Base in keeping with General Swing's desires to train the entire Division to enter combat either by parachute or glider; and prepares for overseas movement—destination unknown;

20-28 April 1944: 11th Airborne moves by train to Camp Stoneman, CA;

2 May 1944: 11th Airborne starts overseas movement by sailing under the Golden Gate Bridge and out into the Pacific—destination still unknown to the majority of the troopers;

June 1944: 11th Airborne closes on Dobodura, New Guinea and starts to set up a base camp with pyramidal tents and palm tree day rooms and company masses amid the kunai grass and along the abandoned taxiways of an Air Corps base;

June-October 1944: Training in the heat and humidity of New Guinea; jump school continues; Division alerted but not committed for an operation in western New Guinea; parachute training jumps continue;

12 October 1944: Southwest Pacific Command orders 11th to move "administratively" to Leyte to prepare for a future operation on Luzon;

11 November 1944: 11th sails for Leyte in nine Navy APAs and two AKAs;

18 November 1944: 11th joins the KING II Operation (Code name for the Battle of Leyte) by landing unopposed on Bito Beach;

0700 21 November 1944: Colonel "Hard Rock" Haugen leads Lt. Colonel Ernie LaFlamme's 1st of the 511th from Bito Beach north to Dulag and Burauen for the initial commitment of an 11th Airborne Division unit to combat;

23 November 1944: Lt. Colonel Ed Lahti leads his 3d Battalion of the 511th to Dulag and Burauen;

24 November 1944: General Swing moves 11th CP from Bito Beach to San Pablo, near Burauen; he assigns Colonel Soule and his 188th the mission of securing the Division's and the 511th's southern flank from La Paz to Bugho; he assigns Colonel Hildebrand and the 187th the mission of protecting the Division base Bito Beach;

26 November 1944: Lt. Colonel Norman Shipley moves his 2d of the 511th to vicinity of Burauen; 511th assembles on Daguitan River bank not far from Burauen;

28 November 1944: Colonel Haugen and 511th relieve 7th Division in place and send "feeler" patrols into the mountains to the west of Burauen; the monsoon rains drench fields, roads, carabao paths, mountain trails and bivouac areas;

28 November 1944: Lt. Colonel Jim Farren moves 152d Abn AA-AT Battalion to San Pablo airstrips;

Late November and early December 1944: 511th PIR moves west from Burauen into the treacherous, unmapped Japanese-infested mountains which form a terrain block across the waist of Leyte; mission: dig out and kill the enemy in the mountains and reach the 7th Division on the west coast near Ormoc;

2 December 1944: Lt. Charles J. Kozlowski and his platoon from C Company of the 187th makes the division's first combat jump onto Manarawat in the hills of Leyte by parachuting successfully from six L-4s and L-5s;

4 December 1944: Lt. Milton R. Holloway jumps his A Battery of the 457th in 13 trips from one Air-Sea Rescue C-47 onto Manarawat;

6 December 1944: Japanese 3d Parachute Regiment drops between 300-350 paratroopers onto the San Pablo airstrips near 11th Airborne Division Headquarters;

December 1944: 511th PIR fights and slogs its way across the rain-soaked, muddy, jungle-covered mountains of Leyte; the 187th GIR, 674th GFAB, 127th AEB, Division troops clear Sa Pablo strips of the Japanese who jumped on the fields and others who came out of the mountains to link up with the Japanese paratroopers; 187th then joins 511th in the mountains; 188th fights along the Division's southern boundary; all Division units fight the rains, the enemy; some units, cut off, fight hunger and lack of supplies;

25 December 1944: 511th and Lt. Colonel Harry Wilson's 2d of the 187th reach Ormoc;

26-28 December 1944: 1st of the 187th, 2d of the 188th, A Battery of the 457th and Division recon. platoon defeat the stubborn, deeply dug-in Japanese in the battle of Purple Heart Hill;

9 January 1945: Sixth Army invades Luzon in Lingayen Gulf area;

15 January 1945: 11th's battle of Leyte is over; all units of the division reassemble at base camp on Bito Beach for rest, recuperation, retraining, rearming and re-equipping;

21 January 1945: 11th Airborne Division review in khaki's on Bito Beach;

January 1945: Division and unit commanders and staffs prepare for next combat operation—the invasion of Luzon amphibiously at Nasugbu by the majority of the glider unit of the division and by parachute on Tagaytay Ridge by the parachute units;

27 January 1945: 11th amphibious units (and part of the 511th RCT) sail from Bito Beach; 511th RCT units drop off on Mindoro;

31 January 1945: Lt. Colonel Ernie LaFlamme lead his 1/188th ashore at Nasugbu and begins the 11th's combat operations on Luzon; Lt. Colonel Tommy Mann and 2/188th, Lt. Colonel Harry Wilson and 1/187th, Lt. Colonel Norman Tipton and 2/187th, Lt. Colonel Luke Hoska and 674th, Lt. Colonel Ernie Massad and 675th, Division troops follow ashore at Nasugbu;

1-2 February 1945: Colonel Soule leads the 188th and 1/187th in Battle of Aga Defile along Highway 17 between Mt. Batulao and Mt. Cariliao;

3 February 1945: Colonel Haugen leads the 511th Infantry parachute assault on Tagaytay Ridge in two serials, one in the morning and one in the afternoon;

4 February 1945: Colonel Nick Stradtherr and the 457th PFA Battalion jump on Tagaytay Ridge; 511th moves out by truck and foot to southern Manila and advances rapidly to the Paranaque River Bridge; 11th Airborne now strung out from Nasugbu to Manila and Eighth Army's "reconnaissance in force" turns into the 11th Airborne's full-scale commitment in the battle for Manila; division beachhead now a few hundred yards wide and 69 miles long;

4-21 February 1945: 511th in the west and 187th and 188th in the center and the east, backed by the Division Artillery and Division support troops, attack the Genko Line, a formidable, 6,000-yard deep bastion of 1,200 concrete and steel pillboxes, studded with five-and six-inch guns, 150mm mortars, 20-, 40-, and 90mm AA guns, leveled against the attacking troops, manned by 6,000 Japanese of the Southern Unit, Manila Defense Force and stretching from Manila Bay east across Nichols Field to an anchor on the high ground of Mabato Point along Laguna de Bay;

11 February 1945: Colonel "Rock" Haugen mortally wounded by a low-trajectory Japanese 20mm. AA round at a battalion CP south of Manila; Colonel Ed Lahti assumes command of the 511th;

17-21 February 1945: 511th and 188th take Fort McKinley; Pierson Task Force (later Pearson Task Force) reduces fanatically-held Japanese position at Mabato Point;

23 February 1945: Major Henry Burgess and 1/511th, Division Recon. Platoon, Filipino guerrilla forces, D/457th, by parachute and amphibious assault, attack Los Banos Internment Camp 40 miles from Manila and 25 miles behind the enemy lines; Colonel Soule and a task force from the 188th move overland to block potential Japanese reinforcements; Burgess and his men kill and rout Japanese guards at the camp, rescue 2,122 starving civilians of many nationalities and return them by amtrac and trucks, before nightfall, to the safety of American lines at New Bilibid Prison in Manila;

24 February 1945: 11th Airborne starts campaign to clear Japanese from Southern Luzon; Plan:

187th GIR: attack and seize Tanauan; 511th PIR: attack toward San Tomas; 188th GIR (2d Battalion): reduce Ternate; 2/188th GIR: protect division supply lines; 675th GFA Bn: direct support of 187th; 674th GFA Bn: direct support of the 511th; 457th PFA Bn: direct support of 188th at Ternate; 472 FA Bn: reinforce fires of the 457th; XIV Corps Artillery: reinforce 6784th and 158th RCT;

1 March-3 April 1945: 1/188th reduces Pico de Loro and Ternate area;

15 March 1945: Colonel George Pearson assumes command of the 187th; Colonel Norman Tipton assumes command of the 188th; Colonel "Shorty" Soule promoted to brigadier general and becomes assistant division commander of the 38th Division;

March-April 1945: 511th seizes Lipa and Malaraya Hill; 511th, 188th and Ciceri Task Force, reinforced by the fires of seven artillery battalions, wipe out Japanese on Mt. Malepunyo; 187th, in a bitter, unorthodox battle, reduce Mt. Macolod, working from the top down rather than from the bottom up;

10 May 1945: Division commences regrouping at a base camp near Lipa; patrols and outposts harass remaining enemy in Southern Luzon; 6,000 replacements for 11th arrive;

11 May 1945: First day in 101 days on Luzon—since 31 January 1945—that the 11th did not kill an enemy soldier; preceding average, 93.8 Japanese per day;

May-June 1945: 11th Airborne Division operates jump school at Lipa;

May 1945: 11th Airborne transfers to a new TO and E: three battalions in 187th and 188th; 188th and 674th become parachute outfits; 472d FA Battalion becomes organic to division; strength of the division goes from 8,505 to 12,000 men;

23 June 1945: Gypsy Task Force—1/511th, G and I Companies; 511th, C/457th, Platoon of C/127th Engineers; 2d Platoon, 221 Medics; 511th Signal Team—under the command of Lt. Colonel Henry Burgess, jump at Aparri at the northern end of Luzon to seal off escape route of remaining Japanese in northern Luzon; included for the first time in the Pacific area were seven gliders, six CG 4As and one CG 13;

26 June 1945; Gypsy Task Force links up with 37th Division;

June 1945: Colonel Ducat McEntee and 541st Parachute Infantry Regiment join the 11th; 541st deactivated and its men sent to units throughout the 11th;

0430 11 August 1945: 11th Airborne alerted to move to Okinawa and Japan within 48 hours;

11 August 1945: 48-hour alert rescinded and 11th starts to move by air to Okinawa;

15 August 1945: bulk of the division closed on Okinawa; 11,100 men, 120 jeeps, 580 tons of equipment flown to Okinawa in 99 B-24s, 351 C-46s, 151 C-47s; monsoons and rains pelt Okinawa;

28 August 1945: Advanced party of 11th lands at Atsugi, Japan to begin the occupation;

0600 30 August 1945: General Swing lands in first plane of the main body of the 11th; on the first day of the move, 4,300 troops of the 11th in 123 C-54s land at Atsugi;

1400 30 August 1945: General MacArthur arrives at Atsugi and is greeted by General Swing and the 11th Airborne Division Band;

2 September 1945: Japanese sign the surrender document aboard the USS *Missouri* in Tokyo Bay;

30 August-7 September 1945: 11th Airborne Division makes the longest (1,600 miles) air transported movement in the war flying 11,708 men, 640 tons of equipment and 600 jeeps and trailers from the Philippines to Japan;

14 September 1945: 11th moves to occupation zones on northern Honshu and Hokkaido;

June 1946: Division zones of occupation: Division Headquarters, 187th and 511th Signal Company in Sapporo, Hokkaido; 152d at Muroran; 511th at Hachinoe, Morioka and Aomori; Division Artillery and 472d at Yamagata; 674th at Jimmachi; 457th at Akita; 675th at Yonezawa; 188th, 127th Engineers (-), 408th QM and Special Troops at Sendai;

24 June 1946: First dependents arrive in Japan;

January 1948: General Swing leaves the 11th and is succeeded in command by Maj. General William M. Miley;

23 March 1949: Major General Lemuel Matthewson assumes command of the 11th;

May 1949: 11th moves from Japan to Fort Campbell, KY; 188th inactivated;

1 September 1950: 187th and 674th formed into a separate airborne RCT and shipped to Korea; 188th reactivated;

September-December 1950: 11th trains 13,000 recalled enlisted reservists for the Korean War;

December 1950: Maj. General Lyman Lemnitzer assumes command of 11th;

2 March 1951: 503d Abn. Infantry Regiment reactivated and becomes part of the 11th;

November 1951: Brig. General Wayne C. Smith assumes command of the 11th;

15 January 1952: Brig. General Ridgely Gaither assumes command of the 11th;

April 1953: Maj. General Wayne C. Smith reassumes command of the 11th;

1954: 11th takes part in atomic tests at Camp Desert Rock, Nevada;

May 1955: Maj. General Derrill Mc Daniel assumes command of the 11th;

May 1956: 11th moves from Fort Campbell to Germany on Operation GYROSCOPE and replaces 5th Division in the Augsbury-Munich area;

October 1956: Maj. General Hugh P. Harris assumes command of the 11th;

1956: 11th goes pentomic—five infantry battle groups of five companies each supported by other branches in a similar five-sided configuration;

April 58: Maj. General RC Cooper assumes command of the 11th;

1 July 1958: 11th Airborne Division inactivated;

7 February 1963: 11th reactivated as the 11th Air Assault Division (Test) at Fort Benning, Georgia under the command of Maj. General Harry W.O. Kinnard to test the air mobility concept which would be used extensively in Vietnam in later years;

30 June 1965: 11th Air Assault Division inactivated and replaced by the 1st Cavalry Division (Airmobile).

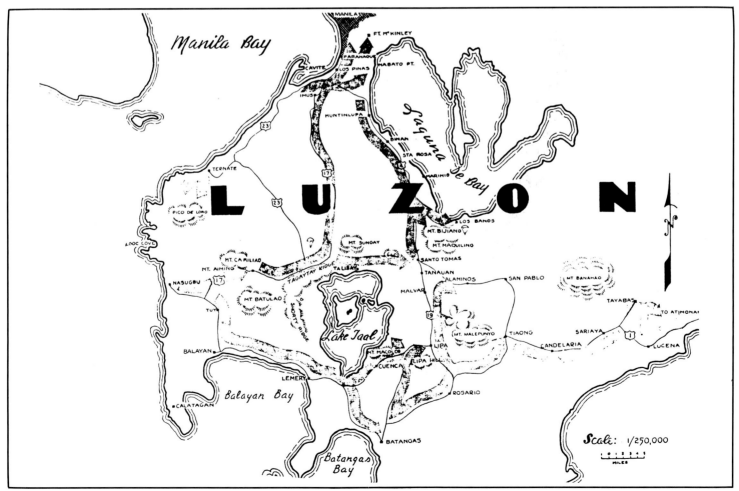

Operation Mike VI – 11th Airborne Division

COMBAT HISTORY

Two score and seven years ago, our Army fathers brought forth on this continent a new division, conceived in liberty and dedicated to the proposition that all men are created equal and that the division would fight—and die, if necessary—to insure the continuity of those profound and perfect propositions. That division was the 11th Airborne Division.

The division was born on 25 February 1943 at Camp Hoffman, soon to become Camp Mackall, North Carolina. Some units, the 511th Parachute Infantry Regiment and the 711th Ordnance Company, for examples, were activated before the 25 of February and thus qualified for unusual legitimacy—or at least priority at parades.

The original commander of the 11th Airborne Division, the man who would train it and fight it and lead it for six years, to include the occupation of Japan, was a tall, 49-year-old, white-haired, ramrod straight, slender professional Army officer and artilleryman named Joseph May Swing, Class of 1915 at West Point. Within a very few weeks, every man in the division would have seen General Swing in unlikely places, checking the training, the administration, the morale, the barracks, the messes, the motor pools, the sick calls, the general condition of "his" men in all aspects of their military lives. The Army had entrusted him with the health and welfare of some 8,000 young, untrained men, and he would see to it that they would be combat ready when the time came for their certain commitment to battle. General Swing was a firm believer in and practitioner of the truism that the harder one trained his men, the more of them would survive on future battlefields. The raw trainees at Mackall might not have thought it at the time, but General Swing was vitally interested in their welfare, well-being and survival on the field of battle. And he was convinced that his airborne division, one-third paratroopers and two-thirds glider men, should be cross-trained and capable of entering combat by either mode. He also correctly foresaw that his division, in manpower and in firepower only about half the size of a standard World War II infantry division, would be required to fight as if it were full size. He planned for these eventualities and when they came to pass, the 11th Airborne was ready.

No one who was with that green and untrained and even un-uniformed group of draftees who first detrained at the railroad station at Hamlet will ever forget the first ride in the backs of 2 1/2 ton trucks into camp—a camp still in the throes of completion, with bulldozers, trucks and graders swarming over the sandy terrain, leaving in their wake clouds of dust, denuded fields and gravel-coated roads. The streets, such as they were, were alternately muddy or dusty, the buildings were unfinished and the training areas were, for the most part, uncleared pine groves. And as the days wore on, the new soldiers, barely into their first set of fatigues and leggins, monstrosities which even the paratroopers of the 11th wore under protest during training, became more and more impressed with the mediocrity and inadequacy of the warped, tarpaper-covered shacks that the Army euphemistically called barracks, mess halls, supply rooms and headquarters buildings. Some men, used to more slightly elegant World War II type barracks, even thought that Mackall's "theatre of operations" type of construction was temporary and that the more permanent structures would be built later.

The original cadre and the fillers remember in detail some of the highlights of the days at Mackall. The stupidity of the battle of the coal boxes. The terror of the first parachute jump. The inherent queasiness of any ride in a glider. The regularity and steadily increasing length of the Friday afternoon "Swing Sessions" for all officers. The incessant and demanding training, running, shooting, training, artillery firing, maintenance, inspections, running, training. The well-attended boxing matches with Lt. Colonel Luke Hoska trying to separate over-zealous pugilists. The emphasis on organized team athletics. The smell of the theaters. The paucity of items in the PXs. The battalion messes. The beer halls. The hitch-hiking to the nearby towns and the race to get back for bed-checks. The maturity and the knowledge of the cadre. The youth of everyone, officers and senior NCOs included. The well-supervised Saturday night dances, at the All Purpose Recreational Halls (APRs), with young ladies in evening gowns and white gloves trucked in by Special Service officers from the local villages around Mackall.

It was during this period of the division's life when it acquired the nickname "Angels," not, as many thought later, at Los Banos when one of the interned nuns was alleged to have said that John Ringler and his B Company of the 511th, descending by parachute next to the camp, looked "like angels coming from Heaven to deliver us." That makes for a good story, but the nickname started much earlier.

Doug Quandt commanded the 457th Parachute FA Battalion from its activation until he left it in New Guinea in the summer of 1944 to become the division G-3. One Monday morning in the summer of 1943, Quandt, at a battery commanders' call, said to one of his BCs: "Well, Godsman, are all of your angels back in camp or are some of them in jail?" And to John Conable he said: "And what about your angels?"

The name persisted in the 457th and later spread to the division headquarters. At Polk, when the paratroopers were wont to remove unauthorized jump boots from "leg" tankers, bundle them up and deposit them in the tankers' orderly rooms, the 11th troopers' reputation for mischief and deviltry began to spread. In New Guinea when the 11th was called to work on the docks unloading equipment, it rapidly became known as "Joe Swing and his 8,000 thieves." General Swing is alleged to have defended his troopers to a superior by saying that "My angels would never get involved in such illegal shenanigans." And so the nickname was born.

The climax of the training at Mackall was the Knollwood Maneuvers in the black of the night of 6 December 1943 when the entire division, by parachute and glider, descended on drop and landing zones around Southern Pines, the Knollwood Airport, Fort Bragg and Pinehurst after a four-hour flight over the Atlantic and back. The troopers of the division did not know it at the time, but the War Department staged the Knollwood Maneuver to prove or discount the validity of the airborne division as an entity.

The value of an airborne unit the size of the division was in jeopardy after the debacle of the 82d in Sicily, where the Navy and ground troopers' ack-ack shot down scores of transports and scattered men and equipment over hundreds of square miles of Sicily's terrain. The War Department had grave doubts about retaining the airborne division organization. General Eisenhower, in fact, went on record as favoring the abandonment of the airborne division concept and using only battalions of paratroopers scattered in wide areas ahead of an invading force against limited objectives. The 11th's success at Knollwood, dropping in and maintaining itself for three days entirely by air, proved to the very high ranking War Department officials present at the drop, to include the Under Secretary of War, Robert Patterson and General Lesley J. McNair, the commanding general of the Army Ground Forces, that the concept of an airborne division was worthy and that the airborne division should be retained in the force structure.

The Knollwood Maneuver reinforced General Swing's belief that, eventually, the 11th Airborne Division should be totally airborne—glider units parachutable and parachute units gliderable. He reasoned that this dual capability of the entire division would give him a valuable tactical flexibility. Consequently, General Swing established jump schools adjacent to the division area wherever possible.

After Mackall came Polk, the Calcasieu Swamps, more tests, a new jump school and, at the end of the Polk sojourn, final exams and graduation. During final training maneuver in the swamps outside Polk, the umpires from higher headquarters found the division combat ready. The men of the division were not surprised at their success. They simply wondered why it had taken so long to send such a superb bunch of troopers into combat.

The train ride to Stoneman, the cruise under the Golden Gate Bridge, the monotony of the 27 days across the Pacific on various Merchant Marine ships—the heat, and boredom, the thirst and the grim food—took the division to Dobodura, New Guinea. Bamboo-framed pyramidal tents; Kunai grass on the drop zones; long jungle marches and hot and humid unit training and combined firing exercises; USO shows with Jack Benny, Lanny Ross, Judith Anderson, Carole Landis in the Division Bowl; bully beef; dehydrated potatoes of the consistency of gravel; non-meltable butter; thatched roof chapels and day rooms; swamps; "The Thing"; company and battery barbers; six cans of warm beer per week, cold once a month; softball; volleyball; football with the championship still undecided between the 511th and Division Artillery; a division review in khakis on the steel-planked runway at Dobodura—all of these events punctuated life for the "Angels" in the heat of New Guinea. But it was soon over.

On 11 November 1944, the division sailed "administratively" (i.e., not expecting to be committed to combat immediately and actually not slated for battle until Luzon) for Leyte aboard nine APAs and two AKAs escorted by nine destroyers. This time the food was better and the crap and poker games longer and more expensive to the losers. On 18 November, the 11th landed at Bito Beach and unloaded. At 0700 on the 21st of November with little time for settling down, Colonel "Hardrock" Haugen, the indomitable, hard-charging, tough commander of the 511th, led Ernie LaFlamme's 1st of the 511th out of the base camp to Dulag and Burauen to relieve the 7th Division in the mountains across the waist of Leyte. On the 23d, Ed Lahti and his 3d of the 511th moved out to join the regiment in Burauen. On the 24th, General Swing moved his CP to San Pablo airstrip. On the 26th, Norman Shipley moved his 2d of the 511th to Burauen. This move united the 511th on the Daguitan River not far from Burauen. Colonel Soule and his 188th took positions to protect the southern flank of the division and the 511th. Colonel Hildebrand and the 187th set up defensive positions and patrols around the base camp at Bito Beach. The rest of the division was scattered from Burauen to Bugho.

The 11th, by now committed in its entirety along the mountainous waist of Leyte, fought wet in the torrential rain, the thick, slippery mud and the almost impenetrable jungle with few and inaccurate maps against a hidden and wily enemy with supply support, medical aid and evacuation, command and control and even platoon drops of men supported mostly by the fleet of Division Artillery's dozen or so Cubs, two-place aircraft.

The Japanese caused some excitement and the temporary diversion of the division's main effort with a battalion-sized parachute attack on San Pablo on the night of 6 December 1944. Lt. Colonel Luke Hoska and his 674th and Lt. Colonel Davis and his 127th Engineers, armed with carbines and pistols, charged across San Pablo airstrip and cleared it of the Japanese. Lt. Colonel George Pearson and his 1st of the 187th then fought a long battle to wipe out not only the paratroopers but also the elements of the Japanese 16th Division which had come out of the hills to assist the paratroopers. Then the 187th and the 188th moved into the hills to fight the Battle of Purple Heart Hill. The 511th and the 2d of the 187th fought along the jungle trails to places like Anonang and Mahonag, Hacksaw Ridge and Rock Hill. Supplies were short because the low clouds and monsoons would not permit aerial resupply with any regularity. The fighting was bloody, hand to hand, grenade to grenade, bayonet to bayonet. Some companies, notably Pat Wheeler's G Company of the 511th on the west end of the mountain chain, were cut off for days and lived on what they could scrounge from the mountain side.

The 11th's troopers who fought in Leyte remember names and places like Dual, Anonang, Takin, Lubi, Patog, Manarawat, Mahonag, Mt. Majunag and finally, Ormoc, Albuera and Bay Bay. They also remember A Battery 457th's jump into Manarawat on 4 December from one Air-Sea Rescue C-47; the paratroopers of B Company of the 187th parachuting into Manarawat from L-4s and L-5s; the surgical tents at Manarawat; the bloody fights at Purple Heart Hill and Hacksaw Ridge; the 511th hacking and bludgeoning its way through the Japanese and the jungle paths; the heroics of Private Elmer Fryar, E Company of the 511th, who fought and fought and finally sacrificed himself for his buddies and won the Medal of Honor; and the incessant rain, the serious lack of food in the hills, the dug-in Japanese, and the long hikes along the slippery, muddy, jungle entangled mountain trails. By the end of December, the 511th and the 2d of the 187th had emerged from the mountains to the beaches of Ormoc Bay near Albuera on the Talisayan River.

By 15 January 1945, the division was back at its base camp on Bito Beach ready for money and money, a bath and a shave, dry clothes, new boots, a division review on the beach and constant speculation on its next mission.

On 31 January, the 188th, followed by the 187th and the rest of the division except the 511th RCT, made an amphibious landing at Nasugbu, in southwestern Luzon. On the 3d of February, the 511th PIR dropped on Tagaytay Ridge ahead of the 188th and 187th moving up from the beach. On the morning of the 4th, the 457th followed by parachute onto Tagaytay Ridge. By the 5th, the 511th was well on its way to attacking Manila from the south, and the division now had a beachhead 60 miles long and 250 yards wide.

After their amphibious landing, the 188th, 187th and the glider elements of the division fought against dug-in enemy positions up the hills toward Tagaytay Ridge through the Mt. Cariliao-Mt. Batulao defile near Aga and across "Shorty Ridge." The 511th fought its way into Manila with hand-to-hand fighting from building to building through Las Pinas, Paranaque and into the outskirts of Manila. Here, on 11 February 1945, the gallant and rugged commander of the 511th was mortally wounded by a round from a Japanese 20mm anti-aircraft gun which the enemy used as a ground weapon with a low trajectory. Colonel Haugen, General Swing, Colonel Tripton and others were in a Spanish style house, the CP of the 2d of the 187th then attached to the 511th, near Libertad Avenue in the southern part of Manila when a number of 20mm rounds hit the house. One came through a small hexagonal-shaped window high on the wall. It exploded in the room, and a fragment hit Colonel Haugen in the chest. He was the only one wounded. He died on a flight to a hospital in Hollandia, New Guinea. Colonel Ed Lahti immediately took over the regiment.

Then came the fight across the open areas of Nichols Field and toward Fort McKinley; the rugged combat against the fanatical defenders dug into the depth and redoubts of the machine-gun-artillery-mine-studded Genko Line; then came Ternate, Cavite and Mabato Point.

In the fight for Fort McKinley, Private Manuel Perez, A Company of the 511th, was a fighting machine, a one-man rifle platoon. A Company was approaching Fort McKinley among the heavily defended Nichols Field-Fort McKinley Road. Perez killed five Japanese in the open and had blasted others in dugouts with grenades. One last strong point blocked the company's route into Fort McKinley. Perez circled around the dugout that contained two .50-caliber dual-purpose machine guns and moved up to within 20 yards, killing four more Japanese on the way. He lobbed a grenade into the position and then shot and killed four more of the enemy who were escaping through a tunnel to the rear. He reloaded and killed four more. One of the escaping Japanese threw his bayonet-tipped rifle at Perez, knocking Perez's rifle out of his hands. Perez picked up the Japanese rifle and killed two more of the enemy. He rushed the remaining enemy soldiers in the pillbox, killed three of them with the butt of the Japanese rifle and then bayoneted the one remaining defender. Single-handedly, Perez killed 18 of the enemy and neutralized the emplacement that had been holding up his company. He was awarded the Medal of Honor but did not live long enough to know it—he was killed a week later in another action.

Later, on 23 February, came the "mission impossible," the rapidly planned, eminently successful, almost casualty free, combined parachute, amphibious, overland rescue of 2,122 civilian internees incarcerated and starving 25 miles behind the Japanese lines at Los Banos. This operation was in four parts: the division Recon. Platoon, reinforced with guerrillas, all under the command of Lt. George Skau, infiltrated around the perimeter of the camp during the hours of darkness (before the attack, Colonel Soule led a reinforced battalion Task Force overland into the area to block any Japanese units who might react to the attack); Henry Burgess led the 1st of the 511th (-) across Laguna de Bay in 51 amptracs; and Lt. John Ringler jumped his B Company of the 511th onto a DZ a few hundred yards from the perimeter of the camp. At 0700 on the 23d, H-Hour for the attack, John Ringler's parachute blossomed beneath a C-47, the signal for the recon platoon and the attached guerrillas to attack the guards on the perimeter and in a calisthenics area in the camp. At the same moment, the first amptrac carrying the rest of the 1st of the 511th waded ashore on the beach at Mayondon Point two miles from the camp. The recon. platoon, the guerrillas and B Company eliminated the guards and started to round up the prisoners. In less than an hour, the amptracs arrived in the camp, and loading of the internees began. The amptracs made two round trips to transport the internees and the paratroopers back behind our own lines. By nightfall that evening, the internees were free and eating a hot meal in the New Bilibid Prison area.

In March and April, the division

Naval Guns in the Manila Genko Line. (Courtesy of Richard N. Loughrin)

Drop Zone, Tagaytay Ridge. (Courtesy of Richard N. Loughrin)

moved into an area south and east of Manila and fought slugging matches with a heavily dug-in and determined enemy around Los Banos, Ternate, Lake Taal, Tanauan, Mt. Bijang, Hill 660, St. Sungay, Mt. Mataasnakahoy, Pico de Loro, Mt. Macolod, Mt. Malepunyo. The 11th's troopers fought like the professional soldiers they had become, giving no ground, using smart and aggressive tactics and gaining their objectives with the fewest casualties possible. General Swing's insistence on hard training at Mackall, Polk and Dobodura paid off in the hard test of combat—the battle was indeed the payoff.

One of the division's last battles and last airborne operations of World War II—and the first in which gliders were used in combat in the Pacific—was the Gypsy Task Force's jump south of Aparri on 23 June 1945, an operation to seal off any Japanese escape routes from the northern part of Luzon. By the 26th of June, Henry Burgess, the commander of the Task Force, had linked up with the commander of the 37th Division moving north.

Following Aparri the Division returned to base camps and began planning their participation in Operations Olympia and Coronet, the programmed plans for the invasion of the Japanese home islands. However, the atom bombs and the advent of the nuclear era intervened and peace and the end of the war were at hand!

But, the division's war time history was not yet over. General MacArthur selected the 11th to lead the post-Hiroshima and Nagasaki invasion of Japan, to furnish his Honor Guard and to occupy Japan from Sendai north to and including Hokkaido.

The 11th Airborne's discipline, esprit, morale, comradeship and fighting spirit gave it an enviable record and history. Its fighting soldiers earned two Medals of Honor, 10 Distinguished Service Crosses, over 500 Silver Stars and innumerable Purple Hearts.

The 11th Airborne Division had been a lot of things to a lot of men. It was a structure, a testing ground, in which, in a very short time, boys, just "off the farm," became highly trained soldiers with a will to fight and who gave vitality, life, and blood to what was initially an inanimate, amorphous non-entity. It was a unit in which men, inured to the hardships of training and combat, became closer than brothers and fought the hard and good fight, not so much because of discipline and fear of punishment but because they did not want to let their buddies down. As Leo Crawford put it: "Although I have been retired from the Army for 29 years and it has been 35 years since I left the 11th, I still feel an attachment for men I served with, like Gamble, Lussier, Ringler and dozens of others that I feel for no others."

It was a small division, half the size of a standard infantry division in men and firepower, which, nonetheless, took on the missions of a full-sized division and proved that heart and courage and training and camaraderie and esprit and loyalty, not only up but down, engender self-confidence and invincibility, making giants of ordinary men. Then they became extraordinary. It was a combat division, a unified group of men cemented together with soul and pride, perhaps even arrogance and swagger. No matter. They were troopers of the 11th Airborne, a division tested on the most awesome of proving grounds, the bloody field of prolonged battle under abominable conditions of weather and hunger, which stretched men's durability beyond expectancy, against a cunning and ruthless enemy who fought under no rules of "civilized" land warfare.

The 11th Airborne Division graduated magna cum laude.

13TH AIRBORNE DIVISION

By James E. Mrazek
Col., USA (Ret)

The 13th Airborne Division was constituted in the Army of the United States on Jan. 9, 1942. It had cadres of the 189th and 190th Glider Infantry and the 513th Parachute Infantry Regiment and supporting units. On Aug. 13, 1943, the Division was activated at Fort Bragg, NC., and Maj. Gen. George W. Griner, Jr. assumed command.

In December 1943, two glider regiments from the 1st Airborne Brigade, the 88th from Fort Meade, SD. and the 326th from Alliance, NE., joined the Division. The 189th and 190th were then inactivated and their personnel transferred to the 88th and 326th Regiments respectively. At this time, Gen. Eldridge G. Chapman, a decorated veteran of World War I, and long the head of the Airborne Command, took command. Col. Hugh P. Harris, who was to win four stars during his career, became chief of staff.

The history of the Division goes back to World War I. Constituted in the Regular Army on July 5, 1918, at Camp Lewis, Washington, it consisted of four infantry regiments. It trained and was to join the American Expeditionary Forces in France. The war ended in November before it was sent to the front, and it was demobilized March 8, 1919. Among its commanders were Brig. Gen. Cornelius Vanderbilt and Maj. Gen. Joseph D. Leitch.

The 515th Parachute Infantry Regiment was activated as a cadre May 31, 1943, at Fort Benning, GA., to process and to administer the officers and enlisted graduates of The Parachute School. In December the regiment was fully activated and went into training to ready for overseas shipment. In February 1944, the 515th replaced the 513th which was reassigned to the 17th Airborne Division.

The 326th had been in the 82nd Division during World War I serving with distinction in historic battles at St. Mihiel, Meuse-Argonne and Lorraine. It became known for having been in active operations longer, continuously and without relief than any regiment in the American Expeditionary Force. It was reactivated with the 82nd Airborne Division at Camp Claiborne, Louisiana, early in World War II, but was soon transferred to the First Airborne Brigade.

The 88th Glider Infantry Regiment had its origin in the 88th Airborne Infantry Battalion, the Army's first air loading unit, activated Oct. 10, 1941 at Fort Benning, GA. During two years it trained the 11th, 17th, 82nd and 101st Airborne Divisions in air transportability and later did so for the 84th and 103rd Infantry Divisions in preparation for their joining in the massive airborne assault on Berlin by the First Allied Airborne Army.

In mid-1944 the 13th suffered a disheartened loss of all its privates and many of its officers. They were rushed to Europe to replace combat losses in the 17th, 82nd and 101st Airborne Divisions. After months of training of replacements, the 13th finally embarked for France in January 1945, months after first scheduled. On arrival it was assigned to the XVIIIth Airborne Corps and went into assembly areas at Sens, Joigny and Auxerre. Soon the 88th was deactivated and its personnel transferred to the 326th Glider Infantry, which was then increased from two to three battalions. On March 1, 1945, the battle-seasoned 517th Parachute Infantry Combat Team, which had fought with distinction in Italy, Southern France, Belgium and Germany, joined the 13th at Joigny, France.

Assignments to combat missions came quickly but with frustrating results. Alerted along with the 17th Airborne Division for an assault against the Nazis at Wesel, Germany, the 13th's participation was cancelled for lack of enough aircraft to airlift both divisions. Next the division geared up for Operation "CHOKER," the landing across the Rhine at Worms. The day before the division was to take off, the 13th's paratroopers and glider troops again moved out of the barbed wire enclosed assembly areas. Paratroopers marched to the airfield, found the C-47s, climbed in the ones they were assigned and secured drop loads. Glidermen loaded and lashed ammunition, pack howitzers, Jeeps and trailers into the gliders ready to take off at dawn. They woke up the next morning to the news the mission had been cancelled while they slept. Gen. Patton had captured Worms while they were loading up the day before!

Next came Operation "EFFECTIVE," which was to deny part of the Alps to the Nazis to prevent them establishing a last ditch stronghold there. New intelligence, however, indicated that this operation was no longer necessary, and it was cancelled. Finally, as the days of the Third Reich were drawing to a close, elements of the 13th were scheduled to land at Copenhagen, Denmark, on a classified mission. It, also, was cancelled. Shortly thereafter, First Allied Airborne Army Headquarters announced that the division would be redeployed to the Pacific to participate in the invasion of Japan after a stop over in the United States. The Division arrived at the New York Port of Embarkation on Aug. 23, 1945 and moved to Fort Bragg, NC. Shortly thereafter, Japan surrendered.

At the end of the war the Division consisted of the following units:
326th Glider Inf. Regt.
458th Prcht. FA Bn. 75mm
153rd Abn. AT Bn.
Hq. Special Troops
409th Abn. QM Co.
515th Prcht. Inf. Regt.
676th Glider FA Bn. 75mm
222nd Abn. Med. Co.
13th Prcht. Maint. Co.
713th Abn. Ord. Maint. Co.
517th Prcht. Inf. Regt.
460th Prcht. FA Bn. 75mm
677th Glider FA Bn. 105mm
129th Abn. Engr. Bn.
513th Abn. Signal Co.
Headquarters Co.
MP Platoon

In addition to six generals that served in the 13th during WWII, 11 others were to become generals. Two wore four stars, three rose to three stars. One of the latter had been a private in the 517th Parachute Infantry Regiment.

On Feb. 25, 1946, the Division was inactivated, and its personnel transferred to the 82nd Airborne Division at Fort Bragg, NC. The 13th was again a matter of history.

Troopers pile out of a C-47 on a practice jump. Note poor body positions guaranteed to produce "strawberries" from the webbing.

No one ever heard of Hoffman, North Carolina before the Airborne built Camp Mackall nearby. The 11th, 17th and 13th Airborne trained there.

17TH AIRBORNE DIVISION

By Bart Hagerman
LTC, AUS (Ret)

ELEMENTS
193rd Glider Infantry Regiment (absorbed into 194th after Bulge)
194th Glider Infantry Regiment
507th Parachute Infantry Regiment (assigned after Holland)
513th Parachute Infantry Regiment
550th Airborne Infantry Battalion (assigned into 194th)
464th Parachute Field Artillery Bn. (assigned after Bulge)
466th Parachute Field Artillery Bn.
680th Glider Field Artillery Bn.
681st Glider Field Artillery Bn.
17th Division Artillery
139th Airborne Engineer Battalion
155th Airborne Anti-Aircraft Battalion
224th Airborne Medical Company
411th Airborne Quartermaster Company
517th Airborne Signal Company
717th Airborne Ordnance Company
17th Parachute Maintenance Company
17th Division Headquarters, Recon Platoon, Military Police Platoon and Band

ACTIVATION
15 April 1943 at Camp Mackall, North Carolina

CAMPAIGNS
Ardennes
Rhineland
Central Europe (Airborne Assault)

UNIT CITATIONS
1st, 2nd and 3rd Battalions, 513th PIR

MEDAL OF HONOR
194th GIR- T/Sgt. Clinton M. Hedrick
507th PIR- Pvt. George J. Peters
513th PIR- Sgt. Isadore S. Jachman
513th PIR- Pvt. Stuart S. Stryker

COMMANDING OFFICERS
Commanding General- MG William M. Miley
Asst. Div. Commander- BG John L. Whitelaw
Div. Artillery CO- BG Joseph V. Phelps
193rd GIR- Col. Maurice G. Stubbs
194th GIR- Col. James R. Pierce
507th PIR- Col. Edson D. Raff
513th PIR- Col. Albert H. Dickerson/ Col. James W. Coutts

CASUALTY SUMMARY
KIA/DOW- 1,314
WIA/IIA- 4,904

TRAINING
Camp Mackall, North Carolina
Camp Forrest, Tenessee

MOTTO
"Thunder from Heaven"

ACTIVATION AND TRAINING

The 17th Airborne Division was activated on 15 April 1943 at Camp Mackall near Hoffman, NC with a cadre selected from the 101st Airborne Division. The 17th's commanding officer from the very inception and the only one the division was ever to have was Major General William M. "Bud" Miley, one of the Airborne's earliest pioneers.

After some nine months of rigorous training in the Sandhills Area of North Carolina, the Division moved in January 1944 to Tennessee for winter maneuvers. Later, in March of that year, the Division returned to garrison at Camp Forrest, located near Tullahoma, TN.

There, for the next five months, the troopers fine-tuned their battle skills. Most of the men qualified for the Expert Infantryman's Badge and 2,150 glidermen qualified as parachutists at a division-operated jump school.

In August the word was that the Division was ready and it was moved to Camp Myles Standish at Taunton, MA to prepare for overseas deployment. Finally, the Division sailed from Boston aboard the USS *Wakefield*. Eight days later, the main body of the Division disembarked at Liverpool, England on 26 August 1944.

Closing at Camp Chisledon, England on 30 August, the Division was assigned to the XVIII Corp (Airborne). Together with the American 101st and 82nd Airborne Divisions, the Polish Brigade and the British 6th and 1st Airborne Divisions, they became part of the First Allied Airborne Army.

COMBAT: THE BATTLE OF THE BULGE

When the Germans began their Ardennes offensive 16 December 1944, the 17th Airborne was soon alerted for movement. On 23 December, the Division was air-transported to the continent landing near Rheims. The mission was clear: Stop the German breakthrough aimed toward Antwerp!

On Christmas Day the Division was attached to the U.S. Third Army and moved into position to defend the Meuse River along a 30-mile line in the vicinity of Charleville, France. Shortly thereafter, the German offensive began to loose its momentum and the threat to the Meuse lessened. It was time to go back on the attack.

The Division was then relieved and moved by truck to the southern flank of the bulge. Operational headquarters for the Division was established in Morhet, Belgium and on 4 January 1944, the Division relieved the 11th Armored Division and went on the attack.

With the 101st Airborne on the right and the 87th Infantry Division on the left, the 17th troopers moved to halt the Germans who were attacking northwest to close the Bastogne corridor.

Colonel Bob Pierce's 194th Glider Infantry Regiment with LTC Edward Sach's 550th Airborne Battalion attached, along with Colonel James Coutts' 513th Parachute Infantry Regiment, made the attack. Colonel Maurice Stubbs' 193rd Glider Infantry Regiment and Colonel Edson Raff's 507th Parachute Infantry Regiment were in reserve.

The Germans hurled dozens of tanks and a heavy barrage of artillery at the attackers and many casualties were inflicted on the lightly-armed troopers. The 2,250 yards of narrow, high-rimmed

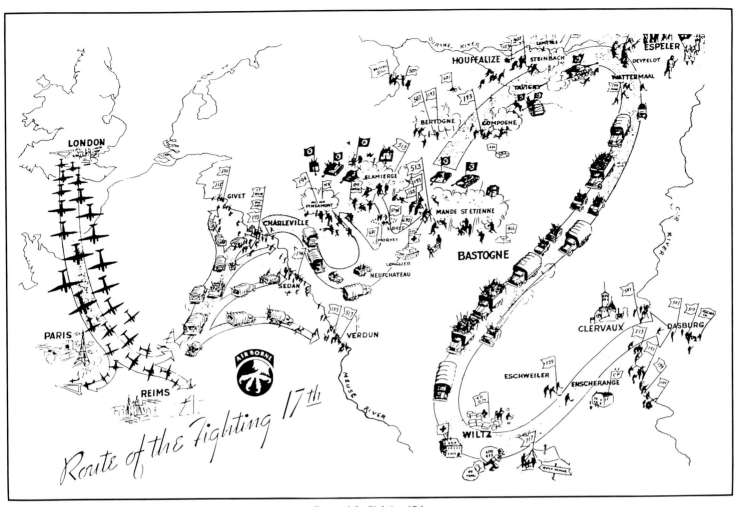
Route of the Fighting 17th.

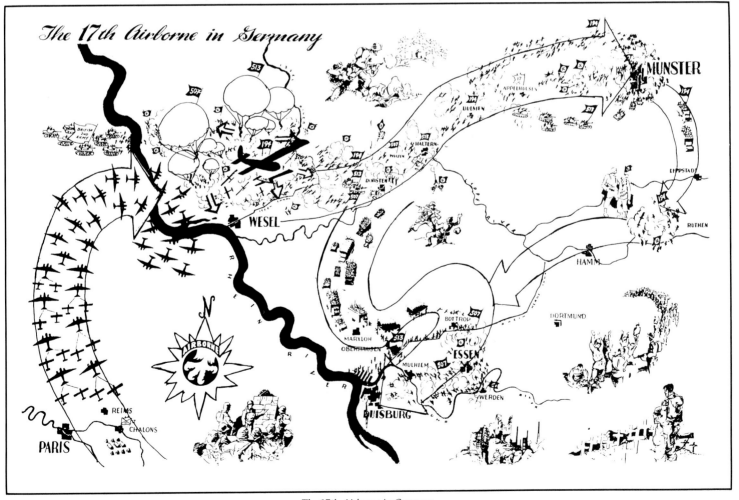
The 17th Airborne in Germany.

road northeast of Bastogne rightfully earned its nickname of "Dead Man's Ridge."

Attacking in a driving snowstorm, the Division battled for control of the ridge. It was a bitterly fought battle that saw the 17th suffer close to a thousand casualties during the three-day battle. Noted military historian S.L.A. Marshal later wrote that no other American division suffered as brutal and as high a casualty rate in their baptism of fire.

On 7 January, it was an all-out attack with the 193rd and the 507th leading the way as the Division stormed the heights to take the objective. At last the German line was broken, and they started to withdraw with the 17th troopers hot on their heels.

Despite blizzard weather and the resulting lack of air support, the Division pushed on to seize Flamierge, Flamizoulle, Renaumont and Heropont. Patrols of the 507th finally reached the Orthe River and made contact with the British 51st Highland Division advancing from the north. The "nose" of the Bulge had been pinched off and a large bag of prisoners fell to the advancing troops.

Attached to a special armored task force, the 193rd drove to take Houffalize, while the rest of the Division moved on through Flamizoulle to take Gives, Bertogne, Compogne, Limerle, Watermal, Hautbellain and Espeler.

After the capture of Espeler, the Division was again relieved and this time trucked to Luxembourg where they closed in an assembly area in the vicinity of Eschweiler. This action marked the end of the Ardennes Campaign and heralded the beginning of the Rhineland Campaign. The 17th's mission now became that of driving the enemy back into Germany east of the Our River and organizing the high ground overlooking the border area.

With the 507th and the 193rd leading the assault, the enemy was forced to withdraw east of the Our River and to move to the protection of the vaunted Siegfried Line. From defensive positions west of the Our, the 507th, 513th and 193rd launched aggressive patrols across the swollen river. Finally, the 507th became the first unit of the 17th to establish positions on the German soil.

Patrols from the 17th continued to probe the east banked enemy positions, but as the snow began to melt and the river ran wide and swift, the launching of an attack had to be delayed. On 10 February 1945, the Division was relieved by the 6th Armored Division. The 17th then moved back to France at Chalons-sur-Marne for rest and refitting and to prepare for the coming airborne assault across the Rhine River into Germany.

OPERATION VARSITY: THE AIRBORNE ASSAULT INTO GERMANY

A new table of organization instituted during early March 1945 saw the 193rd GIR and the 550th Airborne Bn. deactivated. The 550th was designated as the third battalion of a newly-constituted 194th Regimental Combat Team. The badly-decimated 193rd was absorbed into the Division with most of the men going to the 194th as replacements. The 464th Parachute Field Artillery was also assigned to the Division.

Operation VARSITY, the airborne assault into Westphalia near Wesel, began the morning of 24 March 1945. The mission was "to seize, clear and secure the Division area with priority to the high ground east of Derfordt and the bridges over the Issel River, protect the right (southern) flank of Corps, establish contact with the British 1st Commando Brigade Nebraska of Wesel and with the 15th British Division and the British 6th Airborne."

VARSITY, with the 17th Airborne and the British 6th Airborne participating, constituted the largest airborne operation

Fighting alongside the 101st Airborne in Bastogne, 17th Airborne troopers captured these Nazi Panzer Grenadiers.

These 513th PIR troopers make last minute checks as their C-47 crosses the Rhine and they prepare to jump into Germany.

After the Wesel jump, these troopers take shelter in this ditch. Note the parachute canopy hanging in the tree in the background.

These three photos were taken from the gun camera of a P-51 making a strafing run on a canal in the 17th Airborne's landing area east of the Rhine River.

ever staged in one single day. The 507th PIR began landing at 0948 hours and all drops and landings were completed by 1210. Simultaneously, British forces began an amphibious assault over the Rhine River designated to link up with the Airborne forces before nightfall.

A smoke haze hung over the drop and landing zones causing some of the units to be dropped off their designated areas. The new C-46s, used for the first time in combat, proved to be easily set afire by the murderous flak sent from the enemy positions.

In particular, the 513th troopers were dropped some two to three miles northeast of their DZ. Many of the 507th were also scattered. In true Airborne fashion, however the troopers quickly sized up the situation and began attacking the objectives in their area and fighting toward their areas. Everywhere the Germans were found, they were routed from their positions.

In all, some 8,801 Allied paratroopers jumped into the hell below, followed some 15 minutes later by 8,196 glidertroops. They rode in 1,545 planes and 1,305 gliders and were escorted by 543 fighter aircraft—a terrifying sight for the enemy defenders.

By late afternoon all the Division's objectives had been taken, and by the close of the day, 2,873 prisoners had been captured and hundreds of other Germans killed. The stunning victory didn't come without a cost, however. In the one day's action the 17th Airborne had 393 killed, 934 wounded and 164 missing.

THE ADVANCE INTO THE RHUR VALLEY

On March 26, the 17th Airborne, with the 1st British Commando Brigade attached, attacked eastward. The Germans were on the run now and the prisoner bag mounted rapidly. Astride the tanks of the British Guards Armored, the troopers went on to capture the towns of Dorsten, Wulfen and Haltern in rapid succession.

Munster fell to the 17th on 2 April. Then, relieving the 79th Infantry Division, the 17th charged across the Rhine-Herne Canal to take Essen on 6 April. Continuing the advance, the Division then took the industrial cities of Mulheim, Duisberg and Werden.

Troopers from the 194th captured Franz von Papen who had been Chancellor of Germany when Hitler came to power. Von Papen was on the wanted list for war criminals and his capture saw to it that he would be brought to trial at the war's end.

When the German city of Duisberg surrendered to a 507th reconnaissance patrol on 12 April and the Rhur pocket was deemed closed, the combat saga of the Division soon came to an end. Although planning was already complete for the 17th to make an airborne assault in the Berlin area, the decision had been made to let the Russians storm Hitler's last bastion. The 17th was moved from Essen to Marxloh where the administration and care of displaced persons became the Division's mission.

With the surrender of Germany on 7 May 1945, the Division continued its occupational duties until relieved by British forces on 15 June 1945. Many troopers were then transferred to the 82nd Airborne in Berlin and later came home with them to participate in the victory parade in New York City. Others joined the 13th Airborne which was being returned to the States to be readied for shipment to the Pacific and the expected airborne invasion of the Japanese mainland. When the atomic bomb brought the swift surrender of Japan, these troopers were soon discharged and on their way home.

The final echelon of the 17th Airborne returned to Camp Myles Standish and there, on 16 September 1945, the Division was deactivated. Although the 17th was reactivated for the period 6 July 1948 to 10 June 1949 as a training division, it no longer appears on the Army's roles. In September 1945, the odyssey of the 17th Airborne Division had come to a close.

The Division participated in three campaigns: Ardennes, Rhineland and Central Europe. It suffered a total of 1,314 men killed in action or who died of their wounds and 4,904 wounded or injured in action. Officially the Division saw only 65 days of combat during the two years and five months of its existence, but the brief history of the 17th Airborne Division touched the lives of many of America's finest—the "Airborne Soldier!"

"THUNDER FROM HEAVEN"

The 17th Airborne Division insignia seems to bear out its motto, "Thunder From Heaven." The eagle's golden talon on a black background symbolizes the airborne's wartime mission—a surprise strike from the darkness of the sky and the grasping of golden opportunities in the battle below.

BIBLIOGRAPHY

"Paratrooper" by Gerard M. Devlin
"Ridgeway's Paratroopers" by Clay Blair'
"Thunder From Heaven" by Don Pay
"The History of the 17th Airborne Division" by Bart Hagerman and Gardner Hatch

This CG-4A that crashed during Operation Varsity, probably killed the pilot and most if not all of his 17th Airborne troopers.

When the 17th Airborne gliders landed they were looking right into the German artillery. The artillery men then fought like the infantry.

One hour after the 17th Abn. landed over the Rhine, Maj. Gen. William Miley (second from left) gets information from company commander on enemy deployment.

The fraility of the CG-4A glider can readily be seen in this photo of a crash on the Wesel landing zone.

82ND AIRBORNE DIVISION

By Steven Mrozek
SSG/MIARNG

82ND INFANTRY DIVISION WORLD WAR I (1917-1919)

I. Elements:
325th Infantry Regiment
326th
327th
328th
319th Field Artillery Regiment
320th
321st
307th Engineer Battalion
319th Machine gun Battalion
320th
321st
307 Trench Mortar Battery
307th Ammunition Train (ordnance)
307th Field Signal Battalion
307th Sanitary (medical) train (evacuation)

II. Activation:
Camp Gordan, GA (near Atlanta), August 25, 1917. Originally organized with men from Alabama, Georgia and Tennessee

III. Campaign Credits:
Marbache Sector
St. Mihiel Offensive
Meuse-Argonne Offensive

IV. Unit Awards:
None (not done in World War I)

V. Medal of Honor:
LTC Emory J. Pike, September 12, 1918, St. Mihiel; Corp. (later Sgt.) Alvin C. York, Co. G, 328th Inf. October 8, 1918 Meuse-Argonne Offensive.
Number of DSCs - 75
Number of DSMs - 3

VI. Commanders:
Major General Eben Swift (8/25/17-11/23/17)
Brig. General James Erwin (11/24/17-12/16/17)
Brig. General William P. Burnham (12/27/17-10/3/18)
Major General George B. Duncan (10/4/18-5/21/19)

VII. Casualty Summary:
1,413 Killed In Action
6,664 Wounded

VIII. Training:
Camp Gordan, GA, October 7, 1917-April 1918
Divisional Rifle Range, Norcross, Georgia
Divisional Artillery Range, Marietta, Georgia

WORLD WAR II
March 25, 1942-January 3, 1946

I. Elements:
Division HH Company
82nd MP Platoon
325th Glider Infantry Regiment
504th Parachute Infantry Regiment
505th Parachute Infantry Regiment
Division Artillery HH Battery
319th Glider Field Artillery Battalion
320th Glider Field Artillery Battalion
376th Parachute Field Artillery Battalion
456th Parachute Field Artillery Battalion
80th Airborne Anti-Aircraft Battalion
307th Airborne Engineer Battalion
407th Quartermaster Company
307th Airborne Medical Company
82d Airborne Signal Company
782d Airborne Ordnance Maintenance Company
82d Airborne Recon Platoon
82d Parachute Maintenance Company

Attached: 508th Parachute Infantry Regiment (Normandy, Holland, Ardennes, and Rhineland)
507th Parachute Infantry Regiment (Normandy)
401st Glider Infantry Regiment, 2nd Battalion to 36/325th
666th Quartermaster Truck Company (Holland, Ardennes and Central Europe)

II. Activation:
March 25, 1942, Camp Claiborne, Louisiana

III. Campaign Credits:
Sicily, Naples-Foggia, Rome-Arno (504th RCT only), Normandy (minus 504th RCT), Rhineland, Ardennes and Central Europe

IV. Unit Awards:
Belgian Fouraguerres of the Croix de Guerre, 4 October 1945
French Fouraguerres of the Croix de Guerre, April 6, 1946
Netherlands, Military Order of William, October 8, 1945 (attached to the Division's colors)
Netherlands Orange Lanyard, Military Order of William, October 8, 1945 (to be worn by Holland veterans)
Presidential Unit Citation- Awarded to individual units in recognition for outstanding accomplishment on the battlefield (by the end of the war, every unit was entitled to wear this award).

V. Medal of Honor:
PFC Charles N. DeGlopper, Co. C, 325 Glider Infantry, June 9, 1944, La Fiere, France, Normandy
Pvt. John R. Towle, Co. C, 504th Parachute Infantry, September 21, 1944, Oosterhout, Holland
1st SGT. Leonard Funk, Jr., Co. C, 508th Parachute Infantry, January 29, 1945, Holzheim, Belgium

Distinguished Service Crosses: 78

VI. Commanding Generals:
General Omar N. Bradley, March 23-June 25, 1942
General Matthew B. Ridgway, June 26, 1942-August 28, 1944
General James M. Gavin, August 28, 1944-March 26, 1948

VII. Casualty Summary:
Killed In Action - 3,228
(Includes those who died of wounds)
Wounded in Action - 12,604
Missing In Action - 106

Injured In Action - 2,782

Total - 18,720

VIII. Training:
Camp Claiborne, Louisiana, March 25-October 1, 1942
Fort Bragg, North Carolina, October 14, 1942-April 20, 1943

IX. Crest, Logo, Motto, Colors:
Motto: The All-American; America's Guard of Honor

X. Individual Decorations:
DSC - 78
DSM - 1
Legion of Merit - 32
Silver Star - 894
Bronze Star - 2,478

Dominican Republic
April 29, 1965-Summer, 1966

I. Elements:
1st Brigade, 1/504th, 2/504th, 2/508th Airborne Inf. Bn.
2nd Brigade, 1/325th, 2/325th, 3/325th Airborne Inf. Bn.
3rd Brigade, 1/505th, 2/505th, 1/508th Airborne Inf. Bn.
DIVARTY, 1/319th, 2/320th, 2/321st Airborne Art. Bn.
DISCOM, Division H H Company and Band
782 Maintenance Bn.
307th Medical Bn.
407th Supply and Transportation Bn.
82nd Administration Company
Separate Units:
82nd Aviation Bn.
307th Airborne Engineer Bn.
82nd Signal Bn.
1/17 Cavalry
82nd Military Intelligence Company
82nd Military Police Company

II. Activation:
The 82nd Airborne Division has been on active duty since 1942.

III. Campaign Credits:
Dominican Republic, 1965-66

IV. Unit Awards:
None for this operation.

V. Medal of Honor:
None

VI. Commanding General:
Major General Robert H. York

VII. Casualty Summary:
30 killed and wounded

VIII. Training:
Fort Bragg, North Carolina

IX. Crest and Motto:
The All-American, America's Guard of Honor

X. Individual Awards:
Unknown

PROGRESSIVE COMPOSITION OF THE 82d AIRBORNE DIVISION										
YEAR	INFANTRY	ARTILLERY	AAA	AR	ENG	SIG	OD	QM	MAINT	MED
1942	325G, 326G, 504P	319P, 320G, 376P, 456P	80		307	82	782	407	(Parachute Maintenance) 82 PM	307
1943	325G, 326G, 504P, 505P	319P, 320G, 376P, 456P	80		307	82	782	407	82 PM	307
1946	(Airborne Regts) 325, 505, 505	(Airborne Battalions) 98, 319, 376, 456	80	788	307	82	782	407	82 PM	307
1957	(Airborne Battle Groups) 1-325, 2-501, 2-503, 1-504, 1-505	(Airborne Batteries) A, B, C-D, E-319, B-377 (Long John)	44 (50-54) 714 (50-57)		12	82		S&T 408	782 Maint 82 PM	82
1964	(Airborne Bns) 1, 2, 3/325, 1,2/504 1,2/505, 1/508	(Airborne Battalions) 1/319, 1/320, 2/321, B-377 (Long John)	7/60		307	82		S&S 407	782 Maint	307
1970	Division Support <u>1st Brigade</u> (Devils Bde) 1, 2/504, 2/508 <u>2d Brigade</u> (Falcon Bde) 1, 2, 3/325 <u>3d Brigade</u> (Golden Bde) 1, 2/505, 1/508 <u>4th Brigade</u> 4/325, 3/504, 3/505	1/319 1/320 2/321 3/320	7/60	4/68	307, 618 C-307 596	58		82 & 582 S&S 407	782 Maint	307 D-307
1974	Division Support <u>1st Brigade</u> (Devil's Bde) 1, 2, 3/504 <u>2d Brigade</u> (Falcon Bde) 1, 2, 4/325 <u>3d Brigade</u> (Golden Bde) 1, 2, 3/505	1/319 2/319 3/319	3/4* (*1972) 3/73** (**1984)		307	82		82 & 582	782	307

VIETNAM
Feb. 14, 1968-Dec.12, 1969

I. Elements:
   3rd Brigade, 1/505th, 2/505th, 1/508th Airborne Infantry Bn
   2/321st Airborne Artillery Bn.
   C Co. 307th Airborne Engineer Bn.
   58th Signal Company
   82nd Military Police Company
   Company O, 75th Infantry (Ranger)
   1/17th Cavalry, B Troop
   37th Scout Dog Platoon
   82nd Support Bn.
   82nd Military Intelligence Detachment

II. Activation:
   The 3rd Brigade of the 82nd Airborne Division has been on active duty since 1942.

III. Campaign Credits:
   Tet Counteroffensive
   Vietnam Counteroffensive (Phase IV)
   Vietnam Counteroffensive (Phase V)
   Vietnam Counteroffensive (Phase VI)
   Tet 69/Counteroffensive Campaign
   Vietnam Summer/Fall 1969
   Vietnam Winter/Spring 1970

IV. Unit Decorations;
   Cross of Gallantry with Palm (3rd Brigade only)
   Valorous Unit Streamer
   For service in Vietnam: 1/508 Airborne Infantry Bn. and A, B and C Batteries, 321st Artillery Bn.

V. Medal of Honor:
   None

VI. Commanding General:
   3rd Brigade Commander, Brigadier General Alexander R. Bolling
   Brigadier General George W. Dickerson

VII. Casualty Summary:
   212 Killed in Action
   Wounded in Action

VIII. Training:
   Fort Bragg, North Carolina

IX. Motto:
   The Third Brigade was known as the "Golden Brigade."

X. Individual Awards:
   Not Complete- 2-68 to 3-69
   DSC- 2
   Silver Star- 67
   DFC- 5
   Soldiers Medal- 25
   Bronze Star V Device- 240
   Air Medal V Device- 7
   Army Commendation V Device- 82

GRENADA
Oct. 25, 1983 - Nov. 4, 1983

I. Elements:
   2/504th, 2/325th, 3/325th Airborne Infantry Battalions
   1/505th, 2/505th, 1/508th Airborne Infantry Battalions
   Divarty, 1/319th, 1/320th, 2/321st Airborne F Arty Bns
   3/4th Air Defense Artillery Battalion
   307th Airborne Engineer Battalion
   307th Medical Battalion
   407th Service and Supply Battalion
   82nd Signal Battalion
   82nd Combat Aviation Battalion
   1/17th Air Cavalry
   313th Military Intelligence Battalion (CEWI)
   82d Military Police Company
   782d Maintenance Battalion

II. Activation:
   The 82nd Airborne Division has been on active duty since 1942.

III. Campaign Credit:
   Grenada, 1983

IV. Unit Awards:
   None

V. Medal of Honor:
   None

VI. Commanding General:
   Major General Edward L. Trobaugh

VII. Casualty Summary:
   Information not known

VIII. Training:
   Fort Bragg, North Carolina

IX: Crest/Motto:
   The All-American, America's Guard of Honor

X. Individual Awards:
   Not Known

Parts of chapters on Italy, Anzio, Normandy, Holland, The Ardennes, Germany and Post-War and "Power Pack" reprinted from *The 82nd Airborne Division: America's Guard of Honor*, Mrozek, Steven J., Taylor Publishing Company, Dallas, Texas, 1987.

Maps used in the 82nd Airborne Division Section reprinted with General Gavin's permission from his book, *On To Berlin*.

82ND AIRBORNE DIVISION "THE ALL-AMERICANS"

"IN THE BEGINNING..."

Also known as the "All-American" division and "America's Guard of Honor," the 82nd wasn't always Airborne. Originally formed during the First World War on August 25, 1917 at Camp Gordon, Georgia, the 82nd Infantry Division was composed of men from Alabama, Georgia and Tennessee. Later in October the same year, the division received replacements largely from the New England and Mid-Atlantic states. The 2nd was organized into two infantry brigades of two regiments each (325, 326, 327 and 328) and an artillery brigade of three artillery regiments (319, 320 and 321). A regiment of engineers (307), and three machine gun battalions (319, 320, and 321) and an ordnance battalion and medical battalion provided the division's support.

Perhaps the most famous American soldier of the first world war was Sergeant Alvin C. York. Belonging to Company G, 328th Infantry, York was a quiet man from the hills of Tennessee. It was on October 8, 1918, during the Meuse-Argonne offensive that York earned his famous reputation.

After killing a total of 32 of the enemy, York accepted a German major's surrender. Returning back to American lines, York and the survivors of his patrol turned over 132 prisioners. He is creditied with the single-handed destruction of a German machine gun battalion and was personally awarded the Medal of Honor, by General John J. Pershing.

In the course of its service, the 82nd Division spent 105 days in the front lines and advanced over 11 miles and captured 845 prisoners. The division suffered 1,035 killed and 6,387 wounded during the same period. In the service to their country, two soldiers of the "All-American" division won Medal of Honors and 75 won the Distinguished Service Cross.

The United States managed to maintain a fragile neutrality from the wars raging in Europe and in Aisa since 1936. It was the Japanese attack on the American fleet at Pearl Harbor on December 7, 1941 that catapulted the United States deep into the midst of the global war. No sooner the United States declared war on Japan, when Germany announced its declaration of war on America.

Almost overnight, make-shift military camps materialized in obscure parts of the country to accomodate the recent recruits, soon to be soldiers. It was in one such Army installation in Louisiana, Camp Caliborne, that an infantry division with an illustrious past was reactivated. The 82nd Infatnry Division was destined to greatness from the start. Chosen to organize and lead the 82nd was the former commander of Fort Benning, Brigadier General Omar N. Bradley. His second in command was none other than Brigadier General Matthew B. Ridgway.

On June 26, 1942, General Bradley was reassigned as the commander of the 28th Infantry Division, then training at Camp Livingston, Louisiana. General Ridgway was promoted to division commander. Shortly after the change in command, the Army began changing the division's designation. For a brief period in July, the 82nd became a motorized division. At a full-dress review of the division on August 15, 1942, General Ridgway informed the men that the 82nd was to become the first American Airborne division and to supply the cadre for the second, the 101st Airborne Division.

The newly designated 82nd Airborne Division was assigned the 504th Parachute Infantry Regiment. The 325th and 326th Infantry Regiments became the divisional glider infantry. The 504th, formed on May 1, 1942, was first commanded by Colonel Theodore L. Dunn and later by the regiment's executive officer, Lt. Col. Ruben Henry Tucker.

Primarily due to a shortage of CG-4A Waco gliders, the 82nd underwent another organizational change. The previous "one parachute-two glider" regimental structure was reversed on February 12, 1943. The 326th was replaced by the 505th Parachute Infantry Regiment. To balance this change and permit the organization of regimental combat teams, Company B, 307th Airborne Engineers joined Company C of the 307th as parachute troops rather than as glider troops. The 456th Parachute Field Artillery Battalion joined the 376th which was already with the division and attached to the 504th Parachute Infantry.

On April 20, 1943 the division began its movement from Fort Bragg to Camp Edwards, Massachusetts. On the 27th of April the troopers boarded the trains that would take them to a pier in the Port of New York. Loaded down with heavy barracks bags, they boarded the transport. Early on the morning of April 29, 1943 the troopship slipped its moorings and began the journey that would take the division

Elements of the 505th PIR, 82nd Airborne Division advance into Naples, Italy. (Courtesy of Steven Mrozek)

82nd trooper "standing in the door" over Sicily, July 1943. (Courtesy of Steven Mrozek)

to nine countries, from Fort Bragg to Berlin.

After 12 days on the high seas, the men of the 82nd found themselves in the harbor of the Moroccan city of Casablanca. It was on the afternoon of May 10, 1943 where the division disembarked then set foot on African soil. Before anyone could settle down in the temporary camp just outside of Casablanca, the troopers boarded the trains and headed for Oujada. The division spent a total of six weeks at Oujada.

On Friday night, July 9, 1943, the 505th Parachute Combat Team (505th Parachute Infantry, 456th Parachute Artillery, and Company B, 307th Airborne Engineers) reinforced with the Third Battalion of the 504th Parachute Infantry, dropped from the skies over Sicily.

The invasion was long in the planning; D-Day was scheduled for July 9, 1943. A major element of the invasion plan was the preliminary Airborne drop by both American and British paratroopers. Under the codename of LAD BROKE, the "Red Devils" of the British 1st Airborne Division were to strike targets in the vicinity of Syracuse in the Southeastern area of the island. The 82nd Airborne was assigned the Gela area further to the West. Due to the shortage of C-47 transports, the 52nd Troop Carrier Wing could drop what amounted to a reinforced regimental combat team in the first lift.

The first wave was given the designation of HUSKY I. This honor was given to Colonel James M. Gavin's 505th Regimental Combat Team reinforced with the Third Battalion of Colonel Reuben Tucker's 504th Parachute Infantry. The following night, Tucker's other two battalions would be dropped in along with the 376th Parachute Field Artillery Battalion and Company C, 307th Airborne Engineers. The second drop was codenamed HUSKY II.

The drop zone for HUSKY I was an egg-shaped area a few miles east of the town of Gela. From this DZ it was determined by planners that the 505th troopers would have little trouble reaching their objectives, "X" and "Y." The objective at point "X" was near the town of Nescemi located northeast of Gela. This objective was assigned to the 504's Third Battalion. The 505th was to take and hold the key road intersection east of Gela. The intersection was well guarded by a complex of concrete pillboxes armed with machine guns.

The 266 planes laden with paratroopers of the 82nd cleared the fields at 11:15 p.m., July 9, 1943. To avoid "friendly fire" from the hundred of ships of the seaborne invasion force, the troop transports were to fly a long, circuitous route from Africa. They were to go first to the island of Malta, then turn north to the southwest corner of Sicily. The C-47s were then to head back out to sea to avoid fire from the shore batteries. Upon reaching the mouth of the Ocate River, they were to head inland to the designated drop zones. The total distance of this route was 415 miles. To avoid detection by the Germans, the Airborne force would skim over the water at 200 feet, rising to 600 feet over the DZs to allow the troopers to jump. July 9th was selected because there would be a full moon to illuminate the drop zones.

Almost immediately things began to go wrong. The British gliders on the way to their targets near Syracuse were fired on by "friendly" batteries. Ninety loaded gliders crashed into the sea. Army meteorologists estimated the wind speed over the drop zones to have picked up to about 35 mph; (training jumps would be cancelled if the wind exceeded 15 mph). Blown off course, the C-47s never saw the Malta checkpoint, some turned back but most estimated the route and pressed on. Through errors and the wind speed, troopers of the 82nd Airborne were scattered all over southern Sicily. Colonel Gavin and the men in his formation landed about 20 miles east of their assigned drop zone. The last of the paratroops from Colonel Gavin's command had landed by 0100 hours on July 10. Only about 15 percent were in their correct areas.

Once on the ground, the troopers tried to assemble with other members of their stick or unit. Some succeeded in forming groups of several hundred, but a lot found themselves alone behind enemy lines. Of the 3,405 troopers of the reinforced combat teams, Gavin could locate only 20 of them four hours after the jump. Of the many units dropped during the first night, only the 2nd Battalion of the 505th Parachute Infantry, led by Major Mark Alexander, had landed all together in the same spot. Although this feat was a major accomplishment, the 2/505th landed 21 miles outside the airhead, east of the small coastal town of Santa Croce.

The unit that was assigned the northern most objective was Lt. Colonel Charles W. Kouns's 3rd Battalion of the 504th Parachute Infantry. Elements of the 3/504th did land close to its drop zone, but most were scattered over a large area southeast of Niscemi. In spite of the disorganization caused by the drop, all primary objectives were taken. Objective X (Niscemi) was eventually taken by men of the 3/504th and objective Y was neutralized by Captain Syare's A Company of the 505th.

Whether alone or in small groups, the troopers set out to accomplish their mission, that being to disrupt enemy communications and prevent them from mounting a counter-attack on the troops of the 1st Infantry Division, 2nd Armored Division, 3rd Infantry Divison and the 45th Infantry Division landing on the beaches. In one instance, less than 100 troopers of Lt. Colonel Arthur Gorham's 1st Battalion, 505th Parachute Infantry, unknowingly took on advance elements of the elite Herman Goering Panzer Fallschrimjager Division rushing south from its assembly area near Caltagirone to drive the Americans into the sea. Armed with two 60mm mortars and two light machine guns, Gorham's troopers attacked a heavily fortified German outpost. In the ensuing fight the Americans forced the surrender of 40 Italians and 10 Germans. Later inside, 15 dead enemy soldiers were found.

Using captured German weapons and valuable intelligence obtained from prisoners, the 505th men prepared to meet the vanguard of the Herman Goering Division. After destroying two companies of Panzer grenadiers, several vehicles and two tanks, the first battalion men were forced back to new positions where they continued to offer resistance. In this effort, these 100 men bought several hours for American troops coming ashore.

In another desperate action on July 11, 1943, Colonel Gavin and about 180 troops of Lt. Colonel Krause's 3rd Battalion stopped a battle group of the Herman Goering Panzer Division. The small group of the 3rd Battlion men held positions on Biazza Ridge. It was there that the German battle group collided with the All-American paratroopers. Although outnumbered by a superior force, Gavin's men refused to be driven. The combined efforts of the paratroopers' determined and courageous stand and armor and artillery support not only stopped the German flank attack cold, but insured the security of the advancing American troops.

On the night of D plus 1 (July 10th-11th) the 504th Regimental Combat Team (minus the 3rd Battalion), led by Colonel Reuben H. Tucker, loaded their C-47s and took off from the airfields around Kairouan, Tunisia. The night was quieter than two days before and lighted by a quarter moon. The drop zone was behind the 1st Division and everyone had the highest hopes for a safe crossing. Then it happened—one of the war's greatest tragedies. The following account taken from the Division's World War II history SAGA OF THE ALL-AMERICAN describes what happened that night.

> *Nearing the Sicilian coast, the formation of C-47s were fired upon by a naval vessel. Immediately, as though upon a prearranged signal, other vessels fired. Planes dropped out of formation and crashed into the sea. Others, like clumsy whales, wheeled and attempted to get beyond the flak which rose in fountains of fire, lighting the stricken faces of men as they stared through the windows.*
>
> *More planes divied into the sea and those that escaped broke formation and raced like a covey of quail for what they though was the protection of the beach. But they were wrong. Over the beach they were hit again—this time by American ground units, who, having seen the naval barrage, believed the planes to be German. More planes fell, and from some of them men jumped and escaped alive; the less fortunate*

were riddled by flak before reaching the ground. (Twenty-three planes were shot down and 318 504th men were killed.)

Fired upon by our own Navy and shore troops, the 504th Parachute Infantry was scattered like chaff in the wind over the length and breadth of Sicily Island. Col. Tucker's plane, after twice flying the length of the Sicilian coast and with over 2,000 flak holes through the fuselage, reached the DZ near Gela; however, few others were as fortunate and by morning only 400 of the regiment's 1,600 men (excluding the 3rd Battalion) had reached the regimental area.

Other plane loads of 504 men dropped in isolated groups on all parts of the island, and although unable to join the regiment, carried out demolitions, cut lines of communication, established inland road-blocks, ambushed German and Italian motorized columns and caused confusion over such extensive areas behind the enemy lines that initial German radio reports estimated the number of American parachutists dropped to be over 10-times the number actually participating!

Despite the initial castastrope the main force landed successfully and made rapid advances catastrophically. Phase I of the 82nd Airborne's mission on Sicily was completed. Phase II was to be a drive by the division up the southern coast of the island. This drive would encounter light opposition and cover about 150 miles in six days, and result in the capture of 15,475 Italian and German prisoners. During this period, the 82nd was under the command of General George Patton's 7th Army.

The 82nd captured a total of 23,191 prisoners during the entire campaign, but this number does not include several thousand turned over to other units for a processing in the early stages of the operation before the division had assembled as a tactical unit.

So ended the 82nd Airborne's first campaign on the long road to Berlin. Sicily was in Allied hands. Messina fell August 16, 1943 and the 82nd returned to North Africa to prepare for Italy.

There were six different plans to drop American paratroopers in Italy. One such plan called for a combat team to be dropped in the Volterno River sector with the rest of the division in floating reserve. This plan was discarded when the 82nd was given four days notice to prepare for an assault on Rome. The Division's artillery commander, Major General Maxwell D. Taylor, has slipped into Italy near Rome to discuss with Marshall Bandoglio arrangements for the Italian capitulation and armistice. In his conference with the Italian marshall, General Taylor realized that the Italian government was in no position to provide the necessary assistance for the proposed drop on Rome. General Taylor ordered the mission postponed at the last moment. When the cancellation order came, 82nd Troopers had begun to load on the C-47s.

On September 13, 1943, Lt. General Mark W. Clark urgently requested 82nd Airborne Commander General Matthew B. Ridgway to make an emergency Airborne landing behind American lines on the endangered Salerno beachhead. The landing, code named AVALANCHE, was in a precarious situation. The Germans held the jagged mountains which overlooked the Allied positions and had stopped the inland advance cold. Reinforcements were urgently needed in the Paestum sector to prevent the invasion forces from being driven into the sea. Just eight hours later, the 1st and 2nd Battalions of the 504th with two platoons from C Company, 307th Airborne Engineers attached, were dropped on the designated drop zone marked by the new Airborne pathfinder detachments, who were used for the first time. Before morning, both the 504th Battalions were on line.

This mission is still regarded as

Normandy – The much fought over bridge over the Merderey. (Courtesy of Steve Mrozek).

history's greatest example of the mobility of Airborne troops—in exactly eight hours from the time the division had been notified of its mission, the 504th Combat Team was briefed, loaded in the planes and dropped over their assigned DZ. Landing at night, the 504th were in the front line, ready by the morning of the 14th.

The following night Colonel Gavin's 505th jumped into the Salerno beachhead on the same DZ. By daylight the 505th with Company B, 307th Airborne Engineers attached was in line on the right of the 504th. The 325th Glider Infantry with the 3rd Battalion 504th attached had sailed from the harbor of Licata, Sicily at 2000 hours on September 13th. The seaborne element of the 82nd Airborne landed at Salerno late on the 15th. The 325th Glider Regiment went into action on the Sorrentine Peninsula and the 3rd Battalion 504th Parachute Infantry occupied the town of Albanella.

The 509th Parachute Infantry Battalion was attached to the 82nd Airborne during this phase of the operation. The 509th was assigned the mission of dropping in the vicinity of Avellino. Jumping with the 509th on this mission was the first platoon of C Company, 307th Airborne Engineers. In spite of poor visibility conditions, the 509th jumped during the night of September 14, 1943. The drop was badly scattered. It was difficult for troopers to assemble. In one instance the battalion commander, Lt. Colonel Doyle R. Yardley, landed in a German tank park where he was engaged in a fire fight, only to be wounded and taken prisoner. This tragedy was typical of the entire operation. Of the 640 men dropped that night, only 510 were able to work their way back to Allied lines.

During the three days that the 82nd occupied the several hills of Altavilla, approximatley 30 troopers were killed and 150 were wounded.

The 82nd pushed northward by boat and landed in the tiny town of Maiori. Fighting with Colonel Darby's Rangers and supported by British heavy guns, the 325th RCT with the Third Battalion of the 504th attached pushed the Germans from the commanding mountaintops at Mt. San Argela and Chiunzi Pass.

On September 30, 1943, troopers of the 505th Parachute Infantry and Division Reconnaissance elements became the first Allied soldiers to enter Naples, the first major European city to fall to the Allies. On the following day, October 1st, the 82nd Airborne Division led the Allied Army into Naples. The Division was assigned the duty of "policing" the city.

It wasn't long before the "All-American" Division was back in action, this time pushing toward the Volturno. From October 3rd to the 9th, the 505th with British Armor, cleared the Germans from the flats and canals near the southbank of the river. Patrols from the regiment became the first to cross it.

In November the 82nd Airborne, less the 504th Regimental Combat Team (504th Parachute Infantry, 376th Parachute Field Artillery Battalion and Company C, 307th Airborne Engineer Battalion) left Naples for a new assignment in Northern Ireland. Before this division left Naples, Colonel Tucker's Regiment launched its attack through the mountains of Central Italy. The 504th pushed 22 miles ahead of the 5th U.S. Army and the British 8th Army. After crossing the Volturno River, the 504th entered the rail and road center of Isernia. One by one the Italian towns in front of the Regiment fell: Colli, Macchia, Fornelli, Cerro and Rochetta.

On the cold rainy evening of December 10, 1943, the regimental CP was established in Venafro. Two 3rd Battalion companies, G and I, moved to relieve elements of the 3rd Ranger Battalion, who were on Hill 950. The men of Company I took their assigned positions while in the midst of a German counterattack. During the next 12 hours the 504th position was attacked in force by the Germans seven times. By noon the following day, I Company sustained 46 casualties, but still held its positions.

The next morning, troopers belonging to the 2nd Battalion finished the difficult climb up Mt. Sammucro (Hill 1205) and took positions formerly occupied by the 143rd Infantry.

The 1st Battalion, supposedly in reserve, was used for litter-bearing details and to carry food, water and ammunition up the rocky, heavily shelled trails to the 2nd and 3rd Battalion troopers stubbornly clinging to the position on the heights.

By December 20, the men of the 504th RCT were holding not only Hills 1205, 950, 954, 710 and 687, but had patrols operating on Hills 877 and 610. The fighting in this sector of the line consisted of the assaulting of one hill after another.

During the 19 days that the 504th was in action near Venafro, they suffered a total of 54 dead, 226 wounded and two men missing in action. These figures are exclusive of the 367th PFA Battalion and Company C, 307th Engineers, both of which suffered dead and wounded. Most of the casualties were the result of intense enemy artillery fire. Information revealed by the 51 German prisoners taken indicated that German dead and wounded were at least five times greater than those suffered by the 504th.

On December 27th, the regiment was relieved of duty in the Venafro sector and was moved to a new camp in the vicinity of Pignatoro. Orders were received on January 4, 1944, directing the 504th Combat Team to Pozzuoli a suburb of Naples. Unknown to the troopers of the 504th, they had been selected for another operation. This time an amphibious landing in the vicinity of a popular coastal resort area; the code name of the operation was SHINGLE; the name of the town, Anzio.

D-Day for Operation SHINGLE was early January 22, 1944. Essentially the Anzio operation was an end run around the German's troublesome Gothic line. SHINGLE would create a threat to the Italian capital and hopefully break the deadlock further south. The 504th Regimental Combat Team was personally selected for this operation by the British Prime Minister Winston Churchill. It was Churchill who requested General Mark Clark to hold on to the 504th which was to be used initially for an Airborne drop.

When the parachute mission was cancelled, "Rube" Tucker's men were loaded aboard 13 LCIs (Landing Craft-Infantry) and became part of the assaulting force. The 504th was designated to land on Red Beach, south of the coastal town of Nettuno. The landing craft were ordered to land the paratroopers at 0930. As the craft approached the beach they were stopped approximately 70 yards from shore by sandbars.

As the troopers unloaded, six German dive bombers came from out of the sun. When they left, LCI Number 20 was settling in the water after receiving a direct hit. This was G Company's boat. In spite of the occasional appearances of the Luftwaffe, the landing craft continued putting troopers ashore. The landing had been a complete success; the enemy was taken totally by surprise. Unfortunately, the Allied command did not know that and, in turn, did not exploit its early advantage.

The troopers of the 504th were assigned positions on the right flank of the beachhead, along the Mussolini Canal. Shortly after moving into position, the combat team pushed the Germans across the canal to the southern side and began a series of active patrols into the German held sector.

After holding the sector south of Bridge No. 5 to the sea for about a week, the 504th was relieved for reassignment elsewhere on the line. The 1st and 2nd Battalions marched north to support the 3rd infantry Division in an attack to the west while the 3rd Battalion was temporarily attached to the 1st Armored Division in reserve. After several days the 3rd Battalion was committed in the Carreceto sector near the center of the line. It was in this area that the Germans launched their main offensive designed to push the Allies into the sea. The drive was proceeded by an unusually intense artillery barrage early on February 5, 1944. After several days the 3rd Battalion was cut off from supporting British units and forced to withdraw to the war-torn town of Aprilia, known to most Anzio veterans as "The Factory." These troopers had suffered heavy casualties, many companies reduced in strength to between 20 to 30 men.

During the struggle, H Company was ordered to attack and attempt to rescue a captured British general. After a desperate fight, the general was recaptured, but the company found itself cut off. Next, I Company was ordered to make contact with H Company. With only 16 men left in I Company, this mission was carried out successfully. It was for this outstanding performance during the period 8-12 February, that the 3rd Battalion was given one of the first Presidential Unit Citations in the European Theatre of Operations.

For the remainder of their eight-week stay on the Anzio beachhead, the 504th men became involved in defensive operations. On March 23, 1944, the 504th Regimental Combat Team received orders to embark from Anzio and to proceed to Bagnoli. During the eight-week period that the 504th fought in the beachhead, it had lost 120 men killed, 410 wounded and 60 missing in action. After a brief stay at Batnoli, the Combat Team boarded the ship *Capetown Castle* on April 10, bound for England to rejoin the rest of the 82nd Airborne Division.

England provided a welcome and well deserved break from the fighting the division had experienced in the Mediterranean operations. The organizational structure of the division was changed to accomodate three parachute infantry regiments instead of just two. The 504th had just arrived in England on April 22, from Italy and was in need of refitting, replacements and rest. Two other parachute regiments were attached to the 82nd for the next operation. The 507th Parachute Infantry Regiment, commanded by Colonel George Millet (later by Colonel Edson Raff), and the 508th Parachute Infantry Regiment, commanded by Colonel Roy Lindquist, both composed the 2nd Airborne Brigade.

Finally the mission was disclosed late in May of 1944. It would be called Operation NEPTUNE. The operation called for the pre-invasion drop of three Airborne Divisions: the British 6th Airborne Division on the British left flank and the American 82nd and 101st Airborne Divisions to secure the American right flank. Each division had individual objectives which had to be secured before the main invasion force came ashore. The mission of the 82nd Airborne on D-Day was to capture and control the strategic town of Ste. Mere Eglise, control the bridge crossing over the Merderet River and secure the exit routes from the beach area to facilitate the rapid advance of the seaborne troops.

The first C-47s of Force "A" took off at 11:15 p.m. on June 5, 1944. The main flight was preceeded by three regimental Pathfinder teams which dropped 30 minutes prior to the first group.

The 505th Parachute Infantry Regiment landed generally in the vicinity of its drop zone. The 507th Parachute Infantry was scattered. One element dropped in the vicinity of Monteboug; another, south of Carentan; and the remainder, astride the Merderet River, East of the drop zone. The 508th Parachute Infantry Regiment was likewise widely scattered. The bulk of this regiment's paratroopers landed east of the drop zone and some as far away as nine kilometers south of Cherbourg.

Complicating the landing problems were the thousands of hedgerows that subdivided the Norman countryside. Used by the farmers instead of walls or fences, the hedgerows were earthen dikes about four feet high with bushes planted on top. These barriers aided the Germans in the defense and slowed the Allied advance. As a means of defense against gliders, the Germans installed thousands of wooden poles, 8-10 feet high, throughout, fields suspected as being future glider landing zones. Several of the CG-4A Gliders of the 325th Glider Infantry and other glider outfits of the 82nd were destroyed by these formidable obstacles. Many of the paratroopers found many of the fields near the drop zones had been purposely flooded and many of the jumpers were drowned under the weight of their equipment.

In spite of the landing mix ups the resourceful men of the 82nd Airborne managed to accomplish many of the division's objectives. The strategically important town of Ste. Mere Eglise was captured by dawn by the 505th's 3rd Battalion. This town was the first to be liberated on the western front.

With Ste. Mere Eglise in American hands, attention was focused on the other primary division objectives, the two key bridges over the Merderet River at La Fiere and near Chef-du-Pont. The assistant division commander, Brigadier General James M. Gavin, assembled about 500 troopers representing all three parachute regiments. Gavin divided the group into two detachments to secure the two bridges. The bridge at La Fiere changed hands twice during a two-day struggle before the 82nd Airborne achieved final control. The detachment sent to secure the Chef-du-Pont bridge were principally from the 507th Parachute Infantry. These troopers also assumed control of this bridge after much fighting.

The 82nd Airborne Division's second assault wave (Force "B") of the invasion troops came by glider. Before dawn on June 6th, these gliders appeared over their landing zones. Under German anti-aircraft fire, the gliders' landings became scattered. The 325th Glider Infantry Regiment went into action upon landing, reinforcing the division's paratroopers. The men of the 319th and 320th Glider Field Artillery Battalions worked to recover as many of their howitzers and as much ammunition as possible. Initially the 320th was only able to put two guns into action of the 12 flown in. By June 8th, eight howitzers were in action and coordinated offensive action was taken in support of the glidermen of the 325th. Both Glider Field Artillery Battalions fired in support of the crossing of the Douve River by the 508th Parachute Infantry on June 13th.

On June 9th, Private First Class Charles N. DeGlopper of Company C, 325th Glider Infantry Regiment, became the first member of the 82nd Airborne Division to win the Congressional Medal of Honor in World War II. Although wounded several times, DeGlopper stood up in plain sight firing his BAR, covering the withdrawal of his commrades. DeGlopper fired until he was fatally wounded. He had sacrificed his life so that the others would have a chance to escape.

The paratrooper and glider troops of the 82nd Airborne Division were officially relieved of combat duty on July 8, 1944,

"All-Americans" boarding their C-47s for the trip to Holland, Sept. 17, 1944. (Courtesy of Steven Mrozek)

Next stop Holland! – CG-4A Waco Glider is lifted into the sky by a C-47, Sept. 19, 1944. (Courtesy of Steven Mrozek)

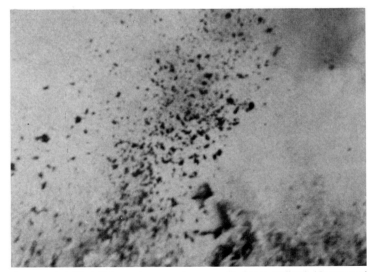
A near miss! 82nd Airborne troopers advance under fire in Holland. (Courtesy of Steven Mrozek)

C-47s drop 82nd Airborne paratroopers near Groesbeek, Holland. (Courtesy of Steven Mrozek)

C Company 707 Airborne Engineer Bn. jumps over Holland, Sept. 17, 1944. (Courtesy of Steven Mrozek)

82nd Airborne landing zone "T" near Groesbeek, Holland, Sept. 19, 1944. (Courtesy of Steven Mrozek)

440th TCC carrying elements of the 508th PIR to Holland, Sept. 18, 1944. (Courtesy of Steven Mrozek)

after 33 days of action. The Division was returned to England for a well-deserved rest and furlough. The Normandy campaign had extracted a terrible cost from the 82nd. Of the nearly 12,000 men that were committed in Normandy, 5,245 were listed as casualties, about 46 percent. Of these, 1,282 were killed in action and 2,373 were seriously wounded.

In the months that followed the Normandy operation the 82nd Airborne experienced some major changes. General Matthew B. Ridgway received command of the newly created 18th Airborne Corps in August 1944. Along with the new command, Ridgway was promoted to lieutenant general. On August 15, 1944, Major General James M. Gavin assumed command of the Division. With the proliferation of Airborne organizations in the armies of the Allied forces on the Western Front, the 1st Allied Airborne Army was organized on August 21, 1944 under the command of General Lewis H. Brereton. The 1st Allied Airborne Army consisted of the 18th Airborne Corps and its three American Airborne Divisions: the 17th, 82nd and the 101st. Also included were the British 1st and 6th Airborne Divisions, the 1st Polish Independent Parachute Brigade and the U.S. 9th Troop Carrier Command.

The 82nd Airborne Division retained its three-parachute regiment structure, which included, at this time, the 504th, 505th and 508th. The 507th was reassigned to the 17th Airborne Division.

In early September, British Field Marshall Bernard Montgomery proposed a plane to secure several of the bridges across the Maas, Waal and Rhine Rivers, and to rapidly advance his British 30th Corps over the Rhine at Arnhem and thrust deep into the heart of Germany. The operation was given the code name MARKET GARDEN. The first phase involved the dropping of three Airborne divisions in key locations. The 101st Airborne Division would land furthest south near the city of Eindhoven; the "Screaming Eagles" would be the first to contact the British armor. Next the 82nd Airborne would take the bridges near Grave and the city of Nijmegen. The northernmost objective was the bridge across the Rhine River near the city of Arnhem. The mission was assigned to the 1st British Airborne Division and the 1st Polish Independent Parachute Brigade.

On the morning of Sunday, September 17, 1944, a total of 1,545 troop transports and 478 gliders took off from 24 airfields in England. The streams of C-47s were protected by 1,130 Allied fighter aircraft.

The invasion aircraft took two routes to their objectives. The British 1st Airborne Division led the 82nd Airborne and travelled by the northern route. The 101st Airborne took the southern route. The 82nd Airborne dropped 7,250 paratroopers that day. General Gavin travelled with Colonel Ekman's 505th in the leading element of the Division's column of transports. Following the 505th came Colonel Bedell's 307th Airborne Engineer Battalion that would jump as a complete unit for the first time. The Engineers were followed by Colonel Tucker's 504th. Colonel Lindquist's 508th was next and bringing up the rear was Colonel Griffith's 376th Parachute Field Artillery Battalion. The 82nd was borne to battle on the wings of 480 transports of the 50th and 52nd Troop Carrier Wings.

Upon landing, the "All-American" troopers rushed to take advantage of the element of surprise and to capture their objectives, which were primarily the bridges over the Maas River, the Maas-Waal Canal and the Waal River. The Grave Bridge was taken by the 2nd Battalion, 504th in about two hours after landing. The 504th's First Battalion was charged with the responsibility of taking possession of the four bridges over the Maas-Waal Canal. The southernmost bridge was captured intact by elements of Company B, 504th and troopers of the 505th. Men of the 504th, 505th and 508th arrived at two bridges near the villages of Malden and Hatert in time to see them blown to pieces by the retreating German soldiers. Early attempts to take the Nijmegen bridges (the highway and the railroad) were foiled by a superior German force.

General Gavin ordered Colonel Shields Warren's 1st Battalion of the 508th to "get the bridge (highway) as quickly as possible."

During the morning hours of September 19, advanced elements of the Guards Armored Division entered the 82nd Airborne sector near Grave. With a clear road up to Nijmegen, units of the British 30th Corps moved up to aid in overwhelming the defending German garrison clinging to the bridge.

Less than an hour before the 82nd Glider Troops were scheduled to land, a strong German force attacked from the Reichswald Forest, east of Groesbeck and overran the landing zones. Elements of the 505th and 508th were soon involved in a desperate struggle for the strategic high ground. The LZ was cleared just as the gliders appeared. During this action, First Sergeant Leonard Funk of Company C, 508th, led a small group of troopers in silencing four German 20mm guns and three anti-aircraft guns and killing 15 of the enemy. For this deed, Funk would be awarded the Distinguished Service Cross.

The situation at the bridge was still deadlocked by September 20. The 505th had made heroic efforts to overpower the German defenders with no success. General Gavin sent for Colonel Tucker of the 504th and discussed a plan to take the highway bridge. Tucker was to cross the Waal River in British Engineer boats and outflank the north end of the bridge. British tanks would supply fire support for the amphibious operation. Gavin further explained that H-hour would be 1400 (2:00 p.m.) the next day, September 20th. Over a hundred artillery pieces would shell the northern shore and British aircraft would strike known German positions.

The Waal River measured 400 yards wide and had a fast current that ran about 10 miles per hour. The 3rd Battalion of 504th would spearhead the operation with the 1st Battalion in support.

The 19-foot engineer boats arrived only 20 minutes before H-hour. The sight of the flimsy canvas boats was far from encouraging. The smoke screen had been blown away as the boats entered midstream. Having never made a river crossing of this nature before and being totally unfamiliar with the canvas boats, many had trouble getting the boats to the opposite shore. In small groups, the troopers eliminated German positions with rifles and grenades; few prisoners were taken. The 504th men flanked both bridges as the 505th, with the aid of several British tanks, overwhelmed the defenders on the southern end. Trapped on the bridge, 267 dead Germans were found after the battle subsided.

Some 200 troopers from the 504th were killed in the river crossing and in subsequent fighting. With the bridge in possession of the Allies, the paratroopers of the 82nd expected to see British armor rolling across to relieve the British paratroopers clinging to hope in Arnhem. The rumbling of armor did not materalize. British-supporting infantry failed to come up in time, so the armor waited. At this point, a wave of bitterness and resentment swept over the men of the 82nd who had made superhuman efforts to take the bridge that day, as all the suffering and loss appeared to be for nothing.

The British in Arnhem were never relieved. After holding out four days longer than planned, the 120 survivors of Colonel John Frost's 2nd Para Battalion broke into small groups in an effort to escape.

On November 11, 1944, elements of the 82nd Airborne began to be relieved by British units. By D+57, two days later, the last of the "All-American" Division was pulled out of the line for a much needed rest. Put aboard trucks, the two American Airborne Divisions were sent to camps at opposite sides of Rheims, France.

During the fighting, after the river crossing, Private John R. Towle of Company C, 504th Parachute Infantry had distinguished himself by attacking advancing German armor single-handedly. Towle succeeded in stopping the armored assault with only a bazooka until he was mortally wounded by a mortar fragment. In recognition of his unselfish act of extreme bravery, Congress posthumously awarded him the Congressional Medal of Honor. Towle was the sceond 82nd Paratrooper to win the coveted award.

The Division's stay at Sissonne, near Rheims, France, provided an opportunity to train replacements and allow the veterans of Holland a chance to unwind after nearly two months on the line.

On December 17, the still understrengthed 82nd Airborne received marching orders for somewhere in Belgium. The division was order by SHAEF (Supreme Headquarters Allied Expeditionary Force) to be on the road in 24 hours. The 82nd was on the move by one hour after daylight on the 18th and trucked in the direction of Bastogne.

The commanding general of the 1st U.S. Army decided to attach the 82nd Airborne Division to V Corps. The Division was to assemble in the vicinity of Werbomont.

From its location in Werbomont, the 82nd Airborne soon began offensive operation against German Field Marshall Von Rundstedt's surprise attack. The story of the Division operations during the Ardennes campaign is best told by its commander, Major General James M. Gavin. The following is taken from his personal report to the War Department.

"I arrived in Werbomont at approximately 2000 hours and about that time the first large group of 82nd vehicles started arriving.

"All during the night the staff worked on closing the vehicles into Werbomont area. About two hours after daylight, December 19th, the division closed in that area.

"At daylight, December 19, it was learned that the North-South road from Bastogne to Werbomont had been cut by the Germans in the vicinity of Houffalize. The depth of this penetration was unknown, but there were rumors from truck drivers that the Germans were on the road in the vicinity of Hotton.

"During the afternoon of December 19, information and orders were received from Headquarters XVIII Corps (Airborne), which had been established about one mile north of Werbomont, that the 1st Army was to hold along the general line of Stoumont-Stavelot-Malmedy and counterattack in the direction of Trois-Ponts to halt the enemy's advance to the northwest. The XVIII Corps (Airborne) assumed command of the sector generally south of the Ambleve River to include Houffalize.

"In compliance with instructions received from Corps Headquarters, the 504th Parachute Infantry advanced and seized the high ground northwest of Rahier, and the 505th Parachute Infantry advanced and seized the high ground in the vicinity of Haut-Bodeax. The 508th Parachute Infantry sent one company to the crossroads one mile east of Bra. The regiment, less one company, occupied the high ground in the vicinity of Chevron. the 325th Glider Infantry remained at Werbomont, having sent the 3rd Battalion to the vicinity of Barvaux and one company to the croosroads at Manhay. Those dispositions were consolidated during the night of December 19-20, and patrols pushed to the front to gain contact with the enemy.

"Shortly after daylight, December 20, I met Colonel Reuben Tucker, 504th commanding officer, in the town of Rahier, at which time he had just received intelligence from civilians to the effect that approximately 125 vehicles, including approximately 30 tanks, had moved through the town the afternoon before moving in the direction of Cheneux.

"If this were the case, the seizure of the bridge over the Ambleve River at Cheneux was imperative if their futher movement was to be blocked.

"Initial contact was made at the western exit of Cheneux by a patrol which had been sent from Rahier by the 1st Battalion of the 504th. They were engaged at once and a heavy fight took place, lasting all day long. This German force, we know now, was the advance guard of a reinforcement battalion of the first SS Panzer Division. The 1st Battalion of the 504th drove them back into Cheneux.

"I went to the 505th Parachute Infantry where I found that they had contacted some engineers who remained in Trois-Ponts. They had occasionally been under fire, but no major German force had moved through the town. All civilians in these northern regimental areas reported that many Germans and much armor had passed through. The situation south of the 505th in the direction of Vielslam was vague. Reconnaissance was pushed in that direction.

"On the afternoon of December 20 at about 1600 hours I was called to Headquarters XVIII Corps (Airborne) to receive orders for an advance to the Vielsalm-Hebronval line. In the meantime, contact had been established with a German SS force, later identified as the 1st SS Panzer Division at Cheneux.

"At Corps Headquarters I received information that they were advancing to the southeast and establishing an active defense along the line Vielsalm-Hebronval-Laroche; that this division, 82nd Airborne, would establish a defensive line from contact with the 30th Division, in the vicinity of La Glieze, to Cheneux - Tois-Ponts - Grant Halleux - Vielsalm - Salmchateau - Hebronval.

"Orders to accomplish this were issued at the division CP at Hablemont shortly before dark, December 20. Units moved promptly and by daylight were organized and prepared to defend. Regiments were in the line in the order, left to right: 504, 505, 508, 325. One battalion of the 325 was held in division reserve in the vicinity of LaVaux.

"In Vielsalm, contact was made with General Hasbrouck who had established the CP of the 7th Armored Division in the town. The division was then fighting around St. Vith, west of Vielsalm. General Jones had established the CP of the 106th Division at Renceveaux.

"Further south at Trois-Ponts, and extending down to Grand Halleux, determined, apparently well planned and executed attacks were being made with increasing strength against the very thinly held front of the 505th. On the south, the 508th and the 325th had no contact with the enemy. The Division Reconnaissance Platoon was pushed south. Information available indicated that the Germans were moving towards the Meuse River.

"The situation in the vicinity of St. Vith appeared to be critical. The town was being over-whelmingly attacked in several directions, and there appeared to be little prospect of preventing its being cut off. On this date, December 21, however, only the narrow neck of land from Vielsalm to Salmchateau, held by the 82nd Airborne Division, connected the St. Vith forces with remaining forces

Headquarter 3rd Bn., 508th PIR in Belgium, December 1944. (Courtesy of Steven Mrozek)

Company D, 504th PIR in Belgium with Father (Cpt.) Kozak. (Courtesy of Steven Mrozek)

of the First Army. Its retention would be decisive.

"The fighting at Cheneux was increasing in bitterness. On this date the first battalion of the 504th, assisted by a company of the 3rd battalion of that regiment, made a final, all-out assault on the Germans. In this attack we destroyed a considerable amount of armor and killed and captured many Germans from the 1st SS Panzer Division.

"Farther to the south and east, the 505th Parachute Infantry was having very hard fighting with the remainder of the 1st SS Panzer Division. The 505th had initially sent a covering force east of the Salm River in the vicinity of Trois-Ponts. Through sheer weight of numbers this small force was finally driven to the river line where it held. Being very much overextended, the regiment managed to hold by diagnosing or estimating the point of German main effort from time to time and then marshalling all available infantry as quickly as possible, beating off the attack at that point. This process was repeated where necessary, day and night until finally the German attacks waned in their intensity about December 23.

"The 508th Parachute Infantry on the Vielsalm-Salmchateau front was without enemy contact except for patrols. The 325th Glider Infantry, aided by the Division Reconnaissance Platoon, had established contact with enemy forces several miles south of their front lines.

"On December 21, I was instructed by the corps commander to make a reconnaissance of the divisional area with a view to withdrawing, after the extrication of the St. Vith forces, to a suitable defensive position that would tie in with the divisions on my right and left.

"I objected to the withdrawal, but the corps commander explained that, regardless of my wishes in the matter, it might be necessary to require the division to withdraw.

"A reconnaissance was undertaken and at its completion it was quite clear that there was but one reasonably good defensive position and that it was the Trois-Ponts-Basse-Bodeau-Bra-Manhay line. At the direction of the corps commander, a reconnaissance was also made of a position farther to the rear, generally along the Cheneux-Rahier-Chevron-Werbomont line.

"From my point of view, it was obvious that the loss of Regne-Lierneux ridge would result in the complete neutralization of the defensive capabilities of the right portion of the division sector. This ridge dominated the entire road out from Vielsalm to Bra. This was the only road not south of the Trois-Ponts-Werbomont road. In addition, all of the division's installations and division artillery were located in the Lierneux-Goronne-Vielsalm valley. Accordingly, orders were issued to the 325th Glider Infantry to extend its right flank and seize and hold Regne and the ridge extending north therefrom. This ridge had to be held at any cost.

"It became clearly evident that the Germans could not bring armor to bear against the sector anywhere between Salmchateau and the Fraiture crossroads, except by bringing it up the Petite-Langlir road. If the Petite-Langlir bridge could be blown, he would be incapable of bringing armor to bear anywhere within this 10,000-yard gap without approaching up the main road toward Salmchateau, which was well-covered.

"The possibility of canalizing his armored attack was obvious, and steps were taken to take advantage of this. Early on December 22, orders were issued to the Engineer Battalion to move without delay and prepare the Petite-Langlir and to destroy the bridge while it was actually being used by German vehicles.

"I ordered the release of the Division reserve battalion of the 325th to the Regimental Commander of that regiment and ordered one battalion of the 504th, and 2nd Battalion, to move at once to the ridge 5,000 yards southwest of Lierneux. These troops went into position during daylight of December 23. On this date the enemy attacked in considerable strength and overran the town of Regne. The 325 was ordered to counterattack and retake the town. The retention of this ridge was most vital if the Division was to accomplish its mission of extricating the St. Vith forces. Supported by attached armor, and with unusual gallantry and elan, the 325 attacked and retook the town and held it until later ordered to withdraw.

"It was becoming increasingly evident that the Germans was determined to ultimately reach Werbomont and move north toward Aywaille and Liege. Colonel Billingslea, commanding officer of the 325th Glider Infantry, was ordered to extend his right flank to include the Fraiture ridge.

"From my viewpoint, its loss would mean that German armor which we had successfully turned back from Trois-Ponts to Regne, with the aid of both terrain and a very active defense, would bypass the division and occupy the Lierneux-Regne ridge mass, thus preventing us from accomplishing our present mission of covering the withdrawal of the St. Vith troops. I accordingly ordered Colonel Billingslea to again extend his right flank and to include in his defensive organization the crossroads southwest of Fraiture. This he did by sending Company F under the command of Captain J. R. Woodruff, to the area. The situation all along the southern front was becoming critical when I visited the battalion commanders of the 325th several times during the period December 22-24.

"On the afternoon of December 23, at about 1730 hours, I arrived at the CP of Captain Gibson in the town of Fraiture. It was then under heavy mortar fire. A considerable volume of small arms fire could be heard to the south and west. SCR-300 contact was made with Captain Woodruff at the crossroads. He stated that he was under terrific attack which was completely engulfing his small unit.

"As it developed, it was the attack of a regiment of the 2nd SS Panzer Division supported by attached armor, attacking with the mission of driving up the main highway to Werbomont. The one company was soon completely overrun. During the hours of darkness, in desperate, close-quarters fighting, Captain Woodruff managed to extricate about 40 men.

"At this point it was evident that there was nothing to prevent the German forces from entering the rear of the division area, which was now closely engaged along its entire 25,000 yard front.

"By telephone, Colonel Tucker was told to be prepared to move the 504 Regimental Headquarters and one battalion to the vicinity of Lansival where he would take over the sector to the right of the division.

"At about daylight, XVIII Corps (Airborne) made available to me Combat Command B of the 9th Armored Division under the command of General Hoge, which had been withdrawn from the St. Vith area. General Hoge reported to my CP at about 0700. At about 0545, December 24, Colonel Tucker was ordered to leave the smallest possible force in the northern sector and to move south to Bra by motor without delay.

"At 0820 verbal orders were issued to General Hoge to hold Malempre until further orders, to contact the 504 on his left and the 7th Armored on his right. The 7th Armored had been recommitted by XVIII Corps (Airborne), down the main road toward Manhay.

"At 1315 hours, General Hoge reported to me that he was holding Malempre. Between Malempre and Fraiture, the 2nd Battalion of the 504th Parachute Infantry was fighting in the woods. There was much close, bitter figthing and the Germans were very roughly handled by Major Wellem's battalion. He finally succeeded in stabilizing his position and containing the Germans, although his frontage was very great, particularly for the wooded sector in which he was fighting.

"During the day of December 24, Colonel Tucker brought up his full regiment, less one battalion which he had left at Cheneux to contain the forces north of the river. That battalion was charged with holding the Ambleve River line from immediately north of Trois-Ponts to where contact was established with the 30th Division in the vicinity of LaGlieze, a frontage of approximatley 12,000 - 15,000 yards, much of it closely wooded country and broken up terrain. The 505 appeared to have all it could do to continue to hold the Trois-Ponts-Grand Halleux line and the 508 was becoming heavily engaged on the Vielsalm-Salmchateau-Joubieval line.

"A conference was held at Headquarters XVIII Corps (Airborne) at about 1330 hours, December 24th, at which time orders were issued for the voluntary withdrawal to the Corps' defensive position. Division plans were completed and orders issued during the afternoon to effect the withdrawal starting after darkness.

"I was greatly concerned with the attitude of the troops toward the withdrawal, the division having never made a withdrawal in its combat history.

"In all of the operations in which we have participated in our two years of combat, and they have been many of multitudinous types, I have never seen a better executed operation than the withdrawal on Christmas Eve. The troops willingly and promptly carried into execution all the withdrawal plans, although they openly and frankly criticized it and failed to understand the necessity for it. But everybody pitched in, and the withdrawal went smoothly.

"On December 25th, we realized that we had just succeeded in withdrawing through a hostile withdrawing force, which was a rather novel maneuver.

"About two days after occupying our new position, an attack was made by the 62nd Volks-Grenadier Division on our left and 9th SS Panzer Division on our center.

"They launched an attack up the main Axis from Lierneux to Hablemont, hitting the 508 and 504 in a coordinated effort that was characterized by great dash and courage. The 3rd Battalion of the 508 was completely overrun. The men remained, however, manning their positions in the houses and foxholes. The battalion commander, Lt. Colonel Mendez, obtained the use of the reserve company of the 2nd Battalion of the 508 on his left, counter-attacked with great gallantry and determination, and drove the 9th SS Panzer from his positions restoring his MLR. The Storm Troopers' losses were extremely heavy. From one field alone, 62 bodies were later removed.

"This ended all offensive efforts of the German forces in the Battle of the Bulge. About a week later the Division attacked, completely overrunning the 62nd V.G. Division and the 9th SS Panzer Division, and capturing 2,500 prisoners, including five battalion commanders. It regained its former position on the Thier-du-mont heights.

"From here, the division withdrew to a rest area from which it was later committed to the attack east of St. Vith, attacking through deep snow over thickly wooded mountains and overrunning a considerable group of German defensive forces in a constant day and night attack lasting for six days. Ultimately they drove into the Siegfried Line to seize Udenbreth and the ridge extending south."

During the final phase of the Ardennes, there occured an incident made possible by the snow conditions of the winter war. On January 29, 1944, First Sergeant Leonard Funk of C Company, 508th Parachute Infantry Regiment, led elements of his unit into the attack on the German held town of Holzeim, Belgium. During the attack, Company C, of the 508th, captured 80 of the enemy. Leaving the prisoners under the guard of four troopers, the main body continued to advance to clear all resistance in the town.

The German prisoners and American guards were approached by three Germans with their captives, members of a patrol from 2nd Battalion, 508th, sent to contact Company C. Because of the similiarity between the enemy's snow suit and our own, the approaching Germans were able to surprise and disarm the American guards. About this time, First Sergeant Funk returned with several other troopers. Sensing that something was wrong, the 508th troopers hesitated a few seconds which gave a German officer time to close in on Sergeant Funk. Once again, owing to the similarity of the snow suits, the German officer was able to get the drop on the Americans. Shoving his Schmeisser machine pistol into Sergeant Funks stomach, the German officer demanded his surrender. Sergeant Funk pretended to comply with the demand and reached up as if to surrender his "Tommy Gun," but with great speed, he quickly reversed the situation by shooting the enemy officer, killing him instantly. This courageous act and quick thinking on Sergeant Funk's part started a battle between the remaining German soldiers and the 508th paratroopers which ended with the troopers in complete control again.

Sergeant Funk was not through with just one of the enemy, he continued firing from the center of the road at the Germans who were making final break for freedom. Many more of the enemy were killed by the fire of Sergeant Funk's Thompson sub-machine gun.

For this encounter, First Sergeant Funk received the Congressional Medal of Honor from President Truman. He became the only living trooper in the 82nd Airborne Division to receive the Medal of Honor in World War II. In addition to the Medal of Honor, Leonard Funk received the Silver Star in Normandy and the Distinguished Service Cross for heroism in Holland. He became the most highly decorated paratrooper in the 82nd Airborne and one of the highest in the Army.

The welcomed rest in the Sissone-Swippes area ended on April 1, 1945. The division was directed to the vicinity of Cologne where the "All-Americans" became part of the operation attempting to encircle German Field Marshall Model and 350,000 of his troops in what would be called the "Ruhr Pocket." The 82nd's role in this operation was a new one - Static Defense. The division held a section of the line on the western bank of the Rhine River which included the city of Cologne. Except for occasional artillery and mortar barrages and small patrols sent across the Rhine River (which separated the opposing forces), the division's duty in Cologne was relatively quiet.

On April 6th the Division ordered a reinforced company to cross the Rhine in the neighborhood of Hitdorf. The job was assigned to A Company of the 504th and a squad from C Company, 307th Airborne Engineers.

The force crossed the river at 0230 in assault boats in waves of platoons. The enemy was contacted immediately upon crossing and a heavy fire fight ensued. The men of A Company found themselves in a minefield and under attack by large numbers of the enemy reinforced with

Company D, 504 marching through Ranier, Dec. 1944. (Courtesy of Steven Mrozek)

armor. In the face of this resistance, the 504th troopers managed to destroy several machine gun nests, establish a roadblock and infiltrate through the enemy and enter Hitdorf. By 0830, the village was cleared and 68 prisoners had been taken.

The Germans made several counterattacks on the village in order to drive out the Americans. The first few were repulsed with heavy losses. After several attempts, the enemy, supported by armor and under cover of smoke, broke through, overrunning several platoon positions. In spite of supporting artillery fire the A Company troopers could not hold their positions and fell back toward the river. Here two platoons established a horseshoe defense, the open end of which faced the water.

By 0130 the next morning, two platoons from I Company were transported over to reinforce what was left of A Company. After repulsing another German attack by over 200 infantry and a platoon of tanks, both companies withdrew back across the Rhine with 13 more prisoners. It was later determined that the 504th troopers had captured 80 Germans and killed or wounded an estimated 350 others. The assaulting force suffered a loss of nine known killed, 24 wounded and 79 missing in action, many of whom are believed to have been killed.

It is hard to say if the river crossing succeeded in its purpose of diverting German troops from a more important location of the pocket to this attack, but the A and I Company force was hit hard by an estimated regiment of infantry plus a platoon of tanks supported by a battalion of artillery. Later the men of A Company, 504th, were awarded a Presidential Unit Citation for the outstanding manner in which they carried out their mission.

Soon after Hitdorf, German resistance in the "Ruhr Pocket" collapsed and the division enjoyed a brief, well-deserved break as occupation troops in Cologne. On April 30, 1945, the 82nd Airborne crossed the northern Elbe River at Bleckede. This was the division's 11th bridgehead-beachhead assault of the war and the last. Making the initial assault were the men of the 505th Parachute Team who also made the initial jump into Sicily two years earlier. Within 24 hours the 505th had advanced over 10,000 yards.

The 504th and 325th Combat Teams moved through the 505th and advanced a total of 52,000 yards by the second evening with the support of Combat Command B of the 7th Armored Division and the 740th Tank Battalion. On May 3rd the 82nd Airborne Division made contact with the Russians; unofficially the war was over. The offical end would come five days later on May 8th, 1945.

Meanwhile on May 2nd another important and unprecedented event had taken place: the surrender of the entire German 21st Army to a single American division—the 82nd Airborne Division. Lieutenant General Von Tippelskirch, commander of the 145,000-man 21st Army, chose to surrender to the Americans rather than the Russians. At the division's CP located in the Marble Palace in Ludwigslust, the German commander signed the surrender document which not only dissolved his command but made his men prisoners. A survey of the equipment captured revealed 2,008 trucks and cars; 109 halftracks, 17 tanks, 197 miscellaneous vehicles, including tractors, motorcycles and buses, 89 trailers and seven eight-inch howitzers. No attempt was made to tally small arms and light equipment, including horses from two Hungarian cavalry divisions.

May 8th, 1945, it was all over. The day the world had waited for all these years had come at last. Elements of the 82nd

Airborne had participated in four airborne assaults: Sicily, Salerno, Normandy and Holland; a seaborne entry at Anzio, and the important river crossing assaults of the Volturno in Italy, the Douve and Merdert of France, the Maas and Waal of Holland and the Rhine and Elbe of Germany. During its 371 days of combat, the division had fought in six countries and had been assigned or attached to every British, American or Canadian army in the ETO except the British 8th Army, which the division fought beside of throughout Italy.

Because of its long service, the 82nd Airborne was not to be redeployed to the Pacific. About the same time, the "All-American" Division was chosen to represent the United States by occupying the American sector of the German capital of Berlin. The 508th Parachute Infantry which had served with the 82nd Airborne since Normandy and which had been detached for special duty on March 1, 1945, was selected by SHAEF to guard its headquarters in Frankfurt-am-Main. The 508th "Devils" would remain in Frankfurt long after the 82nd was sent home and would finally be released from duty in November 1946.

Between May and July, 1945, all of the division's "high point" men were transferred to the 17th Airborne Division for redeployment stateside and discharge.

After the end of the war the 82nd was moved to Epinal in eastern France. Here the division received replacements to fill the gaps made by high point men who were sent home. Many of the replacements had seen combat with the 101st or 17th Airborne Divisions. In July, the division moved to Berlin where it began five months of occupation duty. While stationed in Berlin, the 82nd received scores of dignitaries from all of the Allied nations. The division played host to important military leaders usch as General Dwight Eisenhower, General George Patton, Marshall Zhukov and British Field Marshall Bernard Montgomery. The 82nd created an Honor Guard company composed of some of the most decorated troopers in the division. All of the members selected were over six feet tall and wore white bootlaces, white gloves and white parachute silk scarves. The "All-American" Honor Guard was drilled to precision. During his visit to Berlin, on V-J Day, as it happened, General George Patton complimented the Honor Guard by saying "In all my years in the Army and of all the Honor Guards I've seen, the 82nd Berlin Honor Guard is the best." Since 1945 the 82nd Airborne has been referred to as "America's Guard of Honor."

It was in Berlin that the 82nd Airborne Division was awarded the Orange Lanyard of the Order of William and Belgian Fourragere.

In November the troopers of the 82nd learned that not only was the 82nd going to remain active, but that the division was to march down Fifth Avenue in New York City and receive a hero's welcome. In December the 82nd packed up and left Berlin for the French coast and home. On January 3, 1946, General "Jim" Gavin led his proud division down Fifth Avenue for all of America to see. The 82nd "All-American" Division was home.

With the war over, the United States began to reduce the size of its Army. Only a few of the regular Army divisions were allowed to remain on active duty. The 82nd Airborne was one of these and was the only one on jump status during the post-war era. The revised Army doctrirne designated the "All-American" Division as strategic reserve. The division was to be in a high state of readiness to meet any challenge, anywhere, anytime. To maintain this constant preparedenss, the 82nd participated in a series of tactical field exercises in varieties of weather, terrain and situations.

In 1947, elements of the 82nd Airborne made an air assault against "invading forces" from Florida in Operation COMBINE. That winter, the 505th made a jump into northern New York loaded with snow equipment to train for combat under freezing conditions. Three years later, in 1950, the 82nd Airborne would participate with the reactivated 11th Airborne in the largest peacetime operation, exercise SWAMER.

The 82nd Airborne continued to constantly test new equipment and techniques in an effort to put more troops on future battlefields better equipped ready to meet and defeat the enemy.

In 1958, the 18th Airborne Corps, to which the 82nd Airborne was attached, was designated as the Strategic Army Corps responsible for rapid strike "first in" missions to worldwide trouble spots. Under this doctrine, the 82nd intensified its training to carry out its specialized mission.

During the Cuban Missile Crisis in 1962, the need for such a force became apparent; but it wasn't until 1965 that a unit of the force was deployed in an actual emergency. For the first time in 20 years, the 82nd was called out for a real mission.

On April 29, 1965, the division was assigned the task of restoring peace in the Dominican Republic. The code name for the mission was "Power Pack."

Early the next day the first C-130 transports touched down at San Isidro airfield in the Dominican Republic. Within hours, 33 plane loads were landed and on the move to Santo Domingo, 15 miles away.

The division's 3rd Brigade came in first, followed by the 2nd Brigade and the 1st Brigade. The "All-American" paratroopers had to fight their way through sniper fire across the Duarte Bridge to get into Santo Domingo. That afternoon the division had taken casualties for the first time since the end of World War II. As in future conflicts, the troopers were not permitted to unlease the division's full power; diplomacy and negotiations were given priority. The First Brigade served as part of an inter-American peacekeeping force for over a year, returning in the summer of 1966.

THE "GOLDEN" BRIGADE AND VIETNAM

A major offensive by the NVA/VC during the traditional Tet holiday truce erupted throughout South Vietnam on January 25, 1968. At every point, Allied forces were hard pressed to contain this all out effort by the enemy. With nearly all available reserves having been committed, additional reinforcements were needed to put the Allied forces back on the attack. The unstable situation created by the communist offensive prompted General William C. Westmoreland to make an urgent request for additional troops.

As the backbone of the American Rapid Deployment Force, the 82nd Airborne Division was the only immediate source of combat ready troops. As the Army's strategic reserve force, the division had not been previously committed to the Vietnam conflict.

The situation in Vietnam was critical, and the decision was made to deploy the 82nd's Third Brigade. Notified early in February 1968, urgent efforts were employed to bring the brigade up to strength. Badly needed airborne personnel were reassigned from other units of the division and even from Special Forces units at Fort Bragg.

Led by Colonel (later Brigadier General) Alexander R. Bolling, the advance elements of the Brigade departed for Vietnam on February 14. Traveling by way of Alaska and Japan, they arrived at Chu Lai Airbase at 1100 hours, February 15. Several days and 140 sorties were required to transport the entire 3,650 troopers of the brigade to Chu Lai.

Once "in-country," it was discovered that a majority of the brigade personnel did not meet the requirements for overseas combat deployment. Over 2,500 troopers accepted the option to return to Fort Bragg. Once again replacements were desperately sought. Being less particular this time, most of the new replacements were non-airborne qualified.

Designed to operate as a separate unit, the "Golden" Brigade was composed of 15 individual elements once it was finally organized. The following were assigned to the Third Brigade during its service in Vietnam:

1st Bn. (Abn), 505th Infantry
2nd Bn. (Abn), 505th Infantry
1st Bn. (Abn), 508th Infantry
2nd Bn. (Abn), 321st Artillery
Troop B, 1st Squadron (Abn), 17th Cavalry
Company C, 307th Engineer Bn. (Combat) (Abn)
58th Signal Company (Forward Area Support)
82nd Support Bn. (Abn)
Company A, 82nd Aviation Bn. (Abn)
Company O, (Ranger), 75th Infantry
52nd Chemical Detachment
408th ASA Detachment
518th MI Detachment
37th Infantry Detachment (Scout Dog)
3rd Platoon, 82nd MP Company (Abn)

Coming under the control of the 101st Airborne Division on March 1, the brigade initiated Operation CARENTAN I by incorporating hunter-killer patrols against the enemy infiltrators. Shortly after the conclusion of CARENTAN I, the Third Brigade joined the 101st Airborne and the 1st ARVN Infantry Division in Operation CARENTAN II. Ending on April 1, both operations took place in Quang Tri and Thua Thien provinces. Control of the Brigade was transferred on May 1, 1968 from the 101st Airborne to U.S. Army, Vietnam. By May 17, the Brigade's casualties totaled 43 dead and 270 wounded, while during the same period, the Brigade claimed 727 enemy killed and 18 captured.

In September the Brigade was relocated from I Corps, south to the Saigon area. Under the command of the Capital Military Assistance Command, the Third Brigade was charged with the protection of Tan Son Nhut Air Base and the area northwest of Saigon.

Actively engaged in civic actions as well as aggressive patrolling, the Brigade established and maintained tight control over its area of responsibility. So effective were the Third Brigade's operations, no enemy rocket attacks were launched from its area of responsibility. During this period the Brigade often conducted operations with the 1st Cavalry and the 25th Infantry Divisions.

In September of 1969, the Brigade learned that it would be redeployed back stateside in December of that year. The Brigade's final operation began the same month. During Operation YORKTOWN VICTOR, elements of the Brigade operated in the "Iron Triangle" north of Saigon.

After 22 months of combat in Vietnam, the "Golden" Brigade's job ended on December 12, 1969. A total of 212 "All-Americans" were killed during the Brigade's service in the conflict.

During the course of its service in Vietnam, the Third Brigade received many awards. The entire Brigade was awarded the Vietnamese Cross of GalLantry. The Presidential Unit Citation, Valorous Unit Award, Meritorious Unit Commendation and several Vietnamese commendations were awarded to many individual Brigade units.

Editor's Note: For information regarding the 82nd Airborne's participation in Grenada and Panama, see "Airborne Operations."

In Sicily troopers used the native mode of transportation–donkeys–to carry equipment and weapons. (U.S. Army photo)

101ST AIRBORNE DIVISION

By Robert Jones
Col., AUS (Ret)

DIVISION FACTS

Campaigns
- Normandy
- Rhineland
- Ardennes-Alsace
- Central Europe
- Counteroffensive, Phase III
- Tet Counteroffensive
- Counteroffensive, Phase IV
- Counteroffensive, Phase V
- Counteroffensive, Phase VI
- Tet 1969, Counteroffensive
- Summer-Fall 1969
- Winter-Spring 1970
- Counteroffensive, Phase VII
- Sanctuary Counteroffensive
- Consolidation I
- Consolidation II

Decorations

World War II
Presidential Unit Citation (Army), Steamer embroidered NORMANDY
Presidential Unit Citation (Army), Streamer embroidered BASTOGNE
French Croix de Guerre with Palm, Streamer embroidered NORMANDY
Belgian Croix de Guerre with Palm, Streamer embroidered BASTOGNE
Belgian Fouraguerre
Netherlands Orange Lanyard

Republic of Vietnam
Republic of Vietnam Civil Action Medal, First Class, Streamer embroidered VIETNAM 1968-1970 (HHC)
Republic of Vietnam Cross of Gallantry with Palm, Streamer embroidered VIETNAM 1968-1969 (HHC)
Republic of Vietnam Cross of Gallantry with Palm, Streamer embroidered VIETNAM 1971 (101st Airborne Division)

Note: The campaigns and unit awards listed above are for Headquarters and Headquarters Company, 101st Airborne Division (Air Assault). Individual units within the division have more campaign credits or unit awards.

WWII STATISTICS

Chronology
Activated-15 August 1942
Arrived ETO-15 Sept. 1943
Arrived Continent (D-Day)- 6 June 1944
Entered Combat-6 June 1944

Days in Combat- 214

Casualties (Tentative)
Killed- 2,043
Wounded- 7,976
Missing- 1,193
Captured- 336

Individual Awards
Distinguished Service Cross- 47
Legion of Merit- 12
Silver Star- 516
Soldiers Medal- 4
Bronze Star- 6,977
Air Medal- 46

Campaigns
Normandy
Ardennes
Rhineland
Central Europe

Organic Units
501st Prcht Inf Regt
502nd Prcht Inf Regt
*506th Prcht Inf Regt
**401st Gli Inf Regt
325th Gli Inf Regt
327th Gli Inf Regt
101st Prcht Maint Bn
326th Abn Engr Bn
326th Abn Med Co
81st Abn AA Bn

101st Abn Division Artillery
321st Gli Field Artillery Bn
907th Gli Field Artillery Bn
377th Prcht Field Artillery Bn
*463d Prcht Field Artillery Bn

@Special Troops
426th Quartermaster Company
801st Ordnance Company
101st Signal Company
Military Police Platoon
Headquarters Company
Rcn Platoon
*Band

* Assigned in 1 March 1945 reorganization
** Disbanded in 1 March 1945 reorganization
@ Hq Special Troops activated 1 March 1945. Units previously directly under division.

Shoulder Patch- Black shield with black tab streaming above. On the shield is the head of a bald eagle; lettered inside the tab, in yellow, is the word "AIRBORNE." The shoulder patch dates back to 1923 when the 101st Division was an organized Reserve unit assigned to the state of Wisconsin. The head of the bald eagle alludes to "Old Abe," a live bald eagle carried by the 9th Wisconsin Regiment during the Civil War. (Although the Adjutant General's Office had stated that there was no connection between the 101st Infantry Division and the 101st Airborne Division, the adjutant general on 26 January 1950 stated that the 101st Infantry Division was reconstituted, withdrawn from the Organized Reserves and consolidated with the elements of the 101st Airborne Division (Regular Army) effective 15 August 1942.

Nickname- Screaming Eagle. (The Division gets it nickname from the Division shoulder patch.)

Chronology
15 August 1942- First activation for World War II, Camp Claiborne, Louisiana.
October 1942- Unit went to Fort Bragg, North Carolina, for training under the Second Army Command.
28 August 1943- Unit participated in two phases of the Tennessee Maneuvers.
5 September 1943- Departed the United States for European duty from New York City.
15 September 1943- Arrived in England to receive additional training at Berkshire and Wiltshire.

6 June 1944- Division entered World War II combat. Division was dropped in Normandy behind Utah Beach in the Allied invasion of Northern Europe.

13 July 1944- Returned to England for rest of training.

17 September 1944- Division participated in one of the largest airborne invasions ever attempted. It dropped into Holland for 72 days of continuous action, following 33 days of action in Normandy.

28 November 1944- Division returned to France for further training.

18 December 1944- The unit moved to Belgium to stop the German breathrough. In a short time, the Division was surrounded and cut off from other units. Ammunition and blood plasma had to be dropped from the air.

26 December 1944- The 101st, surrounded at Bastogne, was finally reached by the 4th Armored Division. The stand of the 101st was credited with being one of the greatest single factors which blunted the enemy's drive. For this heroic action, the entire 101st was awarded the Distinguished Unit Citation at Mourmelon, France (March 15, 1945).

17 January 1945- Division moved to Drulingen and Pfaffenhoffen in Alsace and engaged in defensive harassing patrols along the Moder River.

31 January 1945- The Division crossed the Moder in a three-company raid.

26 February 1945- The Division assembled at Mourmelon, France, for training and was placed under the Seventh Army Command.

31 March 1945- Division moved into the Ruhr Pocket.

8 May 1945- VE Day (101st Airborne Division remained on occupation duty until shortly before its inactivation in France.)

30 November 1945- Inactivation, post-World War II, European Theater of Operations.

6 July 1948- Reactivation as a Training Center, Camp Breckinridge, Kentucky.

27 May 1949- Second inactivation, Camp Breckinridge, Kentucky.

24 August 1950- Reactivation as a Training Center, Camp Breckinridge, Kentucky.

1 December 1953- Inactivated again.

15 May 1954- Activated at Fort Jackson, South Carolina, as a training division with personnel and equipment for the 101st coming from the 8th Infantry Division.

16 March 1956- 101st Airborne Division transferred, less personnel and equipment, from Fort Jackson, South Carolina, to Fort Campbell, Kentucky. Use of division designations to identify certain Army training centers was discontinued. The Training Center at Fort Jackson, South Carolina, identified as the 101st Abn Div (Tng), was redesignated as the United States Army Training Center, Infantry.

21 September 1956- Official reorganization ceremonies for the 101st were held at Fort Campbell, Kentucky. The division was organized under the new pentomic concept with five combat groups, each self-contained.

October 1956 through January 1957- The capabilities of the 101st were tested in a four-phase operation: Troop Test JUMPLIGHT. The pentomic concept proved so successful, it was adopted by the Infantry Army-wide.

September 1957- President Eisenhower ordered the division into Little Rock, Arkansas to quell civil disturbances arising out of the federal order to integrate the public schools.

April 1958- In Exercise EAGLE WING, the biggest training exercise yet for the 101st, a division-sized airborne assault was mounted on Fort Campbell drop zones. Because of freak winds, five jumpers were killed and 155 injured.

May 1958- Strategic Army Command (STRAC) was formed with the 101st as one of its spearhead units.

November 1958- Exercise WHITE CLOUD was conducted at Fort Campbell and Fort Bragg involving 17,000 troops. The operation tested the heavy drop, a new technique at this time.

The Regimental Commanders of the 101st Airborne Division being decorated with the Silver Star in Carentan, France on June 20, 1944. L to R: LTC John H. Michaelis, COL. Howard R. Johnson, COL. Robert F. Sink, COL. Joseph H. Harper. (Courtesy of Robert Jones)

The combat-ready trooper of the 101st Airborne Division. Circa: 1958-60. (Photo by SP4 Wm. E. Hook, courtesy of Robert Jones)

September 1959- During Operation RANGER BULLDOZER, an airstrip 2,500-by-60 feet was built in 48 hours under simulated combat conditions.

March 1960- In Exercise PUERTO PINE, 20,000 troops and 40,000 tons of equipment were airlifted to Puerto Rico.

August 1961- Exercise SWIFT STRIKE I gave the 101st the chance to evaluate the new M-14 rifle. The 101st was the first unit to receive this new rifle.

September 1961- The 327th Infantry went to Europe during Exercise CHECKMATE II, a joint exercise in which the armed forces of the United States, Greece, Turkey, Italy and Great Britain participated.

August 1962- The 101st participated in SWIFT ACTION II in the Carolinas. There were 60,000 troops involved in the maneuver.

September 1962- The 101st performed another civil disturbance mission in Mississippi where they were ordered when rioting erupted as a result of the federal order to admit a black student to the University of Mississippi.

July-August 1963- The 101st played a major role in Exercise SWIFT STRIKE III, the biggest peacetime maneuver (involving 100,000 men) ever held.

February 1964- Division reorganized under Reorganization Objective Army Division (ROAD), with three brigade headquarters, each capable of controlling a varying number of combat elements.

April 1964- Over 2,500 paratroopers from the 1st Brigade participated in Exercise DELAWAR, conducted in Iran.

May 1964- Entire division went to California to test and become thoroughly familiar with ROAD concept in DESERT STRIKE.

6 July 1965- 1st Brigade moved via air from Fort Campbell to San Francisco and embarked by ship for Vietnam.

29 July 1965- 1st Brigade arrived at Cam Ranh Bay where they were greeted by two former commanders: Maxwell D. Taylor, ambassador to Vietnam and General William C. Westmoreland, commander of MACV.

August-September 1965- In Operation HIGHLAND and Operation GIBRALTAR, 1st Brigade secured base at An Khe and opened the way for arrival of the 1st Air Cavalry Division.

September-November 1965- In Operation SAYONARA, the 1st Brigade provided security for the incoming Korean Tiger Division.

November 1965- 1st Brigade established a home base at Phan Rang. Because of its mobility, it earned the nickname "Nomads of Vietnam."

November 1965-May 1966- 1st Brigade was involved in Operations CHECKERBOARD (Ben Cat-Lai Khe), VAN BUREN (Tuy Hoa), SEAGULL (Phan Rang), HARRISON (Tuy Hoa), FILLMORE (Phu Yen Province), AUSTIN II (Phan Thiet), AUSTIN VI, (Nhon Co-War Zone D).

June 1966- In Operation HAWTHORNE, the 1st Brigade blunted the NVA offensive in Kontum Province in one of the most viciously contested battles of the Vietnam War. (In September 1968, the Brigade received the Presidential Unit Citation for extraordinary heroism during this operation.)

June 1966-September 1967- 1st Brigade involved in Operations BEAUREGARD (Dak To), DECKHOUSE II and NATAHAN HALE (Tuy Hoa), FARRAGUT (Phan Rang-Song Mao), SUMMERALL (Phu Yen Province), MALHEUR I (Duc Pho) and MALHEUR II, HOOD RIVER, BENTON and COOK (Quang Ngai Province).

September-November 1967- In Operation WHEELER, started near Chu Lai and carried into Song Tranh Valley, the 1st Brigade was involved in its largest single operation in Vietnam.

November-December 1967- At Fort Campbell, the 2d and 3d Brigades began Operation EAGLE THURST and made military history as the only Army division to be completely airlifted into combat. In 41 days, 10,536 troops and 5,118 tons of equipment were airlifted to Vietnam.

13 December 1967- Division headquarters was established at Bien Hoa and division reported to be ready for action.

31 January 1968- Action began with the start of Tet. Within one day, battles were raging throughout the country from the DMZ to the Delta.

February 1968- A 35-man platoon from the 101st defended the Saigon Embassy and successfully repelled attackers.

February-March 1968- In Operation JEB STUART the 2d Brigade fought for control for Quang Tri near the DMZ and fought to reclaim Hue. Two men of D Company, 2/501st were later awarded the Medal of Honor for action in the battle to reclaim Hue.

March-May 1968- Division fought near Hue in Operation CARENTAN I and CARENTAN II.

April-May 1968- Division cut enemy's supply line in the A Shau Valley during Operation DELAWAR.

17 May 1968- Operation NEVADA EAGLE, which lasted 288 days, followed by Operation KENTUCKY JUMPER. Mission was to keep the enemy from harvesting rice crop and to drive them out of the lowlands of Thua Thien Province. Over 4,000 enemy troops were killed or captured and 668 tons of rice captured.

1 July 1968- Division was redesignated 101st Air Cavalry Division, a name change that lasted 11 weeks. It became airmobile with helicopters partially replacing 15,000 parachutes.

August 1968-July 1969- In Operations SOMERSET PLAIN, MASSACHUSETTS STRIKER, APACHE SNOW and MONTGOMERY RENDEZVOUS, the division drove the enemy from A Shau Valley, built an airstrip and opened the valley for the use of armor.

May-August 1969- 1st Brigade fought in Tam Ky.

August 1969-March 1970- Fighting resumed in Thua Thien Province in Operation RICHLAND SQUARE, REPUBLIC SQUARE, SATURATE and RANDOLPH GREEN. A vigorous civic affairs program was also started. The division assisted in the resettlement of more than 8,000 people.

23 May 1970- Division receive the Vietnamese Civic Action Medal for its involvement in civil affairs. This was the first unit award received by 101st since its arrival in the country as a full division.

June 1970 to Present- Participated in unnamed campaigns (civic actions).

6 April 1972- 101st Airborne Division returns from Vietnam action.

COMBAT HISTORY

After the activation of the 101st Airborne Division on 16 August 1942 at Camp Claiborne, Louisiana, Major General William C. Lee observed that, "The 101st...has no history, but it has a rendezvous with destiny." Time and time again, the 101st has kept that rendezvous and, in so doing, has acquired a proud history.

The 101st Airborne Division traces its lineage to World War I with the formation of the 101st Division on 23 July 1918. It was demobilized 11 December 1918 as a result of the armistice. In 1921, the 101st Infantry Division reconstituted and reorganized as a reserve unit with headquarters in Milwaukee, Wisconsin. On 15 August 1942, the division disbanded as a reserve unit and activated in the United States Army as the 101st Airborne Division.

Following the suggestion of Major William C. Lee, the War Department formed the first parachute unit, a test platoon, at Fort Benning, Georgia. This platoon, consisting of volunteers from the

Horsas landing during June 6, 1944 evening flights. (Courtesy of 101st Airborne Division Association)

29th Infantry Regiment, made its first jump on 16 August 1940. The success of this unit led to the establishment of several parachute battalions and regiments during the following two years.

In 1942, Major Lee, by now brigadier general, endorsed the concept of airborne divisions after studying the British Army use of parachute troops during the early years of World War II. The commander of Army ground forces, Lieutenant General Lesley J. McNair, concurred with Lee's recommendation and ordered the formation of the 82d and the 101st Airborne Divisions. The initial personnel and equipment for both divisions came from the 82d Motorized Infantry Division which inactivated at the same time. Brigadier General Lee assumed command of the 101st Airborne Division on 19 August 1942.

Originally, the 101st had one parachute regiment (the 402d Parachute Infantry), two glider regiments (the 327th and the 401st Glider Infantries), and three artillery battalions (the 377th Parachute Field Artillery, the 321st Glider Field Artillery and the 907th Glider Field Artillery). Additional support units were the 326th Airborne Medical Company and the 426th Airborne Quartermaster Company.

Organizing and training the new division was a challenge. In October, the 101st began rigorous training at Fort Bragg. Throughout the fall and winter, General Lee helped to establish a whole new tactic of warfare—the use of airborne troops in battle.

In June of 1943, the 101st received a second parachute regiment, the 506th Parachute Infantry from Camp Toccoa, Georgia. The 506th had trained in the shadow of Currahee Mountain and had adopted the name "Currahee" as its motto. That summer, the division proved itself during the "Tennessee Maneuver" and the Army deployed it to England.

The 101st Airborne Division boarded ships in New York Harbor and arrived in England 10 days later. They spent 10 months in the counties of Berkshire and Wiltshire, training six days a week. Units worked on close combat, night operations, street fighting, combat field exercises, chemical warfare and a number of other military subjects in addition to demanding physical training, which included hikes of 25 miles. They also trained in the use of German weapons. In October, the 101st began its own jump school to train over 400 new personnel and key members of non-jump units of the 101st.

In January 1944, the newly nicknamed Eagle Division received a third parachute regiment, the 501st Parachute Infantry. Two months later, the 401st Glider Infantry Regiment detached one battalion to be a part of the 82d Airborne Division. Major General Lee suffered a heart attack in February and returned to the United States. Major General Maxwell D. Taylor became the new commander.

Meanwhile, the Supreme Headquarters Allied Expeditionary Force planned an invasion of Northern France and called it Operation OVERLORD. The mission of the 101st was to jump in before the waterborne invasion forces landed on the beach. The paratroopers would secure exits from the beachhead and prevent these areas from receiving German reinforcements.

In preparation for its mission, the 101st participated in three Army-wide exercises: BEAVER, TIGER and EAGLE. In May, elements of the division began leaving their training areas for the airfields and marshalling areas. They would not assemble again until they met on the drop zones of France.

At 15 minutes after midnight on 6 June 1944, Captain Frank L. Lillyman led his team of 101st Pathfinders out of the door of a C-47 transport and landed in occupied France. Behind the Pathfinders came 6,000 paratroopers of the 101st

Airborne Division in C-47s of the IX Troop Carrier Command. D-Day began. Running into heavy German flak as they approached the drop zones, many of the troop transports took evasive action and scattered the jumpers over a wide area. By nightfall, only 2,500 men could assemble in their units.

Struggling to carry out the mission of the 101st which was to clear and secure the exits from Utah Beach for the arrival of the 4th Infantry Division, small groups of soldiers valiantly did the best they could. Major General Taylor could assemble only a little over a hundred men, most of them officers, before he set out to secure one of the causeways leading to Utah Beach. Referring to his brass-heavy group, Taylor remarked, "Never were so few led by so many."

On the night of 6 June, the assistant division commander, Brigadier General Don F. Pratt, led 52 gliders during the invasion. Although all of the pilots managed to put down within two miles, only six of them landed in the designated zone. Intelligence reports had not mentioned that most fields were bordered with hedgerows, four-foot earthen fences covered with a tangle of hedges, bushes and trees. Because of the darkness and the hazard caused by these hedgerows, five soldiers were killed in the landings. One of them was Brigadier General Pratt.

Glidermen played an important role during the Normandy operations. As counterparts of the airborne infantrymen, they delivered personnel, equipment, vehicles and weapons to the 101st Airborne Division.

The first daylight glider operation occurred on the morning of 7 June. Using a heavier cargo glider, the pilots delivered 157 personnel, 40 vehicles, six guns and 19 tons of equipment to the 101st Airborne Division. This equipment was crucial to the success the division had in carrying out its objectives.

After the seizure of the causeways, the 101st proceeded toward a new objective, the capture of the town of Carentan, which was the junctional point for the two American forces from Utah and Omaha Beaches and a key to the success of the invasion. For five days the 101st waged a bitter fight and finally managed to dislodge the German 6th Parachute Regiment out of the town and held Carentan until American armor units arrived from the beachhead. During the attack on Carentan, Lieutenant Colonel Robert G. Cole led the 3d Battalion, 502d Infantry in a successful bayonet charge that wiped out a strategically important pocket of enemy resistance. For this action, Lieutenant Colonel Cole became the first member of the 101st to win the Congressional Medal of Honor.

After 33 days of continuous fighting, the 101st Airborne was relieved and allowed to return to England for further training. Elements of the division received the Distinguished Unit Citation and the division commander earned the Distinguished Service Cross.

Throughout the rest of the summer, the Allied airborne forces prepared for several major operations, all of which were cancelled because of the speed of the Allied advance. Logistical problems and stiffening German resistance slowed the Allies short of the German border. On 17 September, the 101st took part in the largest and most daring airborne operation of the war, Operation MARKET GARDEN. Three airborne divisions, the British 1st and the American 101st and 82d, would jump into a narrow corridor in Holland and seize a series of important bridges. At the same time, a British army corps would drive out of Belgium, quickly cross the captured bridges, finally cross the Rhine at the town of Arnhem and then sweep into the German Ruhr in order to end the war earlier.

On 17 September, the 101st jumped

Members of the 502nd Parachute Infantry Regiment hold the flag which flew over Carentan, France. L to R: LTC Robert G. Cole, unknown, SGT. O. Reilly, Major John P. Stopka. (Courtesy of Robert E. Jones)

A Dutch girl and her father greet American airborne troops with the Stars and Stripes as the soldiers of the First Allied Airborne Army advance down a highway after landing behind the German lines in Holland. (Courtesy of 101st Airborne Division Association)

into four drop zones between the Dutch towns of Son and Veghel and set out to seize their objectives. Heavy opposition from elements of several German divisions around the town of Best presented a serious threat to the division and the entire MARKET GARDEN Operation. During this battle, Private First Class Joe E. Mann of the 3d Battalion, 502d Parachute Infantry, became the second member of the division to win the Congressional Medal of Honor. Private First Class Mann shielded the men of his squad from an exploding grenade at the cost of his own life.

Two days after the 101st landed in Holland, the first elements of the British Guards Armored Division reached the Americans at Eindhoven, the first Dutch city to be liberated. While the British continued their unsuccessful drive to capture Arnhem, the American paratroopers fought a series of engagements against superior German forces that were trying to cut the corridor along a 16-mile front. After 72 days in combat, the Eagle Division received relief and, at the end of November, they went to a base camp at Mourmelon-le-Grand, France for a long and well-deserved rest.

The glider operations associated with MARKET GARDEN were among the most extensive in the war. American glider troops of the 101st and 82d Airborne Division departed from 17 different airfields. The 101st alone used a total of 933 gliders. Over 750 of these made landings either on the landing zone or within one mile of it. The men, material and weapons brought in by gliders once more played a decisive role in the success of the mission.

While the 101st rested in France, Adolf Hitler prepared a surprise attack involving 13 German armored and infantry divisions. He hoped to paralyze Allied forces in the west and defeat the Soviet army in the eat. Some 68,822 men of the United States VIII Corps occupied a 40-mile front in the Ardennes region of Belgium. On 16 December, the Germans attacked. The American front began to collapse, and the entire northern wing of the Allied armies in the west was threatened. At 2030 hours, 17 December, the 101st received orders to proceed north to Bastogne.

Brigadier General Anthony C. McAuliffe, the acting commander (General Taylor was on leave), led the 11,840 soldiers to the strategically important Belgian town of Bastogne. They traveled 107 miles in open, 10-ton trucks, most of which had been hurriedly gathered from Rouen and Paris. As the German forces were overrunning the lightly protected approaches to the town, McAuliffe directed the 501st Parachute Infantry Regiment east in the direction of the town of Longvilly, an offensive mission that temporarily disorganized the Germans and gave the 101st time to set up the important defense of Bastogne.

Bastogne was in the center of a highway network that covered the eastern portion of the Ardennes, an area that required mechanized forces to use roads rather than field for rapid movement. It was the mission of the 101st to hold Bastogne and to disrupt the German line of communication. During the battle, Combat Command B of the 10th Armored Division, the 705th Tank Destroyer Battalion and the 969th Field Artillery Battalion were attached to the 101st. These attachments played influential roles in the outcome of the battle.

On 20 December, German troops isolated Bastogne and the 101st by seizing the last road leading out of the town. The success of their offensive in the west depended on the defeat of the 101st and the capture of Bastogne. Strong German armored and infantry forces tried to break through the American lines north, then south and finally west of the town and

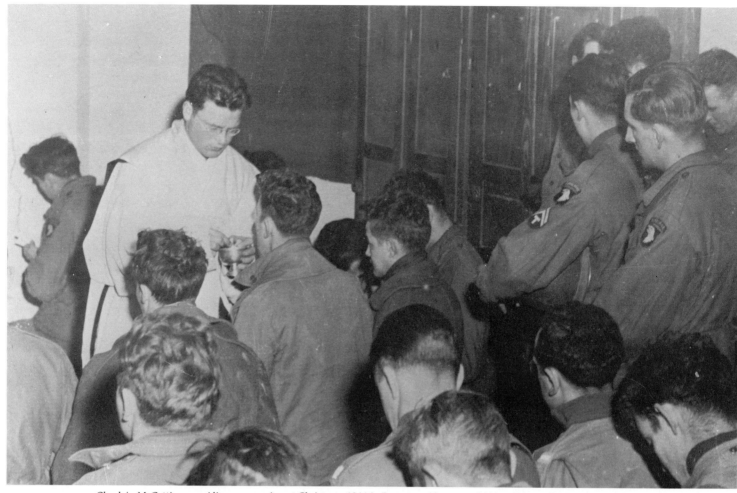

Chaplain McGettigan presiding over services at Christmas, 1944 in Bastogne. (Courtesy of 101st Airborne Division Association)

were beaten back each time. On 22 December, the German commander, Lieutenant General Heinrich von Luttwitz, issued a demand for surrender. General McAuliffe gave his now-famous reply, "Nuts!" Although outnumbered by units from five German divisions, the 101st continued to resist until 26 December when the American 4th Armored Division broke through to Bastogne.

Although no glider assault was planned in connection with the battle for Bastogne, glider troops again played an important role in the war. Encircled and dangerously low on ammunition, the 101st had over 400 wounded housed in civilian facilities and without medical aid. Early in the morning of the 26th, cargo gliders managed to deliver much needed supplies, litter jeeps, aid men and surgeons.

During the next three weeks, the Screaming Eagles encountered some of the hardest and bloodiest fighting of the Bastogne campaign. Teamed with the United States Third Army, they reduced the German pocket in the Ardennes and ended German resistance in the area. For its heroic defense of Bastogne, the 101st was awarded the Distinguished Unit Citation by Supreme Allied Commander General Dwight D. Eisenhower, the first time in the history of the United States Army that an entire division received the award.

On 18 January 1945, VIII Corps relieved the 101st of its task of defending Bastogne. Upon departing, the division received a receipt from the VIII Corps command that read: "Received from the 101st Airborne Division, the town of Bastogne, Luxembourg Province, Belgium. Condition: Used, but serviceable."

At the end of March, the 101st went to the Rhur Region of Germany without the 501st Parachute Infantry Regiment. The 501st remained in reserve for a proposed, but never conducted, special raid to free Allied prisoners of war. After the collapse of the Ruhr pocket, the rest of the 101st moved to southern Bavaria. Crumbling German resistance was only delaying, not stopping, the American advance across southern Germany. The combat mission of World War II for the Screaming Eagles was the capture of Berchtesgaden, Hitler's vacation retreat. Once again teamed with the 3d Infantry Division, the 101st completed its mission and spent the remainder of the war at Berchtesgaden, with some elements in Austria. Battery A of the 321st Field Artillery fired the last combat round for the division in this operation.

While at Berchtesgaden, the 101st received the surrender of the German XIII SS and LXXXII Corps. Several prominent Nazis were also captured. The 506th Parachute Infantry Regiment captured Field Marshal Albert Kesselring, commander-in-chief of the Nazi party. The 502d Parachute Infantry Regiment captured Julius Streicher, the anti-Semitic editor of Der Sturmer, and Obergruppenfuhrer Karl Oberg, the chief of the German SS in occupied France. Colonel General Heinz Guderian, a leading armor expert, was also captured.

On 1 August, the 101st Airborne Division left Germany for Auxerre, France to begin training for the invasion of Japan. When Japan surrendered two weeks later, the operation became unnecessary. The 101st inactivated on 30 November at Auxerre.

During the next 11 years, the 101st activated and then inactivated three times as a training unit, including periods at Camp Breckinridge, Kentucky, from July 1948 to May 1949 and from August 1950 to December 1953. In May 1954, the division activated at Fort Jackson, South Carolina, remaining until March 1956 when Fort Campbell, Kentucky became the new home.

Throughout early 1956, Fort Campbell received new units and reorganized and trained them under the auspices of Headquarters, 101st Airborne Division (Advance). Then on 21 September, the 101st became the Army's first pentomic division. The uniquely equipped and reorganized division comprised five self-contained battle groups: 2d Abn BG, 187th Infantry; 1st Abn BG, 327th Infantry; 1st Abn BG, 501st Infantry; 1st Abn BG, 502d Infantry; and 1st Abn BG, 506th Infantry.

In September 1957, elements of the 101st Airborne went to Little Rock, Arkansas, to assist in maintaining order during a series of civil disturbances. The integration of Central High School in Little Rock proved to be a major milestone in the drive for racial equality. In the midst of the tension and potential violence that surrounded the operation, the courage, tact and discipline of the troops of the 101st prevented a possible tragedy.

Geared to fight on the nuclear battlefield, the 101st participated in a number of important exercises beginning in 1958 with a division-sized exercise called EAGLE WING. In 1960, the 101st conducted exercise QUICK STRIKE which placed the division in a nuclear battlefield situation. In the following year, the Screaming Eagles joined the 82d Airborne Division in a series of large-scale exercises in North and South Carolinas called SIFT STRIKE I, II and III.

The division underwent a second major reorganization in early 1964, shedding the pentomic concept for the Reorganization Objective Army Division organization. The basic components of the new Airborne ROAD Division were the nine infantry battalions, a cavalry squadron and three artillery battalions. This new structure increased the firepower of the 101st, improved ground mobility and facilitated better command and control. Several local operations and a major joint service maneuver, involving 100,000 troops in the Mojave Desert, successfully tested the new organization. The next spring, the Screaming Eagles prepared to send a brigade of infantry and support troops to the Republic of Vietnam.

The 1st Brigade of the 101st Airborne Division landed at Cam Ranh Bay on 29 July.

As the third United States Army unit to arrive in Vietnam, it was to engage in 26 separate operations occurring in three of the four tactical zones of Vietnam. Called the "Nomads of Vietnam" they captured enough weapons to equip eight enemy battalions and took 2,000 tons of rice from the Viet Cong. Medical personnel provided treatment for over 25,000 Vietnamese, and the brigade relocated 15,000 refugees. First Lieutenant James A. Gardner, Specialist Four Dale E. Wayrynen and Sergeant First Class Webster Anderson each received the Congressional Medal of Honor.

While the 1st Brigade participated in KLAMATH FALLS, its last combat operation as a separate brigade, the remainder of the division moved in December 1967 from Fort Campbell to Bien Hoa in operation EAGLE THRUST. The operation made military history as the largest and longest airlift directly into a combat zone. Established at Bien Hoa on 13 December, the Screaming Eagles were ready for action.

The enemy launched the largest single attack of the war, the Tet Offensive, on 31 January. Throughout the assault, the 101st engaged in combat operations ranging as far south of Saigon and as far north as Quang Tri. One platoon from the 2d Brigade battled on the rooftop of the United States Embassy in Saigon, which was under attack by Viet Cong commandos.

Operation NEVADA EAGLE was the largest single campaign ever fought by the 101st Airborne Division. This operation, designed to secure the coastal lowlands (Thua Thien Province) in I Corps from the

Maj. Gen. Matthew B. Ridgway presents unit citation to regiments and companies of the 101st Airborne in recognition of their participation in the initial Normandy invasion, when they were dropped behind enemy lines before the dawn of D-Day. (Courtesy of 101st Airborne Division Association)

Following a night skirmish near Bastogne, Belgium members of the 101st Airborne set out to find and rejoin their unit. Shown in front L to R: Pfc. M. L. Dickman, Pvt. Sunny Sundquist, Sgt. Francis B. McCann.

Viet Cong and North Vietnamese, began on 17 May 1968 and lasted 288 days until 28 February 1969. Thua Thien Province was captured and enough rice was removed to feed 10 enemy battalions for the next year.

One of the most important Viet Cong and North Vietnamese supply and staging areas was the A Shau Valley, which ran along the western edge of the Thua Thien Province. Before NEVADA EAGLE, the 1st Brigade had made a 17-day raid into the valley in an operation called SOMERSET PLAIN. Upon the completion of NEVADA EAGLE, the 101st again attacked the A Shau Valley. In a series of operations known individually as MASSACHUSETTS STRIKER, APACHE SNOW and MONTGOMERY RENDEZVOUS, the Screaming Eagles cut North Vietnamese supply lines, destroyed base camps and seized tons of supplies. The division cleared the way for the first friendly armored vehicles to enter the valley and reopened temporary airstrips abandoned years earlier.

During the APACHE SNOW, the 3rd Battalion of the 187th Infantry assaulted Dong Ap Bia Mountain in one of the most famous and controversial battles of the war. These operations pitted the division against some of the best-trained and equipped North Vietnamese units in South Vietnam. The success of these operations decimated the enemy and forced him to place more reliance upon supply bases in neighboring Laos.

While NEVADA EAGLE was going on, the 101st Airborne Division changed its name to the 101st Air Cavalry Division on 1 July. This change lasted until 29 August 1969 when the Screaming Eagles became the 101st Division (Airmobile), creating the Army's second airmobile division and symbolizing the transition from parachutes to helicopter.

Field action in 1969 and throughout 1970 centered around support of the United States civil operations in the pacification program. Operation RANDOLPH GLEN was a departure from the more conventional use of combat forces in South Vietnam. The 101st provided technical assistance to government officials of Thua Thien Province. Elements of the 101st worked with the Army of the Republic Vietnam (ARVN) 1st Infantry Division providing security against outside communist pressure. Operations TEXAS STAR and JEFFERSON GLEN followed with increased emphasis placed on the Vietnamization of the war effort. Using a network of fire support bases and aggressive patrolling, the Screaming Eagles thwarted enemy thrusts into Thua Thien Province. For its involvement in the many civil affairs programs, the 101st Airborne Division (Airmobile) received the Vietnamese Civic Action Medal on 23 May 1970.

Although a small task force from the 101st participated in a limited incursion into Cambodia in April through June 1970, the most important test of the airmobile concept came in February 1971 during operation LAM SON 719. During this operation, the 101st supported Vietnamese forces in their attack across the Laotian border. Designed to cut enemy infiltration routes and to destroy North Vietnamese staging areas in Laos, the operation began on 8 February when the 101st and other American aviation units airlifted South Vietnamese troop into Laos. For many years, the enemy had controlled the area of Laos adjacent to South Vietnam and had built extensive defenses. The operation ended on 9 April 1971 and despite the increased enemy use of anti-aircraft weapons, artillery and armor, less than one Allied aircraft for every thousand sorties was lost.

South Vietnamese soldiers of the

101st Paratroopers use captured German equipment, Carantan, France, 17 June 1944. (Courtesy of 101st Airborne Division Association)

ARVN 1st Infantry Division, supported by the 101st, invaded the A Shau Valley in Operation LAM SON 720. From April to 13 August 1971, enemy supply lines were cut which cost the enemy both men and equipment. For its actions, the 101st received the Vietnamese Cross of Gallantry with Palm.

In late 1971 and early 1972, the Screaming Eagles withdrew and redeployed to the United States. They were the last United States Army division to leave the combat zone in Vietnam. Vice President Spiro T. Agnew and Army Chief of Staff General William C. Westmoreland welcomed the 101st home during official homecoming ceremonies on 6 April 1972 at Fort Campbell.

Upon arrival back in the United States, the 101st was at 20 percent of its authorized strength on homecoming day. This resulted from an early separation date program, soldiers transferring to other units to complete their tour of duty in Vietnam, and an extensive leave policy. A recruiting program called "Unit of Choice" enabled the 101st to reach 65 percent by December. Rebuilding combat readiness became the major goal of new training programs. On 24 January 1973, elements of the 3d Brigade participated in the largest airborne operation held by the 101st since their return from Vietnam. Exercise QUICK EAGLE I tested the combat readiness of the 3d Brigade, and subsequent QUICK EAGLE exercises tested the rest of the division. By June 1973, the 101st was again combat ready.

Significant identity changes for the 101st took place in 1974. On 1 February, the 3d Brigade announced the termination of its parachute status, and Major General Sidney B. Berry, the 101st commanding general, authorized the wearing of an airmobile badge. When the airmobile designation was dropped on 4 October that same year, the division added the air assault parenthetical designation. Graduates of the Air Assault School each received a newly designed air assault badge, which officially became an Army qualifications skill badge on 20 January 1978 retroactive to 1 April 1974 for any soldier in an air assault unit who had demonstrated the necessary professional knowledge and skill.

Throughout 1975 and 1976, the three brigades of the 101st participated in a number of readiness exercises both at and away from Fort Campbell. The most important of the exercises in 1975 occurred when 2d Brigade and support units departed Fort Campbell for Fort Bliss, Texas to take part in GALLANT SHIELD 75. In the desert environment, the air assault task force engaged large mechanized forces. This exercise served as a forerunner of the largest training exercise the division undertook in the post-Vietnam era. For the first time since the Second World War, the division returned to Europe to participate in REFORGER 76.

The training for REFORGER 76 began late in 1975, followed by the actual processing of men and equipment for departure in the summer of 1976. In July, equipment, vehicles and 318 of the division's aircraft were ready to leave from Norfolk, Virginia. On 7 August, the advance party left for the Federal Republic of Germany, and by 29 August, the deployment was complete. The first tactical exercise began the following week. Using the air assault tactics tested at Fort Campbell along with a German mechanized brigade and an American armored cavalry regiment for ground support, the Screaming Eagles pushed the aggressors back. Within 48 hours, the division, disengaged from the first exercise, moved to its major unit assembly areas to refit and deploy to a new area of operation. The scenario for a second exercise paralleled that of the first. Both times, the 101st fought in a mid-intensity warfare environment, and each time air assault tactics proved effective. The Secretary of the Army praised the leadership, training and professionalism of the division during REFORGER 76, calling the 101st "a great credit to the United States Army."

After the second exercise ended, the Screaming Eagles participated in partnership training with North Atlantic Treaty Organization (NATO) units from Belgium, Britain, Germany and the Netherlands. In addition, the 101st hosted an "Air Assault in Action" demonstration for NATO personnel and made commemorative visits to the division's World War II battle sites at Bastogne and in the Netherlands. Upon redeployment to the United States, a full division review at Fort Campbell on 22 October celebrated the end of REFORGER 75 and safe return of the soldiers who participated in the exercise.

In following years, a "one army" concept developed and received great emphasis. During 1977 and 1978, the 101st led all other United Army Forces Command (FORSCOM) active duty divisions in training assistance for National Guard and reserve units. At the same time, the division continued its major mission of preparing for war while helping to assure the peace. Elements of the division trained at Fort Campbell and in remote training areas such as the Northern Warfare Center in Alaska and the Jungle Operations Training Center in Panama. During 1978, elements of the 229th Attack Helicopter Battalion and the 2d Squadron of the 17th Cavalry again represented the 101st on European soil, participating in REFORGER 78.

In June 1979, the division received the first of its new UH-60A Blackhawk helicopters and integrated them into the air assault concept. The division commander, Major General Jack V. Mackmull, accepted the 101st production helicopter in January 1981.

Fulfilling its readiness role under the "one army" concept, the division played a vital part in helping the United States meet its commitment in the Middle East. Between 7 November and 25 November 1980, elements of the division participated in the Rapid Deployment Joint Task Exercise BRIGHT STAR near Cairo, Egypt. The contingent was a battalion combat team of 900 men from the 1st Battalion of the 501st Infantry along with support units. The exercise gave the division experience in overseas movement, desert warfare and coordination with other branches of the United States armed forces and with foreign allies. Since then, several other exercises involving the Rapid Deployment Force have been carried out.

In late March 1982, the XVIII Airborne Corps designated the 1st Battalion of the 502d Infantry as the replacement unit to be sent to the Sinai peninsula in Egypt for a six-month tour of duty with the Multinational Force and Observers (MFO). Supporting the American commitment to the peacekeeping force established under the terms of the 1979 Egypt-Israeli peace treaty, the Screaming Eagles and the 82d Airborne Division from Fort Bragg alternated six-month tours of duty.

Also during 1982, the division received two Cohesive Operational Readiness and Training (COHORT) Companies. Under this system, soldiers could be associated with a specific unit throughout their service career. After recruitment, the company received initial entry training and advanced individual training at one post. Upon arrival at their first duty station, these first-term soldiers remained with their initial battalions for at least one year and then rotated as a unit to their first overseas assignment, concluding their enlistment abroad. The first division COHORT company was Bravo Company of the 2d Battalion, 502d Infantry, arriving at Fort Campbell in June 1982. Alpha Company of the 3d Battalion, 187th Infantry, arrived as the second COHORT company at the beginning of August.

Reorganizing once more in 1983, this time under a new regimental system, the 327th, the 502d and 187th Regiments became the brigades for the 101st Division. The 327th and the 502d were two of the original units assigned to the 101st at its activation in 1942. The 187th's distinction stems from being the only airborne unit to serve in three wars: World War II, Korea and Vietnam.

Throughout 1984, the division participated in 15 major exercises in the United States, Germany, Honduras and Egypt, helping to maintain the readiness needed to fulfill its assigned mission to deploy rapidly worldwide using the unique capabilities of an air assault division.

In 1985, tragedy struck the 101st after a seemingly routine MFO tour of duty for the 3d Battalion of the 502d Infantry. Returning to Fort Campbell from the Sinai on 12 December, 248 Screaming Eagles perished in an airline crash near Gander, Newfoundland.

In 1987, a division-wide exercise called GOLDEN EAGLE validated the operational and logistical capabilities of the new "Army of Excellence." Spanning an area of operation from Smyrna, Tennessee to Madisonville, Kentucky, the 101st tested the concept of light, flexible, rapidly deployable forces being available for worldwide contingency missions.

Another milestone was reached in September 1989 when the 101st participated in the first full-scale inland water deployment since World War II. As part of the Joint Readiness Training Exercises taking place at Fort Chaffee, Arkansas, the 101st loaded 8 million pounds of equipment and transported them down the Cumberland and Mississippi Rivers.

The air assault concept is continuously being challenged in exercises all over the globe, from Alaska to Egypt. Places such as Bastogne, Carentan, Endhoven, Saigon and Quang Tri will not be forgotten. The 101st Airborne Division (Air Assault) awaits its next rendezvous with destiny.

In their attempt to stem the German breakthrough, the 101st marches out of Bastogne toward Houffalize, Belgium. (Courtesy of the 101st Airborne Division Association)

SECTION I
MEDAL OF HONOR RECIPIENTS

SFC Webster Anderson A/2-320 FA 15 Oct 67
CPT Paul W. Bucha D/3-187 IN 16-19 March 68
*LTC Robert G. Cole 3-502 IN 11 June 44
SP4 Michael J. Fitzmaurice D/2-17 CAV 23 March 71
*CPL Frank R. Fratellenico B/2-502 IN 19 Aug 70
*1LT James A. Gardner 1-327 IN 7 Feb 66
*SSG John G. Gertsch E/1-327 IN 15-19 July 69
*SP4 Peter M. Guenette D/2-506 IN 18 May 68
SP4 Frank A. Herda A/1-506 IN 29 June 68
SSG Joe R. Hooper D/2-501 IN 21 Feb 68
PFC Kenneth M. Kays D/1-506 IN 7 May 70
*SP4 Joseph G. LaPointe, Jr. HHT/2-17 CAV 2 June 69
*PFC Milton A. Lee B/2-502 IN 26 April 68
*LTC Andre C. Lucas 2/506 IN 1-23 July 1970
*PFC Joe E. Mann Co H, 502 PIR 18 Sept 44
SGT Robert M. Patterson B/2-17 CAV 6 May 68
SGT Gordon R. Roberts B/1-506 IN 11 July 69
*SSG Clifford C. Sims D/2-501 IN 21 Feb 68
*SP4 Dale E. Wayrynen B/2-502 IN 18 May 67

SECTION II
DISTINGUISHED SERVICE CROSS
RECIPIENTS

LTC Edmond P. Abood HHC/2-327 IN
2LT Walter G. Amerman
SGT Eldon L. Baker HHC/1-327 IN
*MAJ Jack L. Barker B-101 AVN BN
Raymond F. Barton
*PFC Christopher Bell C/2-502 IN
*PFC William Blakely D/1-501 IN 14 Feb 68
Elton E. Brooks
SP4 Herman Lee Brown L-75 IN
CPT Walter R. Brown A/2-502 IN
*SP5 Lonnie R. Butts HHC/1-327 IN
PFC Jerry A. Cain HHB/2-320 FA
CPT William S. Carpenter, Jr. C/2-502 IN
LTC Patrick F. Cassidy 1-502 IN 6 June 44
SP4 Robert T. Catherman A-326 ENG BN
LTC Steven A. Chappuis 502 PIR
*SGT Paul H. Cline C/3-506 IN 6 Feb 68
*SGT Alfred P. Cofforth A/2-506 IN
1LT Kenneth G. Collins A/1-327 IN
PFC DeForest S. Conner C/1-501 IN
*SGT Albert Contreras, Jr. F-58 IN
*PVT Herman J. Cordes 506 PIR 10 June 44
CPL Virgil E. Danforth
PFC Kenneth J. David D/1-506 IN
SP5 Danny Dennard HHC/2-502 IN
*1LT William L. Dent B/1-502 IN
*MAJ Herbert J. Dexter HHC/2-502 IN
LTC Frank L. Dietrich HHC/2-502 IN
SP4 Clifford Dinkins B/2-502 IN
SGT Michael E. Dorch A/2-502 IN
*CPT Fred O. Drennan 502 PIR 18 Sept 44
George Evule
LTC Julian Ewell
*1LT Robert C. L. Fergusson HHB/2-320 FA
SSG Larry A. Fletcher C/2-502 IN
Edward E. Ford
CPT Jerry R. Fry B-101 AVN BN
CPT Robert L. Friedrich B/3-187 IN
*2LT Michael L. Gandy B/1-501 IN 29 March 68
SP4 James P. Glemser C/1-502 IN
*1LT Harry M. Godwin A/1-327 IN
CW2 Gerald D. Green A/2-17 CAV
Francis L. Harbough
*1LT Ernest O. Harris 502 PIR 6 June 44
SGT Bailey Harrison 502 PIR 6 June 44
SGT William L. Hofatrour C/2-501 IN
SP4 John Hogan C/2-502 IN
LTC Weldon F. Honeycutt HHC/3-187 IN
SGT Robert J. Houston 12 June 44
SFC Jesse A. Issac B/1-506 IN
SGT Kyle D. Jones A/2-506 IN
1SG Donald L. Joubert A/3-187 IN
SGT David R. Kasun D/3-187 IN
1LT Leslie D. Kenndey A/2-502 IN
SP4 Alan Kent B/2-502 IN
LTC Harry W. O. Kinnard
*SGT John W. Kreckel A/2-506 IN 22 July 68
PFC Alfred Kurg A/2-327 IN
Robert E. Langen
PSG William Lawrence, Jr. C/2-501 IN
CPT John P. Lawton A/2-327 IN
Loyd J. Leino
*PFC Orel H. Lev 506 Pir 23 Sept 44
CPT Frank L. Lillyman 502 PIR
1LT James R. Magonyrk C/3-506 IN
CPT John S. Maloney 506 PIR
CPT Gordon A. Mansfield C/1-501 IN
WO1 Marion L. Mark 176 AVN CO, 1-327 IN
Thomas E. Martin
PVT Arthur C. Mayer 9 June 44
CPT Robert G. Mayor A/2-501 IN
BG Anthony C. McAuliffe DIV HQ
1LT Timothy P. McCollum D/1-502 IN
1LT Michael A. McDurmott C/2-327 IN
Tildon S. McGee
Fred R. Metheny
LTC Raymond D. Millener DIV HQ
Jachin Mitchell
CPL George Montilio 506 PIR 6 June 44
*SP5 Dennis F. Moore D/3-187 IN 18 March 68
SGT Kenneth E. Murray HHC/2-502
MAJ James T. Newman C/2-17 CAV
SP4 Philip L. Nichols B/2-327 In
CPT Ralph A. Northrup B/2-17 CAV
1SG Hubert Odom 11 June 44
1LT Daniel L. O'Neill C/1-501 IN
Norman A. Osterberg
PSG George W. Parker E/2-501 IN
SP4 Jesse J. Parker C/2-327 IN
CPT Lloyd E. Patch 506 PIR 6 June 44
SGT Leon Peoples B/2-502 IN
SGT Michael P. Perry C/2-502 IN
Robert G. Pick
LTC Ralph Puckett, Jr. HHC/2-502 IN
Knut H. Randstein
CPT David F. Rich B/2-319 FA
*1LT Roy L. Richardson A/2-502 IN 22 Feb 70
St. Julian F. Rosemond
MAJ Joseph H. Rozelle B/2-17 CAV
T/5 Jack Rudd

1SG Walter J. Sabalauski C/2-502 IN
CPT Francis L. Sampson
2LT Charles J. Santarsiero 7 June 44
SP4 Nicholas W. Schoch B/3-187 IN
*SSG Burnell Simmons A/2-320 FA
*T/SGT James C. Single 327 GIR 7 Oct 44
*CPT Robert M. Snell HHB/2-320 FA
PVT Andrew Sosnack 506 PIR 6-10 June 44
1SG Kenneth N. Sprecher 502 PIR 11 June 44
SFC James D. Spitz A/3-187 IN
SSG Harrison C. Summers 502 PIR 6 June 44
MG Maxwell D. Taylor DIV HQ 6 June 44
Charles J. Timmes
1SG Herman L. Trent B/2-506 IN
Herbert A. Tubbs
*LTC William L. Turner 506 PIR 7 June 44
SGT Dale Urban D/2-501 IN
1LT Jerry T. Walden D/3-187 IN
SGT Billy B. Walkabout F-58 IN
SFC Thomas E. Warran A/3-187 IN
LTC Joseph Wasco, Jr. HHC/2-327 IN
*LTC Carlton G. Werner 327 GIR 9 June 44
1LT Leon Mack Wessel, Jr. HHB/2-320 FA
PVT George A. Whitfield 326 MED CO 26 Sept 44
1LT Michael J. Williams A/2-506 IN
1LT Richard D. Winters 506 PIR 6 Une 44
1LT Jeffrey Wishik D/3-187 IN
PSG Grover Wolford C/2-506 IN
CPT Luther L. Woods C/2-327 IN
SP4 Robert L. Wright D/2-501 IN
SP4 Ronald J. Wright B/2-502 IN
PFC Donald E. Zahn H/3-506 IN 6 June 44

*Award made posthumously

Division Commanders

MG William C. Lee August 42 - February 44
+*MG Maxwell D. Taylor March 44 - August 45
BG WilliaM M. Gillmore August 45 - September 45
BG Gerald St. Clair Mickle September 45 - October 45
BG Stuart Cutler October 45 - November 45
MG William R. Schmidt July 48 - May 49
MG Conelius E. Tyan August 50 - May 51
MG Ray E. Porter May 51 - May 43
MG Paul DeWitt Adams May 53 - December 53
MG Riley F. Ennis May 54 - October 55
MG F. S. Bowen October 55 - March 56
MG Thomas L. Sherburne May 56 - March 58
+MG William C. Westmoreland April 58 - June 60
MG Ben Harrell June 60 - July 61
MG Charles W. G. Rich July 61 - February 63
MG Harry W. Critz February 63 - March 64
MG Beverley E. Powell March 64 - March 66
MG Ben Sternberg March 66 - July 67
*MG Olinto M. Barsanti July 67 - July 68
*MG Melvin Zais July 68 - May 69
*MG John M. Wright May 69 - May 70
*MG John J. Hennessey May 70 - February 71
*MG Thomas M. Tarpley February 71 - April 72
MG John W. Cushman April 72 - August 73
MG Sidney B. Berry August 72 - July 74
MG John W. McEnery August 74 - February 76
*MG John A. Wickham March 76 - March 78
MG John N. Brandenburg March 78 - June 80
MG Jack V. Mackmull June 80 - August 81
MG Charles W. Bagnal August 81 - August 83
MG James E. Thompson August 82 - June 85
MG Burton D. Patrick June 85 - May 87
MG T. G. Allen May 87 - August 89
MG J. H. Binford Peay III August 89 - Present

*Combat Commanders
+Later served as Army Chief of Staff

Note: BG Don F. Prat (6 February 44 - 14 March 44) and BG Anthony C. McAuliffe (5 December 44 - 26 December 44) are frequently listed as division commanders when actually they were acting division commanders. They, like other acting commanders, are not carried on this list.

Immediately following their landing behind the German lines in Holland, Airborne troops disperse to begin the taking of various objectives. Jeep at right also brought in by glider, 101st Airborne, Zon, Holland - Sept. 18, 1944. (Courtesy of 101st Airborne Division Association).

Men of the 101st Airborne are cared for in an aid station set up in a barn. They are the men who held Bastogne for 10 days until they were relieved by Gen. Patton's Third Army, Belgium, Dec. 27, 1944. (Courtesy of 101st Airborne Division Association).

Airborne troops of the 7th Army, Co. C, 1st Bn., 506th Parachute Infantry, 101st Airborne VI Corp, occupy several small towns in their area. The troops inspect each house and barn for enemy troops. Landsberg, Germany, 30 April 1945. (Courtesy of 101st Airborne Division Association)

1ST AIR CAVALRY DIVISION (AIRMOBILE)

By Kenneth D. Mertel
Col. AUS (Ret)

1ST AIRBORNE INFANTRY BRIGADE – 1965-1966

The 1st Airborne Brigade of the 1st Cavalry Division, Airmobile had one of the shortest peacetime and wartime histories of any of the Airborne units of the United States Army.

The Brigade was activated at Fort Benning, Georgia 1 July 1965, sailed to Vietnam in August and September of that year as part of the 1st Air Cavalry Division, Airmobile. The division and the Brigade engaged in its first action against North Vietnamese Army Divisions in the Highlands, based out of An Khe in the 2nd Corps Area. The 1st Air Cavalry Division was the first full division-size unit deployed to Vietnam.

One year after activation, the 1st Airborne Brigade ceased to exist as an Airborne unit, continuing in action as an Airmobile unit with the remainder of the division. All personnel on jump status within the brigade continued in that status and drew jump pay throughout their tour. The following historical facts come from the book *Year of the Horse, Vietnam* published in 1968 by Exposition Press and re-published in 1989 by Ballantine Books as a pocket book. Its author is Colonel Kenneth D. Mertel, U.S. Army, retired, who served as the first commander of the 1st Battalion, 8th Cavalry, Airmobile, Airborne and later as deputy brigade commander of the 1st Airborne Infantry Brigade.

Original plans called for the 1st Air Cavalry Division to be both Airmobile and Airborne in its entirety. Chief of Staff of the Army Harold K. Johnson opposed any airborne for the division, however Chairman of the Joint Chiefs Wheeler insisted on at least one brigade, thus only a brigade side of the division became Airborne.

The first mission for the brigade was to become Airborne. All straight legs in the brigade were encourage to volunteer for Airborne training. Most of the men volunteered and a few of the younger sergeants as well as most of the 2d and 1st lieutenants. Special airborne classes were set up by the Jump School at Benning and the three Infantry battalions of the brigade (1/8, 2/8 and 1/12, engineer, aviation, artillery and the division slice of Division Support and Combat Support Units). The brigade went through training as a unit, the first time since World War II, under supervision of its own officers. Physical and disciplinary training was conducted by respective battalion commanders of the brigade, thus the Jump School was in charge only of the airborne training. All troops were housed in their own respective billets.

During the training, company commanders were received from respective advanced courses. Senior non-coms, to include most of the platoon sergeants, first sergeants and sergeant majors, came from the 101st Airborne Division. Battalion commanders and other senior officers were former paratroopers already part of the former 11th Air Assault Division, now the 1st Air Cavalry Division, Airmobile.

The 8th of July 1965, marked the first parachute jump for those paratroopers already qualified as jumpers. This was a brigade-level jump from C-130s, spilling out over the wet fields on the Alabama side of the Chattahoochee River at 1845 hours. There were no injuries in this first jump, with all participants becoming charter members of the 1st Airborne Infantry Brigade. Several other jumps were made by all members of the brigade as the budding new paratroopers made each of their qualifying jumps. By the end of the month, the 1st Brigade was truly an Airborne Infantry Brigade, ready for whatever combat role would be required in Vietnam.

Incidentally, the first jump by members of the Airborne Brigade was made from a HUEY D Model on 1 July 1965 by the brigade commander, Colonel Elvey Roberts (now three stars retired); Robert Shoemaker, commander of the 1st of the 12th Airborne, Airmobile Battalion (now four stars retired) and Kenneth D. Mertel, commander of the 1st of the 8th Airborne, Airmobile Battalion.

Architect, father and commander of the 1st Cavalry Division, Airmobile was Maj. Gen. Harry W.O. Kinnard, (now three stars, retired). The two assistant division commanders were Brig. Gen. Jack Wright, (now three stars retired) and Brig. Gen. Richard T. Knowles, (also three stars retired).

The 1st Airborne Brigade sailed (most of it) on 20 August 1965, on the USN *Geiger* from Savannah, Georgia. Up anchor at 1745 hours, down the Savannah River to the Atlantic Ocean, around through the Panama Canal, with stops in Hawaii and Guam, it finally debarked in Qui Nhon, Vietnam, 30 days later. The rest of the division sailed on other ships including the aircraft carrier to carry the 435 helicopter and other aircraft organic to the division.

The brigade engaged in maximum weapons training, including firing off the ship's fantail, constant and continuous physical conditioning and squad-and platoon-level skull sessions. Thus, upon arrival in Vietnam, the paratroopers of the 1st Airborne Brigade was ready for killing communists or whatever the mission might be.

The troopers were ferried by helicopter from ship-side to An Khe to join the rest of the division at the division base. Their first missions were defending the base and securing the area from any prowling Viet Cong or regular troops of the North Vietnamese Army.

Numerous battalion level skirmishes occurred over the next few weeks as the 1st Airborne Brigade improved its combat ability and professionalism. The first major engagement of the brigade and the 1st Air Cavalry Division, Airmobile, was the Ia Drang Valley Campaign, 23 October 1965, to 26 November 1965, where the first Presidential Unit Citation was earned in the initial use of airmobile warfare to destroy the major portions of two or more regular North Vietnamese Army Divisions and to drive them from II Corps

Tactical Zone. All elements of the 1st Airborne Infantry Brigade participated in this action, with A Company, commanded by Captain Ted Danielsen, 1st of the 8th, making the first combat night air assault in the history of airmobile warfare.

In November and December of 1965, the 1st Airborne Brigade participated in a series of search and destroy missions clearing the area east of An Khe to Qui Nhon, previously a safe haven for both Viet Cong and regular North Vietnamese army units. This was a most important mission for the 1st Air Cavalry Division Base as An Khe was primarily supplied overland via Highway 19. In addition, the same highway was cleared to Pleiku, through Mang Yang Pass to the west of the division base, the scene of the destruction of French Mobile Group 100 in earlier years. The Airborne Brigade initially opened its pass, committing the 1st of the 8th Airborne, Airmobile Battalion.

During this same time frame, A Company, 1st of the 8th, was selected as the typical rifle company in Vietnam for filming of the famous ABC Television Production "I am a Soldier," one-hour in length, a most pro-Army and pro-Vietnam video film, later designated an Army training film. Aired in March of 1966 in the U.S., this film focused much attention on the Airborne Brigade, and especially the 1st of the 8th. It also covered the first and only jump of the brigade in Pleiku, when a series of training jumps were conducted by most elements of the brigade. While not conducted as an airborne assault, they were used to positioning the units for a series of search and destroy missions.

In January of 1966, it was widely believed by the U.S. media, that the 1st Air Cavalry would make an attack into Cambodia, long used as a safe haven for supply points, hospitals, training areas and headquarters of major elements of the North Vietnamese Regular Army. The brigade was inundated by reports and photographers. The 1st of the 8th alone was overwhelmed with over 30 media people, four or five in each assault rifle company. Although the media and the troops were ready for the air assault into Cambodia, which might have ended the war early on, this was not to be due to timidity on the part of U.S. political leaders and fear of further Russian and Chinese involvement. The attack into Cambodia would not take place until 1970, when the 1st Air Cavalry Division led the assault into Cambodia capturing tens of thousands of tons of vital supplies that had been brought in over past years from China and Russia via the long haul by elephants, pack bicycles and trucks.

The 1st Airborne Brigade continued a series of search and destroy and clearing operations in the area along the China Sea Coast. It was here that the Airborne Brigade won its first Medal of Honor (second for the division in Vietnam) by Dave Dolby of B Company, 1st of the 8th, commanded by Captain Roy Martin. In this same action, Captain Martin won a Distinguished Service Cross and Captain Jerry Plummer won a Silver Star. These were examples of the many demonstrations of heroism in combat shown by members of the Airborne Brigade and the 1st Air Cavalry Division.

Although no longer on airborne status as of 1 July 1966, the brigade continued to lead the way in the many other famous battles in which Sky Soldiers of the 1st Air Cavalry Division, Airmobile participated. All were deemed highly successful operations in airmobile warfare.

AIRBORNE HISTORY
By William C. Roll

3rd Battalion (Airborne), 187th Infantry, 11th Air Assault Division (T). 1st Battallion (Airborne), 187th Infantry, 11th Air Assault Division (T). 1st Battalion (Airborne), 12th Cavalry, 1st Cavalry Division, Airmobile, May 1963 - July 1966.

In February 1963, the 11th Air Assault Division (TEST) was formed at Fort Benning, Georgia, to test the concept of moving men and equipment around the battlefield using the helicopter as the only means of transportation. The Commanding General of the 11th Air Assault Division (T) was Brigadier General Harry W. O. Kinnard. General Kinnard later became the first Commanding General of the 1st Cavalry Division (Airmobile). The 11th Air Assault Division (T) adopted the shoulder patch of the 11th Airborne Division, substituting an airmobile tab for the airborne tab of the 11th Airborne Division patch.

The first battalion formed in the 11th Air Assault Division (T) was the 3rd Battalion, 187th Infantry and its first commander was Lieutenant Colonel John J. Hennessey. On October 1, 1963, the battalion was officially converted from a straight-leg infantry battalion to an airborne battalion, the 3rd Battalion (Airborne), 187th Infantry, "Rakkasans". Many of the officers and men of the battalion were not airborne qualified, however, most volunteered for airborne training to remain with the unit. Special airborne training courses were held for these soldiers at Fort Benning in the summer of 1963. Key non-com and officers' vacancies that did exist in the battalion were filled by experienced, jump master qualified personnel, mainly from the 101st Airborne Division.

For the next twenty months, the 3rd Battalion (Airborne) 187th Infantry trained at Fort Benning, Fort Stewart and in numerous off-port training sites. Over 40 battalion and company size tactical airborne training exercises were conducted. During this period, the 3rd Battalion was redesignated the 1st Battalion (Airborne) 187th Infantry as the 3rd Battalion colors were returned to the 101st Airborne Division at Fort Campbell, Kentucky. Also, command of the battalion was transferred from Colonel Hennessey to Lieutenant Colonel Harlow G. Clark who commanded the battalion through the remaining test days of the 11th Air Assault Division (T). (LTC Clark was later killed in a freak training accident in an OH-13 helicopter that he was piloting while touring the Division defensive perimeter in Ankhe, Vietnam. He was flying with Colonel Hennessey, who escaped the accident uninjured).

The training of the battalion during the final stages of the airmobile testing was highlighted by two Brigade-size, mass tactical airborne operations in the vicinity of Camden, South Carolina. The battalion was joined for these airborne drops by two battalions of the 82nd Airborne Division. The "Rakkasans" subsequently maneuvered against troopers of the 82nd in a realistic training exercise that convinced the decision makers in the Pentagon of the viability of the airmobile concept and the potential lethality of a unit that could move rapidly, with surprise across the battlefield.

In June 1965, a momentous decision for the future of the Army was reached in the Defense Department and announced by the Secretary of Defense. The testing of the 11th Air Assault Division (T) was concluded and authority was granted to organize the Army's first Airmobile Division at Fort Benning. MG Kinnard was given the mission of making the Division "Combat Ready" in just eight weeks. The colors of the 2nd Infantry Division at Fort Benning were transferred to the Republic of Korea and the colors of the 1st Cavalry Division, for the first time in 21 years, were returned to American soil. The 1st Cavalry Division

(Airmobile), to become known as the Air CAV, was born and a new era in modern warfare began.

As were the division colors of the 11th Air Assault Division (T) retired so were the colors of the 1st Battalion (Airborne) 187th Infantry. The officers and men of the battalion now were members of the 1st Battalion (Airborne) 12th Cavalry "Chargers".

An airborne unit in the CAV was unique to a division that had transitioned from horse cavalry to tanks and heavy equipment, to a highly maneuverable unit to be deployed by both helicopter and parachute.

With only eight weeks to prepare for battle, the 1st Battalion (Airborne) 12th Cavalry, then commanded by Lieutenant Colonel Robert M. Shoemaker, launched a concentrated combat training program. The training was almost non-stop with both day and night live-fire exercises the predominant training vehicle. On July 28, 1965, President Lyndon B. Johnson, in a nationwide address, announced to the world what the cavalry troopers already knew; the entire 1st Cavalry Division was to deploy to Vietnam.

In August 1965, the 1st Brigade of the CAV, consisting of the 1st Battalion (Airborne) 12th Cavalry, the 1st Battalion (Airborne), 8th Cavalry and the 2nd Battalion (Airborne) 8th Cavalry and their Direct Support Artillery Battalion, boarded the USNS Geiger in Savannah, Georgia for a month long voyage to Vietnam. Highlights of the deployment, included passage of the Panama Canal and brief stops in Honolulu and Guam, and the extensive training program conducted mainly by squad and platoon leaders to further prepare their soldiers for combat. The daily training regiment included a rigorous ship board physical training program complete with the daily dozen on the hatch covers and open decks of the Geiger. On September 20, 1965, the battalion with its sister units landed in Qui Nhon, Vietnam and subsequently was flown by Chinook helicopter of the Division's Heavy Lift Battalion to the 1st Cavalry Division's base camp in Ankhe.

Upon their arrival in Vietnam the troopers of the "charger" battalion fully expected to conduct airborne as well as airmobile combat tactical operations. Shortly following their arrival the battalion conducted what was to be their only parachute drop. In the rice paddies east of the village of Ankhe an airborne jump was made. This jump was highlighted by the eagerness of the local children from Ankhe to assist the airborne soldiers out of their parachute harnesses and to rigger roll the chutes in record time. The troops were amazed at the ability of these youngsters to show this skill and wondered where it had been learned. Following their support, the children were rewarded individually with C Ration gum, John Wayne bars or a few Piasters.

The battalion subsequently prepared for a battalion combat drop in the vicinity ot Tuy Hoa, which was later cancelled by General William Westmoreland. Though the operation never occurred the Airborne spirit was rekindled in the troopers as they made their morning runs and conducted their pre-jump training complete with hundreds of PLFs.

In October 1966, a decision by General Westmoreland ended the Airborne status of the 1st Battalion (Airborne) 12th Cavalry. The soldiers and officers of the battalion remained on jump status until they concluded their one-year tour in Vietnam. These officers and men are authorized to wear the Airborne Tab over the 1st Cavalry Division patch on the right shoulder of their uniform.

For a more detailed history of the combat action of the 1st Battalion, 12th Cavalry through 1969, please refer to the 1st Air Cavalry Division, Memoirs of the FIRST TEAM, Vietnam, August 1965 - December 1969.

1ST BATTALION, 8TH CAVALRY REGIMENT

By Glenn H. Sheathelm

The 1/8th Cavalry may not have been the earliest Airborne unit, but they were not without a long history. And they were to make their mark in history during the Vietnam War.

The 8th Cavalry started as a horse cavalry in 1866 in California which accounts for the bear on the coat of arms although it is not on the unit crest. The unit served during the Indian Wars from the Canadian border to the Mexican border which was a north to south distinction they would repeat during the Vietnam War. Campaign participation credits during that time included Comanches, Apaches, Pine Ridge, Arizona in 1867, Arizona in 1868, Oregon in 1868 and Mexico in 1877.

During World War I the 8th Cavalry patrolled the border with Mexico after serving at most of the Western posts including Ft. Robinson, Nebraska where an 8th Cavalry guidon and several company rosters are in the museum. Patrolling the Mexican border may not seem significant but there was a fear that Mexico would join sides with Germany during the war and take back lands in Texas and California that they had lost to the United States.

In 1921 the 8th Cavalry became part of the 1st Cavalry Division of which the 8th is still a part. At the time they were still using horses extensively.

During World War II the 8th Cavalry served with the 1st Cavalry Division in the Pacific Theater of Operations and added campaign credits for New Guinea, Bismarck, Archipelago, Leyte and Luzon. They had two Presidential Unit Citations for Luzon and Taegu with A Troop collecting a third Presidential Unit Citation for Manus Island. The 8th Cavalry also collected a Philippine Presidential Unit Citation.

At the end of World War II the 8th Cavalry was stationed in Japan as part of the U.S. occupation forces. They stayed there helping to make changes in Japan under the command of General Douglas MacArthur.

The 8th Cavalry served during the Korean War and added two Korean Presidential Unit Citations and the Chryssoun Aristion Andrias from Greece. The campaign credits included UN Defensive, UN Offensive, CCF Intervention, First UN Counter Offensive, CCF Spring Offensive, UN Summer-Fall Offensive, Second Korean Winter, Korean Summer-Fall 1952 and Third Korean Winter.

During 1963-early 1965, the Air Assault Division Theory was being tested at Ft. Benning, Georgia. Ideas came from the Howze Board and General James Gavin, who had an outstanding reputation in World War II as an Airborne commander, and was very receptive to new ideas. The person selected to test the ideas was H.W.O. Kinnard who was a brigadier general at the time. General H.W.O. Kinnard had served with the 101st Airborne during World War II and had suggested that General McAuliffe's informal response when he heard about the German's surrender demands was appropriate. Thus we have one of the shortest replies to a surrender demand: "Nuts."

One of the units of the 11th Air Assault Division (Test) was the 1st Battalion (Airborne) of the 188th Infantry and some of the cadre of that unit stayed on after it was changed to the 1st Battalion, 8th Cavalry.

Gen. H.W.O. Kinnard sent out word in May and June of 1965 that a new unit was forming at Ft. Benning that had great

D/1st/8th Cavalry takes a break while waiting for helicopters to pick them up near Santana Draw and take them to a battle already in progress near Bong Son. Notice the variety of ways in which the troopers are carrying equipment. Also notice that they are all generally carrying light loads. The 1st Air Cavalry tried resupply runs at least every two days. (Courtesy of Glenn H. Sheathelm)

opportunities for good officers, NCOs and enlisted men. This gave him some of the best from which to pick. Although he wanted the whole division to be Airborne, he had to settle for 1st Brigade along with a scattering of airborne-qualified leaders dispersed through the rest of the division.

One of the officers selected was Lt. Col. Kenneth Mertel who had enlisted in the Army in 1942, finished OCS in 1945 and finished jump school in 1948. He had proved himself as a company commander and battalion S-3 with the 25th Infantry Division during the Korean War and also held a pilot's rating.

Lt. Col. Kenneth Mertel took command of 1st Battalion, 188th Infantry in late June of 1965, and when it became 1st Battalion (Airborne) 8th Cavalry on July 1st he considered it to be the proudest moment of his life. He was going to build an airborne battalion and take it into combat.

The 1/8th was airborne in name and a battalion in name, but actually there were only 55 airborne-qualified people in the unit. Undaunted, Lt. Col. Mertel took them up for their first jump together on July 8th from a C-130. Some of the captains like Roy Martin, Bill Mozey and Theodore Danielson, would also make their marks on history in Vietnam.

While the jump was going on there were others at Ft. Benning who had trained in air assault tactics and wanted to become part of 1/8th. They were going through a fast jump school course which was completed on July 20th.

Others who became part of 1/8th came from other airborne units. One of those was 1st Sgt. Ray Poynter who left STRAC Forces at Ft. Campbell to become 1st Sgt. of B Company. The 101st and 82nd both reluctantly surrendered personnel to 1/8th.

One month after completing jump school, the new people found themselves leaving Savanna, Georgia on the USN *Geiger* bound for Vietnam. When they left on August 20th, 1/8th was "Airborne All The Way" and at full strength. Amazingly, it had been only earlier July when they received the colors from Korea, and they, along with the rest of the 1st Cavalry Division (Airmobile), were going to change the way war was fought.

An advance party from 1/8th had gone over with Major Guy Eberhardt by plane on the 18th of August and while one platoon was on the USN *Patch* and the rest of the battalion was on the *Geiger*, they would start building a base camp at An Khe. Captain Gerrell Plummer and Lt. Roger Talmadge were in that advance party. Lt. Talmadge remembers the soldiers of the 101st getting their laughs at what was supposedly the most modern unit in the Army, attacking the jungle with machetes, scythes and shovels. There was a reason for the apparent madness. They wanted green vegetation to hold down the dust, but that vegetation couldn't interfere with helicopter operations. Thus the "Golf Course" was born.

On September 22nd, one day after leaving ship in Qui Nhon, they arrived at An Khe and the 1/8th was again a full battalion. They spent the rest of the month improving the camp and the perimeter. It was rumored that Lt. Col. Mertel replied, "Oh, that's marvelous" many times during that period as Airborne ingenuity came up with things such as blue generators from undisclosed sources! In any case, Chaplain Spears soon had a chapel and a walk-in outdoor theater that had movies that were generally audible over the roar of the generator to improve the morale of a unit that was already filled with enthusiasm.

Offensive operations were started in October and a film crew went with Captain Theodore Danielson's A Company resulting in the documentary "I Am a Soldier". During that time the operations were routine and the film crew left with good copy even if there were no major battles.

On November 3, 1965, Alpha Company, 1st Battalion, 8th Cavalry made the first night air assault in history to rescue a unit that was already heavily engaged. There were some mistakes. However, considering the obstacles the unit faced, it was a remarkable success.

Early in the evening of November 3rd, C/1/9th Cavalry's Blue Platoon commanded by Capt. Chuck Knowlen ambushed part of a large NVA unit from 8th Battalion/66th NVA Regiment and beat a hasty retreat to LZ Mary where Major Zion was located with the rest of the CP. Because the ambush was very successful, it really stirred up the rest of the NVA. They surrounded LZ Mary and may have overrun it if the platoon had not been reinforced. They had not dug in—even the attached mortar platoon under the command of Stuart Tweedy was not dug in. The mortar platoon had been borrowed from A/1/8th Cavalry because, except for Delta Company, the 1/9th Cavalry had no mortars.

Alpha Company, 1/8th was the 1st Brigade reaction force and was based at Camp Holloway. Lt. Col. John Stockton probably overstepped his authority when he committed A/1/8th without the permission of the 1/8th commander or 1st Brigade. Whether authorized or not, it is now history.

Third Platoon of Alpha Company commanded by Lt. John Hanlon was the first committed. Lt. Hanlon had some new people in the platoon including his Platoon Sergeant Ken Riveer who had previously been with 2nd Platoon. There were some fortunate similarities between the two. They both had a strong but quiet religious faith and both considered Chaplain Spears a friend they could count on. They also both had a sense that they would do their duty to their best and had confidence that the men in the platoon would also do their best. The prayers of both were not for their own protection but for guidance to make the right decisions so that none of the men in their command would die needlessly. 1st Battalion/8th Cavalry had a lot of leaders like that within the battalion.

The flight from Camp Holloway to LZ Mary took abut 20 minutes and it was obvious from a distance that the LZ was hot with flares and tracers lighting it up. Capt. Theodore Danielson didn't know the actual name of the LZ, but the tracers suggested it could be called "LZ Spider Web."

The infantrymen, in typical fashion, had their feet dangling from the open helicopter doors in anticipation of the battle, and they were out on the skids and jumping off as the helicopters crossed the LZ at about a four-foot altitude. The helicopter pilots were quite pleased the troops were unloading so fast because all of them on the first lift were hit, but none so seriously they couldn't continue to operate.

As Alpha Company hit the ground they ran toward the south side of the

1st Platoon of D Company 1st Bn. 8th Cavalry, 1st Air Cavalry Division was well known for being a bit unorthodox. At 1st Brigade and Battalion headquarters they were known as Lt. Reed and his bandits. Lt. Reed, left, and one of his squad leaders are both carrying their rifles with something other than the issue sling. "Lerch" is carrying an issue style frame with two small combat packs made for use on the web belt attached. A carabiner is attached to the retainer for the canteen cap. The carabiner was used for rappelling out of helicopters. 1st Platoon also had the ability to inflict heavy casualties on the enemy troops with very few casualties to themselves. (Courtesy of Glenn H. Sheathelm)

perimeter because the heaviest fire seemed to be coming from that direction. Sgt. Riveer didn't remember any specific order to go in that direction, but the platoon stayed together.

They had difficulty deciding what was the edge of the perimeter because there were no holes so they just dropped in an arc in the grass. They did know the NVA had several advantages in that they were firing from the shadows along the edge of the trees. They had cover from the trees and they had room to maneuver.

One of the first problems was establishing a linked perimeter and Lt. John Hanlon had trouble with that because the only person from 1/9th Cavalry he found was an officer who was hiding under a shattered tree insisting that everyone was going to die.

In the confusion Sgt. Riveer had lost his RTO so he returned to Lt. Hanlon. By then, the lieutenant had been shot and was unable to move his legs because of a spinal cord injury. He then turned command over to Sgt. Riveer along the line and although seriously wounded, passed on encouragement to the other wounded and all his other troopers.

Although Lt. Hanlon didn't want to leave the platoon, he had Sgt. Riveer try to get some of the wounded out when the lift ships came in with 2nd Platoon. Unfortunately, the LZ was still getting heavy fire and although Riveer and his troopers managed to shove one man on, by the time they lifted the second man up, the helicopter was gone. Sgt. Riveer was disappointed at the lack of success getting the wounded out, but he could understand the reluctance of the helicopter pilots to land or hover for any period of time.

The squad leaders Tom Freels, James Fowler and William Teach, performed very well and both Lt. Hanlon and Sgt. Riveer were proud of the performance of the Privates and Specialists Four who stayed together and fought the NVA.

Sgt. Bess from 2nd Platoon was one of the other NCOs that Sgt. Riveer encountered. Riveer explained that he needed more ammunition. Just as Sgt. Bess dropped down, swinging his pack to the ground, he was shot fatally in the neck. It stunned Sgt. Riveer because Bess was a good friend of his, but he took the extra ammunition and went back to his own platoon.

Sp4 Raymond Ortiz, a medic, was wounded when he first got off the helicopter but continued moving around treating the wounded and dragging them to whatever meager cover he could find.

Paratroopers from Co. B, 2nd Bn. (Abn) 8th Cavalry of the 1st Air Cavalry Division, hunker down in the cleared area just north of the Plei Me Special Forces camp following an air assault to relieve NVA pressure on the camp. The ground had been plastered by Air Force bombs for more than three days and the stench of rotting NVA corpses made this first venture into war by the young outfit a nauseating experience. (Courtesy of Steve Wilson)

During that time he was wounded three more times and eventually received a Distinguished Service Cross for his actions which probably saved the lives of several platoon members.

The flares were a problem to the defenders of LZ Mary because they illuminated the Americans while the NVA were in the shadows. When Ken Riveer tried to contact Captain Danielson, the CO of Alpha Company, he found that the radio of Lt. Hanlon's RTO was on the wrong frequency. He finally got it on the right frequency and Captain Danielson got the flares stopped. He also wanted Captain Danielson to alert the next lift not to fire when coming in because he was afraid his men would be hit.

Aerial rocket artillery pounded the tree line and relieved a lot of pressure around the perimeter and Lt. Tweedy's mortars did a fantastic job even though their exposed positions and the darkness forced them to work without plotting boards.

As daylight approached, Sgt. Riveer had 3rd Platoon fix bayonets with the idea that the bayonet charge was preferable to dying in exposed positions after daylight. Third Platoon reached the woodline and found that the NVA had, for the most part, pulled back before daylight. The platoon was exhausted when Captain Danielson called them back to work out a plan with Bravo Company which had just flown in under the command of Captain Roy Martin.

1st Sgt. Gonzales wanted a casualty report from Sgt. Riveer, but instead sent Riveer to Capt. Danielson and got the casualty report himself. Third Platoon was put under the command of the platoon sergeant from 2nd Platoon since he was the senior platoon sergeant in the company and Riveer was an E-6 at the time.

Alpha Company didn't get a break for the next 48 hours. They patrolled during the day, set up ambushes at night and patrolled the following day before being pulled back to provide airfield security at Pleiku for the night. The troopers managed to accomplish this with practically no sleep. Sgt. Riveer guessed 3rd Platoon went 48 hours with no sleep and was proud of the fact they were still able to perform their duties.

Sp4 David Dolby was the first person in the battalion to receive the Congressional Medal of Honor. It was awarded for action on May 21, 1966 when he organized the withdrawal of his platoon after they had walked in front of a line of enemy bunkers. He attacked the enemy bunkers and carried severely wounded Americans to safety.

The battle lasted for about nine hours. Air strikes were called in, but the two pinned-down platoons of B/1/8th weren't actually rescued until Captain Roy Martin took his headquarters group

Elements of the 2nd Bn. (Abn) 8th Cavalry, of the 1st Air Cavalry Division struggle up Chu HO (Ho Mountain) just south of the Plei Me Special Forces camp in an attack on OBJECTIVE CHERRY. It was the first offensive action by the Cav in the young campaign and was where the battalion sustained its first man killed in action, Staff Sgt. Charles Rose of Company B. (Courtesy of Steve Wilson)

over the hill from the back. They charged down on the NVA bunkers dropping grenades in as they went. Capt. Martin received the Distinguished Service Cross for this action.

1st Sgt. Ray Poynter was with Lt. Robert Crum when he was killed and his platoon sergeant wounded. He shot several NVA who tried to rush them and was forced to take cover until Captain Martin's group took the pressure off.

1st Battalion, 8th Cavalry was to fight in a lot of major battles as the war progressed, but one of the most successful was June 22, 1966, when Captain Gerrell Plummer's Bravo Company held a "Mad Minute" at LZ Eagle just as the NVA were launching their attack. Over 100 NVA were killed in that short time and they added another 30 before the day was over. Bravo Company was in the area because they were rescuing C/2/327th which had suffered heavy casualties earlier.

The second Medal of Honor for the battalion went to James H. Monroe who was killed jumping on a grenade after it was thrown near the wounded he had carried to what he thought was safety on February 16, 1967.

On December 2, 1967, Delta Company's 1st Platoon made a significant intelligence haul. While checking with villagers on reports of an NVA battalion moving into the mountains west of Tam Quan, one of the soldiers noticed a reflection on a mountainside. Lt. Reed took his platoon in the opposite direction for awhile then, using concealment of trees and a stream bed, turned back toward the location where the reflection had been seen. He spread his platoon at the base of the mountain and had his Artillery Recon. Sgt. Glenn Sheathelm stay with the left flank squad. Here they would be in the best position to observe fires. They reached the enemy position with no friendly casualties, and the enemy turned out to be a meeting between NVA and VC planning what was later called the Second Battle of Tam Quan. There were several cubic feet of important documents captured and some flags including one that is in the USMA Museum at West Point. It was a major intelligence haul with no friendly casualties because the NVA felt secure enough that security forces were not in the immediate area. Lt. Reed was justified in the pride he had in the platoon. Everything worked perfectly.

1/8th opened the Second Battle of Tam Quan by itself because the rest of 1st Brigade was returning from a rescue mission at Dak To.

1/8th also went into the A Shau and captured a large amount of Russian-made NVA heavy equipment. A truck captured there was to be shipped to the Patton Museum at Ft. Knox, but it was still in country when South Vietnam fell. Sgt. Hillary Craig found that it was possible to kill an NVA tank with an M-72 law if you got close enough. Also, 1/8th found that 57mm flak was very frightening when you were riding a helicopter into the A Shau.

The third Medal of Honor for a member of the battalion went to Donald R. Johnston of Delta Company who threw himself on a sachel charge that the NVA had thrown into bunker. The action, which cost him his life, saved six others in the bunker. The date was March 21, 1969.

The fourth Medal of Honor went to Lt. Robert Leisy who on Dec. 2, 1969 stood in front of an RPG round to protect his radio operator. He continued to lead his platoon and directed the evacuation of the unit's wounded. He later died from his wounds.

1st Battalion, 8th Cavalry returned from Vietnam on March 21, 1971, and, although no longer an airborne unit, they are still carrying on the traditions and a proud history that goes back almost 125 years. They are now part of the unit they joined in 1921, the 1st Cavalry Division, stationed at Ft. Hood, Texas.

It appears that 1st Battalion, 8th Cavalry had about 140 of their own killed in Vietnam, although the figures may not be complete. In honor of those killed a monument will be dedicated during the 50th Anniversary of the Airborne. The monument will be on the walkway below the Tomb of the Unknown Soldiers at Arlington.

Bibliography

Coleman, J.D. (ed.) *The 1st Air Cavalry Division: Vietnam*, Tokyo, Dia Nippon, 1970.

Coleman, J.D. *Pleiku: The Dawn of Helicopter Warfare in Vietnam*, New York, St. Marin's Press, 1988.

"Jumping Mustangs Association"— letters, photographs and conversations with members of the organization. For information, contact "Jumping Mustangs Association," c/o Ray E. Poynter, Route #3 Box 754, Berryville, Arkansas 72616.

Marshall, S.L.A. *Battles in the Monsoon*, New York, William Morrow and Company, 1966.

Marshall, S.L.A. *The Fields of Bamboo*, New York Dial Press, 1971.

Mertel, Kenneth D. *Year of the Horse-Vietnam*, New York, Exposition Press, 1968.

Stanton, Shelby L. *Anatomy of a Division: 1st Cav. in Vietnam*, Novato, CA, Presidio Press, 1987.

Paratroopers of the 2nd Bn., (Abn) 8th Cavalry sweep over the crest of Chu Ho (Ho Mountain) while consolidating their positions on OBJECTIVE CHERRY.

2ND BATTALION, 8TH CAVALRY REGIMENT

By Steve Wilson

Regarding the 1st Cavalry Division General Douglas MacArthur said,

"The First Team...First in Manila, First in Tokyo and First in Pyongyang."

EARLY YEARS

Out of the west they came. Into Ft. Concho, Texas in 1866 rode hard-bitten gold miners, disgruntled settlers and the remnants of the solitary frontiersman. These men rallied to be the West's 8th Cavalry Regiment. The 8th Cav. campaigned in Arizona Territory and in Oregon; then, in 1868, they rode the plains keeping peace between Indians and settlers.

After Cuba, two separate trips to the Philippines and patrol duty along the Mexican border, the 8th Cav. joined the 5th and 7th Cavalry Regiments for the Mexican Punitive Expedition. Then, they joined the 1st Cavalry Division in 1921.

From World War II, the 8th proudly flies campaign streamers for their heroic efforts in New Guinea, Bismarck Archipelago, Leyte and Luzon. Then, once again they were called upon to serve in Korea where they distinguished themselves at the frozen Chosin Reservoir.

REORGANIZATION

On 16 June 1965, Secretary of Defense McNamara formally announced the authorization of an airmobile division. That division would be the First Cavalry. Brigadier General Harry W.O. Kinnard, the division commander of the original 11th Air Assault Division (Test), would be the Cav's division commander with the rank of major general.

He originally wanted the entirety of the Cav. to be parachute qualified. To his dismay, the Pentagon would only permit him a single brigade of paratroopers and various sections of his division support command to be parachute qualified.

On 03 July, in an elaborate ceremony, the 11th Air Assault colors were retired and the First Cavalry Division colors were unveiled and the division was now officially the First Air Cavalry Division, Airmobile.

In a recent interview, General Kinnard, now retired, said, "Though a division commander cannot have favorites, the 1st Brigade was the apple of my eye because it represented what I felt all the combat arms of a division should be- Airborne."

To command the 2nd of the 8th, an extremely capable commander was chosen, LTC James Nix. His challenge was to prepare his new battalion for combat and to instill both the concepts of Airborne and the new airmobile tactics devised, tested and proven by the 11th Air Assault Div (Test). Long hours were spent by Nix and his staff planning and executing intense training programs to ready the new battalion to prepare them for eminent deployment in the jungles, mountains and rice paddies of Vietnam.

TRANSIT

In early August of 1965, the Herculean task of organizing and preparing for transit to Vietnam had been completed. Nowhere in the annals of military history had such a feat ever been accomplished with such panache. With their underwear and bright yellow Cav. patches dyed a jungle green, Nix's "Cavalairs" were ready.

With the battalion transported to the Port of Savannah, nothing remained but to board the USNS *Geiger* for the 31-day sail which would take the Cavalairs through the Panama Canal, half way across the Pacific with brief stops in Hawaii and Guam. The Cavalairs would travel 12,000 miles before they reached their destination at Qui Nhon harbor in the Republic of South Vietnam.

AN KHE-CAMP RADCLIFF

Nix wasted no time in issuing orders that would render the battalion's bivouac area suitable for human habitation. Troopers immediately began preparing their company areas and preparing preliminary defensive positions.

Soon after their arrival, the Cavalairs were called to perform base camp security by preparing and manning the defensive perimeter affectionately known as the "green line." Those first nights were uneventful with the exception of brief outbursts of friendly fire as shadowes fooled tired eyes with illusory images of unseen dangers. Nightly ambush patrols and listening posts were frequently placed outside the perimeter with everyone taking their turn.

PLEIKU CAMPAIGN-PLE ME SPECIAL FORCES CAMP

Until the middle of October, the battalion performed patrolling and picked duties from Qui Nhon through the Deo Mang Pass to An Khe and through the Mang Yang Pass to Pleiku. Their mission: to secure the vital road link, Highway 19, between the three South Vietnamese cities.

Then on 22 October, General Kinnard pulled the 1st Brigade from the Binh Khe area and moved them to Camp Holloway, II Corp Headquarters. Upon arrival at II Corp, LTC Harlow Clark in Colonel Robert's absence, and LTC Nix learned of the ongoing operation by an ARVN relief column trying to relieve the intense pressure on the Plei Me Special Forces Camp south and west of Pleiku.

Alpha and Charlie Companies executed an air assault into LZ South on the 24th. This action secured a tenable position for artillery from which the NVA took heavy casualties.

On the 27th, Nix's troopers again made an airmobile assault but this time on the perimeter of the beleaguered camp. As the trooper's boots hit the ground, the smell of putrefying flesh insulting their nostrils and vomit flowed. The NVA had left a tremendous number of dead. The scene was grim to behold but the camp was, for the time being out of direct danger. The Cavalairs suffered only one wounded trooper and one killed from sniper fire. The 1st Brigade was then given the mission of searching out, fixing and destroying any enemy forces which may provide a threat to Plei Me, Pleiku and the entire AO of the central highlands.

By Halloween, the battalion had assumed operational control of Search Area JIM and found themselves at the wrong end of the longest logistical supply line the Army had ever maintained. The most critical shortage being the life blood of the slicks, Avgas. Food and ammunition were also coming up short. Fortunately, by nightfall the supply pipeline was again open and needed stores were arriving at 1st Brigade's forward base at the Cateka Tea Plantation.

MUER RIVER CROSSING

Soon after daybreak on 4 November, Captain Manley Morrison and his Charlie Company made an air assault near Duc Co along with an artillery battery to support 1/8th Cav. which had secured the patrol area across the Il Drang River.

The remainder of the battalion was left to continue search and clear operations in the general area of an NVA hospital. Nix ordered Captain Linton, CO of Delta Company, to send a recon patrol to investigate a suspected enemy encampment to the north of LZ Cavalair.

Platoon Leader William Ward divided his platoon into two separate elements, leaving half at Cavalair and taking the remaining 17 men.

After working about half a click, Ward's patrol found and captured a sick NVA soldier. Then, crossing a stream and moving up a twin crested-hill mass, they began taking light fire from their right front.

Moving further up the hill, the patrol began taking more and more fire. Then, an estimated 35 to 40 NVA bolted in all directions. Ward ordered Sgt. Harold Rose to continue up the hill and to secure the high ground while Sgt. Jones was to continue in the direction taken by the majority of the NVA soldiers. After traversing some 100 meters, he found, hidden in a wooded depression, 12 more NVA, 10 of whom were sick and wounded. Jones secured these men and sent half of his men back up the hill to aid Sgt. Rose who was not receiving heavy sniper fire and was taking casualties. Sgt. Rose, with the help of the men from Jones, formed into a scrimmage line and continued to assault with grenades up the hill.

After reaching the top of the hill, Rose's group had suffered two KIA and two WIA. One of the wounded was Sgt. Robert Wilson who had destroyed a sniper's nest.

Sgt. Richard Coffee wasn't as lucky. He was to take a fatal round while inspiring his men forward to their objective.

By 11:00 a.m., Linton had informed battalion of the situation and had ordered the remainder of the Recon. Platoon and 10 more men from Bravo Company under the command of Sgt. Marshall to reinforce Ward's force. He and his reinforcements arrived just in time. The NVA had regrouped and moved back into the camp area. Marshall's men killed an additional 10 NVA. He then sent some of his squad on up the hill to reinforce Rose.

While this was taking place, Lt. Trapnell took 2nd Platoon, four men from Alpha Company and a medic to also provide reinforcements. With them went Captain Richard Slifer, Battalion S-2 and his interpreter to interrogate the prisoners.

As this group crossed the stream, it began receiving sniper fire from the western slope of the hill. The platoon deployed on line and began moving up the slope where they encountered even heavier fire. Again the NVA had regrouped and now had placed themselves between Trapnell and Ward.

The enemy forces had placed themselves in such a way as to negate any indirect fire and neither of the Skytrooper elements had sufficient fire power to make a determined movement. So, the Skytroopers and the NVA remained, trading shots.

As the afternoon continued, Nix ordered two platoons from Captain McElroy's Alpha Company to form and attack from the southeast. As they crossed the stream bed, they, too, came under heavy rifle fire. By 2:00 p.m., the battle raged from three separate locations along the ridge line.

Alpha's two platoons under Sgt. Welch continued to labor under heavy fire to reinforce the Skytroopers already in position on the slope until enemy fire became so intense, his advance was stalled. His radio operator, PFC Ronald Luke, provided cover fire as Welch retrieved wounded until Luke saw an NVA soldier draw a bead on Welch. Luke rose to his feet and began firing rapidly drawing fire to himself so Welch could finish pulling the wounded into a small ravine. Luke's radio was hit 11 times, but he continued to draw the enemy's fire until an enemy machine gun cut him down. Both men won Silver Stars for their efforts that day.

Finally, Nix was able to plot and call supporting fire from his mortars and 105mm howitzers, and he was able to identify targets sufficiently enough to call in Huey gunships concentrating fire on the northern, eastern and western sides of the battle area.

By 5:30 the American effort was sufficient to cause the NVA to break contact and move out to the north and west toward Anta village. They left behind many dead, sick and wounded.

The battalion spent the balance of the 4th and on into the 5th putting itself back together again. Nix was quick to rally his troops and prepare for what lay ahead. Charlie Company had reswept the battle area and Bravo Company was sweeping north along the Tae River drainage.

As the night of 5 November began to darken the jungle, Bravo Company had moved to an area of high ground along a trail that led between an abandoned Montagnard village and the Tae River. There they would spend the night with ambushes set out to intercept any NVA that may try to move in the area during darkness.

At dawn, Bravo's CO, Captain Richardson, split his command into a two-pronged sweep. About 10:00 a.m. the first platoon reported receiving fire and taking casualties. As Lt. Meyer reoriented his platoon to the south, second and third platoons began taking fire. It became quickly evident the two maneuvering elements had caught something or someone between them. Nix then ordered Charlie Company to change its direction and begin to close the three clicks separating them from Bravo.

By 10:15 a.m., Bravo was taking even heavier volumes of fire from well-prepared positions and the second platoon had effectively been isolated. Within moments, Bravo Company was surrounded and taking casualties from a battalion-sized force.

While organizing his second platoon to withstand a massive attack, Lt. Castle, though wounded, placed himself in a position where he could continue to direct his platoon and call for napalm runs by Air Force F-104s. His platoon sergeant, SFC Payton Watson, while reorienting one of the platoon's machine guns, killed at least 10 NVA with his M-16 before the North Vietnamese pulled back.

At the same time, Third Platoon, under Lt. Felix King, moved across the fire and swept 200 meters between his position and that of Castle. After making contact with the wounded Castle, he then attempted to return to his previous position but was brought down by a burst of machine gun fire.

It was then that Sgts. Robert Tadilla and John Baer and PFCs James Crafton and Rodrigo Gonzolez disregarded the incoming fire and left the safety of the ravine to retrieve their fallen comrades. Once all were sheltered in the safety of the ravine, there was little to do but wait for Charlie Company.

Just before 1:00 p.m., Charlie Company's first platoon crossed the densely jungled Muer River and came on line. They moved out of the underbrush onto open and upward sloping terrain and moved ahead 75 meters. Before the remainder of the company crossed the river, enemy machine gunners opened up, dropping over half of the platoon in a matter of seconds.

Morrison continued to bring the remainder of Charlie across the river and attempted to bring second platoon on line with the first not knowing that the first platoon had lost its leader and radio operator in the first seconds of the battle. It would be later reported that "the effects were devastating,"

Young PFC Thomas Maynard distinguished himself that afternoon by throwing himself on top of a grenade, sacrificing his life to save that of a buddy. That day also claimed the life of PFC Tony Pendola, the first 17-year-old to die in Vietnam.

Now Charlie Company would repeat Bravo's action in retrieving their wounded. PFCs Donald Pond, Willie Pierce and others left the safety of the shallow ravine to pull back wounded friends. The platoon medic, James Allen, refused to leave his wounded and died for the effort. Pond and Pierce received Bronze Stars with "V" Devices for valor and Allen received the Silver Star for his dedication to saving lives on the field of battle.

The small group of survivors huddled in the ravine for what seemed an eternity not knowing what had become of the rest of Charlie Company. Morrison, unaware that anyone in the platoon was still alive, called in fire support of more than 40 sorties of tack air during the afternoon.

Then PFC Steve Wilson, with little to loose at that point, retraced the platoon's path to see if he could make contact with the main body of Charlie Company. Leaving the comparative safety of the ravine, he made his way under severe sniper fire to the edge of the river where he found Morrison had been wounded and had turned over command to Lt. Weiss. He also received a Bronze Star with "V" Device.

Weiss, surprised to learn that there were survivors from the first platoon, directed men of the third platoon to accompany Wilson back to the ravine to reinforce the remnants of 1st platoon. Meanwhile, the second and third platoons began removing the threat from the company's flanks. And by 6:00 p.m., both Bravo and Charlie Companies were able to link up.

On daybreak of the 7th, with Bravo securing the LZ, Charlie Company conducted a sweep of the battle area. The NVA positions, so deadly the afternoon before, were strewn with enemy dead and the accouterments of war. LTC Nix and Sgt. Coulson from Bravo Company, as they surveyed the area, found and took a prisoner, one last NVA regular still at his position.

It took three Chinooks to out-load the enemy equipment taken from the battlefield. It consisted of three heavy machine guns, three light machine guns, two automatic rifles, 23 AKs, 100 hand grenades, a B-40 rocket launcher, 45 individual packs and about a ton of ammunition. This time the NVA left some bodies on the battlefield to count. The original body count was 198 NVA killed but later changed to 460 NVA killed after two large grave sites were found.

The 2nd of the 8th losses were also high. The battle resulted in the largest number of casualties yet taken in the campaign. The battle had 26 killed, 17 from Charlie Company, 7 from Bravo, and 2 from Headquarters Company with an additional 53 men wounded.

SEARCH AND CLEAR I

With the dawning of 1966, LTC John Hemphill took command of the battalion and led them through Operation MATADOR along the Dak Adrai River near the Cambodian border and Operation WHITE WING in the An Loa Valley and Crow's Foot area near Bong Son.

Next came Operation JIM BOWIE near Kon Truc. Then, back to Pleiku for Operation LINCOLN in March 1966.

LZ HEREFORD

As Operation DAVY CROCKETT began winding down, a civilian irregular defense group from Vinh Thanh ambushed an enemy force and captured interesting intelligence material.

At 11:00 a.m. on 16 May 1966, Captain

J.D. Coleman, now commanding Bravo Company and his men, made a combat assault into a one-ship LZ named Hereford.

After climbing to the ridge line of a mountain, the company began moving eastward along a razor back ridge. Near 2 p.m., the third platoon spotted a single Viet Cong and opened fire. The fire was instantly returned from readied positions. The platoon leader, Lt. Heaney radioed Coleman that he was encountering firm opposition and was maneuvering to his left.

The flanking movement met with instantaneous and violent counterattacks by an enemy platoon. The men fought courageously but were overrun with all being killed except for one man.

Under insurmountable odds, Heaney pressed for superiority, but each time he tried, he received more casualties. Then, the enemy began probing the flanks of the company inflicting heavy casualties.

As mid-afternoon approached, so did torrential monsoon rains cutting the available light under the triple canopy to near darkness. The bad turn in the weather would also limit air reinforcements and resupply efforts. So, Coleman decided to chuck any plans for offensive movements and concentrate on establishing a defensible position on the high ground he now held.

The company had barely formed into a defensive perimeter when the enemy launched a resolute attack from the west. This told Coleman he was up against a force much larger than his own, and obviously, they were well trained.

By 4:30 p.m., Bravo had established a defensible perimeter and had recovered many of its wounded and dead.

The rain continued unabated into the evening. Tube support was unavailable to place sufficient fire near enough to Coleman's perimeter, and only two birds were able to fly within reach of the stranded company. But, as night fell, the enemy threat ended, and by 8:00 p.m., all was quiet.

Coleman now was faced with 20 dead and 40 more wounded and had little choice but to await the arrival of reinforcements from LZ Hereford. The company, though in bad shape, voiced their willingness to "stand and die on that piece of ground!"

The situation was bleak and Coleman knew it. He had only 45 men capable of manning the perimeter and three of his medics were badly wounded. Medic Danny Die was killed as he retrieved wounded from outside the perimeter. He was awarded the Silver Star.

Then, just after 10:00 p.m., 130 men of Alpha Company, 1/12th Cav., under the command of Jackie Cummings, made their way to Coleman's perimeter. In the pitch blackness, Cummings' men were fed into Bravo's line.

At first light on the 17th, Coleman pulled in its listening posts and the combined element initiated a "mad minute" which brought a hail of return enemy fire.

It was apparent at that point that the enemy had placed a battalion-sized element against what was left of Bravo Company and its reinforcements. The men fought bravely for two hours, stacking up the bodies of the enemy one on top of the other outside the small perimeter.

With ammunition all but gone and the order to fix bayonets given, another relief column consisting of Charlie Company, 1/12th caused the enemy to break their strangle hold on Coleman and Cummings, routing them back into the triple canopy.

The end result of the battle was grim for Bravo. They had suffered 25 dead and 62 wounded, and Alpha 1/12th had 3 dead and 37 wounded during the bitter 24 hours on the mountain top. The enemy had suffered as well with 38 NVA bodies actually within the perimeter and 200 additional dead and dying in the surrounding jungle.

Captain Coleman's Bravo Company was the first company-sized unit to receive the Valorous Unit Citation for its heroic actions.

SEARCH AND CLEAR II

June and July of 1966 were taken up with Operations NATHAN HALE in the Tuy Hoa Province and HENRY CLAY in the Trung Phan Province.

In August 1966, LTC Thomas Tackaberry took command of the battalion and led the men back into Tuy Hoa Province on search and clear operations until September when he lead them into Binh Dinh Province on a pacification campaign called Operation THAYER.

October found Tackaberry's battalion participating in Operation IRVING in the Nui Mieu mountains making successful air assaults into LZs Ebony, Playboy, Adam and Esquire.

As the month of October wound down, so did the 2nd of the 8th Cavalry as an airborne unit. Though the men of the 2nd of the 8th continued to fight on bravely, they did so without the benefit of prized silver jump wings.

2ND BATTALION 19TH ARTILLERY

By Glenn H. Sheathelm

In August of 1965, a unit left Ft. Benning, Georgia for Vietnam which was unique in the U.S. Army. It was an artillery battalion that was both airborne and airmobile qualified. The unit was the 2nd Battalion, 19th Artillery.

Although the qualifications of the unit were new, the unit itself was not. It was formed in 1916 and went to Europe during World War I as part of the 5th Division. Service in that division resulted in the unit crest which includes the diamond of the 5th Division, the eagle of Amnedes which was a major battle for the unit during World War I, and the crown on the eagle's head to denote artillery as the "king of battle."

During World War II the 19th Artillery was part of the 5th Division again, and fought across Europe providing support for that division and other units that were part of the 1st Army. During that time period the motto "On the Way" was adopted although it didn't become part of the unit crest.

The 2nd/19th Artillery was part of the 1st Cavalry Division in Korea in June 1965 when the colors were sent to Ft. Benning along with other colors from the 1st Cavalry Division to be the new colors of the 6th Battalion, 81st Artillery which had been in training with the 11th Air Assault Division (Test). What had been the 1st Cavalry Division in Korea became the 2nd Infantry with the 1st Cavalry colors going to the new unit.

Like other units in the 1st Brigade of the 1st Air Cavalry, there was a wild scramble to get people from the 81st Artillery who were not already airborne qualified and other inspired artillerymen through jump school. There were also NCOs and officers who came from other airborne units to be part of the formation of a history making unit.

Lt. Col. Francis Bush took command of the unit that was really breaking tradition. Officials in the U.S. Army had difficulty with the concept of artillery moving other than by truck or tracked vehicles. Some even joked about horse drawn artillery. They weren't even very impressed when CH-47 Chinooks flew by carrying 105mm howitzers.

Gen. H.W.O. Kinnard finally had a 105mm striped down and flown by hanging below a Huey! Since the helicopter was unable to carry the crew for the artillery piece or any ammo, the maneuver was just for show. The 105mm howitzer was one of the M-101 models with all the armor removed and even the much lighter M-102 howitzer was not carried on a regular basis by Hueys in Vietnam. It wasn't practical because, to get the howitzer in action, a crew and ammo were necessary which meant the Chinook was the normal mode of transportation.

2nd/19th Artillery, like the 1st/8th Cavalry, went to Vietnam aboard the USS *Geiger*. With LTCs Bush and Mertel in charge, the respective units joked that they "ran all the way across the Pacific." Exercises included running in place and up and down the stairs on the ship.

2nd/19th Artillery supported the 1st Brigade with different batteries claiming battalions in the brigade as their own. Alpha Battery in particular became very attached to 1st Battalion, 8th Cavalry.

Although they were too far away to support the two companies at LZ Mary on Nov. 3, 1965, they did set up a fire base near enough to support B/1/8 sweeps the day after the battle. Bravo Battery drove off a NVA ambush that was sprung on a relief column heading to the Special Forces camp at Plei Me, and Charlie Battery supported 2/8 Cavalry during clearing operations around the NVA hospital complex.

In early 1966, the battalion received the new M-102 howitzer which increased their ability to provide support for the rapidly moving cavalry units in 1st Brigade. The M-102 was different from other artillery that had been used. It had a base plate that was staked to the ground and instead of the standard split trails, had a one piece trail with a large roller. It was capable of going 6400 miles without being relayed. Aiming stakes were put out in one direction and an infinity collimater was put out in the opposite direction and close to the howitzer. It allowed the 2nd/19th to have rounds in the air within about one minute even though the tubes were pointed in the opposite direction when the call from the forward observer arrived.

In addition to the Presidential Unit Citation which went to the whole 1st Cavalry Division for the October-November 1965 actions, portions of 2/19th added Presidential Unit Citations for October 2-3 and Bravo Battery collected a PUC for the action on LZ Bird at the end of the Christmas Truce in 1966.

At the time of LZ Bird, LTC Wilbur Vinson had just turned over command to LTC Culp. LTC Culp was very interested in new developments in artillery and some of the new ammunition. His arrival was very fortunate because he called a meeting of officers and key NCOs for each of the three firing batteries and the HQ to explain his philosophy of the importance of the new technology to artillery. One of the items covered was the "Beehive" round which could be set to explode into flechettes right at the muzzle if necessary.

When the 22nd NVA Regiment launched its attack on LZ Bird it probably made a mistake because Bird was an established firebase and nearby LZ Pony, which was home for the 2nd/19th Headquarters Battery, didn't even have wire out and was only 7-8 kilometers away. The NVA had put a lot of careful planning into the attack on LZ Bird and wouldn't change their plans even though there was a much more lucrative target nearby.

B/2/19th was on LZ Bird along with C/6/16th Artillery which was a 155th towed unit assigned to 2/19th as medium artillery. 2nd/19th was responsible for clearing all targets and checking data for both units through the Headquarters Fire Direction Control at LZ Pony. Also on LZ Bird was C/1/12th Cavalry which was providing perimeter security and short range patrolling out of the LZ.

Although intelligence information processed by Capt. James Weber's crew at HQ/2/19th on LZ Pony indicated that the NVA were assembling for an attack on LZ Pony or Bird, the NVA were able to get almost up to the wire at night without being seen. The NVA hit LZ Bird with mortar fire and followed it quickly with a large scale ground attack which killed or wounded a lot of the C/1/12th troopers early in the battle. Lt. Campenella who was the forward observer for the company found himself running the company after the other officers were killed or wounded. Captain Leonard Schlenker who was the CO of Bravo Battery didn't even have time to lace his boots before racing to the battery CP. He arrived there pursued by NVA who had already come through the wire. The last word received from Bravo Battery's CP was, "They're right outside the tent." There was gunfire in the background and the artillery net went dead which temporarily left HQ/2/19th in the dark about what was happening.

The NVA had not only breached the wire in several places, but had taken over all of the 155mm howitzers and it was hand-to-hand fighting in some of Bravo Battery's gun pits. To further add to the bizarre scene, some of the fighting was taking place amid Christmas wrappings from the holiday.

The 2nd/19th had lost most of their positions to the NVA and were rallying in the two remaining positions. At one of the guns, Lt. Piper made the decision to use Beehive because the NVA were trying to turn the captured artillery pieces on the defenders. Unable to find the proper red flare in the confusion he started screaming "Beehive" and when he heard "Shoot it!" relayed from forward of the gun, he did and cut a large swath in the NVA forces. He then had the tube traversed to other large groups of NVA and swept them down with additional rounds.

After the Beehive rounds were fired the NVA attack faltered and stopped completely. About the time the NVA attack stalled, Capt. Weber's crew from the HQ/2/19th Artillery on LZ Pony was trying to figure out what had happened. One of his intelligence specialists picked up weak signals on the infantry frequency and started dire adjusting with the 175mm guns on LZ Hammond and 2/17th's 105mm on LZ Pony.

Capt. Weber got into the air with HQ/2/19th's H-13 and pilot and was over LZ Bird surprised to see artillery fire coming in that was obviously being directed from somewhere but not on the normal artillery frequencies. As if the artillery rounds sailing through the captain's same dark airspace were not enough, helicopter gunships almost rammed him. He then took over control of the 2nd/17th Artillery Battery at LZ Pony and directed its fire and told SP4 Sheathelm to move the 175mm to targets of opportunity south and west of the LZ so he would have less chance of an unfortunate meeting between the 175mm round and a helicopter.

The NVA pulled back off the LZ, but not before they wrecked several of the artillery pieces with explosive charges. LTC Culp flew to the LZ as soon as possible to assess the damage and see that everything was being done to get the wounded taken care of and the artillery back in action.

After landing on the LZ, LTC Culp found that Lt. Piper was wounded but didn't want to leave the LZ. He convinced Lt. Piper that he had to go back and brief the S-3, Major Hay, about the battle at LZ Pony, knowing that Major Hay would get Piper to the very dedicated battalion surgeon Captain Reasa. The picture of the battle was getting clearer by the minute at LZ Pony, but when Lt. Piper arrived he wouldn't let the doctor work on him until he had told Major Hay all about the battle. Lt. Piper sat in a metal folding chair on the dirt floor in a tent called headquarters, surrounded by radios, maps and plotting boards, while Capt. Reasa removed shrapnel. Lt. Piper excitedly told Major Hay everything he could about the battle seemingly unaware that it was his body Capt. Reasa was poking and probing.

S/Sgt. Delbert Jennings of C/1/12th earned a Congressional Medal of Honor at LZ Bird for leading assaults on groups of NVA and rescues of wounded Americans on the LZ.

In early 1967, 2nd/19th Artillery added another unit to its control which was a composite battery of 3rd/18th Artillery. It came overland from Highway 1 through the mud to LZ Pony. In some places palm trees had to be used for a corduroy road to keep the tracked 175mm and 8-inch artillery pieces from sinking over the decks.

The arrival also created a new supply problem which resulted in a combat drop of supplies. The C-130s dropped pallets of 175mm and eight-inch ammunition with parachutes. Since the drop had to be out of small arms fire range, the target area was the rice paddies north and east of LX pony.

The pallets disappeared into the muck and the artillerymen from 2nd/19th and 3rd/18th had to go out to recover the chutes and ammo. After disconnecting the parachutes and securing them so they wouldn't get blown around in the rotor wash, Hueys were flown in to pull the pallets out of the muck and shuttle them back to the LZ. The VC fired at the recovery teams but not in significant volume because that would have meant attracting heavy retaliatory fire.

The parachute drop, while an interesting operation, was not very practical. Further resupply of 3rd/18th was done with CH-47 Chinooks and sling-loaded ammo.

After moving the HQ to LZ English, LTC Culp was pleased to learn of a village sick call program that some enlisted men had started while at LZ Pony. He gave the unofficial program his full support and the program grew to the point that he went along on occasion, and Capt. Reasa and his medics got involved.

The arrival of Major Vernon Gillespie as battalion CO was a shock to some of the artillerymen because he differed sharply in style from that of LTC Culp. Major Gillespie was a "Charger" with combat experience in Special Forces (see "National Geographic," January 1965) who obviously was really going to take some aggressive action. He did it with great style, bringing back artillery raids and personally going out to fight battles most commanders would leave to junior officers and COs.

During the time Major Gillespie commanded the 2nd/19th Artillery, it provided one of the most sustained support missions of the war for the unit. It became know as the Battle of Ta Quan II.

It started with the D/1/8th Cavalry's discovery of extensive intelligence information on Dec. 2, 1967, and it got into full swing on the 6th and 7th with other 1/8th units getting into the action near a downed helicopter. By the time the battle was over, they had fired 42,000 artillery rounds either cleared by 2nd/19th or fired by 2nd/19th.

One of the units heavily involved was Capt. Robbin's Alpha Battery of LZ Geronimo. They were a short distance north of the battle area and fired almost continuously for a week. The loyalty of Alpha Battery to 1st/8th Cavalry was such that when offered the chance to go to LZ Mustang for a couple of days rest while another battery worked their guns, they refused. "It is our battalion (meaning 1st/8th) out there and we will stick with them." By the third day the strain was beginning to tell on the artillery crews. In order to give them a break everyone on LZ Geronimo helped carry ammo from the helipad to the gun pits and infantrymen from D/1/8th Cavalry who were being held in reserve in some cases found themselves being trained in artillery. The experience was good for both the 2nd/19th and the 1st/8th.

In January 1968, the 2nd/19th went north with 1st Brigade to I Corps where they became involved in the Tet Offensive and the relief of the Marines at Khe Sanh. The commander during that time was LTC Arnold Boykin.

The 2nd/19th also went into the A Shay in 1968 and provided artillery support for the 1st Brigade. Helicopter resupply was severely tested during that operation because of the 57mm anti-aircraft fire by the NVA and the poor weather conditions.

In the fall of 1968, the 2nd/19th Artillery moved south near Tay Ninh and in March of 1969, fired their one millionth round since arriving in Vietnam. While in the III Corps area the 2nd/19th used its direct fire capability on several occasions to drive back enemy attacks. The first occasion was Charlie Battery at LZ White on March 23rd which was followed by a major attack on LZ Carolyn adding another Presidential Unit Citation for Alpha Battery.

LZ Carolyn was located near the abandoned Prek Klok Special Forces Camp and like the former camp, was a base to support interdiction of NVA and VC supplies. The NVA 95th Regiment attacked the position after an extensive rocket and mortar barrage. In addition to Alpha Battery, the U.S. force at LZ Carolyn included the B/1/30th's position and in all, it exchanged hands three times with the Americans using weapons as varied as Beehive rounds and Bowie knives. Alpha Battery 2nd/19th fired over 800 rounds on or inside the perimeter during the battle. To further add to the confusion, both artillery ammo dumps were hit and for four hours over 600 rounds exploded. It appeared that the whole LZ was on fire. The loss of one ammo dump meant that remaining artillery ammunition often had to be moved from one gun to another under heavy fire. One section chief who went to get more ammo found a wounded NVA hanging onto the rope handle on the other end of the wooden ammo box he had started to lift. The sergeant quickly let go of the box, drew a .45, shot the NVA, jammed the pistol back in the holster and dragged two boxes back to his gun pit. The NVA finally gave and withdrew from the LZ, leaving behind 198 dead.

On May 12, Bravo Battery 2nd/19th drove off an attack on LZ Jamie with the NVA leaving 70 dead behind.

In 1971, when the 1st Air Cavalry Division returned to the United States it went to Ft. Hood, Texas as a TRICAP Division. The 2nd/19th Artillery was placed on inactive status with personnel being assigned to other units.

Bibliography

Coleman, J.D. (ed.) *The 1st Air Cavalry Division: Vietnam*, Tokyo, Dia Nippon, 1970.

Coleman, J.D. *Pleiku: The Dawn of Helicopter Warfare in Vietnam*, New York, Stg. Martin's Press, 1988.

Marshall, S.L.A. *Bird: The Christmastide Battle*, Nashville, Battery Press, 1968.

Ott, David E. *Field Artillery 1954-1973*, Washington, D.C., Department of the Army, 1975.

Reed, David. *Up Front In Vietnam*, New York, Funk & Wagnalls, 1967.

Stanton, Shelby L. *Anatomy of a Division: 1st Cav. in Vietnam*, Novato, CA, Presidio Press, 1987.

Tolson, John J. *Airmobility: 1961-1971*, Washington, D.C., Department of the Army, 1973.

173RD AIRBORNE INFANTRY BRIGADE

By Ellis W. Williamson
Maj. Gen., USA (Ret)

Although the number 173rd has a long history in the U.S. Army, it picked up an entirely new meaning when early in 1963 the Department of the Army ordered the implementation of a new concept. Brigadier General Ellis W. Williamson was called into the office of the Deputy Chief of Staff of the Army for Operations. General Harold K. Johnson, along with several staff officers, explained the new concept. General Williamson was to move to the island of Okinawa where he would organize, train, command and commit to combat if necessary, a unique organization. He was told that this unit would be the only one of its kind in the U.S. Army. It would be extremely mobile and flexible. It would be prepared to respond to emergencies in any of the countries around the "Pacific Rim." The unique aspect of this specific unit was that it would be an especially-tailored separate airborne brigade. It would have all the elements of a complete division, except in lesser numbers. Its capability to project throughout its vast area of responsibility would be close coordination with both the U.S. Air Force and the U.S. Navy.

General Williamson had several units under General Johnson before and stated that maximum flexibility could be obtained only through good, well-trained junior leaders in whom the overall commander had complete trust. General Johnson remarked, "You have my personal support in assembling your team of good people."

During the next two years, the 173rd Airborne Brigade redefined the word "innovation." The command was encouraged to try every new idea that the people of the organization could think of. They were to throw aside and forget those things that did not show promise of real improvements. Those that worked well were to be refined, written and sent back as reports to the United States for consideration. Many of the ideas, particularly in the use of small unit tactics, communications and helicopter operations were perfected and ready for use by the time the Vietnam experience came along.

The "SKY SOLDIERS," as the Nationalist Chinese paratroopers called the 173rd, made thousands of parachute jumps in a dozen different Pacific area countries. They experimented with the use of all types of aircraft, submarines, aircraft carriers and assault boats. They even had their own private jungle training island far south of Okinawa. Within the parameters of safety, all junior leaders were told to be rapid, flexible and pragmatic, i.e. "try most anything that just might work better." Often platoon-sized units were left on the island, completely by themselves, with no interference of a higher commander looking over the leader's shoulder.

As junior leaders became proficient in the basics, they were accorded a level of trust and authority that the Army teaches in the classroom but seldom practices in the field. This philosophy paid off, often to the surprise of senior U.S. officers and our Allied observers.

With its top-notch personnel, its high adventure training opportunities and its 100 percent re-enlistment month after month, the 173rd was the logical draw to be the first U.S. Army ground combat unit to be deployed to Vietnam in May of 1965. The major portion of the brigade landed at Bien Hoa Airfield and found an area that had been battered frequently by enemy raids and shelling attacks. By nightfall of the first day, the SKY SOLDIERS had moved into the surrounding jungles and prepared fighting positions and patrol bases.

They had begun the process of disrupting the Vietcong (VC) operations and plans. For one year thereafter, not a single round of enemy fire landed on or near the Bien Hoa Airfield.

The 173rd Brigade was given some interesting missions:
1. Protect vital installations and personnel.
2. Keep the enemy off balance so that he cannot mount a major attack before additional units arrive from the United States.
3. Clear the enemy from certain areas and lead stateside arriving units into their new positions so that they may get settled before becoming involved in a fight.

An additional mission was later added. "Orient and indoctrinate newly arriving units."

After the brigade had been in country only a short while, the Australian and the New Zealand forces arrived. All of them were placed under General Williamson's command. These forces remained an integral part of the command until the summer of 1966.

In the combat operations to follow, the paratroopers made their superb training pay off. They were the first to go into War Zone D to destroy enemy base camps. They introduced the use of small, long range patrols. They fought the battles of the Iron Triangle, conducted the first major combat parachute jump in the Tay Ninh area, and blocked North Vietnamese Army (NVA) incursions during some of the bloodiest fighting of the war at Dak To during the summer and fall of 1967, culminating in the capture of Hill 875. Elements of the brigade conducted an amphibious assault against NVA and VC forces as part of an operation to clear the rice-growing lowlands along the Bong Song littoral.

From April 1969 to early 1971, the 173rd primarily supported pacification operations. The paratroopers conducted small unit patrols around the hamlets and supported Vietnamese Army combat operations, denying population and rice-growing centers to enemy forces. The sizeable number of enemy troops killed in action, the capture of many weapons and tons of supplies, the enormous amounts of foodstuffs uncovered and the failure of the NVA to mount a major operation in the brigade's tactical area of responsibility were prime indicators of the SKY SOLDIER'S combat success.

The 1st, 2nd, 3rd and 4th Battalions of the 503rd Parachute Infantry Regiment, along with the 1st Mechanized Battalion, 50th Infantry, Troop E, 17th Cavalry, Company D, 16th Armor plus the First Battalion of the Royal Australian

Regiment formed the principal foxhole strength of the brigade. They were well supported by the 319th Airborne Artillery Battalion, a battery of the 161st New Zealand Artillery and the Brigade Support Battalion.

The troopers of the 173rd Airborne Brigade (Sep) wear their combat badges and decorations with pride. The Army records tell the story. During more than six years of nearly continuous combat, the 173rd earned 14 campaign streamers and four unit citations. Its troopers earned 12 Medals of Honor.

As an indication of the reputation the brigade earned, after being in country about six months, the brigade was ordered to transfer 23 of its experienced junior officers from the brigade to other units in Vietnam. These transfers were completed, and within one month, every single one of those officers was given command of a company-sized unit.

Through all its existence the brigade was blessed with outstanding commanders who could develop and trust subordinates. Well over a hundred enlisted men rose to the highest rank of sergeant major, and 49 of the brigade officers went on to wear general officer stars. Two of them wore four stars.

The price paid for this dedication was not easy. The names of 1,534 SKY SOLDIERS are carved into the marble of the Vietnam Memorial, and thousands of Purple Hearts were awarded to members of the brigade.

When the unit's colors were furled at Fort Campbell, Kentucky, in January 1972, a unique chapter in the Army's combat history was closed. Created to "quickly snuff out small brush fires in the Pacific," the 173rd Airborne Brigade (Sep) spent its combat operational life at the cutting edge of the Army's Vietnam Campaigns.

The Society of the 173rd Airborne Brigade is quite active with its national chapter and 12 regional chapters that cover the entire United States plus Australia. The chapters get together about once each three months and the entire group comes together once each year in such places as Fort Bragg, NC; Fort Campbell, Kentucky; Washington, D.C.; Fort Benning, Georgia; Orlando, Florida; Chicago, Illinois and Santa Rosa, California. In 1990, we join the large group of troopers in Washington, D.C. The next year it is Australia, then Fort Benning again, and then Philadelphia, PA. The former members of the 173rd Airborne Brigade enjoy the camaraderie of just getting together; however, that is not all. We honor our fallen, assist and encourage our veterans who need help, and, through our scholarship and incentive fund, we promote the advancement of our children as they strive for excellence. The spirit that created that wonderful organization is still alive and well.

MEDAL OF HONOR RECIPIENTS

Sgt. Larry S. Pierce, * 20 September 65
PFC Milton L. Olive, * 22 October 65
SP5/SFC Lawrence Joel (now deceased), 8 November 65
SGT/CSM Charles B. Morris, 29 June 66
SP4 Don L. Michael, * 8 April 67
PFC John A. Barnes III, * 12 November 67
Maj (Chaplain) Charles J. Watters, * 19 November 67

PFC Carlos J. Lozada, * 20 November 67
SSG Laszlo Rabel, * 13 November 68
CPL Terry T. Kawamura, * 20 March 69
SP4 Michael R. Blanchford, * 3 July 69
SSG Glenn H. English, Jr., * 7 September 70

* awarded posthumously

MOH-RECIPIENTS SUBSEQUENTLY VOLUNTEERING FOR SERVICE WITH 173D

MSG/COL Ola L. Mize, 10-11 June 53 (with 3d Inf. Div.)   SP4/SSG David C. Dolby, 21 May 66 (with 1st Cavalry Div.)

COMBAT OPERATIONS OF THE 173D AIRBORNE BRIGADE

Name of Operation	Date	Location
1. OPORD 5-65	7-8 May 65	Bien Hoa & Vung Tau
2. OPORD 6-65	12 May 65	Bien Hoa Area
3. OPORD 7-65	14 May 65	Bien Hoa Area
4. OPORD 8-65	15 May 65	East of Bien Hoa
5. OPORD 9-65	19-20 May 65	Northeast of Bien Hoa
6. OPORD 10-65	26-27 May 65	Northeast of Bien Hoa
7. OPORD 11-65	31 May-3 June 65	East of Bien Hoa
8. FRAG ORDER 1-65	7 Jun 65	Bien Hoa Area
9. FRAG ORDER 2-65	21 Jun	Bien Hoa Area
10. FRAG ORDER 3-65	22 Jun 65	Bien Hoa Area
11. FRAG ORDER 4-65	23-24 Jun 65	East of Bien Hoa
12. OPORD 15-65	25-26 Jun 65	Southeast of Bien Hoa Highway #1
13. OPORD 16-65	27-30 Jun 65	North of Bien Hoa War Zone "D"
14. FRAG ORDER 7-65	4-5 Jul 65	Bien Hoa Area
15. OPORD 17-65	6-9 Jul 65	North of Bien Hoa War Zone "D"

Name of Operation	Date	Location
16. FRAG ORDER 9-65	16 Jul 65	Bien Hoa Area
17. FRAG ORDER VOGG	19 Jul 65	Bien Hoa Area
18. FRAG ORDER 10-65	21-22 Jul 65	Bien Hoa Area
19. FRAG ORDER 11-65	24-25 Jul 65	Bien Hoa Area
20. OPORD 19-65	28 Jul-2 Aug 65	Phuoc Tay
21. "PLEIKU"	10 Aug-5 Sept 65	Pleiku-Kontum
22. Operation "BIG RED"	7 Sept-8 Oct 65	Ben Cat, Phuoc Ving, Di An, Phu Loc
23. OPORD 24-65	1 Oct 65	Ben Cat
24. FRAG ORDER 15-65	4-6 Oct 65	North of Bien Hoa War Zone "D"
25. OPORD 25-65 (IRON TRIANGLE)	8-14 Oct 65	Ben Cat
26. OPORD 26-65 (NEW HOPE)	21-27 Oct 65	Di An, Phu Loi
27. OPORD 28-65 (HUMP)	5-9 Nov 65	Northeast of Bien Hoa War Zone "D"
28. Operation "NEW LIFE"	21 Nov--17 Dec 65	La Nga River Valley
29. OPORD 29-65 (SMASH)	17-23 Dec 65	Phuoc Tuy
30. OPORD 30-65 (MARAUDER))	1-8 Jan 66	Hua Nghia Province
31. OPORD 1-66 (CRIMP)	8-14 Jan 66	Binh Buong, West of Ho Bo Woods
32. OPORD 2-66 (ON GUARD)	17-21 Jan 66	Di An, Phu Loi
33. Operation "PHOENIX"	26 Feb-22 Mar 66	Binh Duong and Bien Hoa Province
34. OPORD 3-66 (ILVER CITY)	0-22 Mar 66	Long Khanh Province North of Song Be River
35. Operation "DENVER"	10-25 Apr 66	Song Be, Phuoc Long Province
36. Operation "DEXTER"	4-6 May 66	Tan Uyen
37. Operation "HARDIHOOD"	16 May-8 Jun 66	Phuoc Tuy Province
38. Operation "HOLLANDIA"	9-18 Jun 66	Phuoc Tuy Province
39. Operation "YORKTOWN"	24 Jun-9 Jul 66	Long Khanh Province
40. Operation "AURORA I"	9-17 Jul 66	Long Khanh Province
41. Operation "AURORA II"	17 Jul-3 Aug 66	Long Khanh, Binh Tuy Lam Duong Provinces
42. Operation "TOLEDO"	10 Aug-7 Sep 66	Phuoc Tuy & Binh Tuy Provinces
43. Operation "ATLANTIC CITY"	13-22 Sept 66	Dau Ting Airfield
44. Operation "SIOUX CITY"	26 Sept-9 Oct 66	Xom Cat
45. Operation "ROBIN"	10-17 Oct 66	Phu My to Rear Cat
46. Operation "ATTLEBORO"	5-7 Nov 66	Minh Thanh
47. Operation "WACO"	25 Nov-2 Dec 66	Bien Hoa Area
48. Operation "CANARY/DUCK"	7 Dec 66-5 Jan 67	Phu My to Bear Cat
49. Operation "NIAGRA/CEDAR FALLS"	5-25 Jan 67	Cau Dinh Jungle & Iron Triangle
50. Operation "BIG SPRING"	1-16 Feb 67	War Zone "D"
51. Operation "JUNCTION CITY (ALTERNATE)"	22 Feb-15 Mar 67	Tay Ninh Province
52. Operation "JUNCTION CITY II"	20 Mar-13 Apr 67	Minh Thanh
53. Operation "NEWARK"	18-30 Apr 67	War Zone "D"
54. Operation "FORT WAYNE"	1-4 May 67	War Zone "D"
55. Operation "DAYTON"	5-17 May 67	Phuoc Tay Province
56. Operation "CINCINNATI"	17-23 May 67	Bien Hoa/Long Binh Area
57. Operation "WINCHESTER"	23-31 May 67	Pleiku
58. Operation "FRANCIS MARION"	1-18 Jun 67	Pleiku
59. Operation "STILWELL"	18-22 Jun 67	Dak To/Kontum
60. Operation "GREELEY"	18 Jun-14 Oct 67	Dak To/Kontum
61. Operation "BOLLING"	19 September 67-31 Jan 69	Tuy Hoa/Phu Hiep
62. Operation "MACARTHUR"	1 Nov-14 Dec 67	Dak To/Kontum
63. Operation "WALKER"	16 Jan 68-31 Jan 69	An Khe
64. Operation "COCHISE"	30 Mar 68-31 Jan 69	Bong Son
65. Operation "DARBY CREST"	1 Feb-15 Apr 69	The Crescent of Hoai An District
66. Operation "DARBY TRAIL"	1 Feb-16 69	Bong Son
67. Operation "DARBY MARCH"	1 Feb-6 Mar 69	Tuy Hoa
68. Operation "STING RAY"	6-10 Mar 69	An Khe
69. Operation "DARBY PUNCH III"	10 Mar-24 May 69	An Khe
70. Operation "WASHINGTON GREEN"	15 Apr 69-1 Jan 71	Binh Dinh Province
71. Operation "GREENE LIGHTNING"	1 Jan 71-	Binh Dinh Province
72. Operation "GREENE STORM"	6 Feb -15 Mar 71	Binh Dinh Province
73. Operation "GREENE SURE"	17 Mar-21 Apr 71	Binh Dinh Province

187TH AIRBORNE INFANTRY

By Frederick J. Waterhouse

THE "RAKKASANS" 1943-1990

In the early 1940s paratroopers and glidermen of the 187th Glider Infantry Regiment were given the name "Rakkasan" by the Japanese. Loosely translated at best to mean "falling down umbrella," the troopers liked the name. They adopted it, and took the name into combat in three wars: World War II, the Korean War and the Vietnam War.

The battlefield valor of the Rakkasans in those three wars has earned the Regiment 11 decorations and 21 battle streamers, a record for an airborne infantry regiment. The 187th Infantry holds eight Presidential Unit Citations—five from the United States (four U.S. Army, one U.S. Navy). It holds one Presidential Unit Citation from the Philippines and two from the Republic of Korea. The 3rd Battalion, 187th Infantry returned from Vietnam with two of those U.S. Presidential Unit Citations, two Valorous Unit Awards and a Meritorious Unit Commendation. Numerous other units of the 187th hold battle honors and citations including 674th Airborne Field Artillery Battalion, Company E, 3d Platoon, Company A, 127th Airborne Engineers, Anti-tank Platoon, Support Company and Parachute Maintenance Detachment.

Four Rakkasan troopers of the 187th have been awarded the Medal of Honor for gallantry above and beyond the call of duty. Twenty five Rakkasans have earned the distinguished Service Cross and over 411 Rakkasans have been awarded the Silver Star for valor in combat. In those three wars, 3,840 Rakkasan troopers have been awarded the Purple Heart. Nine hundred and sixteen Rakkasans have laid down their lives in defense of America and the free world.

The Rakkasans of the 187th have always prevailed on the battlefield in the traditions of their motto, "NE DESIT VIRTUS" ("Let Valor not Fail"). In doing so, they have made the "Rakkasan" name a legend, forever synonymous with the immortal fighting spirit of the United States Army Airborne.

A LINK FROM THE PAST TO THE PRESENT

Fate and the passage of time have favored the 187th Infantry, allowing it to endure decades of challenge and change and develop into a lead regiment of our nation's only Air Assault Division, a key element of America's Strategic Rapid Deployment Ready Force.

Over the years the 187th Infantry has remained active in federal service while other Airborne units have cased their colors in quiet retirement. The great 60,000-man Airborne Army of World War II was rapidly reduced and retired in 1945. Airborne operations were seen as costly, obsolete and unnecessary in post-war America. The 187th Airborne Infantry remained while other valiant Airborne units—XVIII Airborne Corps, the 13th, 17th and 101st Airborne Divisions and scores of Parachute Regiments and Battalions–were deactivated. The 187th has flown its colors for 47 consecutive years and links the epic saga of the Army Airborne from the past to the present in an untold story of intrepid glory.

ACTIVATION CAMP MACKALL

On 25 February, 1943 the 187th Glider Infantry Regiment unfurled its colors for the first time at Camp Mackall, North Carolina as part of the 11th Airborne Division commanded by Major General Joe Swing. On that day, the men of the Regiment (the glidermen were nicknamed "Devil Dogs" in those early days) became part of a new elite fighting force in the American Army.

Since that day, troopers of the Regiment—as Airborne soldiers, glidermen, paratroopers and air assault infantrymen have proudly given the nation almost 50 years of service in many places throughout the world: New Guinea, the Philippines, Okinawa, Japan, Korea, Germany, Lebanon, Puerto Rico, Alaska, Vietnam, Egypt and Saudi Arabia. In all that time the Regiment has kept salient the legacies of its birth. It has executed from decade to decade with fidelity to elite standards–intrinsic to the creation of a new breed of fighting man in 1940.

War Department planners in 1943 were wary of large-scale airborne operations. German paratroopers in Crete had taken 40 percent casualties in 1941. The American Airborne Division was still untested. The Army's Airborne operations in Africa and Sicily were seriously flawed. General Eisenhower wanted all division-sized Airborne units disbanded. Army Chief of Staff George C. Marshall agreed, but first he proposed to test an Airborne division.

The 11th Airborne Division, with its 511th Parachute Infantry Regiment and 187th and 188th Glider Infantry Regiments, was given the mission of assaulting and capturing the "Knollwood Airport" in North Carolina. The test would determine if an Airborne Division could fly over water on instruments at night to a target, drop with minimal casualties and then wage sustained combat for a long period and be resupplied entirely by air.

Early on 6 December 1943, three battalions of paratroopers from the 511th and four battalions of glidermen from the 188th and 187th lifted off from numerous airfields in the Carolina complex: Mackall Field, Pope Field at Bragg and Laurinburg Maxton Army Glider Base. The landings were perfectly executed in the dark just before dawn.

Watching the massive assault were Assistant Secretary of Defense Patterson, General McNair and General Matthew B. Ridgway. Knollwood Airport was taken at dawn. And for the next five days, the Division simulated bitter combat throughout the sand hills of North Carolina. The entire operation was a success. Because of the Knollwood maneuvers, the War Department changed its plans and went ahead with its activation schedule for new airborne divisions.

187th Airborne Musani-ni Combat Jump 1st Serial.

JAPANESE PARATROOPERS ATTACK ON LEYTE

In war, the 187th has compiled an unparalleled record of courage and valor. In World War II the unit saw combat in the rain-drenched jungles, mountains and swamps of Leyte and Luzon. The 187th holds the distinction of being the only American Airborne unit in the U.S. Army to have been directly assaulted from the air by enemy paratrooper attack.

Four to five hundred Japanese Airborne soldiers attacked San Pablo, Burauen and Buri airstrips in the Battle of Leyte. With 674 artillerymen and 127th engineers fighting as infantrymen, the 187th Rakkasans not only destroyed the Japanese paratrooper force but also three months later, seized Lipa airfield on Luzon a staging area for the Japanese Airborne.

At Lipa the Rakkasans established a jump school and used the control tower as a 34-foot jump tower. Years later the Rakkasans would, in a training jump, parachute into Kanoya airfield in Southern Japan, the former site of the paratrooper school for the Japanese Imperial Army.

The Battle of Leyte was the Rakkasans' first action, and they executed brilliantly under the most difficult combat conditions. At times, there was no food. They lived off the land, scaled mountains and inched along jungle trails under heavy fire. The 187th fought continuously from November to January 1945 on Leyte. They suffered heavy casualties taking Purple Heart Hill at Anonang.

It took them almost two months to fight their way through the jungle and over the mountains from the east coast to the west. In the end they accomplished their mission. They did so with only two battalions. There wasn't a third battalion reserve.

AMPHIBIOUS ASSAULT MADE ON LEYTE JUNGLE

In the fierce jungle battles of Leyte, 187th infantrymen, artillerymen, engineers and medics parachuted, under fire, into the jungle from tiny Piper Cub L-4 aircraft at Manarawat mountain plateau. Initially dropping a 75mm pack howitzer into the area to support infantry, the "Mud Rats," another nickname, expanded this site into a critical base with division headquarters, supply area, field hospital, signal center and airstrip.

By modifying the L-4 aircraft, they were able to evacuate the wounded one at a time. After destroying all Japanese forces on Leyte, the 187th participated in a combined parachute-amphibious assault on Luzon. At Nasugbu Bay, the 187th, accompanied by the 188th, blasted its way into the beach under shore bombardment and slammed into the jungle through the Aga Pass to volcanic Tagaytay Ridge linking up with the 511th Parachute Infantry Regiment which had dropped in serials from Mindoro. In this action the 187th received a Presidential Unit Citation. The 674th Airborne Field Artillery Battalion was cited for its heroic support of the Regiment in the Aga Pass.

NICHOLS FIELD SEIZED; MANILA LIBERATED

After the battle at Tagaytay Ridge, the 187th joined the 511th and 188th attacking toward the city of Manila. Engaging in house-to-house fighting, they moved up Dewey Boulevard and assaulted Nichols Field, Mabarto Point, and Fort McKinley. On seizing the fort, the Rakkasans helped capture Cavite naval base.

After the bloody battle of Manila they moved to southern Luzon and engaged in a bitter three-day battle for Hill 660. The Regiment, now under the command of Colonel George O. Pearson (Big George), then entered what was to become its bloodiest battle of the war—the battle for

Paratroopers of the 187th Abn. RCT jump at base of Mt. Fuju, Japan during Airborne - Marine exercise, "Fuji Blue" 1954.

Mount Macolod. The Rakkasan infantryman struggled for weeks to scale the mountain.

At Macolod, 127th Airborne engineers performed miracles under fire constructing tank roads which allowed the armor to support the infantry with direct fire into caves and bunker complexes. Finally, after 27 days, the Rakkasans, supported by the tanks and heavy artillery, overran the 2,700-foot mountain peak in brutal hand to hand fighting.

After Macolod the Rakkasans deployed into a Lipa airstrip rest area. On 23 June 1945, they participated in their last action of the war. A small force of 187th troopers made glider landings in support of a 511th Battalion parachute assault at the seaport of Appari in Northern Luzon which ended any chance of remnants of the Japanese army escaping by sea. During the Pacific war, the two-battalion 187th Para-Glider Infantry Regiment lost 119 men killed in action. Forty one additional Rakkasans died of battle injuries and 693 others were wounded in action.

U.S. ARMY REVITALIZED IN 1950

The Korean War and the deployment of the 187th Airborne Regimental Combat Team to Korea, with its twin record-setting parachute assaults in that war, gave new life to the tactical importance of "Vertical Envelopment" in the U.S. Army. Thousands of World War II paratroopers saw the chance to rejoin an elite airborne unit in combat. In both Korea and later in Vietnam the 187th Airborne Infantry went to war with a cadre of professional airborne soldiers who possessed the combat experience of the previous war. The cyclic leadership of these officers and enlisted men has always enabled the Rakkasans to achieve in war and in peace.

Over the years the regimental colors have consistently edged to the right of center in deference to military custom and tradition. Today the colors of the 187th Infantry stand in honor, to the "Right of the Line" among the Regiments, by virtue of their longevity, battle honors and service to the nation.

TWO COMBAT JUMPS MADE IN KOREAN WAR

In Korea the 187th Airborne Regimental Combat Team, composed of the 187th Abn. Inf. Regt., 674th Abn. FA Bn., A/127th Abn. Engr. Bn., A/88th Abn., AAA Bn. and other attached units, under the command of Colonel Frank S. Bowen landed at Kimpo airfield in September 1950 reinforcing the Inchon amphibious landings. Within hours the 187th attacked north up the Kimpo-Han river peninsula where it destroyed a North Korean security regiment and cleared the peninsula. In this action the 3rd Battalion, 187th was awarded a Navy Presidential Unit Citation. After this the Rakkasans fought on the flanks of the Army and Marines moving into the south Korean capitol of Seoul.

The 187th Airborne made two classical parachute assaults in Korea: 20 October, 1950, Sukchon-Sunchon, North Korea and 23 March, 1951 Munsan-ni, South Korea. Some historians call these combat jumps the finest parachute operations the U.S. Army ever conducted. The jumps and the heavy drops (300 tons, Sukchon-Sunchon; 174 tons, Munsan-ni) were textbook assaults, records for the U.S. Army at that time. Within 90 days, the technique and lessons learned in these operations were used to modify training at the Parachute School and Infantry Center at Fort Benning and to improve Airborne combat training tactics at Forts Bragg and Campbell.

HOWITZERS, JEEPS, TRUCKS DROPPED FOR FIRST TIME

The 187th Airborne was the first parachute unit to ever drop howitzers, Jeeps, trucks and heavy guns in their initial parachute attack using Fairchild C-119 aircraft. Superbly designed for paratrooper operations, the C-119 (Flying Boxcar) was created for World War II Airborne forces, but only five were available at the end of the war in 1945. The 187th proved the worth of these planes in Korea and developed a unique relationship with staff and crews of the Air Force 315th Air Division which would continue until the end of the war. The "Joint Airborne Planning Team," put together by the 187th Airborne and the 315th Air Division, and the lessons learned would became a valuable teaching aid throughout the U.S. Army and Air Force at all levels of command.

In both combat jumps, the Rakkasans successfully achieved their objectives and fought alone as light infantry for protracted periods against overwhelming enemy forces, defeating them in every battle. In the parachute assault at Sukchon-Sunchon, North Korea, the Regiment received another Presidential Unit Citation for valorous performance in annihilating the North Korean 239th Regiment, inflicting over 800 enemy casualties and capturing the enemy's Regimental Colors.

The 3rd Battalion, 187th with Support Company Anti-Tank Platoon and 127th Airborne Engineers attached, received additional battle honors. On D+1 at Sukchon, PFC Richard G. Wilson was posthumously awarded the Medal of Honor after returning, unarmed, to enemy-held positions to administer medical aid to a wounded Rakkasan after the position was overrun. When the Rakkasans retook the position, PFC Wilson was found dead, shot while shielding the Rakkasan he was trying to help.

THREE DEPLOYMENTS SENT TO KOREA

Deployed to Korea on three separate occasions (twice from Japan) the Rakkasans occupied the North Korean capitol of Pyongyang in 1950. They decimated enemy forces in rearguard actions while covering the Eighth Army withdrawal in the Chinese intervention and held their ground magnificently in the big battles at Wonju in 1951 where they destroyed thousands of attacking enemy troops.

In the battle of Wonju, Company E received Battle Honors after attacking and overwhelming a Chinese Battalion on Hill 255. In savage close combat, E Company Rakkasans hit the Chinese with rifles, grenades and bayonets. After inflicting appalling casualties, the Rakkasans took Hill 255.

On two separate occasions (Uijongbu and Inje) the 187th spearheaded an infantry-armor plunge, Patton-style, through enemy lines over a 30-mile distance. Each time they slugged their way through, they suffered heavy casualties but achieved their objectives. In May 1951 Corporal Rodolfo P. Hernandez, Company G, while severely wounded, launched a one-man counterattack against Chinese hordes with an inoperable M-1 rifle and bayonet. Cpl. Hernandez was awarded the Medal of Honor for his bravery in this action.

PARATROOPERS ORDERED IN TO PRISON CAMP RIOTS

During the Korean War, paratroopers of the 187th Airborne had the mission of restoring order on the Koje Do island when 70,000 rioting prisoners took an American general prisoner while he was attempting to speak with POW camp leaders. Under the command of General Thomas J. Trapnell, the Rakkasans prevented a mass escape, broke up the over-crowded POW compounds, seized the communist ring-leaders and dispersed the 70,000 rebellious prisoners into smaller compounds built by Rakkasan engineers and infantrymen. One Rakkasan was killed on Koje and 31 Rakkasans were wounded by hostile prisoners who had secretly manufactured weapons: zip guns, sheet metal spears knives, barbed wire flails and Molotov cocktails.

REGIMENT MOVED TO THE FRONT AT KUMWHA VALLEY

After cleaning up Koje-do, the Regiment moved to the front with a new commanding officer by the name of Colonel William C. Westmoreland. It was outpost battles, trench raids, artillery duels and nightly ambush patrols into the valley. On a daylight reconnaissance patrol at Kumwha Valley, Corporal Lester Hammond, radio operator, A Company, was posthumously awarded the Medal of Honor after calling in artillery fire on top of his own position after an enemy ambush. In doing so, he saved his patrol but gave his life in the effort.

The battle escalated into an all-day fight with thousands of Chinese rushing into the valley. The 674th Field Artillery used all available guns including corps artillery to decimate the enemy who finally withdrew in a rout. In the Korean War, the 187th Airborne lost 422 Rakkasans killed in action. Forty nine additional Rakkasans died of wounds; 17 missing in action were declared dead; and four died in enemy prison camps. The Rakkasans had an additional 1,705 troopers wounded in action.

VIETNAM WAR MAKES FOR NEW MISSIONS

On 13 December, 1967, the 3rd Battalion, 187th Airborne Infantry, (Air Assault) arrived in Vietnam. In four years of combat the Rakkasans became known as the "Nomad Battalion" because of their many successful operations alone in enemy areas.

The only Airborne Infantry Battalion to fight in all four tactical Corps areas of Vietnam, the men of the 187th distinguished themselves in many battles: at Hue, in the Tet offensive, in the Ho Bo woods, at Khe Sanh, and in the battle of Dong Ap Bia mountain, sometimes referred to as the "Battle of Hamburger Hill." The Rakkasans had 35 men killed and 290 wounded in action in this bloody hill battle. They destroyed a North Vietnamese regiment on the hill and disrupted the enemy's infiltration routes into Vietnam from Laos for months. For this heroic action, the 187th was awarded another Presidential Unit Citation.

In a fierce three-day battle at Phuoc Vinh in 1968, Captain Paul W. Bucha became the fourth Rakkasan to be awarded the Medal of Honor for extreme bravery and gallantry against the enemy.

On September 6, 1968 the 3rd Battalion earned another Presidential Unit Citation. The battle of Trang Bang was one of the bloodiest of the war. Company A reinforced by Company C repelled a heavy enemy attack which lasted 2 1/2 days. After the battle, 141 NVA dead were counted in the wire. One hundred more were found in the perimeter. Thirty four Rakkasans were KIA in the Battle of Trang Bang.

Throughout the war, the Rakkasans engaged in untold numbers of search-and-destroy missions and hundreds of helicopter assaults and carried out long range reconnaissance patrols while continually interdicting the Ho Chi Minh supply trail.

MINES & BOOBY TRAPS MAKE A GUERILLA WAR

In the last two years of the war, 75 percent of Rakkasan casualties resulted from enemy booby traps. In these later years of the war the 187th played a major role in Vietnamization and pacification programs developed to help the South

Vietnamese restore local government and train the South Vietnamese army and local force militia to achieve tactical proficiency on the battlefield. They did this while continually engaging North Vietnamese army and Viet Cong guerrilla forces in the field. When the South Vietnamese army sent large military units into Laos, the Rakkasans protected the man supply route and fought large concentrations of enemy forces in the rear while keeping the vital highway open from Khe Sanh to the Laotian Border.

During its last year in combat the 187th deployed into the strategic Cam Ranh Bay area where it conducted search and destroy missions, fought NVA and Viet Cong forces trying to infiltrate and repelled enemy sniper attacks on vital installation. At this time the Rakkasans launched strong offensive operations deep into an area known as Rocket Ridge.

During its last three months of combat, Rakkasans returned to the Quang Tri area where they heavily patrolled while the U.S. Army continued its Vietnam withdrawal under orders from President Nixon. Casualty data shows the Rakkasans lost 239 men killed in action with an additional 25 dead of wounds. 526 Rakkasans were wounded in action. No Rakkasans were ever taken prisoner.

REGIMENT'S DISTINCTION FOUND IN VERSATILITY

The 187th Infantry has survived many trials and tests in its long history. As a Glider Infantry Regiment, a Para-Glider Infantry Regiment, an Airborne Infantry Regiment, an Airborne Regimental Combat Team, an Airborne Battle Group, an Air Assault Test Unit and now as part of an Airborne Air Assault Brigade, the Regiment has served in three airborne divisions: the 11th, the 82nd and the 101st Airborne Divisions. The 187th holds the distinction of being the only Airborne unit in the United States Army to have entered combat in all three battle modes of "Airborne Vertical Envelopment"— glider, parachute and helicopter. To that history can be added combat jumps from Piper Cub L-4 aircraft at Leyte and combat assault by seaborne landing craft at Luzon.

WAR & PEACE ALLOW SETTING OF NEW RECORDS

The exploits and achievements of the 187th in war and peace are numerous— most of them lost in the shadowy past. In 1945 the 187th established a record when the entire Regiment flew 1,600 miles in hastily requisitioned B-24, B-25, B-29, C-46, C-47, and C-54 aircraft from the Philippines via Okinawa to Atsugi Airfield near Tokyo. Combat loaded and ready to fight (hostilities had not ended), the Rakkasans were the first battle troops into Japan which had not yet formally surrendered.

On arrival, Major General Joe Swing seized the swords from Japanese generals and the 187th secured the field for the arrival of General Douglas MacArthur. The Rakkasans occupied Japan for four years arriving as conquerors and departing as benefactors and friends of the Japanese in 1949.

SWARMER BECOMES LARGEST AIRBORNE OPERATION

During four years of occupation duty in Japan, the 11th Airborne never made a division-sized parachute mass jump. The Regiments were too widely separated across Japan, and there were not enough aircraft. This all changed when the Division arrived at Camp Campbell in the summer of 1949. Planning for Operation SWARMER began in the fall. Headquarters for the operation was located at Camp Mackall. Lessons learned from the Berlin Airlift were employed in the operation. An aggressor force landed

Col. Joseph F. Ryneska, right, Rakkasan Commander, leads "stick" on routine training parachute jump at Sicily Drop Zone, Ft. Bragg, North Carolina.

on the Carolina coast. And a second force landed in Florida and attacked north.

On D-Day April 28, 1950 SWARMER jumped off with men of the 11th Airborne Division including the Rakkasans of the 187th Airborne Infantry parachuting into drop zones at Camp Mackall. That afternoon, paratroopers of the 82nd Airborne Division dropped in to link up with the 11th. High overhead the U.S. Air Force controlled the skies and brought in tons of supplies and equipment for days. On the ground, the paratroopers fought their conventional war, advancing, attacking, withdrawing, defending and attacking once again. In the end, the aggressors were defeated on that same land fought over years before in the Knollwood Maneuvers.

Out of all the Airborne Infantry Regiments participating in Operation SWARMER, the 187th Airborne received the highest performance scores in the operation. Much of the credit for the superior performance of the Regiment must go to its commander at that time—All-American football player, West Point football coach and inspiring leader Lieutenant Colonel Harvey J. Jablonsky. Jabo put fire into the regiment with his insistence that every paratrooper in his unit be physically fit. The high scores and combat readiness of the 187th was central to the regiment being selected for battle in Korea a few months later.

GYROSCOPE OPERATES FROM JAPAN TO THE U.S.

Back in Japan a second time, after the Korean War, the Rakkasans achieved another flight record in 1955 when, under the command of General Roy Lindquist, they participated in the first Operational GYROSCOPE ever conducted by the U.S. Air Force. After months of advance planning and GYROSCOPE Airborne staff changing places in the United States and Japan, a massive airlift took place. Using giant C-124 Globemasters, the entire Rakkasan Regiment included dependents was flown from Japan to Wake Island, to Travis Air Force Base in California, to Camp Mackall in North Carolina in a historic 12,000-mile exchange that brought the 508th Airborne Regimental Combat Team to Japan and returned the 187th Airborne Regimental Combat Team to the United States.

RAKKASANS KEEP PEACE IN BEIRUT, LEBANON

The 187th Airborne Infantry achieved another first when it became the first American paratrooper unit sent to the Middle East. The Rakkasans as the 1st Airborne Battle Group, 187th Infantry, 24th Infantry Division remained in Germany after the 11th Airborne Division was deactivated. Serving in the Augsburgh, Germany area, the Rakkasans were ordered to Beirut, Lebanon as part of a large task force sent in by President Eisenhower after rebel forces overthrew governments friendly to the United States.

The 187th, combat ready and combat loaded, flew from Germany to Turkey and on into Lebanon on 29 July, 1958 where they established themselves in perimeter defenses in and around the Beirut International Airport while making a show of force with U.S. Marines. The rebels kept them on their toes with an occasional round of sniper fire. The summer passed under a wilting Lebanese sun and the Rakkasans manned roadblocks, gave firepower demonstrations for the Lebanese Army, patrolled the mountains and practiced jumping on a drop zone they personally named the "Sahara."

After defusing the situation, the paratroopers returned to Augsburg, Germany on 5 October, 1958. Many strategic studies of this peacekeeping operation have been done by numerous countries. All of them compliment the Rakkasans for their excellent discipline, soldierly bearing and quiet restraint they displayed in keeping the peace.

ASSAULT HELICOPTER TAKES ON COMBAT ROLE

Initially, interest in the helicopter centered on the U.S. Army's ability to rapidly disperse large ground forces in atomic warfare. Secretary of Defenses Robert S. McNamara believed the helicopter could be also used in a direct combat role. He ordered tests. Early in 1962 the Army conducted a series of 40 tests. Basic flying techniques, small unit and weapons tests were conducted at Fort Bragg; mountain jungle operations were tested in Western Virginia. A small group of 187th Infantry Rakkasans were used at Fort Stewart, Georgia in developing airmobile infantry tactic tests.

On 1 February, 1963 the 11th Airborne Division was reactivated at Fort Benning, Georgia as the 11th Air Assault Division (Test). On that same day Rakkasans of the 3rd Battalion, 187th Airborne Infantry became cadre for the 11th Test Division. Soon two other Airborne Infantry Battalions would join the 187th. No strangers to the Rakkasans, they were 1/511th and 1/188th. The 187th and other battalions began testing the Bell Helicopter in numerous combat roles: command and control, attack formations, scouting and screening, reconnaissance, aerial resupply and air-assault tactics.

The 3rd Battalion, 187th tested and perfected these battle concepts for one year. On 4 February, 1964 the 3rd Battalion was assigned to the 101st Airborne Division at Fort Campbell. On this same day the 1st Airborne Battle Group, 187th Infantry was redesignated 1st Battalion, 187th Infantry and moved from Fort Bragg to Fort Benning where it trained for one year with the test helicopter unit.

Additional infantrymen from the 2nd Division at Fort Benning were moved into the 11th Air Assault Division as it matured. In June 1965 the colors of the 1st Cavalry Division in Korea were put on a plane and flown to Fort Benning. In a simple ceremony the colors were presented to the 11th Air Assault Division Test. The 11th was disbanded; 1/187, 1/188, and 1/511th were deactivated. The 1st Air Cavalry Division went to Vietnam shortly thereafter. The 187th had played a key role in development of the air assault concept.

REGIMENT GROWN WITH 5 NEW BATTALIONS

On its return from Vietnam, the 3rd Battalion, 187th Infantry operated alone without its sister battalions for a long period. On 1 October 1983 the Army activated the 1st, 2nd and 4th Battalions, 187th Infantry. The 1st and 2nd Battalions were assigned to the 193rd Infantry Brigade in Panama. The first Battalion was on jump status continually while in Panama.

The 4th Battalion joined the 3rd in the War Eagle Brigade, 101st Airborne at Campbell. Eleven months later on 16 November 1984 the Army completed reorganization of the 187th Infantry as a Regiment by activating the 5th Battalion, 187th Infantry and assigning it also to the 3rd Brigade, 101st Airborne at Campbell. Another change took place on 16 September 1987 when the Army deactivated the two Panama battalions. The 3rd, 4th and 5th Battalions at Campbell were redesignated 1st, 2nd and 3rd.

PEACE GUARANTEED WITH REGIMENT IN SINAI, EGYPT

In recent times the Rakkasans of the 187th Infantry have made two six-month battalion-sized deployments to the Sinai Desert in Egypt as part of the

"Multinational Force and Observers" in support of the Treaty of Peace between the Arab Republic of Egypt and the State of Israel actuated on April 25, 1982. This rotating peacekeeping observer force of three battalions—one each from Colombia, Fiji and the United States perform observation duties in what is called Zone C which has six sectors in the remote barren area of the Sinai peninsula.

The 187th Battalion Headquarters is located at the extreme southern tip of the Sinai peninsula at Sharmel Sheikh in what is known as the South Camp at the juncture of the Gulf of Suez and the Gulf of Agaba. About 250 miles north in Zone C, 20 miles from the Mediterranean coastline, lies the North Camp occupied by the Colombian and Fiji battalions.

The Rakkasan mission is precise. They observe, report and verify implementation of the peace treaty by operating 14 observation posts in four sectors of Zone C. Rakkasans man these isolated outposts for 21 days, then rotate back to the South Camp for another 21 days. The squads on remote sites observe the task force sector for violations of the protocol. Reports from each site are rolled up to the Company Sector Control Center, (SCC) where they are consolidated and verified. Each SCC in turn reports to the Task Force operations center at Sharm el-Shiekh.

Although working in an extremely harsh environment for an extended period of the time, the Rakkasans of Task Force 1-187 Infantry have had many opportunities to visit some of the more aesthetic sights of the region. Trips have been made to Cairo, Egypt where Rakkasans have toured the Great Pyramids and the Cairo Museum, and cities within Israel including Tel-Aviv, Jerusalem and Bethlehem, which are some of the most historical places in the world. Saint Catherine's Monastery on Mount Sinai is a 2 1/2-hour drive from the South Camp.

On 12 December, 1988 Task Force 1-187 held a memorial ceremony in honor of the 247 soldiers of Task Force 3-502nd Infantry who lost their lives in a plane crash at Gander, Newfoundland during their redeployment from the Sinai to Fort Campbell in 1985. In tribute to the occasion, the Legislature of Kentucky passed a resolution honoring the peacekeepers and remembering those "Screaming Eagles" who lost their lives that day. Additionally the resolution recognized the Rakkasans for the sacrifices they have made while serving the MFO. Under the command of LTC Louis D. Huddleston, 1st Battalion, 187th Infantry departed the Sinai and arrived at Fort Campbell on 12 April, 1989.

The 105 howitzer was delivered to the DZ with the blown panel but was not damaged.

Another Rakkasan mission successfully completed.

HONOR TAKES REGIMENT INTO THE FUTURE

The last two decades have seen the 187th Infantry excel in adapting to the revolutionary advances in conventional airborne doctrine. The helicopter has made profound radical improvements in the Army's operational capabilities in the tactical deployment of air assault forces. The 187th has been privileged to advance these new concepts in battlefield air mobility. Today, the 187th Rakkasan Regiment serves in the "War Eagle Brigade" of the 101st Airborne Division—the nation's all-purpose air assault division and lead element of the Strategic Rapid Deployment Ready Force.

The Rakkasan officers and men of the regiment are the finest group of modern soldiers anywhere in the world. Magnificently equipped and superbly trained, these professionals have brought the noble craft of "Soldiering in the Regiment" to new heights of bold elan. The regiment remains young and vibrant while its Rakkasan ranks, proud, aggressive and confident, are the best of America's youth. With a fierce loyalty to the past, they face the future committed to the honor of the regiment. They take the Rakkasan name toward the 21st century with the fire of a noble heritage.

"NE DESIT VIRTUS".

88TH GLIDER INFANTRY

By James E. Mrazek
Col., USA (Ret)

The 88th Glider Infantry was activated as the 88th Infantry Airborne Battalion on October 10, 1941, per General Orders No. 7, Headquarters, The Infantry School, dated September 15, 1941, at Fort Benning, Georgia, with an initial authorized strength of 27 officers and 500 enlisted men. The battalion was attached to the Infantry School for administration.

The battalion was originally commanded by Lt. Col. E.G. Chapman, who continued as battalion commander until his assignment to Headquarters Airborne Command upon its activation in March of 1942. In May of 1942 Col. Chapman was promoted to the rank of brigadier general and assumed command of the Airborne Command and subsequently was in command of all Airborne activities in the United States Army including the training of all divisions and units. In April of 1943, General Chapman was promoted to the rank of major general, and placed in command of the 13th Airborne Division.

Lt. Col. C.L. Keerans, battalion executive officer, was also assigned to Headquarters Airborne Command in March of 1942. Other members of the original battalion staff included Major J.B. Lindsey, S-1, who continued with the battalion until his assignment to Headquarters Army Ground Forces in May 1942, subsequently becoming regimental commander of the 515th Parachute Infantry; 2d Lt.C. Koerschner, S-2, who was assigned to the Army Air Corps in January of 1942; Major R.C. Aloe, S-3, who succeeded Lt. Col. Chapman in command of the battalion and who was subsequently assigned to the Command and General Staff School, Fort Leavenworth, Kansas in December 1942; Capt. Samuel Toth, S-4, to commanding officer of the 88th; and Capt. Hugh P. Harris, to Chief of Staff, 13th Airborne Division.

During the period October 10, 1941 to December 10, 1941, the battalion continued its organization, securing necessary replacements, and one composite company went on an Airborne mission to MacDill Field, Florida, to engage in an air landing problem.

During the period December 10, 1941 to April 1, 1942, a training program was conducted which included a combination of basic infantry training, advanced field problems and initial work in loading and lashing of cargos in C-47 airplanes. Some flying was done and members of the battalion were given orientation flights. During this period the battalion was instructed in the proper use of parachutes and the elements of jump technique.

In February of 1942, Headquarters Airborne Command was activated and the "88th" was relieved from the assignment to the Infantry School and reassigned to Airborne Command. At this time Colonel Chapman, Lt. Col. Keerans, other officers and a cadre of enlisted men were transferred from the battalion to Headquarters Airborne Command, Fort Bragg, North Carolina.

On or about April 1, 1942, the battalion was authorized to expand to full strength (1,000) men and requisition was made for the necessary replacements. For the remainder of this period up until May 4, 1942, the battalion continued training in loading and lashing with C-47, C-53 and C-46 airplanes and mock-ups.

On May 4, 1942, the battalion changed station, moving from Fort Benning, Georgia, to Fort Bragg, North Carolina and during the period following until June 15, 1942, continued training in Airborne tactics and technique. Approximately 600 replacements were received during this period.

June 15, 1942, brought about another reorganization of the battalion. At this time the Battalion was inactivated as the 88th Infantry Airborne Battalion and

Col. Roth reviews Lt. Col. Mrazek's (L) battalion, 88th Glider Inf. (Courtesy of James E. Mrazek)

Members of the 88th Glider Infantry Reg. ready for training.

activated as the 88th Infantry Regiment (less two battalions) and the officers and enlisted men formerly of the 88th Infantry Airborne Battalion became members of the 88th Infantry Regiment.

The reorganization of the unit on June 15, 1942, contained a directive designating the 88th as school troops prepared to teach Airborne tactics and technique to one or more divisions. Consequently an Airborne training instructional unit was organized within the 88th and preparation for this teaching began. The remainder of the organization continued normal training during this period.

During the period June 15, 1942 to September 21, 1942, normal training ensued including another Airborne problem to Myrtle Beach, South Carolina. Reorganization brought about the activation of seven companies in contrast to the four companies of which the organization was previously composed.

During the first week in September the Airborne Training Instructional Team enplaned at Fort Bragg and flew to Fort Sam Houston, Texas where it was stationed temporarily to instruct the 2nd Infantry Division in Airborne tactics and technique. This unit remained in Texas until the first week in November 1942, at which time it rejoined the regiment at Fort Bragg, North Carolina.

On September 21, 1942, the 88th Infantry was reorganized as the 88th Glider Infantry and during the period that followed until February 10, 1943, the regiment engaged in usual training with emphasis on glider training which was taken at the Army Air Base, Maxton, North Carolina, with one battalion at a time being stationed at the air base.

On February 10, 1943, the regiment again changed station moving to Fort Meade, South Dakota, where it was stationed for 10 months. While at Fort Meade emphasis was placed on tactical training of the Infantry Regiment with little emphasis on Airborne instruction. Regimental field exercises and CPXs and a two-week maneuver were held during the period February to August 1943. Early in August 1943, the regiment moved to Fort Robinson, Nebraska, and participated in a brigade maneuver. From Fort Robinson it went to the Army Air Base, Alliance, Nebraska and engaged in battalion, regimental and brigade exercises, returning to Fort Meade on 3 September 1943.

In March 1943, the 9th Army Ground Forces Band was attached to the regiment and remained attached until November 30, 1943.

Prior to April 14, 1943, the 88th was a unit directly under control of Airborne Command. On April 14, 1943, the 88th was assigned to the First Airborne Infantry Brigade. On November 30, 1943, the unit left Fort Meade, South Dakota, and moved by train to Fort Bragg, North Carolina, arriving on the 3rd of December, at which time it was assigned to the 13th Airborne Division.

A brief period of training was undergone at Fort Bragg and on the 15th of January 1944, after a four-day maneuver against the 17th Airborne Division in the North Carolina Maneuver Area, the regiment, together with the rest of the 13th Airborne Division, moved to Camp Mackall, North Carolina.

At Camp Mackall the regiment, under a division training program, underwent first individual training for a period of seven weeks, and thereafter went into small unit and combined training intended to fit the entire organization for combat in a very short period. During this period in the summer of 1944, the regiment participated in a number of field exercises of short duration. One maneuver in the Carolina maneuver area involved the entire division and lasted for a period of about two weeks. For this maneuver the regiment moved by truck to Laurinburg-Maxton Army Air Base, Maxton, North Carolina. The remainder of 1944 was spent in rounding-out training, completion of POM requirements, both training and administrative, for all personnel of the regiment and otherwise completing preparation of the regiment for overseas movement.

At years' end, 1944, the regiment stood ready for departure overseas.

Period 1 through 15 January was spent in making final check of records and equipment for overseas movement. Physical inspections were completed and other administrative arrangements made, and on 16 January 1945 the regiment left Camp Mackall, North Carolina for Camp Shanks, New York, the staging area for the New York Port of debarkation.

On the night of 25-26 January 1945 the regiment boarded the USAT *George Washington*, and at 2611 25 January 1945, left New York.

On the evening of 6 February 1945 the convoy of which the USAT *George Washington* was a part anchored at La Havre, France, and commencing at 1100, 7 February 1945, the troops debarked and moved to Camp Lucky Strike, (Janville) France.

On 19 February 1945, the Regiment left Camp Lucky Strike and moved to Sens, France by train where the unit was billeted together with the 326th Glider Infantry and the 676th Glider Field Artillery.

On the 1st of March in compliance with General Orders No. 4, Headquarters 13th Airborne Division, dated 28 February 1945, the regiment disbanded and all personnel were assigned to the 326th Glider Infantry.

501ST PARACHUTE INFANTRY BATTALION

By Lewis Brown

The 501st Parachute Battalion was activated at Fort Benning, Georgia, on October 1, 1940. It was the first tactical parachute unit to be activated in the United States Army. As early as July 1940, the War Department had determined that a larger unit than the Test Platoon would be required to test and develop the organization, tactics, training procedures and required equipment for parachutists.

Under directive of October 2, 1940, Army commanders were directed to select volunteers to man the 501st Parachute Battalion. These volunteers began arriving early in October 1940 and were housed in tents on the ground overlooking Lawson Field at Fort Benning, Georgia. The unit was organized with the Test Platoon as the nucleus.

Maj. William M. Miley, later to command the 17th Airborne Division, was designated to command this first parachute unit. His initial staff consisted of: Maj. George P. Howell, executive officer; Capt. Roy E. Lindquist, S-1/S-2; Capt. William T. Ryder, S-3; Capt. Richard Chase, S-4; Lt. James A. Bassett, Parachute officer; Sgt. Harris T. Mitchell, Sgt. Maj. The first company commanders were Capt. Robert F. Sink and Capt. William P. Yarborough. The original first sergeants were all from the Test Platoon and were: Sgt. Benedict F. Jacquay- Headquarters and Headquarters Company; Sgt. Hobart B. Wade- Company A; Sgt. Loyd McCullough- Company B; Sgt. John M. Haley- Company C.

Upon arrival the men and officers began an intense physical training program. Parachute jumping and parachute packing groups were established primarily from the Test Platoon to train the incoming personnel. On March 21, 1941, Brigadier Omar N. Bradley and Major Miley presented the parachutist qualification badge to 276 men and officers. Forty-seven members of the Test Platoon and eight officers had previously received the qualification badge that was designed by Captain William P. Yarborough. This group composed the first group of qualified parachutists in the US Army.

The 501st Parachute Battalion furnished cadre for the Provisional Parachute Group activated at Fort Benning, Georgia, on February 25, 1941. Once established, the Parachute Group activated a separate Parachute Training Group for the training of parachute volunteers. These cadres, all from 501st Parachute Battalion, included John Swetish, William N. King, Edward Smith, Charles H. Wilson and others. The 501st Battalion was also directed to furnish cadre for three additional battalions commencing July 1, 1941.

The design and testing of equipment, supplies and weapon containers as well as many other items to be used by the parachutist were carried out by groups composed of both officers and enlisted

Members of the 501st Parachute Infantry Regiment load onto trucks for the movement to Bastogne, Belgium, on 18 December 1944.

Paratroopers of the 501st Parachute Infantry Regiment move through the village of Bastogne, Belgium, on 19 December 1944.

personnel. Parachute training techniques were refined and assistance was furnished the Infantry Board in preparing a Field Manual covering the training and employment of parachutists.

In June 1941, cadre was furnished to form the 502d Parachute Battalion, and in July and August 1941, cadre were furnished for the 503d and 504th Parachute Battalions.

In September 1941, the 501st Parachute Battalion was directed to move to the Panama Canal Zone, to fill a requirement for a tactical parachute unit. Company C, under the command of Captain John B. Shinberger, was dispatched to the Panama Canal Zone as the advance echelon of the Battalion. The battalion, less Company C, arrived in Panama CZ in September 1941.

Upon arrival, the Battalion began filling its ranks with volunteers from the Panama CZ. These men replaced personnel that had been lost in furnishing cadre for the additional battalions. All new volunteers were trained by the battalion as parachutists.

Plans were made and exercises conducted in coordination with other units for the defense of the area. Immediately after December 7, 1941, Lieutenant Colonel Miley was ordered back to the United States to assume command of the newly activated 503d Parachute Infantry Regiment. Lieutenant Colonel Kenneth Kinsler assumed command of the battalion and Major George M. Jones became the executive officer. Shortly thereafter, Lieutenant Colonel Kinsler was ordered back to the United States and Major Jones assumed command of the Battalion.

Orders were received designating the 501st Battalion to become the Second Battalion of the 503d Parachute Infantry Regiment which had lost its second battalion to another mission. The 501st Parachute Battalion joined the 503d Parachute Infantry Regiment aboard a Dutch freighter for a 32-day unescorted voyage to Australia.

Lieutenant Colonel George M. Jones continued to command the battalion until he assumed command of the 503d Parachute Regiment and eventually the 503d Parachute Combat Team. He commanded this unit during most of its wartime action in the Pacific Theatre.

The 501st Battalion, as the second Battalion of the 503d Parachute Regiment, participated in all the wartime action of the 503d Regiment and Regimental Combat Team to include the Markham Valley, New Guinea; Hollandia, Dutch New Guinea; Noemfoor Island, Dutch New Guinea; Leyte, Mindoro, Corregidor and Negros Islands in the Phillippines. It especially distinguished itself in the parachute assault to recapture the fortress of Corregidor. It held a portion of the defense perimeter of the vital topside area during repeated fanatical night attacks by the Japanese during the first night on the island. The loss of this area would have been critical. The battalion held despite numerous casualties and hand-to-hand combat. (See 503rd PRCT History)

The 501st Parachute Battalion was composed of an outstanding group of men and officers, who, in those first days of the parachutist, served as test vehicles and faced the unknown many times. They never backed away from anything. A report from a War Department observer, reporting on Major Miley's first jump stated, "He appeared tense at first, but upon landing, looked around, smiled and said, 'Hell, there's nothing to it.'" That was the attitude of the 501st Battalion. No challenge was too great.

503RD PARACHUTE REGIMENTAL COMBAT TEAM

By Robert Flynn

503RD PARACHUTE INFANTRY REGIMENT
Lineage and Honors

LINEAGE

Constituted 24 February 1942, in the Army of the United States as the 503rd Parachute Infantry (concurrently), First Battalion consolidated with the 503rd Parachute Battalion (constituted 14 March 1941, and activated 22 August 1941, at Fort Benning, Georgia) and Second Battalion consolidated with 504th Parachute Battalion (constituted 14 March 1941, and activated 5 October 1941, at Fort Benning, Georgia) and consolidated units designated as the First and Second Battalions, 503rd Parachute Infantry.

Regiment (less First, Second and Third Battalions) activated 2 March 1942, at Fort Benning, Georgia. Third Battalion activated 8 June 1942, at Fort Bragg, North Carolina. (Second Battalion reorganized and redesignated 2 November 1942, as the Second Battalion, 509th Parachute Inf., hereafter separate lineage; concurrently, new Second Battalion, 503rd Inf., activated in Australia.) Regiment inactivated 24 December 1945, at Camp Anza, California.

CAMPAIGN PARTICIPATION CREDIT
World War II
New Guinea
Leyte
Luzon (with arrowhead)
Southern Philippines

DECORATIONS

Presidential Unit Citation (Army), Streamer embroidered Corregidor cited: WD GO 53, 1945.

Philippine Presidential Unit Citation, Streamer embroidered 17 October 1944 to 4 July 1945, DA GO 47, 1950.

462ND PARACHUTE FIELD ARTILLERY BATTALION
Lineage and Honors

LINEAGE

Constituted 25 February 1943, in the Army of the United States of the 462nd Parachute Field Artillery Battalion. Activated 15 June 1943, at Camp Mackall, North Carolina. Inactivated 21 December 1945, at Camp Anza, California.

CAMPAIGN PARTICIPATION CREDIT
World War II
New Guinea
Leyte
Luzon (with arrowhead)
Southern Philippines

DECORATIONS

Presidential Unit Citation (Army), Streamer embroidered Corregidor cited: WD GO 53, 1945.

Philippine Presidential Citation, Streamer embroidered 17 October 1944 to 4 July 1945, cited: DA GO 47, 1950.

COMPANY C, 161ST AIRBORNE ENGINEER BATTALION
Lineage and Honors

LINEAGE

Constituted 12 September 1942, in the Army of the United States as Company C, 161st Engineer Battalion.

Activated 15 November 1942, at Fort Bliss, Texas, as Company C, 161st Engineer Squadron.

Redesignated 1 May 1943, as the 161st Airborne Engineer Battalion.

Redesignated 2 April 1945, 161st Parachute Engineer Company.

Inactivated 25 October 1945, at Negros Island, Philippine Islands.

CAMPAIGN PARTICIPATION CREDIT
World War II
New Guinea
Leyte
Luzon (with arrowhead)
Southern Philippines

DECORATIONS

Presidential Unit Citation (Army), Streamer embroidered Corregidor cited: WD GO 53, 1945.

Philippine Presidential Citation, Streamer embroidered 17 October 1944 to 4 July 1945 cited: DA GO 47, 1950.

503RD PRCT HISTORY

When the 503rd Parachute Infantry Regiment was formed on 2 March 1942, it incorporated units of the 503rd and 504th battalions. Later it would fill out another battalion with graduates of new classes from The Parachute School. Still later, on the way to the Pacific Theater, it would pick up the 501st Battalion in Panama.

The regiment sailed from San Francisco to Cairns, Australia and set up a permanent camp near Gordonvale where it undertook rigorous training and organizational experimentation. In an effort to find the most effective way to deal with each combat simulation, structural changes were tried, revised and retried. Personnel were assigned and reassigned until the right mix appeared evident.

Each rifle company had three platoons of riflemen, and each platoon consisted of three squads one of which was provided with a light machine gun. Each platoon was assigned one 60mm mortar and one grenade launcher. The grenade launcher was later replaced by the "bazooka," a portable rocket launcher. Mortar platoons of Headquarters Company could very easily be shifted to suit the needs of a combat situation. Medics were assigned to companies on a permanent basis.

Then on 7 August 1943, the 503rd received orders to move to New Guinea. The regiment left Australia by air and by sea and moved into a bivouac area near Port Moresby.

NADZAB OPERATION

The Airborne assault of the airstrip at Nadzab in the Markham Valley was designed to seize the abandoned field and the surrounding area, to clear the airstrip to permit the landing of the 7th Australian Division (which would move to attack Lae about 30 miles to the east) and to intercept enemy forces opposing the amphibious landing at Lae.

On 5 September 1943, the 503rd

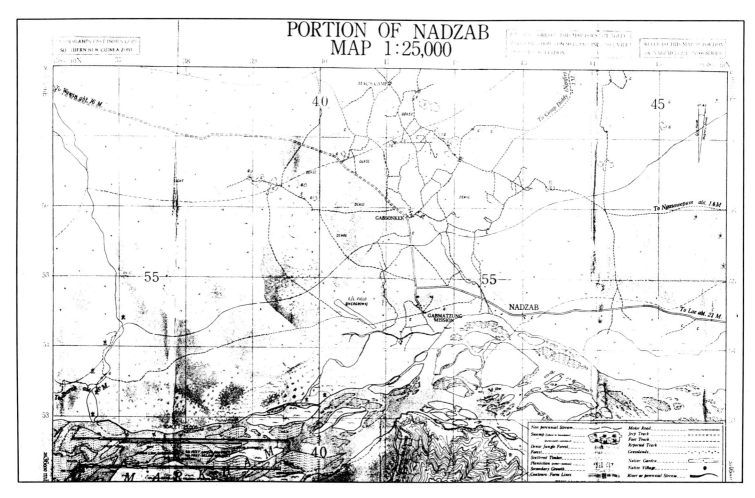

Parachute Infantry regiment put all three of its battalions precisely on the intended drop zone in 4 1/2 minutes while General MacArthur observed the operation from the vantage of a B-17 bomber. It was later reported that General MacArthur had told his aide that this Airborne assault was the finest example of combat efficiency he had ever witnessed.

While the First Battalion seized the airstrip, the Second and Third Battalions sent out patrols to reconnoiter the area surrounding the airstrip and to establish a perimeter. The Australians prepared the airstrip while the 503rd provided the necessary protective cover.

On 15 September, Lt. Col. Tolson sent Capt. John Davis and I Company to look into a report that Japanese forces were evacuating the area to the north of Lae. The lead platoon surprised a large force of Japanese at Log Crossing Village and a bitter fight ensued that lasted through the night. I Company overcame a heavy machine gun emplacement and killed three of the enemy gunners. Lt. Lyle Murphy, leading his platoon downstream, ran into a large Japanese detachment. A fire fight developed that lasted into the night. In the morning, the enemy was found to have moved into the interior of the island.

By 19 September, the 503rd was back at its base at Port Moresby. The airstrip at Nadzab had been completed with two 6,000-feet runways; Lae had fallen; the enemy had retreated into the mountains and their food supplies had been confiscated. Of the 10,000 Japanese troops in the Lae-Nadzab area, 26,000 were killed and more than 600 failed to survive the retreat march.

The Markham Valley parachute operation, because of its precision, reinforced the belief in the vertical deployment of military forces. Due to the unfortunate deployment of Airborne forces in Crete, North Africa and Gela, Sicily, many military strategists were not convinced of the effectiveness of the parachute troops. Markham Valley put to rest the doubts and the concerns. The 503rd Parachute Infantry Regiment returned to camp at Port Moresby.

NOEMFOOR OPERATION

At Port Moresby Lt. Colonel George M. Jones assumed command of the regiment due to the untimely death of Colonel Kenneth Kinsler. General Kruger, Commanding General of the Sixth Army, after observing the 32-year-old commander, promoted him to full colonel in July 1944.

General MacArthur was determined to reach the Philippines before December and had planned to leap-frog island to island, utilizing airstrips along the way to accomplish this objective. Noemfoor Island became a target because of three good airstrips—Kamiri, Namber and Kornasoren.

The 503rd Parachute Infantry mission was to jump at Kamiri and provide support for the 158th Infantry RCT. On 3 July 1944, after two weeks of bombing and shelling by the Fifth Air Force and the Navy, the 1st Battalion, 503rd, commanded by Major Cameron Knox, took off for Hollandia. Colonel Jones was in the lead plane.

The jump was to be made at an altitude of 400 feet, but Colonel Jones made a mental note that the altitude looked less than that. He was correct, but unfortunately, the green light came on over the drop zone and the colonel and his men jumped onto Kamiri Airstrip at an altitude under 200 feet. The pilot had forgotten to adjust the plane's altimeter.

There were numerous injuries due to the jump, some of them so severe that many of the injured would never jump again. Of the 739 jumpers on July 3, 72 were casualties.

The Third Battalion and the remainder of Regimental Headquarters jumped the following day, led by Major John Erickson. This time the planes were to fly in single file at an altitude of 400 feet. The drop zone was cleared of debris and machinery, but the Third Battalion still suffered unusually high casualties during the landing. Of the 1,424 men who jumped at

503rd Parachute Infantry Regiment jumps at Kamiri Airstrip, 3 July 1943. (United States Army Photograph, courtesy of R. J. Flynn)

The 503rd Parachute Infantry Regiment jumps in Markham Valley, New Guinea near Nadzab, 5 September 1943. (United States Army Photograph, courtesy of R. J. Flynn)

Noemfoor, 128 suffered severe injuries. Colonel Jones canceled the proposed jump by the Second Battalion and had them flown from Hollandia to Biak and then moved by landing craft to Noemfoor.

The First and Third Battalions extended a perimeter around the Kamiri Airstrip and aggressively patrolled the area. The Japanese were pushed back into the interior of the island and their food and ammunition supplies were cut off.

The Second Battalion under Lt. Colonel John W. Britten, was ordered to move overland to the village of Inasi. It arrived on the 13th. Meanwhile, the First and Third Battalions were patrolling farther south. On the 13th the First Battalion reached Hill 670 about five miles south of Kamiri. C Company came under fire from the hill. For more than three hours C Company and the Japanese traded gunfire. The enemy appeared to have the heavier weapons so C Company Commander Captain Rucker radioed his position and dug in for the night. At 1845, A and B Companies arrived at Hill 670. The next day patrols made contact with the enemy but found that the Japanese had too much fire power for their light weapons. The artillery of the 147 Field Artillery Battalion was called upon to fire on Hill 670. Companies B and C moved up the hill facing light opposition to find that the enemy had left the area.

On 23 July, while patrolling north of Inasi, a squad from D Company led by Sgt. Ray E. Eubanks encountered a strong enemy position. Sgt. Eubanks directed his squad from a position within 15 yards of the enemy. Sgt. Eubanks used his automatic weapon with telling effect, and when he ran out of ammunition, he used his weapon as a club to kill four more of the enemy before he was mortally wounded. He and his squad had killed 45 of the enemy and routed many others. Sgt. Eubanks was awarded the Medal of Honor.

By the end of August the 503rd Infantry counted 1,087 enemy dead and 82 captured. Many Japanese and Formosan laborers were liberated. The capture of the airfields was completed and fighter planes began operating from the strips. With the airstrips secure at Noemfoor, bombers could attack the petroleum resources on Borneo.

During the Noemfoor operation a new dimension was added. Joining the 503rd was the 462nd Parachute Field Artillery Battalion commanded by Lt. Col. Donald F. Madigan and Company C of the 161st Airborne Engineer Battalion commanded by Capt. James Byer. The combined units were now to be known as the 503d Parachute Regimental Combat Team.

MINDORO OPERATION

The 503rd RCT arrived in Leyte on 18 November to prepare for the invasion of Mindoro. The responsibility of the Combat Team was to effect an amphibious assault landing with the 19th Infantry RCT, to overcome resistance on the island and to secure it for the construction of airstrips to be used in the recapture of Luzon.

The main convoy departed on 12 December 1944, from San Pedro Bay and headed north to Mindoro. The attack force consisted of one light cruiser, 12 destroyers, nine destroyer transports, 30 LSTs, 12 LSMs, 31 LCIs, 16 minesweepers, and several smaller craft. Another group under Rear Admiral Russell S. Berkey was sailing nearby with three cruisers, seven destroyers and 23 PT boats.

On 13 December at about 1500 hours a single kamikaze pilot plunged his plane into the cruiser *Nashville*. In the fires and explosions which followed, 325 casualties—including 135 dead—were counted, including nearly all the task force staff. The *Nashville* was so badly damaged it had to be escorted back to Leyte Gulf. Later that day a large kamikaze attack began. Fifteen enemy planes were intercepted; two escaped to disable a destroyer and killed or wounded most of the men on the bridge.

On 15 December 1944, Navy and Air

Force fighter pilots bombed and strafed enemy airfields destroying many planes on the ground. Naval bombardment forced the Japanese to flee to the hills. By early evening of the first day, the airstrip was captured, a perimeter was established and casualties were light.

Little resistance was met after the invasion except on patrolling operations. Company B was moved over water to Mamburao and marched 20 miles to Paluan where they encountered and ambushed an enemy patrol and killed nine Japanese. The following day they overran an enemy position and both sides exchanged fire. Company B forced the enemy to withdraw and counted 26 Japanese dead. Four paratroopers were killed and 14 were wounded. The company received a commendation from the task force commander.

CORREGIDOR OPERATION

On 3 February 1945, the 503rd Parachute Regimental Combat Team was alerted for a probable mission of seizing Nichols Field on Luzon; but on 5 February the mission was canceled, and the RCT was given the mission to jump on Corregidor Island. The assault troops named ROCK FORCE would combine a parachute jump and an amphibious landing supported by naval and air action. The time was set for 0830, 16 February 1945.

Col. John Lackery and his 317 Troop Carrier Wing dropped the 3rd Battalion Infantry, Battery A, 462nd Field Artillery Battalion; and Company C, 161st A/B Engineers and elements of Regimental Headquarters Company; and a platoon of .50 caliber machine guns of Battery D.

The objective of the first wave was to secure the topside jump fields for the subsequent flights and to secure the high ground overlooking the beach. From this vantage point the paratroopers could provide fire support for the amphibious landing of the 3rd Battalion, 34th Infantry who were to land bottomside at San Jose beach.

The integration of the infantry field artillery and engineers provided the diversity that the parachute regiment had not enjoyed at Nadzab and Noemfoor. The heaviest weapons carried by the parachute infantry had been 81mm mortars, grenades and grenade launchers. With the assignment of the 462nd Parachute Field Artillery, the infantry acquired the potent fire power of the diminutive 75mm pack howitzers and the repetitive destructiveness of the .50 caliber air-cooled machine guns.

On the way to Corregidor, 16 Feb. 1945. (United States Army Photograph, courtesy of R. J. Flynn)

161st Parachute Engineers close one of the many caves on Corregidor. (United States Army Photograph, courtesy of R. J. Flynn)

The 161st Engineers possessed an ambivalent talent to build and to destroy. This small company of 10 officers and 150 enlisted men were assigned to the infantry companies as needed. The engineers placed and deactivated mines and booby traps, operated flame throwers and bazookas and provided the expertise required of demolition teams to root the enemy from the many pillboxes and caves.

At 0830, 16 February 1945, the first plane dropped Col. John Erickson, and the paratroopers of the Third Battalion bailed out over the fortress Corregidor. Thirty-one C-47 planes of the 317th Troop Carrier Wing dropped small sticks of seven or eight troopers as they flew in two columns over the tiny drop zones. Winds gusting at times to 20 miles per hour swept some of the paratroopers over the cliffs of topside. Some were picked up by PT boats circling the island. Others were not so fortunate and were killed by Japanese troops when they tried to extricate themselves from splintered trees and brush.

The jump casualties, although severe, did not prevent the first wave from establishing their position to support the landing of the amphibious forces. The 34th Infantry sailed from Marivels in 25 LCMs toward Corregidor's South Dock. They landed on the beach at 1028 and scrambled through a mine field toward their objective, Malinta Hill.

The second lift of paratroopers

dropped the 2nd Battalion with elements of Service Company 503rd, Battery B of the 462nd and a .50 caliber machine gun platoon of D Battery. The drop zones had been secured, but the enemy had regrouped after the initial assault and became more aggressive. The second lift was subjected to more anti-aircraft, machine gun and sniper fire because the Navy and Air Corps could no longer bomb the topside. Of the 2,050 men who jumped in both lifts the combat team suffered 280 casualties, somewhat less than the planners had assumed.

Two events which happened on the first day of the assault seriously hampered the enemy forces. The first was the death of the Japanese commander Captain Itagaki who was killed by the men of I Company who had overshot the drop zone and came upon the enemy commander and his aides completely by accident and killed all but one aide who was taken prisoner. The second was the destruction of the central communications installation topside. The enemy wire communications from all major locations on Corregidor were obliterated. So, with the island commander dead and the enemy communications destroyed, the Japanese were severely hampered to mount a coordinated attack.

With the topside of Corregidor secure, Colonel Jones canceled the 17 February jump of the 1st Battalion Combat Team and had them flown from Mindoro to Subic Bay and transferred to landing craft for the trip to Corregidor on 17 February.

On the night of 16-17 February the Japanese Imperial Marines assaulted the topside in an effort to gain a foothold against the American paratroopers. These efforts were repulsed but the results were costly on both sides. The Imperial Marines wasted many good soldiers as the paratroopers experienced the first of several enemy banzai charges on Corregidor.

On 17 February, the combat team began to close the caves on Corregidor where the enemy had holed-up during the initial bombing. The exercise which proved most effective throughout this operation was to deploy small groups of five or six men to approach a cave, toss in hand grenades, white phosphorus grenades and a 20-pound satchel charge of TNT. When the demolition charges exploded, a flame thrower operator would aim a short burst of oil/gasoline mixture into the cave entrance. Later, napalm replaced the oil/gas mixture. The constant harassment of the enemy by the 503rd so demoralized the Japanese that they withdrew further into the protective cavern of the fortress and each day the patrols had to root out a fanatical enemy who would not surrender.

On 26 February, the 1st Battalion with Battery A of the 462nd launched an attack on Monkey Point and met strong enemy opposition coming from the tunnel entrance of the hill. Two Sherman tanks moved into position to help the 1st Battalion close the tunnel entrance and drive the enemy into the underground network.

One of the tanks fired into the mouth of the tunnel and apparently hit an enemy ammo dump. A violent underground explosion lifted the top off the hill sending bodies and debris flying through the air. The 35-ton tanks were tossed 35-40 feet down the slope of the ridge. Large rocks and boulders crashed down on many of the men in the area. The explosion killed 54 and wounded 145 troopers. More than 150 Japanese were also killed.

Casualty losses on Corregidor for the 503rd numbered 163 killed and 620 wounded or injured in action. Japanese losses actually counted, numbered 4,509 killed and 20 captured. No reliable figure is available for the hundreds of Imperial Marines who were sealed in caves and tunnels.

At 1000 hours on 2 March, General MacArthur returned to Corregidor. After a tour of the fortress, MacArthur went topside to the parade ground. Colonel Jones called the Honor Guard to attention, saluted General MacArthur and said, "Sir, I present you the Fortress Corregidor." After presenting Colonel Jones with the Distinguished Service Cross, General MacArthur remarked, "I see the old flag pole stands. Have your troops hoist the colors to its peak, and let no enemy ever haul them down."

Military history is replete with battles won by intrepid, clever, decisive commanders against great odds and seemingly insurmountable conditions. But the Rock Force attack against Corregidor on 16 February 1945 must go down as one of the most daring, well-planned and superbly executed operations in the annals of U.S. or any other army. The rugged terrain, the fanatical enemy, the strong defenses, the concrete and steel fortifications stacked the odds in favor of the enemy. Yet the Rock Force achieved success through its daring plan of attack; sound implementation of the battle plan; the bravery, discipline, and training of its soldiers; and the wholehearted support of the other services.

Corregidor, The Rock Force Assault, by Lt. Gen. E. M. Flanagan, Jr. AUS (Ret).

Paratrooper picking his landing between comrades, rubble and brush. (United States Army Photograph, courtesy of R. J. Flynn)

NEGROS OPERATION

Following the Corregidor operation, the 503rd Parachute Regimental Combat Team returned to their base camp on Mindoro. Several hundred replacements, newly arrived from the States, were welcomed to the unit.

HEADQUARTERS
503RD REGIMENTAL COMBAT TEAM
Office of the Commanding Officer
APO 715
28 June 1945
SUBJECT: Historical Report, Negros Island Operation (FO #27).
TO: Commanding General, Eighth Army, APO 343.

1. The assault on enemy-held Negros Island (Field Order #27) was an amphibious landing by 40th Infantry Division (less one regimental combat team and forces required at Panay Island) with the 503rd Parachute Regimental Combat Team, the 164th Regimental Combat Team as reserve. The operation was supported by Air and Naval Forces.

2. The mission of the assault troops was to seize and occupy Northern Negros Occidental, destroy hostile forces and reestablish civil government therein.

3. The forces involved in the operation included:
 a. Ground Troops:
 (1) 40th Infantry Division (less on RCT):
 (a) 185th Regimental Combat Team
 (b) 160th Regimental Combat Team
 (c) 164th Regimental Combat Team
 (d) 40th Division Artillery:
 1. 143rd Field Artillery Battalion.
 2. 213th Field Artillery Battalion.
 3. 222nd Field Artillery Battalion.
 (e) 115th Engineer Battalion.
 (f) 115th Medical Battalion.
 (2) 470th AAA AW Battalion.
 (3) 716th Tank Battalion.
 (4) 239th Engineer Construction Battalion.
 (5) 37th Field Hospital.
 (6) 245th Field Artillery Battalion.
 (7) 503rd Parachute Regimental Combat Team:
 (a) 503rd Parachute Infantry.
 (b) 462nd Parachute Field Artillery Battalion.
 (c) Company C, 161st Airborne Engineer Parachute Battalion.
 (8) Detachment 592 JASCO.
 (9) Detachment 542 EB & SR.
 b. Naval Forces: Detachment Task Force #78.3.2.
 c. Air Forces: Elements 13th Air Force including 322d Troop Carrier Wing.

4. The 503rd Regimental Combat Team was alerted on 25 March 1945 for a probable jump mission vicinity Alicante Airfield, Negros Island, advance to the West and seize and secure Saravia (town), then advance rapidly to the South to effect a junction with the 185th Regimental Combat Team in the vicinity Imbang River Bridge.

Preparations began immediately, which included checking, replacing combat equipment and detailed planning for the movement and mission. Plans for the operation were completed and Field Order #10 was distributed to the lower units 5 April 1945. Later in the day orders were received from Headquarters, Eighth Army, cancelling the jump mission on the recommendation of the Commanding General, 40th Division to the effect that the target area was clear of enemy. Instead, orders were issued for an airborne movement to Panay Island with subsequent waterborne movement to Negros Island for a mission to be designated by the Commanding General, 40th Division. Apply that portion of Field Order #10 as pertained to an airborne movement, the move of the Regimental Combat Team (less one battalion, reinforced) began on 6 April and completed on 8 April with a landing at Pulupandan, Negros Island. (The First Battalion, 503rd Parachute Infantry, C Battery and elements of D Battery, 462nd Parachute Field Artillery remained on Mindoro Island as Eighth Army reserve.) The Regimental Combat Team then entrucked for motorized movement to assembly area as assigned to by Commanding General, 40th Infantry Division. Primary mission and zone of action was assigned to the Regimental Combat Team and issued on 8 April. (Operations overlay to accompany Field Order #17, dated 8 April 1945, Headquarters, 40th Infantry Division).

Paratroopers of the 503rd Parchute Infantry Regiment, part of Gen. MacArthur's forces, land on the bomb-battered terrain of historic Corregidor Island. (U. S. Army photo)

Forward observers radio information to command posts. (United States Army Photograph, courtesy of R. J. Flynn)

The mission assigned to the Regimental Combat Team was to seize division objective within its zone, destroy all hostile forces encountered and protect the left (N) flank of the division. The Second Platoon, Company C, 716th Tank Battalion was attached to the Regimental Combat Team to assist in accomplishing this mission. Leading elements of our assault forces crossed the initial point at 09 0800 April. Included in these elements were demolition sections to disarm the many mines (converted 100- and 250-pound air corps bombs) planted along our approach route. Our approach route followed a series of narrow parallel ridges that sloped upward to our objective. Contact with the enemy was made at 09 1000 April and was never lost during our approach to the objective. Employing rifles, knee mortars, machine guns of all the familiar calibers, including converted air corps types, and occasional dual-purpose anti-aircraft guns, the enemy made his usual fanatical defense from innumerable mutually supporting caves, bunkers and intercommunicating trenches. The terrain was ideally suited to this type of defense. Formerly cultivated fields afforded little cover or concealment of our movements and our uphill attack afforded enemy observation points to observe our progress minutely. The pattern of our attack was to locate enemy strong-points, concentrate supporting artillery, tank, and mortar fire on them and then close with the enemy. Night interdictory and harassing fire by the artillery and mortars effectively prevented concentration of sizable enemy forces for night attacks and the relatively few small attacks were effectively disrupted. Initially, though steady progress was made, it was slowed for several reasons. Lacking the First Battalion Combat Team, protecting our left flank demanded troops that would otherwise have been employed in the assault or as reserves. This, coupled with a rather wide front, thinned our assault forces. However, on the 25th of April the First Battalion Combat Team joined the Regimental Combat Team and our advance accelerated. Coupled with the impetus of fresh troops was the previously unobserved devastating effect the artillery was having on the enemy. Having nothing to combat this, the enemy began retreating into the heavy rain forest and mountains to his rear, leaving only delaying forces. These were quickly overrun and the Regimental Combat Team reached its initial objective 29 April. As the advance to the initial objective had progressed, left flank patrols encountered increasing enemy activity to the North. On 30 April a division order included in its directive to continue along the original line of advance and an order to send sufficient forces to the North to destroy an enemy encountered in the TYAP area. This latter mission was assigned to the First Battalion Combat Team, and the Third Battalion Combat Team remained in position to guard the rear and to patrol laterally. The first battalion met some resistance in the TYAP area; however, the concentration of airstrikes, artillery, 4.2 chemical mortar force coordinated with the ground attack forced the enemy to abandon his positions and scatter into the mountains. Intelligence information indicated the enemy to be withdrawing the bulk of his forces to the South away from the Regimental Combat Team zone of action. Consequently, on 11 May the 40th Division issued orders relocating Combat Team (less one battalion), on the division right flank. The third battalion was attached to the 155th Regimental Combat Team and continued its advance on the division left flank The new Regimental Combat Team mission was to advance to the North and cut the enemy supply and evacuation route to the Southeast; thus containing the balance of the enemy forces in the PATOG HILL 4055 area which the 160th Regimental Combat Team and 185th Regimental Combat Team closed in from the North. Fighting uphill through heavy rain forest, ravines and steep mountainous areas, our forces, supported by aerial and artillery action, succeeded in their mission by emplacing strong forces across the enemy's evacuation route on 26 May. Extensive combat patrolling from this position effectively broke up all organized resistance and forced the remainder to flee deep into the mountains. On 4 June all infantry elements of the 40th Division, including the 503rd Regimental Combat Team, were relieved from the objective area by the 7th Philippine Military District Forces. The mission of the latter was to pursue and destroy the enemy that could be found and contain the remainder in the mountains, denying it access to food or supplies.

At 2400 hour, 9 June, operational control on Negros Island passed to the 503rd Regimental Combat Team. Its mission being to garrison the island maintaining its security, and completing the destruction of the large number of enemy remaining. At the official close of the operation on 20 June, the forces of the 7th Philippine Military District, consisting roughly of 4,000 troops having two combat regiments and one combat battalion, were busy in the accomplishment of the assigned mission.

END OF REPORT

Col. George M. Jones rotated back to the States on 10 August 1945. He was succeeded in command by Lt. Col. Joe S. Lawrie who had served in the 503rd from the outset. On 30 August, Lt. General Kono and eight of his staff officers surrendered the Japanese garrison of approximately 6,500 men to Lt. Col. Lawrie.

The Negros Operation had cost the 503rd 144 men killed and 370 wounded in action.

505TH PARACHUTE INFANTRY REGIMENT

By Ben H. Vandervoort

"I have no doubt that based on its record the 505 was the best Parachute Regiment to come out of World War II."
General M.B. Ridgway

"The one thing that the 505 would always remember...were the parachute lieutenants. Their combat performance was extraordinary."
Lt. Gen. James M. Gavin

The 505 Parachute Infantry fought in every major campaign in the liberation of Europe including four Allied Airborne invasions and Hitler's winter offensive in December 1944. In two years of combat the Regimental Combat Team demonstrated it was one of the most effective fighting regiments in the United States Army. The criteria was battles won and numbers of enemy defeated versus casualties sustained by the regiment. These exceptional soldiers repeatedly defeated numerically-superior Nazi forces.

The regiment was activated in July 1942. All of the enlisted men were "double" volunteers—not draftees. Called to arms by Pearl Harbor they had volunteered for the Army and again for the paratroops. Screened, tested, and jump-qualified by the parachute school, they were the top of the line of America's citizen soldiers. Many would have been in college except for the war. The goal of the regimental cadre was to train them into as tough and as intelligent a group of fighting men as ever pulled on jump boots.

There was no six-and eight-month combat rotation duty tour in those days. A man stayed with his regiment until killed or wounded and evacuated. The regiment was his life for the duration of the war. The esprit-de-corps of the regiment grew with combat experience. His motivation was "mutual faith." Every soldier knew every other man would do his job in combat or would die trying. They would not let each other down. Within the mutual faith all competed to make the play to win the game—like a Super Bowl team. That faith and competitive spirit welded the regiment into a rough tough winning combat team.

In July 1943, 12 months after activation, reinforced by a battalion of our 504 sister regiment, the 505 made its first night Airborne assault into Sicily. The troop-carrying C-47s scattered us 60 miles wide and four miles deep. There had not been enough aircraft or time for sufficient night formation flying and navigation practice without electronic aids. Only one battalion out of the four was delivered to the regimental drop zone. That was the 1st Bn. 505 which accomplished all the missions assigned to the entire regimental combat team. It was a remarkable performance. Our regimental commander, Colonel James M. Gavin, was dropped near Vittoria—about 20 miles from his objective. In the moonlight he rounded up eight combat green troopers. (I was one of them.) He would rather have had a radio. He had no idea of where his regiment was and only a vague idea of where he was. We walked all night.

Mid-morning, we ran into an Italian 35-man anti-paratroop patrol 70 yards in front of us. An intense fire fight ensued. Two of our troopers were hit and lay very still. In the time it takes to fire two dozen aimed shots with a carbine, the Italians were driven to cover behind a stone wall. In the lull, we disengaged straight back, one at a time, the others covering. The colonel was the last man to withdraw from the position. We took temporary cover in a cane brake. We were dirty, sweaty, tired and distressed at having to leave wounded behind. The colonel looked over his six-man command and said, "This is a hell of a place for a regimental commander to be." The next morning the colonel joined his regiment in time to engage it in a fully submerged baptism of fire with the Hermann Goering Panzer Division.

After Sicily came more hard fighting in Italy. Every survivor became a seasoned soldier. Through innumerable small unit actions, the junior officers became knowledgeable, combat-wise leaders. Whatever job the 505 was given, the lieutenants and NCOs made it work. They led like respected older brothers in a fraternity. Their emphasis was on self-discipline rather than discipline applied from above. With never a doubt as to who was in charge, they shared foxholes, rations, hazards and hopes with their troopers. When we left Italy the regiment was lean, mean and battle smart. Units were about full-strength. The 82nd foresightedly brought along its own replacement contingent from the United States. They filled in the gaps caused by casualties. Having met every test the Fifth Army threw at them, the Mediterranean commander tried to hold them in his army, but was outranked by the supreme Allied commander, General Eisenhower.

Bill Mauldin's best <u>Stars and Stripes</u> cartoon of the Airborne commemorates a 505 happening in Italy. When we jumped into Salerno in September 1943 our jump boots were wearing out. There were no replacements because the boots, issued exclusively to the Airborne, were being short-stopped in the supply lines. In those days parachute boots were as sacred to the paratroopers as their wings. In Naples the 3rd Battalion 505 was billeted near the harbor docks. Quartermaster corps longshoremen and supply corps officers showed up wearing shiny new jump boots. That did it! The 3rd Battalion troopers took the boots off their "chairborne" feet and left them on pass, without any shoes. Outraged officers in their stocking feet stormed into the battalion command post. The only sympathy they got there was an interrogation by the battalion duty officer. "Where did you get your jump boots?" In the rear echelons, the word went around. If you weren't jump qualified, don't take your boots to town. The flow of parachute boots to the 82nd Airborne began immediately.

The 505 made the primary Airborne contribution to the Allied invasion of Normandy in June 1944. The 82nd was

Operation Varsity paratroops. (U.S. Army photo)

dropped five hours ahead of the amphibious landings to protect the northwest flank of the American seaborne invasion. The German High Command ordered the strongest mobile force on the Cherbourg peninsula to drive the United States amphibious landings on Utah and Omaha beaches into the sea.

The only full-strength British or American parachute infantry battalion assembled in Normandy on D-Day morning was the 505, 2nd Battalion. With the rest of his division scattered, General Matthew B. Ridgway, the commander of the 82nd, held that battalion on the drop zone. Anticipating the German reaction he released the battalion to defend Ste. Mere Eglise rather than continue with its preplanned mission. Heavily reinforced with self-propelled guns and mobile artillery, the German column was five times the strength of the Americans. For 28 hours that one parachute battalion held off the massive German force, until "line-up" trickled up from the beaches.

Cutting the strategic highway at Ste. Mere Eglise was a regimental achievement. The 3rd/505 took the town and dug in facing south. The 1st Battalion blocked causeways over the Merderet River isolating the west flank. F Company sealed off coastal forces withdrawing from the East. With its rear and flanks thus secured, the rest of the 2nd Battalion counterattacked and made a bloody shambles of the Wehrmacht attack from the north. Later in the war, Maj. Gen. J. Lawton Collins, VII Corps commander said the 505 had conducted "one of the most perfectly coordinated combined attacks laid on in Europe."

Why was the Second Battalion so well assembled? The 505 pathfinder team jumped 30 minutes ahead of the main body. Commanded by First Lieutenant J.J. Smith, they hit the right drop zone. They put all of their electronic homing and D.Z. marking devices in operation on time. It was the only pathfinder team in the two Allied Airborne armies to drop in the right spot and assemble all of its gear properly. After the Sicilian snafu, Jack Norton, our regimental S-3 initiated a theater pathfinder training program in England. J.J. and his people practiced for months with the same pilots and aircrew to get it right. They did. Aircraft from half of southern England released their loads over that lighted drop zone. Glider pilots and stray troopers from two divisions supplemented our strength. We scrounged extra ammunition and weapons from misdropped bundles and jeeps from crashed gliders, including two 57mm anti-tank guns. Our 81mm mortar platoon deliberately left two of its four mortar tubes on the drop zone so they could carry more ammunition. With two tubes they fired 1,000 observed rounds before midnight D-Day.

We thank again the 316th Group of the 52nd Troop Carrier Wing. The 316th put the battalion down right on top of J.J.'s pathfinders. It took cool flying to hold a course wing-tip-to-wing-tip through fog banks so thick the pilots couldn't see beyond their own wings. When they could see, they saw flak. The closer to the flak the stronger the compulsion to turn on the green "Go" light prematurely, drop your load and get the hell out of there.

We crossed the coast of France at 1500 feet altitude, blacked out to protect against enemy fighters and flack. From the door of the lead aircraft I mentally checked off key terrain features through the gaps in the clouds. The crew-chief relayed comments between the pilot and the jump-master via his aircraft intercom head set. Suddenly the green "GO" light came on miles from our destination. That started the jump signal flashing back through the astrodomes of all 36 aircraft in the lead serial. Quickly we had the crew-chief tell the pilot to turn the damn light off, which he did promptly. However, two platoons jumped on the false signal. We didn't see them for three days. Perhaps the poor visibility helped stop the early drop. Breaking out of the fog minutes later we saw the reassuring green glow of the holophane lights marking our drop zone. In the fog it was imprudent to come down to 750 altitude or to slow down to jump speed without risk of a mid-air collision. Now was too late. The 316th came in high and fast, but in a tight disciplined formation right over the heart of the drop zone. That was all we wanted. At 01:41 double-daylight saving time D-Day, we stepped out into the night.

Forty hours later the 505 had made its contribution to the Allied foothold on the continent. Holding the German's principal counterattack north of Ste. Mere Elgise secured the entire American Airborne bridgehead and Utah Beach.

The German Kampfgruppe sustained more than 1,000 casualties in their brief two days against the 2nd 505. In contrast, from an initial strength of 629, the 2nd 505 had 298 casualties during its entire 28 combat days in Normandy. Many of our casualties were merely wounded, evacuated to England and returned to full duty. Some of the paratroopers went AWOL from their hospitals so they would not be left behind during the next invasion.

We spent nine weeks in England refitting and retraining before the Holland campaign. In that brief period the 505 planned and prepared for a dozen different Airborne assaults. The theater planners favored ancient fortifications and capital cities, like Brussels, as strategic objectives. All were cancelled at the eleventh hour—mostly because General George Patton or British ground forces kept overrunning the drop zones. That planning frenzy to get the Airborne back in action prompted our 3rd Battalion commander "Pappy" Hagan, to say, "Hell, it doesn't matter what they think up for us to do. These guys ar so good, they will make it work anyway."

On September 17, 1944, the 505 jumped into the Netherlands in daylight as part of an Airborne armada of 4,000 aircraft and 2,400 gliders. It was the best delivery ever of the 505—except for the 2nd Battalion. Forced to the north by flack in the congested air space, we had to select an alternate drop zone at the last minute. We landed in an open area surrounded by anti-aircraft batteries and dominated by a huge flak tower. The troopers landing nearest each of the ack ack position opened fire. Right now! Catch as catch can—many still in jump harness. Lt. Jimmy Coyle shot one enemy gunner from the air with his 45 automatic pistol. Scrambling off the broken play, we took out the flak tower with a coordinated attack from three sides about 30 minutes after we jumped. The troopers then went to work on their original mission. Next day, I gave a small laudatory critique on their quick reactions under fire and fast assembly on a strange drop zone. That produced the response, "Jesus, colonel; everybody knows you can run faster when you're shot at."

The 2nd Battalion of the 505 fought in the British Army at Nijmegen. It was attached to the British Guards Armored Division and was to help them seize a bridge. The Reconnaissance Battalion of the 9th SS Panzer Division was defending the bridge. The Germans also had dual purpose anti-aircraft/anti-tank 88mm batteries, plus miscellaneous troops in the city at the time of the drop. Every approach to the bridge was covered by 88 cannons firing lethal 22 pound projectiles. The German Recon. Battalion strength was 600 die hard Nazis. In less than five hours of close-quarter city fighting the troopers reduced that fanatic black shirt battalion to 60 men. Working with the Guards Armored Tanks, the 88mm cannon crews were wiped out. We took less than 10 percent casualties. The name of the game was "quick to close, and quick to kill."

Between campaigns, training was rigorous and continuous. Live firing, physical conditioning and equipment maintenance never ceased. Troopers slept with their rifles even in garrison. At any time all one had to do was issue ammunition and field rations and tell them where to go. All that attention to detailed readiness paid off when we were thrown into the Ardennes. Called back off pass without warning they picked up their crew-served weapons and tin hats, entrucked and took off into the unknown a hundred miles away. Each battalion was a formidable instrument of infantry warfare. Each was an experienced, flexible unit of disciplined highly trained professionals with a clear edge over any comparable enemy elements they would meet.

At Trois Ponts, Belgium, in the Battle of the Bulge, Company E, 505, commanded by First Lieutenant Bill Meddaugh, was isolated by itself to stop a couple of thousand enemy soldiers. The Germans were the better part of the 1st SS Panzer Division, the strongest and best equipped enemy division in the German winter offensive of December, 1944. Hitler was counting on them to make a major break-out to the West so his entire army could follow through the gap. The 1st SS were the brutal bastards who shot the helpless American prisoners at Malmedy.

Company E had seven officers, about 135 troopers and one 57mm anti-tank gun and crew. Hit by artillery, mobile flak towers, tanks and infantry, they held Hitler's best for 12 hours. Three times intensive German assaults were thrown back leaving dead Nazis in the E Company foxholes. First SS Panzer Division prisoners were stunned by the way the parachute infantry stood its ground against their armored might. With massed infantry and tanks overpowering their front and both flanks, E Company was almost lost. It cost a shocking 30 percent casualties and culminated the most ill-timed withdrawal ever seen. They had to shoot their way through the woods, down a hundred-foot cliff, across a highway, over railroad tracks and wade an icy river to get back to the battalion.

On E Company's heels came Tiger tanks and infantry trying to force a crossing at the Salm River. They were brought up short by three more tough 505 companies. After 10 more hours the Germans were still on the wrong side of the river. Bruised and bleeding, they slipped away on in the night to find an easier place to cross. They ran into more 505 and never did cross the Salm. Neither E Company nor any of its men received special recognition for that fight. Just another working day in the life of the 505.

The longer we were in the war the harder it was to lose old soldiers. Not only were they key men; they were your friends. We blocked those thought out. Still—we were lucky. No regiment in the European Theater put in as many man-to-man close combat hours and had as many of its original members to survive the war. They came home alive because they fought hard and did their jobs well. Concern for their own safety was secondary. That was not heroics. It was professionalism. They teamed together to make a "lucky" regiment that never lost a battle. They were magnificent and fun. All of us are very proud of our regiment.

The old paratroopers of the 1940s won't fade away. Their mutual faith and competitive spirit lives on today with the Airborne. And the next time America goes for broke they will be there, as invisible pathfinders, to help today's troopers scramble off the broken plays that always have, and always will, come with the Airborne territory.

P.S. *One key to the superior performance of the 505 was Lt. Col. James M. Gavin, the original regimental commander, hand-picked most of the cadre. The original complement included 14 West Pointers. Most of the senior NCOs were Regular Army soldiers. All were experienced in training and motivating men. Whatever the regiment was required to do, Colonel Gavin did it first to set the example.*

2nd Bn., 507th PIR Pathfinder Team and Air Crew Prior to takeoff for Normandy.

The 507th's Regimental Birthday Parade, Alliance Army Airfield, NE. (Courtesy of Hughart).

507TH PARACHUTE INFANTRY REGIMENT

By John Marr
Col., USA (Ret)

Element:
507th Parachute Infantry Regiment.

Activation:
The regiment was activated 20 July 1942 at Fort Benning, Georgia. It obtained the nucleus of its cadre from the 504th Parachute Infantry Regiment and a few officers and men from the 502d and 503d Parachute Infantry regiments.

Campaign Credits:
World War II Campaigns:
Normandy (With Arrowhead) (6 June - 13 July 1944)
Ardennes - Alsace (25 December 1944 - 12 February 1945)
Central Europe (24 March - May 1945)
Rhineland (With Arrowhead) 24 March - 8 May 1945)

World War II Battle Participation

Battle	Dates	Campaign
Battle of the Merderet River	6 - 9 June 1944	Normandy
Battle of Bonneville-Renouf	14 - 17 June 1944	Normandy
Battle of Vindefontaine	17 June - 3 July 1944	Normandy
Battle of La Poterie Ridge	3 - 7 July 1944	Normandy
Battle of Dead Man's Ridge	4 - 14 January 1945	Ardennes
Battle of the Ourthe River Junction	9 - 14 January 1945	Ardennes
Battle of the Our River	27 January 10 February 1945	Ardennes
Operation Varsity	24 March 1945	Rhineland and Central Europe
Battle of the "New York to Paris" Line	26 - 27 March 1945	Rhineland and Central Europe
Battle of Dorsten	27 - 29 March 1945	Rhineland and Central Europe
Battle of the Ruhr Pocket	5 April - 8 May 1945	Rhineland and Central Europe

Unit Citations, Awards and Decorations:

A. Presidential Unit Citation (Army) Streamer Embroidered COTENTIN PENINSULA (War Department General Order 76, 1944).

B. French Croix de Guerre with Palm, World War II, Streamer Embroidered STE MERE EGLISE Department of the Army General Order 43, 1950).

C. French Croix de Guerre with Palm, World War II, Streamer Embroidered COTENTIN (Department of the Army General Order 43, 1950).

D. French Croix de Guerre, World War II, Fourragere (Department of the Army General Order 43, 1950).

Congressional Medal of Honor and Distinguished Service Cross Recipients:

A. Medal of Honor:
Private George J. Peters (Posthumous) Company G.

B. Distinguished Service Cross:
Lieutenant Colonel Arthur A. Maloney
Lieutenant Colonel Charles J. Timmes
Captain Robert D. Rae
Sergeant Dante Toneguzzo

Commanding Officers:
A. Regimental Commanders:
Colonel George V. Millett, Jr. (20 July 1942 - 8 June 1944)
Lieutenant Colonel Arthur A. Maloney (8 - 15 June 1944)
Colonel Edson D. Raff (15 June 1944 - 15 August 1945)
B. Battalion Commanders:
1st Battalion:
Lieutenant Colonel Edwin J. Ostberg
Lieutenant Colonel Benjamin F. Pearson
Lieutenant Colonel Paul F. Smith
Major Roy E. Creek
Major Allen W. Taylor
2nd Battalion:
Lieutenant Colonel Charles J. Timmes
3rd Battalion:
Lieutenant Colonel William A. Kuhn
Lieutenant Colonel Arthur A. Maloney
Lieutenant Colonel Gerhard Boland
Major John T. Davis
Major Roy E. Creek
Major Allen W. Taylor

CASUALTY SUMMARY

CAMPAIGN	Officers	Enlisted Men	Total
Normandy			
KIA	30	273	303
WIA	54	300	354
MIA	29	149	178
Total	113	722	835
Ardennes			
KIA	5	93	98
WIA	27	276	303
MIA	16	280	296
Total	48	649	697
Rhineland and Central Europe			
KIA	12	110	122
WIA	17	219	236
MIA	2	20	22
Total	31	349	380
CUMULATIVE	192	1720	1920

Crest / Logo / Motto / Colors:

Motto: "Down to Earth"
Colors: Orange with Black trim

Individual Decorations:

Congressional Medal of Honor - 1
Distinguished Service Cross - 5
Silver Star - 108
Bronze Star - 351 (Does not include Combat Infantry Badge-related awards)
Purple Heart - 1466
Soldiers Medal - 7

Total - 1938

INTRODUCTION

When the Japanese struck at Pearl Harbor, the U.S. Army had four parachute battalions: one in Panama and three at Ft. Benning, GA. Three months later, in early March 1942, the first parachute infantry regiment was activated. Before the year came to an end, most of the regiments of parachute infantry needed to form the cutting edge of the five airborne divisions brought to the Army rolls in the course of World War II would be constituted. This is the story of one of them. It is the story of the 507th Parachute Infantry Regiment that had a World War II life span of 38 months, functioning for nearly 32 months as a separate regiment before being assigned as an organic regiment of the 17th Airborne Division. The bulk of this historical treatment is devoted to the combat life of the Regiment. Events before and after the combat phase are given much briefer treatment because of space limitations.

ACTIVATION AND TRAINING

The 507th Parachute Infantry Regiment was constituted 24 June, 1942 on the Regular Army inactive list. It was activated 20 July, 1942 in the "Frying Pan" area of Ft. Benning, GA under command of Lt. Col. George Van "Zip" Millet, Jr. As it continued to fill with officers and men it began a 22-week training cycle to prepare for combat deployment.

Training and unit testing continued as the Regiment moved to the "Alabama Training Area" across the Chattahoochee River in September. While training was rigorous and demanding, there was time for leaves and passes. For those, like 2d Battalion HQ Co's T-5 James W. Hyde, Elberton, GA, home visits on a 3-day pass was a breeze.

The regiment left for Barksdale Field in Shreveport, LA, 6 March, 1943 for 12 days of participation in 3rd Army maneuvers along the Sabine River, before moving on to the sand hills of western Nebraska for additional combat training. The speed, precision and disruptiveness of the regiment's Airborne assault brought a halt to the maneuvers, early extraction of the troopers and reconstitution of the simulated battle area so that meaningful maneuvers could continue.

In parallel with its training and maneuvers the 507th was honing its athletes for competition with all takers. Boxing, wrestling and swimming teams were making enviable marks. But it was the basketball team, under the able coaching of 1st Lt. Karl Lillge, now lieutenant colonel, retired, San Jose, CA, that gave the Regiment a wide-spread reputation. It defeated the best college and pro teams in the southeast and capped its 24-2 season with a win over the New York Celtics.

At Alliance Army Airfield, Alliance, Nebraska, 23 March the 507th joined with Air Corps Troop Carrier forces in training. Rigorous advanced unit training to hone the collective skills of the regiment for combat was the daily fare. In the late summer it was put to test in the field under the practiced eyes of a 14-member team of examiners from the Airborne Command, and won high marks which won the entire regiment a vacation bivouac in the Black Hills of South Dakota. There were tactical jumps to test the defenses at distant airfields and fun times, too, for the troopers able to snag a space on one of several jumps made in support of blood bank and bond drives in a four-state area. A new sport was added to regimental repertoire as troopers gathered in riding and roping prizes at rodeos in Alliance and other area towns. The swimming team, with G Co.'s PFC George Tullidge in the lead, made a clean sweep of the regional swim meet. The war in Europe was waiting for the 507th whose skills and readiness were honed to the challenge that lay ahead.

EMBARKATION

Beginning 23 November, 1943, it took a month for the Regiment to get settled into Portrush, Northern Ireland, via Camp Shanks and Ft. Hamilton, NY, the Scottish Ship *HMS Strathnaver* to Liverpool and the liberty ship *Susan B. Anthony* to Belfast. Only small unit training could be done there to keep the edge on trooper skills.

Three months later, 11 March, the regiment crossed to Scotland and took rails to Tollerton Hall in the fabled midland town of Nottingham, England. Again small unit training was the rule. The regiment knew nothing about the Airborne invasion of France. Operation NEPTUNE was now little more than 11 weeks away. Regimental staff would be briefed on 1 April; battalion staffs, 1 May. Pathfinder teams were dispatched to train at North Witham Airfield. Task-oriented training began to take shape as unit commanders, but not their troopers, learned of their missions. Night jumps by battalion and then a regimental mass jump were planned for but not fully implemented because of unusually bad weather conditions.

On 28 May, marshalling for the drop in Normandy began. The 1st Bn. moved to Fulbeck Airfield, the 2d and 3d Bns. to Barkston Heath Airfield and the Pathfinders to North Witham. There, they were sealed into camp behind barbed wire.

Terrain boards with miniature reproductions of the regimental objective area were used for repeated daily briefings of all individuals. Actions to be taken upon landing, assembly and movement to assigned objectives were carefully gone over. Card playing, sports, letter writing and more briefings occupied the passing days until NEPTUNE would be launched. It was to be 5 June, but weather delayed it one day. As daylight on 5 June was drawing to a close, movement toward aircraft began. The 507th was ready and razor-sharp for action. At one minute before midnight the trooper-laden C-47s roared into the air.

THE AIRBORNE INVASION OF NORMANDY NIGHT JUMP

Using 117 aircraft to lift 2,004 troopers, the 507th Parachute Infantry Regiment parachuted into Normandy, France to land on DZ T shortly after 0230 hours, 6 June 1944. Their mission was to: (1) Seize, organize and defend its assigned area; (2) Clear the area within its sector; (3) Assist the 505th Parachute Infantry Regiment in securing the crossing of the Merderet River at La Fiere; (4) Establish and maintain contact with the 508th Parachute Infantry Regiment at Renouf and, (5) Be prepared to advance to the west to the line of the Douve River.

While not achieved as planned the mission was accomplished despite extreme dispersion during the jump and heavy combat casualties. As the heavily loaded C-47s crossed the Cotentin Peninsula coast line, a heavy fog bank and German flak caused many planes to take evasive action and fly at excessive speed and altitudes not used in mass jumps. For example, most of Hq. Co. 3d Battalion landed in the town of Graignes, south of Carentan, 16 miles away.

On drop zone T at Gourbesville, 30 minutes in advance, the three pathfinder teams, comprising 51 troopers under command of Lt. John T. Joseph, now colonel, retired in Monroeville, PA, were making heroic efforts to get equipment set up to guide the oncoming troop carrier aircraft to their correct positions over the drop zone. Units of the 91st German Division, on field maneuvers, reacted promptly and laid down devastating fires on the DZ. So intense and persistent was the enemy response that some crews were unable to set up their Eureka radars and the lights to mark the directional T over which the aircraft would fly.

Thirty-five pathfinders were killed, wounded or captured. Two of the Eurekas, but none of the marker lights, were operating as the oncoming flights crossed the DZ. Nevertheless, Lt. Joseph counted 83 planes passing over the DZ area. Hq. 2d Bn. Sgt. Rolland J. Duff, Ft. Myers Beach, Florida, a Eureka operator for the 2d Bn. team, recalls that his team was engaged almost immediately but was able to assemble seven members of his 17-man team. He operated the radar while the other members provided security under command of 1st Lt. Charles Ames. The team leader, 1st Lt. Ralph McGill was captured soon after landing.

At the time designated for the last flight to cross over the DZ, Duff detonated the highly secret radar equipment and the team moved southward into Chef-du-Pont. A Co. Sgt. Thompson J. Morris, Alpharetta, GA, a member of the 1st Bn. team, remembers the enemy tracers through his parachute, the drop zone covered with German troops and, at daylight, some of his buddies hanging in the trees.

The 2d Bn. was the first element of the main body to jump, starting at 0232 hours. The entire regiment was on the ground in a matter of minutes but widely dispersed over the Peninsula. Most of the troopers landed in the flooded Merderet basin, astride the river. Many drowned under the weight of their equipment.

Some fought valiantly as individuals or members of very small groups before being wounded or captured or both. Others formed into small fighting groups, often with a mix of troopers from different units of the 507th and other regiments. Without exception, these groups performed with unsurpassed determination, courage and effectiveness. Some group, faced with overwhelming enemy strength, resorted to hit-and-run tactics that punished and misled the Germans. Others, pinned into defensive action, held out against staggering odds. Larger groups took the offensive to seize and hold objectives against the best efforts the Germans could muster. The collective effect of these individual and group actions reduced the ranks of the Germans, kept them off balance and confused their leaders as to which display of force would receive counteraction.

On the individual and small group level, reflection of G Co. S/Sgt Joseph Faust of Akron, OH will trigger the memories of other troopers who jumped on D-Day. "It was a night jump into Normandy. I landed in a French pasture. There were a few live cows and a lot of bloated dead cows laying on their backs with their hooves pointed to the sky. Here I am in a strange land. I am a lone human being, surrounded by a herd of living and dead cows. Daylight comes and I join a few troopers and march down a dry highway. In a few hours we see a large line of German tanks headed for our group. In front of the tanks rides a man on a motorcycle. This has to be the most dangerous job in any army. We place four mines on the narrow, tree-lined road. The German rider dismounts and heads for the nearest mine. A BAR speaks and the German will never speak again. The tanks turn around and retreat."

Hq. 1st Bn. radio operator William D. Bowell, St. Paul, MN, remembers dropping 25 miles from the DZ, joining a collection of troopers, evading the enemy for nearly a week, being helped by the French Underground, crossing a river in boats at night, clashing repeatedly with the Germans, being wet and cold at night, despairing breaking through the German lines and, finally, meeting with relief the advancing U.S. infantry and being returned to the Regiment.

Hq. Co. 1st Bn. PFC Ray Ballard, Lake City, FL, dropped with 12 troopers but could link up with only one. Later in the day, they knocked on the door of a French house and asked two French ladies about directions and the presence of Germans. The ladies pointed the way, saying no Germans were in the area. Walking only a short distance, they were fired upon by the Germans, soon captured, stripped of everything they had and beaten. They were taken to the same house they had left and found it to be a German command post. About 1800 hours the Germans brought in H Co. Captain Howard A. "Big Steve" Stephens, now colonel, retired in Tampa, Florida, who was seriously wounded. Stephens was hospitalized in Cherbourg and later liberated. Ballard remained a POW.

Co. C machine gunner, Edward J. Jeziorski dropped near Hebert, about 10 miles southeast of the DZ, on top of a shot-down C-47. He, Dante "Tony" Guzzo of Columbus, OH and Trooper Boyce fought the Germans all day between two Utah

Beach causeways, sometimes in hot actions. They bedded down in a hayloft where they heard excited German voices and the movement of armor throughout the night.

The next evening they made a pact not to ever surrender and Jeziorski positioned his machine gun toward advancing troops they believed to be Germans. They were much relieved when the lead scout turned out to be with a unit of the 4th Division. They briefed the unit's colonel who gave them a jeep ride to an area occupied by the 507th.

E Co. PFC Joseph A. Dahlia, remembers moving along a hedge row, being hit by a sniper bullet ricochetting from his own weapon into his hand. After hiding in a barn, he was found by the Germans, evacuated by horse and wagon to a monastery where he was operated on by a German doctor. He was sent to a prison hospital at Rennes, then liberated when the town was overrun by the 8th Infantry on 4 August.

H Co.'s Clarence Hughart, Arvada, CO, was hospitalized by the Germans in Cherbourg, liberated and because of the severity of his wounds, returned to the U.S. Less fortunate was Regimental S-4, Major Gordon K. Smith, Baton Rouge, LA who was wounded and captured in a close battle, then sent to a POW camp.

Some groups grew larger as they moved about, doing battle. Regimental Commander, Colonel George Van Millett, dropped on DZ T and, with about 40 men, attacked Amfreville to south. German firepower and stubborn resistance forced Millett to skirt around to the west, then to a position southeast of Le Landes. There, he contacted the division. Captain Allen Taylor brought in 250 troops, after slugging it out with the Germans in Fresville. F Co. CO, Captain Paul F. Smith, now major general, retired, Melbourne, Florida. I Co. CO, Captain Allen and Captain Frank brought other troopers and, by D+3 Millett's group numbered 425. Millett was in his assigned defense area, determined to stay and to clear it out. He patrolled and ambushed the enemy and survived heavy pounding by artillery, mortars and small arms fire. The group held out for a period of five days, in the middle of the German 91st Division. D+3 would be a fateful day for them.

Landing east of Amfreville, 2d Bn. CO, Lt. Col Charles J. Timmes, now major general, retired, Fall Church, VA, was able to collect about 30 paratroopers, among them D Co.'s Cpl Arthur L. Sorge, Parma Heights, OH. Timmes could hear the sounds of battle toward Amfreville (probably Millett's attack) and, thinking that F Co could be attacking the town, decided to divert the attention of the German defenders by attacking from the east.

The same overwhelming German strength that repulsed Millett also forced Timmes to fall back to a point northeast of Le Motey where he set up a perimeter with his back to the flooded Merderet River basin. He had no communications equipment with which to reach higher headquarters or other groups. Hoping to gain trooper strength to assault Amfreville during daylight, he sent a patrol led by Lt. Levy to outpost the western approach to the La Fiere causeway at Cauquigny. Levy reported it clear of enemy.

TAKING THE LA FIERE CROSSING

Having landed in the treacherous waters of the inundated Merderet basin and reached the railroad embankment to the east, a group of about 45 officers and men assembled around G Co.'s CO, Captain Floyd B. "Ben" Schwartzwalder, of Syracuse, NY and currently of St. Petersburg, FL. Schwartzwalder (later National Champion Football Coach at Syracuse University and a retired reserve officer) moved south along the railroad to its junction with the road leading to the La Fiere Bridge. Anxious to get to his assigned sector in the Amfreville area before daylight, he peeled off from a much larger force commanded by 508th regimental commander, Colonel Roy Lindquist and headed west, toward the manor at La Fiere and the bridge which lay just beyond.

With G Co.'s 1st Lt. John Marr, now colonel, retired in Arlington, VA, Corporal William Lawton, T/5 Gaspar Escobar of Pico Rivera, CA and Private Marion Parletto, of Harper Woods, MI in the lead, Schwartzwalder's group, which had grown to about 80, ran into heavy fire from the manor. The group moved to the left as Lindquist arrived and mounted an attack on the manor. Marr's advance party, attempting to skirt the manor on the south, came under fire from a German machine gun at the corner of the manor grounds, 15 yards away. They went to ground with Escobar and Lawton hit in the legs. With a hail of hand grenades and tommy gun fire, they quickly silenced the gun and crew. After a pause to get help for the wounded, they continued around to the bridge unopposed as elements of the 505th and 508th broke through to clear the manor grounds.

Linquist directed Schwartzwalder to lead off in the advance over the bridge and the 800-yard causeway which looked to be clear of enemy. Scouts Johnnie Ward and James Mattingly led the way. Ward was 75 yards, Mattingly 10 yards past the bridge and both unaware that Mattingly was abreast of well-concealed MG-42 gun pits on both shoulders of the causeway. A German arose, from the north pit, and fired his rifle but missed Mattingly at point-blank range. Simultaneously sensing the motion of his right, Mattingly whirled and, shooting from the hip, killed the German with an 8-round clip. With lightening speed, he dropped his rifle, flattened to the roadbed and lobbed a grenade into the gun pit. Three severely wounded enemy threw up their hands and, behind Mattingly, was a paratrooper consummately prepared. His entire action consumed not more than 15 seconds.

Schwartzwalder and his group, mostly G Co., reached the Cauquigny churchyard at the western end of the causeway in the early afternoon, certain that Lindquist would follow up with the main body of troopers. Learning from Lt. Levy about Timmes' position in the orchard north of Le Motey, on a straight line to Amfreville, he decided to push on to his objective. His decision had hardly been made when a German ambulance drew up from the south and paused at the crossroads. Its doors were wide open to make visible both American and German wounded. Then it sped off to the west toward Le Motey. Within minutes, German artillery was raining on the junction and the rumble of armored vehicles could be heard in the distant southwest.

There was still no indication of follow-up forces from the east bank. By nightfall a regimental-size German force arrived to battle for the La Fiere causeway and bridge. Schwartzwalder's group and Levy's outpost closed into Timmes' perimeter, helping fend off attacks from two directions and swelling Timmes' ranks to 175. During the night the Germans mounted the first of their vicious attacks to dislodge the Airborne forces holding the eastern end of the La Fiere crossing.

TAKING THE CHEF DU PONT CROSSING

On the east bank of the river basin, Lt. Col Maloney and 1st Bn CO, Lt. Col Edwin Ostberg assembled about 150 troopers near Brigadier General Jim Gavin's Force A headquarters. Among them A Co. Squad Leader Sgt. Halsey H. Dod, Jr. remembers their initial engagement with the enemy was protecting the dawn arrival of glider elements. Gavin ordered them to move southward and seize the La Fiere bridge.

Destroyed German tank on La Fiere Causeway. (Courtesy of Martin J. Tougher)

As they reached the road and rail junction where Schwartzwalder had earlier turned toward the La Fiere Bridge, they received scouting reports that the bridge was under enemy fire.

Gavin ordered Maloney to make a wide, encircling movement to cross the river near Chef Du Pont and drive the enemy from the bridge. Maloney had left with 150 troopers when Gavin learned that the Chef Du Pont bridge was clear of enemy. He, Ostberg and E Co. CO Captain Roy Creek, now colonel, retired, Lake Quivira, KS, hurried directly to Chef Du Pont with about 75 troopers and arrived ahead of Maloney. In a two-hour fight, they drove the enemy from the town. In a quick rush to dislodge the enemy east of the bridge, Ostberg was severely wounded. Maloney arrived and put Creek in command of the 507th elements of his force. They gained the east end of the bridge but could not cross against the withering fires of the German. By 1700 hours, as stalemate seemed likely, Maloney was called to La Fiere with all available troopers, leaving Creek with 34 men and the problem at hand.

Deadly accurate enemy artillery fire came down on Creek's position, killing or wounding 14 of his 34 defenders. This was followed by a message, delivered by a division staff officer, telling Creek to hold the bridge at all costs. It was obvious to Creek that, while it couldn't cost much more, there was some doubt about being able to hold something he didn't have. To add to his woes, an observer in a nearby building reported a company of Germans moving around to his left rear. He turned to the staffer from division and asked for reinforcements. As if by prayers answered, C-47s began dropping much-needed guns and ammunition. Within 30 minutes, 100 men with a 57mm gun arrived.

After putting all available weapons into play, Creek, on a prearranged signal to lift fires, stormed the bridge with the men and rushed to establish a defense at its western approach. The small force held and, at midnight, elements of the 5th Infantry Division, pushing inland from Utah Beach, were in contact. Having lost to Creek's courageous stand at Chef Du Pont, the enemy focused on the La Fiere crossing upstream.

DEFENSE OF THE LA FIERE CROSSING

After seizing the western approach to the La Fiere causeway, the Germans launched a determined tank-infantry attack, supported by heavy concentrations of artillery and mortar fires, along the causeway to retake the bridge. Deadly accurate fire from 57mm antitank guns on the east bank littered the causeway with Mark IV tanks and other enemy vehicles. Dead German infantry were piled up around them making it difficult for German reserves to function.

The attack failed but was renewed on D+1 and D+2. From the start, Lt. Col. Maloney and the three ad hoc companies he formed with 507 troopers, shared the defense of the east bank at La Fiere with the 505th. They took round-the-clock poundings from enemy tanks, artillery, mortars and small arms fire. It was here on D+2 that G Co Squad Leader Sgt. George B. Tullidge, Staunton, VA, was killed in action.

Daily casualties ran high. The experience of G Co Medic T/5 Martin J. Tougher, Pittsburgh, PA is indicative. He was taking a breather with Medic Joe Kolek, Philadelphia, PA on D+2 when the call came from a gliderman who wanted help for his buddy who had a slit throat. Going toward the La Fiere bridge to help the wounded man, Tougher found himself in the midst of battle with shells flying and other casualties to treat. First was a soldier with a buttock wound. Next was Sgt Hohn of G Co. with a large hole in his jaw, a trooper with a head wound, then a gliderman with an arm blown off. He came upon Medic Sgt. Henry Lysek, Gulf Breeze, FL and asked him to go with the gliderman whose buddy was the throat victim. The Germans were still there as Lysek rounded the hedge row. The gliderman killed the Germans but Lysek found the throat victim had died.

German forces battling for the causeway and bridge at La Fiere and against Millett's group near Amfreville, probed and patrolled heavily against Timmes' position near Le Motey on the night of D-Day, D+1 and D+2. Intermittent heavy shelling by artillery and mortars made attack seem imminent. That a major attack to destroy the position did not occur could be attributed to the German preoccupation with Millett and the bridge. Timmes patrolled the Germans at Cauquigny, Le Motey and Amfreville but he still had no way to talk with Chambers, now lieutenant colonel retired, Dublin, OH. Chambers swam the river, reported Timmes' location to division and returned bearing instructions to hold the position.

RETAKING THE LA FIERE CROSSING

Still in need of contact with higher headquarters on D+2, Timmes directed Lt. Marr to make contact with friendly forces across the flooded river basin. Marr and his platoon runner, PFC Norman Carter started at noon and stumbled upon a knee-deep stone road that led them northeast to the railroad embankment. A boat and jeep ride later, they were in the 82d Division CP where it was decided to send a battalion of 325th Glider Regiment across the sunken road at night to attack the rear of the Germans holding Cauquigny and the western end of the La Fiere crossing.

Carter returned to tell Timmes of the plan and Marr stayed to lead the glidermen to Timmes' position. Their arrival gave Timmes his long-needed communications with division but, for a variety of reasons, the night attack failed. Unable to take the position and losing the cover of darkness, the glidermen were severely mauled by the enemy as they fell back to Timmes' flank. Casualties would have been higher had not 1st Bn. Mortar Platoon Leader 1st Lt. Willard "Tex" Young, now colonel, retired, Port Neshes, TX, and his crews delivered 81mm mortar concentrations with surgical precisions to cover the withdrawal.

The 90th Division was pushing in from Utah Beach and the La Fiere crossing was needed for their attack westward. Failure of the night attack meant the Cauquigny approach to the causeway must be taken from the east side. Another glider battalion was tapped to do the job. General Gavin was concerned that any hesitation or loss of momentum might quickly turn into defeat. He alerted Maloney to be instantly ready on signal to carry the attack across the 800 yards of unshielded causeway. Maloney had three 507th composite companies, led by captains Robert D. Rae, Birmingham, AL, Morgan Brakonecke and Roy Creek. Rae would lead, followed by Brakonecke, if needed.

The causeway became an inferno of smoke, fire and deafening noise as the leading company of the 325th moved to the attack. Only a fraction of them made it around the clutter of knocked out tanks and other equipment on the causeway to reach Cauquigny. After a time, the columns of returning wounded seemed to equal the columns of attackers along the coverless causeway. Gavin, Maloney and Rae searched for signs of buckling that might send Rae's 90-man company rushing into battle. Then a Sherman tank with the attack echelon hit a mine and piled into a disabled German tank, nearly blocking passage on the causeway. At that point, Gavin waved and shouted at Rae to get going. Maloney, heedless since his face was covered with blood from a wound, moved onto the bridge to exhort his men forward and insure that Rae would be followed. Maloney's radio operator T/4 Tom Tierney of Gardner, IL states, "I can't describe the braveness of this man. He was the greatest."

The acts of leadership and courage displayed by Captain Rae, G Co.'s Lt. Dave Irwin and other subordinates, inspired glidermen and paratroopers alike. The casualties did not lessen but momentum was gained and the heights at Cauquigny were wrested from the Germans. The attack carried to Le Motey where 1st Lt. Stanley Ardziejewski (Arjay), Hemet, CA led a patrol to contact Timmes. Pushing into the lead, Rae's Company fought through Le Motey and he dug in for the night. As the dawn of D+4 arrived, the 90th Division passed through Rae's positions to renew the attack.

Meanwhile, at 0300 hours D+3, Millett began his Division orders and moved to join up with Timmes. He chose a route that would take him around Amfreville to the north but ran into a strong German position. When the battle was over, Millett and half of his group had been captured. Captain Paul Smith took command of the remainder of the group and proceeded eastward under heavy pressure from the enemy. He reached the railroad and turned south toward Timmes position but was blocked by a powerful German force at La Pesquerie. After heavy fighting, Smith prevailed. That day, D+4, Smith's group of 140 and Timmes' group both assembled with the regiment east of La Fiere to rest and refit.

THE BATTLES OF GRAIGNES

While the 507th groups to the north were battling it out, Hq. Co. 3d Bn. and others that had been dropped at Graignes 16 miles away were focusing on defense of the town. Most of the troopers had landed in flooded areas up to five-feet deep. Many, tangled in their suspension lines, drowned. Survivors, aided by bright moonlight, gravitated to an old 12th century church, outlined against the sky. Stragglers continued to come into the area from all points and by D+1, approximately 180 troopers were positioned in and around Graignes.

As the group grew, activities increased. Defenses were dug, security was posted and patrols were dispatched to locate the enemy. The bridge over the only crossing into the town from the east was prepared for demolition. The acting mayor held a town meeting where it was unanimously determined that complete moral and logistical support would be afforded the troopers. The villagers recovered equipment, supplies and ammunition from the swamp areas, provided food and water to keep the troopers going and supplied important information on German activity. Finally, they contributed greatly in repulsing an attack on 11 June, in which the troopers were outnumbered more than 11:1.

Throughout the period 6-11 June, the Germans initiated light-to-heavy probing attacks. On the 10th, the troopers blew up the bridge just as a large German patrol commenced crossing. After an across-the-water fire-fight, the Germans withdrew. During 10 o'clock mass in the old Roman church, Sunday, 11 June, the Germans made a concentrated attack, then another, followed by probing attacks throughout the day, all of which were repelled. Finally, a fully coordinated, mass assault was launched by the Germans with an estimated 2,000 troops, supported by 75mm and 88mm artillery, heavy mortars and machine guns. The troopers fought back with only light machine guns, mortars, hand grenades, carbines and few rifles. The fires they traded with the Germans were so intense that, prior to midnight, their ammunition was exhausted and the few remaining able-bodied troopers were no longer a match for the German force.

When the Germans penetrated the defenses, an organized withdrawal was not possible. Isolated individuals and even squads slipped into the swamp to make their own way north to the American lines. When Bn. XO Major Charles Johnson was killed it became the task of S-3 Captain Leroy D. Brummitt, now colonel, retired, Satellite Beach, FL to collect as many troopers as possible (80) and lead them back to the 507th. In this endeavor, he was ably assisted by 1st Lt. Francis E. Naughton, now colonel, retired, Norcross, GA and 1st Lt. Earle R. Reed, now lieutenant colonel, retired, Fayetteville, NC.

Sgt. John J. Hinchliff, Coon Rapids, MN remembers it this way: "My machine gun position was behind a hedgerow at the edge of the village. From time to time during our short stay in the village, German patrols came through and were eliminated. On 11 June, we had to withdraw, leaving our dead and wounded behind. I knew we had suffered severe losses when I saw that my assistant and myself were the only two troopers in the vicinity...we made our way to Carentan where we joined up with an American armored patrol which came up from the beachhead. The following day we joined up with elements of our regiment."

Sgt. Charles J. Bucsek of Pittsburgh, PA recounts that he dropped far off the intended drop zone but managed to miss the swamp area. When the Germans penetrated the defense on 11 June, the villagers helped him escape through the flooded area by boat to Carentan, after which he joined the 507th in the vicinity of Vindefontaine. Arthur J. "Rip" Granlund remembers landing in an apple tree and briars after jumping from a very low, fast flying airplane. He led about 36 other troopers to the high ground where they found several equipment bundles. They recovered all of the 81mm mortars, several machine guns and lots of ammunition needed in the defense of Graignes.

LA BONNEVILLE AND THE BAUPTE PENINSULA

After a three-day respite, the 507th was ordered to relieve the 90th Division and continue the attack west toward St. Saveur Le Vicomte. On 14 June, as Captain Brummitt and his Graignes survivors rejoined the Regiment it attacked westward in a column of battalions. The 3d Battalion, in the lead, turned north to block Renouf which was to be taken by the 9th Division. The 3d drew the concentrated fires of mortars, artillery and machine guns, killing G Co. 1st Sgt. Bello, Sgt. Prescott and Lt. Erwin whose courageous leadership had been the key in the La Fiere battle. 1st Plt. Sgt. Burman was severely wounded and returned to the U.S.

The heavy enemy fires shifted to the 1st and 2d Bns. as they continued westward, carried the attack to a small ridge just south and east of La Bonneville where they remained under heavy mortar and artillery attack throughout the night. They pushed forward the next morning and gained 1,000 yards only to find that units to the right and left had not pushed forward to cover the flanks of their attack. Friendly artillery fires fell on their positions to add to the pounding they were taking from the enemy. They held out until 1800 hours when elements of the 505th Regiment passed through to carry the attack westward to St. Saveur Le Vicomte. During the day, Colonel Edson D. Raff took command of the regiment. Losses for the two days were 192.

During the night of the 16th, the regiment moved southeast by foot and truck to relieve the 508th Regiment moved southeast by foot and truck to relived the 508th Regiment in a so-called "quiet sector" and to establish defenses on the line of Francquetot-Coigny-Baupte. It was not quiet for long as the Germans shelled the defenses with artillery and mortars day and night. Enemy positions were patrolled and identified for the attack which was mounted by Major Ben Pearson and his 1st Bn. to seize Vindefontaine on the night of 18-19 June. They were tasked to defend the town until 4 July. The rest of the regiment was passed through by the 508th on 20 June, then relieved on the Baupte Peninsula by the 90th Division on 25 June.

Enemy harassing fires continued to inflict casualties on the battalions in their new areas east of Vindefontaine. The 3d Battalion adjutant, Lt. Wagner, was killed, and the CO. Major John T. Davis was wounded when an enemy shell landed in the CP. Lt. Col. Maloney assumed command of the battalion.

Reconnoitering for enemy presence occupied much of the time. H Co. PFC, Raider E. Nelson, Des Plaines, IL, remembers patrolling in the Vindefontaine area along a hedgerow to reconnoiter a heavy woods with PFC Ulrich Thompson and Sgt. Frederick Hebbe. Short of the woods they took a breather and, looking to their rear, noticed a hug object floating toward them. They identified it was a barrage balloon that had broken loose on the beach. It drifted slowly over them as they grew close to the woods. Enemy in the woods opened fire with rifles and machine guns. The patrol went to ground, then realized the enemy was firing at the balloon. With this enemy disclosure the patrol made a successful return to H Co.

LA POTERIE RIDGE

On 3 July, D+27, 1st Army launched its attack to clear the Cotentin Peninsula. In the vanguard of the attack was the 82d Division with the 507th, minus the 2d Bn in reserve. The 2d Bn, with its own sector in the attack echelon, moved swiftly to its objective and, on the following day, reached Blancelands. While paused there, General Gavin turned them southward to seize the center section of La Poterie Ridge where two other battalions had been repulsed. Here, he attached the 2d Bn to the 508th. With his battalion down to about 200 troopers, Timmes sent a patrol, led by F Co. 1st Lt C.T. Harris to reconnoiter a pass leading to the reverse, south slope of the ridge. Harris returned by 0100 hours and reported no enemy contact, whereupon the entire battalion followed the patrol back through the pass and arrived at the base of the southern slope at dawn on 5 July.

Timmes reorganized and quickly headed his battalion up the grassy slope of the ridge. As they reached the wooded summit the Germans inflicted heavy fire on them. Timmes led E Co. of about 11 men on a charge into the left flank of the German position. He drove them from the ridge and organized a southern slope defense which was later reinforced with a tank company. The response was immediate and prolonged. Intense enemy artillery and mortar concentrations continued to fall on the position for several days. Later in the morning of the 5th, Timmes found the enemy attempting to infiltrate his flanks and suggested to Gavin that other units be dispatched to secure the flanks.

D Co.'s Irvin N. Holtan, El Paso, TX, recalls: "When I raised up above the weeds to see how I was doing, a German raised up to do the same. My M-1 was in my hand on the ground. I snapped a shot at him and missed, and saw my empty clip fly out. Although my heart dropped, I held my empty gun on him, and in my best German told him to put his gun down, and his hands up. His chin about fell to his belt as he knew I had the drop on him, but for some unknown reason didn't shoot again. I didn't tell him the whole story."

Timmes' call for flank protection was quickly translated into orders that sent Lt. Col Maloney's 3d Bn. up the eastern slope of the ridge where, at midnight, they came squarely into a strong German defense. The troopers of G and I companies hugged the earth in the darkness of the deep woods to stay below the German machine gun fires, then were forced to move about as volley after volley of potato-masher grenades came exploding among them. They fought back with rifles, BARs and their own grenades. 1st Lt. Pip Reed kept his machine gun platoon firing from the boundary between G and I companies until their gun barrels were white-hot. Prevented by the thick woods from using his 60mm mortars, G Co. Mortar Squad Leader, Sgt. Gene R. Sylvester, Williamsburg, MA, joined in the grenade battle.

The German position would not yield. With his battalion flanks blinded in the expansive woods and casualties piling up, Maloney sideslipped Captain Ben Schwartzwalder and his G Co. around to the south base of the hill and took the rest of the battalion around the north base into the saddle of the ridge behind the German position. The two forces met at 0830 hours, 6 July, and Maloney contacted Timmes by patrol. An extensive gap still remained between the two battalions. On 7 July, Colonel Raff ordered the 1st Battalion to pass through the 325th Glider Regiment, seize the gap area and protect the flanks of the 2d and 3d Bns. Maloney and Major Pearson moved forward to reconnoiter a position for the 1st Battalion. The Germans opened fire and both commanders were hit. Maloney was badly wounded and evacuated to the U.S. The 1st Bn. seized the gap and, that night, the three battalions dispatched a company each to establish an outpost line (OPL) along the valley road to La Haye Du Puits.

B Co. CO, Captain James F. Brown, Allentown, PA, remembers that his company made the deepest penetration of the 507th's advance into France when it occupied its portion of the OPL. After crawling through field after field that contained enemy dead and hearing the Germans pull out as they advanced, B Co was strafed and bombed at the OPL along the main road into La Haye Du Puits. Brown remembers the difficulties experienced in assisting unseasoned troops as the 8th Division passed through the 507th on 8 July. B Co.'s O'Niel Boe, Reserve, LA, was awarded the Silver Star for gallantry on 10 July while, under fire from the enemy, he recaptured an artillery radio that had been captured from 8th Division elements passing through the B Co. OPL.

The 507th stayed on La Poterie Ridge until 11 July. As it left for the beach and transported to England, the regimental force of 2,004 troopers had been depleted by 938 casualties in 36 days.

ARDENNES CAMPAIGN

The 507th Parachute Infantry Regiment, entered the Ardennes Campaign 26 December, 1944. The purpose of the campaign, more commonly known as the Battle of the Bulge, was to halt the 16 December offensive thrust of the German 5th, 6th and 7th Panzer Armies into Belgium and continue operations to destroy the German army.

Between campaigns, the 507th was assigned to the 17th Airborne Division, under which it refitted, retrained and moved camp from Nottingham to Tidworth Barracks to Barton Stacy in southern England. The troopers had trained hard and were looking forward to having a full-spread Christmas dinner in camp. It was not to be, as Christmas Eve found the regiment encamped at the departure airfield. Christmas Day airlifts to France began, then shut down at mid-day because of wing-icing conditions.

H Co.'s Alton, D. Maharg, now master sergeant, USAF retired Lothian, MD, remembers eating "cold turkey" in Northern France. I Co. PFC Herman R. Neptune, Waterloo, IA, reported surveying what he had received for Christmas as he was awakened at 0630 to pack up for departure: "One Musette bag with toilet articles, rain coat, winter underwear, winter overcoat, light-weight paratrooper coat with lots of pockets, switchblade jump knife, bedroll, wool sweater, gas mask, shovel, six boxes of K-rations, canteen, mess kit, three hand grenades, steel helmet and a duffel bag." After sitting on a plane for three hours, Neptune was rewarded with a hot meal and returned to his tent to await the 26 December shuttle.

The regiment was shuttled to a small airfield in the vicinity of Mourmelon, France. From there they were immediately moved by truck to outpost positions along the Muese River from Nousem to Givet. While in these positions, all units conducted extensive patrolling, searching built-up areas, woods and the like for German parachutists of the special operations force headed by the legendary Otto Skorzney. The story was they had been dropped in American uniforms in the vanguard of the German counteroffensive, to report on and mislead American movements and, otherwise, sow confusion among the Americans. On 2 January, the regiment was ordered to the Bastogne area in the vicinity of Chenet, Belgium.

During the next three days, the 507th, in Division reserve, moved in the direction of Bastogne without making enemy contact. Deep snow, heavy fog and bitter cold had settled into the wide sweep of the hilly, wooded Ardennes. The cold and wet conditions began to test trooper endurance, even in the absence of the enemy contact. Heavy cloud cover and enveloping fog banks created a claustrophobic effect that began to wear on the troopers as the constant harassment fires from well-registered enemy guns and mortars could not be answered by the vaunted Allied air support.

THE BATTLE ON CAKE HILL

On 5 January, the 3rd Bn. occupied Acul and Pinsamont with little or no opposition and halted while the rest of the regiment continued northward to the area of Chenagne. On 8 January, came the first major physical contact with the enemy. In an effort to achieve observation over a major enemy withdrawal route northwest of Bastogne, the 3d Bn. was ordered to move several thousand yards and occupy Hill 450 (later dubbed "Cake Hill") near Chisogne, well ahead of the 17th Airborne Division's forward positions. Beginning at 0200 hours the battalion moved across the snow-covered fields without opposition, arriving on the hill at the first streak of dawn.

Battalion Commander Major John T. Davis and the battalion CP moved into a thick woods on the center of the hill. G Co., commanded by Captain William J. Miller, Bell Glade, FL, occupied the back side of the hill. H Co., commanded by Captain H. A. "Big Steve" Stephens, occupied the left crest. The right crest was occupied by I Co., commanded by 1st Lt. John T. Joseph. It looked out over a deep, wide, wooded ravine. Higher ground beyond that masked the German withdrawal route. Before positions could be dug, the enemy delivered a devastating series of time-on-target (TOT) artillery concentrations, killing Major Davis and many of his CP staff.

I Co.'s Donald J. "Pat" O'Buckley, Horseheads, NY, recalling that Cake Hill was his baptism under fire, thought that not even a rabbit would be able to live through the TOTs as the whole hill lighted. Then an enemy machine gun opened and swept the area in which I Co. was digging its positions. Milt Blackman, Melville, NY, an I Co. messenger, remembers digging in among the trees of the hill when "all hell broke loose" knocking him into his foxhole where he could not move. He remembers being shelled again at the aid station and awaking two days later in a hospital. H Co. Runner Pvt. Oliver C. Foote, Petersburg, VA, also remembers the deadly German artillery attack on the battalion CP in the woods and staying on the hill under continuous enemy fire.

By being out of the woods, checking troop dispositions, Bn. S-3 Captain Leroy D. Brummitt, was spared. He immediately established a new CP in a snow-filled shell crater on the back side of the hill. With only a radio and operator, he kept protective fires coming and with the aid of troopers from the companies, began the evacuation of the wounded and dead from the woods. A day-long blizzard with heavy drifting snows descended on the Ardennes, making it difficult to locate and evacuate the casualties. German probing attacks and their constant pounding of the position with tank, artillery and mortar fires, did not make the process any easier. The jolting loss of Major Davis was magnified a few hours later with the loss of much-revered regimental and Catholic Chaplain Father John J. Verret. He was killed while loading wounded on an ambulance at the rear of the hill.

Shortly after noon Captain Brummitt relinquished command of the Cake Hill force to Bn. XO, Major Allen W. Taylor who arrived from the rear CP. Absent further orders and unable to achieve direct observation over the German withdrawal route, the 3d Bn. battled the blizzard and prepared to give a good account of itself in the event of a major enemy attack. Poor visibility and drifting snow made communication between foxholes and gun emplacements exceedingly difficult.

To the right rear in the hamlet of Laval, enemy tanks behind buildings darted out long enough to fire a few rounds at Cake Hill before darting back to cover. The thought of a tank-led assault from that quarter was ever present. Bn. Mortar Platoon Leader First Lieutenant Auten Chitwood, Osceola, AR, and his 81mm mortar crews, rained down tons of rounds in protective fires to help hold the tanks at bay and repulse the probing attacks.

Just before midnight the German main

attack came. An estimated company-size force in snow caps crossed the ravine, heading directly for the I Co. positions. As the enemy crept closer, I Co. held its fires. As the intruders neared the edge of the position, Joseph gave the signal and his Company loosed devastating fires from their front and flanks that killed or seriously wounded nearly all of the enemy force. The few who gained the position were evicted by hand-to-hand combat.

This well-disciplined action was hardly over when the order came for withdrawal of the 3d Bn. As the force began their silent march to the rear, the sky over Laval suddenly lit up and the clear outlines of a tank on fire were seen. Lt. Chitwood and his men had finally bagged a tank with an incendiary round down the turrey hatch.

Major Roy Creek assumed command of the battalion and Major Taylor resumed his duties as XO. Among those decorated for leadership and courage during the action on Cake Hill were G Co. Technical Sgt, Joseph E. Pawlik, Buffalo, NY (Bronze Star) and 1st Lt. Joseph (Silver Star).

CLOSING ON THE GERMAN FRONTIER

The self-propelled (SP) gun of tank destroyer (TD) units was a new weapon for the 507th in the campaign. Typically, the Germans counterattacked with a company of infantry, supported by four or five tanks. As the 507th continued in division reserve, five tanks attacked the 1st Bn. After some difficulty in communicating with the TD unit, each attached SP gun was put under control of a 1st Bn. officer. Three of the tanks were knocked out and the other two fled.

The 13 of January dawned clear and bright. Allied fighter/bombers worked overtime on the retreating German armor columns as the 507th launched an attack to the east. S/Sgt. Joseph Faust recalls: "The long liens of German Tiger tanks lined up along the edge of the road as far as you could see. Most of the tanks appeared to be in good shape. We thought that the Germans had a choice. They could stay in the tanks and be cremated or they could evacuate and retreat." Before the day ended, Flamizoulle, Frenet, Givery and Gives were under regimental control.

The attack continued the following day and the 1st Battalion reached Bertogne where they were subjected to heavy, day-long enemy shelling and took several casualties. By 1030 hours the 3d Battalion had reached the Ourthe River where they met the British 51st Highland Division on the north side of the river. Pushing onward against the withdrawing enemy the 507th relieved the 193d Glider Regiment southeast of Houffalize on 17 January and remained in positions along Cowan Creek until 20 January. On the 20th the regiment crossed Cowan Creek with the 1st and 3d Bns. abreast and, in its final offensive action in Belgium, attacked northeast and seized Limerle. After a passage of lines by the 513th Regiment on the 21st, the 507th remained in position until the 26th when it moved to Luxembourg.

Throughout the Ardennes combat action, German troops made skillful use of land mines, antipersonnel mines and bobby traps of all kinds. Many tanks, trucks and advancing troops were disabled by these devices as the withdrawing Germans used them to slow down the Allied advance across the frozen terrain. Falling victims to an antipersonnel mine as the 507th advanced to Limerle, was Major Roy Creek who was hospitalized but later rejoined the regiment. He was succeeded in command of the 3d Bn. by Major Allen Taylor.

As the regiment moved into Luxembourg to take up defenses along the Our River, a land mine blew up Lt. Col Ben Pearson's jeep. Critically wounded, he was evacuated to the U.S. Major Paul F. Smith assumed command of the 1st Bn.

As the regiment closed on the Our River the Germans withdrew across the river in the north part of the regimental sector. Here they clung stubbornly to a bridgehead west of Dasburg that extended southward to Rodershausen, about 2,000 yards. On the east bank of the river the enemy prepared extensive defenses along a line of large concrete bunkers.

The sudden thaw at the end of January made crossing the swollen currents of the Our treacherous. It contributed to the strength of the German defenses, and they withdrew the 5th Parachute Division's hold west of the river then blew the bridge. But Lt. Trapp and four men crossed a footbridge that had escaped demolition, shot a German and returned without loss.

As the battalions rotated between the two front-line positions, the river and the reserve position in Clerveaux during the early part of February, they traded shots with the Germans across the Our. The 2d Bn. was then ordered to establish a bridgehead east of the river. With a platoon of the 139th Engineers in support and using boats, E and F Cos. managed to cross the swollen river with much difficulty. Once on the other side, the going was made difficult by mines and booby traps.

After cutting through three barbed wire entanglements, F Co. CO 1st Lt C. T. Harris, led a night combat patrol to destroy a bunker. The moonless night made it impossible to find a proper place for an explosive charge but the patrol captured a prisoner. Harris tried again in daylight, only to be pinned down and received several casualties. Nightfall permitted him to withdraw his prisoner and diminished patrol.

The 3d Bn. relieved the 2d Bn. on 9 February, and repulsed a German infantry attack supported by heavy assault fires. The 3d Bn. and the entire regiment were then relieved by elements of the 6th Armored Division. Having suffered 741 casualties in 45 days of combat in Belgium and Luxembourg, the 507th moved to a bivouac area at Chalons, France.

AIRBORNE INVASION OF GERMANY VARSITY ASSAULT

As the 507th withdrew to the tent camp at Chalons, France, along the Marne River, few of the troopers knew that they were slated to spearhead the Airborne invasion of Germany in a few weeks. Suspicions began to grow quickly, however, as little time was given to cleaning and repairing combat gear before field training was underway. While three-day passes to Paris were to be had by a few lucky troopers, hard, realistic training was the rule of daily camp life. New troopers were received to replace the casualties taken in the Ardennes Campaign. Rehearsal jumps were conducted, using the Marne River to simulate the Rhine. Up to this time the 507th had been a separate regiment; attached to the 82d Airborne for the Normandy invasion and to the 17th Airborne for the Ardennes Campaign. Now, in the last of several airborne force reorganizations during the war, it was made organic to the 17th. It was well-seasoned and ready for the next call to combat.

Operation VARSITY employed the U.S. 17th and British 6th Airborne Divisions, under command of XVIII Airborne Corps, in support of a major effort of the British 2nd Army that would spearhead British 21st Army Group's Operation PLUNDER in a push across the lower Rhine, then to Berlin. The plan called for the 15th Scottish Division of British 30th Corps to begin crossing the Rhine at 0200 hours, 24 March. At 2200 hours, 23 March, the 1st Scot Commando Brigade would cross the Rhine then swarm over Wesel, following a heavy bomber attack on the city.

The VARSITY effort was planned to

put the two Airborne divisions down northeast of Wesel and 4-6 miles east of the 15th Division's assault crossings, during daylight, 24 March. This would put the airborne descent behind the German forward river-line defenses and squarely on top of their reserves and supporting artillery. A daylight drop was chosen to facilitate a quick linkup of forces and to avoid the still active night fighters of the German Luftwaffe. The German II Parachute Corps with the understrength 6th, 7th and 8th Parachute Divisions formed the principal defense force in the Wesel area.

The 17th's mission was to seize, clear and secure its area with priority to the high ground east of Diersfordt and bridges over the Issel River; to protect the Corps right flank; and to make contact with the two river crossing forces and British 6th. The 507th was made a combat team (CT) by the attachment of the 464th PFA Bn. and Btry. A, 155th AA Bn. Its mission was to seize its objective; to assist the crossings of the 15th Division by fire; and to assemble along the east edge of the woods when junction with assault units was made. The 507th CT would lead the XVIII Corps assault. D-day: 24 March. Drop time (P-hour): 1000 hours.

Within four weeks of arrival at Chalons, the regiment was on its way to be sealed into marshalling areas at airfields A-40 at Chartres, southwest of Paris and A-79 at Reims, east of Paris. For several days, map and aerial photo studies of the Airborne invasion area east of the Rhine were updated. Detailed terrain boards of the VARSITY drop area and plans for the assault were studied and restudied by all troopers. Equipment was checked and rechecked. The time between briefings, eating and sleeping was filled with physical conditioning, sports, movies and jesting about Axis Sally's latest radio broadcast prediction that the jump they were about to make would end in bloody failure.

D-Day was clear and balmy as the three C-47s transporting Col. Raff and his Regimental Command Group lifted off A-40 at 07171. In the next plane, leading 1st Bn. assault elements, were Bn. CO Major Paul Smith and B Co. CO Captain John Marr, radio operator Charles Walthall, Creve Coeur, IL, other command group troopers, and Major Ben Schartzwalder who would be spearheading the 17th Division military government efforts. Completing the first serial with 41 planes was the remainder of the 1st Bn., slated to go into regimental reserve north of the DZ upon landing. Next in line with 45 planes

Medics Worthington, Truman and Tougher in Normandy. (Courtesy of Martin J. Tougher)

from A-79 was Lt. Col Timmes' 2d Bn., slated to take and hold positions west of the DZ and assist the British river crossing. With 45 planes from A-79, Major Taylor's 3d Bn. lined up behind the 2d Bn. with the mission of seizing and holding the high ground northwest of the DZ at Diersfordt. These 136 planes, all headed for DZ W, would require two hours and 33 minutes to reach the DZ. By the time Raff bailed out to be the first trooper to touch down, the sky armada behind him measured two hours and 18 minutes in length.

A panorama unfolded as the skytrain moved toward the showdown on German soil. The peaceful French and Belgian landscape was greening and farmers were already in the fields. The zig-zag pattern of old trenches came into view—mute testimony of a war that engulfed the world a generation before. Fighter planes, protecting the air column, cavorted about, doing barrel rolls and other acrobatics. Brussels became visible, then a thick haze of smoke from generators, used to screen the British river crossing, enveloped the air column as it crossed over the release point for its final leg to the drop zone.

Suddenly, the air was filled with hundreds of explosions from flak guns, positioned behind the German river line defenses to fire at slant ranges into the low-flying troop carriers columns. Axis Sally's claim to detailed knowledge of the Allied airborne plan and dire predictions about the outcome seemed to have more credibility as flak-damaged planes could be seen spilling out troopers while veering sharply, losing altitude and burning viciously.

The green light for the 1st Serial came on at 0950. The Regimental Command Group and the 1st Bn. left their planes, confident that Col. Joe Crouch's Pathfinder Group in whose planes they rode had brought them over DZ W. En route to the DZ, Crouch had bet Raff a case of champagne that the drop would be made squarely on the DZ. Raff won easily as Crouch's Group dropped all 45 planeloads to the northwest of Diersfordt 2,500 yards from DZ W, atop a host of German artillery positions, and in the 513th CT area of responsibility. The German gun crews inflicted some casualties during the descent but collapsed quickly in the face of return fire from organized groups of troopers. The story was different on DZ W where the 2d and 3d Bns. and the 464th PFA Bn. landed in the midst of heavy fire from machine guns, mortars and small arms that killed many troopers in the air, at touchdown, during assembly and while attacking German strongpoints.

For the first several minutes of a parachute assault virtually all activity is that of individuals. Some DZ actions, while dangerous at the time, end on a note of humor. Others spell out individual acts of unsurpassed courage and sacrifice. The latter characterizes the acts of G Co.'s Private George J. Peters, Cranston, RI. He and 10 other troopers were fired upon while landing and pinned down by a German machine gun crew supported by riflemen. Without orders and armed with only rifle and grenades, Peters stood up and charged the gun position with rifle fire. He was knocked to the ground by a burst from the German gun but got to his feet to continue his charge. Hit again, and unable to rise, he crawled directly toward the enemy fire. In his dying moments, he hurled a grenade that destroyed the gun crew, caused the riflemen to flee and saved the lives of the other troopers. He was awarded, posthumously, the Congressional Medal of Honor.

D Troops' Paul N. Peck, Galesburg, Illinois, had some early, anxious moments after touching down. He recalls hearing a buzzing, reminiscent of bees, right after touchdown. He started to run and fell with his chute still affixed. Observing bullet holes in his jacket and trousers, he quickly learned that the buzzing was the sound of a German machine gun tracking him across the DZ. Without a helmet, rifle or grenades, he ran across the DZ, then helped a mortar crew get some rounds in the air then got a carbine and took up the duties of scout.

Also caught up in the DZ scramble was D Troop's Sgt. Leo J. Bier, Rockford, IL. Immediately behind Col. Raff in the first plane was A Co. Sergeant and Bazooka-man, Harold E. Barkley. He lowered his bazooka bag on its tether, slipped to avoid an equipment chute and saw himself headed, bazooka-bag and all, toward a hole in a barn roof. Helpless to avoid it, he uttered a prayer and came to an unexpected landing in a hay mow. When the door burst open below he was certain it was the enemy but heard someone shout, "Hold it, that's Barkley!"

H Co.'s PFC Francis A. Maksin was wounded during the DZ scramble, sent to several hospitals in France and, several weeks later, got back to H Co. in Essen, Germany. A Co.'s PFC John H. Nichols remembers the skill and courage of the troop carrier pilots as they brought his plane in for the drop at 400-600 feet as it was being buffeted by flak. He recalls with sadness some of his buddies, still in their chutes, hanging dead in the trees at the edge of the woods. This note of sadness was echoed by H Co.'s PFC Gordon Nagel, Tulsa, OK. Blinded by blood in his eyes, caused by the crash of his machine gun across his face when he received the opening shock of his chute, Nagel did not see his landing. As he regained his sight on the ground, he recognized in quick succession, three dead troopers that had been his close buddies.

H Co. PFC, Bob Baldwin, South Plainfield, NJ, reported seeing his platoon sergeant disintegrate in midair when his bag of explosives was hit by a German bullet, and then, someway, pulling himself over a seven-foot fence while under the fire of German gunners. G Co.'s S/Sgt. Joe Faust was a victim of the white flag ruse so often used by the Germans. Faust saw about 50 flag-waving Germans in the woods. As he advanced to take their surrender, the flags dropped and guns came up, with a hail of fire that sent Faust back to the U.S. with a bullet in the thigh.

Having landed in the wrong area, the 1st Bn. began to battle the enemy in three groups. One group under Cololnel Raff captured a number of artillery gun crews en route to the Diersfordt area. Another group of about 150 troopers assembled with Capts. Harvey of Hq. Co., Marr of B Co. and Joseph of C Co. and headed north to assemble in the woods only to discover that they were clearing German artillery crews from the objective of 2d Bn., 513th CT. Reversing direction, they turned southeast and, by 1100 hours, arrived at the northwest edge of the Diersfordt Castle grounds. Here, they and the Raff group joined Major Smith who had collected about 200 troopers, the bulk of them being A Co. commanded by Captain Albert L. Stephens, Gulfport, FL. Smith's group was locked in battle with a powerful, well-supported German force.

Smith and his group had reduced some well dug-in enemy positions in the Diersfordt area when they were fired on by tanks. They launched an attack toward the castle and knocked out two Mark IV tanks in their first use of the new 57mm recoilless rifle. With about 90 percent of this battalion available and under orders from Raff to continue the attack on the Diersfordt stronghold, Smith kicked off his attack just as the 3d Bn. arrived with I Co. in the lead. Raff halted Smith's attack. He retained A Co. as a maneuver base for a 3d Bn. attack on the castle and dispatched the rest of the 1st Bn. southeastward to assemble in regimental reserve, as originally planned.

Meanwhile, the 2d Bn. had to fight its way off the DZ, reducing enemy strongpoints one by one. Range cards on captured 82mm mortars showed the mortars had been firing on DZ W. Houses used for gun emplacements had to be taken. Once the woods southwest of Diersfordt were reached, however, the Bn. had little difficulty organizing its positions. By 1434, F Co. had contacted the British river-crossing forces. The 464th PFA followed the 2d Bn. onto DZ W and, to the surprise of the Germans, began reducing points of resistance on the DZ with direct fire from their howitzers. With steady accumulation, 10 of its 12 howitzers were in support of the 507th within 3 1/2 hours.

The most determined enemy and the last to fall in the 507th area of responsibility were those entrenched in and around Diersfordt Castle. The 3d Bn. fought its way to the castle and then room by room in an hour of heavy fighting secured all but the turret of the old fortress. By late afternoon, G Co. began its assault on the turret where a large group of officers held out.

Wounded and unable to continue in the fight, Captain Bill Miller passed to S/Sgt Bill Consolvo a faded flag. Earlier that morning, before takeoff, Miller had displayed the flag to the G Co. troopers and vowed that it would be flying high over Germany before day's end. Now it was up to Consolvo. When the German colonel in the turret was severely burned by a trooper's toss of a phosphorous grenade, the holdouts came rushing forth. The castle fight yielded nearly 500 prisoners, five tanks and a host of other weapons. Consolvo climbed the turret, unfurled the flag and secured it to the highest tower. G Co troopers, busy consolidating their positions, stopped long enough to give a proud salute to the victory flag.

The hoisted flag was a fitting punctuation for the last major skirmish of D-Day for the 507th CT. Against a powerful German force that had every advantage of terrain and choice of position, the CT had accomplished all of its missions with distinction. It had converted the adversity of a mis-dropped battalion into the opportunity of an early assault on its main objective. In addition to the five tanks bagged by the 1st and 3d Bn. the regiment corralled 1000 prisoners and captured or destroyed several batteries of artillery, plus countless other weapons, vehicles and equipment, in less than six hours. In the 17th Division area only minor points of resistance were left to be mopped up, a firm link with the British forces had been forged and bridges across the Issel River had been taken for the drive eastward. Operation VARSITY's airborne phase was all but over.

D+1 (25 March) the 3d Bn. was in reserve as the 1st and 2d Bns., with TD guns attached, cleared enemy pockets from the area between the 507th, Scot Commandos at Wesel and British forces making a new crossing of the Rhine, south to north, about 5,000 yards west of Wesel. This put the regiment on Phase Line (PL)

London, facing east. The airborne drop had collapsed German defenses east of the Rhine. To forestall their building a defense elsewhere, speed would be of the utmost importance.

A series of phase lines (PLs)—London, New York, Paris, Boston, etc.—was laid out to the east toward the ancient cultural city of Munster to coordinate the Allied advance. The British 2nd Army and the U.S. 9th Army would attack eastward with the Lippe River between them. Contact between the two Armies would be the Ninth Army's 30th Infantry Division and the 17th's 507th CT. With a company of tanks added to its arsenal, the CT stood on PL LONDON ready for the upcoming sprint.

THE DASH TO MUNSTER

At 0903 on the 26th the 507th CT jumped off in its advance to PL NEW YORK with the 1st and 2d Bns. leading. They advanced against sporadic resistance in what was to become the most prevalent form of delay tactic; one or two SP guns, occasionally tanks and a platoon or so of infantry equipped with machine guns and mortars. These small, highly mobile teams, often supported by artillery and mortars, would open fire from concealed positions, force deployment of the advancing forces, then scoot to the rear to repeat the tactic at another critical point. They operated in a new work of strongpoints, supported by round-the-clock artillery and mortar fire of all calibers. Difficulty in bypassing or destroying these points occasioned further delay.

The German tactics were effective but they were overwhelmed by the speed of the Allied advance. As the 507th CT kept up the pressure through Peddenberg, Schermbeck and Wulfen the battalions leap-frogged from reserve to front line duty, using trucks on occasion to speed up the advance. At 1104 hours on the 27th the 17th Division CG, Major General William M. Miley, called for pursuit tactics which meant that units should step up the pace, bypass points of resistance and go for deep objectives. In keeping with pursuit tactics, the British 6th Guards Armored Brigade, with 513th CT troopers riding their tanks, passed through the 507th, 28 March and sped off to Haltern.

The push was relentless, day and night, with only short respites at key coordination points. It became difficult for commanders to keep up with the forward progress of their front line units. On one occasion Colonel Raff and his bodyguard, Cleo Crouch, Boca Raton, FL went forward to find H Co.'s Captain Stephens who was reported out front trying to locate a German SP gun. They moved forward to the top of a hill but failed to find Stephens. As they turned about to retrace their steps, a dud from the unseen but close by SP gun crashed at their feet. As they ran for the woods the SP gun spit out three more rounds that exploded harmlessly in the trees.

On 29 March the 507th CT, having fought its way over 25 miles in four days, was detached from the 17th Division, attached to XIX Corps and sent to Haltern to relieve the Guards Armored Brigade. By this time German soldiers in uniform were streaming to the west by the hundreds, many in unit formation. At Haltern the CT fought off counterattacks, mainly from the south, dug out pockets of resistance and searched German installations in the city and outlying suburbs and flushed out pockets of enemy. With the fall of Munster, 2 April, XIX Corps relinquished control of the 507th CT back to the 17th Division and it was trucked to Munster where battalion sectors were assigned for clearing and protecting.

Not all the German troops had surrendered, however, as B Co.'s PVT Edward A. Swiski, Hatboro, PA, learned when his patrol entered the city. Following a firefight, Swiski noted a Nazi flag flying over a city building. He climbed a spiral staircase and pulled it down. In the building, the patrol captured and disarmed several civilian-cloaked SS troops that were mingling with the police.

The CT continued its mission of rooting out enemy, collecting weapons, securing facilities and guarding food stores until 1115 hours, 4 April, when all battalions were ordered into a regimental assembly area. After its 12-day, 50-mile fighting trek since touching down on German soil, the CT reversed course and, by 1625 hours, 5 April, was en route by truck to battle against 350,000 troops encircled in the German Ruhr.

BATTLE OF THE RUHR POCKET

Early on 6 April, the 1st and 3d Bns. did a night relief of the 315th Regiment and a battalion of the 314th Regiment of the 79th Infantry Division, along the Rhine-Herne Canal on the south edge of Bottrop. The CT was given four factories to guard in the Bottrop area. Curfew and a policy of minimum troopers on duty were adopted. This gave time for bathing, delousing, clothing change, equipment repair, restoring basic ammunition loads and rest from the rigors of the past 12 days. Combat was not over, however, as the enemy dished up a fairly constant menu of sniper, machine gun, artillery and mortar fires. The troopers along the canal returned the fires and, using assault boats issued to each battalion, systematically patrolled the German side.

At 0300 hours, 7 April, the 79th Division attacked south across the canal from positions to the left, while the 507th conducted a successful fire demonstration to draw German fires away from the 79th. As that action got underway, 3d Bn. forged a small bridgehead across the canal on the right side of the 507th zone with two platoons of I Co. By 2155 hours they had captured 11 prisoners and dispatched a combat patrol, led by H Co.'s First Lieutenant Bartley E. Hale, now Captain retired, Marietta, GA and accompanied by Regimental S-2 Captain George J. Roper, Lower Mills, MA.

The patrol clashed with a strong German force. Roper was killed and the patrol disappeared. Efforts to reach Hale and his men by contact patrols were unsuccessful. The next day, Bn S-2 First Lieutenant Donald C. O'Rourke, later lieutenant colonel, retired, Jacksonville, FL, flew over the point of last contact and seeing 12 guarded troopers lined up against a wall, presumed them to be part of the patrol. When the patrol was freed after the German collapse in the Ruhr Pocket, it was learned that they had exhausted their ammunition and, thus had to surrender.

The 2d Bn. attacked across the canal on 8 April and while reaching its objective on the right flank of the 79th Division, captured 38 Germans. The 3d Bn. hung onto its bridgehead under increasing fires and counterattack activity as the 1st Bn. reverted to reserve. The Germans still had plenty of fight left as the 1st Bn. passed through the 2d Bn. at 0950 hours on the 9th. C Co. came under heavy small arms and artillery fire.

When PFC Dante Toneguzzo's platoon was pinned down by fire from two mutually supporting pillboxes, Toneguzzo arose on his own initiative, in the face of heavy automatic weapons fire, and stormed the nearest pillbox. His grenade caused two kills and nine surrenders. Still under intense machine gun and sniper fire and in disregard of his own safety, he moved on to knock out the second pillbox with a grenade that killed one and caused five surrenders. For this selfless act of heroism and courageous action, Toneguzzo was awarded the Distinguished Service Cross.

By 10 April, the tempo of operations began to step up as the Germans began to withdraw to the south and west, leaving cities open in their wake. All 507th Battalions were linked up south of the Rhine-Herne Canal. At 1430 hours, two five-man jeep patrols entered Essen, home of the renowned Krupp munitions works and returned at 1700 hours to report no resistance in the city. By 0230 hours the next morning all three battalions were in assigned sectors of Essen. They were on the move again at 0700 hours and in Mulheim with a small bridgehead west of Ruhr River by noon.

At 1150 hours on the 12th, orders were received to attack Duisburg. The orders were cancelled but too late to stop two jeep patrols that reached the edge of the city without resistance. They brought back a German medical captain who reported the city ready to surrender. At 1500 hours, the city surrendered to 2d Bn S-2, Lt. Bennett. Colonel Raff signed a formal acceptance of the surrender in the 507th CP at 1610 hours and sent the German commander of the city back home. At midnight, B Co. was dispatched on foot to Duisburg and arrived at 0630 hours to disarm and establish control over the police. It was relieved by a battalion of the 194th Glider Regiment on the 14th and trucked back to the 1st Bn. area.

As the 194th moved into Mulheim, the 507th was ordered to positions along the Ruhr River south of Essen from Kettwig to Dalhauser. The pressure of American forces pushing up from the south put a heavy squeeze on the Germans. Curiously, they seemed to focus on retaking bridges over the Ruhr, especially the one at Werden where the 2d Bn. had established a bridgehead. Time after time, during 14-16 April, the Germans delivered concentrations of heavy caliber fires and launched counterattacks to wipe out the bridgehead, but each time they were repulsed.

LIFE AFTER COMBAT

While open hostilities were over for the 507th, it still had urgent military duties to perform on the ground over which it had been fighting for the past 12 days. For the next eight weeks, until a more permanent occupation structure could be put in place, the mission of the regiment was two-fold: (1) To administer military government to more than 1,000,000 Germans in the Essen Zone, and (2) to provide care for about 25,000 displaced persons (DP). Colonel Raff was the zone commander. Regimental Executive Officer, Lieutenant Colonel William A. Kuhn, now colonel, retired, St. Petersburg, FL, was deputy commander and major, and later Lieutenant Colonel Morgan A. Brakonecke, was camp administrator.

The Zone was divided into subzones that were administered by battalions. The almost insurmountable task was to provide life-sustaining services to the entire population. The DPs were representative of nearly all of the nationalities of Europe, each with unique problems to be solved. Most of them were slave laborers for the huge industrial base of the German war machine and their animosities toward the Germans presented special problems. To facilitate care and repatriation, self-governing DP camps were set up in the subzones.

Military governance was not an endeavor on which the regiment had spent a great deal of training. The disciplined efficiency which marked the success of the 507th battle, however, seemed to provide the ingredients needed for effective and compassionate administration of this mix of the newly-vanquished and the newly-freed. As expected the regiment performed these final military tasks with distinction.

The 507th stood in review for Field Marshall Viscount Bernard L. Montgomery in Essen. Montgomery stood on the hood of a jeep for an informal chat with the troopers. He thanked them for their outstanding performance and told of his personal desire to participate in bringing the war in the Pacific to a close. Memorial Day services were held to honor fallen comrades in arms. Awards ceremonies were held to honor those who stood out in battle.

Sports played a large role in the daily life at Essen. As part of the "GI Olympics" the regiment sponsored intramural competition in most sports. Teams were entered in the 17th Division leagues. At the Duisburg Stadium where the 1936 Olympic Trials were held, the 507th swept the division first round track and field meet. One of the team standouts was Service Company's Harmon "Tex" Walters, also a key player on the 1942-43 regimental basketball team that bested the professional New York Celtics.

On 15 June, the 507th departed Essen for an encampment on the outskirts of Rambervillers, between Nancy and Epinal in eastern France. There, the sports and recreation programs so carefully organized in Essen were revived. Intramurals resumed as company teams vied for regimental championship honors. Another round of play in Division leagues was scheduled. Again the 507th dominated the track and field events, under the able tutelage of majors Ben Schwartzwalder and Roy Creek. Teams in other sports continued to turn in strong showings. Beer gardens were favorite evening haunts where the retelling of war stories reflected the fine art of embellishment that was begun when the last shot was fired in the German Ruhr. The regiment said farewell to its newer "low point" troopers and hello to a tide of "highpoint" troopers joining for the home-bound trip to the USA.

On the eve of the Hiroshima A-bomb,

G Co.'s officers at Essen, Germany. Front, L to R: 1st Lt. Hussey, Capt. Marr. Back: 1st Lt. Jamison, 1st Lt. Arjay, 1st Lt. Bynum, 1st Lt. Wirtz, 1st Lt. Cromwell.

the regiment was alerted to move a staging camp near the port of Marseille, France. The good news was that there would be no further combat deployments. The bad news was the waiting that was still ahead for the anxious troopers. But after a brief stay at the staging camp, the Regiment boarded the liner Mariposa for the eight-day trip to Boston and Camp Miles Standish.

All but the company commanders and a few other key personnel administrators were given immediate orders to reception stations near their homes. They would be released to pick up lives that could never be the same. Shortly, the key persons followed and, on 16 September, 1945, the 507th Parachute Infantry Regiment, once a proud and highly skilled combat regiment with 21 months overseas, four campaigns and two combat jumps to its credit, became a paper-filled collection of footlockers consigned to a records depot.

EPILOGUE

Despite its inactivation at war's end, the 507th Parachute Infantry Regiment was not to be denied making future contributions to the Army and the Airborne community. Many of its officers and men opted for Regular Army careers, made significant contributions to the future of military parachuting and airborne operations. Colonel George Van Millett, first commander of the Regiment, played a key role in the revamping training as a postwar architect in the Airborne School, as advisor to the first Reserve Airborne Division (100th) and as a key staffer on the Joint Airborne Troop Board in the 1950s. He met an untimely death from leukemia in 1955.

Colonel Edson Raff pioneered in development of the Special Forces that have become prominent in the Army's organization and strategy. Long retired, he spent many years in private flying and charting air routes in French Polynesia and now resides in Garnett, KS. Retired Colonel William Kuhn, St. Petersburg, FL, authored the postwar airborne operations field manual and commanded a battle group in the pentomic 101st Airborne Division, designed to fight on the atomic battlefield. Retired Major General Paul F. Smith, Melbourne, FL, then a brigadier general, commanded the 173d Airborne Brigade in the Vietnam War.

The regiment was reactivated 6 July 1948, redesignated an Airborne infantry regiment, and given the mission of training infantry replacements. Corporal Neil Young, now major, retired, Chillicothe, IL, then fresh from jump school, volunteered for duty with F Co. as a member of the cadre of five training instructors to commence the company's initial training cycle. The regiment was inactivated again 25 May 1949.

As part of the Army Regimental System, the 1st Bn. of the Regiment was reactivated on 23 October 1985 and presented its colors by Honorable John O. Marsh, Jr., secretary of the Army. The ceremony took place on Alvin York Field at Ft. Benning, GA, not far from the point where it was first activated in 1942. The 1st Bn. is the U.S. Army's Airborne School. It provides command, training and support to students of all the U.S. Armed Forces and selected foreign nationals in basic airborne, jumpmaster and pathfinder techniques. Its challenge is to qualify and graduate as many as possible of the 25,000 students it receives annually to prepare for participation in unit airborne operations without further training.

Veterans of the 507th have created an association that convenes annually for reunion activities. It has also created a trainer-of-the-year award that goes annually to the NCO who is selected by the 1st Bn. as the top trainer of airborne soldiers at the School. It is named the George J. Peters Trainer-of-the-Year Award in honor of G Co.'s Private Peters, a superbly trained parachutist and Congressional Medal of Honor winner. He made the supreme sacrifice while protecting his fellow troopers on the Operation Varsity drop zone.

Service Co.'s S/Sgt Jack Fessler, Pine Grove, PA, professional baseball player and star basketball player on a great 507th team was killed in Normandy. He was buried there for four years, then brought home. Veteran troopers Andrew Bazaar, Ernest Beck, William Freed, George Smudin, Guy White and Roy Zerbe showed up in Pine Grove to serve as pallbearers and say another farewell to a precious comrade in arms.

A permanent shrine stands in the town of Graignes, France, built from the ruins of the 12th century Roman Church, destroyed by the Nazis during the six-day battle there. At the back wall stands a granite tablet on which is chiseled the names of the troopers and citizens of Graignes who were killed in the battle. One name on the tablet is Arnald J. Martinez. His younger brother Samuel and 507th veterans looked on, 6 July 1986, as Secretary of the Army Honorable John O. Marsh, Jr. lifted a veil from the sign "Rue Du 507 PIR" on a road leading up to the shrine. Mr. Marsh also lifted the veil from a 32" x 48" painting, commissioned by Adolph Coors Company, depicting French-American cooperation in the Battle of Graignes. It now hangs in the Graignes City Hall. In yet other ceremonies on that day, Mr. Marsh bestowed on 96 French citizens, Department of Defense and Department of the Army medals or certificates for corroborated acts of courage of patriotic civilian service.

On D+2, 8 June, G Co. Squad Leader Sgt. George B. Tullidge was killed in the battle for La Fiere Bridge. His last letter, unfinished and a pamphlet of quotes from great people of history, sent by his mother to inspire and sustain him, were found on his body and returned to his mother in Staunton, VA. She had them published under the title, "A Paratrooper's Faith." For many years she provided, at her expense, a copy of the pamphlet to each new trooper joining the 82d Airborne Division. The originals are encased in the entrance hall of the 82d Division Headquarters at Ft. Bragg, NC, where a street also bears his name.

Born on the hot turf of the Ft. Benning "Frying Pan," put through the crucible of training in Alabama, Louisiana, western Nebraska and the British Isles, tested on the battlefields of Normandy, Ardennes, Rhineland and Central Europe, now masters of the art of training parachutists, the 507th Parachute Infantry Regiment lives!

G Co. S/Sgt. Joseph Faust in Normandy. (Courtesy of Faust).

508TH PARACHUTE INFANTRY

By William G. Lord, II

PRELUDE TO COMBAT

excerpts from **History of the 508th PI**

On the 20th of October, 1942, the 508th Parachute Infantry was born at Camp Blanding, FL. This date, however, in no way marked the beginning of the formation of a new regiment in the United States Army, for since early in September, Major Roy E. Lindquist had been laying plans for the activation of the unit he was to command.

The cadre for the 508th came almost entirely from three sources: the 502nd Parachute Infantry, the Parachute School at Fort Benning and the 26th Infantry Division. Every officer and man who became a part of the cadre was personally screened by Major Lindquist.

On the 20th of October at Blanding troop trains began to arrive bringing the regimental commander, now a lieutenant colonel, and his first recruits—men who had been in the Army only a few weeks and who had volunteered for parachute duty. The average age of the new arrivals was low, under 20. Most were in excellent physical and mental shape, and those that weren't were immediately transferred. For six weeks the processing of the new men went on, and the regiment was built up to full strength, battalion by battalion. By the middle of December the regimental strength was 2,300 officers and men, but 4,500 had to be processed before this number was accepted.

After these first active days, life in the regiment settled down to a steady grind of hard work. From six in the morning till six in the evening the men of the 508th trained.

When the move to the Parachute School was initiated by the 1st Battalion on the 3rd of February, 1943, the physical and mental alertness of the men could properly be called superior. Many men have remarked since that they had never seen a unit in such good shape as the regiment was when it left Blanding. Twenty-three hundred civilians had been transferred into good soldiers in a few short months.

The 1st Battalion qualified on the 26th of February, the 2d on the 5th of March, and the 3d on the 12th of March. After 10 days everyone returned to Benning and the move to Camp Mackall, North Carolina, was initiated by battalions. By the first of April the entire regiment had closed at the new station. Lieutenant Lindquist was promoted to the grade of colonel.

By the beginning of September, the entire unit was once again assembled at Camp Mackall when an alert order came down and the 508th moved to Lebanon, Tennessee, to engage in Second Army Maneuvers.

Higher commanders were pleased, and after one more ground problem, Colonel Lindquist received orders to move the regiment back to Camp Mackall to prepare for overseas shipment. While the rest of the maneuvering forces were left to play their games, the 508th made active preparations for shipment to a zone of operations.

The regiment closed in Camp Shanks, New York, on the evening of the 20th of December, 1943. Just 45 minutes from Broadway, Shanks was near the big city, and yet far for the men of the 508th.

Just before dark on the 27th of December full equipment was shouldered and the officers and men of the 508th began the march through camp to the train station. Loaded in a matter of minutes, the regiment began the ride along the banks of the Hudson to the Weehawken ferry. The name of the boat and the location of the pier were still unknown to all but a few of the key officers.

At eight in the morning on the 28th of December, 1943, the ship, its name now revealed as USAT *James Parker,* slid away from the pier and headed out through the Narrows to the Atlantic Ocean.

After more then 11 days at sea the *James Parker* slid unpretentiously into the harbor at Belfast, Northern Ireland, before dawn on the 9th of January, 1944.

Moving by battalions the 508th marched through Belfast to the railroad station, where it entrained for the resort town of Port Stewart, Northern Ireland.

By the end of January the 508th was once again in top physical shape and the plan of training was announced. February was to be spent in Ireland working on small-unit problems and firing weapons. When this ground training was completed, the regiment would move to England where it would receive airborne training. After that the fate of the regiment lay in the hands of the warlords.

On the 10th of March the 508th boarded trains for Belfast. After arriving in the city, the troopers reversed the procedure of the 9th of January and loaded onto a ship bound for Scotland. The boat sailed early in the morning and steamed into the Firth of Clyde, discharging the 508th onto lighters at Greenock before dusk. Entraining once more, the regiment headed east to Glasgow and then south to Nottingham, arriving about midnight. A 10-minute ride ended at Wollaton Park near the ancient home of the Sheriff of Nottingham. The troopers lived in tents located on the ground where legend had it Robin Hood had hunted.

NORMANDY

"I hope that a month or so from today we will be back here preparing ourselves for our second mission. Until then, a happy landing on the Continent, good hunting and good luck." With these words Colonel Lindquist sent the 508th Parachute Infantry from its base camp at Wollaton Park, Nottingham, to the airfields for its first combat jump.

A very decided increase of tension could be noticed as D-day approached.

The mission of all airborne troops was to prevent the enemy from reinforcing his coastal divisions.

The 82nd Airborne Division, with the 508th attached, was to drop from eight to 10 miles inland from the east coast of the Cotentin Peninsula, just west of Ste. Mére Eglise. An all-around defense was to be established in this vicinity, with the Red Devils responsible for the southwest portion of the division's sector.

More specifically, the 508th's mission was to seize, organize and defend its area, destroying the crossings over the Douve

River at Etienville and Beuzeville-la-Bastille, and patrolling the area to the front aggressively. One battalion was to assemble without delay near the center of the defensive area as the reserve battalion of the Division's parachute element, which was to be known as Force A. The glider element, Force B, was to start landing immediately after Force A and was to continue landing different elements until the morning of D plus 1.

Within the regiment, the 3d Battalion was to organize the defensive sector, the 2d Battalion was to destroy the bridges across the Douve and remain in regimental reserve, and the 1st Battalion was to constitute Force A reserve. Each battalion and the Regimental Intelligence Section was assigned an area to patrol.

By suppertime on the 5th of June, all last-minute changes in equipment had been made, bundles had been loaded into the para-racks on the C-47s, and final arrangements had been made between jumpmasters and air crews as to when the bundles would be released. Although the final meal was worthy of kings, the cooks' talents had been wasted for the most part. Preoccupied minds were oblivious to good food. Immediately following the meal the entire unit went blackface, using soot from the huge blackened stoves in the kitchens. After coffee and doughnuts had been consumed, the officers and men of the Regiment waddled out to their planes, fitted their chutes and said their last good-byes.

Since the parachutist is supplied almost entirely by what he carries on his person and what can be safely dropped from an airplane, special clothing had been designated for him. Trousers with large patch pockets were adopted to facilitate carrying large quantities of ammunition and rations. For the drop on Normandy this uniform was impregnated to offer protection against gas attack. In the pockets were carried one complete K ration consisting of three meals, several D ration chocolate bars, two fragmentation grenades, one smoke grenade, one antitank Gammon grenade and other articles to suit the individual.

Over this jumpsuit was worn a belt supported by suspenders. On the belt were hung canteen, shovel, first-aid packet, bayonet and compass. A gas mask was secured to the left leg, a trench knife was strapped to one boot and another aid packet was strapped to the other. These boots, too, were processed to resist gas. Over both shoulders were slung bandoleers of ammunition. Some of the men carried binoculars.

Next, the parachute back-pack and harness fitted, and from the harness was suspended a musette bag containing, in addition to clean socks and extra ammunition, a 10-pound anti-tank mine. The reserve chute was strapped across the chest to secure all this equipment. After putting on his camouflage-covered helmet, adjusting his chin cup and picking up his rifle, the paratrooper was set to go. For the trip across the Channel, a Mae West life preserver was placed over the head.

Under each of the squat C-47s were secured six bundles containing light machine guns, mortars, ammunition and mines. These were to be released by the jumpmaster just before he jumped.

As the formations passed between Guernsey and Jersey islands, the presence of the fighter escort was apparent by the efficient way in which enemy search-lights were neutralized. Before making landfall, the formations flew through thick cloud banks and became hopelessly separated. Although previously jumped pathfinder teams were to set up their radar equipment on the DZ, only one plane in each lift was equipped with a homing device. Consequently after the planes became split up, it was up to each individual pilot to carry his cargo to the proper place.

The first machine-gun fire seemed unreal, the glowing bullets rising very slowly at first and then snapping by with a rush. To the men in the planes, now standing up and pressing for the door, every shot seemed to be headed directly for them. The night seemed to be headed directly for them. The night seemed filled with the snapping of 20mm shells as they burst nearby. The heavier crunch of flak bursts could also be heard. Several of the low-flying craft were damaged by this fire.

When at last the jumpmasters shouted "Let's go!" there was a brief shuffle for the door, a moment of suspense and the helpless feeling of floating calmly and uncontrollably into the German fire. A quick, hopeful glance at the terrain below was enough to tell most of the regiment that they were not in the proper place, but were lost several miles into enemy territory.

The section of France into which the regiment parachuted, relatively flat and only slightly above sea level, is traversed by two rivers, both flooded by the Germans to create invasion obstacles. The Douve River flows roughly west to east a few thousand yards south of the proposed drop zone, and the Merderet River flows north to south, joining the Douve two miles south of the DZ.

The north-south road crossings of the Douve are located at Etienville and Beuzeville-la-Bastille. Two causeways over the Merderet are located at La Fiere and Chef-du-Pont, directly east of the DZ, the Paris-Cherbourg railway crossing farther north.

Although the planes carrying the 2d and 3d Battalions were widely separated, the 1st Battalion lift kept formation fairly well until two minutes from drop time when they encountered extremely heavy fire. Off course for a time, the plane leading this lift picked up the radar signals before the drop and changed direction, dropping the battalion on the radar sets. However, since pathfinder teams themselves had been dropped south of the drop zone, some of the 1st Battalion men landed in the Douve, while the majority landed just north of the river.

In general, the 2d Battalion dropped in an area which straddled the Merderet, east of the DZ. The 3d Battalion was widely scattered. The battalion commander, Lieutenant Colonel Louis G. Mendez, Jr., assembled a small group near the DZ, while one of his companies, Company G, commanded by Captain Frank J. Novak, assembled almost 100 percent near the beach one-eighth miles to the east.

Major Shields Warren, Jr., executive officer of the 1st Battalion, sent numerous patrols through the area, recovering as many equipment bundles as could be found. After dawn this group, now more than 200 strong, started northeast toward the area near Geutteville that had been previously designated for Force A Reserve. After reaching this area, the Red Devils cleared it of enemy and spent the rest of the day fending off German attacks.

Lieutenant Albright with two men moved south, and, climbing a tree, was able to locate the church steeple in Picauville. He could see the bridge at Etienville being dive-bombed. The patrol then returned, made its report and moved north to locate any friendly troops which might be in the area. Lieutenant Albright met First Lieutenant Norman MacVicar, also of E Company, with about 200 men and stayed with them until nightfall. Then, under cover of darkness, he led the group to Colonel Shanley's position.

A patrol from Major Warren's group reported to Colonel Shanley, asking for help near Geutteville. At this time radio contact was established with Regimental Headquarters across the Merderet and orders were issued for both groups to proceed to Hill 30—a small knoll on the west bank of the Merderet from which both the crossing at La Fiere and the one at Chef-du-Pont could be controlled. Colonel Shanley sent word to Major Warren to join him east of Picauville, and at about 1900 on 6 June, the two groups met. They moved in a column of twos to the hill, arriving about 0200 the following morning. Preparation of defensive positions was started immediately, and the 3d Battalion men were sent out to establish a roadblock near the west end of the Chef-du-Pont causeway. Vigorous patrolling went on during the night, both to keep contact with the enemy and to recover any equipment bundles in the area.

After dawn an aggressive enemy attack of company strength failed to penetrate the positions of the paratroopers. Later a Mark III tank attacked and was set afire and driven off by a .50 caliber machine gun manned by Corporal John Kochanic.

The men landing farthest east were those of G Company. Assembling before dawn, complete almost to the man, on a patch of high ground near the coast, about a thousand yards north of Utah Beach, they had ringside seats for the invasion. These G Company men were able to see the LCIs form in waves for the assault, were able to see and hear the initial softening-up barrage, but luckily were in a position where they were not subjected to the fire of our own artillery. During the morning Captain Novak decided to head west in an attempt to cross the Merderet River and reach the regimental defensive position.

The company then moved south toward the highway joining the beaches with Chef-du-Pont. Not being able to move very rapidly through the sniper-infested hedgerows, the company was forced to dig in defensively after dark.

Continuing on their route the next morning, G Company contacted elements of the 4th Infantry Division which had fought its way in from the beach at the main crossroads south of Ste. Mére Eglise.

Two platoons of Sherman tanks, undoubtedly among the first ashore in France, arrived at the crossroads and part of the troopers climbed aboard. Advancing west and then east, this group attempted to flank enemy troops in the Ste. Mére Eglise area, but before it had closed with the enemy, General Gavin came up and ordered all 508th men to Chef-du-Pont, where Colonel Lindquist had established the regimental CP.

The other large group of the regiment east of the Merderet River was composed of Regimental Headquarters Company and part of the three battalions under Colonel Lindquist's direct command. They landed on the banks of the Merderet west of Ste. Mére Eglise. Encountering harassing enemy fire while reorganizing, these men moved southwest with the intention of crossing the river at La Fiere, and then moving to the proposed regimental defensive area.

Stiff resistance was met at La Fiere causeway. The Jerries had plenty of artillery on the other side of the river, and they threw aerial bursts over onto the troopers. The Germans sustained several casualties, but proved to be of too great a strength for the small group of Americans. The east end of the Causeway was cleared, but no one could cross until the Germans were driven from the other side.

Colonel Lindquist made contact with General Gavin, and the 508th group was relieved by one battalion of the 505th. The Colonel then took his men to a railway viaduct east of La Fiere and organized this position for the night.

The next day these men moved north, clearing snipers and small groups of enemy out of the area northwest of Ste. Mére Eglise and preventing the enemy from flanking the town from this otherwise undefended quarter.

By the evening of June 7, D plus 1, the major part of the 508th was assembling in three groups. One, under Colonel Shanley on Hill 30, was composed of men and officers from all three battalions. The second was made up of 1st Battalion men, with Captain Adams commanding south of Hill 30. The third group was Colonel Lindquist's at Chef-du-Pont.

Since elements of the 101st Airborne Division were to occupy the area south of the regimental CP, a patrol was sent from Chef-du-Pont to the small settlements of Le Port and Carquebut to make contact. Both towns were found to be occupied by the enemy, and no contact was made with friendly forces. In the early afternoon Captain Taylor took part of his provisional battalion to Carquebut where most of the Jerries had withdrawn into the buildings. In particular, a great number was crowded into the village church. After a short fire fight the enemy surrendered. Taken prisoner were six officers, seven NCOs, and more than 100 privates, a larger force than Captain Taylor had brought with him. While some of the men marched the prisoners to the vicinity of the CP, the rest of the force cleared Le Port with only slight resistance, taking more prisoners.

Lieutenant Albright, arriving at one of his OPs on Hill 30, found enemy infantry milling around a few hedgerows away, apparently an indication of an impending attack. Taking advantage of the recently established radio contact with regiment, he called for the guns of the 319th Glider Field Artillery to fire a concentration for him. Delivering effective fire, these guns broke up all semblance of organization among the enemy in the area. Lieutenant Albright thus became the first member of the 508th to direct artillery in combat.

At 1130 on D plus 3, Colonel Lindquist decided to move his group across the La Fiere causeway to join the group under Colonel Shanley. Moving out before noon, the men proceeded up the railway running north from Chef-du-Pont. Equipment chutes were visible everywhere in the swampy areas around the railroad tracks. Arriving at the causeway, the men spread out well before crossing, for although the 325th Glider Infantry had already cleared the area, long-range machine-gun fire harassed any attempted crossing. The effectiveness of our artillery could be seen at the west end of the causeway, where a small group of buildings was almost obliterated. Dead Germans and Americans lying a few feet from each other testified to the bitterness of the battle that was fought here by the 325th. Fighting their way south from the causeway against small isolated groups of Germans, the men soon arrived in the vicinity of Picauville, overrunning from the rear the enemy who were attacking Colonel Shanley's positions.

At this time the battalions were commanded by Major Warren, Lieutenant Colonel Shanley and Lieutenant Colonel Mendez. Lieutenant Colonel Batcheller, still missing, was later found to have been killed on D-day. The depleted size of the regiment was testimony to the fact that the success with which the Red Devils had harassed the enemy had not been without cost.

June 11 and 12 were spent in regrouping the regiment and in preparing for the next mission. The overall situation on the peninsula at this time was good. All the beaches had been taken, and a sizeable beachhead had been formed. Although the British on the left flank were running into much armor, in the American sector the Germans had been incapable of launching a large-scale coordinated attack. A bridgehead south of the Douve River was the next task of the airborne divisions. After the peninsula had been cleared the Americans would attempt a breakthrough to the south. An area had to be prepared from which this attack could jump off.

At midnight on the 12th Lieutenant Goodale led Company F across the river at Beuzeville-la-Bastille in assault boats. On reaching the south shore, he radioed the regimental commander for artillery on the town to disorganize the enemy. A 15-minute barrage by not only the 319th Field,

BATTLE CASUALTIES

Type of Casualty	Normandy	Holland	Ardennes	Total
Killed in Action	307	131	101	539
Died of Wounds	26	15	33	74
Died of Injuries	3	0	0	3
Wounded in Action	487	389	398	1274
Injured in Action	173	80	273	526
Totals by Campaign	1161	681	828	2670

Figures for Missing in Action are taken from records at the end of each campaign. Many men so listed were later reported as prisoners and some were later returned to military control.

COMBAT AWARDS

Decorations	Number
Medal of Honor	1
Distinguished Service Cross	14
Legion of Merit	3
Silver Star Medal	118
Soldier's Medal	7
Bronze Star Medal (Issued in Orders)	378
Foreign Decorations	19
TOTAL	540

but also by heavier artillery, was climaxed by F Company's attack on the town. While the artillery barrage was being fired, men of the 307th Airborne Engineer Battalion began to bridge a gap in the causeway leading across the river. By the time enemy resistance had been neutralized, including the destruction of two German tanks, the regiment began to move across the river. By 0500 in the morning the entire regiment had completed this move, and all battalions were on the way to their objectives.

The 1st Battalion led the regiment across the causeway and immediately started toward its objective. By 1600 in the afternoon the 1st Battalion was located at Coigny and had established an all-around defense. Major Warren then sent Companies A and B with a 57mm anti-tank gun from B Battery of the 80th Airborne AA Battalion attached to each company, to clear the area south of the Douve through which the Regiment did not pass, but for which it was responsible. In this maneuver Company A, commanded by Captain Adams, ambushed and destroyed five light tanks and routed a tank CP. Both companies returned to Coigny by 2300 that night, their mission completed. Because they had not protected their armor with sufficient infantry, the Germans lost 12 tanks in one day to the 1st Battalion.

The 3d Battalion crossed the causeway at Beuzeville-la-Bastille after the 1st, and followed by Regimental Headquarters Company, they moved southwest along the hedgerows. Although they did not have to leave the approach-march formation, small enemy outposts were encountered from time to time, and the area was spotted with snipers.

By early afternoon 3d Battalion was in position with its front to the south between Pont Auny and Hotot. The regimental command post was located in Taillerfer.

The 2d Battalion's objective, Baupte, proved to be by far the most difficult of those assigned to the regiment. Moving across the Douve at 0500 in the morning, the battalion was joined by the F Company force which had cleared Beuzeville-la-Bastille. While still a mile and a half northeast of Baupte, the battalion ran into extremely heavy small-arms fire. Colonel Shanley drew his command into a perimeter defense and sent patrols to reconnoiter to the front.

At shortly after 1600 in the afternoon after the 319th had laid down a preparatory barrage, the 2d Battalion attacked Baupte. Companies D and F were in the assault echelon, while Company E remained in reserve. Company F cleared the southern half of the town after a fire fight that lasted more than an hour. Company D fought its way to the outskirts of the northern end of the town against bitter opposition. In the wake of these two companies were several groups of dead Germans who had decided to fight to the end of the hedgerow strong points.

Northeast of Baupte, D Company encountered a strongly defended vehicle park. Colonel Shanley sent his reserve, E Company, through the section of the town already cleared by F Company. Then D and E Companies joined in a coordinated attack that soon had the Germans reeling. The 2d Battalion bazookas accounted for 10 tanks in seizing the motor pool, and German matériel of all kinds was found by the troopers, including 50 vehicles, gasoline and rations. By dark the battalion had reorganized and was in complete control of the town and the causeway. A railroad bridge east of the town and a nearby culvert were demolished by the attached platoon of Company D, 307th Airborne Engineers.

On the causeway south of Baupte, contact was made with the 101st Airborne Division. The bridgehead south of the Douve now extended from the vicinity of Carentan in the east to a point near Pont Auny. The battle for Baupte was the fiercest offensive action in which the regiment had participated, and the 2d Battalion did a job of which it was rightly proud.

At 0730 on the 14th the 3d Battalion was attacked by a small force in the vicinity of Pont Auny. With effective use of 81mm mortars which obtained some bursts off buildings, the attack was quickly repulsed.

At 1030 the 1st Battalion moved from the Baupte area to clear the ground in front of the 3d Battalion's sector.

During the early evening the regimental CP moved to Château-Fracquetot, and the 1st Battalion went into regimental reserve at Fracquetot. Leaving Company D behind, the 2d Battalion pulled out from the Baupte area and moved to Coigny, a reserve position. Company D remained in defense north of the town with combat groups on the north end of the causeway, on the approach to the demolished railroad bridge southeast of the town, and near the culvert west of Baupte.

At break of day on the 15th a force of enemy cyclists estimated to be about 50 in number, attempted to cross the bridge west of Baupte and were repulsed swiftly by the combat group of Company D located at this point. Two of the Jerries surrendered and nine dead were left on the road, while the rest retreated in disorder, abandoning most of their cycles. This constituted the last bit of action for the regiment on this push south of the Douve, for late in the morning orders were issued to move north of the river. At 1500 the 3d Battalion was relieved by the 2d Battalion. Beginning at midnight the rest of the 508th was relieved by the 507th Parachute Infantry.

On the 16th of June the regiment was in division reserve northwest of Etienville. The 3d Battalion moved two miles north of St. Sauveur where it relieved the 3d Battalion of the 505th. Colonel Mendez's men were to protect the right flank of division and corps, covering the main highway running through St. Sauveur to Cherbourg in the north and Paris in the south. Shortly after the 3d Battalion was in position, the remainder of the 508th moved west of St. Sauveur-le-Vicomte in preparation for relief of the 505th Parachute Infantry. This relief was not merely a trade of positions, but amounted to a push through the 505th in order to occupy an extended defensive position. While this was taking place, many small groups of Germans were neutralized. By 2300 the relief was completed, and the regimental CP had been located at St. Sauveur.

This defensive position was located on the peninsula bordered on the north and east by the Douve River, on the west by the Cherbourg-Paris highway and on the south by the Prairies Marécageuses. Although actual troop positions did not cover all this area, it was cleared of enemy by constant aggressive combat patrols. The highway bridge across the swamp was prepared for demolition and blown on Division order by Company C. Company A then joined with Company C in establishing a strongpoint to cover this bridge. The regiment with the 3d Battalion detached remained in position until the morning of the 20th of June.

On the 18th Colonel Mendez received orders to move his battalion back to the vicinity of Etienville. Early the next morning the 3d Battalion marched to an area east of Etienville, prepared to cross the Douve River in assault boats.

Once again 307th Airborne Engineers proved their worth to the Red Devils by the efficient manner in which the crossing was accomplished. Although there was no enemy fire to meet the troopers, almost everyone had a few tough moments when they realized their vulnerability in the small rubber boats, sitting on the river like so many ducks.

At dawn of the 20th the battalion attacked toward Pretot to protect the left flank of the 325th Glider Infantry.

It soon became apparent that the position in Pretot was untenable.

In the early afternoon, the battalion commander issued the order to withdraw to the high ground 600 yards north of Pretot. Later in the evening the 507th Parachute Infantry relieved the Red Devils and they moved back to a reserve area on the south bank of the Douve. Here they met the rest of the regiment which, after the 90th Division had passed through, had moved there by truck from the St. Sauveur area.

The Cherbourg Peninsula had been cut by a sudden sweep to the west coast. The enemy in and around Cherbourg was cut off from reinforcements and American troops were fast mopping up the peninsula. In fact, only one obstacle now lay in the way of General Bradley's First Army before it could effect breakthrough into France proper. This obstacle was a series of hills and ridges on the south edge of the peninsula which afforded the enemy excellent defensive positions and enabled him to observe American activity. To the 508th was assigned the mission of clearing a part of this ridge, along with the rest of the 82nd Division.

Having been alerted for an attack the evening before, the regiment was making last-minute plans when the operation was postponed on the 21st of June. Instead, defensive positions were to be established two miles south of Etienville. During the remainder of June the three battalions of the regiment alternated positions in the regimental sector, allowing everyone to get equal rest.

The long-expected attack on the hills and ridges to the south was set for the 3rd of July. Reconnaissance patrols, notably three made by Corporal Ellis and Private First Class Kennedy of the regimental S-2 Section, procured information of the enemy on Hill 131. The 1st Battalion remained in a defensive position while the 2d and 3d prepared to move through them at H-hour, 0630.

When the 508th began to advance, it was nearly at double time for the preparatory barrage had devastating effect.

By 0900 the 2d and 3d Battalions had reached their initial objectives and before noon 2d Battalion was at the base of Hill 131, the highest ground on the peninsula, while the 1st and 3d Battalions had reached secondary objectives. As the attack advanced deeper and deeper into German territory, the reeling Germans seemed less and less capable of resistance, a factor which resulted in more and more prisoners.

Ordered to move to Hill 131, the 3d Battalion helped the 2d Battalion hold this objective and then pushed on to the south in preparation for the attack on the next ridge of which Hill 95 was the major peak. The 2d and 3d Battalions moved, into an assembly area in the woods west of Blanchelande while the 1st Battalion assembled north of the woods. By this time less than half of the more than 2,000 men and officers dropped on D-day were present for duty.

Shortly after midnight, battalion commanders received an attack order from Colonel Lindquist.

The 1st Battalion was to advance on the left and clear a finger of high ground thereby preventing the enemy from setting up killing lanes to annihilate troops moving up the adjacent draw. However, this ground was under friendly artillery fire from the unit on the left, and both 1st and 3d Battalions were forced to advance down the draw. Evidently the slight let-up in the attack during the night was sufficient for the Germans to effect a hasty reorganization, for fire, both small-arms and artillery, from the objective was extremely intense.

This situation threw the bulk of the burden on the 2d Battalion's shoulders. At this time the 2d Battalion was not of sufficient strength to clear Hill 95 in one swoop. Consequently Captain Graham sent D and F Companies around the base of the hill, establishing interlocking automatic fire in front of the hill cutting the high ground off from the rest of the Germans. This was accomplished, but only because of a remarkable display of leadership on the part of First Lieutenant Floyd Pollette, commanding Company F, and because of the absolute refusal of his men to surrender an inch of ground despite the intense artillery to which they were constantly subjected.

Now that both flanks of the hill were secured, combat patrols began to operate toward the wooded crest with the intention of clearing Hill 95 of enemy. When this was accomplished, 2d Battalion troops could move onto the hill under cover of darkness.

During the afternoon Captain Royal R. Taylor arrived at the 2d Battalion CP to assume command. In 24 hours the 2d Battalion was commanded by four different officers.

Although the 2d Battalion had cleared Hill 95, the enemy, taking advantage of sunken roads, draws and hedgerows, refused to break contact or to give up any more ground than absolutely necessary. As a result all 2d Battalion patrols which started to operate to the front got only a few yards before drawing enemy fire. Despite this the patrol leaders were able to give accurate reports as to enemy troop concentrations, and both the 319th and the

81mm mortar platoon were firing almost constantly with remarkable effect.

At 1000 on the 7th of July, the regimental commander received Operations Memo No. 4 from the 82nd Airborne Division, instructing the regiment to assembly near the 3d Battalion's position in division reserve. By 1145 all units were closed in the reserve area.

Contact with the enemy was now broken and, lying in rain-filled slit trenches, the troopers began to sweat out the much-rumored trip to England.

On the 12th of July the entire regiment entrucked for Utah Beach where, after the usual period of stumbling around with all equipment packed on weary backs, the troopers were able to lie down on top of the ground, between blankets and out of range of artillery for the first time since D-day.

Late in the evening on the 13th the Red Devils marched onto the beach proper, and like the thousands of German PWs who looked across the water toward England, they waited for LSTs to come in on the high tide. Before midnight two ships were loaded by the regiment, and the trip to Southampton began.

The men of the 508th had established a record in Normandy of which they could be rightly proud. Many Germans had been killed by the Red Devils hours before the first troops hit Utah and Omaha Beaches. Though split into several small groups, the fighting men showed such aggressive spirit that a coveted Distinguished Unit Citation was won for their first two days of combat. Of the 2,056 men who had dropped on D-day, 1,161 had become casualties—307 of them buried on French soil.

HOLLAND

As had been promised, immediately on returning to the Wollaton Park Base Camp half the regiment left on five-day furloughs and leaves.

Late in August the entire 82nd Airborne Division to which the 508th was still attached, held a review at an airport outside of Leicester, England, for General Eisenhower. The general commended the division on its fine appearance and its record and shocked a few by referring to bigger and better airborne operations in the future. About this time it was announced that all airborne and troop-carrier outfits had been placed under a single command, Lieutenant General Lewis H. Brereton's First Allied Airborne Army. Major General Ridgway, who had commanded the 82nd since June of 1942, became commander of the American contingent, the XVIII Airborne Corps, while Brigadier General Gavin assumed command of the 82nd.

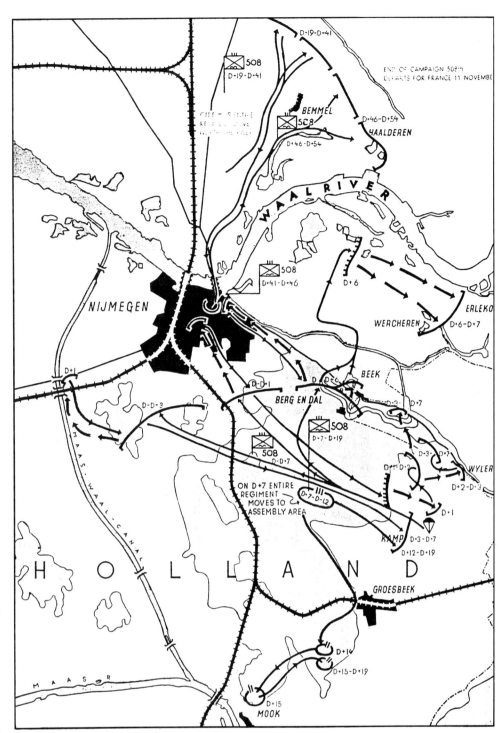

Nijmegen and the Crossing of the Waal. (Map from History of the 508th PI by William G. Lord, II)

The unexperienced rest was almost as short as the traditional 10-minute break, however, for on September 14 the entire 82nd was once again at the airports. Constant practice had made the process of preparation almost a matter of second nature, and by the evening of the 16th, everything was ready for the next day's operation. Since it was to be a daylight jump, the formations would fly over some of the heaviest flak concentrations in Europe to reach the DZ south of Nijmegen, Holland. It was anticipated that opposition in the air would be heavier than in Normandy, but that none of Normandy's confusion after the jump would be present, as the pilots should have no trouble finding the drop zones.

Lieutenant General F.A.M. Browning, of the British Army, was to command the corps made up of the British 1st Airborne Division, plus the 82nd and 101st American Airborne Divisions. These units were to land along the main highway joining the Dutch city of Arnhem with the British Second Army's positions along the Escaut Canal. If successful, the British would be able to drive to the Zuider Zee and cut off an entire German army in western Holland.

This zone of operation was divided into three sectors with the British element taking the northernmost sector in the vicinity of Arnhem. The 82nd was to be responsible for the area south of Arnhem, its main objectives being the cities of Grave and Nijmegen. The 101st was assigned the southern area centering around the city of Eindhoven. The 82nd's sector was further divided so that the 508th's zone of responsibility lay just south of Nijmegen, the 1st Battalion having the initial objective of De Ploeg, the 2d Battalion De Hut, and the 3d Battalion Berg-en-Dal.

By midnight the hangars were silent, as all the troopers slept except the three or four on duty. A breakfast of fresh eggs and bacon just after daybreak was not fully appreciated, and by 0830 everyone was moving out to the planes. Last-minute checks were made, the planes were loaded, and by 1100 the 508th was airborne.

At a few minutes before 1330 in the afternoon of September 17, the C-47s dumped their loads south of Nijmegen and headed back for England. The flight had been nearly uneventful, although much of the beauty of the scene below was not enjoyed by the jumpers. The aerial convoy left England at North Foreland and proceeded due east across the North Sea until landfall was made on the Dutch coast. At this point heavy flak met the invaders, but casualties were extremely light. The flight across Holland to the objective was almost entirely over flooded fields with only an occasional building interrupting the surface of the mirror-like water. It was a cloudless Sunday afternoon with the sun shining bright and hot upon the Dutch countryside. Made up of lifts of 42 planes flying three minutes apart, the long sky train stretched from England across the North Sea, through Holland to the drop zones, and then back to England by the same route. So huge was this armada that as the first planes were returning after the drop, others were just taking off from English airfields. Around this long column snarled angry fighters, while in all directions could be seen groups of medium bombers returning from the missions which had prepared objectives for the airborne assaults.

The drop zone was occupied by most of an enemy antiaircraft artillery battalion, but the sight of the hordes of descending skytroopers scared most of the Germans away from their weapons, and the regiment met very little resistance during the assembly. In slightly more than an hour the 508th was assembled and the battalions were moving towards their objectives.

The terrain on which the regiment dropped was nothing like the ground it had fought over in Normandy. About five miles north of the drop zone the Rhine River separates on its course into the North Sea. The upper branch is known as the Neder Rijn by the Dutch, while the southern branch, the main artery, is called the Waal. Nijmegen, a city of nearly 100,000, lies to the south of the Waal, while Arnhem, about 15 miles north, is situated on the right bank of the Neder Rijn. Ten miles south of Nijmegen is the Mass River, and to the west of the city flows a canal joining the Maas and the Waal. A main highway runs from Eindhoven in the south, through Nijmegen and on up to Arnhem. This highway crosses the Waal over one of the largest and most modern bridges to be found in Europe.

The area immediately south of Nijmegen contains some of the highest ground in Holland, and it was here that the 508th dropped. Only two kilometers to the east was the Holland-Germany frontier. In order to insure that this area was completely secured, it would be necessary to control all the high ground, pushing the enemy out onto the flats.

By 1830 Lieutenant Colonel Shields Warren, Jr., commanding the 1st Battalion, had led his men to the initial objective, De Ploeg, and had occupied it. A and B Companies, reinforced with the 81mm mortar platoon and machine-gun squads from Headquarters Company, were sent as a strong combat patrol to feel out enemy defenses in the city. Their limit of advance was to be the highway bridge over the Waal.

The two companies advanced rapidly north into the city. At a traffic circle about eight or ten blocks from the southern approach to the bridge, the Germans finally appeared in force and a heavy fire fight started.

The forward movement of the two companies had been almost completely halted, and it would be very difficult to maintain control of a large unit at night through the unfamiliar streets. Knowing that a control tower for the demolition of the bridge was nearby, Captain Adams took A Company's 2d Platoon, led by Lieutenant George D. Lamm, and set out to find and destroy this building while the rest of the force remained at the traffic circle.

It was estimated that the building was defended by a reinforced platoon armed with at least four machine guns and a light artillery piece, probably a 40mm AA-AT gun. In the fight that followed both Lieutenant Lamm and Captain Adams distinguished themselves. Sergeant Alvin Henderson, who had already earned a reputation in Normandy for being a fierce fighter and who had escaped from the Germans after spending some time as a prisoner of war in France, rushed across the street to the control tower and started heaving grenades in the windows. The building was set on fire and the enemy withdrew without attempting to use the bridge demolition equipment. Sergeant Henderson was subsequently killed and was awarded the Distinguished Service Cross posthumously.

During the morning of the 18th the Germans made a strong counterthrust on the southern edge of the regimental sector near Wyler. By 0800 in the morning Krauts were beginning to pour over the drop zone, which D Company was struggling to hold as a landing zone for the gliders. D Company was greatly over-extended and was forced to give some ground. Consequently C Company was sent from De Ploeg to the high ground near Groote Flierenberg, about two miles to the southeast with the mission of securing a line of departure for a later attack.

After dark on the 17th, Lieutenant Colonel Louis G. Mendez, commanding the 3d Battalion, sent G Company into Nijmegen by a different route than that taken by the 1st Battalion.

By early morning of the 18th, G Company was within 400 yards of the bridge but was unable to seize it.

The 2d Battalion had the quietest time on D-Day with only occasional skirmishes during the night on some of the roadblocks on Nijmegen's southern outskirts. At 0330 on the 18th, Lieutenant Lloyd L. Pollette led his E Company platoon with two squads of machine gunners from Headquarters Company attached, toward the rail and highway bridges over the Maas-Waal Canal on the Regiment's left flank. Seizing these bridges was not originally assigned to the Regiment, but E Company was now assigned to take and hold them until relieved by the 504th Parachute Infantry. The location of the highway bridge corresponded with Check Point 10 on the secret Division overlays, and so the name Bridge 10 was adopted by the men assigned to take it.

The small force reached a point only a few hundred yards from the bridges when the Krauts saw them and opened fire. Because their left flank was exposed, the troopers received fire not only from the objective but from the opposite bank of the canal. In a matter of minutes the German small-arms fire was reinforced with extremely accurate mortar and artillery fire. Eight of the parachutists were killed almost immediately and several more were wounded, cutting the effective strength of the small force nearly in half. A strongpoint in a house just short of the objective was neutralized by bazooka and machine-gun fire. Realizing that it would be necessary to have supporting fire to take the objective, Lieutenant Pollette called battalion for the 81mm mortar platoon which could not deliver fire on the bridge from the positions they were occupying. The force moved back to a small group of buildings to reorganize and evacuate their wounded.

By this time the defenders had blown the railway bridge and had slightly damaged the highway bridge. Supporting by very effective mortar fire, Lieutenant Pollette's men seized the bridge and held it until the 504th Parachute Infantry sent relief in the early morning. Lieutenant Pollette later received the Distinguished Service Cross for his actions during the morning.

While Bridge 10 was being attacked, D Company was having plenty of trouble on the glider landing zone. The Germans launched one of their first coordinated counterattacks of the operation across the landing zone. D Company had organized a series of small strong-points to defend the area and, being over extended, were forced to fight a delaying action. C Company, meanwhile, was clearing a line of departure on the high ground to the northwest. By noon Company B had followed Company A out of Nijmegen and the 1st Battalion was ready to launch its attack to clear the LZ.

At 1230 the 1st Battalion emerged from the woods northwest of the landing zone, with C on the right, B on the left and A in reserve. The Germans were firing 20mm shells point blank into the woods as the men debouched, and some of the troopers were pinned down. However, the attack as a whole went very smoothly, almost at double time. The first high ground was taken easily, and from commanding positions, supporting fire was delivered for the rest of the push. By 1400 in the afternoon the 1st Battalion was on its objective, and while more than 200 casualties were inflicted on the enemy, less than 15 were sustained by the troopers.

Company A pushed up to a small group of houses just short of Wyler to climax the attack, and B and C Companies neutralized 16 20mm guns. Just as the battalion reached its objective the first of the gliders landed behind them. For once the men could see the results of a hard attack. Where only a few minutes before the enemy was waiting, many gliders were now safely landing.

For the remainder of D plus 1, September 18, and during the night all was quiet in the regimental area. The 1st Battalion was located from Wyler to Kamp along the edge of the DZ. The 3d Battalion was farther north in the vicinity of Berg-en-Dal, and the 2d Battalion was located just south of Nijmegen.

The tactical situation as it stood on the morning of the 19th demanded that the enemy be denied the use of the highway running southeast from the city through Wyler, since this could be used to supply German units still fighting for the city. This meant that the 3d Battalion would have to roadblock the highway in its sector, while Hill 75.9, overlooking the highway between Beek and Wyler, would have to be taken.

Roadblocks were established on the afternoon of the 19th by the 3d Battalion which was now reinforced by F Company at the main intersection in Berg-en-Dal.

Company A, commanded by Lieutenant John P. Foley in the absence of Captain Adams, was alerted to seize Hill 75.9 with an attached platoon of Company G. Company A by this time had reduced to two officers and 42 enlisted men.

Company A approached the hill from the south, since it had been reported that the enemy was on the other three slopes. The attached platoon of Company G had been repulsed three times during the morning by what was believed to be a company of German paratroopers. Two hundred yards south of the crest, still undetected, Lieutenant Foley's force deployed and charged up the hill. The Krauts were momentarily surprised, and they left their holes to withdraw down the northern slope.

Armed very heavily with automatic weapons, the Germans counterattacked several times, but in ferocious hole-to-hole fighting, A Company pushed them down the slope.

Enemy dead and wounded were scattered all over the hillside, but A Company had lost 17 men, 10 of whom were killed.

Company B took Wyler on the 19th at the same time that A Company made its assault on Devil's Hill, and once secured, the town was roadblocked. After inflicting heavy casualties on the enemy, B Company withdrew from Wyler and set up a defense around a roadblock southwest of the town.

The 3d Battalion was also busy during the 19th and 20th of September. I Company's roadblock at Beek had been forced back on the 18th and reestablished on the 19th of September. When all the officers and most of the non-coms present became casualties, Corporal Robert Chisholm of I Company took command of the roadblock—a total of 83 men. For the exemplary leadership Chisholm showed in this and the ensuing action, he later received the Legion of Merit.

At about 1500 on the 20th of September, the enemy began to systematically shell Beek. After 15 minutes of concentrated artillery and mortar fire, the Germans made a swift, strong attack on the town, driving the outposts. When Corporal Chisholm saw that his force was not capable of holding back the German tide, he ordered and personally conducted rear-guard action, withdrawing his command to the high ground south of Beek.

It was necessary to clear Beek, so H Company attacked late on the night of the 20th. The town proved to be much more heavily fortified than expected, and the platoon on the right was unable to proceed farther north than the main highway.

It became apparent that at least 300 Germans, reinforced with several armored cars, were defending Beek. When H Company attacked, the enemy moved north of the highway, splitting the town, and plastered with artillery an area they had evacuated.

Shortly after dawn of the 21st another attack on Beek was launched, but once again, the Germans withdrew to their stronghold and proved too much for the lone company. About noon H Company, reinforced with a platoon from F Company and one from G Company, again attacked. F Company's men advanced through a draw to the southeast of the town in the face of withering fire. As the Germans reinforced their positions, Lieutenant Toth ordered his men back to the high ground just in time to contain a strong counterattack.

As H Company prepared for a fourth attack, the Germans began to stream from the town. H Company moved in, pushing the Krauts in front of them. By 1800 on the 21st Beek was secured.

The battle for this small town was a decisive one and a tough one for H Company. Fewer than 50 men remained after the battle, although a full-strength company of 120 men had been committed 24 hours before. The high cost of the town was caused by the fact the maneuvering was impossible since Beek was bordered on three sides by the flats, which offered virtually no cover.

On the 20th of September the 2d Battalion left its defensive positions on the southern outskirts of Nijmegen and moved to positions south of the drop zone in the vicinity of Kamp and Voxhil. Except for the usual exchange of artillery and patrol actions, this sector was quiet until the battalion was relieved by the 504th Parachute Infantry at midnight on the 24th.

On the 22nd of September, Colonel Mendez ordered Company I northeast from Beek across the flats as a combat patrol with the mission of clearing the area between the 3d Battalion's positions at Beek and the Waal River.

The company moved out of Beek on the main road across the flats. When it reached the road junction 1,000 yards northeast of the town, the company split, one platoon holding the road junction, one moving up the left-hand fork and one up the right-hand fork. Lieutenant Delbert C. Roper led his platoon to the left and initially encountered no resistance. While crossing a field near one of the dikes, Lieutenant Roper's men were surprised to see a German rise up behind the dike, waving the troopers on. With just enough time to hit the ground before the Germans opened up, the platoon was pinned down and sustained several casualties. One machine gun on the left flank was able to go into action against the enemy, and under this covering fire, the platoon withdrew.

During the early morning of the next day, the 23rd of September the 3d Battalion moved from positions near Beek and Berg-en-Dal and moved north to prepare for an attack across the flats. The line of departure was in the vicinity of Polden, three miles east of Nijmegen on the south shore of the Waal. At 0730 the battalion jumped off, with a troop of four Sherman tanks of the Sherwood Rangers Yeomanry attached to each company.

On the northern flank I Company reached its objective, the brick kilns at Erlekom within the hour. H Company met no resistance in taking its objective, the brick kilns at Erlekom within the sector.

Company G, on the right, reached its objective, Thorensche Molen, in the face of heavy resistance, but was forced to withdraw by the accurate enemy artillery. Four times during the 23rd and 24th G Company attacked and reached its goal, but each time found the objective untendable, as the area was extremely open and the men were fighting under the very eyes of enemy observers. Finally they secured positions 600 yards to the north at Wercheren Lake and covered the flats with fire. The cause of the repeated attacks on this position was to secure the northeast end of the dike that ran from the base of Devil's Hill to Thorensche Molen. If this dike could be captured, it would offer cover to its defenders and would afford a continuous line of defense from below the DZ to the banks of the Waal east of Nijmegen.

The remainder of the campaign in Holland was strictly defensive. During the week following the jump the regiment had been constantly on the move, and by continually carrying the fight to the enemy, numerous casualties were sustained. Only a handful of reinforcements was to arrive before the regiment was to move back to base camp in the middle of November, and for the remaining month and a half of the campaign, all three battalions would be fighting a little more than half-strength.

British tanks from the Guards Armored Division entered our division area over this bridge at 0820 on the morning of the 19th, and at 1830 on the 20th, the first tank crossed the highway bridge in Nijmegen and moved north toward Arnhem. The battle for the bridge, one of the toughest in Holland, had been climaxed earlier in the day with an assault across the Waal by the 504th to seize the north end of the bridge while the 505th reinforced by British tanks assaulted the southern approach.

On the 29th of September the 2d Battalion relieved part of the 504th Parachute Infantry in the Voxhil area. At the same time the 1st Battalion dug in breakthrough positions to the rear of the 2d Battalion. The 3d Battalion remained in division reserve behind the 2d.

The 2d Battalion patrolled the area to the front vigorously during the last few nights of September.

A systematic shelling of the 2d Battalion's positions began at first light the next morning, the 30th. All day long, light medium, and heavy artillery kept falling throughout the battalion area.

During the evening of the 1st the enemy artillery began to fall again. This time the Germans laid down their barrage in great depth. In fact, the 3d Battalion in division reserve near the regimental CP was raked by shells.

The barrage was climaxed with an attack by an estimated battalion of Panzergrenadiers on E Company's position.

E Company on the left had its right platoon overrun and forced back about 600 yards. One squad of E Company on the roadblock at the right of the company zone of responsibility withdrew slightly in excellent order and cut down several Germans as they approached the roadblock. D Company, on the right, and the left-hand platoon of E Company stood fast.

By midnight the German attack had almost spent itself, and F Company, commanded by Captain Martin, was ordered to counterattack. In a brilliant night attack which hit the enemy on the flank where they had pushed through E Company the line was soon restored.

During the next day 28 prisoners were cleared from the 2d Battalion to regiment. At least 50 German dead were in the area, and four armored cars and one Panther tank were knocked out. The 2d Battalion casualties numbered 48, including 12 men killed. This counterattack was the last fierce fighting to take place in the 508th sector south of Nijmegen.

On the 6th of October the entire 508th loaded into amphibious trucks and crossed the highway bridge over the Waal to the Arnhem-Nijmegen island, a peninsula bounded by the Waal on the south, the Neder Rijn on the north and the junction of the two rivers on the east. By last light the regiment had closed in a defensive area north of the town of Bemmel.

From the 6th until the regiment was relieved from the Nijmegen area on November 10, the war in Holland was static.

Twice during the period of slightly more than a month the regiment returned for a rest in Nijmegen, and although the town itself was off limits to all American troops, the men of the 508th were billeted quite comfortably in a large schoolhouse. Showers, movies and hot food were enjoyed by everyone.

On the 11th of November, the 26th anniversary of Armistice Day, the men of the 508th marched 22 miles to Oss, Holland, preparatory to returning to a new base camp in France.

On November 14 the 508th closed at Camp Sissonne, France, a former artillery post in the Reims area.

THE ARDENNES

On returning from Holland the men and officers of the regiment were looking forward to a long rest, comfortable quarters and furloughs in England and Paris. The amount of rest the regiment would get was of course up to General Eisenhower and the gods of war.

Passes to Paris started at the end of November. The first quota for the regiment was very large and for awhile it looked as though every one would see the city before Christmas.

On December 16th the relatively good life came to an end. At dawn on that fateful day, approximately 12 enemy divisions pushed through the lightly held Ardennes in Belgium.

At 2000 on the 17th the regiment, still attached to the 82nd Airborne Division,

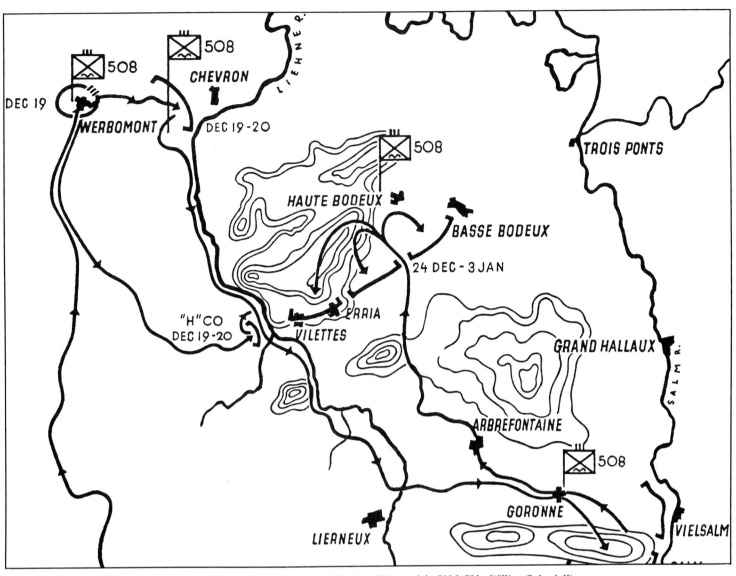

The Regiment Arrives in the Ardennes. (Map from History of the 508th PI by William G. Lord, II)

was alerted for immediate movement to Belgium.

By 0900 on the 18th the 508th, loaded in huge tractor-trailer trucks, joined the division convoy as it cleared Sissonne and headed for Werbomont, a small Belgian village nestled around the junction of two of the most important roads in the sector. Here, because the situation was so fluid, the division established an all-around defense and awaited orders.

At 1800 on the 19th, 12 hours before the regiment had arrived at Werbomont, orders were issued to move to the hill mass in the vicinity of Chevron, 2 1/2 miles to the east. The move was made on foot without incident. By midnight the regiment, less H Company, was at Chevron. H Company was outposting about five miles to the south at Bras.

When the 508th was alerted for another move late in the afternoon of the next day, the troopers still had not made contact with the enemy. By first light on the 21st, despite the fact that no one had had any sleep, the 508th, still out of contact with the enemy, was tactically disposed along Their-du-Mont, a ridge a thousand yards south of the Belgian village of Goronne.

The 508th was not the only unit to make this move. Rather, the entire 82nd Airborne had changed position and now was pushed out like a long finger into the middle of the north side of the wedge the enemy had driven into the American lines. At the tip of this finger was the 508th, supported as usual by the 319th Glider Field Artillery Battalion. On the 21st of December the 1st Battalion was detached from regimental control and placed in division reserve. This seriously affected the defensive set-up of the regiment, since a secure defense could not be maintained in the heavily wooded area and around Goronne with only two battalions. On the next day, December 22, the 1st Battalion returned to Colonel Lindquist's control and was placed on the regimental left flank. Early in the morning on the 22nd snow began to cover Thier-du-Mont, and the troopers received their first taste of winter warfare.

Soon the enemy began to sweep east across the 508th front toward Provedroux.

The enemy armor moved northeast toward the regimental area and attacked the town of Salmchâteau, several miles in front of the 2d Battalion's positions.

At 0200 on the 24th, 40 vehicles and more than 300 men passed the 3d Battalion's positions on their way to the rear, and all the bridges over the Salm as well as many of the small culverts to the front of the 508th area were destroyed by demolition to hold back the enemy tide.

On the afternoon of December 24 a field order came down from Division headquarters. "At H-hour on D-night 82nd Division withdraws to new defensive positions...Covering force will consist of one platoon per rifle company."

The withdrawal of the main body of the regiment was accomplished without incident, and in fact was later commented upon as being masterfully done. By 0415 the main body was closed along the line Basse Bodeux-Erria-Villettes.

The critical area for the 508th covering force therefore centered around the bridge over the Salm River in Vielsalm.

This was the only place the Germans could get across without constructing new roads and a new bridge.

With the river and both banks shrouded by smoke, the advance elements of the Germans effected a crossing. B Company's machine guns split the enemy formation with deadly bands of steel, and the troopers held.

Farther south, in the vicinity of the bridge, A Company's covering force commander was having even more trouble. Lieutenant George D. Lamm's platoon positions straddled the road from Vielsalm to Goronne, a few hundred yards west of the bridge. The main weight of the German attack fell upon the 24 A Company men. Two LMGs and two BARs were the only automatic weapons these men had to fire against the Germans.

Lieutenant Lamm realized that withdrawal of A Company's covering force would have to be effected now or never. He signaled the squad leaders to have their men begin falling back, covering each other as they came. Within 15 minutes Lieutenant Lamm had assembled all the men from his platoon except a few who had been overrun on the right. When the group began to move out, these men made their way through the enemy to rejoin the outfit.

The other two battalions were not his as fiercely as the 1st Battalion was, but the enemy was streaming into the regimental area and along the ridge.

The withdrawal of the covering force could now be carried out with some semblance of control. The platoons from all three battalions reported to Lieutenant Colonel Shanley, covering force commander, and the seven-mile trek to the new positions began.

By 0800 on Christmas morning the entire regiment was assembled on the hill mass overlooking Villettes, Erria and Basse-Bodeux, Although not participated in by the whole Regiment, the delaying action of this covering force was one of the best pieces of fighting in the 508th's history.

Christmas Day for the 508th was neither a day of rest nor joy. There was no turkey dinner, no presents, not even mail.

The next days were spent in strengthening positions. Defensive warfare in Holland had proven that a defensive position is never perfect; it must be continuously improved.

The end of December and the beginning of January were spent improving defenses and feeling out the enemy positions to the front in preparation for an attack.

During the first few days of January it became apparent that is was not the American plan to stay on the defensive in the Ardennes. The German counteroffensive was halted just east of the Meuse, and Supreme Headquarters was faced with the problem of regaining all the lost ground. On January 2, 1945, the Division field order for a large-scale attack arrived at the regimental CP. The attack jumped off on January 3, ending the defensive phase of the Battle of the Bulge for the 82d Airborne. At the beginning of the push, the 508th played a reserve role, backing the 504th and 325th.

The 508th Parachute Infantry remained in reserve until the evening of January 6, at which time the three battalions were in separate assembly areas north of Odrimont, awaiting orders from the Division commander.

At midnight on January 6 the battalion commanders returned from a meeting with Colonel Lindquist and issued the attack order to their staffs and company commanders. The objective, ironically enough, was to be Thier-du-Mont, the ridge from which the 508th had withdrawn on Christmas Eve. Each company would take the area with which it was most familiar—the area which it had defended previous to withdrawal.

The early hours of the 7th were busy ones for the entire outfit, for ammunition had to be drawn, and rolls had to be made and spotted for supply personnel to pick up.

The worst part of the attack on January 7 was not that so many men were killed in the assault, for this is to be expected in war, but rather than many men who were wounded in the attack died that night from exposure. The thermometer hovered around zero all day and then drooped much lower at night. The entire ridge was buried in two feet of snow, and the few paths that ran up the side of the hill were snowbound. Not even a jeep could get near many of the wounded. Search parties combed the thick woods all night in hope of finding some of the wounded, but many were not found until too late. G Company, which had led the assault, arrived on position with 33 men in fighting shape, though it had jumped off that morning with more than 100 men.

After three days in defense on the ridge, the regiment was relieved by the 75th Infantry Division.

By dawn on January 11 the entire 82nd Airborne Division had been relieved and was resting in billets in and around Chevron, Belgium, about 20 miles from the front.

Although alerted several times to go back in the line, the 508th was not committed to action until January 21. At this time motor movement to the Deidenberg area, on the northeastern edge of the Bulge, was initiated. By shortly after noon the regiment had relieved part of the 23rd Infantry, 2d Infantry Division and elements of Combat Command A of the 7th Armored Division.

Relieved from this position on the 24th of January during an intense enemy barrage, the regiment moved into Corps reserve for two days. On the evening of the 26th the 508th again moved to the front, this time in the vicinity of St. Vith, the first town of any size to be seized by the enemy at the outset of the battle on the 17 of December.

On January 28 the attack jumped off, and once again the 508th was held in reserve as two of the other regiments moved off in the assault echelon.

In the early afternoon of the 29th the 1st Battalion pushed to Holzheim where a pitched battle was fought for the town. First Sergeant Leonard A. Funk, who had been awarded the Distinguished Service Cross for heroism while helping to clear the Holland DZ, personally led the assault and C Company moved into town. The battalion immediately set up a defense around Holzheim. At this time approximately 90 Germans who had been taken prisoner in the assault were confined under guard in one of the buildings.

A short time later Sergeant Funk approached the building to check the guards. His tommy gun was slung over his shoulder, with bolt back and safety off. As Funk approached the house the German officer poked a Luger in his stomach and demanded surrender. The sergeant stepped back to get shelter from the corner of the building and reappeared with submachine gun in hand, nearly cutting the German in half with .45 slugs. Funk emptied his gun into the group of enemy and then reloaded, firing burst after burst at the fleeing enemy. Funk was joined by troopers who had been attracted by the firing. When the first sergeant put his smoking weapon back on his shoulder, both sides of the road and the area surrounding the building were covered with fallen enemy. Estimates showed that about half of the 90 Germans were killed and the rest wounded. None escaped. For his quick thinking, aggressive action and his absolute lack of fear, Funk was later presented the nation's top award, the Medal of Honor, by President Truman.

The 3d Battalion broke out of the woods in sight of Lanzerath. Hastily, the men deployed in an assembly area while the battalion staff and unit commanders received final orders from Colonel Mendez. The attack began.

Lanzerath fell after only slight resistance.

Before dawn on the 31st the regimental CP displaced forward to Lanzerath as the 3d Battalion prepared to leave town on its way east to German territory. The 2d Battalion was to move on the 3d's right. As the columns began to twist slowly forward, Lanzerath came under direct tank fire.

By early afternoon the entire regiment was on its final objective, a ridge overlooking a northwest-southeast road along the German border. By this time the regiment was well into the Siegfried Line, as evidenced by the dragon's teeth, pillboxes and heavy artillery pieces encountered.

It was on January 31 that the final objective was reached and the regiment remained in position until February 2 when a slight shift was made to include some of the area formerly held by other units of the division that were taken out of the line and sent north 50 miles to the Aachen area.

On the 4th of February the 508th moved into old barracks near Rencheux, the scene of the 1st Battalion covering force's struggle on Christmas Eve.

On the night of February 7 the regiment moved from the rest area in Rencheux to the vicinity of Aachen. The small German village of Hahn became the new regimental assembly area.

During the evening the 508th relieved the 517th and part of the 505th Parachute Infantry in the little town of Bergstein, about three miles west of the Roer River.

The morning after the 508th took positions in Bergstein the troopers jumped off in an attack to the river. The 2d Battalion led the rest of the regiment. Advancing about 1,400 yards in the face of enemy fire and extensive antipersonnel minefields, the battalion was held up by a killing lane of fire from the south.

Consequently, the 2d Battalion took up defensive positions to hold the ground already gained.

In the early evening of the same day, February 9, orders came down from the 82nd to seize and hold the high ground on the west bank of the river. At 0200 on the 10th the 1st Battalion attacked. The move was successful and the enemy was dislodged. By 0850 the 1st Battalion was on position.

The 2d Battalion meanwhile pushed on to its objective on the right of the 1st Battalion. By midmorning the entire regiment was in position with the 1st Battalion on Hill 400 and the 2d Battalion on the adjacent ridge to the south. The 3d Battalion remained in reserve 1,000 yards to the rear. These positions were kept until February 18. During this period extensive patrolling went on to the front. On the evening of the 17th the battalion commanders were called to a meeting at the regimental CP. When they returned and assembled their officers, everyone felt sure that at last the attack order for a difficult mission was being given. However, the actual fact of the matter was that the entire 82nd Airborne was being returned to theater reserve near Sissonne, France.

By February 20 the regiment had arrived from Aachen at Sissonne. The mode of travel had been the uncomfortable but westward-bound 40-and-8s. Though no one really suspected it at the time, the fighting days of the 508th Parachute Infantry in World War II had come to a close.

OCCUPATION DUTY

On April 4 the long attachment of the 508th to the 82nd Airborne Division came to an end. The regiment was placed under the direct control of First Allied Airborne Army. Immediately the regiment packed up and moved by truck to the railroad station at Laon. Once again boarding the 40-and-8s the troopers wee jostled westward across France, arriving the next day at Chartres, southwest of Paris. Here the regimental CP was established and the regiment went to nearby airfields, prepared to jump on 48 hours' notice to liberate prisoner-of-war camps if the Germans restored to atrocities. East of the Rhine the Allied armies were cutting Germany to ribbons.

In the latter part of May the regiment once again moved back to Sissonne to gather up all the equipment and await assignment. On June 8 movement was initiated to Frankfurt-am-Main, Germany, where the 508th was to be stationed with the Army of Occupation. On the 10th of June the 508th arrived at its new station and immediately began occupation duties.

Frankfurt-am-Main was designated as the location of General of the Army Dwight D. Eisenhower's Supreme Headquarters, Allied Expeditionary Force. The 508th Parachute Infantry was chosen to guard the headquarters and to form honor guards for all visiting dignitaries. Billeted not in Frankfort itself, but in the little suburb of Heddernheim, the troopers were able to live very comfortably.

The primary mission of the regiment was the security of the big headquarters. This was accomplished with two battalions, one being assigned to guard the wired-in inclosure where the headquarters was located and another being given the job of guarding the nearby towns of Oberursel, Bad Homburg, and Königstein.

Immediately after VE-Day, a point system was announced to determine eligibility for discharge. With one point for each month in the Army, one for each month overseas, five for each battle star or decoration, and twelve for each child, 85 points was set as the initial figure to let men out of the Army.

With the 82nd and 101st Airborne Divisions slated for shipment to the States early in 1946, the 508th was the only parachute unit to stay in the theater. Consequently by the end of the year most of the men in the regiment were either re-enlistees or low-point men. Three hundred volunteers for the Regular Army left the regiment, in November for re-enlistment furloughs.

This is the story of the 508th Parachute Infantry from its activation on October 20, 1942 to January 1, 1946. The story has concerned itself in the main with the fighting units of the regiment, the nine rifle companies and their supporting units. However, it was not intended that the rest of the regiment, or the units which helped the regiment, be overlooked.

Behind the battle-weary trooper who carried his M1 through France, Holland, Belgium and Germany were many helping hands. Parachute maintenance men packed chutes almost continually through the training period, just previous to the two combat jumps and immediately after the jumps for resupply. The supply personnel of the regiment burned the midnight oil in garrison getting the 508th ready for its missions and toiled long hours in combat to make sure that those missions would be fulfilled. The truck and jeep drivers in Service Company drove continually, often over rocky, muddy or icy roads to insure that the men at the front received all available supplies. The personnel section back at the base camp worked long hours compiling records, writing letters of condolence and keeping track of personal effects.

In garrison, the men and officers of the Medical Detachment were called "pill rollers," but when German steel started landing nearby, the aid man and the surgeon were the doughboy's best friends. Not soon will be forgotten the quiet services the chaplains held a few hundred yards behind the lines whenever possible. Not soon will be forgotten the cheerful smiles and words of Chaplains Elder and Kenney on their frequent trips to the front.

In three years more than 10,000 men passed through the regiment, though the tables of organization called for roughly 2,500 officers and men. In three campaigns the regiment won five unit decorations for heroic action. Members of the regiment have won every comb at decoration authorized by the United States Government and decorations from the governments of Great Britain, France and the Netherlands. The 508th left more than 600 men buried in foreign soil.

On January 1, 1946, the primary mission of the regiment was strategic reserve for the European Theater.

After returning from Frankfurt as a unit, the 508th Parachute Infantry Regiment passed into history at Camp Kilmer, New Jersey, on November 24, 1946—four years, one month and four days after it had come into existence.

508th ARCT

By Bob Murray

The 508th Parachute Infantry Regiment (PIR) was inactivated on 25 November 1946, at Camp Kilmer, New Jersey. However, with the escalation of the Korean War, a need was perceived for another Airborne unit for possible use in that theater as well as to serve as part of the national mobile reserve.

The lineage of the 508th PIR was passed to the new unit which was designated as the 508th Airborne Infantry and was activated at Fort Bragg, North Carolina on 16 April 1951. COL Joseph P. Cleland was selected as the regimental commander and served until 28 September 1951. He was followed by COL Joe S. Lawrie who served in that position until 30 June 1952. He was succeeded by COL George O. Pearson.

The new 508th was at Fort Benning, Georgia during the period May through June 1951 for Airborne training. From there, it was infantry training at Benning from June to December 1951.

Phase one of infantry unit training was completed as the 508th acted as an aggressor force maneuvering against the 47th Infantry Division at Camp Rucker, Alabama. Then followed a number of night training exercises and night jumps during the period January through March 1952.

Joining the 508th in this training program were the 320th Airborne Field Artillery Battalion and the 598th Airborne Engineer Company, both of which were activated on 1 August 1951. The 519th Airborne Quartermaster Company, activated on 25 August 1953, was also assigned to the growing force. The 427th Airborne Anti-Aircraft Battery, which was activated later on 8 April 1955, was also assigned, and these five airborne units comprised the 508th Airborne Regimental Combat Team.

In April of 1952, the 508th, along with its then assigned units, began training as an ARCT in the massive Exercise Longhorn with the 82nd Airborne Division in the San Angelo-Lampasas-Fort Hood area of Texas. The ARCT in its first year as an active Airborne unit exceeded the 19 regimental jumps of its wartime predecessor, the 508th PIR.

The ARCT provided troopers for the testing of new equipment for Airborne units. Included was the testing of special desert warfare gear in the Yuma, Arizona area.

Troopers were also used in the support of Ranger training in Northern Georgia and in Florida near Eglin Air Force Base. The unit was rapidly acquiring a reputation of being ready to deploy anywhere at anytime.

Moved from Ft. Benning to Ft. Campbell, KY, the ARCT was ready for its next operation when in early July 1955, they were tasked to participate in Operation GYROSCOPE with the 187th taking up the 508th's station at Fort Campbell. The 508th remained on duty in Japan through 1956 when it was returned to Ft. Campbell. On 22 March 1957, the 508th ARCT was deactivated. The five units comprising the ARCT were either deactivated or redesignated and reassigned to other Airborne units. The saga of the 508th ARCT had finally come to an end.

SHOULDER INSIGNIA

The shoulder insignia worn by the 508th ARCT was an adaptation of that worn by the 82nd Airborne Division. The same shape and same colors of red, white and blue were used, but the "AA" in the 82nd's insignia was replaced with a blue wyvern. (The 508th ARCT troopers also often wore the circular insignia of the 508th PIR as a patch on the left pocket of their jacket.)

COAT OF ARMS

508th Airborne Infantry:

SHIELD: Azure, on a bend argent a lion passant guardant gules, armed and langued of the first.

CREST: On a wreath of the colors argent and azure, a pheon with 14 barbs, divided per pale or and sable, in front of a wyvern statant gules.

MOTTO: Fury From the Sky

The two principal colors of the shield, blue and white, are the current and old colors of infantry. The lion on the coat of arms is the same as the French leopard used in the arms of Normandy and commemorates the organization's landing and campaign in that province. The silver bar, called a bens, is in honor of the organization's service in the Rhineland. The wyvern in the crest is taken from the shoulder sleeve insignia of the 508th Airborne Regimental Combat Team in which the regiment was the primary elements in the 1950s. The color red alludes to the unit's unofficial nickname, "Red Devils." The arrowhead is divided into two colors, yellow and black, in reference to the two assault landings made by the 508th in Normandy and in the Rhineland. The 14 notches in the arrowhead allude to the regiment's overall honors in World War II—four campaigns and 10 decorations.

320th Airborne Field Artillery Bn.

SHIELD:L Gules, on a palmetto tree eradicated or a Lorraine cross azure.

CREST: None.

MOTTO: Volens et potens (Willing and Able)

The shield is scarlet for artillery. The palmetto tree refers to South Carolina, the home area of the organization after World War I. The Lorraine cross commemorates service in France during World War I.

BATTLE CREDITS

NORMANDY
6 June to 24 July 1944

RHINELAND
15 September 1944 to 21 March 1945

ARDENNES-ALSACE
16 December 1944 to 25 January 1945

CENTRAL EUROPE
22 March to 11 May 1945

NORMANDY
6-7 June 1944

509TH PARACHUTE INFANTRY BATTALION

By Morton N. Katz

Elements: None

Activation:
Activated at Fort Benning, Georgia, October 5, 1941 as the 504th Parachute Battalion.

Campaign Credits:
Algeria, French Morocco, Tunisia, Naples, Foggia, Rome, Arno France, Rhineland and Ardennes.

Unit Citations, Awards and Decorations:
The Distinguished Unit Badge and Oak Leaf Cluster (Two Clusters for Company C); given two Army and two Corps Commendations and the regimental insignia of the 3rd (French) Zovave Regiment; given honorary memberships in the 1st British Airborne Division.

Medal of Honor Recipient:
S/Sgt. Paul H. Huff- Anzio Beachhead, February 29, 1944. First to be given to an American parachutist.

B. Distinguished Service Cross:
Captain William P. Moir, November 28, 1942
1st Lt. Henry F. Rouse, March 25, 1944
Sgt. Mike Baranek, March 25, 1944
T/Sgt. Carl R. Clegg, March 25, 1944
Pvt. Leslie B. David, May 11, 1944
Cpl. Boggs C. Collins, May 11, 1944
Pvt. Edwin C. Hicks, May 11, 1944
Captain Harry J. Stone, May 10, 1944
Captain Carlos C. Alden (Doctor), August 8, 1944
Captain Roy E. Gaze, December 1, 1944

Commanding Officers:
Lt. Colonel Edson D. Raff- Retired Full Colonel
Colonel Doyle R. Yardley- Retired Full Colonel
Lt. Colonel William P. Yarborough, Jr.- Retired Lt. General
Major Edmund J. Tomasik- Retired Full Colonel

Casualty Summary:
Killed In Action: Approximately 208
Wounded: Approximately 2,600

Crest:
Gingerbread Man- Black and gold, representing a trooper standing in the door of transport.

Individual Decorations:
Congressional Medal of Honor- 1
Distinguished Service Cross- 10
Silver Star- 62
Soldiers Medal- 3
Bronze Star- 38
Legion Of Merit- 5
Croix De guerre with Silver Star- 6

The 509th Parachute Infantry Battalion was originally known as the 504th Parachute Battalion. When the first parachute regiments were formed, this unit became the 2nd Battalion, 503rd Parachute Infantry. The first parachute unit to go overseas, it sailed June 4, 1942 under command of Lt. Col. Edson D. Raff, landing in Gouroch, Scotland on June 10, 1942. Stationed at Hungerford, England, the Battalion trained with Major General Sir Frederick A.M. Browning's 1st British Airborne Division. The American troopers were made honorary "Red Devils" and accorded the right to wear the red beret or tam of the British parachutists.

To help spearhead the North Africa invasion, the unit took off on the night of November 7, 1942 from Land's End, England. The first American parachute or Airborne outfit to enter combat, it was bound for Tafaraoui Airport, Oran, Algeria. The C-47 transports arrived November 8 after a flight of 1,600 miles, the longest Airborne invasion in history. Three planes were forced down in Spanish Morocco, others in the Sebkra d'Oran, a salt waste, where they were strafed.

Among the first to die was Private John C. Mackall of Company E. In his memory is named Camp Mackall at Hoffman, North Carolina, the Army's largest Airborne training center.

Here also Capt. William J. Moir became the first parachute medic to be decorated, receiving the DSC for his treatment- while under fire- of the wounded. Following defense of Tafaraoui from the French Foreign Legion coming from Sidi bel Abbes, the troopers reorganized and went on to Maison Blanche Airport outside Algiers. From here they took off on November 15, 1942 to jump at Youcks les Bains near Thelepte airport, on the Tunisia border. Here the French awarded the men of the Battalion the regimental insignia of the Third (French) Zouave Regiment. This airfield became base for the almost legendary Colonel Phil Cochran, immortalized as Milt Caniff's "Flip" Corkin.

First into Tunisia to contact the Africa Korps and first into Gafsa, the unit fought at Kasserine Pass, Feriana, Sheitla and in the first battle of Faid Pass, taken by a combined parachutist, infantry, tirailleur and TD team.

Their third combat jump was Lt. Dan A. DeLeo's demolition mission at El Dejm to blow the bridge on Rommel's main supply line. Faulty air navigation and inexperienced men untrained as a unit combined to keep this jump, made in December 1942, from full success, but DeLeo and his men destroyed much track, blew up an armored train and caused much confusion among the enemy.

Following redesignation as the 2nd Battalion, 509th Parachute Infantry (to avoid confusion with the reorganized 503d Regiment back in the United States), the unit moved from Maison Carree to Boufarik for training. By request of General Mark W. Clark, the Battalion was assigned to the newly-formed Fifth Army, and moved to Oujda, French Morocco in March 1943. There was built Camp Kunkle, named for the first parachute officer killed in combat. Battalion men staffed the new Fifth Army Airborne Training Center and gave Air Force glider pilots valuable infantry training.

The 82nd Airborne Division left the United States on April 29, 1943, landing at Casablanca on May 10. Veteran troopers of the 509th were on hand to welcome the new arrivals to the Oujda area. With the division, the battalion moved to Kairouan, Tunisia in June, and remained in Seventh Army reserve with the 325th Glider Infantry during the Sicily campaign.

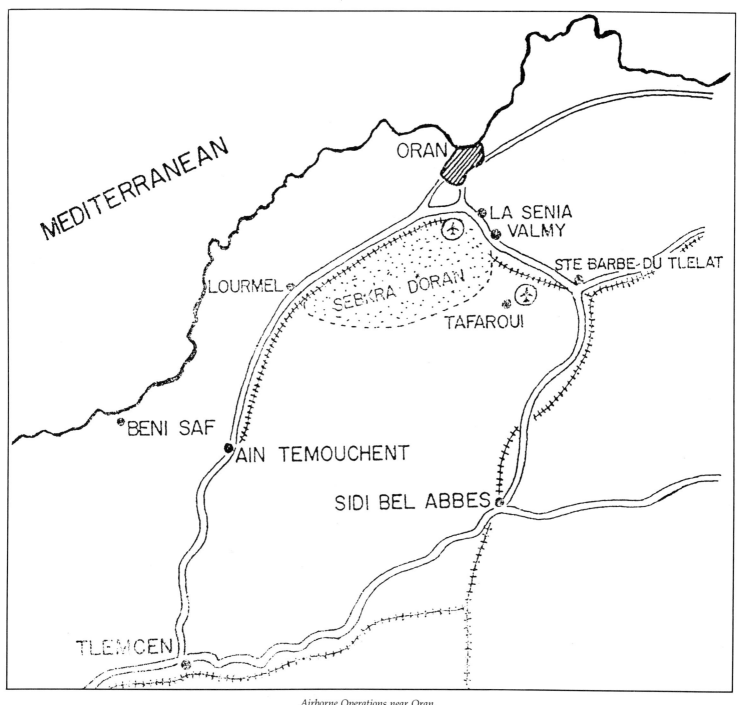

Airborne Operations near Oran.

The most unusual exploit was that of the Parachute Scout Company. On September 9, 1943, led by the late Capt. Charles C.W. Howland, the company left Bizerte as part of a joint Army-Navy task force. In an amphibious operation in Naples harbor, the islands of Ventotene, Ischia, Procida, Capri and Ponza were captured, the troopers missing Benito Mussolini by seven hours on the last-named island. A radar station and hundreds of the enemy were captured without loss of a man.

From Comiso, Sicily, the Battalion took off on the night of September 14-15, 1943 for its fourth parachute mission. Jumping at Avellino, Italy in a desperate attempt to relieve pressure on the Salerno beachhead, the troopers cut German supply and communication lines to successfully disorganize the enemy, despite heavy losses including the commanding officer, the late Colonel Doyle R. Yardley.

Lt. Col. William P. Yarborough, Jr., formerly with the 504th Parachute Infantry, took the Battalion from Piscinola, Naples to Macchia in November for a short tour with the 504th. Crossing the Volturno River, the unit joined the Rangers in Venafro and on November 11, 1943, captured Mount Croce, overlooking Venafro. This position and Mount Corno alongside, were held until December 14, when the 45th Infantry Division took over.

On November 29, 1943, the name of the Battalion was again changed, this time to the 509th Parachute Infantry Battalion, by Fifth Army orders.

Training at Baia Bay in Pozzuoli preceded the D-Day landing on the Anzio-Nettuno Beachhead at H-Hour plus 2, on January 22, 1944. The battalion landed with the Rangers, and subsequently on the beachhead suffered its heaviest casualties. Here S/Sgt. Paul H. Huff of the 509th became the first parachutist to win the Medal of Honor.

For stopping the desperate German drive of February 29, 1944 at Carano, near Garibaldi's tomb, which was designed to split and destroy the beachhead, the battalion became the first American parachute unit to win the War Department (Presidential) Unit Citation. On the night of March 15-16, 1944, Company C captured two houses used later as a base for the Sixth Corps break-

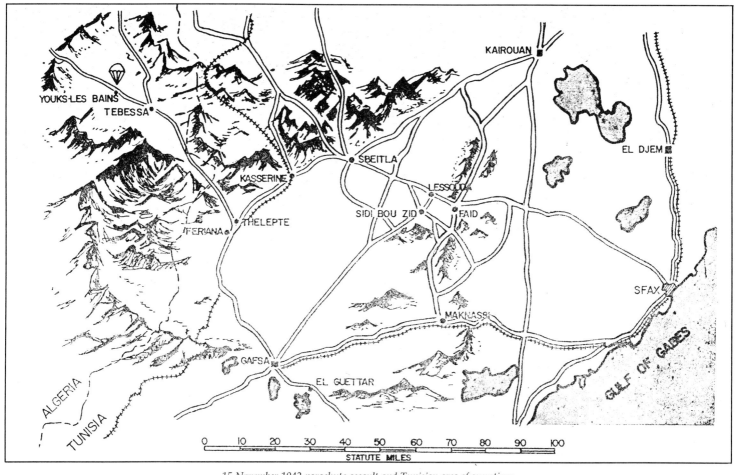

15 November 1942 parachute assault and Tunisian area of operations.

out, and became the first parachute company to be cited, its troopers adding the Oak Leaf Cluster to the Distinguished Unit Badge.

After Rome's fall, the outfit trained at Lido di Rome and on August 15, 1944, made its fifth combat parachute mission, spearheading the Seventh Army invasion of Southern France. The battalion, leading the 1st Airborne Task Force, was first to land at Le Muy and St. Tropez, and to liberate Cammes, Juan les Pims, Antibes and Nice, receiving enthusiastic welcomes from the FFI and the people. Troopers fought at La Courbaisse, Lantosque, Peira Cava and Fort Mille Fourches, and were withdrawn to La Gaude on November 23.

Moving to Villers-Cotterets in Northern France on December 8, 1944, the battalion was attached to the 101st Airborne Division, but not for long. When the 101st went to Bastogne, Major Edmund J. Tomasik led the battalion to Manhay, Belgium, to join the 3rd Armored Division. In fighting at Soy and Hotton elements of Hq. 3rd Division were rescued. In bloody fighting at Sadzot the troopers accomplished their mission of holding the Manhay-Grandmenil-Erezee supply line at all costs, and thoroughly defeated the 1st and 2nd Battalions of the 25th SS Panzer Grenadier Division. This fighting from December 22-30, 1944, resulted in the unit's second citation. The men hung an Oak Leaf Cluster on their Distinguished Unit Badges, and the men of Company C put on their second cluster.

Attached to the 7th Armored Division, the Battalion captured Born on January 20, 1945, and later captured Hunningen and cleared the woods north of St. Vith, to let the 7th Armored Division roll through unopposed. After this action, seven officers and 43 men came down the hill on January 29, 1945, from this last action. All others were dead or hospitalized. Men released from hospitals came to the final CP in Trois Fonts to face the bitter news that the 509th was to disband. Units were to be consolidated for the final push on Germany, and special units were no longer needed.

On March 1, 1945, the 509th Parachute Infantry Battalion was officially disbanded. Officers and men of this great outfit were assigned to the 82nd Airborne Division, which was then at Stavelot, and some to the 13th Airborne Division.

In its overseas history the unit earned battle honors for Algeria-French Morocco, Tunisia, Naples-Foggia, Rome-Arno, Southern France, Rhineland and Ardennes. Invasion arrowheads were included in the first five. The 509th was a Combat Infantry Battalion, won the Distinguished Unit Badge and Oak Leaf Cluster (two clusters for Company C), was awarded the French Croix de Guerre with Silver Star, given two Army and two Corps Commendations, plus the Belgian Croix de Guerre for the action at St. Vith, the regimental insignia of the 3rd (French) Zouave Regiment, and given honorary membership in the 1st British Airborne Division.

Individuals earned a Congressional Medal of Honor (first to be given an American parachutist), 10 Distinguished Service Crosses, 62 Silver Star Medals and two Oak Leaf Clusters, five Legion of Merit Awards, 38 Bronze Star Medals and two Oak Leaf Clusters, three Soldier's Medals and six Croix de Guerre with the Silver Star Decorations.

Its unit records include being first parachute or Airborne outfit overseas and in combat, making the longest airborne invasion in history; most number of combat parachute jumps in EAME Theatre: first into Tunisia, first to fight alongside and be commended by the French, first to fight the Afrika Korps, first into Southern France, first to jump with war correspondent. While in England members of Hq. and Hq. Co. made the lowest altitude mass parachute jump of all time: 143 feet.

517TH PRCT

*Condensed from **Paratroopers' Odyssey: A History of the 517th Parachute Combat Team** (Military narrative by LTC Charles E. LaChaussee, AUS Ret.) and **Chronicle of the 517th PRCT** (Compiled by Clark Archer)*

ELEMENTS
517th Parachute Infantry Regiment
460th Parachute Field Artillery Battalion
596th Parachute Combat Engineer Company

ACTIVATION
Activated as elements of 17th Airborne Division
Declared an independent unit and shipped to Italy (May 44)

CAMPAIGN CREDITS
Rome-Arno
Southern France
Rhineland
Ardennes-Alsace
Central Europe

UNIT DECORATIONS
517th PRCT - French Croix de Guerre (Draguignan, France)
517th PRCT - Belgian Croix de Guerre
1/517th PIR - Belgian Croix de Guerre (Soy-Hotton, Belgium)
2/517th PIR - Belgian Croix de Guerre (Saint Vith, Belgium)
1/517th PIR - Presidential Distinguished Unit Citation

MEDAL OF HONOR
PFC Melvin E. Biddle B/1/517th PIR (Soy-Hotton, Belgium)

DISTINGUISHED SERVICE CROSS
Lt. Col. William J. Boyle
S/Sgt. Albert P. Deshayes
T/Sgt. Wilford C. Anderson
T/5 Sporos Gogos (P)
T/Sgt. George W. Heckard
PFC Nolan L. Powell

COMMANDING OFFICERS
517th PRCT Col. Rupert D. Graves
1/517 PIR Lt. Col. William J. Boyle/ Lt. Col. Robert E. McMahon
2/517 PIR Lt. Col. Richard J. Seitz
3/517 PIR Lt. Col. Melvin Zais / Lt. Col. Forest S. Paxton
406th PFAB Lt. Col. Raymond L. Cato
596th PCEC Capt. Robert W. Dalrymple

CASUALTY SUMMARY
Killed in Action, 247
Wounded, 1,576

TRAINING
1/517 PIR/17 Airborne Division, Camp Toccoa, Georgia
2/517 PIR/17 Airborne Division, Camp Toccoa, Georgia
3/517 PIR/17 Airborne Division, Fort Benning, Georgia

MOTTO
ATTACK

INDIVIDUAL DECORATIONS
Medal of Honor, 1 (PFC Melvin E. Biddle)
Purple Heart, 1,576
DSC, 6
Legion of Merit, 5
Silver Star, 131
Air Medal, 2
Bronze Star, 631
French Croix de Guerre, 17
Soldiers Medal, 4

COMMENTS
The 517th PRCT accumulated over 150 combat days.
The 517th PRCT was the only independent Combat Team in the ETO.

ACTIVATION AND TRAINING
The story of the 517th Parachute Regimental Combat Team begins with the activation of the 17th Airborne Division on March 15, 1943. The Division's parachute units were the 517th Parachute Infantry Regiment, the 460th Parachute Field Artillery Battalion and Company C, 139th Airborne Engineer Battalion. The 517th was at Camp Toccoa, Georgia; the 460th and C/139 were at Camp Mackall, North Carolina.

For the next several months all men volunteering for parachute duty at induction stations throughout the United States were sent to Camp Toccoa. The 517th was charged with screening the volunteers and assigning those qualified to either infantry, artillery or engineers. Officers of the 460th and C/139 were placed on temporary duty at Toccoa to help with the screening, and men assigned to those units were sent to Camp Mackall.

As units filled up they were to be given basic training at their home stations and then sent for parachute qualification to Fort Benning, Georgia. After jump training, all units, including the 517th would join the 17th Airborne at Camp Mackall.

Receiving and screening one to two hundred men a day was a pretty big order for the 517th. On activation the regiment had a total strength of nine officers, headed by newly-appointed commanding officer Lt. Col. Louis A. Walsh, Jr. They were joined three days later by the "cadre" under command of Major William J. Boyle, bringing the regiment's strength to about 250.

Through the spring of 1943 trains arrived at Toccoa daily with contingents of 50 to 150 men; each group was met at the station and trucked to the parade ground where a 34-foot-tall parachute "mock tower" had been erected. Lieutenant John Alicki, favored by fortune with a rugged appearance, greeted them with a blood-and-guts speech intended to scare off the timorous.

"In" and "Out" platoons were formed, those who survived the mock tower went to the "In" platoon for further screening. This consisted of a medical examination by Regimental Surgeon Paul Vella and his staff, followed by an interrogation by their potential officers as to why they has applied for parachute duty. Many answers were interesting and some hilarious. A few had been advised by doctors to take up parachuting to help overcome their fear of heights. Some with criminal records had been told their slates would be wiped clean. Those failing the screening process were sent to the "Out" platoon and the balance assigned to units. As men assigned to the artillery and engineers moved to Camp Mackall the infantry began basic training.

Military organizations are strongly influenced by the character of their commanders. Because of its isolation and greenness this was particularly true of the 517th. At age 32, Louis Walsh was young, cocky and aggressive. He had been with the Airborne since its earliest days and had spent three months as an observer with U.S. forces in the Southwest Pacific. Having seen combat in its most primitive form under atrocious conditions, he was determined to prepare the 517th to survive, fight and win under any circumstances. To reach this goal Colonel Walsh set extremely high standards. Physical conditioning was paramount.

Each trooper was required to qualify as "expert" with his individual weapon, "sharpshooter" with another and "marksman" with all crew-served weapons in his platoon.

It had been planned to fill the battalions in numerical sequence. By the end of April, Major Boyle's 1st Battalion was almost complete. At the end of the following month Major Seitz' 2nd Battalion was pretty well on its way. By late June or early July, while Major Zais' 3rd Battalion was still waiting for its first recruit, the flow of volunteers to Toccoa was suddenly turned off. It was announced that the 3rd Battalion would be filled with Parachute School graduates who had already completed basic.

In late summer an advance detail staked out a claim at Camp Mackall and the regiment moved to Fort Benning for parachute training. The 517th breezed through jump school with no washouts, setting a record that has endured to this day. School Commandant General Ridgely said that the 517th's Battalions were without equal in discipline and effectiveness—which says a great deal for Colonel Walsh's selection and training methods. The 517th troopers were the first to wear the steel helmet in jump training; until then a modified football helmet had been used. On completion of jump training the 1st and 2nd Battalions moved on to Mackall while the 3rd remained at Benning to complete fill-up.

Camp Mackall was not much different from Toccoa, but bigger on level ground. Everyone was quartered in the same one-story, uninsulated "hutments" heated with coal stoves. The 17th Airborne was big on athletics, and the 517th shook it up a little by fielding football and boxing teams that won Division Championships.

One day an inspection team from Headquarters Army Ground Forces arrived at Camp Mackall to test the regiment's physical fitness. Using more-or-less scientific statistical sampling methods, men and units were selected and put through their paces. Individuals took the Physical Fitness Test consisting of pull-ups, push-ups, and other weird calisthenics done against time. Platoons and companies were chosen to run and march, with and without equipment, for various distances. When all was done the results were analyzed and announced. The 517th had taken first, second and third place in all tests and events, scoring higher than any unit tested before or since.

Through the fall the regiment conducted unit training—tactical exercises for the squad, platoon, company and battalion. Effort was made to conclude each phase of training with a parachute jump. Sometimes jumps had to be cancelled because of weather or lack of airplanes, but men and units averaged one per month.

In February the regiment moved to Tennessee to take part in maneuvers being conducted by Headquarters Second Army. The "Tennessee Maneuvers" were a sort of little practice war that went on year-round. Participation in the Tennessee Maneuvers was supposed to be the final test before a unit could be pronounced combat-ready.

One cold day in March when all were shivering and knee-deep in mud, it was announced that the parachute elements of the 17th Airborne Division were being pulled out for overseas shipment as the 517th Regimental Combat Team. So, from the mud of Tennessee, the 517th PRCT emerged. The parachute units were hastily shipped back to Camp Mackall to prepare for overseas movement.

The 460th Parachute Field Artillery Battalion, with an authorized strength of 39 officers and 534 enlisted men, consisted of a headquarters and four firing batteries, each with four 75mm pack howitzers. The 75 threw a 13.9-pound shell for a maximum range of 9,650 yards. The 75 broke down into seven pieces for parachute drop.

Company C, 139th Airborne Engineer Battalion, was redesignated the 596th Airborne (Parachute) Engineer Company. The 596th had a company headquarters and three platoons with an authorized strength of eight officers and 137 enlisted men. The engineers were lightly armed and equipped, but highly trained in their missions of construction and destruction.

The 517th RCT received no special augmentation to allow it to function as a separate unit. It was expected to operate as a small division.

On return to Camp Mackall all efforts were concentrated on preparation for overseas movement. In the midst of this activity the word spread one day that Colonel Walsh had been relieved. It was a real shock to 517th troopers. But in the Army, as elsewhere, life must go on. Colonel Walsh's successor was Lt. Col. Rupert D. Graves, USMA '24, who came from command of the 551st Parachute Infantry Battalion.

In early May the RCT components staged through Camp Patrick Henry near Newport News, Virginia. On May 17th the troopers climbed the gangplanks for their great adventure. The 517th boarded the former Grace liner *Santa Rosa*, while the 460th and 596th loaded onto the Panama Canal ship *Cristobal*.

ITALY

One dark night the ships slipped through the Straits of Gibralter and it became obvious that the destination was Italy. This idyll came to an end when the *Santa Rosa* and *Cristobal* docked at Naples on May 31st. The troopers filed down gangplanks into waiting railroad cars and were carried to a staging area in the Neapolitan suburb of Bagnoli. Enroute, Colonel Graves was handed an order directing the RCT to take part in the attack from Valmontone to Rome the next day. The 517th was ready to go, but since crew-served weapons, artillery and vehicles had been loaded separately it would have to be with only rifles. After this was pointed out, the order was cancelled and the RCT moved on to "The Crater."

Gradually weapons and vehicles arrived. On June 14th the outfit struck tents, stowed away extra gear and moved to a beach to wait for LSTs to carry it to Anzio. The troopers filed aboard, were handed C-rations, and told to make themselves comfortable anywhere they could find space on the crowded decks. In the evening the ships raised ramps, backed out into the channel and headed north. During the night the RCT's destination was changed. At midday the LSTs put in at bomb-wracked Civittavecchia, dropped ramps and the troops marched off to bivouac several miles inland.

The RCT was attached to Major General Fred L. Walker's 36th Infantry Division, which under IV Corps was operating on the left of Fifth Army. A long truck ride and a short foot march on the 17th of June brought the units south of Grosseto. Colonel Graves was handed an overlay marked with zones, objectives and phase lines. The regiment was to join the division's advance north from Grosseto the next day.

At daylight on June 18th, the rifle battalions filed through Grosseto heading north-east on Highway 223. Mechanized cavalry had reportedly been through the area and found it clear, but the leading company of Major Boyle's 1st Battalion ran into a storm of machine gun fire as it entered the Moscona Hills. The troopers fanned out, took cover and returned fire. The Germans held a group of farm buildings in a small valley. With a platoon of B Company attached, C Company moved to the ridge overlooking the farm and opened fire. Enemy machine gun fire clipped leaves from a hedgerow; within a few minutes 10 C Company men were hit.

Colonel Graves had received no word from the 1st Battalion, but its predicament was obvious. He committed Lt. Col. Dick Seitz' 2nd Battalion to envelop the enemy from the right and sent I Company from the 3rd Battalion to protect the western flank. Battalion 81mm mortars and 460th guns opened up. Under this fire and with pressure on their front and flank, the Germans pulled out.

In the early afternoon the advance was resumed. At twilight the battalions took up rough perimeters and halted for the night. On the east I Company had become trapped in a minefield under machine gun fire. It was extricated after dark.

In its all-important first day of combat, the regiment suffered 40 to 50 casualties but inflicted several times that number upon the enemy. The next seven days were spent in almost continuous movement. The Germans tried to make an orderly withdrawal while the Americans pressed them hard. For the 460th the period was a continuous, 24-hour-a-day operation. Gun batteries continually leapfrogged each other; usually two batteries were in position while the other two were moving forward. The principle chore of the 596th Engineers was road reconnaissance and mine-sweeping.

On June 19th the 2nd Battalion captured the hilltop village of Montesario. On the left the 3rd Battalion moved through Montepescali against light resistance, going on to take Sticciano with 14 prisoners. The RCT bivouacked overnight June 22-23 on a ridgeline south of Gavarrano. Next morning the RCT moved across the Piombino Valley and closed into an assembly area behind the 142nd Infantry. On June 24th the 2nd Battalion entered the eastern outskirts of Follonica under heavy artillery and Nebelwerfer fire.

During the night of June 24-25 the 3rd Battalion made a long infiltration, emerging next morning on high ground overlooking the dry stream bed of the Cornia River. At 0800 the 1st Battalion passed through the 3rd to seize Monte Peloso, dominating a broad valley with the town of Suvereto about a mile north on the far side. The attack was preceded by a heavy artillery barrage fired by 36th Division Artillery under 460th direction. Moving in column along the dry stream bed, 1st Battalion met minor delays as skirmishers with "burp guns" fought to slow the advance. Under cover of a smoke screen laid down by 1st Battalion's 81mm Mortar Platoon, one company moved west in a shallow envelopment to the left. PFC Carl Salmon silenced a machine gun with rifle fire, and troopers rushed the hill. The enemy force had been a detachment of the 29th SS Panzer Grenadier Division. The remainder of the battalion came forward and the position was consolidated.

Enemy artillery fire continued heavy on Monte Peloso through the night. A haystack on the crest had caught fire during the afternoon. After dark it became an aiming point for the German artillery. While the 1st Battalion had been taking Monte Peloso, Colonel Graves had been studying the terrain to the north. It was ideal for defense, with steep hills overlooking broad open fields. In the distance he saw Tiger tanks moving around. Graves estimated that there would also be minefields with which to contend. The colonel was planning a night attack to Suvereto. However, the 517th went into IV Corps reserve and remained in that status until early July.

The 517th had been sent to Italy in response to a Seventh Army request for airborne troops for ANVIL, the invasion of Southern France. Troops had been withdrawn from the line (including 517th's) and air and naval forces were assembling.

On July 2nd the Combined Chiefs of Staff issued a directive to the CINC Mediterranean to go ahead with ANVIL (renamed DRAGOON) on 15 August. As a by-product of this directive the 517th RCT was released from IV Corps and moved to join the First Airborne Task Force in the Rome area.

The German Nineteenth Army was along the Mediterranean coast. Four divisions and a corps headquarters were west of the Rhone. East of the Rhone the LXII Corps at Draguignan had a division each at Marseilles and Toulon and one southwest of Cannes. There were an estimated 30,000 enemy troops in the assault area and another 200,000 within a few days march.

The planners decided early that an airborne force of division size would be needed. Since there was none in the Mediterranean, a force of comparable size would have to be improvised. In response, the 517th RCT, 509th and 551st Parachute Battalions and the 550th Airborne Battalion were provided. Other units in Italy were designated "gliderborne" to be trained by the 550th and the Airborne Training Center. By early July the concentration of airborne forces in the Rome area was almost complete. Two additional troop carrier wings totalling 413 aircraft were enroute from England.

H-Hour and D-Day were tentatively set for 0800, 15 August. The 517th RCT had been allocated 180 C-47 aircraft in four serials. The Combat Team was sealed off on August 10th. Maps, "escape kits" and invasion scripts were issued. During the last hours of daylight on the 14th, equipment bundles were packed, rigged and dropped off beside each plane. Around midnight the paratroopers formed by sticks and marched to their planes. After slinging the pararack bundles they fitted parachutes, adjusted weapons and equipment and climbed aboard. At 0100 on August 15th, 396 C-47 aircraft began turning over their engines. At 10-second intervals, planes taxied down dirt runways, lifted off and circled into formation.

DRAGOON - THE INVASION OF SOUTHERN FRANCE THROUGH D+4

Radio beacons would guide the serials from Elba to the northern tip of Corsica. From there, radar and Navy beacon ships would lead them to Agay, where each serial should descend to 1,500 feet, slow to 125 miles per hour, and home-in on its drop zone by beacons and lights to be put out by pathfinder teams. Each plane carried six equipment bundles in pararacks beneath its belly.

Most of the pathfinders missed their drop zones. The 517th team dropped early at 0328. North of La Ciotat the aircrews dropped 300 parachute dummies and a large quantity of "rifle simulators" which went off in firecracker-like explosions as they hit the ground.

The four serials bearing the 517th RCT began drops at 0430. First to arrive was Lt. Col. Dick Seitz' 2nd Battalion in Serial 6 flown by the 440th Group from Ombrone. Lt. Col. Mel Zais' 3rd Battalion was due next in the 439th Group's Serial 7 from

Orbetello. The 460th Field Artillery (less Battery C) in Serial 8 with the 437th Group from Montalto fared better than the 3rd Battalion but not as well as the 2nd. Twenty plane loads jumped early and were spread from Frejus to the west. Last in was Serial 9 at 0453, flown by the 438th Group from Canino with Major Boyle's 1st Battalion and Battery C of the 460th. One platoon of the 596th had dropped with the 509th. One platoon had dropped with the 2nd Battalion and one with the 3rd Battalion.

All told, only about 20 percent of the 517th RCT landed within two miles of the DZ. Regardless of where they landed the 517th troopers went to work with the tenacity and aggressiveness that characterized parachute outfits. The Germans were not anxious to tangle with the Allied paratroopers but nevertheless put up a stiff fight.

Actions throughout the next three days threw the Germans into a state of chaos. Enemy convoys were attacked, communication lines severed and German reinforcements were denied access to the beach landing areas. Towns and villages were occupied as troopers fought toward their objectives. Le Muy, Les Arcs, La Motte and Draguignan became names to remember.

Part of the 3rd Battalion had proceeded toward Fayence shattering enemy lines and installations as they moved. Remaining troops of the 3rd Battalion assembled from Seillans, Tourettes and Callian. Those troops landing to the east of Tourettes were joined by troops of the British 2nd Independent Parachute Brigade. The combined force annihilated a large German convoy speeding reinforcements to defensive positions near the beach.

Lt. Col. Boyle and a handful of 1st Battalion men made a gallant stand at Les Arcs. Remaining elements of the 1st Battalion captured assigned objectives.

The 460th Field Artillery, under Lt. Col. Ray Cato, had a bulk of its guns deployed and ready to fire by 1100.

The 2nd Battalion pushed through to join with the 1st Battalion as Germans began massing their forces on the outskirts of Les Arcs for an all-out counterattack.

The 3rd Battalion completed a 40km forced march as the RCT consolidated. The team attacked all assigned German positions clearing the way for Allied beach forces to push toward the north.

The 1st Platoon of Capt. Bob Dalrymple's 596th engineers had joined assault operations with elements of the 509th Parachute Battalion near Le Muy. The 2nd Platoon conducted operations south of Les Arcs. The 3rd Platoon had joined attack operations with 3rd Battalion.

By D+3, German opposition within the airhead had ceased. The 517th RCT was given a new mission.

The Airborne operation was a remarkable performance, considered by many military historians the most successful of the war. Within 18 hours 9,099 troops, 213 artillery pieces and anti-tank guns and 221 vehicles had been flown over 200 miles across the Mediterranean and landed by parachute and glider in enemy-held territory. Despite widely-scattered landings, all missions assigned had been accomplished within 48 hours. Airborne task force losses included 560 killed, wounded and missing, and 283 jump and glider casualties. 517th PIR losses included 19 killed, 126 wounded and 137 injured through D+3.

THE CAMPAIGN IN SOUTHERN FRANCE

As VI Corps moved west, the Airborne Task Force reverted to Seventh Army control and was assigned to protect the Army's eastern flank, while the main forces moved up the Rhone Valley. The British 2nd Parachute Brigade returned to Italy and was replaced by the First Special Force. Protection of the Army's eastern flank meant moving as far east as practicable and then protecting the best ground available. The initial Task Force objective was the line Fayence-La Napoule. The 517th RCT was assigned the left, the Special Service Force the center and the 509th/551st the right in a narrow strip along the coast.

The 2nd and 3rd Battalions were charged with the capture of Fayence and Callian. This was accomplished by August 21st. Saint Cezaire fell to Companies G and I on the 22nd. During the attack, Company G had been pinned down. Company I surged through heavy fire up the mountainous slope to take the objective. For this action, it earned a commendation from Task Force Commander Maj. Gen. Robert T. Frederick.

Saint Vallier, Grasse, Bouyon and La Roquette fell in quick succession. In the attack on La Roquette, Company E distinguished itself and received a commendation from General Frederick.

The RCT's momentum was slowed by a line of enemy fortifications extending from the Maritime Alps to the sea. The Germans attempted to hold a series of forts at all costs. On September 5th, Company D succeeded in taking some high ground near Col de Braus. Heavy fighting ensued. Companies G and H were successful in capturing Col de Braus. A step closer to the heavily defended Sospel Valley.

The 1st Battalion, supported by 460th fire, pressed into Peira Cava. A red-letter day of the campaign occurred when Ventebren and Tete de Lavina were captured by the 2nd and 3rd Battalions.

The remainder of September was spent digging defensive positions in and around Peira Cava. The 517th RCT now held a thinly manned 15-mile front, using mines and booby-traps to take the place of troopers. Attacks on Hill 1098 ended the month with the roar of artillery duels echoing through the Maritime Alps.

Despite heavy artillery fire, a patrol from Company F pushed into Sospel on September 29th. The Germans withdrew as Company B moved up to occupy Mount Agaisen. The siege of Sospel was over after 51 days of continuous fighting. Troopers fanned out in pursuit of the enemy. 517th involvement with the campaign was terminated on November 17, 1944. The RCT marched 48km to La Colle. On December 6th the RCT moved from La Colle to entrain at Antibes for movement to Soissons and assignment to XVIII Airborne Corps.

The 517th RPCT suffered over 500 casualties and had 102 men killed in action. On July 15, 1946, the President of the Provisional Government of the French Republic issued Decision Number 247 awarding the French Croix de Guerre to the RCT.

ARDENNES-ALSACE

All elements of the RCT were quartered in Soissons by December 10th. Every American airborne unit in Europe was now part of General Matthew B. Ridgway's XVIII Airborne Corps. This included the 82nd and 101st Airborne Divisions just back from Holland and the 517th and other separate units up from the Mediterranean. Additionally, the 17th Airborne Division was now in England and was scheduled to come across to France in the near future.

During the night of December 15-16 the German army launched its last great offensive of World War II, striking with three armies against weak American positions in the Ardennes region of Belgium and Luxembourg. The Allies were taken totally by surprise. The Germans made their main effort with the Sixth SS and Fifth Panzer armies, while their Seventh army on the left made a limited holding attack.

Movement orders came for the 517th and 1100, December 21st. One Battery of the 460th and a platoon of the 596th were attached to each rifle battalion for movement.

Orders were received through XVIII Airborne Corps which directed the 1st Battalion to the 3rd Armored Division sector near Soy, Belgium. Pressure from German armor had made the situation so fluid that it was impossible to tell exactly where the front began. Company D was immediately attached to the 3rd Armored's Task Force Kane. This unit held the key point on which the front hinged. Companies A and B detrucked northeast of Soy and was ordered to attack along the highway leading from Soy to Hotton.

The mission of the 1st Battalion was to take the commanding ground around Haid-Hits, then remove the enemy from the high ground at Sur-Les-Hys. The object was to facilitate a breakthrough and free surrounded elements of the 3rd Armored in Hotton.

Company B led the attack until forced to hold a line due to heavy tank and automatic weapons fire. It became necessary for Company A to bypass the planned route to Hotton. While this maneuver saved casualties, it was necessary to fight for every foot of ground along the entire route. Fighting on the return trip from Hotton to Soy was as heated as on the trip in. The Soy-Hotton mission was so well executed despite fanatic resistance that the 1st Battalion was awarded the Presidential Distinguished Unit Citation. The cost: 150 wounded and 11 men killed.

While the 1st Battalion was attached to the 3rd Armored, the balance of the RCT was kept busy. The morning after arrival in Belgium, Company G was detailed as a security force for the XVIII Airborne Corps CP. The RCT (less 1st Battalion and Company G) was attached to the 30th Infantry Division near Malmedy. The RCT Headquarters opened at 1000, December 23rd, at Xhoffraix. On Christmas Day the RCT was released from attachment to the 30th and returned to XVIII Corps control.

When the RCT was attached to the 30th Division, the 460th tied in with 30th Division Artillery and fired 400 rounds in missions south and east of Malmedy. During the nine days in December, the 460th fired more than 30 TOTs.

The fall of Manhay to the 2nd SS Panzer Division on Christmas Day sent shock waves throughout the Allied Command. From Manhay the Germans could continue north toward Liege or turn against the flank of the 3rd Armored and the 82nd Airborne. Urgent directives descended upon General Ridgway demanding that Manhay be retaken at all costs.

The directive to recapture Manhay arrived in RCT Headquarters at 1400 on December 26th. The 517th was to attach one battalion to the 7th Armored Division for the mission.

The 3rd Battalion (less Company G) under Lt. Col. Forest S. Paxton was given the assignment. One platoon of the 596th Engineers and a section of the Regimental demolitions platoons was attached. The battalion would have to cross two miles of terrain covered with snow and underbrush, in darkness, before reaching the line of departure. The attack would jump off at 0215 after a 10-minute TOT by eight battalions of artillery.

The attack proceeded as planned after 5,000 rounds were fired in four concentrations. By 0330 the last pocket of resistance was eliminated. A counterattack at 0400 was driven off. The 3rd Battalion suffered 36 casualties, including 16 killed.

Early on New Year's Day, the RCT was attached to the 82nd Airborne and alerted to go on the attack. On January 3rd, the RCT, acting as the left flank of the 82nd, attacked south along the Salm River. The 551st Parachute Infantry, as an attached unit, fought through Basse Bodeux, while the 2nd Battalion captured Trois Ponts. The southerly attack continued to Monte Fosse where advance elements were subjected to intense shelling.

The 1st Battalion moved through ground already taken to seize Saint Jacques and Bergeval. The 3rd Battalion continued its attack across the Salm River and moved to the east. On January 9th, they circled around the 551st and closed on the bank of the Salm at Petit-Halleux. That night, advance details of the 75th Infantry Division arrived to make arrangements for relieving the 82nd in the area. To get them off to a good start, 3/517 under direction of the 504th crossed the Salm and seized Grand Halleux.

Colonel Graves received orders on January 11th that the RCT (less 2nd Battalion, attached to the 7th Armored Division) was attached to the 106th Infantry Division. The immediate job was to relieve the 112th Infantry at Stavelot and along the northern bank of the Ambleve. This was accomplished by the 1st Battalion on January 12th.

A new attack was launched at 0800 on January 13th, to seize a line running from Spineux, north of Grand Halleux, to Poteaux, eight miles south of Malmedy. The 1st and 2nd Battalions moved to the south capturing Butay, Lusnie, Henumont, Coulee, Logbierme and established blocks at Petit Thier and Poteaux. The RCT had now reached the limits of the prescribed advance.

While most of the RCT had been involved with the 106th and 30th Infantry Division, the 2nd Battalion moved from Goronne to Neuville for assignment to the 7th Armored Division. Colonel Seitz and his men were assigned to Combat Command A at Polleux. On January 20th, Task Force Seitz attacked south from an assembly area near Am Kreuz to capture Auf der Hardt woods and formed defensive positions on the southern edge. On reaching the objective, a patrol was sent to the village of Hochkreuz.

At 1500 Company F was detailed to join a tank company for an attack on Born.

On January 22nd, the task force led CCA through In Der Eidt Woods and closed in attack positions a mile northwest of Hunnange. At 1700 TOT concentrations were fired on Hunnange and the attack moved out. By dark Task Force Seitz had overrun Neider Emmels and Hunnange and was in contact with other 7th Armored Division forces.

Defensive positions were taken facing south and southwest. A road block was established at Lorentswaldchen and patrols were sent to the outskirts of Saint Vith. At 1400 on January 23rd, Combat Command B passed through Task Force Seitz and completed the capture of Saint Vith.

On January 24th orders were given to clear the Saint Vith-Ambleve road that remained in enemy hands. At 0600 on January 25th, the Battalion moved out for its attack position. By 1400 the objectives were secured.

On February 1st the 517th PRCT joined the 82nd near Honsfeld. Next day the 1st Battalion took up a blocking position to protect the northern flank of the 325th Glider Infantry while the 3rd Battalion moved into position to support if required. All objectives of the attack plan were met, and on February 3rd, the RCT received orders attaching it to the 78th Infantry Division at Simmerath.

The 78th was to attack east on February 6th to seize Schmidt and the Schwammenauel Dam. The 517th RCT was to move north to the Kleinhau-Bergstein area, relieve elements of the 8th Infantry and attack south from Bergstein during darkness on February 5th to seize the Schmidt-Nideggen Ridge. The Germans had prepared the strongest defenses of the western front in this area.

By 0600 on the morning of February 5th, all units had closed at Kleinhau. The German line ran from Zerkall west and south of Hill 400 to the Kall River. After dark the 2nd and 3rd Battalions moved into attack positions. Five to six hundred yards below Bergstein, both battalions hit minefields and concertina wire. The troopers attempted to move forward by crawling and probing, but all efforts proved futile. Men were blown up by Schu mines, Tellermines and "Bouncing Bettys." In Bergstein the troopers found some protection from small-arms fire but little else.

In mid-morning the 596th Engineers began working in relays to clear a lane through the largest minefield encountered by the Allies in World War II while under direct enemy observation and fire. For 36 hours the 596th continued this genuinely heroic effort. In the 1st Battalion area, Company A sent a patrol from Hill 400 to Zerkall.

In the early afternoon of February 7th, Colonel Graves was informed that the 517th was released from the 78th Division and attached to the 82nd Airborne in place. Task Force A had been formed, consisting of the 517th and the 505th Parachute Infantry. The 517th was to continue its planned attack.

During darkness on February 7th, the 1st and 2nd Battalions prepared to go on the attack. At 2145 the 2nd Battalion moved down the lane through the minefields. By 0100 Company E and the remains of Company F were at the edge of the Kall Ravine. At 0145 the 1st Battalion was 400 yards southeast of Hill 400. North of the Kall, the 2nd Battalion troopers came under savage machine gun and mortar fire. The 1st Battalion rearranged to Hill 400. At noon a 3rd Battalion patrol was sent west to contact the 505th at the predesignated point on the Kall. Three efforts to reach the point were turned back by machine gun fire.

The rifle strengths of the 517th Battalions, now reduced to company size, would be relieved by the 508th Parachute Infantry that night.

December and January casualties were 653: 565 wounded and 78 killed. February casualties in Germany were 287: 235 wounded and 52 killed. These numbers do not include evacuations attributable to disease and frozen extremities.

EPILOGUE

The 517th Parachute Regimental Combat Team accumulated over 150 combat days during five campaigns on battlefields in Italy, France, Belgium and Germany.

The battalion casualty rate was 81.9 percent. The Team suffered 1,576 casualties and had 247 men killed in action.

PFC Melvin E. Biddle B/1/517th PIR was awarded the Congressional Medal of Honor for heroic actions during the Soy-Hotton engagement.

On February 15, 1945, elements of the RCT were assigned to the 13th Airborne Division. The 13th was deactivated in February of 1946.

In addition to the one Medal of Honor, troopers of the 517th PRCT earned 131 Silver Stars, 631 Bronze Stars, 1,576 Purple Hearts, 6 Distinguished Service Crosses, 5 Legion of Merits, 4 Soldier Medals, 2 Air Medals and 17 French Croix de Guerres.

Demostration jump, August 1943 - 541st PIR, Ft. Benning, Georgia. (Courtesy of Harry Chase).

541ST PARACHUTE INFANTRY REGIMENT

By Harry Chase

Although never to taste combat under its own colors, the 541st Parachute Infantry Regiment nevertheless played an important role in the development of the concept of American Parachute Infantry.

The Regiment was activated at Ft. Benning, GA on Aug. 12, 1943. Reportedly, its ranks were filled with top graduates of the jump school and men who had scored exceptionally well on their army entrance tests.

The unit became part of the country's strategic reserve and was utilized to test new doctrine and tactics as well as new delivery methods and weaponry. The well-trained and well-disciplined 541st served the Airborne well as they consistently proved the value of airborne warfare.

Colonel Ducat M. McEntee, a 1935 graduate of West Point, commanded the 541st during its entire time in the roll of U.S. Armed Forces. His executive officer was Major Harley N. Trice.

One mission performed by the 541st had a profound effect on airborne warfare by American forces. Known as the Knollwood Maneuvers in North Carolina during early December 1943, the exercise involved the 17th and 11th Airborne Division and the 541st Parachute Infantry Regiment.

Washington had become disenchanted with the use of airborne troops and, fearing high casualties, was on the verge of disbanding all airborne units in favor of line infantry. The two-week maneuver was scheduled to test the value of continuing the Airborne concept.

Little suspecting the consequences of the maneuver, the troopers of all the involved units pitched in with typical Airborne enthusiasm and despite some pretty horrid weather conditions, proved beyond a doubt the value of the Airborne concept. The 541st was particularly outstanding as they accomplished their mission with high grades and the Airborne was saved.

In addition, the 541st was continually called upon to demonstrate airborne tactics for congressional visitors and other VIPs. There is really no way of telling how many times their professional skills were the difference between funding and personnel cuts for all the Airborne.

From the ranks of 541st came many of the officers and men that formed the cadres for other new airborne units. The 541st also conducted a 13-week basic Airborne Infantry Training Cycle and from this program came many of the men assigned as replacements to overseas airborne units.

The 541st was stationed at Ft. Benning on two separate occasions and at Camp Mackall, NC on two occasions. In July 1945, the 541st finally was shipped overseas, arriving at Manila in the Philippines with a pending assignment to the 11th Airborne Division.

The 11th was then preparing for a combat assault on the Japanese homeland and the prospects for casualties were expected to be tremendous, especially among the airborne forces that would lead the way.

Much to the disappointment of the men of the 541st, the unit was then deactivated and the personnel absorbed into the 11th Airborne Division's parachute regiments. The 541st, after three years of vaunted service, was no more.

The rest is history. The atomic bomb brought about a quick end to hostilities in the Pacific Theater and the 11th Airborne deployed to Japan as an occupational force.

Although the 541st never saw combat as a unit, "graduates" of the regiment served in all the major airborne divisions and saw combat in every World War II Theater. Three major cadre shipments were sent to the European Theater to augment the troops' losses there.

The 541st consequently has no combat record as a unit. But, from its ranks came many of the top officers and NCOs who wrote Airborne history. It deserves and has a unique place in the 50 years of Airborne warfare in the United States Army.

541st PIR, August 1943 - Ft. Benning, Georgia. Demostration jump through smoke into infiltration course. (Courtesy of Harry Chase).

542ND PARACHUTE INFANTRY REGIMENT

By John C. Grady

The first event related to the birth of the 542nd Parachute Infantry Battalion was the activation of the 542nd Parachute Infantry Regiment 1 Sept. 1943 at Ft. Benning, GA per G.O. #58 issued by the Airborne Command.

The designated commanding officer was Lt. Colonel William T. Ryder per Special Order 206 dated 28 Aug. 1943 at Camp Mackall, NC, this being the headquarters of the Airborne Command.

Historical Note: The 542nd Parachute Infantry was structured as an Infantry Regiment, but did not use the title of "regiment." The first staff officers were per General Order #1: 1 Sept. 43: executive officer, Major George E. Bailey; adjutant, Captain Robert E. Speer; intelligence (S2), 1st Lt. Charles R. Barrett; plans and training (S3), 1st Lt. Alex L. Wallau; supply (S4), Captain James C. Givens.

The 542nd Parachute Infantry Regiment was designated as a "training regiment" for advance infantry training of newly graduated students from the Parachute School before their movement overseas. Field commanders indicated that they were receiving many troopers that did not have sufficient basic and advanced infantry training before arrival at their theater of operations.

6 March 44: The foundation for the 542nd Parachute Infantry Battalion was formed with Special Order #56, issued by headquarters, 542nd Parachute Infantry Regiment at Ft. Benning, GA (Alabama area). Paragraph 4 reads: "Lt. Colonel Wood G. Joerg, 020793, Inf. is hereby relieved from duty as regimental executive officer, and assigned to HQ & HQ Company, 3rd Battalion and designated duty as battalion commander."

This same special order transferred 20 officers to the 3rd Battalion along with Lt. Colonel Wood G. Joerg. They formed the initial company officers structure, designated as Companies G, H, I & HQRS.

17 March 44: The 542nd Parachute Infantry Regiment was designated the 542nd Parachute Infantry "Battalion" per General Order #22, Headquarters, The Parachute School, Ft. Benning, GA.

The battalion was to be commanded by Major Arthur D. Henderson and staffed by transfers and volunteers from the "Regiment," both officers and non-commissioned officers. The basic framework of men being from the 3rd Battalion of the "Regiment." This unit was first quartered in the Alabama area of Ft. Benning, GA, then moved to Camp Mackall, NC, 1 July 44.

During the short period of time that Lt. Colonel Wood G. Joerg commanded the 3rd Battalion, he set the tone and inspired and motivated the small group of officers, cadre and enlisted men. He expected them to excel in their duties with a smile! He expected them to be mentally and physically tired at the end of each day of duty. The office and service personnel were ordered to take field training with the line-company troops. The tactical mission was rumored to be the demolition of an electrical power dam located in the industrial Rhur Valley of Germany and training was directed for such an operation. However, the unit was never committed to such an operation. When the tactical assignment was canceled, Lt. Colonel Wood G. Joerg returned to his former unit, the 551st Parachute Infantry Battalion, located in Europe.

The first separate 542nd Battalion special order was issued on 21 March 1944 at Ft. Benning, GA. This order assigned 27 company officers and most of the non-commissioned officers to the companies. This first special order was signed by order of Major Arthur M. Henderson.

Major Henderson would remain as the commander of the Battalion and Airborne Center Training Detachment at Camp Mackall, NC until 25 May 1945.

1 July 44: The 542nd Parachute Infantry Battalion remained at Ft. Benning, GA (Alabama area) until it was relocated to Camp Mackall, NC.

The advance party moved north per Special Order #73 dated 23 June 1944. The move was completed by 7 July 1944.

At Camp Mackall the battalion was assigned to the Airborne Center Command Headquarters. The unit remained at this assignment for one year, still retaining the designation of 542nd Parachute Infantry Battalion.

July 1, 1945: the unit changed its name to the "Airborne Center Training Detachment" and it was reorganized by the removal of Companies G and H. These two "line" infantry companies were dissolved and many were transferred to Company I, which was renamed the Infantry Company. G and H Companies were replaced by a glider company and a battery of 467th Parachute Field Artillery.

During the life span of the 542nd Parachute Infantry Regiment, the 542nd Parachute Infantry Battalion and the Airborne Center Training Detachment furnished our overseas airborne divisions and regiments with several thousand highly trained and motivated fighting paratroopers and glidermen.

Also, the 542nd Parachute Infantry Battalion provided the Airborne Command with paratroop units for equipment testing and development, airborne firepower and operations demonstration teams, "War Bond" sales promotion teams and U.S. Air Force troop carrier command training schools.

The Airborne Center (Airborne Command) was transferred to Ft. Bragg, NC along with all of its support troops during the month of November 1945. Camp Mackall, NC was closed shortly thereafter.

The Airborne Center Training Detachment at Ft. Bragg, NC was transferred to Ft. Benning, GA per movement order dated 17 January 1946, effective 21 January 1946. After arriving at Ft. Benning and while at Ft. Bragg, NC, most of the personnel were processed for discharge or re-enlistment. The remaining troops were redesignated as members of a training regiment.

The senior commanding officers of the 542nd Parachute Infantry Battalion and the Airborne Center Training Detachment were: Lt. Colonel Wood G. Joerg- Battalion Commanding Officer; Lt. Colonel Arthur M. Henderson- Battalion Commanding Officer; Major Ronald F. Thomas- Battalion Executive Officer; Major Robert H. Miller- ABCTD Commanding Officer; Major David Rosen- ABCTD Executive Officer.

RECOLLECTIONS OF THE 542ND PARACHUTE INFANTRY REGIMENT

By Brig. Gen. (Ret) William T. Ryder

BACKGROUND

In early April 1943, while serving as Parachute Training Officer on the Airborne Command at Camp Mackall, I was assigned to Temporary Duty as Liaison Officer with the 52nd Troop Carrier Wing, then at Pope Field, North Carolina.

The 52nd TCW was preparing to fly to North Africa in support of U.S. airborne operations planned for Sicily (Operation HUSKY).

My assignment was to provide liaison between the 52nd TCW and the 82nd Airborne Division during combat planning and training, accompany the paratroopers into combat and, on completion of the action, return to the Airborne Command with an appropriate report and training recommendations. Additionally, I was to carry in and set-up for testing and subsequent troop carrier use, one of the early model homing devices for guiding troop-carrying or supply aircraft to drop zones.

On 13 July, I jumped into Sicily with the 505th Parachute Infantry from the plane carrying its regimental commander. Seven years earlier, I'd served with then Lt. Jim Gavin. Little did I then envision that good fortune was to grant me the opportunity for an intimate, close-up observation in combat, of one of World War II's truly great military commanders.

Following the 505th's night assault, another night drop was made on 12 July by the 504th Parachute Infantry. On that fateful night, 23 aircraft were shot down by anti-aircraft fire. The fire was mainly from supply and troop ships laying offshore that had been subjected to punishing attacks by German night bombers, just before our troop carriers arrived over their convoys.

Despite this mishap, Operation HUSKY was successful, and on completion of its assault and mopping-up phases, I rejoined the 52nd TCW in North Africa. Soon after I was sent to appear before a joint services board, called by Air Marshall Tedder in Algeria, to investigate and determine corrective measures for the tragic mishap of the 504th's follow-up night drop. On completion of that assignment I was ordered back to the Airborne Command at Camp Mackall.

During my absence, the Airborne Command had been authorized to activate another parachute regiment...the 542nd Parachute Infantry Regiment. To my great surprise, and greater delight I was designated to command that regiment!

ACTIVATION

Paragraph 10 of Special Orders No. 206, Hq Airborne Command dated 28 August, 1943, relieved Lt. Col W.T. Ryder from duty with Hq A/B Comd and assigned him to "542d Prcht Inf, Ft Benning, Ga effective upon activation 1 Sept. 1943, and atchd Prcht Sch, Ft Benning, Ga pending activation of the 542nd Prcht Inf."

With eager compliance, I proceeded to Benning where I learned that the 542nd had been allocated a regimental tent encampment area just across the Chatahoochee River in Alabama. I learned also that 15 officers and 50 enlisted men, most of whom held non-com ratings, were enroute to serve as a cadre to help organize and complete activation of our 1,500-man regiment.

Although the cadre contained few "old timers," all of them, officers and men alike, were highly-spirited, completely dedicated and eager to get on with the demanding job of shaping, moulding and training a parachute regiment that would truly be their very own!

There was a special added incentive to boost everyone's morale. The Airborne Command had indicated that <u>if</u> the 542nd did well in its Basic and Unit Training Phases, there was a strong probability that it would be sent overseas early for completion of its Combined Training!!

By late September, sufficient officers and enlisted men were on hand to flesh out the regiment. It was austerely, but adequately encamped under canvas in Alabama, fully-equipped and eager to embark on a vigorous, all-out, full-time training program. Officers and men were charged-up with hopes and aspirations that the turn of the year would see us well on our way to join in the war <u>overseas</u>!

TRAINING

The Basic Training Phase, as the name implies, is designated to instruct and condition the soldier so that, in combat, he can well-care for himself and his equipment and well-execute his assigned role within the basic unit of which he's a part. Starting with untrained, but healthy and fit men, this phase normally takes 12 to 14 weeks.

In addition to special physical development necessary for men to perform effectively in combat, they must acquire the skills and knowledge required for proper care and use of assigned weapons and equipment. Finally, they must learn how to perform their individual roles as supportive team players within the basic squad unit.

Although obviously biased, I believe that the 542nd worked especially hard and performed exceptionally well throughout the arduous basic training period. I sensed that officers and men alike were inspired by an awareness of prospects of early overseas deployment.

I vividly recall two rather unique attitudes manifested by the troops during this initial training phase. One had to do with night training and the other our men's aggressive pride in their status as paratroopers.

Doctrine emphasized that a dark, night-time environment favored the paratrooper, once committed into action. Although the enemy might be defensively entrenched, he could not be certain about the numbers, locations, or concentrations of the parachutists deployed against him. In short, while the enemy might not know where the paratroopers were...the paratroopers did know where the enemy was! Consequently the paratrooper could move and attack more effectively than the enemy at night.

One preliminary exercise used to train individuals in night movement, was to send them on a short, half-mile night course with only a luminous compass and a heading to an initial station, about 250 yards distant from the starting point. From that station, he'd be given a new heading to the next station...and so on, until the course was completed. The station consisted of an instructor, well-hidden under a shelter half, from which a weak gleam of light from a lantern or battery lamp was permitted to escape along the prescribed approach route. The approaching soldier had to be reasonably accurate on the assigned compass heading, or he would not likely find the initial, or succeeding stations as he proceeded through the night. He was of course, admonished to move stealthily and quietly in the dark and avoid any outcries or noisome movements.

Results of such preliminary night training revealed that the "country boy" was generally more at ease and could move more effectively and efficiently at night than the "city boy." The former was evidently more familiar with, and at ease in, working and moving within a "nighttime" environment. The latter, accustomed to lighted streets and thoroughfares at night, found exposure to the dark a somewhat bewildering and often scary experience, by virtue of a lively imagination trying to sort out strange sights, sounds and shapes in the dark.

But persistent drilling in these and other night-time exercises soon successfully developed the necessary confidence and aptitude for night operations.

There was another form of "night operations" in which 542nd troopers appeared to excel. These were periodic fist fights in nearby Columbus, Georgia that, too often, erupted when airborne "paratroopers" and armored forces "tankers" encountered each other in local bars or clubs.

Both elements were convinced that they were the truly _elite_ members of the Army's fighting forces. Both were ready, at the drop of a hat, to prove their convictions. Should a member—or members—of one group make any disparaging remarks or gestures toward the other, immediate and intense unarmed combat almost certainly followed. Invariably the local police had to be called in, in numbers, to restore order, separate combatants and put them under safe constraint, until they could be returned to their units.

Many's the time I, or another regimental officer, had to hurry down to Columbus, placate police and owners of damaged premises, and return to camp with a bedraggled, battered but unbowed group of troopers. Routine military justice was, of course, administered, but the defense was invariably the same, viz, no self-respecting paratrooper could reasonably be expected to tolerate passively such unwarranted slurs from tankers!

However, those high-handed and high-spirited shenanigans were not permitted to interfere with the serious and all-important job of getting on with our training. By early December, I was convinced that when Airborne Command inspectors came to administer Basic Training Proficiency Tests, the regiment would be passed with high marks. But, circumstances unknown to us and beyond our control, were to disrupt our training and end our aspirations for early regimental deployment overseas.

In the latter part of December, the 542nd was directed to provide about a hundred men as Pacific area replacements, and I was directed to accompany them, by troop train, to the Ft. Ord, California embarkation center. Regulations, at that time, required airborne replacements shipping out from the West Coast to divest themselves of wings, jump boots and insignia that identified them as airborne, while awaiting shipment overseas.

The purpose of such constraints was to preclude any local espionage agents who might be in the area from noting and reporting on movements of airborne troops.

Not only were there frequent violations of those orders, but often "violent" demonstrations with duly constituted local military personnel, charged with their enforcement.

The Airborne Command directed me to deliver our replacements in good order, investigate embarkation processing problems with the Ft. Ord staff and recommend troop indoctrination and other measures that might alleviate the situation.

While at Ft. Ord I received urgent, secret, telegraphic orders to return immediately to Benning. The orders authorized "first priority" air travel enabling me to get the first available homeward commercial or military air space.

Although mystified, I was elated by anticipations evoked by the "urgent" nature of the order. I felt certain the orders had something to do with movement overseas. My anticipation was valid, but my elation was unjustified!

On arrival at Benning, I was greeted with the disheartening and morale-shattering news that the 542nd had been directed to select and process, without delay, about a thousand men, for immediate shipment to England. The troopers were needed to fill out airborne units, then begin preparations for our invasion of Europe. The Airborne command orders specified not only that we expedite the replacements processing, but that they be the best conditioned, trained and qualified troops the regiment could provide. No men with injuries on the sick lists or under any current or pending courts-martial action were eligible for shipment.

With heavy hearts, but in a determined and disciplined manner, our officers and non-coms compiled with the orders' letter and spirit. By mid-January less than a battalion strength of men remained in the 542nd encampment.

And...there were no indications from higher headquarters when, or if, the regiment would be reconstituted to full strength. Even should that occur, those of us remaining were faced with the unhappy thought of having to start at Basic Training Phase, all over again.

Prior bright hopes for early regimental deployment overseas were dissipated. They were replaced by doubtful uncertainties regarding our respective personal and organizational futures.

During this disheartening period, I'd been promoted to the rank of full colonel. The exultation from the "eagles" was sobered by the awareness that I'd lost a regiment. There is no question in my mind that, given a choice, I would have willingly and gladly swapped the former for the latter!

Soon after my promotion, by fortunate coincidence for me, the Airborne Command was directed to dispatch an officer, with the rank of full colonel, to the South West Pacific Theatre to serve as Airborne Officer on General MacArthur's staff.

Aware of my keen disappointment and perhaps as a compensation for the disastrous loss of "replacements," the Airborne Command offered to relieve me from the 542nd and to accept the assignment to General MacArthur's staff.

Here, at last, was a certain and immediate chance for shipment overseas. Naturally, I gratefully accepted the offered assignment and by mid-February 1944, had joined General MacArthur's staff in Brisbane, Australia. And, it was to be my good fortune to serve under General MacArthur from the day I joined his headquarters in Australia until he departed from Tokyo, Japan to end his long and distinguished military career.

This is a brief summation of my recollections. I must leave to those who remained behind with the 542nd Parachute Infantry Regiment to carry on the saga of its history.

550TH AIRBORNE INFANTRY BATTALION

By John A. Wentzel

The 550th was activated July 1, 1941 in the Panama Canal Zone at Howard Field.

The unit was made of regular Army volunteers from the 33rd Inf., 5th Inf., 14th Inf., 2nd FA Signal Corps and engineers based in the Canal Zone.

The U.S. Army Center of Military History has acknowledged that the 550th was the first "Airborne Infantry Bn," so constituted.

We trained in B-18 bombers before the CG-4A gliders came down to Panama.

The 550th departed from Panama on 8-20-43 and arrived 8-30-43 in San Francisco. Departed from there by troop train 9-3-43 and arrived at Camp Mackall, NC 9-9-43. Members of the 88th Airborne Infantry Bn. were then assigned to our unit. The 88th Airborne Infantry Battalion was formed at Ft. Benning, GA. It consisted of all volunteers, regulars or draftees. Moved to Ft. Bragg and was stationed at 9th Div. and Log St. roads. They trained the Airborne units that were stationed at Ft. Bragg. Later an instructional team was selected and sent to South Dakota. The balance of the unit was to be assigned to the 550th unit at Laurinburg Maxton Air Base in Laurinburg, NC.

Departed (POE) to Newport News, VA, 4-3-44. Arrived in Oran, Africa 5-10-44 with C Co., under Capt. Sumpter in command arriving in Barri Italy. Later to be joined with the Bn. in Sicily. Security mission in Oran 5-18-44. Left by boat for Naples, Italy. Arrived 5-21-44. Moved by rail to Reggio Calabria and then by ferry to Messina, Sicily early in June. Trained three weeks in Sicily. Returned to Italy, vicinity of Rome, early in July. Prepared for the invasion of Southern France. Proceeded to Follonica Airfield 8-14-44.

On Aug. 15, 1944, took off from Italy by glider in the a.m. Our LZ was near LeMuy which was attacked and secured. Moved to town between Frejus and LeMuy 8-29-44 to take up defensive positions.

Moved to vicinity of Grasse and from that area we moved to Barcellonette, 8-29-44. Assumed defensive position and patrolled Condamire and the Condamire Valley. Many small unit attacks.

Moved to rest area at Latesque, France, Created roadblocks and security patrols. Contacted 551st PIB at St. Martin Vesuble. Left Lantosque 12-7-44 moving through Reims and LaHarve to Aldbourne, England by boat. Arrived around 12-10-44. Moved to Oxford 12-22-44.

Left by plane for France 12-25-44. Landed at Dreux and Reims. Moved by truck to vicinity of Sedan. Left for Bastogne 1-1-45 moving through Nevfchateau and arriving at Laveselle 1-3-45. Offensive and defensive combat in the Battle of the Bulge. Moved into attack positions. Heavy fighting at Renvamont and in the woods south of Hillmont. Began mopping up woods near Bethomont, Belgium. 1-17-45 and moved to Raisonkneul 1-23-45.

Battalion moved to rest area. Arrived at Huldangs and remained until 2-3-45 then moved to Chalons-Sur area until 2-10-45.

Inactivated effective 3-1-45, transferred to the 17th Airborne Division and formed the 3rd Bn. of the 194th GIR in time for the mission across the Rhine.

Lineage

Constituted 10 June 1941 as the 550th Inf. Airborne Inf. Bn. Activated July 1, 1941 at Howard Field, CZ. Disbanded March 1, 1945 in European Theater.

Reconstituted and redesignated 550th Airborne Inf. Bn., 11-23-48 at Camp Picket, VA.

Campaign Streamers (WWII)
1. Rome - Arno
2. Southern France (with arrowhead)
3. Ardennes - Alsace
4. Rhineland

Deaths in Action

Fifty-four men lost in action. Mostly were casualties in the Battle of the Bulge as we were on the southern part and were fighting to break through from the South to free the troops.

Deaths since Discharge

There has been 34 known deaths since the association was formed 37 years ago.

The 550th made the Southern France invasion and then joined the 17th Airborne for the Varsity Operation.

551ST PARACHUTE INFANTRY BATTALION

By Douglas C. Dillard

ELEMENTS
Hq. & Hq. Co.; A, B, C Companies and Medical Detachment.

ACTIVATION
This was a separate unit *(The Army Lineage Book, Volume II, Infantry,* page 833): Department of the Army, 1953: "Constituted in Army of the United States 30 October 1942 as 551st Parachute Infantry Regiment. Elements of 1st Battalion activated 26 November 1942 at Fort Kobbe, CZ 1st Battalion inactivated in Belgium and Regiment concurrently disbanded 10 February 1945."

Actually, the personnel of the 551st Parachute Infantry Battalion (there never was a regiment) were selected at Fort Benning, (the Frying Pan), Georgia, in early December 1942 (less C Company, which was added to the Battalion at Fort Kobbe, CZ in early January 1943).

CAMPAIGN CREDITS
World War II: Rome-Arno; Southern France (with arrowhead); Rhineland; Ardennes-Alsace.

UNIT CITATIONS
All companies of the 551st are entitled to French Croix de Guerre with silver-gilt star embroidered "Draguignan" (DA GO 43, 1950).

"An elite parachute unit of magnificent bravery and well-tempered morale. During the landing operations in Provence (Southern France), it found itself, on 16 August 1944, at Draguignan, completely cut off from the main body of the Allied forces. Limited to its own arms for five days, it endured the furious enemy attacks which were supported by a rain of artillery and mortar fire. In spite of heavy losses, it proceeded to attack, freeing Draguignan and taking numerous prisoners, among them several of high rank in the German army".

DISTINGUISHED SERVICE CROSS
PFC Milo C. Huempfner
Narrative of action: Headquarters Company, 551st Parachute Infantry Battalion, for extraordinary heroism in connection with military operations on 23 and 24 December 1944. When his truck was immobilized and encircled by the enemy, Private Huempfner, after destroying his vehicle, remained in a Belgian town for two days waging single-handed warfare against a hostile force. During this period, he destroyed two armored vehicles, neutralized a machine gun position and killed three of the enemy. On three occasions, he left his place of concealment to warn friendly convoys of the presence of the enemy, thereby saving many lives and much vital equipment. The extraordinary heroism and individual courage of Private Huempfner reflect great credit upon himself and are in keeping with the highest traditions of the military service. Entered military service from Wisconsin.

COMMANDING OFFICERS
The only graduates of the U.S. Military Academy:
LTC. Wood G. Joerg, 0-20793 (26 November 1942 to 30 August 1943 and 2 February to 7 January 1945)
LTC. Rupert D. Graves, (1 September 1943 to 1 February 1944)
Major William N. Holm, 0-23282 (8 January 1945 to 10 February 1945).

CASUALTY SUMMARY
In operation DRAGOON, the 551st Battalion was in continual action from the D-Day drop on 15 August 1944 until 17 November 1944 for a total of 94 days: approximately 30 KIA and 150 WIA. Casualty rate: 21 percent.

In the Battle of the Bulge, in which the Battalion was engaged from 21 December 1944 to 8 January 1945 for a total of 18 days: approximately 175 KIA and 615 WIA (plus two captured while guarding 12 of our wounded). Casualty rate: 94 percent.

On 8 January 1945, the Battalion strength was 14 officers and 96 men (45 of these were walking wounded who remained on the line for duty.

TRAINING
The 551st Parachute Infantry Battalion, upon arrival in the Panama Canal, was organized into lettered companies of H & HQ, A, B and C. (Note C Company was formerly C Company of the original 501st Parachute Infantry Battalion). The other elements of the 501st had joined the 503rd Parachute Infantry Regiment as it passed through the Panama Canal enroute to the Pacific.

The jungle training expertise was utilized by the new battalion to support specialized small unit operation, such as Special Forces of the 1970s. All elements of the battalion were permitted to elect a specialty such as: demolitions, light aircraft, operation of armored vehicles, light boat handling, communications, etc. Conceptually, the battalion could field any type specially-tailored force for any type operation, in a jungle environment.

Additionally, development and experimentation with all types of airdrop activities continued and helped in the eventual acceptance of squad-size air bundles, nylon net cargo containers for large, bulky items and the canvas MI rifle containers the Airborne used throughout World War II. The Rigger Section, led by, then 1st Lt. Bud Schroeder, personally tested each development before the line troops. Several low altitude drops were made in Panama from altitudes as low as 500 feet. Significantly, no unusual number of casualties resulted from these tests.

The objective of this arduous and extensive training was to make a parachute assault followed by a glider landing to capture the island of Martinique from the Vichy French. The island was being used as a resupply base for German submarines in the Caribbean.

Before the scheduled operation, the island government changed over to the Free French Forces.

While under the Airborne Command at Camp Mackall, NC, the battalion conducted nine months of company and battalion level exercises as a separate battalion. Several tactical jumps at company level were made at Alachua Army Air Field, Gainesville, Florida. Such operations were to test long range flight and organizational airborne concepts.

During this same period, the battalion was selected by the Airborne Command to test the use of gliders as transports for parachutists. Several individual and company-sized jumps were made from gliders. There was no significant casualties from these tests.

The jumps were made from gliders (CG4A) and resulted in a mass battalion jump on October 31, 1943 at Camp Mackall, NC The troopers jumped from C-47s towing two gliders each transportation paratroopers. The ground pattern was very close due to rapid delivery from these transports simultaneously.

For these test jumps, the battalion was commended by M.G. Chapman, CG, Airborne Command, Camp Mackall, NC.

Upon arrival in Sicily, near Trapani, the battalion conducted company-and battalion-level exercises for a 60-day period. Such training was oriented toward Italian campaign-style combat versus the jungle training for which the battalion was noted.

During these exercises, two members of Company B were killed, due to short rounds of 81mm mortar fire used in the exercise. One facet of training development was the placement of 81mm mortars in sand bagged jeep trailers to facilitate mobile fire support since the 81mm mortar was the only organic fire support elements the battalion directly controlled.

CREST/LOGO

Crest was designated by the enlisted men in Panama in 1943 via a battalion-wide contest. The silver eagle with blue shield on breast containing a green palm tree, symbolic of activities in Panama, with the exposed machete, the weapon of the jungle and a red bolt of lightening symbolic of the battalion motto, in letters "ATTERICE Y ATAQUE"—"LAND AND ATTACK."

MISCELLANEOUS

The battalion has been honored by a memorial erected by the French and Belgian citizens at the following locations:

FRANCE
  A. Draguignan- The Avenue of the 551st Parachute Infantry Battalion
  B. Valbourges- Dedication of Family Chapel- Stevens Estate.
  C. Marble Tablet dedicated in St. Martins ve Subie, Maritime Alps.
  D. Bronze Plaque dedicated in Isola, Maritime Alps.
  E. Bronze Plaque dedicated in St. Entenne d' Tinee, Maritime Alps.
  F. Bronze Plaque dedicated in La Turbie.

BELGIUM
  A. La Chappel- Odrimont - Noire-Fontaine.
  B. Liernaux- Trois Ponts.
  C. Rochelinval.

In Rochelinval the battalion dedicated a marble marker in remembrance of the Belgian people.

COMBAT OPERATIONS
CODE NAME: DRAGOON/ANVIL

D-DAY: 15 August 1944

AIRBORNE UNITS INVOLVED
  551st dropped as a separate unit, subordinate to 1st Airborne Task Force
  Other units: 517th Reg. Combat Team; British 2nd Indep. Para Brgde.; 509th Prcht. Inf. Battalion; 460th, 463rd and 596th Arty/Engineer Support Units, 550th (ABN) Glider Bn.
  Note: These other units except for the 550th, which followed the 551st) dropped some 13 hours prior to the 551st drop.

TROOP CARRIER ELEMENTS
  437th Group, 53rd Wing, Army Air Corps

MISSION STATEMENT
  To draw German defensive forces back from the coast of South France and thus aid the U.S. 7th Army ground forces' amphibious assault.

OBJECTIVES
  GEOGRAPHIC: The city of Draguignan was the first designated objective following consolidation.
  TACTICAL: Disrupt rear-area communications, destroy enemy material, aid the French Resistance.

ENGAGEMENT DATA
  U.S. KILLED: 3 IIA on landing; total: 19 KIA and 60 WIA during first 8 days. 11 KIA and 90 WIA during remaining 86 days of action in South France.
  ENEMY FORCES: Captured approximately 700 during 94 days of Southern France Campaign.

SYNOPTIC ACCOUNT
  The 551st made an eminently successful on-target landing at 1806 hours on 15 August 1944, occupied the city of Draguignan on D+1 (including the capture of Maj. Gen. Bieringer, commander, German 62d Corps, with his staff). The Bn. then moved south through the lines of the approaching U.S. 7th Army forces and turned east, fighting its way along the coast of the French Riviera. Together with the 509th Prcht. Inf. Bn., it then liberated Cannes on 24 August. On 29 August elements of the 551st liberated the city of Nice. The Bn. then continued driving the German forces in a northeast direction, proceeding up into the French Maritime Alps.

For the next 12 weeks, the 551st, with the 509th on its left and the 550th Glider Bn. on its right, fought a successful holding action against the Austrian 5th Hochgebirgesjager Division, in an effort to protect the right flank of U.S. 7th Army. The 551st was relieved by the U.S. 100th Infantry on 17 November 1944.

Significantly, on D+2, the tactical plan changed for the 551st from defense to offense. The Bn. S-2 (Captain Hartman) was advised by members of the French Resistance that the German troops had pulled out of Draguignan and that French flags had been unfurled in the city. Now, the S-2 was told, the Germans were returning to the city and the French patriots would be placed in a precarious position.

After advising HQ. 1st ABTF, the Bn. was ordered to "hold present position with minimum force. Attack and seize Draguignan." The major element of the battalion began a night infiltration of German positions to secure the high ground over Draguignan. The battalion attack by all companies surprised the German garrison of approximately 750 men. By noon D+3, the battalion had secured the town, destroyed the element of the German LX11 Corps, and captured M.G. Bieringer.

Draguignan was entered and effectively secured early on the morning of August 17, 1944. It was the first major city in southern France to fall to Allied Forces. It should be noted that the 551st held the city for three days before the link-up occurred with the forces that landed on the beach—a scheduled 24-hour operation. For this action, the battalion was awarded the French Croix de Guerre with silver gilt star.

COMBAT OPERATIONS
CODE NAME: BATTLE OF THE BULGE

D-DAY: N/A

AIRBORNE UNITS INVOLVED
551st Prcht Inf. Bn. (Attch'd to 30th Inf. Div.)
HQ & HQ Company
Companies- A, B and C
Medical Detachment

TROOP CARRIER ELEMENTS
N/A

MISSION STATEMENT
Intercept the German drive through the Ardennes ("Bulge"); deplete German combat resources as heavily as possible in the process.

OBJECTIVES
GEOGRAPHIC: 7 Jan 1945, Rochelinval; 24 Dec, Schmidt; 4 Jan, St. Jacques; 27 Dec, Noirefontaine; 5 Jan, Dairomont.
TACTICAL: Turn back and destroy German thrust through Ardennes

ENGAGEMENT DATA
U.S. KILLED- KIA, approximately 175; WIA, approximately 615.
ENEMY FORCES- Captured, approximately 425

SYNOPTIC ACCOUNT
The battalion entered the Battle of the Bulge on 21 December 1944 with an augmented strength of approximately 860 officers and men. Initially attached to the 30th Infantry Division, en route to Francochamps, Belgium. The battalion came under fire from German SS Forces fording a stream at Hotton, Belgium. By rerouting the convoy, the German spearhead was bypassed by the battalion. During the period 080021 Dec 1944, the battalion reinforced 30th Infantry Division position in and around Francorchamps, Ster and Stavelot, Belgium. On 24 December 1944, the battalion was ordered to attack and capture Schmidt, where the 30th Infantry Division casualties were trapped by German forces. After moving into attack positions, the 30th Infantry Division realized the 551st was only one battalion of paratroopers, not a full regiment, so the attack was cancelled. On 25 December 1944, the battalion was attached to the 82nd Airborne Division and moved to Rahier, Belgium in Division Reserve, but in support of the 508th Parachute Infantry Regiment.

On order of General Gavin, the battalion executed a reconnaissance in force. The battalion moved through the 508th lines 110027 Dec 1944 to throw the Germans off balance and to capture prisoners and identify the German units in contact. Secondary mission was to place fire on La Chapelle, Odrimont and Amcomont. The objective Noirefontaine was 4,000 yards from the line of departure. The battalion crossed the LOD at 230027 Dec 44, in columns of companies in the following order: B, C and A—Light Machine Gun Platoon attached to Company A and Company C, 81st MM Mortar Platoon in support with mortar observer and an SCR 300 Radio attached to Companies A and C. Both A and C Companies had an artillery forward observer attached. XVIII Airborne Corps Artillery elements were in direct support.

After considerable close-in in contact with German Forces, Company C swung around Company B to isolate the village. After 30 minutes of fighting and 400 rounds of HE light ammunition expended on the area, the position was taken. A and B Companies repulsed two determined German counterattacks. Fighting ensued in all directions and became rather confused. At 09230, the battalion was on its objective Noirefontaine; due to timing the battalion had to begin its withdrawal by 0500. The mission was accomplished, despite the fact that the Germans maintained contact with the battalion for approximately 1000 yards after the withdrawal began:

Score for the Raid:
A. 30 German KIA (including a company commander).
B. Six prisoners.
C. Destroyed one American, captured halftrack.
D. U.S. casualties: 1 KIA, 2 MIA and 15 WIA.

By 020028 Dec 44, the battalion recrossed the lines of the 508th Parachute Infantry Regiment.

On 1 January 1945 this unit was in defensive positions, in the vicinity of Rahier, Belgium. On 2 January 1945 the organization was attached to the 517th Regimental Combat Team, for an assigned mission on 3 January 1945. During the entire day of 2nd January, preparations were made for the attack of the following day.

On 3 January 1945 the battalion moved out as an assault unit in the northern offensive. The Battalion Command Post was established at (630978) Coordinates. The leading element left the line of departure at 0935. Enemy was immediately contacted. From 3 January 1945 to 7 January 1945 the battalion was heavily engaged against the enemy and seriously depleted by casualties, suffering both from exposure and wounds. The enemy units which were engaged, during this period, were initially, the 190th Regiment, 62nd VG Division and then the 183rd Regiment of the same division.

On 5 January 1945 the unit was relieved from the 517th Regimental Combat Team and attached to the 504th Parachute Infantry Regiment, effective 061312 January 1945. The Battalion Command Post was now at 678957 coordinates.

On 6 January 1945 the companies of this battalion moved out to consolidate positions west of the Salm River (690952). Contact was made with the 517th Combat Team on the left flank.

On 7 January 1945 the battalion was given a mission to take the town of Rochelinval, Belgium. Company A was to take the town while Company C set up a base of fire at the nearby woods and knocked out the road block. Company B took the high ground to the south of the town. One the town was taken the forward command group continued to direct the operation from in and around the town. The final objective (the Salm River) was reached. Companies sent patrols across the river, while the unit prepared defensive positions to hold the ground taken.

The final day of commitment, personnel of the 18th VG Division that had moved from east of the Salm River were killed, captured or wounded and large amounts of enemy equipment was captured, i.e., MG-34 and 42 machine guns, mortars and all types of small arms.

During the operation, between 400 and 410 prisoners were taken. Enemy dead was estimated to be 400. This unit lost approximately 400 men as battle casualties and non-battle casualties.

On 9 January 1945 the unit was relieved by the 517th Combat Team and moved to Juslenville, Belgium, where it remained until 27 January 1945.

The battalion, now less than company size, was to be inactivated and all personnel were transferred to the 82nd Airborne Division.

Special 555th Parachute Infantry Company unit test jumping the C-46 (Dakota) at Camp Mackall in September 1944. (Courtesy of The Shoe String Press, Inc.).

Major General James M. Gavin and First Lt. Roger S. Walden watch a stick of jumpers.

555 PARACHUTE INFANTRY BATTALION (THE TRIPLE NICKLES)

By Bradley Biggs
LTC, USA (Ret)

From day one of this nation's existence, Americans have seen a distorted picture of the American black serviceman. But in 1947, the Triple Nickles became the locomotive in moving the country out of the woods of discrimination and into the light of becoming a fairer and less hypocritical society. This contribution cannot be measured nor described in terms adequate to the moral and mature growth of our society and with all of the advantageous economic opportunities for millions—then and now.

In the wartime winter of 1943-44, many years before "black pride" became a popular slogan, a small group of black American soldiers gave life and meaning to those words. In an army that too often relegated blacks to menial jobs, they succeeded in becoming the nation's first all-black parachute infantry test platoon, company and battalion. They were the 555th Parachute Infantry Battalion of the 82nd Airborne-the Triple Nickles.

They were the first airborne soldiers to mass jump from a glider, testing this concept under the Airborne Center; the first airborne unit to pioneer and train as "Smoke Jumpers," the unit that executed one of the most difficult missions in one of the best kept secrets of World War II—combating the Japanese balloon invasion in the northwestern area of the United States; the first Army unit to train in and to demonstrate airborne tactics under atomic concepts; the first airborne unit to demonstrate joint Airborne-Navy-Air Force combat operations concepts; and for years, the largest airborne battalion in the U.S. Army.

The history of the Triple Nickles is an untold chapter in black social history. The high professional standards of the men of the 555th, their skill in airborne operations, fearlessness in testing new concepts and effectiveness in training other service personnel marked them as black pioneers whose regular Army unit was to be integrated. The 555th showed how this major goal could be achieved without incident. Their vision opened the way for the present-day military.

The story of these airborne soldiers is that of pioneering courage and one that needs retelling as a reminder and a lesson. It is a story of black men beating down humiliation with the indomitable will to succeed. It was this handful of soldiers—black soldiers—that dealt white military racism one of its most humiliating setbacks. They survived, they endured.

In so doing, they kept in the forefront of their every effort and day of service that a paramount requirement and obligation in being first is to make the path easier for those who follow.

The legend of their achievement will endure; the memory of having been a part of this great milestone in history that will never be repeated will be a lasting pride forever.

The 555th Parachute Infantry Battalion was the ultimate result of the 555th Parachute Infantry Test Platoon of seven officers and 20 enlisted volunteers. These men then constituted the cadre for Company A, 555th Parachute Infantry Battalion which was expanded to the 555th Parachute Infantry Battalion. This unit later became the 3rd Battalion, 505th Airborne Infantry Regiment when it was integrated into the 82nd Airborne Division.

Briefly the activations and deactivations were:

- Test platoon recruitment and selection by Major Hoover, S-1, The Parachute School, Ft. Benning, Georgia took place during the period late November to early December, 1943.
- Company A, 555th PIB was activated 31 December, 1943, The Parachute School, Fort Benning, Georgia.
- Redesignated the 555th Parachute Infantry Battalion on 25 November 1944 at Camp Mackall, NC per General Order No. 13, Hq., Airborne Center, Army Ground Forces, Camp Mackall, NC dated 2 November, 1944. Absorption of the 555th Parachute Infantry Company provided the nucleus of the battalion's strength.
- At redesignation as a battalion, it was under the operational control of the Airborne Center, (Gen. Darby, commanding) and later attached to the 13th Airborne Division for administration, training and supply.

Chronicle of Significant Activities: A special company-size unit was formed for operations during the Battle of the Bulge. However, the Japanese balloon attack on the west coast cancelled the European assignment and the 555th was ordered to Pendleton Air Base, Oregon for training and operations as the Army's first parachute "Smoke Jumpers." The primary unit under Operation FIREFLY, the 555th executed scores of smoke jumps successfully in combating the Japanese balloon attack on the U.S.

Returning to Camp Mackall, NC for a short period, the battalion was moved to its permanent home at Ft. Bragg, NC in late 1945.

Attached to the 13th Airborne Division for a few months, it was detached from that unit and attached to the 82nd Airborne Division in the early spring of 1946.

The battalion was designated as the demonstration unit for planning pre-jump marshalling area operations and dispersion techniques under atomic warfare concepts.

Selected to participate in several parades, it presented to the public the elan, pride and all that is best in airborne units.

In 1947 and again in 1948, as the 3rd Battalion, 505th Airborne Infantry Regiment, 82nd Airborne Division, it was selected to perform the airborne and group troop mission in several outstanding combined arms demonstrations, namely: Operation COMBINE I, Ft. Benning and Operation COMBINE II, Eglin AB, FL.

In November 1947, Gen. James M. Gavin welcomed and personally conducted the ceremony which integrated the 555th into the 82nd Airborne Division as the 3rd Battalion of the 505th Airborne Infantry Regiment, then commanded by Col. William E. Ekman.

With integration, the battalion, then the largest airborne battalion in the Army, played a significant role in convincing the military hierarchy and the public that the black Airborne soldier is second to none.

The glider tow on a Combine I demostration. Gliders were loaded with reinforcements for the parachute drop operation.

Officers of the test platoon that paved the way for the world's first all-black parachute infantry battalion (wings awarded 4 March 1944) L to R: First Lt. Jasper E. Ross, Second Lt. Clifford Allen, Second Lt. Bradley Biggs, Second Lt. Edwin H. Wills, Second Lt. Warren C. Cornelius, Second Lt. Edward Baker.

The number of black paratroopers swelled into the thousands with them playing significant combat roles in both Korea and Vietnam. The 2nd Company was the first all-black unit to make a combat jump (Korea). Many of those troopers were former Triple Nickles.

Commanders were:

555th Parachute Infantry Company- 1st Lt. James H. Porter.

Company A, 555th Parachute Infantry Company- Capt. James H. Porter.

555th Parachute Infantry Battalion- Capt. James H. Porter, Capt. Joseph Gates.

(Capt. Gates was the CO when the battalion was integrated.)

The 555th produced for the 80th AA, 503rd AA, 187th, 2nd Ranger Co., plus other airborne assignments.

OPERATION FIREFLY

The sordid history of the treatment of the Japanese Americans in World War II will forever remain a black spot on the conscience of this nation. During Operation FIREFLY, the 555th Parachute Infantry Battalion was meticulous in adhering to a fault the regulations of secrecy concerning its operations against the Japanese balloon invasion of the northwest United States in 1945. Had it not done so, the plight of the Japanese Americans would have been exacerbated to an ever greater degree of humiliation and hardship. The motion picture industry took deliberate pains to depict the Japanese as brutal and vicious; the liberation of Japanese-occupied territories in the Pacific exposed brutality by the Imperial Japanese forces; the release of American prisoners of war with stories of horrible treatment, plus accounts from our allies, added to the hostile feelings against Japanese Americans.

If it were known that Japanese balloons, the first unmanned intercontinental ballistic missiles, had been successful in reaching our shores, the Japanese military machine would have strengthened its effort in this area. If the secrecy of the 555th operation had been broken, there is no telling what additional maltreatment would have befallen the Japanese Americans in concentration camp internment.

Hence, the conduct of The Triple Nickles during this heretofore highly secret and untold story contributed immeasurably to the well-being of most Japanese Americans already incarcerated in concentration camps in the west.

For the first time in the annals of military history of any nation, a military organization of paratroopers was selected to become "Airborne Firefighters." The Triple Nickles became not only the first military firefighting organization in the world but pioneered methods of combatting forest fires still in use today.

This first, late to be recognized, is just now being honored in such famous places as the Smithsonian Institution and some national archives.

The potentially disastrous impact of this Japanese attack on continental U.S. by the first unmanned intercontinental ballistic missile can be seen in the instance when one of the balloons descended in the Hanford, Washington nuclear plant area. The balloon became tangled in electrical transmission lines near the plant making the plutonium slugs. This caused a temporary short-circuit in the power for the nuclear reactor cooling pumps. Backup safety devices restored power almost immediately, but if the cooling system had been off much longer a reactor might have collapsed or exploded, a granddaddy to Chernobyl.

Not mentioned publicly at the time was the possibility that Japan might have equipped the balloons with the capacity to carry out some form of chemical-biological warfare. Their experiments with prisoners of war in the notorious unit 731 were not known until much later—but they began in 1937 and pointed to the existence of a Japanese program to develop deadly biological agents. Such agents quite possibly could have been delivered in quantity to the United States' mainland by balloons.

TRAINING OTHER ARMY, AIR FORCE AND NAVAL SERVICE PERSONNEL

The 555th was the first military unit to organize and demonstrate how marshalling area activities for a military operation should be conducted under atomic warfare conditions. Several Army organizations viewed and learned from these demonstration activities.

Even before then, and in the mid 40s, members of The Triple Nickles trained white pilots in glider jump techniques and other aerial delivery-type operations. They were so successful that General Hap Arnold, then chief of staff of the Army Air Corps, congratulated members of the 555th training teams.

While in the northwest, on a special operation, one company helped to train U.S. Naval and Marine Corps pilots in air-to-ground operations in preparation for combat in the Pacific Theatre.

The important post World War II Operations, COMBINE I and II, employing joint combined Army, Navy and Air Force units, enabled the 555th to demonstrate its airborne and organizational skills to thousands of observers—military, civilian and foreign. The selection of this assignment could have gone only to a full-strength, highly proficient and skilled airborne battalion. In each of these two assignments (COMBINE I in 1946 and COMBINE II in 1947) The Triple Nickles trained Air Force and Naval service personnel in the use of crew-served weapons and individual small arms musketry.

Accolades and praise abound from all circles.

Service Company drill field, June 1944. Second Lt. Edward DuBois Baker putting the second platoon of the 555th Parachute Infantry Company through the second favorite paratroop exercise after running–pushups. They are in the famous leaning rest position. (Courtesy of The Shoe String Press, Inc.).

75TH INFANTRY RANGER REGIMENT

By Robert F. Gilbert

The 75th Infantry Ranger Regiment is linked directly and historically to the thirteen Infantry Ranger Companies of the 75th that were active in Vietnam from February 1, 1969 until August 15, 1972. The longest, sustained combat history for any American Ranger unit in more than three hundred years of United States Army Ranger history.

The 75th Infantry Regiment was activated on Okinawa during 1954 and traced its lineage o the 475th Infantry Regiment, thence to the 5307th Composite Provisional Unit popularly known as Merrill's Marauders. Historically, Company I, (Ranger) 75th Infantry, 1st Infantry Division and Company G, (Ranger) 23rd Infantry Division (Americal) produced the first two U.S. Army Rangers to ever earn the Medal of Honor as a member of and while serving in a combat Ranger company.

Specialist Four Robert D. Law won the first Medal of Honor with I/75th while on a long range patrol in Tinh Phoc Province Vietnam. He was from Texas.

Staff Sergeant Robert J. Pruden won the second Medal of Honor with G/75th while on a reconnaissance mission in Quang Ni Province Vietnam. He was from Minnesota.

Conversion of Long Range Patrol Companies of the 20th, 50th, 51st, 52nd, 58th, 71st, 74th, 78th and 79th Infantry Detachments and Companies and Company D, 151st Infantry Long Range patrol of the Indiana National Guard, to Ranger Companies of the 75th Infantry began on 1 February 1969. Only Company D, 151st retained their unit identity and did not become a 75th Ranger company, however, they did become a Ranger company and continued the mission in Vietnam.

Companies, C, D, E, F, G, H, I, K, L, M, N, O, and P (Ranger) 75th Infantry conducted Ranger missions and operations for three years and seven months every day of the year while in Vietnam. Like the original unit from whence their lineage as Neo Marauders was drawn, 75th Rangers came from the Infantry, Artillery, Engineers, Signal, Medical, Military Police, Food Service, Parachute Riggers and other Army units. They were joined by former adversaries, the Vietcong and North Vietnamese Army Soldier who became, "Kit Carson Scouts," and fought alongside the Rangers against their former units and comrades.

Unlike Rangers of other eras in the 20th Century who trained in the United States or in friendly nations overseas, Rangers of the 75th Infantry Vietnam were activated, trained and fought in the same geographic areas in Vietnam. It was a high speed approach to training.

Training was a combat mission for volunteers for the 75th (Ranger) Infantry, Vietnam. Volunteers were assigned, not accepted in the various Ranger Companies, until, after a series of patrols, the volunteer had passed favorably, the acid test of a Combat Ranger and was accepted by his peers. Following the peer acceptance the volunteer was allowed to wear the black beret and wear the red, white and black scroll shoulder sleeve insignia bearing his Ranger company identity.

All Long Range Patrol and 75th Ranger Companies were authorized Parachute pay. Modus Operandi for patrol insertion varies, however, the helicopter was the primary means for infiltration and exfiltration of enemy rear areas. Other methods included foot, wheeled and tracked vehicles, airboats, Navy Swift Boats and stay behind missions where the Rangers remained in place as a larger tactical unit withdrew. False insertions by helicopter was a means of security from ever present enemy trail watchers.

General missions consisted of locating the enemy, bases and lines of communication. Special Ranger missions included wiretap, prisoner snatch, platoon and company sized raid missions, ambush and bomb damage assessments following B-52 bomber arc-light missions.

Staffed principally by graduates of the U.S. Army Ranger School, paratroopers and Special Forces trained men, the bulk of the Ranger volunteers came from the soldiers who had no chance to attend the schools, but who carried the fight to the enemy. These Rangers remained with their units through some of the most difficult patrolling action(s) in Army history and frequently fought much larger enemy forces when compromised on their reconnaissance missions.

Rangers from Company M/75th were known to patrol in two man reconnaissance teams, however the six man Ranger team was standard and a twelve man heavy team was used for combat patrols in most instances.

Rangers remaining on active duty following Vietnam became instructors at the Army Ranger School, HALO and HAHO Schools, the SCUBA School and were mountain climbing instructors at the Army Cold Weather School, Alaska. Some became tabbed Rangers, by attending Ranger School after serving from one to five years in a Long Range Patrol or 75th Ranger Company in Vietnam.

The Vietnam Rangers of the 75th Infantry were awarded the title Merrill's Marauders by Secretary of the Army Stanley Resor during 1969 for having lived up to the standards set by the original Marauders during World War II. Army Chief of Staff Creighton Abrams who observed the 75th Ranger operations in Vietnam as Commander of all US Forces there selected the 75th Rangers as the role model for the first U.S. Army Ranger units formed, during peacetime, in the history of the United States Army.

On 31 January 1974, the 1st Battalion (Ranger) 75th Infantry was formed followed by the 2nd Ranger Battalion on 1 October 1974. Both of these highly trained and equipped Battalions made the combat parachute assault on the island of Grenada during operation Urgent Fury on 25 October 1984. The combat jump was performed to rescue American students there and restore democracy.

Using their bodies as shields, the Rangers led the students to safety during the evacuation while Cuban communists shot at them. Earlier, Company C, 1st

Staff Sergeant James Moran, Ranger with L Company, 75th Infantry, 101st Airborne Division (Airmobile), uses a mirror to signal a helicopter in the jungles of I Corps. (U. S. Army photo by Spec. 4 David Volk, courtesy of 101st Airborne Division Association)

In Vietnam Lt. Lynn Moore (foreground), Snuffy Anderson and Muggs Mullis open fire during a chance contact with the enemy.

Ranger Battalion was involved in Desert One, the attempt by Delta Force to extract the Americans held hostage by Shiite Moslem fanatics in Iran.

To meet the worldwide contingency missions, training for the 75th Rangers is being conducted in the jungle, arctic and desert continuously.

On February 3, 1986, Secretary of the Army John O. Marsh Jr., redesignated the 75th Infantry Ranger Regiment as the 75th Ranger Regiment. On 17 April 1986, Marsh presided over the transfer of colors and lineage back to the Rangers.

During World War II and Korea, company and battalion commanders or specially trained sergeants controlled fires for tactical units. Rangers in grades E-4 to E-6 controlled fires from the USS New Jersey's 16 inch guns, a two thousand pound projectile, in addition to helicopter gunship, piston engine and high performance jet aircraft while operating, frequently, far beyond conventional artillery and infiltrating enemy base camp areas and capturing prisoners or conducting other, covert, operations.

Ranger Team leader, Paul Morguce fires a M79 grenade launcher during H/75th Ranger chance contact. (Photo by Doug Stearn).

Bob Law, the first Ranger in U. S. Army history to win the Medal Of Honor while serving in a U. S. Ranger unit in combat.

James A. Champion, only known Ranger MIA from Vietnam following a three day battle in the Ashau Valley. (Courtesy of Robert F. Gilbert).

Korean Era Airborne Ranger Company Lineage

World War II	Korean Conflict	Supporting in Korea
None	Eighth Army Rgr Co (8213th ArmyUnit)	25th Infantry Division
None	Rgr Tng Cmd/Ctr (3440 Army SvcUnit)	US Army Infantry Center
Co A, 1st Ranger Inf Bn	1st Ranger Inf Co	2d Infantry Division
Co A, 2d Ranger Inf Bn	2d Ranger Inf Co	7th Infantry Division
Co A, 3d Ranger Inf Bn	3d Ranger Inf Co	3d Infantry Division
Co A, 4th Ranger Inf Bn	4th Ranger Inf Co	1st Cavalry Division
Co B, 1st Ranger Inf Bn	5th Ranger Inf Co	25th Infantry Division
Co B, 2d Ranger Inf Bn	6th Ranger Inf Co	1st Infantry Division
Co B, 3d Ranger Inf Bn	7th Ranger Inf Co	USA Infantry Center
Co B, 4th Ranger Inf Bn	8th Ranger Inf Co	24th Infantry Division
Co E, 2d Ranger Inf Bn	9th Ranger Inf Co	31st Infantry Division
Co F, 2d Ranger Inf Bn	10th Ranger Inf Co	45th Infantry Division
Co C, 3d Ranger Inf Bn	11th Ranger Inf Co	40th Infantry Division
Co D, 3d Ranger Inf Bn	12th Ranger Inf Co	28th Infantry Division
Co E, 3d Ranger Inf Bn	13th Ranger Inf Co	43d Infantry Division
Co C, 2d Ranger Inf Bn	14th Ranger Inf Co	4th Infantry Division
Co D, 2d Ranger Inf Bn	15th Ranger Inf Co	47th Infantry Division

NOTE: As the 9th Ranger Infantry Company, activated on 5 January 1951, initially completed training at Fort Benning, Georgia, assigned personnel were used as filler replacements for deployed Rangers companies in the Far East. A second 9th Ranger Company completed the required training, and was attached to the 31st Infantry Division in November.

Rangers inbound to patrol box with Cobra gunship flying cover for the Rangers and helicopter lift ship.

Back L to R: Michael Plunkett, Stanley C. Stryker, "Chief" Bill McCabe. Front L to R: Franklin Anderson, "Chief" Alfred Lee, James L. Peterson. (Courtesy of Greg Krahl)

RANGER INFANTRY COMPANIES (ABN) — KOREA

By Emmett E. Fike

The Korean War was fought from 25 June 1950 to 27 July 1953, over a largely mountainous peninsula shaped somewhat like Florida and approximately the size of Minnesota. It was, with the exception of General MacArthur's brilliant end run at Inchon, largely a head-to-head slugging match over mountains and razor back ridges. The winters were a sub-zero icy hell and when the summer monsoons came, all was mud, and the heat sometimes reached 100 plus degrees for days.

American artillery and the air support performed brilliantly.

Armor, though severely hampered by terrain and bad roads, added its weight. But, in the final analysis, it was like all the other wars of history: an infantry war.

There were no Ranger units in the United States Army when the North Koreans launched their massive attack across the 38th parallel. On 30 June 1950, President Harry Truman ordered U.S. ground forces committed. Elements of the 24th Infantry Division were in combat action by 5 July.

The enemy began a guerilla-type war in the American and South Korean rear areas at the very beginning by infiltrating their forces to the rear. There they would reassemble to conduct raids and establish road block positions, inflicting on the UN Forces heavy damage and many casualties.

It soon became evident that the American and U.N. Forces badly needed the same capability. A call was made for volunteers to form a Ranger company at Camp Drake, Japan, on 25 August 1950. From service units throughout the Far East Command volunteers were accepted and the 8213th Army unit was formed. It is known today, as it was then, as the 8th Army Ranger Company. The Ranger company was trained on the spot in the Pusan Perimeter.

The 8th Army Ranger Company was on Hill 205 near the Yalu River when the Chinese Communist Forces entered the war during Thanksgiving week 1950. The Chinese ran head on into Task Force Dolvin which was made up of several units including the 8th Army Ranger Company. The Rangers, although they suffered heavy losses, made a very good account of themselves.

Meanwhile, General MacArthur, U.N. commander, agreed that these special units were needed. General J. Lawton Collins ordered the establishment of a Ranger Training Center at Ft. Benning, Georgia to be activated and for the Army to train such units. Colonel Van Houten was put in command of the Ranger Training Center on 15 September, 1950. His instructions were to begin training of Ranger-type units immediately.

On 29 September 1950, The Ranger Training Center, (Airborne) was organized as the 3340 ASU.

When the call went forth for volunteers to step forward, the 82nd Airborne Division was in formation, and it is reported that the whole division stepped forward. The selection process was long and difficult, and only the best paratroopers were chosen.

The first training cycle arrived at Ft. Benning on 20 September, 1950 and began training on Monday, 2 Oct., 1950. Three companies of airborne qualified officers and enlisted volunteers, a total of 19 officers and 314 enlisted men, began training. These first three companies were composed of white soldiers.

On 9 October 1950, another company of Airborne qualified officers and enlisted volunteers arrived for training. These were black paratroopers from the 3rd Bn., 505th and the 80th Airborne Anti-Aircraft Battalion of the 82nd Airborne Division. Initially designated the 4th Ranger Company, they were redesignated the 2nd Ranger Infantry Company (Airborne), the only Department of the Army-authorized all-black Ranger unit in American military history.

The first training cycle was completed on 13 November 1950. The 1st, 2nd and 4th Ranger Companies departed for shipment to the Far East while the 3rd Ranger Company remained to train again with the second cycle, consisting of the 5th, 6th, 7th and 8th Ranger Companies.

The 1st Ranger Company arrived in Korea on 17 December, 1950 and was attached to the 2nd Infantry Division. The 2nd and 4th Ranger Companies arrived in Korea on 29 December, 1950. The 2nd Company was attached to the 7th Infantry Division and the 4th was attached to Headquarters, Eighth U.S. Army and the 1st Cavalry Division.

Later, when the 3rd Ranger Company arrived in Korea, it was attached to the 3rd Infantry Division. The 5th Ranger Company went to the 25th Infantry Division; the 8th Ranger Company went to the 24th Infantry Division; and the 6th Ranger Company went to Germany where it was assigned to the 1st Infantry Division.

The 3rd training cycle consisted of the 9th, 10th, 11th and 7th Companies, the 7th having been held over again like the 3rd Company was in the 2nd cycle. This cycle completed its training and graduated from Ranger School on 13 April, 1951. The 7th, 10th and 11th Ranger Companies departed for Camp Carson, Colorado where they underwent additional mountain training in the rugged Rocky Mountains of Colorado.

The 9th Ranger Company was broken up and went into the pipeline as replacements to the six Ranger Companies already in Korea, which had already suffered heavy losses.

It must be noted here that there were eventually three 7th Companies. The first 7th Company returned to Ft. Benning from Colorado and became cadre and demonstration platoons.

There were two 9th Companies. The first one, as before mentioned, became replacements to the six Companies in Korea. The second 9th Company was composed of volunteers from the 31st "Dixie", Division.

The 10th and 11th Ranger Companies went to the 45th and 40th Infantry Divisions that were in the final phases of combat training in Japan.

The fourth Ranger training cycle commenced training on 30 April, 1951. Making up this cycle was the 12th, 13th, 14th, 15th and Companies A and B. The 12th Ranger Company upon completion of Ranger training, was assigned to the 28th Pennsylvania, "Keystone" Division. The 13th Ranger Company was assigned to the 43rd Division. The 14th Ranger Company went to the 4th Division, the 15th Company went to the 47th in Minnesota, the "Viking" Division. Companies A and B were assigned to the 3rd Army.

The Ranger companies in Korea distinguished themselves in combat. Their battle flag is adorned with campaign streamers for the UN Defensive, UN Offensive, the CCF Intervention, First UN Counter-Offensive, CCF Spring Offensive, UN Summer Offensive, a combat jump at Munsan-ni, battles at CQ4594, Chipyong-ni, 299 Turkey Shoot, Bloody Nose Ridge, Majori-ri, Hill 205, Hill 628, Hill 832, Objective Sugar, May Massacre, Sangwiryang, raids at Changmal, Hwachon Dam, Hill 383, Toopyong-ni and Rogers, Zebra and Croft.

In July 19561, an order was sent to all commands of the U.S. Army to deactivate the Ranger Infantry Companies (Airborne). The Ranger companies in Korea were deactivated first and the others followed suit within a few months. The Order stated that personnel being qualified parachutists should then be transferred to the 187th Abn RCT, then stationed in the combat zone.

In November 1951, the last Ranger company was deactivated. Men of the 6th Ranger Company stationed near Stuttgart, West Germany, made the last parachute jump as a Korean War Airborne Ranger Company. They then had to travel halfway around the world to reach the 187th ARCT and combat. It was never their choice, and a bitter pill to swallow when they had been assigned to the European Theater after volunteering for the Rangers and combat in Korea.

In 1983, the survivors of this small band of "Forgotten Warriors of the Forgotten War" after 32 years came together again and formed their association, The Ranger Infantry Companies Airborne, Inc. Today they are a close brotherhood, gathering for regional meetings all over the country every year and every two years for a grand national reunion.

They have marked the gravestones with a Bronze Ranger Scroll of 80 percent of the 148 Rangers who died during the Korean War. Twenty-two graves can never be so marked, as these Rangers were missing in action and their remains never recovered from behind enemy lines.

This "Elite Brotherhood of the Elite" quietly watches and assists each other and the young active duty Rangers of today in true Ranger Spirit, for they set the standards which the Airborne Rangers of today are required to meet.

Elements of an Airborne Ranger Infantry Company in Korea, 1951. (U.S. Army photo)

THE FIRST SPECIAL SERVICE FORCE

By Bill Story

"THE STORY OF THE NORTH AMERICANS: 1942-1944"

At the start of World War II, after France had been defeated and the whole of Europe overrun by the German armies, and after Britain had suffered at Dunkirk, there was a need to continue to carry the fight to the enemy on the Continent. It was at this time in early 1940, that the British Commandos were born. Their success in small unit actions drew favorable attention to this form of military tactics. They were called Special Service troops.

After Pearl Harbor, President Roosevelt was anxious to have American troops committed as quickly as possible to the European Theater, rightly seeing this as being the hardest nut to crack. He supported the concept of the Rangers, and they were the first to train in commando-style tactics with the Commandos. A select few of them were the first to be committed to battle in Europe at Dieppe with the Canadians.

Thus, when Roosevelt's aid, Harry Hopkins, and Chief of Staff George C. Marshall came back from meetings in Britain with Prime Minister Winston Churchill and Combined Operations Chief Lord Louis Mountbatten with an idea to form a hard-hitting, ski-and parachute-trained multi-nation unit for major raids on the Continent in winter, Roosevelt enthusiastically adopted it and told Marshall to proceed with speed. The Canadians agreed to provide troops, when the British were unable to, and it was decided not to have Norwegians.

Thus the First Special Service Force came into being. The U.S. lieutenant colonel who had examined the idea and recommended against it, Robert T. Frederick, was told to undertake it and given the opportunity to write his own orders. They were broad—almost carte blanche as Army orders go. Time was of the essence. No paper pushers could stand in the way.

Seeking remote, snowy, hilly, tough terrain for training, Ft. William Henry Harrison, a National Guard camp in Helena, Montana was selected from a number of locations. Meanwhile, volunteers from the U.S. and Canadian armies of the time were being chosen. Men had to have rugged, outdoor backgrounds and better than average intelligence.

They came from virtually every U.S. Army unit and every Canadian unit at home and aboard. There was no separation between soldiers of the two nations—two Canadians and two Americans found themselves tenting together in the dusty Montana camp as it literally expanded around them. It was an interesting experience for both groups: sorting out seniority—for many were sergeants and above; learning what not to say about the other's country; examining uniforms and U.S. weapons and exchanging personal histories.

There were games with marching orders—since all were mixed regardless of nation, commands were given Canadian or U.S. Army style, depending on who was giving them. Eventually a composite drill was worked out. In the meantime, everyone did what came naturally, with the attitude, "Who Cares? Let's get on with it." Those hung up on this and other problems of the time were privileged to leave, if they weren't thrown out bodily for their own failures. What ensued over the long term was a well-knit, highly-trained body of soldiers with an extraordinarily high esprit de corps.

Force training was intense. There were tough calisthenics followed by runs over the Scratch Gravel Hills outside Fort Harrison. There were walks, no marches, but rambling walks lasting the entire day, as Forcemen were taught the right technique for walking up hills and mountains. Later, these were followed by day-long marches with full rucksacks and weapons along roads and highways.

Parachuting was taught quickly, with the aid of plywood mock-ups of C-47 cabins for exit training; harnesses hanging from 4x4 structures, the rudiments of control and landing. Then, wearing football helmets, jump boots and coveralls, Forcemen went up and jumped from the C-47s which made up the Force's Air Section. There was nothing fancy about the training; it was quick and dirty.

In the meantime, Americans helped Canadians familiarize themselves with U.S. weapons, including the M-1, far better than the bolt-action Canada Lee-Enfield. For Canadian machine gunners it was no problem catching on to the .30-caliber, belt-fed Browning. It proved to be a far superior weapon than the water-cooled 301 Vickers on which they were trained. Then came the Browning Automatic rifle praised by Americans because of its long-barrelled accuracy and damned by Canadians because of its slow pace of fire, its seemingly forever-collapsing bipodal legs and its length, which always seemed to catch in the brush just when one wanted to be quiet. "Can't compare to the Czech-designed Bren gun," they said. "It has a larger magazine; it's air-cooled with interchangeable barrels; it's lighter; it has a higher rate of fire." This kind of comment went on until the Supply Department obtained Johnson light machine guns (called "Johnny Guns" by the Forcemen) from the Marine Raiders for a swap of part of the Force's exclusive supply of the new and mighty powerful explosive—RS.

The basic element of the force was the Section—a 12 to 15 man group—armed with a .30-caliber, a Johnny Gun, a mortar, a bazooka, M-1s, 45 pistols, grenade launchers and, if need be, flame throwers. The result: awesome firepower for an infantry-size unit of the time. A platoon, which had two sections, could lay down a real curtain of fire.

One bit of training unique to the Force was Canadian Army battle drill. It was training in basic fire and movement tactics. Movements were marked out initially on the parade ground, hence the "drill." Right flanking, left flanking, pincer—all were susceptible to this approach which gained fast acceptance by the Americans once they understood its value.

Forcemen also trained under live fire—again unique to the U.S. Army of that particular time. The result was that when the first combat action came, Forcemen knew what the situation called for and how to execute it.

They had demonstrated this for the inspector generals of both nations who graded them for overseas capabilities. Col. Bill Walton, co-author of the popular book THE DEVILS BRIGADE, was on the U.S. Army Team and said many years later that the Force scores went right off the sheets—higher than ever before recorded for any unit.

Demolitions, mountaineering with ropes, picks, pitons and rings, ski training with Norwegian army experts; amphibious training with the U.S. Navy at Norfolk, plus the usual unit and larger force tactics rounded out the training phase while the Force leadership sought an action command. The original European targets had been scrubbed for a variety of reasons.

First call was for the Aleutian Islands, a far-flung chain off the coast of Alaska. Japanese-held Attu had been taken bloodily for both the U.S. and the Japanese. Now it was time for Kiska, the second occupied island. Two regiments of the Force rubber-boated in for the landings, while another waited on Amchitka next to the C-47s, waiting to jump in. But, the Japanese had fled leaving equipment, weapons and the like along with booby traps. The only casualties came when green troops fired on each other on adjacent hills while a Force patrol dug safe foxholes in the valley between.

A quick call from the Mediterranean Theater in late August placed the Force aboard Navy transports back to the U.S. After a delay en route for the troops, orders came for the Forcemen to embark for Italy via North Africa. They trucked into their new base camp in Italy, south of Cassino, by late November 1943 and were committed to their first real action after the Kiska dry run by December 3. A fast turnaround for that time and an unusual one.

This was a rugged job—taking a hill mass called Monte La Difensa (so named because it resembled a fortress) after two U.S. divisions had failed and a British division had blunted itself on the adjoining Monte Camino. Casualties had been heavy in these prior attempts. Something new was needed, and the Force provided it.

Careful scouting of possible routes up the imposing and well-defended mountain led to a daring plan. First, a move would have to be made secretly around the German flank during the night. The Force would remain concealed during the following day in the forests that extended part way up the mountainside and then advance during the next night to the base of an escarpment which soared 40 or more feet. This still left another 300 plus yards before they could reach the

Ft. William Henry Harrison, Helena, Montana, winter of 1942-1943. Men of the First Special Service Force train with the 30 caliber on a clear winter's day. (Courtesy of Lew Merrim, Force Photographer).

very top where the 3rd Battalion, 129th Panzergrenadier Regiment had its bunkers and emplacements. It was an awesome task for the Force's Second Regiment. Left sitting on their chutes waiting to jump into Kiska, they had been promised first crack at the Krauts, and they got it.

Debarking in front of Corps' artillery, rucksacks piled high with food and ammunition, the Second Regiment marched all night through off-again, on-again cold Italian winter rain, to arrive at the base of La Difensa and up into the trees where they would lie concealed during the daylight hours. The last Platoon of the last Company made it as the skies began to lighten. Fortunately for the wet Forcemen, the clouds broke and the sun shone on the day that followed. Sterno stoves came out to heat K-rations, last minute orders were discussed, and when night fell, the advance up the slope through the trees to the escarpment began in total silence.

Meanwhile, the entire 5th Army artillery worked over the mountain tops in one of the greatest concentrations of artillery fire yet seen in Italy. The entire mountain top seemed to be on fire. Freddie's Freighters, so named because of their commander and the enormous-looking loads they carried in their rucksacks and packboards (both items new to the Italian front), moved calmly up under the barrage until the enemy escarpment loomed ahead. Then it was the turn of a small group of climbers to call on their mountaineering skills and rope their way up the cleft in the rock to the top in absolute silence.

Soon, ropes snaked down, and 1st Company made its way up, followed by the Second and Third until all of First Battalion was on top and spread out to either side of the cleft before moving ahead for the final ascent. It had taken two hours for First Battalion to complete its climb. Other than a German burp gunner at the top who was guiding mortar fire on to the regular Difensa trails with his tracer, nothing else stirred as the First Battalion moved out in skirmish line to complete the final move.

As they neared the crest at about 4:30 a.m., 1st Company flanked left and 2nd Company came abreast with 3rd Company further back. As 1st Company moved into assault formation, loose rocks around the German position were dislodged, and the Panzergrenadiers awakened to the fact that they were under direct attack from an enemy which had come from an unexpected direction in complete silence. Red and green signal flares went up, followed by illuminating flares which silhouetted Forcemen as they began to advance. All hell broke loose as the battle started. Battle drill training paid off as the platoons went into the attack against the various strongpoints around the saucer-shaped top of La Difensa. Soon the Panzergrenadiers were fleeing down the back side of the mountain to safety. Incoming mortar fire brought casualties, including the death of the battalion commander, the wounding of the executive officer and the death of more Forcemen.

On the rocky top, there were few places to take shelter except those which the enemy had already marked on his map. This included the log-covered bunker where the wounded lay, attended by a young German corpsman. Counterattacks were mounted and thrown off; snipers were everywhere and had to be wrinkled out.

Essentially, the main battle was over. The Force had achieved in about two hours what the 5th Army planners had figured would take almost a week. Then it was just a matter of holding on, reinforcing with 2nd Battalion, Second Regiment and with elements of 1st Regiment. The Service Battalion and the Third Regi-

Members of 2nd Regiment, First Special Service Force line up to board C-47 to parachute into Kiska Island in the Aleutians, August 14, 1943. The jump was cancelled when the Force's 1st and 3rd Regiments found the Japanese had fled. (Courtesy of Bill Story)

MG Robert T. Frederick presenting awards to officers, other ranks enlisted men of the First Special Service Force, Menton, France. Note Force colors–Force flag, Canadian flag (of the time) and U.S. flag. Gen. Frederick commanded the 1st Airborne Task Force on the landings in S. France in August 1944. (Courtesy of Bill Story)

Ft. Wm. Henry Harrison, Helena, MT. Forceman is helped by Service Bn. man to wrap up his chute and pack after jump. Note football helmet. (Courtesy of Bill Story)

The Anzio Beachhead. Fight by night, farm by day. Note foxholes dug into Mussolini Canal bank in back. (Courtesy of Bill Story)

ment were pressed into supply haulers—no mules could make it up that first night or even later. Everything had to be packed on the backs of the men. Six or eight men were required to handle stretchers down the steep slope to the Regimental Aid Station located about half way down near Force Headquarters. Many Force wounded chose to walk down if they could.

The next objective—taken a few days later by a casualty-weakened 1st Battalion–was Monte la Remetanea. Coupled with clearing the saddle and the lower reaches of the mountain on the German side, this enabled the British 56th Division to take Monte Camino, Difensa's near-twin, with its monastary-capped peak. On December 9, a weary but satisfied Force was relieved to return to its base at Santa Maria Capua Vetere, to mourn its dead, to wait for the return of the lightly wounded and to prepare for the next mountain assault.

Christmas Day the Force hit Monte Sammucro—Hill 720—of that mountain mass which towers to 1,305 feet. First Regiment handled that mission successfully and after the battle, could look down on San Vittore in the valley and back to La Difensa-Remetanea on the other side of the valley. The success at La Difensa had breached the German Winter Line and now the entire 5th Army was moving forward as the Germans skillfully fought a rearguard action as they withdrew toward the city of Cassino. Monte Cassino, although not as high as La Difensa, had an equally steep approach.

The Force fought its way through trackless wastes on mountain top after mountain top. Sammucro was followed by Radicosa, then Monte Majo and finally Monte Vischiataro, towering over Monte Cassino.

It was time for the Force to have a real rest, and to integrate replacements from both the U.S. and Canadian Armies who had been training at Santa Maria CV. The Force sought volunteers directly. No one was shipped from the replacement depot to fill vacancies.

The Force's respite from action proved to be short. The Fifth Army's end-run to the Anzio Beachhead, designed to cause the Germans to fall back from their strong defense lines further south, failed to do the job. One major casualty was Darby's Rangers. Surrounded at Cisterna by the German's favorite tactic, an armored pincer movement, it was effectively eliminated as a fighting force. Heavy German reinforcements brought the beachhead advance to a halt.

The Force was needed to shore up the right flank of the circle, along the Mussolini Canal down to the Tyrrenian Sea. The combat echelon had fewer than 1,100 effectives. Second Regiment had been merged for the time being into one battalion. It stayed in reserve while the First and Third went on the line.

Combat patrols of one size or another—on one occasion the full battalion—kept the Germans off balance starting with the first night and day when they were driven back from the other side of the Canal which they considered their territory. The other two regiments kept up their patrolling, either for information (prisoners) or to drive the German outpost line back yet another farm house or two toward their Main Line of Resistance. The "No Man's Land" was widest at the sea, narrowing toward the place where the canal split and the 504th PIR had its positions.

Since this was farm country and the farmers had fled leaving their animals, it wasn't long before Forcemen were milking the cows back of their lines and gathering eggs from "No Man's Land." Brash or unknowledgeable enemy soldiers with the same idea, out on egg patrol, were summarily captured and sent to the rear.

After 99 straight days on the line without relief, the Force was finally pulled back, reinforced and given some rest prior to the break-out from the beachhead. May

10 saw them lead the way on the right flank and dash for the hills. As expected, heavy casualties were encountered the first day as adjoining units failed to keep up with the speed of the Force and Tiger tanks hammering the exposed flank. Hard fighting in the hills helped clear the road to Rome. The Force was the first Allied unit in strength into Rome on June 4, 1944, capturing all its assigned bridges over the Tiber before being relieved.

Then the Force was sent South to another Santa Maria: Castellabate this time, well below Naples, to train for the landings in Southern France and to absorb the new recruit volunteers who had arrived. The Force went ashore by rubber boats on August 15, 1944, one year to the day of their amphibious exercise at Kiska in the far Alutians. Again, it was island landings, designed to take out strong points which could interdict the main landings. And, again, it was hours in advance of those main landings. Once more the Force did its job. Moving to the mainland, the Force fought east toward the Franco-Italian border, where it stopped on orders from Allied Commander General Eisenhower. In behind them lay a host of cities and towns freed of the German oppressor, including Menton, the site of the last Force Headquarters.

On December 5, 1944, this unique Allied fighting unit was disbanded. Canadians stepped out of ranks and formed up as the 1st Canadian Special Service Battalion—the first time this hitherto paper organization had been fleshed out with bodies. They marched off the field. The remaining Americans did not close ranks but left empty the places where their Canadian buddies had stood, until they, too, were dismissed.

Today, battle honors of the First Special Service Force are proudly carried by the U.S. Army's Special Forces Groups. In Canada, the same colors (sic) are carried by the Canadian Airborne Regiment, part of Canada's Special Service Force. Today's Special Forces are now a separate branch of the Army, as the Force essentially was. Special Forces soldiers of all ranks now proudly wear the crossed arrow insignia of the original Force. And the badge of the Army Special Operations Command is a red spearhead similar to the striking USA/CANADA Spearhead worn by Forcemen.

Of the Force, it can be said, "They never failed a mission."

Commanding Officers—Initially, Col. Robert T. Frederick; secondly Col. Edwin A. Walker (both later Major Generals). Executive Officer—Paul D. Adams (later General).

Regimental Commanders—1st Regiment: Col. Cookson Marshall (KIA); later Col. J.F.R. Akehurst (Canadian). 2nd Regiment: Col. D.D. Williamson (Canadian), Col. Robert Moore. 3rd Regiment: Col. Jiggs Mahoney; then, Col. Edwin A. Walker; finally, Col. Wilson Becket (Canadian).

Battalion Commanders—1st Battalion, 1st Regiment: Maj. Gus Heilman, LTC W.S. Gray (KIA), Major T. E. Pearce (KIA), Maj. Ed Mueller (later BG), Maj. Con Tate (Canadian). 2nd Battalion: Maj. Jerry McFadden (Canadian), Maj. S.V. Ojala (KIA). 2nd Regiment 1st Battalion: LTC Tom MacWilliam (KIA), Maj. Ed Thomas (later BG); 2nd Battalion: LTC Bob Moore, Maj. S.W. Waters (Canadian), (later LT Gen). 3rd Regiment, 1st Battalion: LTC T. Gilday (Canadian), Maj. Ray Hufft (later MG); 2nd Battalion: LTC John Bourne (Canadian), (later Colonel).

Decorations: close to 250 awarded to Forcemen. They ranged from the Canadian Mention in Dispatches, Military Medal, Member of the Order of the British Empire and the Distinguished Service Order to the U. S. Bronze Star, Silver Star and Distinguished Service Cross. In addition, some 35 or 36 enlisted men/other ranks were given battlefield commissions. Bear in mind the Combat Echelon strength of the Force was 1,800 plus or minus. Its life was slightly more than two years: July 1942 to December 1944. In the years that followed its dissolution to the present, in both the Canadian and U.S. armies, there were an uncounted number of Colonels, three Brigadier Generals, five Major Generals, one Lieutenant General (head of the Canadian Army) and one four-star General, all of whose warmest boast was: "I was a member of the Force."

Battle Honors—Canada: Monte la Difensa-Monte la Remetanea, Monte Majo, Anzio, Rome, Advance on Tiber, Southern France. United States: Pacific Theater, Aleutians, Mediterranean Theater, Naples-Foggia, Rome-Arno, European Theater, Rhineland, France-Italian Border, Southern France.

In the Mountains in Italy, 1943. Chief Scout of the First Special Service Force, Francis B. Wright of Ottawa pauses to confer with comrade. (Courtesy of Bill Story).

Menton, France-August 1944. Members of 1st Platoon, 4th Co., Second Regiment relax after two days spent in taking Ft. Mont Agel. (Courtesy of Bill Story).

Activation Day Ceremony, 8 May 1963. Major General William Yarborough presents the unit colors of the 6th Special Forces Group (Airborne), 1st Special Forces to LTC Charles Kasler. SGM Charles Petry watches. (Courtesy of C. Kasler Collection).

SPECIAL FORCES

By Clyde Sincere

FOREWORD

Special Operations is, and has been a continuing expanding technology. This chapter deals with, and relates to, a limited portion of the total spectrum of Special Operation/Special Forces. Specific units reflected to varying degrees in this chapter are the 1st, 3d, 5th, 6th, 77th, 7th, 8th, 10th, 11th, 12th, 19th and 20th Special Forces Groups, plus the 46th Special Forces Company (Airborne), later redesignated the 3d Battalion, 1st Special Forces Group (Airborne) and 1st Special Forces located in Lop Buri, Thailand.

Special Forces training never stops. To perform his unconventional warfare mission, the Special Forces soldier must be trained to infiltrate deep into enemy-held territory. He may travel by air, water or land. Detachments are static-line parachute qualified, and some may even infiltrate by parachuting from altitudes in excess of 20,000 feet, day or night. Water transport may require a trip by raft, submarine, surface swimming or swimming underwater with scuba equipment. Land infiltration may require walking, crawling or using existing surface transportation. All means of infiltration are hazardous and the training must be thorough.

Duties in Special Forces for officers and non-commissioned officers primarily involve participation in special operations and interrelated fields of unconventional warfare. These include performing foreign-internal defense and direct-action missions as a part of a small operational detachment. Duties at higher levels involve command, control and support functions. Frequently, duties require regional orientation, including foreign-language training and in-country experience. Special Forces places emphasis not only on unconventional tactics, but also on knowledge of conventional light-infantry doctrine and low-intensity conflict. Missions can include activities in waterborne, desert, jungle, mountain or arctic environments. Special Forces soldiers are professionals in the business of helping others during peacetime and in war. They maintain their high standards of skills through devotion to duty and through rigorous training programs.

Our nation's capability for immediate response to aggression throughout the free world was enhanced considerably by the activation of the United States Army 10th Special Forces Group (Airborne) on 20 June 1952. Time and time again since the formation of that first special forces unit, Special Forces, its personnel and its constant readiness to react quickly and effectively has been a deterrent to communist exploitation and other explosive situations in friendly nations. As President John F. Kennedy once stated, "The Special Forces soldier is the true embodiment of the term 'soldier stateman,' for he is the soldier and statesman of our time."

FROM THE BEGINNING

The Army Special Forces can trace its official origins as a commando strike unit to the unique United States Army-Canadian First Special Service Force, which was formed during World War II on July 9, 1942, in Montana. This initial force, designed for sabotage operations behind enemy lines in Scandinavia and other locations, was commonly referred to as the "Devil's Brigade" and was led by Major General Robert T. Frederick. Instead of operating in this fashion, however, the brigade was used as elite infantry in ground operations and earned a reputation for aggressive mountain fighting in Italy.

The First Special Service Force won the tough six-day battle for Monte la Difensa and broke through enemy positions, clearing the way for the Fifth U.S. Army to move up the Italian peninsula from Naples. The present-day Special Forces derives its modern crossed arrows distinctive unit insignia from the First Special Service Force, which was first authorized the crossed arrows badge by the Secretary of War on February 26, 1942.

The roots of Special Forces' unconventional warfare can be traced directly to the United States Office of Strategic Services (OSS) formed during World War II by Colonel William O. Donovan. "Wild Bill's" organization parachuted and infiltrated well-trained volunteers behind enemy lines in almost every theater of operations. For example, American OSS Jedburgh agents regularly parachuted into German-occupied France in small teams, consisting of two officers with an enlisted radio operator, and linked up with French resistance fighters to supply them with liaison, equipment and technical assistance for guerrilla tasks. The OSS operated successfully in Burma and Northern India as well where Detachment 101 performed valuable reconnaissance and raiding missions against the Japanese.

The Army Special Forces was also based on the lineage of Colonel Frank D. Merrill's renowned "Merrill's Marauders," officially known as the 5307th Composite Unit (Provisional). This rugged jungle-warfare organization fought in Northern Burma where its 3,000 soldiers engaged the Japanese in five major battles and 17 other actions. The most famous epic of the Marauders was completing a wide sweeping movement, virtually unsupported except by air, that placed them several hundred miles behind enemy front lines to capture the vital Myitkyina airfield.

When the Army Special Forces was first formed, it also inherited the lineage and honors of the six ranger battalions of World War II. Theses battalions traced their ancestry to the first American Rangers raised from militia volunteers in New Hampshire during the French and Indian Wars. Led by the notable Major Robert Rogers, these rangers operated with stealth and daring against the Native Americans, and the colonists later used these same tactics to defeat the British in the Revolutionary War. During the latter conflict, for instance, South Carolina "Swamp Fox" Francis Marion cut the English supply lines throughout the southern colonies and stifled attacks by the Loyalists.

The Special Forces was bestowed the honors of the World War II Rangers which operated and fought throughout Europe and the Pacific. Ranger companies also fought in the Korean War and Vietnam conflict.

The modern Army Special Forces was steeped in this rich legacy when it was formed in 1952. That June, Colonel Aaron Bank, a former Jedburgh commander, was recalled from Korea to activate the original 10th Special Forces Group in North Carolina. The Army records that he was probably "a little disappointed" at the size of his new command because merely one officer, one warrant officer, and eight enlisted men were present instead of the 2,500 men authorized for the group. Colonel Bank was destined to take this small nucleus of troops to the newly-built Psychological Warfare Center at Fort Bragg, and eventually mold it into the unconventional war spearhead of all United States military counterinsurgency efforts.

Within nine months of initial formation, Colonel Bank had 1,000 men who were either combat veterans of the Rangers, the OSS, or hand-picked soldiers chosen for their ability to learn all types of warfare. The story of Colonel Bank and the birth of Special Forces can be found in his book, *From OSS to Green Berets* (Presidio Press, 1986).

The new Special Forces warriors were trained in all aspects of warfare in order to survive and operate successfully in hostile territory. The Special Forces were trained primarily to form and lead resistance movements throughout enemy-dominated regions using unconventional techniques in case of war. During periods of low hostilities, they were envisioned as a specialized force that could train Allied countries in every military field from parachuting to basic weaponry. In addition, the Special Forces teams could be dispatched into troubled areas and execute important scouting, rescue or commando missions at the highest national level. For the first 13 months the 10th SF Group trained in basic guerrilla warfare techniques at Fort Bragg and Camp Mackall in North Carolina, and pursued specialized training in mountain, glacier, rock-escarpment and cold weather methods at Camp Carson, Colorado and the Windriver Indian Reservation in the Grand Tetons of Wyoming.

All Special Forces personnel were experts in essential areas such as demolitions or radio communications. They were combined into 12-man teams, which continually rehearsed infiltration methods and combatant skills such as underwater operations and jungle fighting. The thorough training demanded of all Special Forces required volunteers of very high intelligence capable of gaining familiarity with any foreign weapon or other language. These troops also learned how to master their environment, despite sparse or adverse field conditions by thorough knowledge of basic woodsmanship, personal combat and survival principles. As a result, the Special Forces became an exceptionally elite group of physically robust and versatile troops, who were soon constantly engaged in field training and clandestine missions throughout the world.

As the Korean War concluded, and U.S. concerns over East Asia subsided, Special Forces began preparing to wage guerrilla warfare against Soviet-controlled Europe in event of a Russian invasion. In mid-1953, the basic 10th Special Forces Group was alerted to deploy to Germany, where operations could be conducted close to or launched behind the Iron Curtain. On November 11, the group was directed to relocate to Bad Tolz, West Germany. The group was split in half to fulfill this requirement, and the main contingent going overseas to Germany retained the title of the 10th Special Forces Group.

The troops remaining at Fort Bragg in the United States were reformed as the 77th Special Forces Group under Colonel Jack T. Shannon, the previous executive officer of Bank's original group. The new 77th was formed officially on September 25, 1953, and adopted Shannons' motto, "Anything, Anytime, Any Place, Anyhow!" The new group redoubled its planning and active field training in order to provide a versatile unconventional warfare capability for global contingencies.

Meanwhile, in Germany, the 10th Special Forces Group became well-established in its field of European responsibility. Its men began wearing distinctive headgear known as green berets. These berets were used occasionally even during field maneuvers at Fort Bragg, but they were part of a wide assortment of unusual headgear worn by Special Forces troops while training in the field. These distinctive uniform items were worn both for comfort and to distinguish Special Forces troops from conventional soldiers who were limited to wearing issue caps.

The early berets were restricted to field operations and never worn with the dress uniform. The Green Beret, however, quickly became associated with the commando tactics of the Special Forces. During 1954, Colonel William Ekman of the 10th Special Forces Group approved wearing the green beret in Europe, and by the following year every Special Forces trooper was wearing it as a treasured part of his uniform.

The first real effort to get the green beret approved on an Army-wide basis was made at Fort Bragg, the home of the 77th Special Forces Group, by the commander of the Army Psychological Warfare Center, Colonel Edson D. Raff. Although his attempts to gain recognition for this unique headgear were rebuffed time and time again by higher Army authorities, the Special Forces community kept up a heated struggle to gain official sanction for the green beret. While the green beret debate was raging, the Special Forces were awarded their own shoulder sleeve insignia.

In 1956, Captain John W. Frye designed the arrowhead-shaped patch containing an upturned dagger crossed by three lightning bolts. The shape of the arrowhead connected Special Forces to its native American heritage of hunting stealth. The dagger was characteristic of the World War II-era "Devil's Brigade" First Special Service Force. The three bolts of lightning represented the Special Forces ability to infiltrate by air, sea or land. The patch was colored teal blue, the traditional color of Army unassigned branch personnel, which symbolized the fact that Special Forces belonged to no particular branch of the Army, but rather embraced personnel from all areas of the Army. Finally, the black and gold Airborne tab, showing actual parachutist capability, was added to highlight the paratrooper qualification of all assigned Special Forces troops.

During the same year, the Psychological Warfare Center and School was established. The center not only offered specialized instruction in all phases of unconventional warfare, but also served as an institute of higher learning for the research and conduct of counterinsurgency operations. Special Forces troops were trained in five military occupational specialties: American and foreign weaponry; medical techniques; basic field and combat engineering; radio and advanced communications; and intelligence gathering and operational use.

The Special Forces continued to grow. In April 1956, a small number of hand-picked officers and non-commissioned officers from the 77th Special Forces led by Colonel Shannon were chosen for a Far East mission and deployed to Japan, where they formed the 8321st Army Special Operations Detachment. In June 1957, this detachment became the basis for the cadre of the 1st Special Forces Group on Okinawa, an island near and south of Japan. Other personnel of this Asian-oriented group were drawn from Fort Bragg and assisted in the formation of the 14th Special Forces Operational Detachment in Hawaii.

Within two years, the dominant focus of Special Forces affairs was switching from Europe to Southeast Asia. Larger and larger numbers of Special Forces trained personnel were dispatched as military advisors to the Kingdoms of Laos and Thailand, and the Republic of Vietnam. Special Forces A-Detachments were also assigned permanently (one each) to the Republic of Korea and the Republic of China (Taiwan). This transition was complicated by Special Forces requirements in response to the Berlin crisis and various emergencies in other parts of the world, such as the Congo.

In 1960, the Special Forces organization was radically altered by the Department of the Army. The 1st Special Forces was activated as the parent regiment for all Special Forces groups. One of the existing groups was redesignated as a result, and the 77th at Fort Bragg was reduced to a single digit as the 7th Special Forces Group. The 1st and 10th Special Forces Groups remained the same, except that they were now part of the 1st Special Forces (Regiment). The Army also gave the Special Forces the official heritage of the First Special Service Force and all Ranger units, an unfortunate decision that later caused friction when the Army Rangers raised their own regiment during the Vietnam conflict. In September 1961 the Special Forces added the 5th Special Forces Group, with direct responsibility for Southeast Asian contingencies, to its ranks at Fort Bragg.

The Special Forces was heavily committed to Southeast Asian combat advice and support when President John F. Kennedy authorized the distinctive Green Beret for his counterinsurgency warriors. President Kennedy had always expressed his keen desire and faith in special warfare as an effective foreign policy tool of the United States and visited the Special Warfare Center in October 1961. The Center commander, Brigadier General William P. Yarborough, met the commander-in-chief wearing the green beret. Kennedy was pleased with the appearance of the green beret and approved its wear by all Special Forces members. On December 10, 1961, the green beret was designated by the Army as the Special Forces' official headgear.

The Special Forces presence in Southeast Asia, as well as other parts of the Third World was not well reported. Many projects were shrouded in secrecy, and the media had a tendency to concentrate on incidents of combat belligerence. The Special Forces accomplishments in this region were actually premised on completing fundamental pacification tasks such as securing rural villages from terrorist acts, health treatment and civic action. At first, Special Forces temporary-duty mobile training teams sent to Vietnam originated from the 1st Special Forces Group on Okinawa. Special Forces extended its operations to include the diverse ethnic groups and native "Montaguard" hill tribes, and teams were also sent from the 5th and 7th Special Forces Groups in the United States. The main Special Forces project became the Civilian Irregular Defense Group (CIDG) program, which assisted hamlet defense, raised strike forces and formed its own mobile guerrilla units. Much of this history is recounted in Charles M. Simpson's volume, *Inside the Green Berets* (Presidio Press, 1983).

The Special Forces were received by the Vietnamese as caring and concerned professionals who spoke the language and helped local communities with schooling, new farming method, and self-protection. The Viet Cong resented this unwelcome intrusion to their plans of conquest and introduced a systematic program of assassination and selected attacks against Special Forces targets. Special Force casualties inevitably increased among its isolated teams, most of which were deployed in South Vietnam's remote areas of swamp and jungle. The communist guerrillas also tried to intimidate the villagers into resisting Special Forces efforts by telling them that Special Forces medics were witch doctors and that their food was poisoned. These primitive enemy tactics failed as people soon came to respect and even treasure Special Forces assistance.

The central South Vietnamese government, however, remained coup-prone and its main army continued to suffer defeats at the hands of Viet Cong formations well endowed with Sino-Soviet weapons and ammunition. The Special Forces were spread thin as they were ordered to hold more and more areas throughout South Vietnam. Higher-level B and C detachments were staffed to control the numerous A teams stationed from the northern Demilitarized Zone to the tip of the Mekong Delta.

The Special Forces provided valuable intelligence on enemy plans, upgraded defenses and trained hamlet militia. The guerrilla expertise of the Special Forces made them prime anti-guerrilla fighters. During September 1962, Colonel George Morton was appointed to form a provisional Special Forces group in Vietnam. By September 1964 the entire 5th Special Forces Group deployed from Fort Bragg to become the new headquarters for all Vietnam-based Special Forces operations.

The Special Forces was involved in heavy fighting throughout its Vietnam service. The early battles were marked by the heroic defense of Nam Dong—where Captain Roger Donlon earned the first Medal of Honor since the Korean War on July 6, 1964—and the repulse of Viet Cong wave assaults outside Dong Xoai by Second Lieutenant Charles Q. Williams' team. The Special Forces defense of the border highland outpost of Plei Me in 1965 became one of the most significant military actions of the Second Indochina War. Located in Vietnam for over a decade, the Special Forces fought with such valor that this small element—averaging about 3,000 men at any one time—earned 17 Medals of Honor, 90 Distinguished Crosses and thousands of other decorations.

During the Vietnam conflict, the Army Special Forces was instrumental in conducting strategic reconnaissance and raiding operations that extended beyond the Vietnamese boundaries, covering the flanks of Army and Marine divisions and patrolling the border with a series of forts that resisted continual attacks. During the pivotal battles of Tet-68, Special Forces led CIDG forces rendered valuable front line infantry service that recaptured many fortified towns and cities.

At the peak of its service in Vietnam, Special Forces manned over 80 camps and forward operating bases and advised 40,000 CIDG soldiers. The last commander of the 5th Special Forces Group in Vietnam, Colonel Mike Healy, summed up the great success of Special Forces-advised CIDG tribesmen, "We took them out of loincloths and put them into uniforms and now they are elite forces." Shelby L. Stanton's *Green Berets at War* (Presidio Press, 1985) provides the best reference for Special Forces in Southeast Asia.

While Vietnam dominated Special Forces efforts during this period, other Special Forces units were deployed on domestic action and combat assignments throughout the world. In August 1962 one reinforced company of the 7th Special Forces Group was transferred to the Panama Canal Zone and became the 8th Special Forces Group the following April. This group was dedicated to anti-communist operations throughout Latin America. In the spring of 1963 a cadre of officers and sergeants from the 7th Special Forces Group was selected for the headquarters of the newly authorized 6th Special Forces

Group, which was designed to cover troublespots and support for the Middle East.

In December of 1963, the 7th Special Forces Group again parted with a portion of its cadre to form the 3d Special Forces Group, which was officially activated during March 1964 to cover the continent of Africa. At this point, the 7th Special Forces Group itself was the global contingency force of all Special Forces. The last Special Forces element fielded during the Vietnam era was the 46th Special Forces Company, later known as the 3d Battalion of the 1st Special Forces Group, which went to Thailand.

During the years immediately following the withdrawal of Special Forces from Vietnam, and the adverse conclusion of the Second Indochina War once the Americans departed, Army Special Forces sustained severe cutbacks. The 3d Special Forces Group was deactivated in December 1969, the 6th Special Forces Group in March 1971, the 8th Special Forces Group in June 1974. The worldwide strength of Special Forces was reduced over 70 percent, and unconventional warfare expertise at a national level diminished to alarming levels.

During the 1980s, the United States decided to rebuild its Special Forces as part of a revitalized counter-terrorist and counterinsurgency posture. This transition was highlighted when the Special Forces reactivated the 1st Special Forces Group and deployed elements back to Okinawa in 1984. The Special Forces was reshaped to conduct unconventional warfare, strategic reconnaissance, foregoing internal defense, direct-action and other special-operations missions. Special Forces teams were sent to such diverse areas as Lebanon in the Middle East, the Philippines and El Salvador in Central America. Finally, Army Special Forces was granted equal status with all other components of the Army when it was granted its own branch status on September 11, 1987.

The Army Special Forces today is considered the cutting edge of United States' foreign policy in most critical areas around the world. The Special Forces is constituted of the 1st Special Forces Group, headquartered at Fort Lewis, Washington with a battalion on Okinawa; the 5th Special Forces Group located at Fort Campbell, Kentucky; the 7th Special Forces Group situated at Fort Bragg, North Carolina, with a battalion in the Panama Canal Zone; the 10th Special Forces Group at Fort Devens, Massachusetts, with a battalion forward in Germany; the 11th Special Forces Group (USAR) at Fort Meade, Maryland; the 12th Special Forces Group (USAR) at Arlington Heights, Illinois; the 19th Special Forces Group (ARNG) at Camp Williams, Utah; and the 20th Special Forces Group (ARNG) at Birmingham, Alabama. The few selected men of the Special Forces remain a true elite fighting force, dedicated to the voluntary professionalism of succeeding in any mission, no matter how difficult the circumstances or how dangerous the adversary.

A. Special Forces Branch Insignia

The gold, crossed arrow insignia is the same worn by members of the 1st Special Service Force formed in World War II. It reflects qualities of the Special Forces soldier—straight and true—and today is the official branch insignia of the Special Forces.

B. Special Forces Tab

The tab is awarded to those individuals who successfully complete the Special Forces Qualification Course or the Special Forces Officer Course. It can now be awarded retroactively to certain designated groups. These may include the individuals who served in the joint Canadian-American organization, the 1st Special Service Force or past members of the legendary Jedburgh teams of the OSS—both early predecessors of today's Special Forces.

C. Special Forces Patch

The arrowhead shape represents craft and stealth—skills of the American In-

Secretary of Defense Melvin Laird visit to Bad Toelz, Flint Kaserne, Special forces Detachment (Airborne), Europe, 1969. LTC (P) Ludwig Faistenhammer, Jr., Commander and Major Clyde J. Sincere, Jr. B-1 Detachment Commander left of Secretary Laird. LTC Robert Arkley, DCO, with an aide to Secretary Laird to the right rear. (Courtesy L. Faistenhammer Collection).

dian. The upturned dagger represents the unconventional nature of Special Forces operations. The three—lightning bolts symbolize blinding speed and strength as well as to denote the three methods of infiltration—air, land and sea. The gold color of the dagger and lightning bolts stands for constancy and inspiration and is set against a teal blue background. A black and gold airborne tab is worn directly above the patch.

D. Special Forces Crest
Prominent on the black and silver crest is the Special Forces motto "De Oppresso Liber," meaning "liberate from oppression." The crossed arrows symbolize the role of Special Forces in uncentional warfare and reflect the qualities of the Special Forces soldier - straight and true. The upturned knife represents the one issued to members of the First Special Service Force.

FORT BRAGG–HOME OF ARMY SPECIAL FORCES

Today's Special Warfare Center and School is the product of more than three decades of special-operations development which began at Fort Bragg in June 1952, when the Psychological Warfare Center moved from Fort Riley, Kansas. In 1982, two distinct organizations were formed to meet expanding special-operations missions: The Special Warfare Center and the 1st Special Operations Command, which took over responsibility for the conduct of special-operations missions. In May of 1987 the Special Warfare Center came under the operational control of the newly created unified command, the United States Special Operations Command.

The addition of the word "School" to the Special Warfare Center's name in May 1986 reflects the organization's mission of training most elements of U.S. Army special-operations forces. Historically, special-operations forces have been viewed as operating "deep in the enemy's homeland," for strategic objectives and have been considered to have little relationship to conventional forces. However, the JFK Center and School trains to the theme of integration of special-operations forces with all levels of conventional forces, in support of the Air-Land Battle Doctrine.

U.S. ARMY SPECIAL OPERATIONS COMMANDERS

A. Commander-in-Chief, USSOCOM, MacDill AFB, Florida:
General James J. Lindsay 1987-

B. Commanders: (Various "Higher" Headquarters, U.S. Army Special Forces/Special Operations, Fort Bragg, North Carolina, 1952-Present):
Colonel Aaron Bank 1952-
*Colonel Charles H. Karlstad 1952-1953
*Colonel Gordon Singles 1953-1954
Colonel Thomas A. McAnsh 1954-1956
Colonel Edson D. Raff 1956-1957
*Colonel William J. Mullen 1957-1958
Brigadier General George M. Jones 1958-1961
Major General (Ret LTG) William P. Yarborough 1961-1965
*Brigadier General Joseph W. Stillwell, Jr. 1965-1966
Brigadier General Albert E. Milloy 1966-1968
Major General (Ret LTG) Edward M. Flanagan, Jr. 1968-1971
Major General (Ret LTG) Henry E. Emerson 1971-1973
Major General Michael D. Healy 1973-1975
Major General (Ret GEN) Robert C. Kingston 1975-1977
Major General (Ret LTG) Jack V. Mackmull 1977-1980
Major General Joseph C. Lutz 1980-1984
Major General Leroy N. Suddath, Jr. 1984-1988
Major General James A. Guest 1988-1989
Lieutenant General Gary E. Luck 1989- Present

U.S. ARMY SPECIAL FORCES MEDALS OF HONOR WINNERS

The nation's highest decoration for valor was established by an act of Congress for the Navy and Marine Corps in 1861 and for the Army in 1862. The first decoration formally authorized by the government to be worn as a badge of honor, it is awarded in the name of Congress to those who have distinguished themselves in conflict with the enemy by gallantry and risk of life above and beyond the call of duty. Of the 155 soldiers who earned the Medal of Honor in Vietnam, 17 were members of Special Forces.

The following individuals have distinguished themselves and have upheld and preserved the high ideals and standards of the Special Forces.

Recipients are listed chronologically by date of action for which they were awarded. Ranks noted are those at time of action

Rank/Name	Date Awarded	Date of Action
Cpt. Roger H.C. Donlon	5 Dec. 1964	6 July 1964
2nd Lt. Charles Q. Williams	9 June 1966	9 June 1965
*1st Lt. George K. Sisler	6 July 1968	7 Feb. 1967
SSG Drew D. Dix	16 Jan. 1969	31 Jan. 1968
SSG Fred W. Zabitosky	7 March 1969	19 Feb. 1968
*SFC Charles E. Hosking, Jr.	23 May 1969	21 March 1967
*SFC Eugene Ashley, Jr.	18 Nov. 1969	7 Feb. 1968
*SGT Gordon Yntema	2 Dec. 1969	18 Jan. 1968
*SP5 John J. Kendenburg	7 April 1969	13 June 1968
*SFC William M. Bryant	16 Feb. 1971	24 March 1968
SFC Robert L. Howard	2 March 1971	30 Dec. 1968
SSG Franklin D. Miller	15 June 1971	5 Jan. 1970
*SGT Brian L. Buker	16 Dec. 1971	5 April 1970
SGT Gary B. Beikirch	15 Oct. 1973	1 April 1971
SSG Jon R. Cavaini	12 Dec. 1974	4 June 1971
SSG Roy P. Benavidez	24 Feb. 1981	2 May 1968
*1st LT Loren D. Hagen	6 Sept. 1974	7 Aug. 1971

* Awarded Posthumously (8)

TASK FORCE "IVORY COAST" "OPERATION KINGPIN"

Background

By 1970, the struggle in Vietnam had become one of the longest conflicts in our nation's history. Throughout the country, heated debate grew as to how the war should be fought or even if it should be fought. However, regardless of political views, one issue tended to unify every American and that was concern for our Prisoners of War (POWs) held in North Vietnam.

In August of 1964, shortly after the Tonkin Gulf incident, the first American pilot was shot down over North Vietnam. By spring of 1970, over 350 other pilots had likewise been downed and were being held in North Vietnamese prisons along with hundreds of others designated as Missing in Action (MIAs).

Exposed to horrid conditions and frequent torture most American prisoners of war in North Vietnam were never allowed to contact the outside world. The North Vietnamese even refused to release a complete listing of who had been captured or who was injured.

Numerous individuals, including many foreign representatives, tried to intercede on behalf of the POWs and gain at least adequate medical care for the men, but most such efforts failed. A few prisoners had been set free, but the North Vietnamese plainly stated that the prisoners as a group were hostages and would not be released until the last American had left Vietnam.

The United States had been gathering information about the POWs from other sources since the first pilot was shot down. It was essential that the Air Force and Navy know at least the approximate location of prisoners to avoid tragic accidents during bombing missions.

Aerial photographic reconnaissance was one method used to gather information concerning the POWs. Late in 1968, photographs revealed that some American prisoners were being held at a compound some 30 miles northwest of Hanoi. However, the information was not conclusive and aerial surveillance uncovered no further developments until 1970.

In May of 1970, reconnaissance photographs revealed the existence of two camps west of Hanoi: one, Ap Lo, and another on the Son Cong River near Son Tay. A comparison of old photos and those of 1970 showed that the Son Tay compound was being expanded. To add further evidence, the POWs inside the camp had used a complicated code to signal that there were 55 Americans present. One photo even identified what seemed to be a large "K" drawn in the dirt which was a code for "Come get us."

Brigadier General Donald D. Blackburn became intimately involved in what was to follow, as special assistant for Counter Insurgency and Special Activities at the Pentagon. General Blackburn informed General Earle Wheeler, chairman of the Joint Chiefs of Staff, of development at Son Tay. It was at this meeting held on 25 May that General Blackburn suggested a small group of Special Forces volunteers be given the task of rescuing the POWs.

Both General Blackburn and Colonel Arthur D. "Bull" Simons shared the same enthusiasm for unconventional operations and when the Son Tay question arose, the brigadier already knew who he wanted to lead the rescue mission. General Blackburn had long wanted to plan such an operation. He wanted to do something to the North Vietnamese that would really rattle their cage. Colonel Simons also felt the situation called for extraordinary measures. He strongly believed that "people in the rear area should never feel comfortable. The idea is to discomfort the enemy as much as you can."

It was decided that the mission would be a joint Army-Air Force effort with most of the training scheduled for Eglin Air Force Base in Florida. Air Force Brigadier General Leroy J. Manor was chosen to be the overall commander and would see to the administrative aspects of the mission while Colonel Simons would take charge of training and lead the actual raid.

While General Manor was occupied finding his subordinates to pilot the assault helicopters, Colonel Simons was recruiting his team at Fort Bragg. He chose Lieutenant Colonel Elliot P. "Bud" Sydnor to be his deputy and Captain Richard J. "Dick" Meadows to head the compound assault team. Colonel Simons knew and had the utmost confidence in both men.

Colonel Simons had commanded B Company of the Sixth Rangers during the invasion of the Philippines and thus had an opportunity early in his career to form a definite set of opinions concerning leadership and unconventional tactics. Good leadership should see that casualties were kept at an absolute minimum. Soldiers were entitled to leaders that could "smart their way out" of a tight situation. Colonel Simons was such a leader. After the "White Star," he brought every one in his group back to the U.S. alive.

It was said that when Colonel Simons planned for a mission, his planning was meticulous. There was very little ever left to chance. He felt that when planning an unconventional operation, "The more improbable something is, the surer you can pull it off." General Blackburn was sure he had the right man for the job.

It was believed that since the compound was more than 20 miles west of Hanoi, Son Tay was isolated enough to enable a small group to land, release the prisoners and withdraw.

There were trees just inside the compound wall, but the planners felt that enough space was available for a small helicopter to land inside the camp and release an eight-or 12-man assault team.

By doing this, the raiders could get into the cell block before the North Vietnamese could harm the prisoners. Outside the compound walls, two larger helicopters could land with the rest of the Special Forces contingent. One would blow a hole through the wall and aid in rescuing, while the third contingent would sweep through the prison support area and set up roadblocks to prevent NVA reinforcements from reaching the camp.

It was also agreed that a full-sized mock-up of the Son Tay compound be constructed on the training site at Eglin Air Force Base. However, intelligence experts feared Soviet photographic satellites would detect such a large construction and surmise its meaning. As a result, the compound was constructed of 2 x 4 lumber and heavy cloth so that the various structures could be rolled up and put away. Night training was no problem, but daylight training was to be limited to those four hours a day when Soviet satellites were not in position to photograph the area.

For training purposes, a table-sized replica of the Son Tay compound was built by the CIA and equipped with special viewing devices. Thus Colonel Simons' men would be able to see the ground before them exactly as it would appear the night of the raid. Code named "Barbara," the model was precise down to the last bush and tree.

However, time was running out. It would have been impossible to keep such a mission secret for any length of time. Then, on 13 November, it was announced that six more POWs had died. On 19 November, President Nixon gave the go ahead to Secretary of Defense Melvin Laird.

Execution

On 18 November, one day in advance of the president's go ahead, the Son Tay raiders landed at Takhli, Thailand. However, even then only four men in the ground force knew what the target was. Only Simons, Sydnor, Meadows and Cataldo, the group physician, knew what the mission was to be.

The men were finally told details of the raid five hours before takeoff. Colonel Simons simply told them, "we are going to rescue 70 American prisoners of war, maybe more, from a camp called Son Tay. This is something American prisoners have a right to expect from their fellow soldiers. The target is 25 miles west of Hanoi." Colonel Sydnor continued with the briefing, but as Colonel Simons left the room, he was given a loud round of applause.

Time was to be a crucial factor. It was estimated that the whole raid must be completed within 26 minutes or else NVA reinforcements might arrive in great enough numbers to endanger the whole mission.

As the raiding force neared Son Tay, a diversionary mock air strike was launched from the aircraft carriers *Oriskany*, *Hancock* and *Ranger* stationed in the Tonkin Gulf. Only instead of bombs, the aircraft flew over Haiphong and dropped flares. The diversion worked. Attention was drawn to the east while Colonel Simons' force closed in on Son Tay from the west.

The raiders then loaded into a C-130 and flew to Udorn, Thailand, where they transferred to Huey helicopters. The assault force left Udorn led by a C-130 tanker which would later be used to refuel the aircraft en route.

Captain Meadows was to lead the way by setting his 13 men down in the compound only seconds after they had neutralized the guard towers. In this way his small team would be able to reach the POWs before the guards. However, the rotors struck a tree and the craft plunged groundward. Fortunately, only one man was injured (a broken ankle). The men quickly exited the crashed vehicle and dashed for the cell blocks. Captain Meadows used a bull horn to announce, "We're Americans. This is a rescue. We're here to get you out." Three minutes after Captain Meadows' men hit the ground, a blast knocked a gaping hole into the south wall and reinforcements streamed into the compound.

The men pouring through the south wall were supposed to be from Colonels Simons' group. However, it was Colonel Sydnor's men instead. The helicopter carrying Colonel Simons had landed 400 yards to the south at a similar looking compound labeled the "Secondary School." Intelligence officers had been concerned that such a mistake might be made, but contingency plans had been formulated for just such a situation.

Air Force Lieutenant Colonel John Allison, pilot of Colonel Sydnor's men, noticed that the helicopter carrying Colonel Simons' contingent had landed at the "Secondary School" and not the designated target. Realizing that he might have to execute the raid with 22 fewer men than planned, Colonel Sydnor put his contingency plan into effect.

Colonel Simons immediately realized that a mistake had been made, but he was sure that Colonel Sydnor and Captain Meadows could handle the situation until he arrived. He wanted to recall the helicopter, take off and fly the 400 odd yards back to the prison compound, but suddenly a heavy firefight broke out all around him. This was very unexpected. No additional NVA troops were supposed to be this near to the POW compound. Colonel Simons immediately pressed the advantage and attacked. He later remarked, "The attack was pressed with great violence—because surprise doesn't work if you don't use violence and speed."

Colonel Simons immediately ordered the helicopters back and began loading his men. Twenty-seven minutes after Colonel Donohue crashed into the compound, the raid was over. Scores of NVA and foreign troops were left dead and wounded around Son Tay, but Colonel Simons did not lose a man. Fifty-nine men landed and 59 men flew back to Thailand. The operation went perfectly. Colonel Simons had made General Blackburn's ambition a reality. He had really rattled their cage.

Colonel Simons's team broke contact, boarded the helicopter and flew to the POW compound a short distance away. Nine minutes into the raid, Colonel Simons was outside the prison walls and the basic plan was reinstated. Inside the compound, Captain Meadows had eliminated most hostile elements and was still searching for the POWs. Most of the 30 plus guards at Son Tay were dead and wounded, but a disturbing fact was becoming obvious. There were no prisoners. Captain Meadows radioed to Simons," Search complete, negative items." Son Tay was a dry hole.

However, Colonel Simons observed something else during his action at the "Secondary School." It quickly became apparent that his opposition was not North Vietnamese. They were taller, dressed differently and better equipped than the NVA. Colonel Simons' 21 men inflicted several dozen casualties among the strangers in a few short minutes and in doing so perhaps saved his whole raiding force. Only later was it learned that the heavily guarded compound housed a Soviet or Chinese contingent that was training North Vietnamese air defense technicians.

Results

After the raid, U.S. POWs were all moved to the "Hanoi Hilton" where they were able to organize as a POW unit for the first time and get up to date news from recent POWs.

They were also provided better treatment and rations.

The raid was also a morale factor for the POW's and their families, in that an American fighting force actually attempted a rescue.

TASK FORCE IVORY COAST 21 NOVEMBER 1970

Roll of Honor

A. Distinguished Service Crosses: 6
CPT Richard Meadows 1970
MSG Thomas Kemmer 1970
SFC Tyrone Adderley 1970
COL Arthur D. Simons 1970
SSG Thomas Powell 1970
LTC Elliott Sydnor 1970

B. Silver Stars: 40

C. Purple Hearts: 1

1ST SPECIAL FORCES GROUP (AIRBORNE), FIRST SPECIAL FORCES

The First Special Forces Group (Airborne) is descended from the First Special Service Force, a joint Canadian-American unit of World War II. The unit was formed July 5, 1942 at Fort William Henry Harrison. Following intensive preparation including parachute and mountain warfare training, the famous "Devil's Brigade" fought campaigns in several theaters. These included the landings at Attu and Kiska to eject the Japanese from the Aleutian Islands, as well as combat operations in North Africa, Italy and Southern France. The 1st Special Service Force was disbanded in January 1945 after earning distinguished combat record.

The 1st Special Forces draws much of its tradition from many predecessors in the field of unconventional warfare: Rogers' Rangers of colonial days; Francis Marion, the Swamp Fox, and his irregulars of the American Revolutionary War

Mosby's Rangers of the Civil War; and the Office of Strategic Services and Merrill's Marauders of World War II.

The modern history of 1st Special Forces Group commenced on 1 April 1956, with the activation of the 14th Special Forces Operational Detachment (Area) at Fort Bragg, North Carolina. Hand-picked from the 77th Special Forces Group (Airborne), the members of this detachment—along with the 12th, 13th and 16th SFODs—were specifically selected and trained for the purpose of establishing a special operations capability in the Asian-Pacific theater. These units were transferred to the Pacific in two increments. The 14th SFOD (Area), under the cover of "8521st Army Service Unit" was transferred to Fort Shafter, Hawaii in June 1956. Shortly afterwards, the 12th SFOD (Regiment), 13th SFOD (Regiment), and 16th SFOD (District), were moved to Camp Drake, Japan under cover of "8231st Army Unit."

On 24 June 1957, the 1st Special Force Group (Airborne) was officially activated at Camp Drake, Japan, although all its elements were either en route to Okinawa or on temporary duty in South Vietnam on that date. Group activation ceremonies were conducted on 14 July 1957 at Fort Buckner, Okinawa, following arrival of the operational detachments and the 248th Quartermaster Detachment (Rigger). A staff officer from U.S. Armed Forces Far East was initially assigned as Group commander, but broke both his legs on his first parachute jump and was evacuated. Command of the group was then assumed by LTC A. Scott Madding of the 14th SFOD, a highly decorated veteran who had served with Merrill's Marauders and OSS in World War II and with Ranger and partisan units in Korea.

The period 1957-1960 was a time of intensive training for the 1st Special Forces. A large contingent of Republic of Korea Special Forces troops was trained in Okinawa, while the 1st Special Forces Group deployed mobile training teams to conduct missions in Thailand, Taiwan, the Philippines, Indonesia and South Vietnam. Simultaneously, the 1st Special Forces Group also conducted internal training operations to qualify Special Forces volunteers who had not been through training at Fort Bragg, as well as to cross-train qualified personnel in additional team skills. During the same period, the group grew in strength from 55 personnel in July 1957 to 364 personnel by October 1960.

On 30 October 1960, all Special Forces groups were reorganized under the combat arms regimental system. 1st Special Forces Group was redesignated 1st Special Forces Group (Airborne)—"First Special Forces" in recognition of its descent from the First Special Service Force. At the same time, the lineage and honors of B Company, 1st Ranger Battalion (World War II) and the 5th Ranger Company (Korean War) were assigned to 1st Special Forces Group.

The most visible change occurring at this time was the restructuring of the group's organization. To accommodate the growth in size and to streamline control and administration, the detachments were reorganized on 15 December 1961 into four lettered companies: A, B, C and D. Each Company consisted of an Operational Detachment C, functioning as the company headquarters and a varying number of subordinate ODBs and ODAs.

A Company was the largest of the four companies with a strength of 47 officers and 165 enlisted men. B Company was activated with 36 officers and 114 men; C Company with 40 officers and 120 enlisted. D Company was organized with only a cadre of five personnel. A fifth company, Company E (Signal), was activated on 19 February 1964, to provide communications support to deployed detachments. Group strength continued to increase, reaching a peak in 1963 of 232 officers, four warrant officers and 1,026 enlisted men.

With the advent of the 1960s, 1st Special Forces Group's activities increasingly focused on operations in the Republic of Vietnam. The 14th SFOD had conducted the first mission to train Vietnamese Rangers near Nha Trang in the summer and fall of 1957. Commitment of 1st Special Forces teams to Vietnam increased steadily thereafter, with numerous detachments deploying from Okinawa on extended TDY missions to train and lead units of the Vietnamese Special Forces (LLDB), Rangers and Civilian Irregular Defense Group (CIDG). This headquarters element directed Special Forces operations in Vietnam from 1962-1965, although the manpower continued to come from detachments of the 1st Special Forces Group on Okinawa and the 7th Special Forces Group at Fort Bragg.

In early 1965, USASFV (P) was replaced by the 5th Special Forces Group, which deployed from Fort Bragg. Even after the arrival of the 5th Special Forces Group, however, the 1st Special Forces Group continued to dispatch teams to Vietnam, maintaining at least six ODAs in country at all times to participate in Special Operation Group (SOG) reconnaissance missions. These missions frequently involved cross-border operations into neighboring Laos.

Operations in Vietnam were only one aspect of 1st Special Forces Group activities, however. The group simultaneously carried out security assistance and civic action missions throughout Southeast Asia and the Pacific. A Special Action Force Asia, or SAFASIA, organized with 1st Special Forces Groups' commitment in Thailand, eventually grew to such a scale that in 1967 the group's D Company was detached and permanently stationed in country. Redesignated the 46th Special Forces Company, this unit operated in Thailand for the next four years. Among its accomplishments was the afterwards deployment to Vietnam.

Increased experience in supporting civic action and relief operations resulted in the establishment of Disaster Assistance and Relief Teams, or DARTs. The 1st Special Forces Group organized a number of DARTs, each consisting of an A Team augmented by two doctors and four to six medics from the SAFASIA medical detachment. Engineers from Group of the 539th Engineers were attached as needed.

These task-organized DARTs operated successfully in Luzon, Indonesia, the Marshall Islands and even in the outer island of the Ryukyu chain. The greatest successes of the program occurred during the 1971 Pakistan floods, and the 1972 floods and famine in the Philippines. Teams from the 1st Special Forces Group were literally life-savers during both calamities. Operating rescue boats, inoculating civilians, distributing food and directing rebuilding efforts, the DARTs saved lives and salvaged livelihoods and earned America many friends. The Philippines Presidential Unit Citation was awarded to the 1st Special Forces Group (Airborne) for the latter action.

In 1972, just prior to the Philippines Disaster Relief Operation, the 1st Special Forces Group was again reorganized. Companies A, B and D were consolidated and redesignated as 1st and 2nd Battalions, 1st Special Forces Group (Airborne). The change was, for the most part, nominal. A C Detachment remained the command and control element with a lieutenant colonel in command. Operational Detachments B were now designated as lettered companies of the battalions; the name and role of the A Teams remained unchanged.

Even as these organizational changes were occurring, a chapter in the history of the 1st Special Forces Group was coming to a close. The disappointing conclusion to America's Vietnam experience foreshadowed a shift of strategic emphasis away from Asia and the Pacific, back to NATO and Europe. The size of the Army was being reduced to peacetime levels, and decisions on these reductions were being made by leaders more concerned with maintaining armor and mechanized infantry strength than with maintaining unconventional warfare capabilities.

Special Forces, which had grown to a force of seven groups in 1963, was cut severely in strength. In 1974, the 1st Special Forces Group (Airborne) was cut from the force, furling its colors on 28 June. Left behind was a single team, U.S. Army Special Forces Detachment Korea, to provide a special operations presence in the Far East.

During inactivation ceremonies held at Fort Bragg, Brigadier General Michael P. Healy remarked; "When we furl our colors and place them in the hands of the custodian of all combat unit colors, we carefully let it be known that those who remain are quite capable of resurging. The beret flashes which are deposited to the rear of the Special Warfare Memorial will be held in abeyance for that day when we will unfurl those colors and proudly take our place in line." For 17 years of service in theater, the retiring group was awarded the Meritorious Unit Citation with streamer embroidered "PACIFIC AREA."

Ten years later, the resurgence of which General Healy had spoken came to pass. Once again the need for special operations forces in the Asia-Pacific Theater became apparent to the leaders of the Army and the nation, and the 1st Special Forces Group (Airborne) was once again called to duty. One battalion of the group would be forward-stationed at Okinawa, while two battalions and the group's headquarters and separate companies would be organized at Fort Lewis, Washington.

The first element to be activated was Company A, 1st Battalion, which was reactivated at Fort Bragg on 15 March 1984. This company and the remainder of the 1st Battalion was assembled and deployed to Torii Station, Okinawa during the spring and summer of 1984, under command of Lieutenant Colonel James Estep. An in-theater activation ceremony for the battalion was held at Torii Station on 19 October 1984, with Lieutenant General Alexander Weyand present as reviewing officer. Activation of the rest of 1st Special Forces Group (Headquarters Company, 2nd and 3rd Battalions, Service Company, Signal Company and 1st Military Intelligence Company) officially commenced on 4 March 1984 at Fort Lewis, Washington.

That date a"carrier day, or activation of the cadre...was marked by a parachute jump and rucksack march led by Colonel David J. Baratto, the 1st Special Forces Group commander. On 4 September 1984, a formal activation ceremony was held at the Fort Lewis parade ground. The group's colors were presented to Colonel Baratto by Major General Leroy N. Suddath, after which the 1st Special Forces group passed in review. A number of distinguished guests were on hand for the ceremony, including General Maxwell R. Thurman, vice chief of staff of the Army.

COMMANDING OFFICERS OF THE 1ST SPECIAL FORCES GROUP (AIRBORNE)

* LTC Albert S. Madding April 1956 - Nov. 1957
LTC John M. Cole June 1957 - July 1957
Col Marshall Wallach Nov. 1957 - Aug. 1959
Col Francis B. Mills Aug. 1959 - July 1961
Col Noel A. Menard July 1961 - Aug. 1962
Col Robert W. Garrett Aug. 1962 - July 1964
Col Francis J. Kelly Aug. 1964 - June 1966
*Col Jonathan F. Ladd July 1966 - May 1967
*Col Harold R. Aaron May 1967 - May 1968
Col Robert B. Rheault May 1968 - May 1969
*Col Charles M. Simpson, III June 1969 - Aug. 1971
Col John C. Geraci Aug. 1971 - July 1973
Col Elliot P. Sydnor, Jr. July 1973 - July 1974
Col David J. Baratto March 1984 - July 1986
Col John D. Blair, IV July 1986 - July 1989
Col Eddie J. White July 1988 -

*Deceased

COMMAND SERGEANTS MAJOR OF THE 1ST SPECIAL FORCES GROUP (ABN)

MSG Robert L. Voss 1957 - 1958
SGM Lory Bell 1958 - 1960
SGM Robert Hoiser 1960 - 1962
SGM George W. Dunaway 1962 - 1964
SGM Patrick E. Pettingill 1964 - 1965
CSM George W. Odum 1965 - 1967
CSM Myron J. Bowser 1967 - 1968
CSM Duane C. Vierk 1968 - 1969
CSM Dewey C. Simpson 1969 - 1972
CSM Galen C. Kittleson 1972 - 1974
CSM Arthur F. Garcia 1984 - 1986
CSM Jeffrey H. Raker 1986 - 1987
CSM Gregory T. McGuire 1987

1ST SPECIAL FORCES GROUP (AIRBORNE), FIRST SPECIAL FORCES ROLL OF HONOR

A. Distinguished Service Cross: Date Awarded

*CPT Herbert F. Hardy, Jr.	1964
MSG Herman J. Kennedy	1964
CPT Wallace Viau	1965
SFC Henry Bailey	1965
SFC Steven Comerford	1966
*SSG Ronald Terry	1968

*Posthumously

	Number
B. Silver Stars	44
C. Bronze Stars w/ "V"	244
D. Air Medals	499
E. Purple Hearts (Wounded in Action)	293
F. Combat Infantryman Badges	554
G. Combat Medical Badges	88
H. Killed in Action	40
I. Missing in Action	2

Ironically, the first and last Special Forces soldiers to die in Vietnam were members of the 1st Special Force Group, (Airborne), 1st Special Forces:
CPT Harry G. Cramer, 21 October 1957
SGT Fred C. Mick, 12 October 1972

46TH SPECIAL FORCES COMPANY (AIRBORNE)

The 46th Special Forces Company (Airborne) was originally constituted in the Regular Army in 1966 as Company D (Augmented), 1st Special Forces Group (Airborne), 1st Special Forces and activated at Fort Bragg, North Carolina. The unit was constituted and allotted to the Regular Army as the 46th Special Forces Company (Airborne) on 15 April 1967.

June of 1966 saw the first embryonic strings of this reborn unit. The company staff started administrative and logistical planning. A Tactical Operations Center was established which developed and instituted a 10-week pre-mission cycle. A four-week course of instruction on China and Southeast Asia, under the auspices of East Carolina College, was presented to the first group of 75 students from the company. At this time, additional detachments within the company began pre-mission training. The unit was oriented toward Thailand and everyone stayed busy attending classes on history, geography, ethnology and customs of Thailand, down to the most minute details.

On the second day of October 1966, an advance party departed from Fort Bragg and arrived at Takhli Royal Thai Air Force Base on 4 October after 38 hours in the air with only several intermediate stops. The following day, portions of this party went to headquarters, MACTHAI, to meet with both U.S. and Royal Thai army officials, initiating plans for a counter insurgency training program to be presented to Royal Thai Army combat arms units. Commencing 11 October 1966 the bulk of the company departed Fort Bragg in three shipping increments via C-141 aircraft. On 15 October, Company D (Augmented), 1st Special Forces Group (Airborne), was present and accounted for at Camp Pawai, Lopburi, Thailand.

No sooner was this unit in country when detachments were dispatched to establish base camps at various locations. One detachment moved 300 miles to the northeast to set up camp at Nam Pun Dam; another went 125 miles east to Nong Takoo; a third detachment moved almost out of country by going 600 miles south to Trang. A significant event concerning this last detachment is that it parachuted into its training area complete with TOE equipment. In conjunction with an Air Force Air Commando Squadron, this turned out to be the largest airborne operation to date in Thailand and reputed to be the longest non-stop airborne operation executed by Americans in a C-123 aircraft.

The initial training task for all the detachments was the conducting of a five-week three phase counter insurgency training program for selected infantry companies fo the Royal Thai Army (RTA). The program was conducted at three training camps using combined RTA and U.S. Special Forces training teams. In addition, the teams provided advisory support to the RTA Special Warfare Center and Special Forces Group.

During the next several years there were several minor reorganizations within the unit regarding authorized strength. The unit continued with its missions and other assigned tasks until its inactivation on 31 March 1972 at Lopburi, Thailand. The 46th Special Forces Company (Airborne) was redesignated as 3rd Battalion, 1st Special Forces Group, 1st Special Forces, and relocated to Okinawa. In 1974 it was inactivated with the 1st Special Forces Group.

The unit had a brief but admirable history and its presence in the Kingdom of Thailand was felt to such a great extent that it immeasurably reduced the threat of insurgency and most likely avoided a colonized state.

COMMANDING OFFICERS OF THE 46TH SPECIAL FORCES COMPANY (AIRBORNE)

LTC Robert H. Bartelt 1966-1968
COL Stephen R. Johnson 1968-1970
COL Paul H. Coombs, Jr. 1970-1972
LTC William P. Radtke 1972-1973
LTC George Maracek 1973-1974

COMMAND SERGEANTS MAJOR OF THE 46TH SPECIAL FORCES COMPANY (AIRBORNE)

SGM Arthur R. Senkewich
SGM Edward F. McDougall
*SGM Marshall Lynch
*SGM Carlos Leal
SGM Jerry Wareing

*Deceased

3RD SPECIAL FORCES GROUP (AIRBORNE), 1ST SPECIAL FORCES

The 3rd Special Forces Group (Airborne) was constituted on 5 July 1942 in the Army of the United States as Headquarters and Headquarters Detachment, 1st Battalion, Second Regiment, 1st Special Service Force, as a joint Canadian-American organization, activated 9 July 1942 at Fort William Henry Harrison, Montana.

The unit was reconstituted 15 April 1960 in the Regular Army, consolidated with Headquarters and Headquarters Company, 3rd Ranger Infantry Battalion; and designated as Headquarters and Headquarters Company, 3rd Special Forces Group, 1st Special Forces.

The group was activated 5 December 1963 at Fort Bragg, North Carolina and became the fourth such group at the John F. Kennedy Center for Special Warfare (Airborne).

One of the first assignments for the group was to construct a demonstration village, later known as the Gabriel Demonstration Area, which displayed the skills and activities of the Special Forces soldiers and brought many favorable comments from hundreds of visitors, foreign and domestic.

The first group commander was Col James B. Bartholomees, the former director of instruction at the then-named Special Warfare School, and SGM Charles R. Ferguson was the top NCO. The commander and 17 NCOs made their first parachute jump from a CV-2 (Caribou) aircraft over Sicily Drop Zone 18, December 1963.

The three companies, A, B and C, were activated in the spring and fall of 1963-1964, as were the Signal and Aviation Companies, the 1st Civil Affairs Co, the 19th PSYWAR Co., the 534th Engr. Det. and the 705th Intel. Group Detachment. Within 18 months, the group became a fully organized Special Action Force (SAF). Members went on Mobile Training Teams (MTT) to nations in the Far East, Middle East and Africa.

Col. Leroy S. Stanley assumed group command on 1 September 1965, and Col. Bartholomees became the JFK deputy commander. MTTs were conducted in Ethiopia, the Congo, Colombia, Mali, Argentina and Iraq on missions ranging from 45 days to six months. The group also participated in SWIFT STRIKE, WATER MOCCASIN, CHEROKEE TRAIL, BIG OAKIE, TOTOHATCHEE, QUICK KICK AND MAPSTRIKE. The group continued to maintain the "Liberty Village" at the Gabriel Demonstration Area and presented 47 performances during 1965 alone.

During December 1965, men began receiving alerts for Vietnam duty and some left immediately, leaving the group understrength. In 1966, a team went to the Colorado Rockies to study the effects of dysprosium while working at high altitudes with very little oxygen. The group continued to perform a variety of missions until it was inactivated on 1 December 1969 at Fort Bragg, North Carolina.

CAMPAIGN CREDITS
Naples-Foggia
Anzio
Rome-Arno
Southern France (w/Arrowhead)
Rhineland
Aleutian Islands

COMMANDING OFFICERS OF THE 3RD SPECIAL FORCES GROUP (AIRBORNE) 1ST SPECIAL FORCES
Col. James B. Bartholomees 1963-1965
Col. Leroy S. Stanley 1965-1966
Col. Jesse G. Ugalde 1966-1967
Col. John P. Arntz 1967-1968
Col. Elmer Monger 1968-1969

5TH SPECIAL FORCES GROUP (AIRBORNE), 1ST SPECIAL FORCES

The 5th Special Forces Group (Airborne), 1st Special Forces lineage derives from two units of World War II fame: the 1st Special Service Force—a combined Canadian-American organization—and the 5th Ranger Infantry Battalion. The 1st Special Service Force was activated on 9 July 1942 in the Army of the United States as Headquarters and Headquarters Detachment, 1st Battalion, Third Regiment, 1st Special Service Force.

The Headquarters and Headquarters Detachment, 1st Battalion, Third Regiment, 1st Special Service Force was first activated and trained at Fort William Henry Harrison, Montana, in July 1942. The force was first committed overseas in North Africa, conducting the initial assault at Arzew, Algeria. After North Africa, the unit participated in the Italian campaign and also saw action in France. The Group was disbanded in France on 6 February 1945.

The 5th Special Forces Group, 1st Special Forces was reconstituted on 15 April 1960 in the Regular Army and concurrently consolidated with Headquarters and Headquarters Company, 5th Ranger Infantry Battalion and was redesignated as Headquarters and Headquarters Company, 5th Special Forces Group (Airborne), 1st Special Forces.

On 21 September 1961 at Fort Bragg, North Carolina, the 5th Special Forces Group (Airborne), 1st Special Forces was officially activated and subsequently became the "prime" Special Forces unit operating in Vietnam.

However, Special Forces detachments had been serving in Vietnam since the mid 1950s as mobile training teams, under sponsorship of either the Military Assistance Advisory Group (MAAG), Vietnam, or with the Central Intelligence Agency. In 1962, the CIA conceived the Civilian Irregular Defense Group (CIDG) program, initially with the Rhade tribe, in Darlac Province. The success of this initial effort with the Montaguards warranted an expansion of the CIDG effort to other tribal groups which, in turn, led to the requirement for additional Special Forces.

In July 1962, a meeting was held at HQ CINCPAC, and a decision was made under National Security Action Memorandum (NSAM) 57, for the U.S. Army to assume responsibility from the CIA for implementation of the CIDG program. A provisional group, commanded by a colonel, was to be established in Vietnam, and the U.S. Army was to assume complete responsibility for the CIDG program by 1 July 1963. The code name for this transfer of responsibility was Operation SWITCHBACK, and the late Colonel George C. Morton had the honor of being selected as the first commander of U.S. Army Special Forces (Provisional) Vietnam.

U.S. Army Special Force (Provisional), Vietnam, which was approved at

Cardinal Spellman meeting Special Forces CPT Clyde Sincere following the dedication of Special Forces Chapel, Nha Trang, South Vietnam, Christmas 1966. Right is COL Francis Kelly, Chaplain Charles J. McDonnall makes the introduction. (Courtesy of L. Faistenhammer Collection).

CINCPAC in July 1962, consisted of: Headquarters Detachment, 5th Special Forces Group (Airborne), comprised of 15 officers and 61 enlisted personnel, all PCS; one C detachment from the 5th Special Forces Group (Abn) on TDY; four B detachments from the 1st and 5th Special Forces Groups on TDY; and 36 A detachments on TDY from the 1st, 5th and 7th Special Forces Groups.

In view of the fact that all U.S. Army units in Vietnam in 1962 had organizational colors, and since colors were required for honor guard and other ceremonies at the Special Forces Operating Base (SFOB) at Nha Trang (which was visited by the chief of staff, U.S. Army, and many other dignitaries), U.S. Army Special Forces (Provisional) designed a flag made by "Cheap Charlie Tailors" on Tu Do Street in Saigon. This flag served as the organizational colors for the U.S. Army Special Forces (Provisional) Vietnam from September 1962 until the 5th Special Forces Group (Airborne), 1st Special Forces was formally established in Vietnam on 1 October 1964, at which time it absorbed the personnel and equipment of the provisional group. Full deployment of the group was completed in February 1965.

Although young in years, the 5th Special Forces Group is rich in tradition, glory and honors. From the Special Forces operational base at Nha Trang, the group spread throughout the four military regions of South Vietnam. Its operational detachments established and manned camps at 270 difference locations and trained and led indigenous forces of the civilian irregular defense groups.

The CIDG Program proved to be highly successful, being expanded throughout the 1960s to embrace tribal groups such as the Rhade, Ja Rai, Kha, Meo, Khmers from Cambodia and certain groups of ethnic Vietnamese such as the Hoa Hao and Cao Dai, religious groups which had, in previous periods of Vietnamese history, fielded powerful private armies.

THE MIKE FORCES

One of the most significant changes in the organization of the Civilian Irregular Defense Group occurred during the summer of 1965, when the Mike Force was established. The Mike Force was a direct outgrowth of the "Eagle Flight" Detachment formed at Pleiku on 16 October 1964 to react to emergency combat situations at the small Special Forces camps within the western highlands. This "eagle flight" consisted of five 1st Special Forces Groups personnel of Detachment A-334B and 36 Rhade tribesmen. These Montaguards were trained in various Special Forces skills and underwent parachute training. The eagle flight troops were rewarded with higher ranks, a special pay scale and hazardous duty allowances of 1,000 piasters per month. They were armed with M-2 carbines and directly supported by six transport helicopters and three helicopter gunships from Camp Holloway outside Pleiku. The "eagle flight" was used for reconnaissance, search and seizure and camp reaction missions.

As a result of the continued success of this mobile "eagle flight," each C-Detachment within 5th Special Forces Group dedicated an A-Detachment to raise and train reaction forces for its own corps tactical zone, commencing in the fall of 1965. These were known as Mike Forces. However, the country-wide Nha Trang Mike Force and IV former was organized as a battalion of 594 Rhade, Cham and Chinese troops by Detachment A-503; and the latter used Chinese and Cambodians under Detachment A-430, created at Don Phuc with Special Forces cadre selected throughout the Detachment C-4 area.

The Mike Forces were intended as multipurpose reaction units, but instead they were initially utilized primarily as interior guards for the C-Detachments and the group headquarters compound. The value of these troops in large part resulted from the fact that they represented the only combatant formations exclusively under Special Forces command. Authority over other CIDG resources was either shared with the LLDB or technically under LLDB control. With the hiring of more Nung camp security elements and bodyguards, the Mike Forces were freed for a wide range of field duties.

The 184-man Mike Force companies became the reserves of Special Forces, each containing a 34-man "eagle flight" reconnaissance platoon. The Mike Forces differed from the camp strike forces in many respects. Mike Forces were specifically designed for employment under short-reaction time conditions and were not restricted by camp defense responsibilities.

The Mike Forces were conventionally organized with crew-served weapons, such as medium mortars and a range of recoilless rifles, which allowed them to deliver much of their own supporting fire. Because of the higher wages and more strenuous training, the Mike Force troops were better-than-average CIDG soldiers. They were theoretically airborne-qualified, although personnel turbulence and continual combat requirements kept many companies from achieving that status (as late as 1970, for example, only two of the 12 companies in the 4th Mobile Strike Force Command were actually airborne).

The Mike Forces also had limitations. As mobile response forces, they usually lacked intimate knowledge, which the local CIDG possessed, of the terrain and inhabitants in an operational area. In some cases language and ethnic differences created friction with camp strike forces. Logistically, Mike Forces were also subject to the same loss of motivation in extended field operations away from their home bases that plagued the whole CIDG program. Finally, after intense Vietnamese pressure, they were brought under joint Special Forces-LLDB control in December 1966.

The main Mike Force utilization was threefold: reinforcement of camps under construction or attack, performing raids and patrols and conducting small-scale conventional combat operations. During 1966, the Mike Forces became an effective and critical factor on the Special Forces battlefield. In accord with their increase in importance, the number of companies was expanded with each C-Detachment jurisdiction.

The Nha Trang Mike Force was given the two-fold mission of servicing as a country-wide reaction to support besieged camps and deploying into any Special Forces tactical area of responsibility as required.

BLACKJACK FORCES

The 5th Special Forces Group realized that the Viet Cong had numerous secret bases and complete freedom of movement throughout certain areas of South Vietnam. The Special Forces decided to stab at the elusive VC network by using small bands of Special Forces-led CIDG troops operating as mobile guerrilla forces. The Mobile Guerrilla Force was the brainchild of Colonel Francis Kelly, who dubbed its operations with his favorite call sign, "Blackjack." In fact, Blackjack operations became so famous that Colonel Kelly's code word became his nom de guerre, and his other nickname, "Splash" (gained by avoiding the coral drop zones on Okinawa), faded from use. Colonel Kelly's mobile guerrilla task forces were a refinement and amplification of the mobile strike concept.

Each mobile guerrilla force consisted of an A-Detachment that controlled one 150-man Mike Force company trained in extended patrolling and a 34-man combat reconnaissance eagle flight platoon. The mobile guerrilla task forces were theoretically intended to operate independently for as long as 30 to 60 days in remote regions of South Vietnam. The roving mobile guerrilla forces were to be resupplied by aerial means, such as modified napalm bomb canisters filled with foodstuffs and ammunition and dropped by A-1E Skyraiders. The mobile guerrilla forces employed conventional ranger tactics to locate, watch and harass Viet Cong safe havens through ambush, destruction of storage areas and airstrike direction.

The pilot mobile guerrilla mission was BLACKJACK 21. After five weeks of training and planning, the month-long combat sweep was conducted through the mountains and valleys of the Plei Trap Valley in southwestern Kontum Province. Led by the assistant group operations officer, Captain James A. Fenlon, Task Force 777 was composed of 15 Special Forces and 249 Montagnard Mike Force troops and was devoid of LLDB participation. Although Colonel Kelly appointed Fenlon to be in charge of BLACKJACK 21, in actuality the Montagnards responded only to their own Mike Force company commander, 1st Lt. Gilbert K. "Joe" Jenkins.

Future Blackjack operations (and there were many) proved to be effective "combat vehicles" for Special Forces. COMUSMACV, in concert with the senior advisors to the four tactical combat zones, directed the commander, 5th Special Forces to execute a series of Blackjack operations throughout South Vietnam which proved to be highly successful.

Additionally, the 5th Special Forces Group (Airborne), 1st Special Forces supported COMUSMACV in the overall tactical effort in Vietnam by providing personnel to the Military Assistance Command Vietnam, Studies and Observation Group (MACVSOG). MACVSOG was organized in February 1964 by General Paul D. Harkins, COMUSMACV. Its basic mission was to conduct covert operations in the denied areas of Vietnam, Laos and Cambodia. MACVSOG drew its personnel from a number of military sources but primarily from the U.S. Army Special Forces community. The majority of these Army Special Forces personnel was administratively assigned to the 5th Special Forces Group, however, operational control rested with chief, MACVSOG through his OPS-35 commander. OPS-35 was responsible for ground operations such as recon. missions performed while conducting cross-border operations.

Personnel assigned to OPS-35 missions were assigned to one of the following FOBs (Forward Operating Bases): FOB-1 (Phu Bai), FOB-2 (Kon Tum), FOB-3 (Khe-Sanh), FOB-4 (Marble Mountain, Da Nang), FOB-5 (Ban Me Thuot) and FOB-6 (Ho Ngoc Tao). In 1968, SOG's FOBs were consolidated into Command and Control Bases, i.e., FOB-1, -3 and -4 were designated as CCN (Command and Control-North) (Da Nang), FOB-2 became CCC (Command and Control-Central) (Kon Tum) and FOB-5 and -6 became CCS (Command and Control-South) (Ban Me Thuot).

Chief MACVSOG commanders reporting directly to COMUSMACV during the Vietnam conflict were:

*Colonel Clyde Russell 1964-1965
Colonel Donald V. Blackburn 1965-1966
Colonel John K. Singlaub 1966-1968
Colonel Stephen Cavanaugh 1968-1970
Colonel John "Skip" Sadler 1970-1972

*Deceased

MACVSOG'S OPS-35 GROUP MISSIONS OVERVIEW:

In September 1965, the worsening situation in the Republic of Vietnam caused the United States to undertake limited ground reconnaissance actions in Laos. These operations, initially named "SHINING BRASS" but subsequently known as "PRAIRIE FIRE," involved small reconnaissance teams composed of indigenous civilians led by Vietnamese or U.S. Special Forces personnel assigned from the Military Assistance Command Studies and Operations Group (MACSOG). The teams conducted on-the-ground reconnaissance missions in Laos to determine the nature and extent of enemy activities in the assigned areas of operations.

When the enemy later began moving major amounts of supplies through the Cambodian port of Sihanoukville on the central coast of Cambodia and into the sanctuary areas along the South Vietnam border, a limited ground reconnaissance program (initially, DANIEL BOONE; later, SALEM HOUSE) was authorized in May 1967 to gain information on these activities.

A total of 3,683 missions into Laos and Cambodia were conducted prior to the termination of U.S. participation in April 1972. The House Appropriations Committee and the Senate Appropriations Committee were briefed on the nature of these activities, their functions and costs, including casualties. Additionally, the Subcommittee on U.S. Security Agreements and Commitments Board of the Senate Foreign Relations Committee held extensive hearings on U.S. operations in Laos and Cambodia, including MACSOG operations. Detailed information was provided to the committee at that time.

PRAIRIE FIRE OPERATIONS

PRAIRIE FIRE was the code name for MACSOG cross-border intelligence collection and interdiction operations into southern Laos against enemy bases and infiltration routes. The rationale for the operations into southern Laos was based on strong evidence in early 1965 that the Laos corridor was being used as an infiltration and resupply route in support of the Communist effort in South Vietnam. During the period 1965-1972, the code name assigned to cross-border operations into Laos changed from SHINING BRASS to PRAIRIE FIRE to PHU DUNG. For purposes of clarity, these operations will be referred to as PRAIRIE FIRE.

Missions included such intelligence and intelligence-associated activities as emplacing sensors; prisoner apprehension; area, point and linear reconnaissance by small teams; and selected reconnaissance by larger units. South Vietnamese personnel performed PRAIRIE FIRE operations with U.S. Army Special Forces or Army of the Republic of Vietnam advisors/commanders and were supported with U.S. trooplift and gunship helicopters and U.S. tactical air strikes (TACAIR). PRAIRIE FIRE teams were trained in air-control procedures and made considerable use of tactical air and helicopter gunship support in their operations in Laos.

The PRAIRIE FIRE historical records and operations reports do not indicate how many U.S. personnel accompanied each operation nor how many operations were U.S.-accompanied. U.S. personnel were authorized to accompany PRAIRIE FIRE cross-border operations from 20 September 1965 to 8 February 1971. It is believed that during this period, virtually all of the 1,446 reconnaissance team and 203 platoon and multiplatoon operations conducted involved U.S. participation. The operational guidelines for the conduct of PRAIRIE FIRE missions provided that, generally, the organization of a U.S.-accompanied reconnaissance team would include three U.S. personnel. Larger units normally included five to six U.S. personnel with a platoon force and 20 to 22 U.S. personnel with a multiplatoon force.

No U.S. personnel participated in ground reconnaissance in Laos after February 1971. U.S. air support of Vietnamese-led teams was authorized by appropriate U.S. civilian authorities until March 1972. In recognition of the complete Vietnamization of the operations in Laos, the Vietnamese code name PHU DUNG was applied to all operations after 7 April 1971.

PRAIRIE FIRE security guidance precluded advising next-of-kin of actual location of casualties since this information would comprise the area of operation. Generally, the U.S. Armed Forces, notifying next-of-kin, indicated the loss location as either "South-east Asia," "classified" or "along the border." On 9 May 1973 the secretary of defense approved the release of the actual location of PRAIRIE FIRE casualties to the next-of-kin.

In view of the special security precautions protecting these operations, the PRAIRIE FIRE casualty data included in the Office of the Joint Chiefs of Staff (OJCS) data bank could not reflect actual locations. These casualties were grouped with the South Vietnam data. The data submitted to Congress prior to 25 July 1973 also reflected South Vietnam. At that time Congress was advised that there had been 76 U.S. personnel killed in action in Laos in conjunction with PRAIRIE FIRE.

SALEM HOUSE OPERATIONS

SALEM HOUSE was the name for MACSOG cross-border operations in northeastern Cambodia. When the enemy buildup of logistic and base-camp facilities in the border area of northeastern Cambodia created a threat to the safety of U.S. forces in the Republic of Vietnam, selective and reconnaissance interdiction were authorized to access the enemy threat. The names of these operations varied from DANIEL BOONE to SALEM HOUSE to THOT NOT (when the South Vietnamese assumed complete responsibility). For simplicity, these operations will be referred to as SALEM HOUSE.

The missions of SALEM HOUSE operations was basically intelligence collection and verification. The approval to initiate SALEM HOUSE cross-border operations was provided on 22 May 1967. Approval was subject to restrictions such as:

Only reconnaissance teams were to be committed and could not exceed an overall strength of 12 men to include not more than three U.S. advisors.

Tactical air strikes and/or the commitment of additional forces were not authorized across the border into Cambodia. Teams were not to engage in combat except to avoid capture.

No contact with civilians was permitted.

No more than three reconnaissance teams could be committed on operations into Cambodia at any one time.

The total number of missions could not exceed ten in any 30-day period.

By October 1967 appropriate civilian authority approved SALEM HOUSE operations along the entire Cambodia-South Vietnam border to a depth of 20 kilometers. The use of helicopters for infiltration was authorized at a rate of five per month to a depth of 10 kilometers into Cambodia.

In December 1967 with State Department concurrence, the secretary of defense authorized the use of forward air control aircraft over the SALEM HOUSE area to control helicopters and to conduct reconnaissance of landing sites. Only two such flights were authorized per SALEM HOUSE mission.

After the Tet offensive of 1968, SALEM HOUSE cross-border ground reconnaissance operations into Cambodia were modified. The emplacement of land mines with self-destruct features was authorized in October 1968. By December, the depth of these operations was increased to 30 kilometers in the northern part of Cambodia. In the central and southern operation areas—where specific JCS approval was required for any group reconnaissance operations—penetrations were limited to 20 kilometers. While the restriction on numbers of participating U.S. personnel was removed, total team size remained constrained to 12 members, and various additional restrictions reemphasized the intelligence collection and verification nature of SALEM HOUSE.

During Cambodian incursion in 1970, authority to conduct SALEM HOUSE operations to 200 meters west of the Mekong River in the FREEDOM DEAL air interdiction zone was granted and use of tactical air operations support of SALEM HOUSE also was authorized. No reconnaissance teams ever reached the Mekong, however, due to range and lift limitations of the helicopters involved. After the ground operations into Cambodia ended on 30 June 1970, no additional U.S. ground personnel were permitted to take part in SALEM HOUSE. However, use of tactical air and helicopter gunships to support SALEM HOUSE operations conducted by the South Vietnamese in no larger than platoon-sized operations was continued—when such support was required and was clearly beyond Vietnamese capability. Troop lift helicopters were exclusively manned by the Vietnamese after 30 June 1970. In April 1970 in recognition of the complete Vietnamization of the ground reconnaissance operations, the name was changed to THOT NOT. The program continued until 30 April 1972, when all U.S. involvement in the Vietnamese cross-border operations terminated.

SALEM HOUSE historical records and operational reports do not indicate how many U.S. personnel accompanied each operation nor how many operations were U.S.-accompanied. U.S. personnel were authorized to accompany SALEM HOUSE cross-border operations from 22 May 1967 to 30 June 1970. During this period, virtually all of the 1,119 reconnaissance teams, nine platoon and one multiplatoon operations are believed to have involved U.S. participation.

SALEM HOUSE targets, dates, penetration points and landing zones were submitted by message from COMUSMACV to CINCPAC for approval, with information copies to the OJCS, the secretary of defense, and the secretary of state. Approval of the schedule was assumed if no objections were raised.

As with PRAIRIE FIRE operations, SALEM HOUSE security guidance precluded advising next-of-kin of the actual location of casualties since this information would have compromised the area of operation. The U.S. Armed Forces in notifying the next-of-kin generally indicated the loss location as either "Southeast Asia," "classified" or "along the border." On 9 May 1973, the secretary of defense approved the release of the actual location of SALEM HOUSE casualties to the next-of-kin.

In view of the special security precautions protecting these operations, the SALEM HOUSE casualty data included in the JCS data bank did not reflect actual casualty locations. The SALEM HOUSE casualties, for security reasons, were grouped with the South Vietnam data. On 25 July 1973 Congress was advised that there had been 27 U.S. personnel killed in action in Cambodia as a result of SALEM HOUSE operations.

Despite being one of the smallest units engaged in the Vietnam conflict, the 5th Special Forces Group (Airborne), 1st Special Forces colors fly 21 campaign streamers, and its soldiers are among the most decorated in the history of our nation. Sixteen Medals of Honor were awarded, eight posthumously. The Group has been awarded the Presidential Unit Citation (Army), Vietnam 1966-1968; the Meritorious Unit Commendation (Army), Vietnam 1968; Republic of Vietnam Cross of Gallantry with Palm, Vietnam 1964-1969; Republic of Vietnam Civil Action Honor Medal, First Class, Vietnam 1968-1970. On 5 March 1971, the colors of the 5th Special Forces Group (Airborne), 1st Special Forces were returned to Fort Bragg, North Carolina, by a 94-man contingent led by Colonel (now major general, retired) Michael D. Healy.

On 10 June 1988, the colors of the 5th Special Forces Group (Airborne), 1st Special Forces were cased at a ceremony marking its departure from Fort Bragg. The colors were officially uncased by MG Allen, Col Harley Davis and CSM J. Dennison on 16 June 1988 at its new home at Fort Campbell, Kentucky. The 5th Special Forces Group (Airborne), 1st Special Forces continuously maintains a high state of readiness and has deployed teams throughout Africa and Southwest Asia. The 5th Special Forces Group (Airborne), 1st Special Forces stands ready to answer the call of freedom in defense of the United States anywhere in the world.

5TH SPECIAL FORCES GROUP (AIRBORNE) ACTIVATED AT FORT BRAGG, NORTH CAROLINA 21 SEPTEMBER 1961 COMMANDING OFFICERS

*Col Leo H. Schweiter 1961-1962 CONUS (Ft. Bragg, North Carolina)
*Col Lloyd E. Wills 1962-1963 CONUS (Ft. Bragg, North Carolina)
*Col George C. Morton 1962-1963 Vietnam
*Col Herbert F. Roye 1963-1964 Vietnam
*Col Theodore Leonard 1963-1964 Vietnam
Col John H. Spears 1964-1965 Vietnam
*Col William A. McKean 1965-1966 Vietnam
Col Francis J. Kelly 1966-1967 Vietnam
*Col Jonathan F. Ladd 1967-1968 Vietnam
*Col Harold R. Aaron 1968-1969 Vietnam
Col Robert B. Rheault 1969 Vietnam
Col Michael D. Healy 1969-1971 Vietnam
Col Jay B. Durst 1971-1972 CONUS (Ft. Bragg, North Carolina)
Col Earl L. Keesling 1972-1973 CONUS (Ft. Bragg, North Carolina)
Col Audley C. Harris 1973-1974 CONUS (Ft. Bragg, North Carolina)
Col Raymond Maladowitz 1974-1976 CONUS (Ft. Bragg, North Carolina)
Col Clarence L. Stearns 1976-1977 CONUS (Ft. Bragg, North Carolina)
Col Robert A. Mountel 1977-1978 CONUS (Ft. Bragg, North Carolina)
*Col George W. McGovern 1978-1980 CONUS (Ft. Bragg, North Carolina)
Col Holland E. Bynam 1980-1983 CONUS (Ft. Bragg, North Carolina)
Col James Guest 1984-1985 CONUS (Ft. Bragg, North Carolina)
Col Lawrence W. Duggan 1985-1987 CONUS (Ft. Bragg, North Carolina)
Col Harley C. Davis 1987-Present CONUS (Ft. Bragg, North Carolina and Ft. Campbell, Kentucky)

*Deceased

5TH SPECIAL FORCES GROUP (AIRBORNE), 1ST SPECIAL FORCES COMMAND SERGEANTS MAJOR

*SGM Phil Hoffman 1961
SGM Harmon Shelton 1961-1962
SGM C. Wade 1962-1964
*SGM James Tryone 1964
SGM Clyde W. Francis 1964-1965
SGM John F. Pioletti 1965-1966
SGM George W. Dunaway 1966-1967
CSM Robert D. Mattox 1967-1968
CSM George W. Odom 1968-1969
CSM Myron J. Bowser 1969-1971
*CSM Paul Darcy 1971
CSM R. Maddox 1971-1972
CSM Arnold S. Beckerman 1972-1973
CSM James L. Lyons 1973-1975
CSM Marion A. Spicer 1975-1976
CSM Joseph W. Lupyak 1977-1980
CSM Janusz Borkowski 1980-1982
CSM R.S. Rivera 1982-1983
CSM Forest K. Foreman 1983-1986
CSM Joseph L. Dennison 1986-Present

*Deceased

5TH SPECIAL FORCES GROUP (AIRBORNE), FIRST SPECIAL FORCES–ROLL OF HONOR

A. MEDALS OF HONOR
Date Awarded
2nd LT Charles Q. Williams 9 June 1966
*1st LT George K. Sisler 6 July 1968 MACVSOG (CCN)
SSG Drew D. Dix 14 Jan. 1969
SSG Fred W. Zabitosky 7 March 1969 MACVSOG (CCN)
*SFC Charles E. Hosking, Jr. 23 May 1969
*SFC Eugene Ashley, Jr. 18 Nov. 1969
*SGT Gordon D. Yntema 2 Dec. 1969
*SP5 John J. Kendenburg 7 April 1969
*SFC William M. Bryant 16 Feb. 1971
SFC Robert L. Howard 2 March 1971 MACVSOG (CCC)
SSG Franklin D. Miller 15 June 1971
*SGT Brian L. Buker 16 Dec. 1971
SGT Gary B. Beikirch 15 Oct. 1973
*1st LT Loren D. Hagen 6 Sept. 1974
SSG Jon R. Cavaini 12 Dec. 1974
SSG Roy P. Benavidez 24 Feb. 1981

B. DISTINGUISHED SERVICE CROSSES
SFC Michael F. Carpenter 1965
SFC James T. Taylor 1965
SSG Daniel F. Crabtree 1965
SP5 Forestal A. Stevens 1965
SP5 Michael J. Hand 1965
SGT Harold T. Palmer, Jr. 1965
*SFC Maurice A. Casey 1966
SFC Gerald Grant 1966 MACVSOG (CNN)
*SFC Donald L. Lehew 1966
*SSG Billy Hall 1966
CPT John D. Blair, IV 1966
CPT Tennis Carter 1966
CPT Craig R. Chamberlain 1966
*CPT James B. Conway 1966
CPT Paul M. Trees (w/OLC) 1966
2nd LT Louis A. Mari 1966
SFC Bernard G. Adkins 1966
SP5 Wayne H. Murray 1966
*SP5 Phillip Stahl 1966
SFC Victor Underwood 1966
SGT Russell P. Bott 1966
SGT Max D. Speers 1966
CPT Clyde J. Sincere, Jr. 1967
SFC Billy D. Evans 1967 MACVSOG (CNN)
*SFC William G. Ferguson 1967
SFC George H. Heaps 1967
SFC Leonard W. Tilley 1967
SFC Jack L. Williams (w/OLC) 1967
SFC Morris G. Worley 1967
*SSG John E. McCarthy 1967
*SSG Hubert O. VanPoll 1967
MAJ George Maracek 1967
CPT Norman E. Baldwin 1967
CPT Chester Garrett 1967
*1LT James F. Godsey 1967
1LT Willie Merkerson, Jr. 1967
2LT Joseph W. Moore 1967
*MSG Bruce R. Baster 1967 MACVSOG (CCN)
MSG Thomas J. Sanchez 1967
*MSG Samuel S. Theriault 1967
*SFC Domingo R.S. Borja 1967 MACVSOG (CCN)
*SP4 Richard P. Teevens 1967
*SGT Jackie L. Waymire 1967
SGT Timothy W. Clough 1968
SGT Peter M. Stark 1968
*SP4 John F. Link 1968
*SP5 Paul R. Severson 1968
SSG George C.D. Allen 1968
SFC Manuel C. Bustamante 1968
*MSG Robert D. Plato 1968 MACVSOG (CCN)
*1LT Thomas L. Swann 1968
*1LT Peter W. Johnson 1968
CPT Thomas W. Jager 1968

*CPT Stephen Gabrys 1968
LTC Daniel F. Schungel 1968
*SSG Gerald L. Watson 1968
*SSG Lloyd F. Mousseau 1968
*SSG Balfour O. Lytton, Sr. 1968
*SSG Leslie L. Brucker, Jr. 1968
*SFC Leroy N. Wright 1968
*SFC Paul H. Villarosa 1968 MACVSOG (CCN)
*SFC Linwood D. Martin 1968
SFC William Ledbetter 1968 MACVSOG (CCN)
SFC Gilbert L. Hamilton (w/OLC) 1968 MACVSOG (CCN)
*SFC Benedict M. Davan 1968
SFC Jose Rodela 1969
SSG Melvin Morris 1969
*MSG Robert G. Daniel 1969
*1LT John J. McHugh 1969
SGT Michael D. Buchanan 1969
SGT Gary M. Rose 1969 MACVSOG (CCN)
SGT Robert D. Pryor 1969
SGT James N. Pruitt 1969 MACVSOG (CCN)
SFC Antonio J. Coehlo 1970
SFC Walter G. Hetzler 1970
*SFC Otis Parker 1970

* Posthumously

C. **SILVER STARS** 814
D. **SOLDIERS MEDALS** 232
E. **BRONZE STARS W/"V"** 3,074
F. **BRONZE STARS FOR SERVICE** 10,168
G. **PURPLE HEARTS (WOUNDED IN ACTION)** 2,658
H. **KILLED IN ACTION** 562
I. **MISSING IN ACTION** 31
J. **PRISONERS OF WAR** 11

6TH SPECIAL FORCES GROUP, 1ST SPECIAL FORCES

Constituted 5 July 1942 in the Army of the United States as Headquarters and Headquarters Detachment, 2d Battalion, Third Regiment, 1st Special Force as a joint Canadian-American organization.

Activated 9 July 1942 at Fort William Henry Harrison, Montana.

Reconstituted 15 April 1960 in the Regular Army; concurrently consolidated with Headquarters and Headquarters Company, 6th Ranger Infantry Battalion (see Annex) and consolidated unit designated as Headquarters Company, 6th Special Forces Group, 1st Special Forces. Activated 1 May 1963 at Fort Bragg, North Carolina. (Organic elements constituted 30 October 1963 and activated in December 1963 at Fort Bragg, North Carolina.

Annex- Constituted 16 December 1940 in the Regular Army as Headquarters and Headquarters Battery, 98th Field Artillery Battalion. Activated 20 January 1941 at Fort Lewis, Washington. Converted and redesignated 25 September 1944 as Headquarters and Headquarters Company, 6th Ranger Infantry Battalion. Inactivated 30 December 1945 in Japan.

CAMPAIGN PARTICIPATION CREDIT
WORLD WAR II

Algeria-French Morocco (with arrowhead)
Tunisia
Sicily (with arrowhead)
*Naples-Foggia (with arrowhead)
*Anzio (with arrowhead)
*Rome-Arno
Normandy (with arrowhead)
*Northern France
*Southern France (with arrowhead)
*Rhineland
Ardennes
Central Europe
*Aleutian Islands
*New Guinea
*Leyte (with arrowhead)
*Luzon

DECORATIONS
Presidential Unit Citation (Army), Streamer embroidered EL GUETTAR
Presidential Unit Citation (Army), Streamer embroidered SALERNO

Lineage and Heraldic Data
Presidential Unit Citation (Army), Streamer embroidered POINTE DU HOE
Presidential Unit Citation (Army), Streamer embroidered SARR RIVER AREA
*Philippine Presidential Unit Citation (Army), Streamer embroidered 17 OCTOBER 1944 TO JULY 1945 96th Ranger Infantry Battalion, cited DA GO 47, 1950.

6TH SPECIAL FORCES GROUP (AIRBORNE) FIRST SPECIAL FORCES

The 6th Special Forces Group (Airborne) became the sixth United States Army Special Forces Group as Major General William P. Yarborough, then commanding general of the Special Warfare Center, Fort Bragg, North Carolina, passed the colors of the new unit to Lieutenant Colonel Charles L. Kasler at a retreat parade on Wednesday, 8 May 1963.

Colonel Kasler came to the 6th Special Forces Group from the office of Chief of Staff, Special Warfare Center where he performed the duties of Chief of Staff, Special Warfare Center. Assisting Colonel Kasler in forming the 6th Special Forces Group were seven outstanding officers including a former member of the famed World War II First Special Service Force and a Korean Medal of Honor winner. All of the officers were combat veterans, experienced Special Forces men, and most had served tours in the Republic of Vietnam.

Sergeant Major Charles L. Petry was assigned as the group command sergeant major.

The group flash consisted of diagonal stripes representing the three Special Warfare Center groups from which the 6th Special Forces Group drew its personnel. Red, for the men of the 7th Special Forces Group (Airborne); white, for the Special Forces Training Group (Airborne); and black, for the 5th Special Forces Group (Airborne).

At the time of the activation of the 6th Special Forces Group, the Army's three other Special Forces Groups were all stationed outside the continental United States; the 1st Special Forces Group (Airborne) was on Okinawa; the 8th Special Forces Group (Airborne) was stationed at Fort Gulick, Canal Zone, Panama; and the 10th Special Forces Group was stationed in Bad Toelz, Germany.

Twenty-three members of the 6th Special Forces Group celebrated the founding of the 6th Special Forces by conducting a parachute drop at Fort Bragg's Sicily Drop Zone on 9 May 1963.

In World War II, Colonel Kasler fought at Bataan and Corregidor and participated in the infamous "Bataan Death March."

Serving as the group executive officer was Lieutenant Colonel Eldred E. "Red" Weber who had served previously as the S3 of the 5th Special Forces Group (Airborne). During World War II he fought in five campaigns as a member of the U.S.-Canadian First Special Service Force.

Rounding out the remaining original staff members of the 6th Special Forces Group (Airborne) were Major Hampton Dews, S3; Major Lewis D. Allen, S4; Captain George W. Gaspard, S1; Captain Ola L. Mize, S3 training officer and the recipient of the Medal of Honor. Captain Richard A. Clark was the S2 training officer and First Lieutenant Harold P. Skamser, area specialist, S3.

The geographical area of operations for the 6th Special Forces Group (Airborne), 1st Special Forces was the middle-east. Group training was directed toward this mission area to include cold weather and mountain training which was conducted in Alaska. Unconventional Warfare and Counter Insurgency training was conducted in the national forests of North Carolina. Additionally, Mobile Training Teams were deployed to Pakistan and Ethiopia.

Individual members of the 6th Special Forces Group (Airborne) were selected for specific and special missions in South America and the Dominican Republic.

Training provided by personnel of the 6th Special Forces Group (Airborne) produced a generation of Allied Special Forces who have served in key positions throughout the Allied Special Forces community.

The 6th Special Forces Group (Airborne), 1st Special Forces was deactivated in 1971.

COMMANDING OFFICERS, 6TH SPECIAL FORCES GROUP (AIRBORNE) FIRST SPECIAL FORCES

LTC Charles L. Kasler 1963-1964
Colonel Clarence W. Patten 1964-1966
Colonel Joseph McCulloch 1966-1967
Colonel Samuel V. Wilson 1967-1968
LTC Howard D. Kinney 1968-1969
Colonel Francis B. Kane 1969-1970
Colonel Jay B. Durst 1970-1971

COMMAND SERGEANTS MAJOR, 6TH SPECIAL FORCES GROUP (AIRBORNE) 1ST SPECIAL FORCES

SGM Charles R. "Slats" Petry 1963-1964
SGM Arthur Senkewich 1964-1965
SGM Ed Denton 1965-1966
SGM Charles R. Ferguson 1966-1967
SGM Charles Burnett 1967-1968
*SGM Phillip J. Hoffman 1968-1969
SGM Clifton Phillips 1969-1970
CSM Robert D. Mattox 1970-1971

*Deceased

77TH SPECIAL FORCES GROUP (AIRBORNE)

The 77th Special Forces Group (Airborne), was organized and activated in the fall of 1953, with personnel drawn from the 10th Special Forces Group then preparing for deployment of Bad Toelz, Germany. The actual activation of the unit occurred on 25 September 1953, with Lieutenant Colonel "Black" Jack T. Shannon as its first commander. The group's motto was "Anything, Any Time, Any Place, Any How."

The next commander was Colonel Edson Duncan Raff, a colorful combat veteran whose service during World War II was highly praised by General Dwight Eisenhower. Colonel Raff encouraged the wearing of the beret by members of the 77th Special Forces Group; however, official sanction was not to take place until five years later.

The 77th Special Forces Group carried out a very rigorous training and sports program. Training was conducted from the high mountains of Colorado and Wyoming to the semi-tropics of the Virgin Islands. Additionally, the 77th Special Forces Group conducted an annual exercise in the swamps near Camp Lejeune, North Carolina culminating this annual exercise with a four-day survival problem. Also, amphibious training was conducted several times a year near Norfolk, Virginia. It included beach landings, troop carrier submarine exercises and high-speed transport familiarization. Typical of the membership in the 77th Special Forces Group was a private from Poland, a corporal from Shanghai, China and a private from Finland. (It was during this time that a number of displaced Europeans joined the U.S. Army under the "Lodge Act" and added an enviable foreign language capability to the group.)

A major accomplishment in February 1957 was the deployment of Mobile Training Team 1A to the island of Taiwan for a six-month TDY stint for the purpose of training a Republic of China cadre in all aspects of Special Forces operations and techniques. Fifty Chinese officers were selected to be trained. These same officers would subsequently establish the first Republic of China Special Forces Training Center at Lung Tan, located approximately 50 miles south of Taipai.

The Mobile Training Team consisted of six officers and six non-commissioned officers commanded by Lieutenant Colonel Eugene "Mike" Smith. Shortly after its arrival in Taiwan, MTT 1A was augmented by Major Donald "Paddy" O'Rourke, Master Sergeant Henry Furst and Sergeant First Class Everett C. White from the 14th Special Forces Operational Detachment stationed at Fort Shaffter, Hawaii. Little did these personnel realize that, upon completion of their TDY tour on Taiwan, they would find that they had been relocated to the newly activated 1st Special Forces Group (Airborne) on Okinawa. The 14th Special Forces Operational Detachment had passed into history.

Major General F.W. Farrel, while serving as the 82d Airborne Division commander, once told a Special Forces class: "Conventional warfare is outmoded, and we must prepare ourselves for the unconventional in any future conflict." His words were very similar to the views held by the famed General Ord C. Wingate, another strong advocate for "special" warfare.

The 77th Special Forces Group was not destined to move overseas as a unit. The men remained headquartered at the Special Warfare Center—a title adopted in 1956—and in June 1960, became members of the 7th Special Forces Group (Airborne), 1st Special Forces.

77TH SPECIAL FORCES GROUP (AIRBORNE) CREST

The Group Crest combines salient point of the Special Forces Patch, background to the wings and motto. It is teal blue and yellow, the 77th Special Forces colors. The group motto is enscrolled on the bottom. The background of the crest is teal blue with the bar sinister, running through it in yellow. There are three arrows in the upper left hand corner of the crest reflecting the three methods of entry to and exit from an operational area: land, sea or air. There is a globe in the bottom right hand corner which represents the global aspect of Special Forces operations. The eagle surmounting the crest represents eternal vigilance, which is the watchword of a Special Forces unit.

COMMANDING OFFICERS, 77TH SPECIAL FORCES GROUP (AIRBORNE)

LTC Jack T. Shannon 1953
Col Edson D. Raff 1953-1954
LTC Benjamin F. Delameter 1954-1955
*Col William J. Mullen 1955-1956
Col Noel A. Menard 1956-1957
Col Julian A. Cook 1957-1958
Col Irwin A. Edwards 1958-1959
Col Donald D. Blackburn 1959-1960

*Deceased

7TH SPECIAL FORCES GROUP (AIRBORNE), 1ST SPECIAL FORCES

The 7th Special Forces Group (Airborne), 1st Special Forces had earned a distinguished reputation during its history. It was originally constituted as the 1st Company, 1st Regiment, 1st Special Service Force on 9 July 1942 at Camp William Henry Harrison, Montana. It was subsequently disbanded in France on 6 January 1945. It was reconstituted on 15 April 1960 and redesignated Headquarters and Headquarters Company, 7th Special Forces. On 6 June 1960 it was consolidated with Headquarters and Headquarters, 77th Special Forces and designated Headquarters and Headquarters, 7th Special Forces Group (Airborne). 1st Special Forces had established one of the finest records of any element of the United States Army.

In the early 1960s the Group was actively involved in operations in South Vietnam, Laos and Thailand. Examples of this involvement are the Mobile Training Teams of "Project White Star" which provided military assistance to Laos on foreign internal defense and unconventional warfare missions. The missions terminated after a coalition government was formed in 1962.

The first series of Mobile Training Teams to Vietnam were also employed in the 1960s. From March 1963 to August 1964 numerous operational detachments were deployed to Vietnam where they conducted foreign internal defense, civic actions and intelligence missions. Col. (retired), then Capt. Roger Donlon was presented the first Congressional Medal of Honor awarded in Vietnam for actions at Nam Dong on 6 July 1964.

The longest-held POW in the Vietnam Conflict (the Second Indo-China War), Col. (retired) Floyd Thompson was a member of the 7th Special Forces Group when he was captured in 1964. He was not released until 1973. Additional involvement of operational teams in the early 1960s was in several countries of South and Central America, including the Dominican Republic in 1965 and 1966, in order to safeguard American lives.

Although the 7th Special Forces Group (Airborne) is not the oldest regular Special Forces Group on active roles, it is referred to as the building block from which Special Forces expanded during the term of President John Fitzgerald Kennedy. Throughout the early 1960s military assistance requirements for Mobile Training Teams far exceeded the manpower resources within the 7th Special Forces Group. Therefore, the 7th Special Forces Group was called upon to assist in activating other similar units.

In 1961, the 7th Special Forces Group provided the nucleus for the newly activated 5th Special Forces Group (Airborne), which later was given the mission of advising the South Vietnamese Army. In May 1962, the 7th Special Forces Group was tasked, once again, to provide the nucleus of another Special Forces Group, this time oriented toward Latin America. On 14 May 1962, the advance party from Company D, 7th Special Forces Group, departed for Fort Gulick, Panama Canal Zone, to establish what was to be later designated as the 8th Special Forces Group (Airborne) (Special Action Force), 1st Special Forces. The 3d Battalion, 7th Special Forces Group provided officers and enlisted men to form the cadres of two additional Special Forces Groups, the 3d and the 6th, activated at Fort Bragg, North Carolina, with missions oriented to the Middle East and Africa.

The 7th Special Forces Group (Airborne), 1st Special Forces conducts training year-round in almost every country of Central and South America. The training year is divided into two six-month cycles: Deployment for Training (DFT) and the Mobile Training Team (MTT).

During the DFT cycle, detachments will deploy to a Central/South American country to conduct training for the betterment and the benefit of the deployed element. During the MTT cycle, detachments will deploy to a Central/South American country upon the host country's request to train and advise military and paramilitary forces conventional or unconventional warfare.

Today, members of the 7th Special Forces Group (Airborne), 1st Special Forces are subject to being killed in action as they continue the stuggle against opponents of democracy in Latin America:

In El Salvador, the U. S. maintains up to 55 Special Forces military advisor/trainers (President Reagan agreed in 1981 not to exceed that ceiling without Congressional approval). The Special Forces are involved in almost every aspect of the Salvadoran military. MTTs rotate in and out, providing instuction in weapons usage, tactical intelligence, planning communications, logistics, maintenance and more. They teach everything from firing a rifle to flying a helicopter. They have trained several counter insurgency battalions which are now considered the elite of the Salvadoran army. SEALS have trained with the Salvadoran navy at La Union.

SFC Gregory A. Fronius was killed in action on 31 March 1987, when guerrillas attacked the Headquarters, 4th Infantry Brigade, 4th Military Zone, El Paraiso, Department of Chalatenango, El Salvador. He was awarded the Purple Heart posthumously.

The 7th Special Forces Group (Airborne), 1st Special Forces remains ready to perform whatever missions assigned and retains its capability of providing Mobile Training Teams anywhere to allies requesting assistance, continuing in the tradition of excellence which has marked its entire existence.

7TH SPECIAL FORCES GROUP (AIRBORNE), 1ST SPECIAL FORCES COMMANDING OFFICERS

COL Donald D. Blackburn 1958-1960
COL Irwin A. Edwards 1960-1961
*COL Clyde R. Russell 1961-1962
COL William Evans-Smith 1962-1964
*COL Edward E. Mayer 1964-1965
COL Joseph J. Jackson 1965-1966
LTC William N. Jackson 1966-1967
COL Robert W. Hakala 1967-1968
LTC Robert E. Furman 1968
COL Daniel F. Schungel 1968-1970
COL Elmer E. Monger 1970-1971
COL Joseph B. Love 1971-1973
COL Ronald A. Shackleton 1973-74
COL Charles E. Garwood 1974-1976
COL Charles W. Norton 1976-1977
COL Timothy G. Gannon 1977-1979
COL William T. Palmer 1979-1980
COL Edward T. Richards 1980-1983
COL Stuart Perkins 1983-1985
COL J.P. Waghelstein 1985-1987
COL Robert C. Jacobelly 1987-1989
COL Hugh F. Scruggs 1989

*Deceased

7TH SPECIAL FORCES GROUP (AIRBORNE), 1ST SPECIAL FORCES COMMAND SERGEANTS MAJOR

SGM John Pioletti 1960-1961
SGM James L. Hallford 1961-1962
SGM Robert Depue 1962
SGM Loner Westmoreland 1962-1963
SGM Edward McDougall 1963-1967
SGM Edward Denton 1967
SGM Robert Taylor 1967-1968
*CSM Edward Gaydosik 1968-1969
CSM Harmon D. Hodge 1969-1971
CSM Durham 1971
CSM Calvin Thomas 1971
SGM David Clark 1971-1972
SGM Foster Alexander 1972
CSM Robert E. Shaw 1972-1975
SGM Gerhardt Kunert 1975
CSM James C. Pickles 1975
CSM Galen C. Kittleson 1976-1978
CSM Peter Morakon 1978-1980
CSM J. Roye 1980
CSM George A. Zacher 1981
CSM Kenneth Chavis 1982-1984
CSM Ivan Ivanov 1984-1988
CSM Henry D. Luthy 1988

*Deceased

7TH SPECIAL FORCES GROUP (AIRBORNE), FIRST SPECIAL FORCES–ROLL OF HONOR

A. MEDAL OF HONOR
Date Awarded
CPT Roger H. C. Donlon 5 Dec. 1964

B. DISTINGUISHED SERVICE CROSS
MSG Gabriel R. Alamo (Posthumously)

C. SILVER STARS 4
D. BRONZE STARS W/"V" 26
E. PURPLE HEARTS (WOUNDED IN ACTION) 56
F. KILLED IN ACTION 9

8TH SPECIAL FORCES GROUP (AIRBORNE), 1ST SPECIAL FORCES

The 8th Special Forces Group (Airborne) was constituted on 5 July 1942 in the Army of the United States as Headquarters and Headquarters Detachment, 1st Battalion, First Regiment, 1st Special Service Force, as a joint Canadian-American organization, activated 9 July 1942 at Fort William Henry Harrison, Montana. Disbanded 6 January 1945 in France.

The unit was reconstituted 15 April 1960 in the Regular Army; consolidated with Headquarters and Headquarters Company, 1st Ranger Infantry Battalion and designated as Headquarters and Headquarters Company, 8th Special Forces Group, 1st Special Forces. It was activated 1 April 1963 at Fort Gulick, Canal Zone (organic elements concurrently constituted and activated). Inactivated 30 June 1972 at Fort Gulick, Canal Zone.

The former Headquarters and Headquarters Company, 1st Ranger Infantry Battalion was withdrawn 3 February 1986 and retained its own lineage. This unit was originally constituted 27 May 1942 in the Army of the United States as Headquarters and Headquarters Company, 1st Ranger Battalion. Activated 19 June 1942 at Carrickfergus, Northern Ireland. Redesignated 1 August 1943 as Headquarters and Headquarters Company, 1st Ranger Infantry Battalion. Disbanded 15 August 1944 in the United States.

Reconstituted 1 September 1948 in the Army of the United States as Headquarters Company, 1st Infantry Battalion; concurrently, activated at Fort Gulick, Canal Zone. Inactivated 4 January 1950 at Fort Gulick, Canal Zone. Redesignated 24 November 1952 as Headquarters and Headquarters Company, 1st Ranger Infantry Battalion and allotted to the Regular Army.

Col. Arthur D. "Bull" Simons was the original commander of the 8th SFG (Abn), 1st SF, in Panama. His nickname "Bull" derived from his robust physical appearance and his powerful speech and deep command voice. "The Bull's Horn," a stylized drinking cup formed from the horn of a bull and used for "prop blasts" during Col. Simons's tenure, is still proudly maintained in the 3rd Battalion, 7th Special Forces Group (Airborne), 1st Special Forces Trophy Case. Col. Simons continued his association with Special Forces during subsequent assignments at DA and was the commanding officer and prime moving force behind the famous Son Tay prison raid in 1971.

CAMPAIGN CREDITS

Naples-Foggia
Anzio
Rome-Arno
Southern France (w/Arrowhead)
Rhineland
Aleutian Islands

COMMANDING OFFICERS, 8TH SPECIAL FORCES GROUP (AIRBORNE) 1ST SPECIAL FORCES

*COL Arthur D. Simons 1963-1965
COL Magnus L. Smith 1965-1967
LTC Eldred E. Weber 1967-
LTC Louis F. Felder 1967-1969
LTC George E. Dexter 1969-1971
COL B.J. Pindkerton 1971-1972

*Deceased

10TH SPECIAL FORCES GROUP (AIRBORNE), 1ST SPECIAL FORCES

The history of the United States Army's 10th Special Forces Group (Airborne), 1st Special Forces dates back, so far as its mission is concerned, to the formation of the Office of Strategic Services (OSS) under the command of Colonel William O. "Wild Bill" Donovan. Volunteers for the unit were extremely specialized and carefully selected and their missions took them behind enemy lines in every theater of operations. American, British and French agents of the OSS parachuted into France in small teams which consisted of two officers and an enlisted radio operator.

The OSS undertook a number of highly successful unconventional warfare operations, the most notable being in support of the D-Day Invasion. The OSS dropped 77 men behind the German lines before D-Day to assist and organize opposition; Jedburg Teams, 78 uniformed members of the armed forces, dropped into France on D-Day to coordinate resistance; Operational Groups, 356 U.S. Army volunteers who spoke French, dropped into France soon after D-Day to bolster guerrilla/partisan warfare activities. The OSS was disbanded in 1946. Based upon the success of the OSS, Brigadier General Robert McClure, Col. Aaron Bank, Col. Fertig and Col. Russ Volckman were convinced the Army should have a permanent unit whose wartime mission was to conduct unconventional warfare.

On 19 June 1952, Col. Aaron Bank assumed command of the 10th Special Forces Group at Fort Bragg, North Carolina. Present for duty on that day were seven enlisted men, one warrant officer and Col. Aaron Bank. By the end of June, 122 officers and men had arrived and were assigned to Headquarters and Headquarters Company.

Sprinkled among these initial arrivals were a few former OSS, Ranger and Airborne officers, with an occasional Lodge Bill soldier ("Lodge Bill" soldiers were East Europeans or stateless volunteers in the U.S. Army).

The U.S. Army created the 10th Special Forces Group to conduct partisan warfare behind Soviet lines in the event of a Soviet invasion of Europe. From the very start, the Army planned to deploy the Group to Europe. On 10 November 1953 the 10th Special Forces Group was split in half. One-half deployed to Bad Toelz, Germany as the 10th Special Forces. The remaining troops formed the new 77th Special Forces Group at Fort Bragg, North Carolina.

In 1955, Special Forces received its first publicity: two articles in the New York Times announcing the existence of a U.S. Army liberation force designed to fight behind enemy lines. The Times correspondent noted the distinct "foreign" nature of the Special Forces, as many of its volunteers were refugees from Eastern Europe. Major features in these articles were photographs of 10th Special Forces troops wearing berets and of refugees from Eastern Europe with their faces blacked out in the photos to conceal their identities. Notwithstanding such sensationalism, the articles provided relatively accurate descriptions of the 10th Special Forces Group's soldiers.

As the 10th Group became established in Germany, a new item of headgear, the green beret, appeared in rapidly increasing numbers. The 10th Special Forces Group decided to adopt the wear of the beret with all uniforms. The Group commander, Colonel William Eckman, authorized the wear of the beret and it became

Group policy in 1954. By 1955, every Special Forces soldier in Germany was wearing the green beret as a permanent part of his uniform.

Captain Roger Pezzelle designed the silver Trojan Horse badge for wear on the beret. It remained the unofficial badge until 1962 when the official Distinguished Unit Insignia and green cloth "flash" were authorized.

The differences in mission, organization, manning and modus operandi set the 10th Special Forces Group apart from the conventional Army units. Notable differences in other external symbols began to appear. The 10th Special Forces Group carried the "mountain rucksack" rather than the standard Army field packs. Likewise, the men soon did away with spit-shined jump boots, opting instead for mountain boots, which were the most practical field boot possible for the European climate. Mountain boots became trademarks of the 10th Special Forces Group.

The original A Detachments were called FA Teams and consisted of 15 men. Each FA Team was designed to advise and support a regiment of up to 1,500 partisans. FB Teams (equivalent to the current ODB or Company Headquarter) commanded two or more FA Teams. FC Teams (today's ODCs or Battalion Headquarters) were designed to command and control FA and FB Teams in a single country including Guerrilla Warfare (GW) Area Commands. The Group Headquarters was called the FD Team, designed to command and control the entire group when employed in two or more countries. The fact that this original organization has changed very little over the many years is indeed a tribute to those who devised the first table of organization and equipment (TO&E).

As time passed, the Group commander and staff conducted visits to England, Turkey, France, Norway, Italy, Greece, Iran and Spain. As a result, the idea of exchanging training with foreign soldiers was generally received with great enthusiasm. Soon, A Detachments trained with Western European and Middle Eastern armies. Men of the 10th Special Forces Group trained with airborne, commando, ranger, raider, militia and clandestine organizations in England, France, Norway, Germany, Greece, Spain, Italy, Turkey, Pakistan, Iran, Jordan and Saudi Arabia. A Detachments worked across cultural and linguistic borders, learning how to subsist on native food and establishing and maintaining rapport with the host nation forces.

Based upon the unit mission and the indentities of the soldiers, an aura of secrecy surrounded the 10th Special Forces Group in the early days. Soldier quality standards were also very high. A certain breed of man stayed and subsequently attracted more of his own kind. One group of men who found places in Special Forces that suited their temperaments and special abilities were the so-called Lodge Bill troopers. Many of the Lodge Bill men still had families behind the iron curtain. A few of the more notable Lodge Bill soldiers were Sgt Paul Ettman, a refugee from Poland; Stefan Mazak, a Czech and veteran of the Marquis and the French Foreign Legion; Henry "French" Szarck, a Pole and a veteran of four armies; Peter Astalos, served in the Romanian and German Army during WWII; Martin Urich participated in the largest tank battle in history, the Battle of Kursk in the USSR.

Then there was Larry Throne, one of Special Forces' more illustrious soldiers. He entered the Finnish Army in 1938. About a year later, he began a six-year period of continuous combat against the Soviets during World War II. After various assignments with front line infantry units, he volunteered for commando activities behind the Soviet lines. Lieutenant Thorne's most daring exploits began in 1942 as he conducted numerous deep penetration missions. On one, he personally led a small group of men behind Soviet lines, ambushed and destroyed a Soviet convoy, killed over 300 of the enemy and returned without a single casualty.

In June 1944, Thorne's unit was employed in the front line role as the last reserve available. They conducted a counterattack against the spearhead of an attempted enemy breakthrough operation. This action occurred only a few hours after his unit had returned from a mission behind Soviet lines. The commander of the particular sector described CPT Thorne's boldness during the attack, "My plan was to provide CPT Thorne with some artillery and mortar support which probably would have delayed the start of his counterattack about an hour. However, CPT Thorne's plan was to attack immediately, before the Soviets had a chance to dig in. CPT Thorne counterattacked in daylight through dense forest, surprising the Soviets and resulting not only in destruction of a Soviet battalion, but also in saving the desperate situation."

CPT Thorne repeatedly exposed himself to extreme hazards; his leadership and heroism made him a national hero in Finland and earned him the Mannerheim Cross, Finland's highest military award (equivalent to the Medal of Honor) in July 1944.

CPT Thorne joined the American Army after World War II, volunteered for Special Forces duty, was assigned to the 77th Special Forces Group and later transferred to the 10th Special Forces Group. He received a direct commission as a 1LT in the Signal Corps. He was a HALO parachutist, SCUBA diver, mountaineer and skier. In 1962, as a CPT he led his Special Forces Detachment onto the highest mountain in Iran to recover the bodies of an American Army air crew lost in a plane crash and to secure the classified material they transported. After unsuccessful attempts by others, his detachment's success was largely attributed to CPT Thorne's superb leadership. He volunteered twice for Vietnam. On 18 October 1965, on his second tour with the Studies and Observations Group (SOG), he was declared missing in action.

In Africa, the 10th Special Force Group served without fanfare, often wearing no identification patches, berets or other insignia, sometimes even operating in civilian clothes. This deliberate and commendable low profile should not obscure the story of those missions.

In the summer of 1960, the commanding officer of the 10th Special Forces Group, Colonel "Iron Mike" Paulick, read about the trouble in the Congo. A wave of violence against the remaining white in the former Belgian colony was developing following its independence on July 1, 1960. There was no hint in the news media that the 10th would be involved. Col. Paulick called Lt. Sully Fontaine. Lt. Fontaine had a remarkable record. A Belgian by birth, he worked for the British SOE during World War II and parachuted behind Nazi lines into France because he spoke French better than English. He later held a commission in the Belgian Army and served in the Congo.

Col. Paulick told Lt. Fontaine to put together a team and get ready to go. Lt. Fontaine's choices were: Sobiachevsky, a Russian; George YosicYosich, who led partisans in Korea; "Pop" Grant; Charles Hoskins, who later won the Medal of Honor posthumously; and Stefan Mazak, a tough little Czech and ex-French Foreign Legionnaire. U.S. Ambassador Timberlake ordered a smaller unit to Leopoldville in the Congo to help save American and European lives. This team consisted of three helicopters, three light single-engine airplanes, an Air Force ham radio expert and the Special Forces element from Bad Toelz. Lt. Fontaine selected CPT Clement and Sgt. Stefan Mazak for this mission.

As it started, the mission nearly came to a disastrous end as the team touched down at the wrong airfield. A meeting with Ambassador Timberlake and Belgian paratroopers took place and the mission was defined. At the larger airfields, Belgian paratroopers with larger planes would be in charge. The Special Forces Team would control operations on the small airfields. The mission was to evacuate as many Europeans and Americans as possible and move them to Leopoldville for large scale evacuation. Despite minor contact and a few holes in their aircraft, the mission was accomplished. In fact, nine days following their arrival, the Special Forces Team evacuated 239 refugees without a single casualty. Stefan Mazak exposed himself to hostile forces again and again during this operation, only to be killed in action later during a classified operation with SOG in Vietnam.

As the 1960s began, counter insurgency was a new concept for Special Forces. Although the 10th Special Forces Group (Airborne), 1st Special Forces was not directly involved in Southeast Asia, most Special Forces personnel, by normal rotation, served with Special Forces units in Southeast Asia. Counter insurgency was to be the answer to communist expansion, subversion and so called "Wars of Liberation."

During these years, the 10th Special Forces was handed responsibility for an awesome chunk of new geography. In addition to Europe, it became responsible for North Africa, the Middle East and Southwest Asia, as far east as Pakistan. Two companies, (C Detachments at the time), were designated to remain responsible for the East European Guerrilla Warfare (GW) role. The other C Detachment trained for counter insurgency (CI) missions. Besides the normal Special Forces training, this C Detachment gathered intelligence and commenced language training in Arabic, Urdu, Farsi, Greek, Turkish and Pushto.

In Jordan, Major Joe Callahan and his B Detachment were directed to establish and run a paratroop school. Four weeks later, the school was opened, and a group of singularly uninspired Arabs showed up for the training. Special Forces teams were accustomed to unexpected obstacles, so jump school started for this "ragtag" group. It was rough for both the Arabs and the team, but the mission was a complete success. An enthusiastic King Hussein was on hand for the graduation parachute drop.

In 1963, the Royalist government of the Kingdom of Yemen was overthrown by a Nasser-supported group. A new republican government was installed under President Al Salol. The son of the old King Mohammed Al-badr fled to the hills to fight a guerrilla war, where he was supported by King Faisal of Saudi Arabia. Company C, 10th Special Forces Group, commanded by Maj. William Hinton, deployed to Saudi Arabia and trained 350 selected officers and NCOs of this guerrilla force in the basics of insurgency and counter insurgency.

One B Detachment and three A Detachments later travelled to Iran and trained with the Iranian Special Forces. The Iranian Special Forces was actually only an airborne battalion. The A Detachments trained the Kurdis tribesman, now fighting in the mountains of Iran. Captain Steve Snowden and his A team trained the nucleus of the Turkish Special Forces, including airborne qualification, Special Forces tradecraft and SCUBA diving for selected officers. Never did one A Detachment accomplish so much for one nation, constructing training apparatus for the airborne course, conducting classroom instruction for 350 officers and NCOs and presenting field training in air, on land and in water. Captain Mike Boos and his detachment went to the hills and deserts of Pakistan to train with the Baluch Regiment of Special Warfare Warriors.

Special Forces, as with all armed forces, was severely affected by the withdrawal from Vietnam, the tide of anti-militarism, force-strength cuts and the reduction in overseas deployments. Old antagonists of elite units attempted to do away with Special Forces. In September 1968 the 10th Special Forces Group, minus the 1st Battalion, was redeployed to Fort Devens, Massachusetts. The Special Forces Group was lucky to have survived and not be dropped from the rolls, as were the 1st, 3rd, 6th and 8th Special Forces Groups. Fort Devens was a far cry from Flint Kaserne in the Alps of Bavaria. Nevertheless, Colonel Vernon Green, the commander of the 10th, and Sergeant Major Art Senkewich were able to maintain training standards by moving detachments back to Europe on maneuvers and on joint training with our Allies.

Colonel Jack Isler and approximately 300 personnel were hand-picked and comprised the stay-behind element which formed the new stay-behind unit. On September 3, 1968, this organization was an officially designated Special Forces Detachment (Airborne), Europe, with permanent headquarters located in Flint Kaserne, Bad Toelz, Germany. This detachment was a compact and highly trained group of airborne combat specialists who were ready at a moment's notice to infiltrate by air, land or sea behind enemy lines to establish and train large guerrilla forces. It also had the capability of carrying out vital and strategic missions with or without the help of guerrilla units. It served the free people of the world as a deterrent to aggression and a potent weapon to be used in the event of all-out aggression by any enemy force.

In December 1968, Colonel Ludwig "Blue Max" Faistenhammer, Jr. assumed command of Special Forces Detachment (Airborne), Europe and remained as its commander until July 1973. Few, if any other Special Forces commanders, have enjoyed the privilege and honor of commanding a Special Forces unit of that size for that length of time. During Colonel Faistenhammer's tenure of command, the flash worn by Special Forces personnel in Europe consisted of a green background which represents the color used by the 10th Special Forces Group from which the battalion was originally formed. The black, red and gold (yellow) diagonals represent the national colors of the Federal Republic of Germany, the host country of the battalion.

On August 21, 1972, the unit was redesignated Detachment A, Special Forces Support Battalion, 10th Special Forces Group (Airborne), 1st Special Forces. This change was primarily a name change and did not affect troop strength. The realignment also resulted in the renaming of Company A, 10th Special Forces Group (Airborne), 1st Special Forces, which became the 1st Special Forces Battalion, 10th Special Forces Group (Airborne), 1st Special Forces. Also, personnel assigned to the 1st Battalion removed the Special Forces Detachment (Airborne), Europe flash from their berets and replaced it with the solid green flash of the 10th Special Forces Group (Airborne), 1st Special Forces.

Detachments from Bad Toelz continue to receive mission assignment and assistance in analysis, study and preparation from the Group Headquarters at Fort Devens. Day-to-day command and control, budget and contingency mission assignment functions remain with European-based Army elements. The Battalion continues to maintain close relation with their Bavarian neighbors and to take advantage of the training grounds and facilities that exist in Europe. The Bad

Toelz area of southern Germany is one of the best Special Forces training environments in the world with ideal terrain and with the local farmers, woodchoppers and townspeople being staunchly supportive of Americans.

From 11 May 1983 through 25 October 1985, the 10th Special Forces Group deployed 17 separate mobile training teams (MTT) to support the Lebanese army. The mission was to advise and assist the Lebanese army training centers. The 10th Special Forces MTT and the Lebanese developed a training program for over 5,000 officers, NCOs' basic training, Safra for unit training, Wata Al Jawz for unit combined arms live-fire and Haef Jumayyid for urban live-fire training. Training programs for NCO combat leaders, basic training for over 900 LAF conscripts, long-range reconnaissance training for the Lebanese Rangers and advanced unit training and maintenance for mechanized units were also conducted. Despite the confusion of the current situation in Lebanon, the training programs conducted by the 10th Special Forces Group for the LAF were extremely successful.

On 2 June 1985 an MTT from the 1st Battalion, Bad Toelz, Germany, spent four months in Somalia conducting disaster relief operations.

From 8 May 1986 through 23 December 1986, Company C, 3rd Battalion, 10th Special Forces Group, trained the entire nucleus of the Nigerian Airborne. Initially, parachute riggers were trained to pack the parachutes, followed by Airborne cadre selection and training, and then the 10th Special Forces Group soldiers observed this cadre as they trained the remainder of the Nigerian Army. Mission accomplished, the Nigerian Airborne School was established. In conjunction with the Airborne MTT, one detachment trained the Nigerian Army in riverine operations, maintenance, tactics and patrolling.

Each year, 10th Special Forces Group Detachments from the First Battalion in Germany, and from the Second and Third Battalions at Fort Devens participate in Joint Combined Training Exercises (JCET) with West European armies, learning and sharing training techniques. During 1987, 30 JCETs were conducted with Norway, Denmark, Spain, Italy, Belgium, Federal Republic of Germany, United Kingdom, Portugal, Canada, France, Luxembourg, Netherlands and Greece.

The 10th Special Forces Group's individual and detachment training for 1988 significantly enhanced the capability to execute the full spectrum of Special Forces missions. By direction of the group commander, the group focused on the total UW spectrum of military and paramilitary operations in enemy-held, politically sensitive territory. This training included: GW, subversion, target interdiction, offensive actions and intelligence reporting.

The group participated in 1st SOCOM EDRE/ARTEP evaluation exercise CASINO CAMBIT 1-88 and in JCRX FLINTLOCK '88. These exercises constituted extreme challenges for the entire group and provided true tests of overall combat readiness. Summer of 1988 found the group in an intensive individual and collective training posture highlighted by JCET participation in Belgium, Denmark, West Germany and Italy.

The group also hosted a contingent of Belgian commandos for a 30-day JCET at Ft. Devens, conducted an extremely successful Jumpmaster School for active and reserve component Special Forces soldiers and participated in an expanded West Point summer training program to pass critical skills on to our future leaders. Col. Roger G. Seymour passed the colors and command of the group to Col. Jesse L. Johnson on 24 January 1988. The 10th Special Forces Group (Airborne), 1st Special Forces closed out the summer of 1988 with the 2nd Battalion deploying to Ft. Story, Virginia for three weeks of Maritime Operations training to maintain baseline skills in these critical tasks.

The fall of 1988 and into 1989 found the soldiers of the 10th Special Forces Group in a normal, diverse training posture. The 3rd Battalion conducted a six-week GW exercise in the White Mountain National Forest, while 2nd Battalion executed an 11-week formal cross-trained program in MOS 18C and 18E. 1st Battalion conducted FTX Alpine Friendship in the Bavarian Foothills continuing to hone the skills required in combat tasks. The soldiers of the 10th Special Forces Group clearly demonstrated absolute mission readiness and the level of overall professionalism that enable the 10th Special Forces Group to state emphatically that there is indeed something "special" about Special Forces.

10TH SPECIAL FORCES GROUP (AIRBORNE), 1ST SPECIAL FORCES COMMANDERS

Col Aaron Bank 1952-1954
*Col William E. Ekman 1954-1956
*Col William E. Harrison 1956-1958
*Col Michael Paulick 1958-1961
Col Salve H. Matheson 1961-1963
Col Jerry M. Sage 1963-1965
Col Stephen E. Cavanaugh 1965-1967
Col Robert E. Jones 1967-1968
Col Vernon E. Greene 1968-1970
Col R.W. Woolard 1970-1971
*Col Selby F. Little 1971-1973
Col Robert A. Hyatt 1973-1975
Col George E. Palmer 1975-1976
Col Othar J. Shalikashvili 1976-1978
*Col Edward V. Cutulo 1978-1980
Col John A. "Scott" Crerar 1980
Col Paris Davis 1980-1981
Col Roman Rondiak 1981-1982
Col Richard Potter 1982-1984
Col James Zachary 1984-1986
Col Roger G. Seymour 1986-1988
Col Jesse Johnson 1988-Present

*Deceased

10TH SPECIAL FORCES GROUP (AIRBORNE) COMMAND SERGEANTS MAJOR

MSG Paul Allen 1952-1953
MSG Earl Witcher 1953-1956
*MSG James Hallford 1956-1957
*MSG Ray Miller 1958
SGM Charles Ferguson 1958-1961
SGM John Pioletti 1961-1965
*SGM James Hallford 1965-1969
SGM Arthur Senkewich 1969
CMS John Pioletti 1969-1970
CSM John Vierk 1970-1971
CSM Mac Williamson 1971-1972
CSM Williams 1972-1973
CSM Rodriguez 1973-1974
CSM Donald Wilson 1974-1978
CSM Robert Mulcahy 1978-1981
CSM R.S. Rivera 1981-1983
CSM Steve Holmstock 1983-1985
CSM George Moskaluk 1985-Present

*Deceased

COMMANDING OFFICER AND COMMAND SERGEANT MAJOR, 11TH SPECIAL FORCES GROUP (AIRBORNE), 1ST SPECIAL FORCES

Colonel Paul E. Lima
CSM Hubert M. Doyle

COMMANDING OFFICER AND COMMAND SERGEANT MAJOR, 12TH SPECIAL FORCES GROUP (AIRBORNE), 1ST SPECIAL FORCES

Lieutenant Colonel John G. Townsend
SGM John M. Getzinger

19TH SPECIAL FORCES GROUP (AIRBORNE), 1ST SPECIAL FORCES

The 19th Special Forces Group (Airborne), 1st Special Forces was officially activated in the State of Utah on 1 May 1961, and grew from a number of small Special Forces detachments first authorized to the United States Army National Guard in 1959.

In 1961, Army National Guard Special Forces detachments were expanded and reorganized into four newly designated Special Forces groups within the National Guard structure. Reorganizational changes soon became commonplace in succeeding years. By February 1966, the 19th Special Forces Group (Airborne) was only one of two Army National Guard Special Forces groups still in existence. From 1966 through 1972, the 19th Special Forces Group (Airborne) inherited various Special Forces detachments and supporting elements from the 16th Special Forces Group (Airborne), headquartered in the State of West Virginia, and thus continued to expand.

During the history of the 19th Special Forces Group (Airborne), 1st Special Forces, it has at various times, encompassed elements in Utah, Montana, Maryland, Rhode Island, New York, West Virginia, Colorado and Rhode Island.

The majority of annual training conducted by members of the 19th Special Forces Group has been provided at Camp Dawson, West Virginia or Camp Williams, Utah, where numerous unconventional warfare training exercises have been performed. In 1968, the 19th Special Forces Group (Airborne) focused on desert operations in the mountain ranges around Dugway Proving Ground, Utah. Various other training was conducted at the Jungle Warfare School, Fort Sherman, Panama Canal Zone. In 1977, the 19th Special Forces Group (Airborne) commenced a major orientation toward the Pacific theater, as both the U.S. Army and National Guard Bureau began to stress overseas deployment training exercises for reserve component units in support of the Total Force Concept. From then on, numerous training exercises have been conducted in Korea, usually with Republic of Korea Special Forces personnel. Other detachments of the 19th Special Forces Group (Airborne) have participated in arctic warfare training exercises in Alaska. Heavy snows and adverse weather conditions, (down to -30°F in temperature) provided detachment personnel a number of challenging experiences. In 1986, the annual training for the 19th Special Forces Group (Airborne) consisted of a major overseas deployment to Guam supplemented by detachments from other Special Forces Groups. Training was primarily in guerrilla warfare exercises conducted on the island of Saipan in addition to that training which was performed on Guam. Some individual detachments and selected members have also participated in training in other overseas areas.

Members of the 19th Special Forces Group (Airborne), 1st Special Forces have won their share of individual awards and decorations throughout the years for exceptional service in the National Guard. These include the Meritorious Service Medal, Army Commendation Medal, Army Achievement Medal, Army Reserve Component Medal, Armed Forces Reserve Service Medal, Reserve Component Overseas Duty Training Ribbon and the Humanitarian Service Medal.

During the many overseas deployments, members have participated in foreign airborne training that resulted in the award of the coveted foreign parachutist badge of the particular country in which they happened to be conducting training.

COMMANDING OFFICERS, 19TH SPECIAL FORCES GROUP (AIRBORNE), 1ST SPECIAL FORCES

Col. Jack M. Minnoch 1961-1966
Col. Dana F. Peck 1966-1969
Col. Joe T. Decarria 1969-1973
Col. Donald Spradling 1973-1976
Col. Gilbert H. Iker 1976-1978
Col. Richard B. Hansen 1978-1980
Col. Douglas F. Anderson 1980-1983
Col. James G. Martin 1983-1987
Col. Steven J. Dangerfield 1987-Present

20TH SPECIAL FORCES GROUP (AIRBORNE), 1ST SPECIAL FORCES

Special Forces came into existence in the Alabama National Guard in May 1959, and consisted of the 116th Special Forces Operational Detachment (D Detachment) located at the Homewood Armory near Birmingham. In addition was one C Detachment in Montgomery, one C Detachment in Gadsden, one C Detachment in Huntsville, one B Detachment in Montgomery, one C Detachment in Huntsville, one B Detachment in Birmingham and nine A Detachments at various locations in Birmingham, Tuscaloosa, Decatur, Tallasses and Alexander City. Major General Henry Cobb (retired) was a primary organizer and original commander of the unit.

Altogether, this Special Forces organization initially consisted of approximately 60 personnel. Almost half this number was lost within the first six months of that year. Most quit or transferred because of the airborne training requirement. However, as some were lost, others were recruited, and by January 1960, the total number of personnel had increased to about 100. Approximately 40 were airborne qualified by then.

Initially drill periods were conducted on week nights (Tuesday 1900-2130). It became obvious very quickly that Special Forces detachments required more training time than was allowed by meeting in this once-a-week manner. As a result, training was moved to weekends. The training activities consisted primarily of classes on survival, escape and evasion, raids and ambushes, use of radios, etc. During the first year, there was only one full-time member who kept the property books and performed general administrative duties for the entire unit.

The first unit airborne operations had to be conducted at Fort Benning. Those who had become airborne qualified that first year (1959) would travel to Fort Benning by vehicle, and the jumpmasters at Fort Benning would conduct the airborne operation utilizing U1A aircraft. In early 1960, Graham Drop Zone at Fort McClellan was developed specifically for the unit by the Corps of Engineers. Jumpmasters for unit airborne operations were provided by the 77th Special Forces Group from Fort Bragg, North Carolina.

The first annual training period (summer camp) for members of the newly formed 20th Special Forces Group occurred during the period 5-21 July 1959 at Fort Bragg where they trained with 77th Special Forces Group. In 1960 and 1961, annual training consisted of bivouacs and field training activities at Fort McClellan.

On 8 July 1961 these original Special Forces Detachments were reorganized as the 20th Special Forces Group (Airborne), 1st Special Forces with elements in Alabama only. In September 1963, the 20th Special Forces Group (Airborne), 1st Special Forces was again reorganized to include elements not only in Alabama but also in Florida, Louisiana and Mississippi. In 1964, the 123d Quartermaster Company, located in Birmingham and Pell City, Alabama, was attached to the 20th Special Forces Group for support in packing and maintenance of parachutes for airborne operations.

In 1965, the 20th Special Forces Group

was reorganized to include the 1st, 2d and 3d Battalions. The 1st Battalion initially consisted of Headquarter and Headquarters Detachment in Huntsville and Company A in Montgomery. The 2d Battalion initially consisted of Headquarters and Headquarters Detachment in New Orleans, Company A in Baton Rouge and Company B in Lafayette, Louisiana. In 1966, Signal Company (originally Company G, Graysville, Alabama) was made a part of the 20th Special Forces Group.

In 1968 the 123d Quartermaster Company was disbanded and the members were absorbed in Headquarters and Headquarters Company, 20th Special Forces Group, (Airborne). The rigger personnel of the former quartermaster company began packing parachutes in a shed at the Birmingham Airport. Later, they used the armory facilities in Pell City, Alabama. In 1969, they relocated to new equipment facilities next to the Fort Hanna Armory in Birmingham where they remain currently. Also in 1972, the 4th Support Battalion was added to the 20th Special Forces Group and included the rigger, transportation, maintenance, medical, food service, group supply, aviation and administrative elements.

In 1972 the 20th Special Forces Group (Airborne), 1st Special Forces was again reorganized to consist of elements in Alabama, Mississippi, Florida, Maryland and Rhode Island. In 1976 the Rhode Island element was detached, and the 4th Support Battalion, added in 1972, was redesignated as Service Company, consisting of the same elements as previously mentioned.

During the 1980s the 20th Special Forces Group became actively involved in many overseas deployment training exercises. On every occasion the unit fulfilled its training mission responsibilities with success.

On 1 July 1988, the 20th Special Forces Group (Airborne) was reorganized under a new LTOE. A Support Company was added to consist of what was previously Service and Signal Companies. Components of the Support Company are located at Fort McClellan and Pell City, Alabama. Additionally, as part of this reorganization, the 165th Weather Flight (Kentucky Air National Guard) and the 356th Military Intelligence Company (Florida Army Reserve) became closely affiliated with the 20th Special Forces Group.

Since its conception in 1959, the 20th Special Forces Group has continued to grow in strength, and training activities have continued to expand with that growth. Total personnel now number over 1,300 with 150 full-time positions. With the spirit and capability of its members, the 20th Special Forces Group can now be compared favorably with any Regular Army Special Forces group. This unit has an interesting history and its accomplishments can be reviewed with pride. Men of the 20th Special Forces Group (Airborne), 1st Special Forces look forward to a proud and exciting future as they continue to be an important part of our nation's total Special Operations Force.

The current commanding officer of the 20th Special Forces Group (Airborne), 1st Special Forces is Colonel James L. Horak. Assisting him as the group command sergeant major is Garnett D. Hooper.

10th Special Forces Group (Airborne), 1st Special Forces Staff, January 1988. Front Row L to R: LTC McNamara, XO, Colonel Roger Seymour, CO, CSM George Moskulak, CSM. (Courtesy of 10th Group Collection).

Training Jump. 10th Special Forces group (Airborne), AARON DZ, eight kilometers west of Bad Toelz, Germany, December 1953. (Courtesy of C. Sincere)

NCO Review (Courtesy of 10th Group Collection).

Col. Rucker's Parachute School Personnel, Club De Pine, N. Africa. (Courtesy of Guy Hiestand)

Col. Rucker and Parachute Personnel, Kunming, China 1945. (Courtesy of Guy Hiestand)

OFFICE OF STRATEGIC SERVICES (OSS)

By Harry F. Bailey

When Prime Minister Churchill and President Roosevelt met shortly after Japan's attack on Pearl Harbor, they decided that the first major military action of the war should be a joint British-American invasion of North Africa. William J. Donovan, or "Will Bill" as his friends knew him, suggested to President Roosevelt that American agents infiltrate North Africa "as a concrete example of what can be done in the field of espionage, sabotage, economic and diplomatic warfare." Donovan had been appointed director of the newly established Office of Coordinator of Information (COI) on July 11, 1941. The COI was non-military and was established to collect and analyze all information and data which might affect the national security.

In June 1942, part of the COI was transferred to the Office of War Information (OWI) and the rest was militarized under the Joint Chiefs of Staff to become the Office of Strategic Services (OSS). William J. Donovan was appointed brigadier general (later major general) and recalled to active service as director of America's first secret intelligence and special operations service. The first steps had been taken toward the future concept of Special Forces and a Central Intelligence Agency.

Continental United States (CONUS) Headquarters for the OSS was located in Washington, D.C. and Maryland. The rest of OSS was loosely organized into numerous small detachments scattered throughout Europe, the Mediterranean region, the far East and Burma, India and China.

Shortly after the activation of OSS in 1942, a parachute school modeled after the school at Fort Benning, Georgia was organized solely for OSS personnel. Training of OSS personnel was highly classified. A group of airborne and rigger qualified personnel under the command of then Captain Lucius O. Rucker established a small school known as Area A-3 in Prince William Forest Park just west of the Quantico Marine Corps Base. Equipment was primitive. However, it provided landing training which was most important to a parachutist.

The school was scheduled for expansion but by December 1942, the decision had been made to move the school overseas. Rucker went to the Parachute School at Fort Benning and recruited several instructors to round out the group already assembled. Louise D. Davis, a member of the original Test Platoon, John W. Swetish, James R. Hilty and Edward R. Smith originally with the 501st, were quickly signed up. They were moved to Fort Myer, Virginia to assembly for overseas movement. "Wild Bill" Donovan met with the group in a horse barn at Fort Myer and emphasized the importance of what OSS was doing to shorten the war.

A group of approximately 28 people under the command of Captain Jerry Sage reported to Fort Hamilton, NY for overseas shipment and arrived in Oran, Algieria, January 13, 1943. They traveled by train to Algiers where they were welcomed by Colonel William A. Eddy, USMC, who commanded the Mediterranean Region OSS operations. Some of the group went their several ways to scattered OSS station sites while parachute school personnel went to Club de Pins on the Sidi Ferruch, North Africa Peninsula.

Club de Pins was a former beach resort and empty villas were plentiful. These villas provided quarters for the detachment. The area was occupied and administered by the British who maintained radio communications with their people on special operations. They helped the "Yanks" set up the training areas and practice drop zones.

In a very short time The Parachute School, Company A, 2677th Regiment OSS (Provisional), APO 534, USA Army, under LTC Lucius O. Rucker was in business. The schedule called for ground training to commence Monday mornings followed by a training jump Tuesday afternoon: two more jumps on Wednesday, followed by the fourth and fifth jumps on Thursday, and the course was over. Friday served as a day of rest unless weather conditions required make-up jumps.

Quite often an agent would complete the course on Friday and weather permitting, would climb aboard a specially modified bomber, either British or American, and drop behind enemy lines during the hours of darkness on Saturday or Sunday. Most American and Allied OSS personnel attended jump school either at Fort Benning, Georgia, England or in North Africa. Graduates were presented with American Jump Wings.

Personnel at the OSS Parachute School packed bundles of weapons, ammunition, explosives and money to be dropped to the partisans led by OSS agents. School personnel often flew as jump master (dispatcher on British aircraft) to drop personnel and supplies. The primary departure base was Blida Airfield near Club de Pins, and the aircraft might be an American B-17, B-24 or B-25 bomber or sometimes a British Halifax or Sterling bomber.

The Operational Groups or OGs of the OSS were associated administratively with the 2677th Regiment (OSS) Provisional but acted independently as the 2671st Special Reconnaissance Battalion (Separate, Provisional). General Orders No. 72, War Department, Washington, D.C., 18 July 1946, cites a few of their accomplishments:

Company A, organized into small operational groups, "infiltrated behind the enemy lines by parachute and, maintaining contact with their headquarters by radio, organized extensive partisan forces." The unit was "cited for outstanding performance of duty in action against the enemy in Italy from 15 April to 1 May 1945."

Company B, was "assigned the mission to parachute into Central and Southern France in strategic areas in advance of the invasion forces. "Its mission was to arrange for supply of the Free French Forces by parachute drops of arms, ammunition and money. The OGs together with the FFI "established road blocks, mined roads, ambushed columns, attacked enemy installations and received the surrender of over 10,000 German troops." The Company was "cited for outstanding performance of duty in action against the enemy in Southern France from 1 to 15 August 1944."

Company C infiltrated behind enemy lines, in uniform, by parachute or by sea "leading resistance forces in attacks against the enemy." The unit was cited for outstanding performance of duty in action against the enemy in Greece from 15 August to 1 September 1944.

Detachment 101 (Kachin Rangers) engaged primarily in intelligence and guerrilla activities in Northern Burma from April 1942 to July 1945. General Orders No. 7, War Department, 17 January 1946, signed by Dwight D. Eisenhower, chief of staff awarded the Unit Citation to this OSS unit for action 8 May to 15 June 1945 for leading 3,200 native troops who "met and routed 10,000 Japanese throughout an area of 10,000 square miles, killing 1,274 while sustaining losses of 37"

Jedburghs were three-person teams formed of Allied and American personnel dropped into France, Belgium, and Holland after D-Day to direct the underground army. Many were taught to jump at the Royal Air Force (RAF) School at Ringway near Manchester, England. Those that were not jump trained at Ringway were trained in North Africa at the OSS Parachute School in Club de Pins.

In May 1944, the first Jedburgh teams were sent to North Africa to be parachuted into Southern France. By the end of August 1944, all the remaining Jeds had jumped into France or the Low-Countries to lead the Maquis in hit-and-run forays.

Many of the OSS agents were recruited from enemy occupied countries. They were secretly removed to a safe training location by submarine or surface vessels and trained in special techniques. These agents were then air-transported back to their point of origin to organize resistance groups and sabotage enemy operations. When agents were to be parachuted into an area behind enemy lines or in occupied countries they were first given an abbreviated parachute training course.

In October 1944, many of the OSS secret operational groups commenced phasing out, returning to the United States or going on to new areas. Company B, 2671st Battalion returned to the U.S. and then went to China and trained Chinese Commando units. Company A, 2677th Regiment, the Parachute School, left the Mediterranean area of operations in October 1944. Some of the personnel were assigned to the Parachute School at Fort Benning. A few selected individuals shipped the jump school equipment to Detachment 202 in Kunming, China and the rest of the small contingent arrived in India, April 28, 1945. They reached Kunming in May and set up a Parachute School to give a large Commando force jump training. The Chinese were subsequently dropped into Indo-China. The OSS personnel were awarded the Chinese Breast Order of Yun Hui and Chinese Parachute Wings.

With the end of the war the need for OSS personnel dwindled until President Truman by executive order September 20, 1945 abolished the Office of Strategic Services. On October 31, 1945 the unit was deactivated.

Enough cannot be said for those individuals whose notes and recollections were the basis for this very brief synopsis of OSS/Airborne activities. Among them, president of the Veterans of OSS, Geoffrey Jones and Veterans of the OSS, Guy Hiestand, Edward R. 'Yardbird' Smith and John Swetish.

Bibliography:

Cave Brown, Anthony and Wild Bill Donovan, **The Last Hero**, Times Books, 1982.

Russell, Francis, **The Secret War**, Time-Life Books, Inc., 1981.

Douglas A-26 attack bomber. Both fast and small, this type of aircraft was frequently used to drop OSS agents deep into Germany. (U.S. Air Force photo)

The OSS Detachment of the U.S. Seventh Army was headquartered in this chateau near Lyon, France. (U.S. Army photo)

A force of about 16,000 men and women received a total of 2,005 medals for gallantry and proficiency. These included:

    Distinguished Service Medals, 3
    Distinguished Service Crosses, 51
    Navy Crosses, 2
    Legions of Merit, 209
    Silver Stars, 148
    Distinguished Flying Crosses, 76
    Soldiers Medals, 72
    Bronze Stars, 773
    Air Medals, 34
    Navy Commendation Ribbons, 4
    Medals of Freedom, 9
    Purple Hearts, 206
    Distinguished Service Order (UK), 1
    Orders of the British Empire, 3
    Military Crosses (UK), 3
    Military Medals (UK), 3
    Legions of Honor (France), 4
    Croix de Guerre (France), 108

This list was complete only to January 5, 1946; to it must be added perhaps 300 more American and Allied decorations and awards and several hundred certificates of service from foreign governments. In sum, therefore, the number of medals, certificates and other recognitions awarded to OSS men constitutes a tangible, remarkable testament to the OSS's performance.

Yet it did its work with remarkably few casualties. In all, 143 men and women, excluding subagents, were killed in action, and about 300 men and women were wounded or captured while on active service.

ABOUT THE JEDBURGHS

For those of us in other branches who knew only vaguely about the reason for and the valorous service rendered by this elite group, the following excerption from Carey Ford's **Donovan of OSS** gives a most concise description of the some 300 men who served in the Jedburghs:

At the Quebec Conference in the late summer of 1943, Secretary of War Stimson and Generals Marshall and H.H. Arnold and Admiral Ernest King—with, for once, the firm backing of President Roosevelt—obtained the agreement of the prime minister for a cross-Channel strike, code-named OVERLORD, as the major Allied effort in 1944.

Target date for OVERLORD was set for May, though the actual landings at Normandy, which Churchill had predicted would be determined mainly by the moon and the weather, did not take place until June 6. With D-Day established, OSS/London embarked with British SOE on a joint Anglo-American enterprise, designated Special Force Headquarters (SFHQ), which would organize all underground resistance in France in support of the forthcoming invasion. Their program called for large-scale paramilitary activity, aided by a maximum delivery of arms and supplies, which would culminate in an all-out Maquis attack on the Germans on D-Day. Once the landings had been made, the resistance forces would create chaos behind the enemy lines, disrupting communications, ambushing troops and convoys, blocking all escape routes and conducting anti-scorch measures to prevent the demolition of key installations by retreating Nazis. To integrate this Free French uprising with the advance of the Allies, a number of three-man SFHQ teams called Jedburghs would be trained and dropped in uniform into France and Belgium and Holland after D-Day to direct the underground army.

At Milton Hall, an Elizabethan manor house some one hundred miles north of London, was assembled a heterogeneous group of 240 volunteers: American and British and French and Belgian and Dutch, Army and Navy and Marine, officers and enlisted men, all living and learning the tactics of unorthodox warfare in a staid atmosphere of oak-beamed halls and Cromwellian armor. Para-

trooper boots thudded from a training harness onto the neat lawns, and men practiced silent killing in the sunken gardens. From the croquet pitch came the crackle of small arms or the louder explosion of Sten guns and Enfields, and the acrid smell of burnt powder blended with the traditional odor of boxwood and roses. Demolition charges shook the far end of the golf course, and a hand-powered wireless set—everyone had to master the Morse code—stuttered day and night in the paneled library.

The Jeds were taught to jump at an RAF school at Ringway, located in a sheep pasture in the rolling downs country near Manchester. They dropped from war-weary Sterling Bombers, 10 men to a plane, five forward and five aft of the Joe-hole. The first jump was made in pairs, to avoid fouling; the second was in strings of five; on the third jump all 10 men descended together, while a British jump-instructor on the ground coached them over a bullhorn. "Get your knees together, Number Two!", "Pull down those front risers, Five!", "Always come in for a forward landing." The final lesson, a night drop under combat conditions, concluded their three-day course. Some hardened Fort Benning paratroopers with the Jedburghs were astounded; back at Benning it was six weeks before your first jump.

Gradually the trainees grew better acquainted, and individual teams began to form. At the outset these teams had been assigned by the colonel in charge, but it was decided that the partnerships made by mutual consent were more efficient in combat. After a preliminary courtship period, shy and rather self-conscious, an American and a Frenchman would decide to become "engaged," and their "marriage" was duly solemnized in the following day's orders. A radio operator, usually an enlisted man, would be invited to join the team, the trio would be given their own code name—Dauntless, Harvard, Argonne—and henceforth they would work and eat and live together in a tight-knit comradeship on which their very lives might depend. Sometimes, after a marriage, the members of a team would realize that they were incompatible and apply for a divorce; but it was better to find out at training school than the field.

In May of 1944, the first Jedburgh teams were sent to North Africa, to be parachuted into Southern France. During June, six more teams were dropped behind the lines in Brittany, from which they radioed vital intelligence to the Normandy beachhead. By the end of August, after General Patton's breakthrough, all the remaining Jeds had jumped into France or the Low Countries, to gather about them their own little armies of Maquis and lead them in hit-and-run forays, the basic strategy of guerrilla warfare. "Surprise, mitraillage, evanouissement," the Maquis called it: surprise, kill, and vanish. Of the 82 Americans who participated in the Jedburgh operation, 53 received the Distinguished Service Cross, Croix de Guerre, Legion of Merit, Silver or Bronze Star, or Purple Heart—claimed to be the highest percentage of citations awarded to a single group in the entire war.

OFFICERS

Everett T. Allen; Stewart D. Alsop; Robert M. Anstett; B. McDonald Austin; Aaron Bank; Douglas D. Bazata; William C. Boggs; John H. Bonsall, D; Russell W. Brazelton; Paul F. Brightman; Charles E. Brown III, D; Charles M. Carman Jr.; Philip H. Chadbourne, Jr.; William E. Colby; Lucien E. Comein; Paul Cyr; Conrad C. Dillon; Philip W. Donovan; Harry A. Dorsey; William B. Dreux; Rene Dussaq; Ian Forbes; Ray H. Foster; Horace W. Fuller; Charles J. Gennerich; John J. Gildee; Nelson E. Guillot; Walter C. Hanna, Jr.; Robert E. Heyns; Stephen J. Kherly; Bernard M.W. Knox; Robert A. Lucas; Cyrus Manierre; Henry D. McIntosh; Richard V. McLallen; Robert K. Montgomery; Raymond E. Moore; Robert G. Mundinger, D; Cecil F. Mynatt; John M. Olmsted; William H. Pietsch, Jr.; Jack T. Shannon; John K. Singlaub; McCord Sollenberger; Mason B. Starring; John W. Summers; Lawrence E. Swank, D; Steve W. Thornton, Jr.; Harvey A. Todd; Shirley R. Trumps; George M. Verhaeghe; Lucien H. Wante; George Thomson.

D- Deceased

ENLISTED

William H. Adams; Robert J. Anderson; Albert V. Bacik; Robert R. Baird; Theodore Baumgold; Jacob B. Berlin; Willard Beynon; James R. Billingsley; Lucien Bourgoin, D; John L. Bradner; James J. Caldwell; James J. Carpenter; Francis J. Cole; Roger E. Cote, D; Anthony J. Denneau; Elmer B. Esch; Richard C. Floyd; Norman R. Franklin; Berent E. Friele; Lewis F. Goddard; Arthur Gruen; Frank A. Hanson; Michael F. Henely; Robert R. Kehoe; Lucion E. Lajeunesse; John E. McGowan; Charles P. Mersereau; Howard V. Palmer; Roger L. Pierre; Francis M. Poche, Jr.; Vincent M. Rocca; Carl A. Scott; Don A. Spears; John L. Stoyka; William T. Thompson; Thomas J. Tracy; John Van Hart; Lee J. Watters; John A. White; William L. Zielski.

D- Deceased

DETACHMENT 101 "THE KACHIN RANGERS"

Excerpt from USS Fall 1981 Newsletter

Called the "Kachin Rangers" during World War II, Detachment 101 was commanded first by Colonel Carl F. Eifler, and fought in Burma under General Joseph W. Stilwell (subsequently succeeded by General Daniel I. Sultan). They were the only OSS group awarded the Presidential Distinguished Unit Citation: General Orders No. 7, War Department, 17 January 1946, signed by Dwight D. Eisenhower, chief of staff.

Detachment 101 had been engaged primarily in intelligence and guerrilla activities in Northern Burma from April '42 to July '45 and had led "Merrill's Marauders" into Myitkyina. The award, however, was specifically for an infantry action 8 May to 15 June 1945 against 10,000 veteran troops from the Japanese 18th and 56th Divisions. The battle was fought in the Shan States of Burma in a drive south from the old Burma Road between Lashio and Mandalay to Laihka, Loilem and Lawksawk. The job was to cut off the Japanese escape route on the Taunggyi-Kentung road into Thailand. The 3,200 native troops led by 300 American OSS men "met and routed 10,000 Japanese throughout an area of 10,000 square miles, killed 1,274 while sustaining losses of 37" the Citation said in part.

Today, the 101 Association has about 500 members and associates on their mailing list out of a wartime total American membership of probably 750. Another 250 of the 101'ers were British, Australian and Burmese, and that total of 1,000 OSS people trained and led about 11,000 native troops in battle over a three-year period. More than 100 Detachment 101'ers are members of the VSS now.

The OG's—Operational Groups—were one of the principle branches of OSS whose function was "to train and supply and lead guerilla forces in enemy territory." These groups have also been further described as being "organized by target countries, France, Italy, Greece, etc., consisting of 32 men (sometimes 16, the smaller groups with at least 1 officer and 1 radio operator) with language qualifications, dropped behind the lines in uniform to carry out both espionage and sabotage operations in concert with the resistance."

Here's a copy of the WD Gen. Order No 72, that can give you a sanitized idea of what these heroic comrades of ours accomplished:

BATTLE HONORS - As authorized by Executive Order 9396 (Sec. I, WD Bul. 22, 1943), superseding Executive Order 9075 (Sec. III, WD Bul 11, 1942), citations of the following units, as approved by the Commanding General Mediterranean Theater of Operations, 3 July 1946, are confirmed under the provisions of section IV, WD Circular 333, 1943, in the name of the President of the United States as public evidence of deserved honor and distinction. The citations read as follows:

1. Company A, 2671st Special Reconnaissance Battalion (Separate) (Provisional), is cited for outstanding performance of duty in action against the enemy in Italy from 15 April to 1 May 1943. Company A, composed of officers and enlisted men who volunteered for extra hazardous duty in the conducting of operations behind enemy lines, engaged in extensive operations in Italy under the direction of the 15th Army Group Headquarters. These men, organized into small operational groups, were infiltrated behind the enemy lines by parachute and, maintaining contact with their headquarters by radio, organized extensive partisan forces. In the final phase of the offensive of the 15th Army Group, they led these partisan forces in all-out attacks. The officers and enlisted men of Company A, 2671st Special Reconnaissance Battalion (Separate) (Provisional), despite the constant danger of attack and capture, by their courageous leadership and participation in the operations of these resistance forces were instrumental in causing them to organize and attack the enemy and were a constant inspiration to them, thus reflecting great credit on themselves and the armed forces of the United States.

2. Company B, 2671st Special Reconnaissance Battalion (Separate) (Provisional), is cited for outstanding performance of duty in action against the enemy in Southern France from 1 to 15 August 1944. Assigned the mission to parachute into Central and Southern France in strategic areas in advance of the invasion forces, Company B, composed of officers and enlisted men who had volunteered to perform extra hazardous duty, contacted French Forces of the Interior, arranged for their supply by parachute drops of arms, ammunition and other supplies, and led them in operations as directed by Allied Force Headquarters. These men, along with French Forces of the Interior, established road blocks, mined roads, ambushed columns, attacked enemy installations and received the surrender of over 10,000 German troops. The presence of Company B, 2671st Special Reconnaissance Battalion (Separate) (Provisional), deep in occupied France and their active participation and leadership in dangerous operations was a potent factor in inspiring the French Forces of the Interior to take such an active and important part in attacking the enemy in support of the Supreme Headquarters Allied Expeditionary Forces and Seventh Army invasions, thus reflecting great credit on themselves and the armed forces of the United States.

3. Company C 2671st Special Reconnaissance Battalion (Separate) (Provisional), is cited for outstanding performance of duty in action against the enemy in Greece from 15 August to 1 September 1944. The officers and enlisted men of Company C volunteered for extra hazardous duty consisting of infiltrating behind the German lines, in uniform, by parachute or by sea and leading resistance forces in attacks against the enemy. A total of 15 officers and 159 enlisted men were parachuted into strategic areas of Greece, or entered it by sea, and organized and led the Greek-partisans in a campaign to cut off the lines of retreat of the German forces. The men of Company C 2671st Special Reconnaissance Battalion (Separate) (Provisional), with their partisan bands, destroyed many bridges, blocked roads, attacked German convoys and caused severe loss in enemy personnel and equipment, thus reflecting great credit on themselves and the armed forces of the United States.

The suspended harness at Camp A-3 in national park west of Quantico, Virginia, 1942. (Courtesy of Guy Hiestand).

THE VIETNAMESE AIRBORNE ADVISORY DETACHMENT

By BG John D. Howard, USA

MACV TEAM 162

In the spring of 1975, the Vietnamese Airborne Division, along with most units of the Army of the Republic of Vietnam (ARVN), melted away under an avalanche of North Vietnamese infantry, armor and artillery. The disintegration of ARVN—both elite units and those not so elite—was hastened by gross ineptness at the national command level and the termination of military aid by a hostile U.S. Congress.

When the North Vietnamese Army (NVA) tanks broke through the gates of Independence Palace in Saigon on 30 April, the Vietnamese Airborne Division, the longtime ARVN strategic reserve, no longer existed. Hence, the 30-year war ended, not by the popular uprising that had been predicted by the communist world and its left-wing supporters in the United States, but by massive force of conventional arms that sapped the last strands of moral fiber of the South Vietnamese.

The Vietnamese airborne units had enjoyed a long association with its western allies. It had started in 1951 when the French organized the 1st Vietnamese Paratrooper Battalion (BPVN) to augment the over-extended French Expeditionary Corps (FEC). Five battalions were eventually activated and as the Indo-China War dragged on, the French finally began to train a fledgling officer and NCO corps and integrated these young aspirants into the structure of the BPVN. Some of these officers, Tran Van Don, Cao Van Vien, and Do Quoc Dong, later rose to the highest levels of the ARVN. They earned their spurs during the first Indo-China War in fighting along "la rue sans joie," in the set-piece battles north of Hanoi, and at Dien Bien Phu.

The United States Army filled the vacuum created by the FEC departure from Indo-China following the Geneva Convention of 1954. A Military Assistance Advisory Group (MAAG) already had been established in September of 1950 to coordinate the flow of U.S. aid, but now it became the dominant military presence in the South.

Meanwhile, the new president, Ngo Dinh Diem, quickly consolidated his power in the new republic by using the airborne battalions to destroy the "Cholon Mafia" known as the Binh Xuyen. These gangsters had operated with impunity in the Chinese Quarter of Saigon throughout the colonial period; Diem used his paratroopers to rid the city of this threat. His handling of this crisis reinforced U.S. support of the new government and caused the MAAG chief to make plans to establish an advisory detachment with the Vietnamese Airborne Group.

The creation of the Vietnamese Airborne Advisory Detachment, later known as MACV Team and called the "Red Hats" by most, began an 18-year affair that finally ended on a hot February afternoon in 1973 with a deactivation ceremony

South Vietnamese Airborne Commander confers with his U. S. counterpart on combat operation. (U. S. Army photo, courtesy of Bill Knapp)

A 19-year old North Vietnamese soldier captured north of Tau Son Nhut is transported by chopper for further questioning. (U. S. Army photo, courtesy of Bill Knapp)

prompted by the January cease-fire agreements.

The Advisory Detachment was always an amalgamation of the "best and the brightest" that the U.S. Army had to offer. Even when U.S. units dominated the war (1966-1970), there was always an overabundance of volunteers to serve as advisors with the Vietnamese Airborne. Unlike the bulk of the advisory effort after 1965, no one ever had to be "drafted" into Team 162. This phenomena was a testimony to the professionalism and fighting fervor of the Vietnamese paratrooper and how the Vietnamese Airborne Division was viewed by the U.S. Army as a whole.

Airborne advisory duty was always considered as "career enhancing" service in an American unit—and that couldn't be said for many other MACV billets. Several former Airborne advisors, General Jim Lindsey and General Norm Schwarzkopf, went on to wear four stars while many others have ultimately been selected to be general officers and command sergeants major.

The initial advisory efforts with the Vietnamese Airborne Group were technically oriented rather than tactical. They emphasized planning, training, equipment and logistical issues. A reflection of growing U.S. Army staff influence emerged in 1959 when the Joint General Staff (JGS) redesignated the Group as the Vietnamese Airborne Brigade. In 1961, the MAAG significantly altered the course of war by directing that advisors would be assigned to combat battalions and that a Battalion Combat Advisory Team (BCAT) would consist of a captain, a lieutenant and a senior non-commissioned officer.

As U.S. combat support assets, such as helicopters and USAF aircraft, began to arrive in Vietnam, the influence and role of the advisor also increased. With all these indicators of an expanding U.S. commitment to the Vietnam War came another—U.S. casualties. The first airborne advisor to be killed was Captain Don J. York who died on 14 June 1962 in an ambush of a vehicle convoy north of Saigon. Don York headed the Team 162 killed in action (KIA) list which would eventually number 21 U.S. Army advisors, three Air Force Forward Air Controllers (FACs) and two Army advisors who still are counted as missing in action...the most of any division advisory detachment in RVN.

By December 1965, the brigade was upgraded to a division because two more infantry battalions (2d Airborne Battalion and the 8th Airborne Battalion) and an artillery battalion were activated. Concomitantly, advisors were assigned not only to combat but also to combat support and combat service support elements within the airborne division. MACV Team 162 was the largest advisory effort of its kind.

Advisors were actively engaged in all the division's combat operations, training and logistical upgrades. During this time, the airborne division became the most aggressive in the Republic participating in the critical battles of the war. The most notable were Hue and Phu Cat in 1966; Dak To in 1967; Khe Sanh, A Shau Valley and the Tet Offensive in 1968; the defense of Tay Ninh in 1969; and the Cambodian operation in 1970. In addition, the division had conducted three combat parachute assaults: in Binh Long in 1965 and in Chuong Thien in 1966 and 1967.

During Tet, the division received its

Abn. Bde. Adv. Det., December 1964. 1st Row: MAJ C. East, LTC Casteel, LT Tobin, unknown, SFC J. Maldasapo, COL J. Hayes, LT W. Kopenhavner, Lund, CAPT J. Necgard, LT Murray, SGT MAJ Campisi. 2nd Row: MAJ J. Lindsay, CAPT T. Throckmorton, CAPT B. Golden, unknown, MSG Lewis, CAPT. D. Scholtes, LT Sewall, CAPT. B. Losik, unknown. 3rd row: SFC G. Colvin, SFC A. Combs, unknown, CAPT L. Suddath, unknown, CAPT B. Webb, unknown, unknown.. (Photo by SP 4 John F. Watts, courtesy of Gen. James J. Lindsay)

most difficult assignment of the war. While defending sectors of Tan Son Nhut, other battalions were fighting in Saigon, Cholon, Quang Tri, the Citadel in Hue and Van Kiep. By the end of Tet, airborne units had killed over 2,000 enemy, captured 146 NVA soldiers and accounted for 800 individual and crew-served weapons. The fierce fighting also claimed the lives of two more advisors, SFC John Church and Captain Terry Sage. The division's role in the fighting was summarized as follows in an official MACV account:

"The 8th Battalion hit an NVA regiment head-on to halt the enemy surge across the runway at Tan Son Nhut and drive them off the air base. Meanwhile, two companies of the 8th Battalion were sent to assist the remnants of the 6th Battalion, now reduced to less than 200 men, when they defeated two enemy regiments at Dak To just three days earlier and stopped two NVA battalions in the Joint General Staff compound. In Saigon, the 1st Battalion held off a threat on the Presidential Palace, cleaned out Cholon and regained control of the radio station. The 5th Battalion returned to Saigon to assist in fighting there immediately after operations north of Da Nang while the 3rd Battalion was fighting NVA units at Go Vap and Xom Moi north of the capital. The 2d Battalion and 7th Battalion joined the 9th Battalion to break the siege in the Citadel. Previously, the 9th had fought two NVA regiments in Quang Tri. Still undergoing training, the newly activated 11th Battalion repelled a reinforced VC battalion attack on Van Kiep Training Center."

LAM SON 719, the attack into Laos in February 1971, started an attrition process that eventually resulted in the division being "bled white." For the first time, a MACV order prohibited the advisors from accompanying their units into combat. Command concern over U.S. casualties and growing reliance on "Vietnamizing" the war prompted the directive.

An advisory base was established at Khe Sanh when the units crossed into Laos. While individual airborne battalions fought the North Vietnamese valiantly, Vietnamese commanders were unable to synchronize U.S. fire support assets. Heretofore, this had been accomplished by advisors, and now their absence created a void in the paratroopers' ability to orchestrate the battle.

During the fighting on the way to Tchepone, five of nine battalion commanders were killed or wounded, and when Firebase 31 fell, the 3d Brigade commander and his staff were captured by the NVA. The airborne division's withdrawal from Laos resulted in a long standdown period while the leadership attempted to fill its depleted ranks and began to assume an even greater responsibility for prosecuting the war.

Notwithstanding the deficiencies that emerged during LAM SON 719, U.S. troop withdrawals continued unabated. President Nixon was determined to decrease U.S. involvement in Vietnam. The draw-down affected not only U.S. units but also had impact on the advisory effort. Although Vietnamization was in full swing, HQ MACV issued strength reductions for all detachments.

By the fall of 1971, advisors had been removed from infantry battalions in all ARVN divisions except those in the airborne division. However, Team 162 was not totally protected from the cuts. Advisory positions in the artillery, engineers, and combat service support units were either eliminated or reduced to one individual. But, the decision to protect the BCATs in the airborne division proved to be decisive when the NVA launched their long-anticipated offensive in 1972.

The "Easter Offensive" which began on 31 March 1972 introduced a new face of war inside RVN. North Vietnam's defense minister, Vo Nguyen Giap, directed conventional attacks in Military Regions (MRs) I, II and III involving the commitment of practically all his regular forces. These division-sized formations, well-balanced in armor, infantry and artillery, were oriented on the destruction of ARVN, trapping if possible the remaining U.S. personnel in the country.

The basis of the North's action revolved around the assumption that Vietnamization was a failure and that the U.S. public was so adverse to continued involvement in the Vietnam War that President Nixon would be unable to bolster the Thieu government. However, North Vietnam had grossly underestimated the U.S. response.

Giap's offensive stretched the airborne division even more than the LAM SON 719. During the initial stages, brigades were deployed along Rocket Ridge in MR II, in Quang Tri Province and to reinforce the defense of An Loc. Fighting was intense an casualties were very heavy. Because of its strategic location north of Saigon, the NVA stranglehold on An Loc caught the attention of the international media. The 1st Brigade helped break that siege and stave off what could have been a disastrous defeat.

By late June, with the NVA stopped at An Loc, the airborne division was completely assembled north of Hue and marshalled for an attempt to wrest Quang Tri City from North Vietnamese occupation. The fact that it was the only province capital in NVA hands made it assume an importance far beyond its tactical significance. By frontally attacking this fortified position, both the airborne division and the Vietnamese Marines frittered away the strength of their units.

The airborne battalion advisors earned their pay by coordinating massive amounts of U.S. assets that were sent from the U.S. to stem the invasion. While U.S. firepower, particularly U.S. tactical air support and B52s, provided the margin of victory, it could not replace the experience and elan that was lost on the battlefields of Quang Tri, An Loc and Kontum.

By the fall of 1972, the fighting in I Corps had tapered off and actually degenerated into a stalemate. Both sides were clinging to each other like two heavyweights in the 14th round of the championship fight. The Vietnamese Airborne, the Vietnamese Marines and the NVA were content to look at each other across the Thach Han River near Quang Tri City—neither adversary having the inclination nor the combat power to mount a significant offensive.

Similarly, the progress of the Paris Peace talks was not lost on any of the combatants. Advisors were queried on status of the talks and the rationale behind U.S. negotiating positions. Of course, Team 162 members were not "consulted" on strategic policy matters and hence were not as much in the dark as their counterparts. When the announcement came that an accord had been reached, advisors were assigned the task of checking the status of the territorial claims of each side. At the same time, efforts were made to get battalion advisors out of the field in preparations for returning to the United States.

The loss earlier in January of Major Bill Deane and SSG Elbert Bush as MIAs had heightened MACV's anxiety concerning casualties. While the Vietnamese vio-

U. S. Advisors receive decorations for their actions during the TET offensive, 1968. (U. S. Army photo, courtesy of Bill Knapp)

lently disagreed with the provisions of the Peace Accords, they did not hold the American advisors responsible or transfer their distaste to their counterparts. The Team, by then only numbering 42 officers and NCOs, was ordered back to Saigon and assembled at Tan Son Nhut in February 1973. It was finally over for the Americans. Team CONEX's were cleaned out and after-action reports were written. Good-byes were said and only the lights needed to be turned off.

General Frederick C. Weyand and Colonel Marcus Hansen, the last senior advisor, presided over the deactivation of Team 162. Just like so many other days in Vietnam, it was a hot, sunny afternoon. Orders were read and the dead were honored in manner befitting their service and sacrifice. Then, in true Airborne fashion, there was an advisor party which lasted well beyond its allotted time and became the premier event of the 1973 Saigon social season. It was the only way it could have ended.

For most career U.S. military personnel, the final capitulation of the Republic of Vietnam has been almost too painful to recount. It was especially trying for the ex-members of Team 162 who watched the evening news in March and April 1975 and saw their former counterparts fleeing and units that they had once fought with dissolve in the face of the NVA onslaught.

President Thieu's untimely withdrawal of the airborne division from I Corps, coupled with the catastrophic Central Highlands retrograde, was the catalyst for defeat. Those actions caused the thin veneer of ARVN professionalism to crack, and once it started there was little anyone could do to stop it.

While the press preached the theme that the "agony of Vietnam" was finally over, many Red Hats knew otherwise. For those Vietnamese paratroopers who could not or chose not to escape, there was a long period of "re-education" or re-settlement to the "new economic zones." It proved to be the ultimate sanction for many. Some were fortunate enough to immigrate to the United States. The Society of the Vietnamese Airborne was in the vanguard of the sponsorship drive. The society was extremely active helping its former comrades not only start new lives but also become productive members of their newly adopted communities and country. Probably no other private organization could boast of such an effort. These contacts have remained strong to this date. It proved that the Airborne spirit was, and still is, a state of mind—and that once an Airborne Soldier, always an Airborne Soldier.

MACV TEAM 162 HONOR ROLL

The following personnel gave their lives in the line of duty while serving the United States and the Republic of Vietnam as advisors in the Vietnamese Airborne Division:

CPT Don J. York- July 1962; CPT Thomas W. McCarthy- March 1964; SFC Alfred H. Combs- June 1965; CPT Paul R. Windle (USAF)- June 1965; 1st LT Robert M. Carn, Jr. (USAF)- August 1965; CPT Donald R. Hawley (USAF)- December 1965; SFC John E/ Milender- January 1966; SFC Clifford L. Robinson- June 1966; CPT William T. Deuel- September 1966; CPT Cary H. Brux- October 1966; PFC Michael P. Randall- May 1967; 1st LT Charles L. Hemmingway- June 1967; 1st LT Carl R. Arvin- October 1967; SFC John L. Church- January 1968; CPT Terrance F. Sage- January 1968; 1st LT Arthur M. Parker III- May 1968; SFC Charles J. Moore- September 1968; SP4 Ronald I. Kirkpatrick- July 1969; 1st LT William E. Bonner- December 1969; 1st LT Gerald L. Maseda- October 1970; CPT Lee G. Grimsley- April 1971; SFC Alberto Ortiz, Jr.- April 1972; CPT Glen S. Ivey- May 1972; 1st LT Grady T. Triplett- August 1972.

The following soldiers are missing in action:

Maj. William L. Deane- January 1973; and SSG Elbert W. Bush- January 1973.

War Zone C, May 1966. Major Guy S. (Sandy) Meloy with his counterpart, Lt. Col. Ho Trung Hau, who was the Commander of Task Force I, ARVN Airborne Division.

CPT Blair of the U. S. Advisor Team talks to ABC-TV reporter, Don Baker, about captured VC in background. (U. S. Army photograph courtesy of Bill Knapp)

CPT Wes Taylor (later Brig. Gen.), senior advisor with ARVIN, 8th Airborne Battalion, coordinating clearing operation north of Saigon During February 1968. (U. S. Army photograph courtesy of Wes Taylor)

MARINE PARACHUTE BATTALIONS "THE PARAMARINES"

By Charles L. Updegraph, Jr.

As had been the case with the raider units, the Marine Corps parachute units could trace the impetus for their development to the employment of special purpose forces by the European powers during World War II. Although the Marines had had limited previous experience with parachutists (As early as 1927, a group of 12 Marines parachuted from a transport plane over Anacostia in Washington, D.C.), realization of the parachute concept on an enlarged scale occurred only after the outbreak of World War II.

In May 1940, Major General Holcomb, commandant of the Marine Corps, tasked the Division of Plans and Policy with the preparation of a plan for the utilization of Marine parachute troops. For planning purposes the force would consist of: one battalion of infantry at full strength; one platoon of 75mm pack howitzers (two guns); three units of fire for all arms; three days' rations and water; no vehicles other than hand-drawn; added light antiaircraft and antitank protection as appropriate. The resulting paper envisioned the employment of parachute units in three distinct tactical situations: as a reconnoitering and raiding force with a limited ability to return to its parent organization. This assumed that the objective was sufficiently important to warrant the sacrifice of the force, as a spearhead or advance guard, to seize and hold strategic installations or terrain features until arrival of larger forces and as an independent force operating for extended periods, presumably in a guerrilla role in hostile territory.

The Marine plans for parachute units gained added impetus after inspection of Army training facilities by several Marine officers in the summer of 1940 and after naval attaches began to collect reports on the use of parachute forces by the Germans, Russians and French. By October of 1940, the commandant had decided that one battalion of each infantry regiment would be trained as "air troops" to be transported and landed by aircraft. Each air troop battalion would include a company of parachutists with an estimated requirement for 750 parachute personnel for the entire Corps.

An initial detachment of 40 trainees (two officers/38 enlisted) arrived at the Naval Air Station, Lakehurst, New Jersey on 26 October 1940 where they were quartered while in preliminary parachute training. Using towers in nearby Hightstown, the detachment completed tower training on 6 November and moved to the Marine Base, Quantico, Virginia for added physical conditioning prior to making jumps from aircraft.

Concurrently, the commandant indicated that the training of parachutists should proceed "as fast as facilities and personnel are available" in order to train the estimated 750 men in the shortest time. In implementation of this decision, a second detachment of three officers and 44 enlisted personnel underwent tower training at Hightstown from 30 December 1940 to 15 January 1941. On 26 February, both classes graduated together: the first qualified as parachute jumpers and riggers while the second qualified as jumpers. Training of subsequent groups continued throughout the spring, and by July 1941, a total of 225 jumpers had graduated from the Lakehurst course.

Unfortunately, the facilities at Lakehurst and the available towers at Hightstown were inadequate for the demands of the Marine Corps; therefore, Captain Marion L. Dawson was sent to San Diego in February 1941 to prepare additional facilities there. In March, the entire second class together with six riggers from the first transferred to San Diego, and formed the 1st Platoon, Company A, 2d Parachute Battalion. The third class from Lakehurst was subsequently transferred as well, forming the second platoon of Company A.

Company A, of what would become the 1st Parachute Battalion, was formed at Quantico on 28 May 1941, and the Headquarters Company was formed on 10 July. The battalion itself was officially organized effective 15 August 1941 Captain Marcellus J. Howard commanding. On 20 September, Company A of the 2d Parachute Battalion was redesignated Company B of the 1st Parachute Battalion. On 28 September, the battalion moved from Quantico to New River for further training, and there, on 28 March 1942, the battalion was completed with the formation of Company C.

Company B, 2d Parachute Battalion was formed on 23 July 1941 at San Diego, and the battalion organization became effective on 1 October, Captain Charles E. Shepard, Jr., commanding. A new Company A, to replace that detached in September, was formed 7 February 1942, and the battalion went to full strength on 3 September 1942 when Company C was organized.

There was no shortage of volunteers for parachute training although the qualifications were stiff. An applicant was required to be unmarried, athletically inclined, above average in intelligence, 18-32 years of age, and have no physical or mental impairments. Extra pay amounting to $50 per month for enlisted personnel and $100 per month for officers was authorized in June 1941 and surely resulted in increased numbers of volunteers. Nevertheless, the aura of adventure surrounding the parachutists and the promise of action seemed more important to most applicants.

In May 1942, the Parachute School Detachment was formed within the Marine Barracks, Naval Air Station, Lakehurst, but the war had so increased the Marine Corps' training requirements that it was decided to establish two Marine parachute schools. The commandant received approval from the secretary of the Navy to establish Parachute School, Marine Corps Base, San Diego, effective 6 May 1942 and Parachute School, New River, effective 15 June 1941.

Facilities were so tight at San Diego that 10 quarters at Bern Camp were utilized for initial classes. Even so, the school was able to plan on classes of 36 students at an entry rate of one class per week for the 10-week course. The first group entered on 27 May 1942 and was followed by a second on 1 June. This schedule would be maintained until September when the new training facility at Camp Gillespie was ready. Once established at Camp Gillespie, the school standardized its course at 361 hours (six weeks), with a new class entering weekly. The training was in three phases. The first was ground training which included instruction in parachute tactics, map reading, demolitions, techniques of fire, scouting and patrolling, combat swimming and weapons training. Phase two concerned parachute material and included training in parachute packing, flotation gear and cargo containers. The final phase was jump instruction which started with controlled and free tower jumping, covered the use of suspension lines and led to six actual jumps after which parachutist's wings were presented. The course was difficult: a 40 percent wash-out rate was standard.

Organization of the school at New River was beset by difficulties and delays from the beginning. Initial attempts to locate qualified instructors were unsuccessful and finally resulted in a request to the commandant for permission to select personnel from the 1st Parachute Battalion, assigned at the time to the 1st Marine Division, Fleet Marine Force. Greater problems arose concerning facilities. The camp was built from the ground up and although the parachute towers were scheduled for completion by late August, heavy rains and a shortage of contract labor delayed them until 25 September. Similar delays afflicted the parachute building and the training building, neither of which was ready until November. Of a total of 250 parachutes requisitioned, only 50 had arrived by October and most had to be sent on to the Marine Corps Air Station at Cherry Point for repair prior to use.

The New River facility never did attain the size of that at Camp Gillespie although it did turn out 50 graduates per month. Camp Gillespie produced 70 per month, and increased this to 100 per month by early 1943. This increase, coupled with a relatively static demand for parachutists, permitted the closing of the New River Parachute School on 1 July 1943, after which all training was accomplished at Camp Gillespie.

The 1st Parachute Battalion left New River on 7 June 1942, travelling by train to Norfolk, Virginia where it embarked on the USS *Mizar* (AF-12) on 10 June. It sailed via the Panama Canal, arriving at Wellington, New Zealand on 11 July. One week later, the battalion, less a rear echelon, embarked on the USS *Heywood* (AP-12) and sailed to Koro, Fiji Islands where it participated in rehearsals prior to the Guadalcanal landings.

The 1st Parachute Battalion, together with the 1st Raider Battalion and the 2d Battalion, 5th Marines, made up that portion of the Guadalcanal landing force which, under Brigadier General William H. Rupertus (assistant division commander, 1st Marine Division) was scheduled to land in the Florida Island group, across the Sealark Channel from Guadalcanal. The 1st Parachute Battalion, under Major Robert H. Williams, went ashore on Gavutu at H-plus four hours. The delay resulted from a shortage of landing craft—the parachutists had to wait until the raider landings on Tulagi had been completed and those landing craft available. The landing was made in three waves each of company strength. The first landed against light opposition and made limited progress inland. The next two waves met heavier fire and were pinned on the beach until Company B overran enemy positions on and near Hill 148, Gavutu's lone hill. Reinforced by Company A, Company B then captured the remaining Japanese positions atop Hill 148, and by nightfall, the parachutists were engaged in mopping up the island. Shortly after the initial assault, Major Williams was wounded and relieved as Commanding officer by Major Charles A. Miller. On 9 August, two days after the landing, the battalion moved from Gavutu to Tulagi where it took positions as a security force near the government buildings.

On 8 September, the 1st Parachute Battalion and the 1st Raider Battalion, both under Lieutenant Colonel Merritt A. Edson, carried out a raid in the vicinity of Taivu near the village of Tasimboko, Guadalcanal. The raiders landed at Taivu Point and advanced toward Tasimboko, while the parachutists landed some 2,000 yards east of the village and took positions to protect the flank and rear of the raider advance. After an intense fire fight with enemy forces defending the village, the raiders and parachutists entered the base area and destroyed food, medical equipment and military stores left by the fleeing Japanese. By evening of 8 September all units had reembarked and returned to the Kukom landing area.

Several days after this raid, the 1st Parachute Battalion, again in conjunction with the 1st Raider Battalion, was ordered to occupy the ridge southeast of Henderson Field, Guadalcanal. Enemy activity increased starting on the 12 September and reached a peak during the night of 13-14 September when the Marine lines repulsed strong and repeated assaults by determined enemy units. Known afterward as the Battle of Bloody Ridge or Edson's Ridge, this action scattered the Kawaguchi Force against which the earlier Tasimboko raid had been staged.

On 18 September, the 1st Parachute Battalion was withdrawn from Guadalcanal, proceeding by ship to Noumea, New Caledonia, where it was quartered in a temporary camp while constructing a new camp to be named Camp Kiser. After moving into Camp Kiser on 8 November, the battalion picked up a rigorous two-phased training schedule. The parachutists were reindoctrinated in jump techniques, parachute packing and patrolling/scouting. They also received tactical training in the form of platoon, company and battalion problems. This training period lasted into 1943.

The 2d Parachute Battalion, at full strength after the organization of Company C on 3 September 1942, sailed from San Diego on 20 October 1942 and arrived at Wellington, New Zealand on 31 October. The battalion went into camp at Titahi Bay, south of Wellington, until 6 January 1943 when it sailed to Noumea to undergo further training with the 1st Parachute Battalion.

The 3d Parachute Battalion (Major Robert T. Vance) was organized on 16 September 1942. Initially composed of Headquarters Company and Company A, it was assigned to the new 3d Marine Division, then forming in San Diego. Company B and Compant C was activated on 10 December 1942. The battalion was attached to Amphibious Corps, Pacific Fleet, at Camp Elliott on 4 January 1943, and on 5 and 13 March it departed for Noumea in two echelons. All units had arrived and settled in Camp Kiser by 27 March to continue training with the 1st and 2d Parachute Battalions.

On 1 April 1943, the 1st Parachute Regiment was activated at Tontouta composed of the 1st, 2d and 3d Parachute Battalions, Regimental Headquarters and Service Company and Regimental Weapons Company. Companies A, B and C of the 2d Parachute Battalion were redesignated Companies E, F and G, while Companies A, B, and C of the 3d Parachute Battalion were redesignated Companies I, K and L. Lieutenant Colonel Robert H. Williams, recovered from wounds suffered on Gavutu, was appointed regimental commander.

The 4th Parachute Battalion was organized starting 2 April 1943 with the formation of Company B at Camp Elliott. On 1 July, the New River Parachute Training Battalion was reorganized and designated Headquarters Company, Company A and Company C of the 4th Parachute Battalion. These units joined Combat B at Camp Pendleton where the entire battalion (Lieutenant Colonel Marcellus J. Howard) remained in a training status until disbanded on 19 January 1944.

After an extended period of training, the 1st Parachute Regiment started movement in September 1943 to Guadalcanal. On 18 September, the 2d Parachute Battalion with advance elements of the 1st and 3d Parachute Battalions, sailed from Noumea on the USAT *Noordam*, arriving at Tassaforanga on 22 September. The remaining units of the regiment embarked on the USS *American Legion* (APA-17) and sailed from Noumea on 26 September, arriving at Guadalcanal on 28 September. On 29-30 September, the entire regiment reembarked and sailed to Vella Lavella.

Shortly after the arrival of the regiment on Vella Lavella, Lieutenant Colonel Victor H. Krulak, commanding officer of the 2d Parachute Battalion, was summoned to I Marine Amphibious Corps Headquarters on Guadalcanal. He was advised of the impending (1 November) Bougainville landings and ordered to land with a raiding force on the island of Choiseul, there to create as great a disturbance as possible in order to confuse the enemy and to mask the true location of the major assault. Krulak returned to Vella Lavella to map out a plan for the raid and was joined on 24 October by an Australian coastwatcher who supplied last-minute information on enemy forces and locations. Prior to becoming a parachutist, Lieutenant Colonel Krulak had served in Shanghai as a company commander with the 4th Marines and was with the Amphibious Force, Atlantic Fleet from 1941 until he volunteered for parachute training.

The raiding force consisted of Companies E, F and G, reinforced by a communications platoon, a regimental weapons company with eight light machine guns, and a detachment from an experimental rocket platoon. Total strength was 30 officers and 626 men.

Embarked in four high speed transports (USS *McKeon, Gosby, Kilty* and *Ward*) the force left Vella Lavella in the evening of 27 October, escorted by the destroyer USS *Conway* (DD-507). The battalion landed unopposed near Voza and moved inland a mile and a half to set up a base camp. On 28 and 29 October, patrols were sent out to reconnoiter Japanese positions at Sangigai to the south and on the Warrior River to the north.

The attack on Sangigai started at 1100 on 30 October when Company E opened fire on enemy forces in the village. The Japanese defenders quickly retreated toward the mountains, directly into positions prepared by Company F which had executed an enveloping movement through the mountains in order to attack the enemy flank. On the heels of the retreating Japanese, Company E entered the garrison area and destroyed all buildings and facilities, a barge and about 180 tons of supplies. By 0800, the raiding party had returned to its base camp having suffered six dead and 12 wounded while killing at least 75 of the enemy.

On 31 October, a second raiding party (Major Warner T. Bigger) was sent north to Nukiki and then overland to the Warrior River. This group bombarded Japanese installations on nearby Guppy Island with 143 mortar rounds, starting several large fires. After encountering strong enemy resistance near the Warrior River, the party was withdrawn aboard personnel landing craft (LCPs). On 1 and 2 November, Kurlak continued to send out patrols from the base camp and maintained aggressive contact with the enemy. By 3 November, the Japanese were slowly realizing the limited size of the raiding force and started closing in on the beachhead area. The battalion was taken off Choiseul during the night of 3-4 November after laying extensive minefields and booby traps. The proximity of the approaching Japanese was attested to by the sounds of exploding mines as the last parachutists boarded LCIs.

On 22 November, the 1st Parachute Battalion (Major Richard Fagan) embarked 23 officers and 596 enlisted personnel on four infantry landing craft (LCIs) and departed Vella Lavella for Bougainville. It arrived off Empress Augusta Bay the next day and, after going ashore, was attached to the 2d Raider Regiment in I Marine Amphibious Corps Reserve. In 27 November, the 1st Parachute Battalion, with attached units, was designated the task organization for a raid on Japanese supply facilities near Koiari, south of Cape Torokina. In addition to the parachute battalion, the group included Company M of the 3d Raider Battalion and a forward observer team of the 12th Marines.

Major Fagan's force embarked on landing craft at Camp Torokina early in the morning of 29 November and, after a one hour voyage southward, moved in toward the beach at Koiari. It had been intended that the Marines would land a short distance from known Japanese forces and attack from the rear but, immediately after landing, it was discovered that the raiding force had come ashore in the midst of a major enemy supply area. The Marines dug in after forming a hasty perimeter. They were surrounded on three sides by Japanese and had their backs to the ocean. A fierce battle raged for several hours during which the parachute forces were under almost constant fire from mortars and machine guns. By afternoon casualties were mounting and ammunition was nearly exhausted. Shortly before 1800, three destroyers arrived off the beach after having been diverted from escort duty. They opened fire directly to the flanks of the Marine beachhead while 155mm guns from Cape Torokina fired parallel to the shore. Thus, protected by a three-sided box of fire, the battalion was able to board rescue boats which dashed in to the beach. Total Marine casualties were 15 killed, 99 wounded and seven missing of a landing force which had numbered 24 officers and 505 enlisted personnel.

On 3 December, the 1st Battalion on Bougainville was joined by the rest of the 1st Parachute Regiment (less the 2d Battalion which stayed in camp at Vella Lavella). Two days later the regiment was sent to the 3d Marine Division front, and for the next four days fought off enemy patrols, snipers and ambushes. On 10 December the force was relieved by elements of the 9th and 21st marines and moved to the 9th Marines regimental reserve position. Later, on 22 December, the 1st Parachute Battalion, Weapons Company and a Headquarters and Service Platoon were attached to the 2d Raider Regiment and relieved the 1st Battalion, 3d Marines near Eagle Creek. The unit remained there, generally strengthening the positions, until it was in turn relieved by elements of the 132d Infantry Regiment, Americal Division on 11 January 1944.

This would be the final combat role for the 1st Parachute Regiment. The 2d Parachute Battalion moved from Vella Lavella to Guadalcanal on 2 January 1944 and remained there until 18 January when it embarked for San Diego. On 15 January elements of the regiment on Bougainville started embarkation, also destined for San Diego and reorganization.

The disbandment of the parachute units had been under study for many months at Headquarters Marine Corps. The requirement for trained parachutists had been met fully by Spring 1943 and training quotas were revised and lowered. Earlier the weekly student assignment quota at the Parachute School, Camp Gillespie, had been reduced from 70 to 50 and in mid-April it was further reduced to 30. Later in April, the commandant determined that the Marine Corps then had enough trained parachutists to meet any current or projected requirement. It was suggested that the New River Parachute School be disbanded, an action which subsequently was taken when, on 1 July 1943, the school was closed and the personnel used to form the 4th Parachute Battalion. By the fall of 1943 it became apparent that abandonment of the parachute program would release some 3,000 troops and save $150,000 monthly in payments to parachutists who were unable to jump for lack of aircraft. On 30 December 1943, the commandant ordered the disbandment of all parachute forces. The 1st Parachute Regiment, minus Air Delivery Section, was assigned to Fleet Marine Force, San Diego where it would cadre the then-forming 5th Marine Division. The reorganization was effective on 29 February 1944.

There were no combat jumps by Marines during World War II. During the summer and fall of 1943, consideration had been given to drops at Kolombangara and Kavieng, but plans were never developed. This failure to utilize the parachutists was the result of four factors, three of which were peculiar to the Pacific war. First was a lack of sufficient airlift capability—at no time did the Marines have resources to air-lift more than one battalion (i.e., six transport squadrons). Next, land-based staging areas were not available for mass flights. A third factor was the long distances involved, both from the Continental U.S. and among the island targets. Finally, the objectives assigned to the Marine Corps were generally small, densely defended islands unsuited for large-scale parachute operations. Although determined to be a "luxury" which the Marine Corps could not afford, the parachute units had made noteworthy contributions to the tradition to the Corps, and the esprit and high state of professionalism inculcated in their personnel would be apparent one year later when the 5th Marine Division went ashore on Iwo Jima.

Commanding Officers, Marine Parachute Troops

1st Parachute Regiment
Lt Col Robert H. Williams, 1 April 1943 - 15 Jan 1944; Maj Richard Fagan, 16 Jan 1944 - 29 Feb 1944.

1st Parachute Battalion
Capt Marcellus J. Howard, 15 August 1941-1 Oct 1941; Capt (Maj) Robert H. Williams, 2 Oct 1941 - 7 August 1942; Maj Charles A. Miller, 8 Aug 1942 - 17 Sept 1942; Capt Harry L. Torgerson, 18 Sept 1942 - 27 Sept 1942; Lt Col Robert H. Williams, 27 Sept 1942 - 31 Mar 1943; Maj Brooke H. Hatch, 1 Apr 1943 - 27 Apr 1943; Maj Robert C. McDonough, 28 Apr 1943 - 9 May 1943; Maj Richard Fagan, 10 May 1943 - 10 Jan 1944; Maj Robert C. McDonough, 11 Jan 1944 - 29 Feb 1944

2d Parachute Battalion
Capt (Maj) Charles E. Shepard, Jr., 1 Oct 1941 - 4 May 1943; Maj (Lt Col) Richard W. Hayward, 5 May 1942- 31 Mar 1943; Lt Col Victor H. Krulak, 1 Apr 1943 - 7 Nov 1943; Maj Warner T. Bigger, 8 Nov 1943 - 29 Feb 1944.

3d Parachute Battalion
Maj Robert T. Vance, 16 Sept 1942 - 9 Dec 1943; Maj Harry L Torgerson, 10 Dec 1943 - 29 Feb 1944.

4th Parachute Battalion
Lt Col Marcellus J. Howard, 1 Jul 1943 - 30 Sept 1943; Maj Tom M. Trotti, 1 Oct 1943 - 19 Jan 1944.

Parachute Battle Honors
1st Parachute Regiment
Vella Lavella Occupation, 4-16 Oct 1943; Choiseul Island Diversion, 28 Oct 1943 - 4 Nov 1943; Occupation and Defense of Cape Torokina, 4-15 Dec 1943; Consolidation of Northern Solomons, 15 Dec 1943 - 12 Jan 1944.

1st Parachute Battalion
Guadalcanal-Tulagi landings, 7-9 Aug 1942; Capture and Defense of Guadalcanal, 10 August 1942 - 18 Sept 1942; Vella Lavella Occupation, 4-16 Oct 1943; Occupation and Defense of Cape Torokina, 23 Nov 1943 - 15 Dec 1943; Consolidation of Northern Solomons, 15 Dec 1943 - 12 Jan 1944; Presidential Unit Citation (attached to 1st Marine Division) Solomon Islands, 7 Aug 1942 - 18 Sept 1942.

2d Parachute Battalion
Vella Lavella Occupation, 1 Sept 1943 - 16 Oct 1943; Choiseul Island Diversion, 28 Oct 1943 - 4 Nov 1943.

3d Parachute Battalion
Vella Lavella Occupation, 7-16 Oct 1943; Occupation and Defense of Cape Torokina, 4-15 Dec 1943; Consolidation of Northern Solomons, 15 Dec 1943 - 12 Jan 1944.

4th Parachute Battalion
None—in training status until disbanded.

U.S. Marine paratroopers with their weapons on Vella La Vella Island, 1944.

AIRBORNE TROOP CARRIER COMMAND

By Martin Wolfe

INTRODUCTION

By Colonel Vito S. Pedone, USAF (Retired)

Any documentation on the "**U.S.A. Airborne Fiftieth Anniversary**" must include the participation of the Troop Carrier Units. Certainly the C-47 aircraft is the catalyst that brought the parachutist, the glidermen and the aircrews together to plan and to execute a wide variety of airborne operations in all theaters of operation during World War II.

While the D-Day invasion and the Holland operation received a great deal of publicity, there were other airborne operations such as Nadzab, New Guinea, Corregidor and Laguna de Bay in the Philippines that also played a major role toward victory in the Pacific Theater.

One must also acknowledge the special operations of the Air Command Groups and the Combat Cargo Groups in dropping special airborne units and supplies and in aiding medical evacuation in the China-Burma-India Theater. This section describes the contribution the men of the Troop Carrier made toward the overall success of airborne operations during World War II.

THE AIRBORNE PLATFORM

"The Plane That Changed the World"—that was the title of a TV show in 1985, celebrating the 50th anniversary of the beginnings of commercial aviation. The inaugural flight of the wonderful DC-3, the first profitable airliner and the most spectacularly successful plane ever built, came in 1935. Commercial aviation did change the world—in effect, shrinking time and space, providing ever-increasing numbers of people with easy, long-distance travel.

That same DC-3, given a more rugged floor, wider doors, a bit more power and some electronic and engineering gadgetry, became the Troop Carrier Command's great C-47, the plane that also changed the world by helping liberate it from the most dangerous tyrannies ever known. Without their solid confidence in the abilities of the C-47, our military chiefs never would have authorized the immense numbers of specialized troop carrier pilots, navigators, glider pilots, crew chiefs, radio operators, ground engineering and communication technicians and all the other combat and non-combat specialties needed to man fleets of C-47s in airborne assaults against Hitler's Europe.

Some military historians argue that the German 88mm cannon was the best weapon of World War II; but we troop carrier veterans feel that honor belongs to the C-47. This sweet-flying, rugged, dependable ship was a prime feature of the American "Arsenal of Democracy." Everyone knows the C-47 provided the main platform for the "point of the lance"—our paratroopers—for every major invasion of Hitler's Europe. But also—given a hook in its tail—the C-47 was able to pull fully loaded assault gliders—gliders almost as big as itself.

It was the C-47 that accomplished spectacular resupply missions such as those at Bastogne and several famous paratrooper and glider operations in the Philippines, the South Pacific and Burma. Put to work lugging jerricans of gasoline, transporting ammunition, jeeps and small artillery pieces, moving whole regiments from England to France or mules from one point on the Burma Road to another—there didn't seem to be any task this remarkable plane couldn't accomplish. Maybe some airplane other that the C-47 might have provided the scaffolding on which the Troop Carrier Command was structured; but we Troop Carrier veterans find this hard to believe. "Our plane" is so tightly wrapped up in our thoughts with all we did back then, it is impossible for us to imagine the World War II airborne operations without the C-47.

The Air Transport Command, like the Troop Carrier Command, also flew C-47s—but did not carry paratroopers or gliders into combat. Today this makes for an "image problem" for troop carrier vets. We would like everyone to remember that, unlike the Air Transport Command, our crews had to fly low and slow over enemy-held territory in unarmed, unarmored planes, without self-sealing gas tanks often in situations where a single rifle bullet could bring us down. Most Troop Carrier outfits, it is true, took fewer casualties than those in fighters or bombers; but we too, had to watch while our buddies in near-by planes went down in flames.

"Vertical envelopment": this was by far the most important reason for America's creation of the Troop Carrier Command. Placing large numbers of troopers in "airheads" behind enemy lines was the reason we were separated from the ATC in 1942 and given a separate chain of command within the Army Air Corps. The Twelfth Air Force, in North Africa for the invasion of Italy, got the XII Troop Carrier Command; the Ninth AF in Britain got the IX TCC, and the Tenth; Thirteenth and Fourteenth Air Forces in China, the CBI and the South Pacific also got troop carrier components.

When our military leaders in Washington and London became convinced that the invasion of Hitler's "Fortress Europe" (and perhaps, later, Hirohito's Japan) would demand entire divisions of airborne soldiers, they pushed the growth of Troop Carrier by spectacular rates. In the ETO alone, by the time IX TCC was at full strength, it was flying more than 2,000 of those wonderful C-47s, and its total combat and non-combat personnel was around 25,000.

However, the first few combat operations of Troop Carrier, taken together, disappointed airborne enthusiasts who were promising spectacular military successes through "vertical envelopment." In the invasions of North Africa, Sicily and Italy between November 1942 and August 1943, poor planning, bad weather and inexperienced power and glider pilots made for only partial successes and some outright disasters. Some of these disasters—like the tragic episode at Sicily on July 11, 1943, when our own Navy (and some of our shore artillery) shot down 27 of our Troop Carrier planes, six of which still had paratroopers inside—could

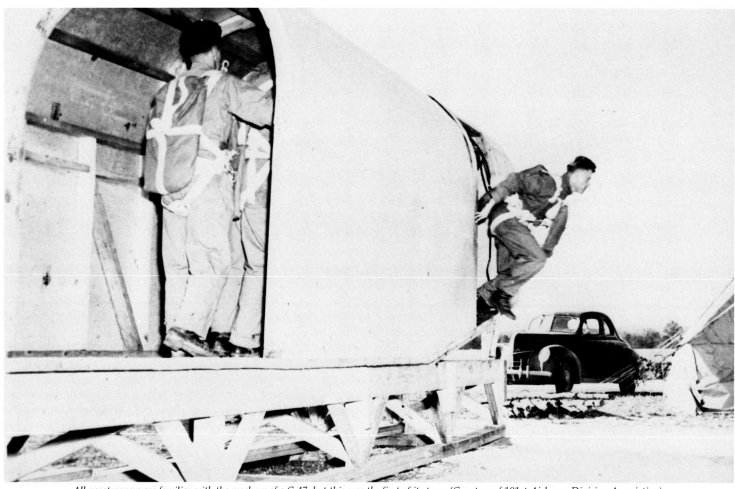
All paratroopers are familiar with the mock-up of a C-47, but this was the first of its type. (Courtesy of 101st Airborne Division Association)

hardly be blamed on Troop Carrier. But for several months, our top generals debated whether they should abandon hopes for "vertical envelopment." Would paratroopers be kept on the ground as elite infantry? And would Troop Carrier be restricted to freighting missions, plus (possibly) a few small-scale commando-type airborne raids?

Fortunately, Troop Carrier units, then training in the U.S., under close scrutiny from Washington, were demonstrating increasingly superior tactical skills in handling huge formations and massed drops and also were absorbing the hard-won lessons of every airborne operation in the Mediterranean area. When our top commanders in Europe, contemplating the up-coming invasion of France, decided that they could not crack Hitler's "Atlantic Wall" without airborne soldiers operating behind the German shore defences, the die was cast. Troop Carrier units in the States were authorized to continue expansion and to get ready for the invasion of Normandy.

Operation NEPTUNE, the invasion of Normandy, called for "vertical envelopment" by three airborne divisions: our 82nd and 101st Airborne Divisions and the British 6th Airborne. Casualty predictions were frightening: Eisenhower's air chief estimated that 50 percent of the Troop Carrier planes and 75 percent of the gliders, and all their troopers, would be shot down. These predictions were wrong but the very first drops of the 82nd and 101st troopers, in the early morning of June 6, were not brilliant successes.

Many Troop Carrier planes dropped their sticks far off target, and tragically, some troopers were released over the English Channel and drowned. An unreported cloud bank lying across the Cotentin peninsula had broken up many of the tight "V-of-Vs" Troop Carrier formations, and some rattled pilots had become further confused by streams of ground fire tracers arcing up at them in that menacing night sky. Even so, the majority of the paratroopers were set down close enough to their DZs so that they were able to reach most of their unit goals. The rest is history—they ripped German communications to shreds, thus totally confusing Hitler as to Allied objectives.

Unfortunately for our Troop Carrier self-image, that far-from-perfect drop in the early morning of D-Day—coming after the disappointing missions in Sicily and Italy—tended to color the evaluation of all Troop Carrier's performances throughout the rest of the war. Veterans of other outfits, and many historians, have assumed that all Troop Carrier operations were characterized by the bad scattering of paratroopers and glidermen. But this is completely wrong. The very next large-scale airborne operation, the glider missions into Normandy that same day (the evening of D-Day), was executed extremely well. So were the remaining smaller Troop Carrier operations during the rest of the Normandy invasion. Furthermore, later Troop Carrier operations—especially the large-scale invasion of Southern France, Holland and Germany—were nearly perfect and were often characterized by airborne commanders as "parade-ground jumps."

Another reason Troop Carrier has an "image problem" is that its responsibilities were so complex and so different from anything the U.S. had ever attempted that even back in 1943-45, any discussion of airborne warfare in newspapers or magazines had to be preceded by an explanation of "vertical envelopment." This remains a problem even today, especially concerning glider missions. The word "glider," to most people, calls up images of a sport craft, not an assault weapon. Troop Carrier veterans often encounter surprise—even disbelief—when they explain that in World War II hundreds of assault gliders, most often with a dozen or more airborne soldiers in each glider, were used in several massive Allied invasions of Europe and some smaller operations in the CBI. Assault gliders—the American "Wacos" (CG 4As) and the British "Horsas" and "Hamilcars"—are the forgotten machines of World War II.

Troop Carrier outfits were organized along strict Army Air Forces lines into commands, wings, groups and squadrons. But aside from combat missions and time spent in combat training, Troop Carrier squadrons remained pretty much on their own. Each squadron was run like a small airline with its own personnel, engineering, communications and managerial functions. Squadrons kept their own squadron records, and they usually ate in squadron messes, had their own quartermaster and tech supply sections, and—in all but higher-echelon engineering—did their own repairs, maintenance and inspections. Each of our section chiefs, in effect, was his own boss. For 19 out of 20 of us, therefore, the only "headquarters" we knew were our squadron orderly rooms and operations offices. Apart from combat we had little to do with group, wing or command HQ. This is the way we liked it, and this system worked.

The 60 highly-trained Troop Carrier squadrons we had in the ETO by June 1944—with their diverse and demanding assignments—and the 36 squadrons in the MTO, the CBI and the Pacific, are spectacular proofs of what organizational miracles America could perform when she had to. Starting from scratch—with no airborne theory, let alone experience—our country was able to turn out a seemingly endless stream of soldier technicians who could pull 100-hour inspections on those Pratt & Whitney engines, or take the gremlins out of complicated radio and radar sets, or patch a C-47 so riddled with flak and bullet holes its skin looked like Swiss cheese—and get that ship ready to fly by 5:30 a.m. the next day.

For the cutting edge of our squadrons, we had literally thousands of extremely capable pilots; and though most of them had never had a plane ride before 1942, let alone any actual piloting experience, they could fly good "V-of-Vs" formations (at night, if required) with their wingtips practically touching those of the nearest planes. By the summer of 1944, our power pilots had become so profitable they could land an overloaded C-47 in a small cow pasture and could take off and fly a good pattern with two Waco gliders hanging on their plane's tail.

The astonishing versatility provided by Troop Carrier C-47s was dramatized in our next large operation. The bulk of Troop Carrier's squadrons in the MTO (which had been in North Africa and Italy) had been flown to England back in early 1944 to participate in the invasion of Normandy. Then, a month after Normandy—with the invasion of France going along fairly well—an entire armada of Troop Carrier planes (half the strength of many ETO groups) was flown to Central Italy with stops at Gibralter or Algeria. This was an immense logistical accomplishment. Plucked out of England, all these Troop Carrier units were set down in Italy two days later, intact and ready for anything.

Together with planes which had remained in the MTO, and with shiploads of WACO gliders hastily delivered to the docks at Naples, the ETO contingents were put under a command apparatus dubbed the "Provisional Troop Carrier Air Division."

To our great surprise and relief, Allied mastery of the skies for this invasion was almost complete and Nazi resistance on the ground was relatively weak—nothing like Normandy. For most of us in the C-47 crews, those Southern France missions were milk runs, with virtually no crew casualties. In the books, this invasion of Southern France (August 1944) is called "The Champagne Campaign," partly because of the riotous welcome the delighted French on the Riviera gave our airborne troopers and glider pilots.

Our paratrooper drops and glider missions in Southern France (part of Operation DRAGOON) were rated as some of the most accurate of the war. In spite of a dense ground fog covering most of the target areas, only a few planes dropped their sticks far off the DZs. It was a different story for the glider landings. These were even more accurate but much more costly. Airborne troopers and glider pilots on Operation DRAGOON were hurt more because of some hectic traffic jams during landings than because of ground fire.

At relatively small total cost and thanks to what has been called "a military administrative miracle" (assembling so quickly all the components of PTCAD), Operation DRAGOON provided a powerful head start for the liberation of the crucial ports of Marseilles and Toulon and for the Seventh Army's rapid invasion up through the Rhone Valley toward Eastern France and Germany.

It was Troop Carrier's very versatility that made its next two major assignments so dangerously controversial. When American infantry and armored divisions broke through the German lines in Normandy and began racing toward Paris and Eastern France, they found themselves outrunning their supply columns. Tanks and personnel trucks were left stranded without gasoline. For our commanders this was especially galling, since the overall military situation seemed to show that little effective Nazi resistance stood between the Allies and the German border.

Troop Carrier C-47s were thrown into the breach. Our planes were loaded down with five-gallon jerricans of gas or drums of diesel or ammunition, and then we would fly from England to designated dump sites in France or Belgium. Because of the C-47's famous ability to land practically anywhere, these gas and ammo dumps could be in any convenient and decently level meadow. Often they were ahead of our infantry advances. On the return trip we would be given wounded personnel to transport back to British hospitals. This meant rigging our planes with litters and picking up flight nurses and med-evac technicians so that we could fly back as hospital ships. We were working harder than ever, sometimes making two flights a day to the Continent.

General Patton, especially, was clamoring for more gasoline. Each and every C-47, he thought, should be turned into a freighter to supply his rapidly-advancing tanks. Several top American commanders urgently backed his demands.

But our airborne division commanders, egged on by British General Montgomery and supported by Chief of Staff George Marshall, were pressing for additional "vertical envelopments" in back of the fleeing Germans. This they saw, was the way to "end the war in '44!"

Even before our planes which had been involved in Southern France could make their way back to England, Troop Carrier was repeatedly put on alert for drops east of the retreating German lines. And each time one of these alerts came down from SHAEF, we would have to stop freighting supplies and begin lengthy preparations for an airborne assault. Sometimes such preparations required practice formations to allow us to test planned flight patterns. Often we would carry along a token number of airborne troopers in our planes and tug a few gliders to try out a projected air assault operation. This, of course, took time away from our ability to mount freighting operations. Days would go by when little or no gas would be delivered by air to Patton.

These invasion alerts proved terribly frustrating for troop carrier personnel and for the airborne units we were slated to deliver. Some of the alerts were pushed forward to the point where the troopers were lined up in front of our planes, ready to board. Then each and every one of them were cancelled. The reason was that while we were still readying for an air assault our plunging armor units would get so near the designated DZs that the assault alert would be cancelled as useless. Then we in Troop Carrier, as soon as possible– but always with costly delays—would get back into the freighting business.

We were being yanked back and forth by commanders with entirely different concepts of how we should be used. Patton was furious, but Eisenhower, after some agonizing debate, came down on the side of those who wanted another airborne assault some time in late August or September 1944. Thus was born the ill-fated invasion of Holland, Operation MARKET-GARDEN, known in movies and books as the campaign that called for going "one bridge too far."

In the midst of these debates over whether troop carrier would be fighters or freighters, SHAEF pulled the wraps off plans for an entirely new sort of command: the First Allied Airborne Army (FAAA). This unprecedented military formation was intended to consolidate tactical control over both airborne infantry and troop carrier units. All the American and British airborne divisions in the ETO and all the American and British combat troop transport planes based in Britain were placed under the control of FAAA. Finally, it seemed, a rational command structure was forged that would weld the separate components needed for "vertical envelopment." To demonstrate that we in Troop Carrier were no longer under the control of the Ninth Air Force, we now sported a shoulder patch with "The Big White 1"—the FAAA—under the words "Allied Airborne."

General Montgomery's scheme for an air assault in back of German lines (Operation MARKET-GARDEN) was entirely original and very, very daring. It encompassed two dove-tailed operations. Operation MARKET was to be a kind of "multiple vertical envelopment," one each near the towns of Eindhoven, Nijmegen and Arnhem, with the three separate airborne objectives to be won on the first day of the invasion. These towns were on a line running north from the Belgian-Dutch border to the Lower Rhine River. Part of the Dutch town of Arnhem lies north and east of the Rhine.

Operation GARDEN called for ground troops to smash north and connect up these three centers of what Montgomery called MARKET's "airborne carpet." Then massive Allied infantry and tank forces could pour this "carpet" into north Germany. From there they could sweep south and encircle the industrial heartland of the enemy. Montgomery assured SHAEF that this would bring the already faltering German war machine to its knees.

But against all expectations, the German commanders around Arnhem performed miracles of military improvisation. They pulled their weary troops together and fought a brilliant defensive battle against the three Allied "airheads." They were able to prevail partly because the British ground forces which were given the task of forging the three "airheads" into one consolidated salient, advanced up the invasion route with agonizing caution. This allowed the Germans to mount increasingly stronger attacks against the hard-pressed 101st and 82nd Airborne troopers around Eindhoven and Nijmegen. Meanwhile, the British 1st Airborne Division was not able to consolidate its hold on Arnhem bridge partly because the British high command had decided to drop their troopers too far from that town.

C-47s and gliders marshalled for Varsity Mission across the Rhine. 95th Troop Carrier Sqn., 440th Troop Carrier Group, France 1945. (Courtesy of George Buckley)

What turned out to be the final blow came when, on the third day of the bloody fighting in Holland—with the British still only inching their way up what was becoming known as "Hell's Highway," the weather turned bad. Impenetrable clouds cut down troop carrier's operations severely, and several ships that tried to bull their way through that soup to resupply the hard-pressed airborne units became hopelessly lost and had to return to England with troopers or supplies still inside or gliders still on their tail.

Airborne troopers who participated in MARKET-GARDEN, and military historians who have examined it, agree that none of the reasons for the defeat can be attributed to either the airborne forces or the troop carrier units. The American troop carrier units gave our airborne near-perfect drops and glider tows. General Matthew Ridgway, who was flying overhead in a bomber as an observer near Nijmegen, said:

"The drop was perfect, the best we'd ever done. Despite the fact that planes were being lost to AA fire, those magnificent pilots of the 52nd Troop Carried Wing held formation perfectly, and hit their drop zones on the nose."

In the Eindhovven area, some of the drops of 101st Division units were so compact the troopers didn't have to go through an SOP assembly ("rolling up the stick"). They were able to grab their equipment and go off immediately to their unit objectives.

When all is said, it is not the monumental size nor the operational intricacies of MARKET that linger longest in the memory. It is the heroism of the men who flew burning, disintegrating plans over their zones as coolly as if on review and gave their lives to get the last trooper out, the last bundle dropped. It is the stubborn courage of the airborne troops who would not surrender though an army came against them. In the sense that both troop carrier crews and airborne troops did all that men could do, there was, as General Gavin said, no failure in MARKET. (John Warren, Airborne Operations in World War II, European Theater, p.155.)

Between the end of September and the last week of December 1944, Troop Carrier in the ETO was taken off combat status and limited to freighting personnel and supplies. A few of our groups moved from England to France to get ready for the expected invasion of Germany. Some of us began to train with another great American airborne division now in the ETO, the 17th Airborne Division. With the start of the German offensive in the Ardennes, this entire division was airlifted to France just before Christmas and rushed on line.

During the last week of December we had the intense satisfaction of being part of the drop of arms and supplies to the surrounded 101st troopers at Bastogne—missions that earned us a bit of credit for helping win the Battle of the Bulge, the German counteroffensive into Belgium. This time, even the severest critics of Troop Carrier had to admit that "flak was heavy, but not a single plane took evasive action." A plane in one outfit, the 81st Troop Carrier Squadron, racked up a record for that unit, coming back to base with 90 flak or bullet holes. General McAuliffe, commanding the 101st at Bastogne, conveyed to General Paul Williams, chief of IX Troop Carrier Command:

(The) admiration of all of us in the 101st Airborne for the grand job of air re-supply you furnished us during the siege of Bastogne....Despite intense flak, the much-needed ammunition and medical supplies were dropped just where we wanted them. Needless to say, Bastogne could not have been held without this excellent support.

The very last combat assignment of ETO Troop Carrier (March 24, 1945) was Operation VARSITY, the invasion of Northern Germany just across the Rhine, near the little town of Wesel. This operation was the greatest single airborne mission ever attempted and, in all probability, the greatest that the world will ever see. It was based on an immensely complex plan. In order to put the maximum number of troopers into one area as quickly as possible, 1,500 planes and 1,300 gliders were sent into the air on the same mission. It was a vast air armada, stretching more than 200 miles from tip to tail and it was dangerously packed with multiple streams of aircraft from side to side—an immense traffic problem. Practically anything that could carry troopers or tug gliders it seemed, was up there. There were converted British bombers as well as American and British transports. Many of the C-47s were tugging two gliders apiece, and the British were deploying their huge Hamilcar as well as their large Horsa gliders. And for the first time in European combat, VARSITY was using the C-46 Curtis Commando transport, already so successful in flights across "the Hump" (the Himalayas) between India and China.

This immense mission succeeded in catapulting some 22,000 troopers and glidermen into the German homeland within a few hours. Our American armor was already across the Rhine in a few salients south of the VARSITY area. There was little the Germans on the western front could do but run or surrender.

Casualties in VARSITY for troop carrier crews and glider pilots were heavy—the worst of the war, for many of our outfits. The C-46 squadrons especially, were hurt very badly. Their planes not equipped with self-sealing full tanks, proved too vulnerable to incendiary tracer bullets. Ever since, in troop carrier histories, the C-46 has been dubbed "the flaming coffin." Nineteen of the 72 C-46s assigned to the 313th Troop Carrier Group were shot down, fourteen of them in flames.

Glider pilots in VARSITY, fighting in support of the airborne troopers after they got their gliders down on the ground, were showered with praise for their brave performance. And we should never forget the large number of Troop Carrier pilots in this operation who sacrificed themselves and their planes in order to get their airborne troopers to the right DZs or LZs. Soldiers already on the ground stared up in awed disbelief as plane after plane, desperately hurt, struggled to get their paratroopers or gliders to their targets and then spun into the ground.

Summarizing VARSITY after the war, top airborne and troop carrier generals agreed that this mission had been performed with text-book perfection. General Gavin called VARSITY "the highest state of development attained by troop carrier and airborne units."

One reason for the excellent performance in VARSITY was Troop Carrier's ability to learn from experience. For example, we had learned that running an operation over several days, as we had in the Holland invasion (Operation MARKET), left you vulnerable to devastating weather changes. VARSITY, therefore, was set up as a one-day operation. And we learned way back during the invasion of Normandy that all our training for dropping troopers at night—aimed at giving them (and us) as much protection as possible—was entirely wrong. Since the Allies had control of the air, airborne assaults stood to lose far less when launched in daytime.

Some Army Air Corps units wound down their operations very quickly in March 1945, once German resistance collapsed. There was little need, now, for Allied bombers and fighters. But we in Troop Carrier were busier than ever. The end of the Nazi empire left hundreds of thousands of starved, sick, and bewildered victims stranded in Germany,

Austria and Poland. These were the concentration camp inmates and "displaced persons"—mostly French, Polish, Russian, Belgian and Dutch—who had been taken from their homelands by the Germans and were now in desperate need of care and transportation back home. Troop Carrier was put to work helping to empty the slave labor settlements, the concentration camps and the prisoner of war camps. It was hard, tiring, and emotionally exhausting work. But our reward was the realization that now we were not in the business of helping kill the enemy. We were coping with the task of helping victims of Nazi crimes against humanity. The road and railroads of Continental Europe were a shambles. We in our C-47s provided a quick and relatively comfortable means of getting these people to hospitals, to rehabilitation centers or to transfer centers as close as possible to their homes. Many of us feel that this work, which took almost all of April and May 1945, was Troop Carrier's finest hour.

In July and August of 1944, most of Troop Carrier in the ETO were shipped or flown back to the States. We thoroughly expected that after a 30-day furlough most of us—especially those without a lot of points—would be sent to the Pacific for the final onslaught against Japan's main islands. But of course, the atom bomb changed those plans.

By the end of 1945, it was already becoming clear that Troop Carrier was not destined to long survive the war. Troop Carrier had been created to respond to the particular needs and opportunities of World War II. Even before the Korean War began (1950), our military chiefs could see that "vertical envelopment" was a tactic whose time had passed. The French, it is true, later tried to keep their hold on North Vietnam through use of troop carrier-style paratrooper operations and through re-supply of their soldiers surrounded at Dien Bien Phu (with the C-47s we had given them), but they failed. Most importantly, it was clear that military technology had advanced to the point where massed flights of slow transports and gliders would be easy targets for the new ground-to-air missiles.

Our Military Airlift Command—in its huge new cargo planes that seemed so gigantic to World War II veterans—will continue to transport and airland troops but not to "airheads" behind enemy lines. For small-scale airborne operations, the military helicopter soon replaced all fixed-wing aircraft, though C-47s continued to serve very usefully in many capacities all during the Korean and Vietnam wars. Rapid Deployment Units certainly will use paratroopers in the future but probably only in commando-type raids in "brush-fire" wars. And, most certainly, the day of assault gliders is over.

Never again will "sky trains" of massed hundreds of C-47s and gliders, carrying whole regiments of paratroopers and glidermen, play a part in our country's armed struggles. But Troop Carrier veterans can feel the pride that comes with knowing that we were given a crucial, complex and demanding job—and did it well.

C-47 work horse of Troop Carriers - WW II.

AIRBORNE PATHFINDERS

By Dennis M. Davies

Known from the early days of the war was the priority requiring an accurate way of delivering airborne troops to their proper drop and landing zones. The Allies had been working on development of tactics and equipment that could be used to eliminate these problems.

One of the better ideas to come out of this partnership was the Rebecca-Eureka radar transmitter/receiver. Eureka was the portable transmitter that was carried by the paratrooper to the drop zone. After assembly, the set would send a preset signal that would be received by the Rebecca set in the lead aircraft of the group heading for that particular drop zone. This system was an English design and was manufactured in the United States.

This and other numerous developments greatly enhanced the chances of large numbers of aircraft finding their respective drop and landing zones. This would save much needed time in assembling the troops, especially on night-time operations.

The Eureka-Rebecca system was still in the experimental stages when it was decided it would be used on the North African invasion, Operation TORCH. A Eureka set had been smuggled into Algeria to help guide the planes carrying the 2nd Battalion of the 509th Parachute Infantry. Their targets were to be the airfields at La Senia and Tafarroui.

The planes took off from Lands End, England on the evening of 7 November 1942. It was to be a 10-hour flight that took them over Spain. Unfortunately, the experiment did not work. Because of unrelated navigational problems, the planes crossed the African coast farther west over Morocco and did not come within range of the Eureka signal. Since the paratroopers did not arrive on schedule, the operator destroyed the set and went into hiding until friendly troops arrived.

On 14 September 1942, the 509th was again having problems with the Eureka signal. This time its target was to be the area around Avellino, Italy. The high hills in the area interfered with the signal making it difficult for the Rebeccas to receive it. Only 10 of 40 aircraft found the drop zone.

The first tactical use of Pathfinders was on 13 September 1943. The 504th Parachute Infantry of the 82nd Airborne Division jumped near Salerno, Italy to help support the beach landings. This was led by a Pathfinder Team from the 504th. This was the only such unit in the division. The following night the 505th Parachute Infantry jumped on the same drop zone guided by its Eurekas. Both of these jumps were made with a fair amount of success.

In June 1944, 26 504th Pathfinders (with two in reserve) volunteered for the Normandy jump. These troops were disbursed into the teams from the 507th and 508th Regiments the reason being that these units lacked combat experience at the time. This knowledge was a great asset to these teams.

The commanders of both the troop carrier and airborne units decided to form a combined service organization to carry out future pathfinder operations. In early 1944, a new unit was formed in England under command of the U.S. 9th Air Force. This new unit was designated the 9th Troop Carrier Pathfinder Group (Provisional). The group was to be commanded by then Lt. Col. Joel L. Crouch, a well-seasoned C-47 pilot.

Many hours were spent training on navigational problems, night parachute jumps and working with the equipment that was new to most of the troops. By the middle of May 1944, the group was finished with its training and was waiting for its first mission. It was at first believed that the group would be disbanded after the upcoming invasion of Europe, but the near future would show the need for a more permanent unit such as this.

It was not long of a wait before the invasion became a reality. On the evening of June 5, 1944, Lieutenant Colonel Crouch was at the controls of the C-47 leading the pathfinder group to its drop zones on the Cotentin Peninsula. Aboard this aircraft was Captain Frank Lillyman of the 502nd Regiment, 101st Airborne Division. Capt. Lillyman was in command of the 101st Pathfinders and, at approximately 12:15 a.m., jumped with his team on to Drop Zone A just east of Ste. Mere Eglise. It is believed that Capt. Lillyman was the first American Paratrooper to land on French soil on D-Day.

After the Normandy Campaign, the 9th Troop Carrier Pathfinder Group participated in the Southern France and Holland combat airborne operations. Two plane loads of pathfinders led by Col. Crouch were parachuted into the surrounded town of Bastogne, Belgium that was being held by the 101st Airborne Division. The purpose was to guide the troop carrier aircraft carrying much needed supples and ammunition to the hard fighting but weary troopers.

On March 24, 1945, the groups' pathfinders were set up on the American held side of the Rhine River to help guide the 17th Airborne Division to its drop and landing zones near Wesel, Germany. This was the last large scale airborne operation in the European Theatre, so with the fighting coming to an end, the need for the large unit ceased to exist. The 9th Troop Carrier Pathfinder Group (Provisional) was disbanded around the end of May 1945, and its personnel returned to their parent units.

This was the end of an elite group, but not the end of the pathfinders as a part of the airborne and air-assault units throughout the Army. The U.S. Air Force has Combat Control Teams that also work in a pathfinder-type function for air-to-ground communications. These units draw their lineage from these elite World War II groups.

The first use of pathfinder insignia came to life in May 1944. The involved troops were volunteers and professionals, so it was decided that a special insignia was warranted. One of the aircraft navigators Lieutenant Prescott, was asked to help with its design. He was well-known around the group to be a very good artist.

The final design was shown to Col. Crouch and approved for use. Thus, the flaming winged torch came to be the mark

Members of the 101st Pathfinders exit a Huey helicopter from Company C, 158th Aviation Battalion, 101st Airborne Division (Airmobile) onto a Helicopter Tree landing Platform at Campbell Army Airfield. (U.S. Army photo by PVT Jeffrey W. Burkey)

of distinction of the American Pathfinders. Lt. Prescott took the design to an insignia manufacturer in London and had the first lot produced.

The first issue of the patch was on 5 June 1944, just in time for the invasion of Normandy. Some personnel managed to have the patches sewn on their uniforms so they could be worn during the invasion. This was not liked by some of the senior officers because, if the person wearing it was captured, German intelligence would want to know its meaning. Since the pathfinder group was a fairly new and secretive unit, they felt the insignia would draw unwanted attention. It was decided that the insignia would be worn only on dress uniforms on the lower left sleeve. There may have been individual cases of them being worn in the field, but there is doubt it was done with authorization.

Both the paratroopers and aircrews alike were entitled to wear the insignia. In certain instances, foreign troops trained in the various pathfinder schools set up during the war by this unit were also considered entitled to wear it. To wear the patch you did not have to be involved in a combat operation; you only had to have completed the training.

The original winged torch insignia was issued on a piece of dark purple wool approximately 2 1/2 X 3 inches. It was then trimmed down to the shape of the wing and torch, leaving a small border. It was then sewn on the lower left sleeve with the torch end on the downward angle.

Besides the purple wool version, there is a later version that appears to be an English, made piece which is identical except it is on blue wool. There have been other slight color differences with the yellow and blue thread on the flame. The blue wool version came in both a flat and padded version. As with other types of insignia there are also in existence hand embroidered and bullion thread variations. These would have been private purchases. One version was hand embroidered on dark red wool but it had no special meaning.

The "Pathfinder Wing" was approved by the Department of Heraldry in 1952 and is still worn by pathfinder trained personnel in the U.S. military. Postwar versions exist on a black cotton twill backing and later versions of the subdued type of black embroidery on o.d. material.

They are now worn on the left breast pocket flap and come in a gold colored metal with the flame detailed in colored enamel. Another metal version painted black is to be worn with subdued insignia and a subdued cloth version on o.d. that is to be folded and sewn like the current subdued parachutist wings are worn. Again, these have many different variations.

This is far from being the final word on the subject; and yet, it is a basic informative identification of the insignia and of the people that wore it and brought the pathfinders into being.

"AIR ASSAULT"

*Excerpted from "Air Assault" by Public Affairs Office,
101st Airborne Division (Air Assault)*

THE ARMY'S ONLY AIR ASSAULT DIVISION

The 101st Airborne Division (Air Assault) is a unique organization, using its helicopters to provide tactical mobility and flexibility on a scale never before realized. This combination enables the division to engage the enemy so fast and with such crushing firepower that the mission of the Army's combat forces to "close with the enemy and destroy him" might well have a divisional postscript—"at 120 knots."

KEY ORGANIZATIONAL FEATURES

The best way to get an idea of the unique capabilities of an air assault division is to review key organizational features. There are three brigade headquarters, each commanding three air assault infantry battalions; an aviation brigade with two assault helicopter (UH-60 "Blackhawk") battalions, one medium helicopter (CH-47 "Chinook") battalion, three attack helicopter battalions (AH-1 Cobra/AH-64 Apache), an air cavalry squadron and a command aviation battalion; a division artillery headquarters with three 105mm howitzer battalions.)A 155mm howitzer battalion is assigned in general support to the division.) The Division Support Command is composed of a maintenance and a supply and transportation battalion, an aviation maintenance battalion and a medical battalion which includes an air ambulance company equipped with UH-60 Blackhawk MEDEVAC helicopters.

The division's primary aerial workhorse is the UH-60 Blackhawk. The two assault helicopter battalions provide over 100 Blackhawks for air assault operations of combat troops and equipment. The CH-47 Chinook constitutes the backbone of the logistical support system, while the AH-1, Cobras and AH-64 Apaches of the attack helicopter battalions and the air cavalry squadron provide the anti-armor punch of the division. The air cavalry squadron includes a mix of UH-60, AH-1, and OH-58 helicopters and performs the cavalry role of reconnaissance and security. The command aviation battalion provides OH-58s and UH-1Hs for command and control within the division.

HIGHLIGHTS OF COMBINED ARMS EMPLOYMENT

The air assault division has been characterized as the Army's "all purpose" division, a term suggestive of its strategic and tactical flexibility and its firepower. The strength of the air assault division resides in the collective mobility provided by its organic helicopters which gives it the capability to mass, disperse and recycle forces rapidly throughout the battlefield. The air assault force is particularly well-suited for screening, covering force and delay operations and reinforcing economy-of-force roles, rear area security operations and offensive operations into the enemy's rear areas. These operations can be conducted in all types of terrain and weather, recognizing that periods of bad weather and reduced visibility enhance concealment of flight routes and degrade the enemy's surface-to-air missile and high-performance aircraft capabilities. The air assault division provides a flexible, highly mobile reserve capable of concentrating forces quickly at the critical time and place. Obstacles which plague ground forces such as river crossings, refugee and traffic congestion and towns and villages, represent no obstacle at all for heliborne troops. The air assault division offers a new dimension of tactical operation to the battlefield because of its relative freedom from such obstacles.

The division can lift simultaneously into battle combat elements of several of its nine infantry battalion task forces. The rapid movement of these forces over long distances to landing zones on or near the objective provides fresh assault troops for battle, eliminating arduous foot or motor marches. Division howitzers can be moved at treetop level by Blackhawk helicopters flying at 120 knots or transported internally in the Chinook helicopter, adding surprise to their employment. Thus, indirect fire support can be massed at the critical point in an extremely short reaction time. Combat power is built progressively by aerial reinforcement and resupply until the total division combat force can be brought to bear against the enemy. This force includes ground TOW missile systems which ideally could destroy enemy tanks or armored vehicles at ranges in excess of 3,000 meters. Meanwhile, AH-1 Cobras and AH-64 Apaches provide an aerial engagement platform capable of massing against enemy armor or defeating it in detail over great distances. The dividends earned in battlefield initiative against, for example, a Soviet T-62 tank whose maximum effective range is 2,000 meters, are obvious. Forward arming and refuel points for the attack helicopters are established to maintain the high level of anti-armor support to deployed forces. On such a fluid battlefield, flexibility, utility and security of the command and control structure would be achieved by frequent shifting, often at night, of the division's assault command posts, keeping well forward in the battle area. The bulk of the tactical operations centers and supporting signal systems of the headquarters would disperse to distances of 30 kilometers or more as required by the tactical situation. The sustained recycling of helicopters to re-position assault units and artillery support calls for a high level of aviator proficiency in nap-of-the-earth (low level) and non-illuminated night flying techniques. These techniques are sharpened and tested by realistic flight training against divisional air defense weapons simulating tactical disposition of enemy systems. The payoff in aviation proficiency is the high degree of coordination and cooperation that exists between the aviation units and supported combat, combat support, and combat service support units of the division.

The division moves by air and fights on the ground as a combined arms team supported by air reconnaissance and attack helicopter striking power. It sustains its fighting momentum and maneuverability by recycling helicopters supported by forward arming and refueling points. It employs air movement to minimize fatigue and moves its command and control structure by air rapidly over long distances. Because of the tactical efficiency and extraordinary firepower achieved by this combined arms approach, the division does not habitually provide for a reserve in organizing for a given operation. As the battle develops, the need for a ground tactical reserve, as such, is met by extracting by helicopter the least-engaged force and airlifting it or by directing attack helicopter units to the critical point of action.

AIR ASSAULT TRAINING

The unique skills required to deploy the division and successfully execute air assault tactics in combat do not come easily. Considerable emphasis must be placed on the specialized training required by air assault soldiers and units. To this end the division operates an Air Assault School which teaches the individual soldier air assault techniques and tactics and such fundamental skills as rigging, repelling, climbing troop ladders and pickup zone and landing zone operations. Aviator skills are refined through continuing emphasis on tactical flying and night operations. The Strategic Deployability School prepares officers and noncommissioned officers from each company and battalion for strategic air and rail movement duties.

The Air Assault School and Strategic Deployability School teach individual skills. Unit skills and mission training are developed through an intensive program of strategic deployability training using Air Force aircraft and the division's 30 rail cars, with each unit required to conduct air and rail load training at least semiannually. Tactical training is enhanced through continuing emphasis on camouflage discipline, night operations, anti-armor tactics and air assault techniques.

ANTI-ARMOR CAPABILITY

Air assault operations conducted against a sophisticated, mid-intensity threat require a formidable armor-defeating capability. The division's ground armor-defeating arsenal includes the TOW and Dragon guided missiles and the LAW/AT-4 rocket. The APACHE is the premier aerial anti-armor weapons system. Each air assault battalion's TOW and Dragon weapons systems constitute a devastating anti-armor capability.

SUMMARY

The 101st Airborne Division (Air Assault) is a one-of-a-kind division and combines a high degree of strategic mobility with an extremely high degree of tactical mobility within the target area. It can be deployed strategically as rapidly as any other division, and once on the battlefield, it can bring to bear powerful ground and aerial firepower at the critical time and place. Its potent arsenal of weapons makes it more than a match for enemy forces. Use of organic helicopters to move combat, combat support and combat service support units on the battlefield gives the division the ability to deploy and redeploy rapidly to engage the enemy and operate over a wide area. The air assault division normally is assigned to a corps. It operates from a position where its mobility permits its use in a variety of missions throughout the corps area.

Air Assault permits today's airborne soldier to be accurately placed within the airhead, eliminates the reorganization problem and avoids the loss of individuals due to landing injuries.

A Screaming Eagle from the 2nd Brigade, 101st Airborne Div. (Airmobile) is silhouetted against the setting sun as his squad prepares to move out to their night position on the coastal plain near Camp Evans in northern I Corps. (Photo courtesy of 101st Airborne, by Lt. Milton Parsons)

Cannoneers of Bravo Battery, 2nd Bn. (Ambl), 11th Artillery of the 101st Airborne Div., utilize their airmobile capability, assault onto Fire Support Base Blaze during a recent artillery raid. (U.S. Army photo by Spec. 5 Daniel Weber, courtesy of 101st Airborne Division Association)

Members of the 3rd Bn, 187th Infantry, 101st Airborne Div, Ft. Campbell, KY disembark from UH-1H "Huey" helicopters to establish a defensive perimeter around the landing zone for the remaining three waves of helicopter landings as part of a training exercise. (U.S. Army photo by Pfc. Steagall, courtesy of 101st Airborne Division Association)

"Airborne All The Way"

The Spirit of Airborne
Original sculpture by William F. Porteous

WORLD WAR II AIRBORNE TROOPERS LEFT LEGACY OF COURAGE, PRIDE AND DARING

By William B. Breuer

Since D-Day in Normandy on June 6, 1944, I have had enormous respect and admiration for the American Airborne soldier. On that epic invasion, I became indelibly aware that the paratrooper and gliderman were special breeds of warriors, and that their jobs were of extreme peril requiring the highest order of courage, daring and resourcefulness.

On D-Day, I made the H-hour amphibious assault at Utah Beach, and although some of my comrades fell, the resistance was amazingly light. As we pushed inland against little more than token opposition, the reason for our incredible good fortune became starkly evident: American airborne men had been there before us, paving the way.

Our parachutists' mission was to disrupt the Germans and prevent them from launching an attack on Utah Beach while we were coming vulnerably ashore. So successful were our troopers that Utah Beach was never threatened, and an entire American infantry division stormed ashore that day with astonishingly minimal casualties—12 dead and 46 wounded—instead of being slaughtered in the bloodbath predicted by some top Allied leaders.

However, paving the way at Utah Beach had not been cheap for our paratroopers. Victory in war is never cheap. Sprawled in the lush green fields of Normandy and dangling from their harnesses in trees were scores of lifeless forms wearing the coveted jumpboots and baggy pants of the American paratrooper.

Near dusk on that D-Day, swarms of American gliders soared in and crash-landed onto fortified German positions only a hundred yards to our front. We looked on in horror and in anguish as the flimsy crates were ripped in flight by enemy bullets and torn apart when they smashed into trees and thick earthen hedgerows, killing and mutilating many of those aboard. Miraculously, some of those glider troopers scrambled from the wreckages and under a hail of bullets charged dug-in enemy machine guns.

It is incredible to recall that, earlier in the war, armchair commandos, safely ensconced in the War Department in Washington, had denied hazardous duty pay to our glider soldiers. Those swivel-chair experts had concluded that soaring into battle in canvas-covered matchboxes over and behind enemy lines, being raked in flight by telephone poles, ditches, trees, boulders, stone walls and earthen hedgerows, did not constitute hazardous duty.

Some day I hope they will build a magnificent monument for those other unsung heroes, the C-47 crews and glider pilots, who shared the airborne peril in equal measure and with equal fortitude.

American paratroopers were (and are) proud members of the world's most exclusive and honorable fraternity. Membership could not be purchased or bestowed due to wealth, social stature or political connections. Rather, entry into the paratrooper fraternity had to be earned by enduring the most grueling training program that diabolical minds could conceive and later by measuring up in the crucible of battle.

Due to the nature of the paratrooper business, there was a unique and closely-knit relationship between officers and enlisted men, one of mutual respect for the confidence in each other, a kinship cemented by the fact that everyone leaping from airplanes into combat shared identical dangers. Bodies floating earthward under billowing white parachutes looked alike to hostile gunners on the ground. Paratroopers, whether generals, privates, cooks, riflemen, surgeons, engineers, artillerymen or chaplains all stood equal chance of being riddled by bullets in mid-air or plunging to their deaths with a "streamer".

The same situation held true for those who pounced on the enemy in gliders. When a fragile Waco crammed with soldiers smashed into a tree at 90 miles per hour or was shot down in flight, no one aboard was spared due to rank or military occupation. One of the two American generals (Don Pratt) killed in action during the war lost his life when his glider was torn asunder on crash-landing in Normandy.

If I had to sum up in one word the reason for the battlefield success of America's elite paratroop and glider outfits, that one word would be pride. Each airborne soldier had, and has, enormous pride—pride in himself, in his country and in his unit. That pride has sustained him, even in the blackest hours. In the mists of long ago, in World War II, pride ignited the paratroopers' torch of honor, and it has been passed along over the years and is still burning brightly in 1990.

A few years ago, I received a letter from a World War II paratrooper who had been grievously wounded in that conflict and has suffered incessant pain since then. This old warhorse wrote:

"Those of us who went through it know that war is horrible. But I would do it all again. Sometimes a man has to stand up and fight for what he thinks is right!"

Where does America get such gallant men?

A Korean War paratrooper, Rudy Hernandez, who was awarded the Medal of Honor and suffered disabling wounds, was asked of late: "If you could be granted one wish, what would it be?" Rudy pondered the question, then replied softly: "I would like to have the chance to serve my country again."

Where does America get such gallant men?

In more current times, teenaged paratrooper Harry Shaw lost both legs when raked by machine gun fire while jumping into Grenada in the mission to rescue some 700 American students from the clutches of the brutal Communist gang that had seized control of the government in a bloody coup. Life would seem to be bleak indeed for a youth of 19 suddenly minus both legs. But from his hospital bed of pain, Harry Shaw thrust out his jaw and declared: "It's going to be tough out there, but I'm going to lick the world anyhow!"

Where does America get such gallant men?

Today, there are cynics who consider it strange that old airborne comrades come back to each other in reunions year after year. With great gusto, they tell the same war stories—well, almost the same stories—allowing for a slight exaggeration each year. They hug each other and slap each other on the back. They call each other by old nicknames largely forgotten. And they laugh uproariously over recollections of romantic escapades of long ago in North Carolina or Australia or France or the Philippines or Japan.

Cynics say: "Surely these grown men can't enjoy going through all that same old nonsense year after year." But the cynics cannot grasp that a group of airborne comrades are linked today by a common experience too incredible ever to be fully comprehended by anyone who wasn't there. The cynics have no conception of how mutually shared fear, suffering, danger and anguish forged such powerful bonds between airborne fighting men that neither the passing of decades nor any force on earth can break. The cynics fail to understand the elation Old Warhorses feel when they have faced the black Angel of Death many times, when they have beaten the odds, and when they find themselves able to celebrate their good fortune with comrades.

However, even today World War II's paratroopers and glider warriors feel an aching void for buddies of long ago who were cut down on far flung battlefields in the sunrise of their lives and had to be left behind. It's tough to die when you're in your teens or twenties. They didn't want to die. They wanted to come home, too. Time will never dim the airborne troopers' visions of their cheerful, boyish faces.

Most of the American paratroopers and glidermen (many of whom were also qualified parachutists) who survived World War II have "taken a little off the top and put on a little around the middle." And they admit: "In those days, we were the young lions—today we're the aging pussycats." And yet, they have lost none of their airborne spirit and deep love of country.

World War II American airborne troopers can be immensely proud of the fact that their courage, daring, suffering and sacrifices contributed enormously toward assuring that their loved ones and those Americans not yet born could live in freedom, dignity, relative prosperity and without fear. That is the priceless endowment that American airborne troopers and all fighting men of World War II have bestowed upon the current and future generations—and let no one forget it.

In 1990, while the 50th birthday of the American airborne is being celebrated from sea to shining sea, this nation stands vigilant in a volatile and dangerous world. However, should an ill-advised aggressor choose to test the mettle of today's airborne young lions, he will find that they will react with the same high standards of fierce fighting skills, courage and dedication set by their predecessors in World War II.

As long as there are venturesome young Americans who are imbued with the flaming Airborne spirit passed along since World War II and who accept the challenge to "stand up and hook up" or drop into the fiery jaws of hell in a helicopter, then the freedoms with which Americans have been blessed for 214 years will continue to endure.

William B. Breuer's 17th book, **Geronimo!, The story of American paratroopers in World War II,** *was published by St. Martin's Press in fall 1989.*

Parachute blossom over North Carolina as the 513th PIR makes a practice jump.

THE NEW AERIAL DOUGHBOYS

By William P. Yarborough
Lt. Gen., USA (Ret)

Editor's note: The following is a reprint of an article written and published in the October 10, 1941 edition of the POINTER at the West Point Military Academy. The author is Lt. General William P. Yarborough (then Captain) and one of Airborne's earliest and most famous pioneers. General Yarborough sent us this article along with his permission to reproduce it. We believe it captures the atmospherre of the time when parachuting as a military weapon was both experimental and highly challenging for the individual.

In our Army, a "Parachutist" is not merely one who parachutes. Parachuting is a great sport. The tedious preparation preceding the actual descent comes closer to being work.

We parachutists are fanatically fond of our particular branch of the infantry. Granted it means weeks of grueling physical conditioning, days spent in folding and re-folding parachutes, and hours hanging in suspended harnesses. There has grown up among us a bond which the layman may or may not understand. Perhaps it is because all of us, officers and enlisted men, have faced danger together. We have seen our comrades "crack-up" all too often; and, on one occasion, we watched one of our pioneer paratroopers plunge to a soldier's death when his parachute was fouled by the propeller blast of the airplane from which he was jumping.

These things bind men with a closer mutual understanding. A paratrooper learns deep lessons in human psychology which other less-fortunate followers of the military profession must often learn only in the heat of battle. Most of the parachutists I know are slow to call another man "yellow." They have seen too many examples of stark courage to doubt that most men possess an ample supply of it.

As I said before, our parachutists are extremely proud of their organization. They have had the privilege of making their own traditions and watching them grow. Take, for instance, the cry, "Geronimo!"

Last September, a confidential letter from the War Department to the Infantry Board at Fort Benning had followed closely the receipt of the news of Holland's invasion by German "fallshirmjaeger" parachutists. Little Finland had already felt the might of the vertical envelopment by Russian chutists. The time had come for the United States to re-examine one of its own developments—airborne fighting men dropped by parachutes.

The test platoon authorized by the War Department letter consisted of two officers and 48 enlisted men, all infantrymen. They knew of airplanes in a general way, but some of them had never flown. Weeks of ground training had brought them up to the eve of their maiden parachute jump. That night was one never to be forgotten. No one was what you might call "scared" but there was not a member of the platoon who did not feel a certain nervous exhilaration over the events which were to transpire on the morrow.

Most of the platoon went to the movies that evening. It was called "Geronimo." It was a fast moving, full-of-action, hair-raising movie and the men enjoyed it because it gave their tired minds a well needed rest.

Bright and early the next morning, a solemn file of silent figures wound down the Lawson Field Road toward an Army bomber waiting on the apron. At the rear of the column marched Sergeant Charles Eberhardt with a steady determined step.

"Say, Eberhardt," a voice called from the middle of the column, "I bet you're gonna take one look at the hangers 1,500-feet down and pass out".

"Could be," replied the six-foot-two non-commissioned officer, "I ain't saying I won't, see, but if I'm okay once I get out into the air, I'll holler—let's see. I'll holler, 'Geronimo!' loud enough for you guys on the ground to hear."

So Sergeant Eberhart's "Geronimo!" was heard by the "guys" on the ground that day, and will continue to be heard whenever members of the 501st Parachute Battalion are aloft. "Geronimo" has come to mean things like "Hell, I ain't scared," and "Let's show 'em how it's done, gang." Some paratroopers have even abandoned the "one thousand, two thousand, three thousand" for a carefully timed "Gee—ron—imo."

The history of any outfit is a composite history of the men who were part of it. The story of Sergeant Grossi falls into such a category. It is somewhat as follows:

I was standing near the Lawson Field hangars one day with a group of other officers. We were all intently watching two silvery C-39 transports as they circled the field 1,500 feet up. Both ships showed gaping holes in their fuselages where the cargo doors had been removed. These open doors meant that the ships were filled with parachutists who would bail out as soon as their "jumpmasters" could spot the red and white marker panel on the ground.

As the first ship passed over the panel, a tiny dark figure leaped into space. It described a short arc and was brought to an abrupt stop by the parachute which popped open mushroom-like above it. The first jumper was followed at rapid intervals by the remainder of the plane's occupants. Each parachute seemed to blossom with mathematical precision. We counted them, "One, two, three." The opening chutes reminded me of smoke puffs made by exploding anti-aircraft shells, "Four, five, six, seve—." I stopped counting abruptly. One figure had dropped from the airplane door, completely through the parachute formation. I could see the trooper turn over and over in the air as he fell, his static line trailing behind him like a cobweb. Closer and closer to earth he kept falling like a meteor, until he was just barely visible over the tops of the hangers. The plummeting figure must not have been more than 200 feet from instant death when we saw a flash of white, and it stopped abruptly. We could hear the resounding crash as Grossi's emergency chest pack took hold." Jesus H Christ" I heard the officer next to me exclaim, "that lad nearly got a short discharge."

Now the notable thing about Grossi was not that he leaped from an airplane without having his static line hooked. Such a thing, if it had been Grossi's fault, would have shown lack of proper training on his part and would have been dealt with accordingly.

The thing that does put Grossi's name

on the roster is what occurred subsequently when Major Miley, the commanding officer careened out onto the field in his official car, looking for Grossi's fragments. He found the latter with his main parachute still on, calmly rolling his reserve into a neat ball.

"My God, Grossi, are you okay?" The commanding officer's eyes searched the trooper for possible fractures.

"Yes Sir," Grossi replied, "my reserve was overdue for repacking anyhow."

We need men like Grossi where the going is tough and odds against us are great.

Another example of cool unassuming courage was the case of Findlay. Findlay just got out of the hospital last week. How he got in there in the first place, is the story:

I was in C Company orderly room making out a training schedule one day, when Sergeant Peters, my charge of quarters poked his head into the office and stated in even tones, "Captain, there's a guy hanging on the tail of an airplane out here."

I was out of the office like a skyrocket. I looked around for the plane whose motors I could plainly hear. Sure enough, there it was, one of our transports, and streaming from its tail was a dark object being towed through the upper air at 100 miles per hour.

"How did it happen, Sergeant Peters?" I asked, my eyes riveted to the plane and its human tow-target.

"I don't know, sir," Peters replied, "I reckon his pack opened before he cleared the door. I was watching the jumpers come out, when suddenly this guy's chute whipped into the stabilizer and stayed there. Must be five more men in the plane from the way I counted."

The five in the plane were most certainly in a bad way, not to mention the unfortunate trooper on the tail. The ship would have a very difficult time landing with a man hanging from the stabilizer. The man, of course, would be killed in such an operation.

The pilot, however, kept his head in the best Air Corps tradition. I could see that he was fighting for altitude and trying to keep the giant ship as close to the field as possible.

The man on the tail was beginning to spin around and around. "He must be dead," I thought. "The risers will twist around his neck like steel cables."

The plane was over Alabama and was now headed for Lawson Field at about 2,000 feet.

Suddenly there was a burst of white. The great airplane seemed almost to halt momentarily as Findlay's reserve parachute bit into the slipstream, but the violent jerk pulled his main canopy free. The tenseness vanished from the watcher's faces. Findlay seemed to be all in one piece in spite of the three-thousand-pound pull it must have taken to tear his main chute loose. The latter now hung straight down in a tattered tangled mass. Findlay hung limply in his harness and drifted toward Alabama.

I ran down the hill from the barracks, trying to keep in sight the drifting chute. I saw it come to earth on a cleared hillside about five miles away. Findlay was down; I believed he must be dead. In any case I was determined to find out. I ran to the parachute shed and snatched my chute. Nor was I alone. Two of my lieutenants had had the same idea. Without saying a word, we all leaped onto one of the battalion's trucks and started for the field. But we were too late. Already an observation plane had circled around Findlay as he was descending and reported that the trooper had waved his hand at the ship as it passed him. As a medical officer was speeding toward the spot, we reluctantly agreed that jumping to Findlay's aid would bring help no sooner, so we abandoned the idea.

Soon after, the transport landed and disgorged a grim faced group of parachutists who had witnessed the whole drama from the air.

"He was as cool as a cucumber," said one, "I seen them risers twisting around and around like they was going to pinch his head off. All the time he kept grabbing them and trying to keep his body from turning."

"Yeah," said another. "He must have been getting plenty dizzy just before he cut loose. I seen his hands creep down to his ripcord sure and steady like he knew exactly what he was doin'. I thought the whole tail of the ship was coming off when he cut loose. He sure made mincemeat out of the de-icer boot in the stabilizer."

Findlay was alive all right. True, he had a broken nose, a fractured hip, a broken collar bone and possible internal injuries, but his spirit was still intact. When the Medico was removing his jumping boot from the injured member, Findlay, through clenched teeth, allowed that he didn't by a damn site want to lose that boot, and "would the doc take care of it, because a guy has to have jumping boots if he's going to be in a parachute outfit."

Findlay's crack-up caused him to miss attending Riggers School at Chanute Field, but it put his name on the list with Sergeant Grossi. Paratroopers don't forget people like Findlay. His kind of pluck is legal tender anywhere.

Parachutists are not a superstitious lot as a whole. In fact, we have consciously tried to keep little phobias and fancied hexes from growing up. We have many times jumped 13 men from a plane. One time we bailed 13 out on Friday the 13th and there was not so much as a sprained toe in the crowd. Being free from "jinx" ideas means that we are not bound to wait for propitious days for jumping. Neither do we refuse to bail out if we find that we have lost our "good luck" penny or that our "lucky" undershirt has been lost by the laundry.

I say, we are comparatively free from superstitious jinx, but there is one idea that is quite persistent among us. It has mainly to do with the law of averages. In short, most of us feel that in spite of all the care we exercise in the packing, inspecting and manipulation of our parachutes, somewhere in the cards lies a malfunction or a crack-up. It may not be fatal, but we are convinced it's there just the same.

I got mine out of my system for a while on my jump. I won't say I was especially nervous that day but I remember being a little apprehensive, for no reason that I could put my finger on. Anyhow here's the yarn:

I was in a C-39 Transport above Lawson Field. My parachute was bulky and awkward. I hardly dared to move about for fear of ripping open the canopy cover and spilling the carefully folded canopy onto the floor. Roy Lindquist, who sat in back of me tapped my shoulder and poked a finger toward the window. His lips formed the words, "There's the panel." Seven hundred and fifty feet below, I could make out the familiar outline of the marker over which we would shortly bail out.

"Hook up!" the command rang out a cut above the roar of the motors. Mechanically my left hand found the heavy snap fastener on the end of the static line, I grasped it and gave it a jerk to break the linen thread with which it was secured to my harness. Then I snapped the fastener over the steel cable which ran the length of the fuselage overhead.

We were very near the panel now, for the pilot was throttling down in order to lessen the effect of the "prop blast" on our parachutes.

I was the last in line. Eleven jumpers, looking like men from Mars were lined up in single file ahead of me. We waited for the command which would send us catapulting into space. This was our first mass jump. We had practiced it a score of times with dummy chutes from a "mock up" door in our training area. Now we were separated by a split second from the real thing. I could feel the tenseness in the air. It wasn't fear—it was more like that feeling I had often experienced before a swimming race when I was on the Academy squad.

The command, "Go!" came like a pistol shot. I was moving toward the door as fast as I could move. The column of jumpers grew shorter and shorter as each disappeared over the side. Now I was on the threshold of the door for a split instant. I followed Roy Lindquist so closely that both my hands were on his parachute pack when we hit the prop blast.

I was descending head down. Below me I could see three parachutes, their chutes in various stages of opening, falling around me.

The familiar jerk which marks the opening of the main canopy seemed a little more violent than usual. When the stars cleared away I looked up into my chute. Something was wrong. In place of the usual broad undulating hemisphere of silk I saw what appeared to be two small parachutes each about one-third the diameter of my main canopy. The two small chutes were tugging furiously at my linen risers, which meant that I was falling fast. The ground was getting closer at an alarming rate of speed. I acted instinctively—my right hand found the rip-cord on my emergency chest pack. I pulled it, expecting to see a burst of white as my secondary chute inflated. Instead, I witnessed a remarkable thing. The cover on my emergency chute was fully opened revealing the folded parachute beneath it. The latter, having no pilot chute to draw it out of my pack, was being held in place by air pressure created by my violent oscillation. I threw the rip cord grip away and grasped the emergency chute, tearing it from the pack and casting it away from my body in an effort to inflate it. Before the air could get to the channel, however, my oscillation threw me into the streaming silk of the emergency canopy. It wrapped around my body, covered by eyes, hampered my arm movements—I caught a glimpse of the ground, now only 50 feet below. For an instant, I felt panicky—but my strangest feeling was indignation. I was going to write to Washington about a ————— ————— chest pack that wouldn't open without handling with kid gloves.

The next thing I remember was the anxious face of a Medical Corps soldier as he raised a fold of the silk which was piled on top of me in layers. The fog slowly cleared away. Major Miley was watching me as I got up, and he looked as relieved as I did when I discovered no broken bones.

My chute had gone through what is known to riggers as a "malfunction." In this case, one half of the parachute had been turned inside-out in some unaccountable manner. This reduced, by about half, the effective lifting surface of my main canopy. I had been lucky—not even a sprained ankle.

Perhaps I have painted a rather dismal picture of parachuting. If I have, please let me hasten to state that such a portrayal is indeed a false one. The unusual occurrences which I have related are, of course, very exceptional. A normal jumping day is one in which the worst casualty is nothing more severe than a moderate bump or a split lip from contact with the reserve pack.

Physically, paratroopers are built to take it. Men chosen for service with our battalions must pass a rigid physical examination before they are accepted. Men who are psychologically unfitted for parachuting are summarily rejected.

Our first desire, is to obtain not parachutists, but good infantrymen. We can make a good parachutist out of a good infantryman. In fact, an A-1 infantryman comes darn close to being good at most jobs anyway.

Our troopers are armed with the Garand rifle, the machine gun, the Thompson sub-machine gun, the 60mm mortar, hand grenades, and the .45 caliber pistol. We hope to obtain heavier weapons, such as the 81mm mortar, anti-tank guns, heavy machine guns and perhaps flame throwers and pack howitzers in the near future.

Progress in experimentation with aerial delivery units for dropping food, arms, ammunition and miscellaneous supplies is being made steadily. We maintain a very close liaison with the great Air Corps laboratories at Wright Field in Dayton, Ohio.

No longer is the 501st Parachute Battalion unique. The 502nd was activated on July 1st. The 503rd came into being on September 1st. From there on, the sky is the limit. We want candidates—all they need in the way of qualification is pride, (for of such, courage is made) youth, enthusiasm, loyalty, physical condition and an adventuresome spirit. Geronimo!

Paratroops entering G-53 Transport plane, Camp Mackall, North Carolina. (U.S. Signal Corps photo)

GLIDER PILOTS: 1942-1945

By George "Pete" Buckley
Former Glider Pilot, 74th
TC Sqdn, 434th TC Group

They were a mixed bag of individuals. All were volunteers, many lacked military discipline; and all were willing to fight friend or foe at the drop of a hat to protect their reputation as the "down and go" boys. Referred to by the Airborne as the "Crazies," they were fiercely independent and quite belligerent at times toward any attempts by the higher ups to organize or shape them. In spite of this negative attitude they had an Esprit de Corps unsurpassed by any other U.S. Army Air Corps outfit in World War II.

In the Foreword of *Silent Wings*, the definitive book...author Major Gerard Devlin (retired), General William Westmoreland described them as intrepid pilots whose duty it was to deliberately crashland their gliders behind enemy lines and then go on to fight as combat infantrymen. They were the only Army aviators during World War II who had no motors, no parachutes and no second chance. They were the least heralded heroes of the World War II. (Permission granted by Major Devlin to use this paragraph, 12/2/89).

The glider program got off the ground in early 1942 with a projected goal of 5,000 active duty glider pilots. These volunteers came from all branches of the Army, from privates up to and including majors; some came straight from civilian status. Almost all of them wanted the opportunity to fly and to get in on the ground floor of this new concept in airborne warfare.

Early in the war, their lack of ground combat experience was more than offset by the determination and sheer guts they displayed in getting their gliders to assigned areas under very hazardous conditions. Their efforts were best described in General Order #212, dated 23 August 1944, awarding battle honors to the 9th Troop Carrier Command. It reads as follows:

These USAAF Glider Pilots flew unarmed and unarmored fragile craft at minimum altitudes and airspeeds and at times highly unfavorable weather conditions over water and land into the face of heavy flak and vigorously defended enemy positions with no possibility of employing evasive action. They brought into battle glider infantry, anti-tank weapons, vehicles, supplies and ammunition which were sorely needed by Airborne troops who had preceded them and were engaged in deadly hand-to-hand combat with the enemy.

Their job as glider pilots became without a doubt one of the most hazardous and demanding jobs both in the air and on the ground that the Army Air Corps had ever created. Theirs was a triple role that no other Army pilots had been called on to perform. As mentioned earlier, they were rated pilots in the air and after making a combat landing they became combat infantry soldiers. Lastly, when the need was there, they were qualified, and many times had to assemble and test fly their own gliders. As the war progressed many flew as co-pilots in C-47s in order to give Troop Carrier pilots who had been flying night and day, seven days a week, a much needed rest.

They served in almost all theaters of operation during World War II. A very small group served in Alaska as part of a downed-fliers rescue team, the theory being that it was much quicker and possibly cheaper to use gliders instead of dog teams for long range rescue operations. Some were stationed in Australia, New Guinea, India and, late in the war, the Philippine Islands. The largest contingent of GPs was assigned to the 1st Air Commando Group in the CBI under the command of Colonel Philip Cochrane. There, missions deep into the Burmese jungles far behind Japanese lines were classic examples of glider operations in vertical envelopment warfare. Due to the jungle terrain and the savage capabilities of the enemy, gliders and their occupants forced down in these areas stood very little chance of survival. This was the least desirable of all theaters to be assigned.

The first baptism of fire of American GPs in the European-Mediterranean Theater was the Airborne invasion of Sicily. Due to a shortage of English Horsa Gliders, the British borrowed a large number of our C-G4As. Approximately 30 American GPs from the 316th Troop Carrier Group volunteered to accompany the British as co-pilots and observers. The glider phase in this operation was just short of catastrophic. Some were forced to ditch at sea because high headwinds kept them from reaching their LZs. Others were shot down by Allied warships who were trigger-happy from previous attacks by enemy planes. Those that did land on the island were widely scattered away from their objectives.

The American GPs upon landing fought side-by-side with their British counterparts and acquitted themselves quite well. They were later made honorary members of the British Glider Pilot Regiment and were awarded and authorized to wear British Glider Pilot Wings. The valuable lessons learned here and in Burma were to be applied to coming airborne invasions.

Normandy involved the first large scale use of Allied gliders in World War II. The lead glider in the U.S. phase was piloted by Col. Michael Murphy, the senior glider pilot officer in the European Theater. Most of the landing techniques used here were developed by Col. Murphy and his staff in the United States and England. Col. Murphy was a world famous stunt pilot and barnstormer from prewar days. He held many national and international championships and was almost as well-known as Gen. James Doolittle. He was not required to fly the Normandy mission but insisted upon leading his boys in, over the objections of the higher brass. Tragically, his glider crashed upon landing, killing the co-pilot, Lt. John Butler and Gen. Donald Pratt, assistant division commander of the 101st Airborne Division. Col. Murphy, severely injured, was evacuated back to the United States.

The successful landings in and around the St. Mere Eglise area were not as concentrated as hoped for. Small fields surrounded by 60-foot tall trees and thick hedgerows, plus heavy opposition on the ground, contributed greatly to many fatalities and injuries. Again, glider pilots performed well and gained much needed ground combat experience which would prove useful in the coming Southern France and Holland airborne operations.

Before pulling out for return to England they had been used to guard division CP and to secure and hold landing areas for incoming glider serials coming in on D plus 2, 3 and 4. Their last job was to escort hundreds of German prisoners through the lines to the beachhead area. After returning to their home bases, they received more ground combat training at Ogbourne, St. George in southern England.

The Southern France airborne invasion, although not on as large a scale as Normandy, went off without serious mishap except for the normal screw-ups that do occur during undertakings of this type. Casualties among glider pilots and airborne personnel were much lower than expected. All glider pilots involved were returned to England in a very short time to make ready for future airborne operations.

Holland, 17 September 1944. This date would go down as the largest Airborne operation in history. Unfortunately, everything that could possibly go wrong, did. Flights were aborted because of weather conditions on D plus 2 through D plus 6. Some flights were cancelled completely. Glider serials on day 3, the 19th of September, were a disaster. Dense fog over the English Channel caused many gliders to ditch. Others stayed on tow and returned to their English bases. Many gliders and their tow planes were shot down between the Channel coast and their LZ areas in Holland. One group of tow planes overshot their objectives and their gliders cut loose 10 miles inside Germany.

In spite of all these difficulties on the fly-in, the glider pilots after landing put the knowledge gained on previous missions to good use. Approximately 200 glider pilots were organized to plug a weak point in the airborne perimeter caused by the withdrawal of the glidermen and the paratroopers to the other sections of the line which was being hammered by repeated German counterattacks. One glider pilot, Captain Elgin Andross of the 313th Troop Carrier Group, distinguished himself when he led a group of fellow glider pilots through an ambush that had been set up by elements of a German parachute regiment. In the ensuing action, while escaping from the ambush, the GPs killed and wounded many of the enemy and Captain Andross was later awarded the Silver Star for this action. The glider pilots had again proved their mettle as combat soldiers.

Within 10 days all GPs accounted for were back in England for more flying and ground combat training. In December 1944, a small group of them were called upon again to assist in resupplying the 101st Airborne Division which was in a bit of trouble in and around the Bastogne area. Their gliders brought in surgical teams, supplies and badly-needed ammunition.

The flak was extremely heavy, causing a large number of C-47s to be shot down in proportion to the small number used. Most gliders reached their proper LZ; and the GPs again became foot soldiers in the coldest winter in Europe in over 50 years.

March 24, 1945, VARSITY (Rhine crossing). This would be the largest one-day drop of Airborne forces to date. It would also be the first time Allied gliders would land in fields unsecured by paratroopers and the first time that American glider pilots would form up after landing and fight as individual units under the control of airborne commanders.

During the first night of action, GPs from the 77th Sqdn., 435th Group, aided by glidermen, fought a pitched battle which raged most of the night. When dawn broke, it was found that they had annihilated a large group of German infantry led by two tanks and a mobile flak wagon that had been trying to break out of the airborne perimeter to get back to their own lines. German dead and wounded were scattered over a large area of the crossroads that the GPs had been holding and well over 75 prisoners were now in their hands.

During the height of the battle at very close range, Glider Pilot Elbert Jella, with a borrowed bazooka, managed to knock out one of the tanks. The GPs from the 77th Sqdn. received a personal commendation from the airborne commanding general and Stars and Stripes dubbed this action "The Battle of Burp Gun Corner."

After all LZs and DZs and the perimeter had been secured, 500 GPs were withdrawn back over the Rhine, bringing with them over 2,000 German POWs who had been captured by airborne personnel and GPs shortly after landing.

This brought an end to glider pilot operations in the ETO. By now they were a well-oiled, experienced team who had proved themselves in air and ground combat. No one ever questioned their flying skill, loyalty or aggressiveness. Flying large combat gliders required exceptional skill and judgment of heights, speeds and sink rates in order to successfully land that first time around. There were no second chances. All these skills were acquired through rigorous training before they were awarded the coveted wings of the rated Glider Pilot.

At this time, many began packing for the long trip to the Pacific and some high point men went home to the United States for discharge. All told, it was a job well done by this small group of dedicated, maverick air soldiers who contributed greatly to the success of World War II airborne operations. Unfortunately over the years they have received very little recognition or publicity for their deeds.

Today in this year of 1990, there are approximately 1,500 survivors that have banded together and belong to the National World War II Glider Pilots Association based in Dallas, Texas. They have their own first-class military airborne museum located in Terrell, Texas. Articles on display include a completely rebuilt CG-4A and hundreds of items of Airborne memorabilia. All Airborne personnel, past and present, are cordially invited to stop by for an interesting and worthwhile visit.

Try as they did, the glider troops had little success at trying to float the CG-4A, as these Camp Mackall troops found out in an exercise in early 1943.

THE GLIDER PILOT

By Douglas W. Wilmer
WW II Glider Pilot

The American glider pilot turned out to be an uncut diamond for the Army Air Corps which for the most part, never perceived that they had a treasure in men who could live on the "edge" or "brink" without falling apart. Every assignment given to them was well done, whether it was flying as co-pilots in the C-47 aircraft or other squadron or flight responsibilities. But the job he was trained for was more than adequately performed under extremely adverse conditions. This is that story.

The world situation of 1941 and '42 has an important bearing on our story. The German war machine was very successful and rumors of great deeds done by German gliders reached the American War Department. One of these rumors—the almost unbelievable account of the mighty underground impregnable fort in Belgium-Eban Emael whose major guns protected the bridges of the Key Albert Canal—was lost to troops and special explosives that had been landed on top of the fort in gliders. A year later (20 May 1941) 75 German DFS-230 combat gliders and paratroops captured the strategic island of Crete.

There was a small group of American generals interested in the German Airborne and was making plans for an American airborne effort. One of these was General Henry H. "Hap" Arnold, who wrote a memo to Maj. Gen. G.H. Brett early in 1941 that he would like to see a two-man jeep constructed so as to take off as a glider, land as a glider, shed its wings and continue on as a jeep. Perhaps he had seen reports of the 1930 Russian project in which wings had been fitted to a light tank and it was flown as a glider and shed its wings when it landed. Gen. Brett who was acting chief of the Army Air Corps, sent Gen. Arnold's suggestion out for bids and the Glider Program was official and underway.

On 5 June 1941 orders were cut sending six flying officers to the Elmira, NY Soaring School and six more officers to the Frankfort Sailplane School in Joliet, Illinois. These officials were to become familiar with flying gliders and to provide leadership to the many students that were to follow.

It had been reported that Germany had over 50,000 trained glider pilots. Early estimates for the American Glider Corps was for 30,000 men. This figure was never met and in a few months was lowered to about 10,000. The best information we have today indicates that 6,000 Glider "G" Wings were issued in World War II.

Within the Air Corps Command there was a great deal of concern that the Glider Program not conflict with either the flying cadet program or the acquisition and manufacturing of regular airplanes. Orders were issued that only non-airframe manufacturers could bid on glider projects and a large number of household names with wood-working shops, beer companies, pickle plants, piano companies, automobile, streetcar and casket plants were heard from. Of these, the Ford Motor Co's stationwagon plant at Iron Mtn., Michigan became a major supplier.

Richard DuPont, one of the country's leading sport glider pilots, was chosen by Gen. "Hap" Arnold to head the program.

For the most part, the early glider pilot fledglings were men who had flying experience, some with just a little and others with a lot. He may have already been in the service or still a civilian. In any case, he was volunteering to fly gliders for the Army Air Corps. He would have rather flown fighters, but for one reason or other he could not pass the physical or educational standards required for the flying cadets. Perhaps a third were unable to make the age requirements.

These prospects were surely the unsung heroes of the Glider Corps because they could have sat the war out working in a defense plant with no fear of being drafted for any service. He had strong patriotic feelings and wanted to do his part for his country with a skill he thought important—flying. If you had flying experience and were in the armed services, you could get an immediate transfer to the Glider Corps and many did just that.

Another source of qualified airmen came from the cadet and enlisted pilot programs of the Air Corps. These were men who were rightly or wrongly "washed out" of those programs and were given the opportunity of applying for further training in the glider program. To be accepted, however, he had to have proficiency in flying and even though these were rejects from the Flying Cadet Program, later events proved that the Army Air Corps gave up too quickly on these men.

To most of the glider pilots in the above category, this was a bitter pill to swallow and this second chance at flying was just the thing they needed to prove their worth to themselves and their country. To my knowledge, all enlisted pilots that were washed out were not given any other options and were automatically transferred into the Glider Program.

The Glider Program of that day seemed to be a "catch all" for anyone with flying experience for whom the Air Corps could not find an immediate assignment. For example, Americans who had gone to Canada before "Pearl Harbor" to join the Royal Air Force and returned to the States, CPT trainees and others.

All glider students with flying experience were "A" students and made up the major part of the 6,000 or so glider pilots that were issued wings. But many glider pilots were "B" students who had no flying experience at all, and these men were trained first to fly power craft and then to fly gliders.

The flying training of the glider pilot was performed largely by civilian instructors and consisted of light airplane flying at civilian contract schools widely spread over the United States. After being checked out in the light airplanes like "Cubs"—which most of the guys already knew how to fly—our GP was introduced to the "dead stick" landings. This was the name given to turning off the switch of your plane and landing it without any power.

He soon became proficient in flying

airplanes without power and was moved to schools that trained in the flying of light two- and three-place gliders. The Army operated the Advanced Glider Schools that flew the large cargo type CG-4A WACO gliders after which he received his "G" wings. This was not the finish of his training, however; if anything, it was just the beginning. Tactical training in a combat glider was a must before assignments to a Troop Carrier Group.

Ground training and weapons training continued whenever he was not flying. Some glider pilots received combat training that approached that of "Rangers," but some received little or none; and these were mostly the early GPs that had been sent to Africa and the later power pilots that were sent over to Europe to fly as co-pilots on the Rhine mission.

Some who knew the glider pilot during World War II thought that the "G" in his wings stood for "guts;" but to the glider pilot these wings meant that he had completed a long and difficult flying course and that this was his badge of graduation. They were awarded once, but they were his to keep forever.

The serious part of training to be a glider pilot was the tactical use of the combat glider. The many different facets of flying his aircraft into different types of terrain and performing slow, small-field landings over barriers with full loads separated the men from the boys.

In combat, the gliders were towed into the most hostile environment that one could imagine. But getting there many times was much more difficult. He had to fly long, exhausting tows without co-pilots or any method of gaining relief from the stress and strain of holding his craft in the proper position in formation. The air had been tormented by hundreds of tow planes with two engines each and the propellers had created unbelievable "prop-wash." There was bad and borderline weather, inept navigation and tow planes that went down both from enemy and mechanical actions. He had to contend with bursting flak above and below and directly in front, small arms fire and poles planted like trees but connected with wires on the LZ (landing zones). These LZs may or may not have been zeroed in for mortar and artillery fire by a determined enemy.

But this glider pilot was also determined. He was prepared at all costs to deliver a very valuable cargo of men, equipment and supplies entrusted to his keeping as close to his predetermined LZ as his tow pilot, aircraft and navigation could get him. Getting there was only part of his responsibility. He was required to defend himself and report to the Command Post (CP) for further assignments.

During the Holland mission (Sept. 1944) 300 glider pilots were called out one night by General Gavin to replace paratroopers who were needed for another important assignment. American military history tells us that this was the first time ever that a unit composed of all-officer personnel solely occupied a stretch of front line positions. The military situation also required that these lightly-armed GPs supply themselves with the necessary weapons from wherever they could be found, usually from dead troopers.

The World War II glider pilots flew many glider missions but the larger combat missions are put at eight and include: Sicily-July 1943, Burma-March 1944, Normandy-June 1944, Southern France-August 1944, Holland-Sept. 1944, Bastogne-Dec. 1944, Wesel (across the Rhine) March 1945, and Luzon P.I., June 1945.

The glider was used mostly to deliver the heavy muscle part of the airborne battle needs: jeeps, artillery pieces, trailers, engineering equipment, radios, land mines, gasoline, ammunition and much more. The following is a ball park figure of what the eight combat missions meant in what was carried and supplied to the Airborne: Over 3,800 gliders were flown in which more than 24,000 troops were put down, over 1,500 jeeps carried, more than 700 pieces of artillery landed plus over 4,000 tons of other equipment and supplies were landed. Many more resupply gliders were flown, but these were not counted as combat missions and their figures are not included.

What was the driving force that cooly placed the glider pilots in the seats of these fragile, motorless craft that were neither crashproof or bullet proof? It was the endearing love for flying and a rare ability to live in a constant environment of "close calls."

In judging the glider pilot and what he did and how he did it; one must remember that at any time the GP had the option of staying on tow or releasing himself; but in spite of heavy flak and small arms fire—so bad at times it seemed like you could walk on it—he had the "guts" to hang on until the last second in order to deliver his badly-needed troops and supplies to a battlefield that was always encircled by an enemy. It also must be remembered that in addition to the glider pilot, the other part of the Troop Carrier team—the tug pilot—also was being shot at by the same enemy and he also had the "guts" to see that his gliders got as close to the LZ as he could get them. These men were unsung heroes in every sense of the word. After all, they were flying unarmed aircraft and for most of the missions their planes did not have seal-proof gas tanks.

One cannot write about the glider pilots without mentioning the glider soldier, the C-47 tow plane or the WACO CG-4A glider. First, the glider trooper, or those "who were roped in." If there was one man of the Airborne that the glider pilot had a great deal of respect for, it was the man behind him that he flew into combat or on non-combat flights. This glider soldier, besides being an excellent "storm trooper type" also had to have his inner-self, his fears and his nerves under control and a runaway imagination. Sitting in the back end of a box-like structure with very little light, he had next to no way of seeing what was happening. This claustrophobic feeling did nothing to lower the fears—both of flying and of going into combat. The GP must have known that his charges were at least 200 percent dependent on him to bring them safely through this flight. It must have seemed like an eternity. Then there was the glider touching down and rolling to a stop, the feelings of relief and then the mad rush to get away from his "skyway" adventure as fast as possible. History should find a special place for these rare and brave men.

No one in Airborne has to be reminded that his airplane was special; now the whole world knows that the C-47 (DC-3) was and still is a wonderful flying machine. Without it there would be a different story about the use of Airborne troops in World War II. Thank you, Donald Douglas, wherever you are!

Every glider pilot I know still has a very fond spot in his heart for the aircraft he flew—the WACO CG-4A glider. It took better care of its pilots and soldiers than it is given credit for. The "old girl" was put together with welded steel tubing and this made all the difference if it took a rough landing to deliver two tons of men, equipment or supplies.

To the glider pilot, she would not win any beauty contest but she flew well; and as long as you watched your down wind turns she performed her assigned tasks very well. She was a real "Queen of the Air."

After 48 years or so, the surviving glider pilots still feel a deep sense of pride for the assignments that they completed. To these same survivors, whether they know it or not, he had been given by his training and experience a combination of internal assets that were invaluable in meeting the challenges of postwar life and endeavors.

As a well-known glider pilot, Earl Goodwin put it when describing the large number of successful glider pilots after the war, "Their postwar drive and motivation was very much like the tempering of steel. His GP life did for him what tempering does for steel. You heat soft iron until it is cherry red then plunge it into cold water." The result is hard steel. Glider pilots will agree with that.

After 48 years or so, the present day GP would like the American people to know more about what he did and how he did it. The Airborne Museum located on the municipal airport of Terrell, Texas has a complete CG-4A Glider and other memorabilia of the World War II Glider Pilots. For further information, write the museum at P.O. Box 775, Terrell, Texas 75160.

While the CG-4A in the background seems to have only a broken tail, the one in foreground is completely destroyed.

THE GLIDER SOLDIER...
THOSE WHO WERE ROPED IN!

By Douglas Wilmer
Reprinted from "The Static Line," October 1989 issue. Used by permission of the author.

It strikes me that the airborne army had three very unusual individuals in its personnel make-up. The paratrooper is well-known, but the glider soldier and his pilots (both glider and power) are not as well-known for their part, and this column deals in a small way with the glider soldier.

First he rode in the back of a glider (usually the CG-4A WACO) that has been called a very "fragile but tough" delivery system from the air. There are still C-47s flying today after all these years. The gliders didn't look like much, but they were real "queens" of the air. The way they were made (welded steel tubing) gave them a toughness that was not realized until they were put to one of the many tests that they were put to during the history of their "hard" use. They were not bulletproof or crashproof, but most of the time they protected their load and occupants enough to deliver them in an effective state to meet the enemy.

Riding in a closed-in container like the backend of a CG-4A glider was not easy. It did not allow for viewing where you were going. Troopers faced each other, and if one got airsick, everyone else usually followed. All of the German gliders had their troops facing forward or backward, and one wonders why our aircraft engineers didn't do the same—no question that it would have been better for the glider troops. As a glider pilot, I always thought that the troopers I was flying were having exceptional problems in flight. Not only was the visibility low, but incidence of air sickness was very high. I remember a large maneuver at Camp Mackall—my glider was one of the last in a serial of 50 or more on a hot day and I could hear one of the troopers in the far back throwing up above the noise of the fabric beating against the side of the glider in flight. When we landed I noted the poor guy falling through the door, completely spent, almost unable to move, and I asked his 2nd/Lt., who was also on board if the Airborne Dept. could help these troopers from getting air sick. He told me that this fellow always got sick, but this time the doctors gave him something to prevent it and it seemed to have made him worse. Of course, these were the days before an effective air sickness drug was available, but I don't think that I ever carried glider troops on a hot summer day that either some or all did not get airsick.

The large wing and a large formation of tug ships usually meant a very rough ride and as most of you know, until you have ridden in a glider under these conditions and on a hot day, you have never really had a rough ride.

There have only been a few times that I carried any glider troopers at night, but what a difference. The ride on tow was usually smooth and breathtakingly quiet in free flight-like floating on a big pillow. This was when you got the true feel of flying and it always reminded me of our sailplane training.

I guess you had to be a glider pilot to realize how special and unique glider troopers really were. I realize that most of the later guys were volunteers and also that many of you at the time would have been happier if you had been in a different line of work. However, in spite of the fears of flying and the fears of going into combat, you had enough pride, love of country and unit, to hang on and deal with your fears in a personal way. I suspect there were as many ways of dealing with them as there were persons.

I believe these were the facts that made you different from all of the other soldiers of World War II. You not only had to be good soldiers (and you were), but you also had to handle unbelievable internal pressures that went with the job in addition to all of the other pressures of warfare. In other words, you had to deal with a job description and conditions unequaled in any other soldiering or specialty in World War II.

Once you were in a glider and it was airborne, you had very little control over your own well-being. The paratrooper had much more control over himself, and if asked, he would tell you he would rather jump than ride.

You can be proud of what you did and how you did it; and I for one am glad and proud that I had a small part in flying you and putting you down as part of that glider effort that was very important to the overall airborne success. Remember, it could not have happened without you.

Two glider troopers of the Airborne raise the tail of the CG-4A glider and prop it up during front end loading of the glider. The tail is also raised for unloading. (Photo from film strip-Abn. HQ.)

GOOD TRAINING NEVER ENDS FOR THE AIRBORNE TROOPER

By Paul Troth
LTC, AUS (Ret)

Training of parachute and glider troops in the early formative days of the Airborne was a continuous process. A constant search was made for closer teamwork in combined training, improved methods, weapons, vehicles and aircraft. Never before had men, weapons and equipment in the American Army moved by air and entered combat from above the battlefield by vertical envelopment. Now, 50 years later, that search still goes on. Airborne troopers are proud of their training.

The Provisional Parachute Group organized at Fort Benning, Georgia, just 50 years ago placed the early emphasis on physical training and know-how. In a four-week course at The Parachute School, training started with a grueling week of push-ups, obstacle courses, forced marches and physical development. Those volunteers who survived went on to spend a week in the packing sheds to learn rigging and how to fold and pack their reserve and main chutes. Next came jump training out of mock-ups and from the parachute towers. Live jumps from a DC-3 or a C-47—where paratroopers learned to stand up and hook up, inspect their equipment, stand in the door, and "GO" with a step into space—led up to graduation day and a "prop blast." Qualified paratroopers had good reason to shout "Geronimo!"

With glidermen, the early emphasis was also on physical development and complete understanding of loading and lashing vehicles, guns, ammunition, equipment and men in airplanes and gliders. The cargo glider CG-4A was most often used by glider-troopers with hours of practice in raising the tail and raising the nose of the glider, followed by loading and lashing of vehicles, guns and equipment. Every glider soldier learned to tie two half-hitches, the baker bowline and the slippery half-hitch. Those three useful knots had to hold fast in flight, stand the stress of sudden stops and be ready for quick release on landing. Glider troopers also learned flight discipline, the use of seat belts, the importance of balanced loads and the center of gravity in aircraft. Orientation flights with troop carrier pilots of the Air Corps was an essential part of tactical training on daylight flights and night maneuvers.

With all of this training for movement by air, basic training of the individual soldier still received top priority. As General James M. Gavin said in the early development of airborne troops: "Once that paratrooper or glider-soldier hits the ground behind enemy lines with only the weapons and supplies he can take with him, he'd better be the best damned fighting man in the U.S. Army."

He was so right.

In October 1942, when the 82nd and 101st Airborne Divisions moved into Fort Bragg, North Carolina, to continue their training for combat as airborne troops, the training program was logically divided into three parts. The three parts included: basic ground training for 13 weeks, airborne training for 13 weeks and the combined training of the various arms and services included in the brigade or division for 12 weeks. With the expansion of the airborne effort from infantry to all elements of a division, other arms and services were developed and trained for airborne operations. The various service schools such as the Infantry School, the Field Artillery School, the Engineer School, the Antiaircraft School of the Coast Artillery, the Signal School and others helped in that training.

Looking back again to the year 1940, William T. Ryder and James A. Bassett were the two young lieutenants who volunteered to command and train the first test platoon for parachute duty. Starting from scratch, they and their eager group of volunteers were the first paratroopers to step into a tactical training problem from above the battlefield. What a beginning and what a 50-year record we celebrate today.

Fifty years ago, Colonel William C. Lee, affectionately known by many even today as "the first airborne soldier," was in command of the Provisional Parachute Group at Fort Benning, Georgia. Colonel Lee had already won his first battle of World War II, which was on the side of those who fought to keep paratroopers in the Army. The Air Corps had made a strong bid for the project. So also had the Chief of Engineers who saw paratroopers as saboteurs and demolition experts. The final decision was strongly supported by the contention of General Leslie J. McNair, chief of infantry at General Headquarters, War Department, that the primary mission of parachute troops was ground action and that air transport was only another means of transportation. Colonel Lee and the chief of infantry won the decisive battle that we celebrate today.

Only a few months later—in October 1941—the 88th Infantry Airborne Battalion was activated at Fort Benning. Lt. Col. Elbridge Gerry Chapman was named commanding officer with Maj. Hugh Harris, Lt. John A. Wallace and others on his staff. They were immediately faced with the need for airplanes and gliders suitable for moving men and equipment. John Wallace, in his training sessions in future weeks, always pointed out with a sly smile that in converting the commercial DC-3 to the hard-working C-47 of the Troop Carrier Command, the Douglas manufacturer had strengthened the floor, widened the door, installed folding seats, included loading rings and ramps, but unfortunately they had removed the attractive airline hostesses from their cargo aircraft. What a blow!

The first air-landing unit of the United States Army was the 550th Infantry Airborne Battalion formed in July 1941. The initial mission of the 550th was the defense of the Panama Canal Zone and its nearby territory using air-landing infantry. Available aircraft was limited.

Less then four months after Congress declared war in December 1941, the Airborne Command was activated by War Department directive. The orders brought together under one headquarters the Provisional Parachute group and the 88th Airborne Battalion. The new headquarters also included the existing parachute battalions and The Parachute School at Fort

Benning where jumpers were trained. Units stationed in the Panama Canal Zone were beyond jurisdiction of the new Airborne Command.

The orders directed that the Airborne Headquarters would move as soon as possible from Fort Benning to Fort Bragg near Fayetteville, NC. Col. William C. Lee, Infantry, was named commanding officer of the Airborne Command. Lt. Col. Elbridge Gerry Chapman, Infantry, was chief of staff. The officers and men of their previous units comprised the new staff and headquarters company. The Airborne Headquarters was moved to some empty barracks buildings at Fort Bragg in April 1942.

History tells us that the Airborne Command, and later the Airborne Training Center at Camp Mackall, NC, activated and trained the 82nd, the 101st, the 11th, the 17th and the 13th Airborne Divisions and the Airborne Brigades. The new Airborne Headquarters was also responsible for the training of separate parachute regiments and battalions, field artillery units, combat engineers, signal corps units, medics and many spare parts trained to move by airplanes, parachutes, and gliders.

On this fiftieth anniversary of Airborne troops in our U.S. Army, we might summarize what happened in the early months of World War II, following Pearl Harbor. It was a tremendous training mission which we celebrate today:

1. Among the initial missions of the new Airborne Command Hq. were these: To activate and to train airborne units, parachute, glider and air-landing, for combat; to train other units and organizations for movement by air; to develop tactical and training doctrines for airborne forces; to collaborate with the Army Air Corps in the development of all types of carriers for troops and equipment. Another important part of the cooperation with the Army Air Forces was the procedure for air support.

2. The Army Air Corps cooperated by the organization of the Troop Carrier Command in June 1942. The primary mission of the Troop Carrier Command designated by Army Air Corps was threefold: (a) To provide for the air movement of troops, including gliderborne troops, parachute troops, air-landing troops, and their equipment; (b) to supply and resupply these ground troops in the combat zone; (c) to evacuate casualties in airplanes and gliders.

3. A third most important group was procurement and supply which developed and delivered the aircraft, parachutes, weapons, equipment and supplies needed by combat units. Constant search was maintained for the effective lightweight weapons and vehicles to increase the firepower and mobility of airborne units and then to keep units on the ground supplied with ammunition, gasoline and other necessary items. It was a very big order and the cooperation on the part of all concerned was outstanding.

Units in training during the early years of the airborne practiced night flights with parachute drops and glider landings at low altitude. The risk was great, but Troop Carrier pilots and airborne troopers performed with courage. Units were organized and equipped as combat teams of infantry, field artillery and combat engineers. Anti-aircraft artillery units, signal communication and pathfinder teams and medical corps units were part of the effort. A principle objective of all airborne training was to give the individual soldier confidence in himself and his equipment and to overcome fear—fear of the unknown.

After the training of the 82nd and the 101st Airborne Divisions at Fort Bragg, both divisions took part in extensive maneuvers before movement overseas. The 11th, 17th and 13th Airborne Divisions were trained at Camp Mackall, NC. Training maneuvers used landing areas, drop zones and the airports at Camp Mackall, Pope Field at Fort Bragg, the Laurenburg-Maxton Airfield and the small airport at Knollwood in Pinehurst, NC.

When General Chapman was transferred from Airborne Headquarters to become commanding general of the 13th Airborne Division, his Chief of Staff Josiah

Airborne troopers were trained on airplane and glider mock-ups to learn the principles of safe loading. The mock-ups were dummy cargo compartments that gave troops ground training without precluding the use of airplane and gliders from necessary flight missions. (Photo from Airborne Training Bulletins)

Major Harold "Beaver" Beaty and Colonel Chester B. DeGavre of the Airborne Board about to make a test jump with the new Hart parachute designed to reduce oscillation. (Courtesy of Paul Troth)

T. Dalbey was promoted to Brigadier General in command of the Airborne Training Center. In effect, the training mission was essentially the same. However, the sizeable reduction of the troop list eased the responsibilities of the headquarters and decreased the demands made on it.

A most important unit of the Airborne Headquarters during all those months and war years was the Airborne Board under command of Colonel Chester B. DeGavre. Colonel DeGavre had completed jump training at The Parachute School. He was promoted and made president of the Airborne Board. He was closely involved in training.

The Airborne Board, with its group of trained paratroopers from Headquarters Company and from the 542nd Parachute Infantry, was constantly at work testing and developing various containers and parachute packs for dropping crew-served weapons, radios, demolition kits, ammunition, rations and other supplies. Many new types of parachutes were tested and considered for standardization to improve control in the air and to reduce oscillation. The daisy chain was developed to hold parachute bundles together. The specialists in field artillery used the daisy chain to drop and hold together the 75mm pack howitzer in seven parachute packs—two from the door and five from the belly of the C-47 from bomb racks. Heavier artillery was tested and flown in airplanes and gliders.

To pass the word to airborne units in training, the Publications Section and the Engineer Reproduction Section were busy parts of the Airborne Headquarters. Their jobs were to write, produce and distribute many airborne bulletins, film strips and training films. The training bulletins, produced with Airborne Command approval, were believed to be preferable to standard field manuals since their subject matter was not applicable to the entire Army and because they could be published more rapidly and changed more readily. All of us knew we were exploring new fields. We were doing, testing, saying and writing things about an airborne effort which was entirely new in the U.S. Army and the nation 50 years ago.

One of the greatest kicks our troopers who had worked on training films received was in the Hollywood film "Objective Burma." Our troopers shouted their delight and surprise when the rescuing units who landed in gliders in the Hollywood version of "Objective Burma" were none other than our own troopers landing among the pine trees at Camp Mackall. Hollywood had clipped our film of 11th Airborne Maneuvers.

Cargo gliders may be obsolete today as new assault forces are trained in helicopters and in other aircraft with clam shell doors for faster loading and unloading and capable of vertical take-off and landing. Larger, long-range aircraft are available today for moving combat troops and heavier equipment. Parachute training methods continue with high standards for combat duty. After 50 years, the individual soldier is still the most important guy in airborne operations.

Our airborne soldiers have helped to write American history in North Africa, Sicily, Normandy, Southern France, Holland, Bastogne, the jump across the Rhine and the drive on Berlin by the 82nd, 101st, 17th and 13th Airborne Divisions. Airborne history tells of the parachute jumps and troop landings in the Pacific Theatre of War on Lae Salamaua, Corregidor, Tagaytay Ridge, the liberation of Manila and the occupation of Japan by the 11th Airborne. In more recent years parachute, and airborne troops have added to the American story in Korea and in Vietnam.

Airborne troopers today are still proud to be "Airborne All the Way!"

As a matter of record, we can refer in closing to a letter of commendation received from the secretary of war when he visited Airborne Headquarters for the purpose of inspecting division training, parachute training and the glider training school in the very early days of World War II. Secretary of War Henry L. Stimson wrote in his letter: "I brought back with me to Washington much of real value from my trip....Your Command shows the results of great care and forethought in planning of training. Your troops are young, tough and enthusiastic, and I know that when the day comes in which the nation will rely upon them, they will perform outstandingly."

The secretary was right: Airborne troopers have performed outstandingly!

*NOTE—Extracts of this report on training are from the book **The Airborne Story** or **An Old Soldier's Sketchbook** written and illustrated by Paul Troth, Lt. Col. AUS retired, and published by Vantage Press. Troth served with the Airborne Command and Airborne Training Center from 1941 to 1946. He was on active duty before Pearl Harbor at the Infantry School, Fort Benning, in the early days of the Provisional Parachute Group and the 88th Airborne Battalion. He was in command of the small airborne detachment to General Douglas MacArthur's headquarters in Manila, Philippine Islands, in late 1945.*

LONG RANGE RECONNAISSANCE PATROL (LRRP) U.S. ARMY NATIONAL GUARD

By Steven Mrozek
SGT, MIARNG

Some of the most important elements necessary for unit commanders to make correct decisions in battle is the availability of accurate, timely intelligence about the enemy, his ability to fight and the terrain over which the battle will be fought. In this age of technical wonders, the most reliable means of acquiring this information is still by extraordinary soldiers equipped with special skills sent behind the enemy's lines to report back what they saw.

In response to this need for dependable battlefield intelligence, in the late 1950s several NATO countries began the development of the Long Range Reconnaissance Patrol (LRRP). Perhaps the most influential proponent was the British Special Air Service (SAS) whose practice of operating in small teams or patrols served as a model. The Army formed the first two American LRRP companies in West Germany on 15 July 1961. These two companies were attached to V and VII Corps.

The LRRP concept was expanded to include elements of the National Guard. In 1967, Companies D and E, 151st Infantry from Indiana became the first reserve component LRRP units organized. The following year, Michigan joined Indiana by organizing its own LRRP companies. On 1 February 1968, Companies E and F, 425th Infantry officially assumed their new mission. These two companies were formed from assets of B and C Companies, 1st Battalion (ABN) 225th Infantry, Michigan National Guard.

Accepting its new tactical responsibility with intensity and vigor, today's F Company (LRS), 425th Infantry has endured the test of time to evolve into the oldest existing LRRP Company in the entire United States Army.

After several reorganizations, the company was reorganized once again on 15 November 1965. This time the unit was known as Company B, 1st Battalion, 225th Airborne Infantry. This configuration would last for three-years before the company's mission became more specialized. Formed from assets of the disbanded airborne battalion, on 1 February 1968, F Company, 425th Infantry became one of the first companies to assume LRRP responsibilities. Company E along with its Detroit-based sister unit Company F, accepted the challenge few units could match. Years of difficult and demanding training lay ahead for the men of the 425th. Training exercises overseas as well as stateside would bring the companies in contact with the elite of our NATO Allies and the latest techniques and thinking of the LRRP doctrine.

A further consolidation on 1 February 1972 brought the two Michigan LRRP companies together in Pontiac. The result became the present F Company, 425th Infantry.

Back on 1 February 1969 all Army LRRP units were designated RANGER while still retaining their original LRRP mission. This authorization entitled F Company (Ranger), 425th Infantry to wear the coveted black beret and its own distinctive ranger "scroll" shoulder patch.

In 1981, F Company was given the responsibility of providing the LRRP capability for the XVIII Airborne Corps. This corps affiliation was transferred to the VII Corps in late 1983. Although the VII Corps was then stationed in Germany, F Company remained stateside but often trained overseas in Europe.

A change in status came on 1 September 1987 which replaced the title of Ranger with that of Long Range Surveillance Unit (LRSU). The black beret was replaced with a maroon one and an airborne tab replaced the ranger scroll. Later that year, F Company (LRS), 425th Infantry became affiliated with the military intelligence brigade of the I Corps stationed at Fort Lewis, Washington. This is the current assignment of the company which can claim to be the oldest existing LRRP company in the entire Army.

Army National Guard Long Range Patrol and Ranger Companies

LRP/Ranger Co	Served	Home Station	Assets Source
Co. E, 65th Inf.	1 Apr. 71 - 29 Feb. 80	San Juan, Puerto Rico	755th Trans. Co. (Md. Cgo. Trk)
Co. G, 143d Inf.	1 Apr. 80 - present	Houston & Austin, Texas	Co. A, 2d Bn. (Abn.), 143d Inf. 36th Abn. Bde.)
Co. D, 151st Inf.	1 Dec. 67 - 1 Mar. 77	Greenfield & Evansville, IN	Co. B & C, 1st Bn. (Abn.), 151st Inf. (38th Inf. Div.)
Co. E, 151st Inf.	1 Dec. 67 - 1 Mar. 71	Muncie, Indiana	Co. A, 1st Bn. (Abn.), 151st Inf. (38th Inf. Div.)
Co. E, 200th Inf	1 Dec. 69 - 1 Feb. 72	Mobile, Alabama	778th Maint. Co. (Lt. Equip.) (Direct Spt.)
Co. A, 259th Inf.	1 Nov. 71 - 1 Jun 74	New Castle, Delaware	1049th Trans. Co. (Acft. Maint.) (Gen. Spt.)
Co. E, 425th Inf	1 Feb. 68 - 1 Feb. 72	Pontiac, Michigan	Co. B, 1st Bn. (Abn.), 225th Inf. (46th Inf. Div.)
Co. F, 425th Inf.	1 Feb. 68 - present	Detroit (Until 72), Pontiac, Michigan	Co. C, 1st Bn. (Abn.), 225th Inf. (46th Inf. Div.)

THE SPECIAL ALLIED AIRBORNE RECONNAISSANCE FORCE

By Les Hughes

As early as 1943, Allied planners were concerned with the possibility that Allied prisoners of war (PWs) might be the target of localized or wholesale violence during a collapse of the Third Reich, or that the chaos surrounding such a collapse would, at the very least, impose considerable hardship on the PWs. The planners suggested that the PW contact teams of the various armies might be expanded to include airborne teams that would parachute into the vicinity of PW camps with which it was desired to make early contact.

It was not until February 1945, however, when the collapse of Germany appeared imminent, that SHAEF was provided with a specific mandate for dispatching troops whose mission would be to secure the safety of Allied PWs and to provide for their early evacuation. As a result of its mandate, SHAEF created in March 1945 the Special Allied Airborne Reconnaissance Force (SAARF). A golf course and its facilities, which formerly served as the Headquarters of the 21st Army Group, at Wentworth, was allocated for SAARF's Headquarters and Training Camp. Although both the Office of Strategic Services (OSS) and the Special Operations Executive (SOE), Britain's counterpart to the OSS, provided training support and personnel, and the 1st Allied Airborne Army provided personnel, SAARF remained under the direct control of SHAEF.

SAARF was genuinely an Allied unit. Brigadier J.S. Nichols, a British officer, was selected to command SAARF, and Col. J.E. Raymond, an American, was appointed deputy commander. The make-up of the operational personnel was even more international: 96 British, 96 American, 120 French, 30 Belgian and 18 Polish personnel. Furthermore, not all operational personnel were men: there were several women who had served as SOE agents in Nazi-occupied Europe. Of the 96 Americans on the operational side, 18 were drawn from the OSS while the balance came almost completely from the 82nd and 101st Airborne Divisions. Most personnel were volunteers (the official record states that everyone was a volunteer), many enticed by the promise of an early return to the States after completion of their SAARF missions, but some of the officers recall they were simply handed orders for SAARF and told to get a move on.

SAARF was structured into three-man teams (a misnomer in some cases, considering the presence of women in the unit) comprising, generally, two officers and an enlisted radio operator, with all personnel being airborne qualified. Because the time available for training was brief, the personnel in each team were of the same nationality (unlike the OSS's Jedburgh teams, the national make-up of which was intentionally mixed). Though time was short, an effort was made to weld the force into a single unit and to establish a unit espirit; and to that end, two distinctive sleeve insignia were created for the unit and issued to all personnel.

The collapse of Germany was more rapid than anticipated, and SAARF was able to mount but a single airborne operation—Operation VIOLET—that would prove to be the last airborne operation of the war in the ETO. On the night of 26 April 1945, six SAARF teams—two British, two American, and two French—were dropped at three sites near Stalag XIA near Altengrabow, Germany. The American team designated CASHBOX was an all-OSS team comprised of Captain J. Brown and Sergeants D. Murphy and E. Porada; the other American team PENCIL, was comprised of two officers from the 504th Parachute Infantry Regiment Captain G. Warfield and Lieutenant W. Meerman and an OSS radio operator Sergeant P. Potter.

Obviously, it was not the intention of a force of this size to secure the release of PWs by force; rather the teams' mission was to gather intelligence regarding conditions in the camp and, if possible, to establish contact with the German forces maintaining the camp. In the hope of encouraging cooperation among those with whom the teams might come in contact, each team member carried a document from SHAEF, written in German on one side and in English on the other, identifying him as a member of SAARF and describing his mission as one of determining whether those holding the PWs had discharged their Geneva Convention obligations and requesting assistance in fulfilling his mission. It is doubtful, however, that this document provided any great feeling of security to those who carried it in their kit.

The teams were scattered during the drop, and within 24 hours the senior officer, Major P. Worrall of the SOE, and his team were captured. To his surprise, Worrall and his team, soon to be joined by three other captured members of the mission, were taken to Stalag XIA rather than turned over to the Gestapo. Worrall proceeded to press his case for cooperation with the camp's commandant, who, with one ear cocked to the sound of Allied artillery in the distance, and the other to reports of Germany's crumbling resistance, treated his new prisoners with kid gloves; and Worrall became a de facto negotiator between the camp commandant, the PWs, and the Allied forces, eventually being allowed to communicate by radio with SHAEF from the commandant's office! The negotiations culminated with an agreement on the part of the Germans to surrender the camp and its 20,000 PWs (about half of whom were Russians) to the American 83rd Infantry Division on 3 May 1945.

Ironically, for one of the teams, it was not the Germans who turned Operation VIOLET into a harrowing experience. The team in question, an all-French team, was dropped into the middle of the German Scharnhorst Division, which it managed to elude only to be captured by advancing Russian troops. After strenuous interrogations and three weeks of being held prisoners, the Frenchmen managed to escape their Russian captors and make their way back to France. The incident presaged the shifting alignments the world would soon witness.

Because of the speed with which the Third Reich was disintegrating, the bal-

SUPREME HEADQUARTERS, ALLIED EXPEDITIONARY FORCE

COPY NO. 9

25 APR 45
DATE

To Whom It May Concern:

1. The bearer of this letter, further identified by personal identification tags or documents is:

MEERMAN / William / A. / Lieut / O-1296236
(Last Name) (First Name) (Middle Initial) (Rank) (Number)

of the regularly constituted military forces of United States
(Allied Country)

and a member of the Special Allied Airborne Reconnaissance Force, operating directly under this headquarters.

2. The mission of this individual is in connection with Allied Prisoners of War now in GERMANY. He comes to determine whether those who have such prisoners under restraint or control, are fulfilling their obligations under the Geneva Convention. In an emergency he is prepared to call for aid in obtaining food and medical supplies as required for their maintenance and well being and has facilities for establishing contact with Allied Agencies for that purpose.

3. All are enjoined to render such assistance to this individual and his party as may be necessary in the fulfilment of this mission, and to facilitate his return to his proper station.

By direction of the Supreme Commander:

T. J. DAVIS
Brigadier General, USA
Adjutant General

Signature of Individual:
William A. Meerman

Issued by: *J.S. Nichols*

Special Allied Airborne Recon. Force

---

ance of SAARF's teams were conveyed via surface transportation to their assignments. The teams found themselves assisting in the pursuit of war criminals and in the repatriation of those held in PW camps and Nazi jails (the latter being a difficult task as some of the civilian prisoners were criminals who had committed acts that would have led to their incarceration under any regime).

SAARF was formally disbanded in July 1945. A short-lived and obscure unit, SAARF was a strange note, most of the hardened veterans of airborne and special operations who served in it would have agreed, on which to end one's wartime career. And yet, one must wonder whether some of these men have not looked back and felt an increasingly sense of satisfaction that their last mission was a humanitarian one.

THE U.S. ARMY PARACHUTE RIGGERS

By Thomas R. Cross
Col., USAF (Ret)

The original World War II Parachute Maintenance Sections and Companies were made of combat arms personnel from the separate Regiments, Combat Teams and the Airborne Divisions. The TO&E for the Parachute Maintenance Company had either a -1 or -11 identification that designated its organizational personnel and equipment. It was not until the early 1950s that the Parachute Maintenance Companies were transferred to the Quartermaster Corps.

After World War II the 517th Parachute Regiment was absorbed into the 505th Parachute Regiment at Fort Bragg, NC. In 1946 and 1947 many were transferred out of the 82nd Airborne Division and were on their way to Japan where most ended up in the 11th Airborne Division.

This was a "no joy" period for the Airborne as the U.S. Army took upon itself to down-play the airborne units by making them take off their airborne patches etc. The 82nd lost its airborne tab over the AA patch; and the airborne cap patch was also a casualty of this short-sighted policy. The 11th Airborne Division was able to retain the airborne tab on its divisional insignia as it was an integral part of the patch. The 11th was permitted to wear the overseas cap while in Japan, but upon its return to the States in 1949 it was ordered to wear the "flying saucer" garrison cap. We thought this a low blow and that we looked funny as hell in our suntan uniforms with the bloused parachute boots and the silly looking "flying saucers"—we did too.

Upon my reporting to the 11th Airborne Division in November 1947, Major General Swing assigned me as commanding officer Special Troops (Rear) and as the 11th Airborne Division Parachute Maintenance Officer. Special Troops (Rear) consisted of the 408th Airborne Quartermaster Co., the 711th Airborne Ordnance Maintenance Co. and the 11th Airborne Parachute Maintenance Co.

I had no idea what a Parachute Maintenance Company did other than to pack and repair parachutes. I found out very quickly that it was engaged in other activities of an airborne nature in addition to running a Parachute Riggers School—supporting the 11th Airborne Jump School and the 11th's Flight Section at Lanier Field which even then consisted of some 21 L-4 aircraft (most of which were illegal). All of the above plus supporting the 11th Airborne Division's Airborne activities was unique in itself as the 11th was spread out in five separate military camps extending from Sapparo to Sendai, Japan. I quickly obtained the 4820 MOS for a parachute rigger so I could fathom my newly assigned duties.

What I found out in November of 1947 was that I had inherited an outstanding and a very unique organization from Major Percy Lisk, the former 11th Airborne Div. parachute maintenance officer. The company turned out to be of battalion size with a strength of 350 U.S. Army personnel and about 75 or more Japanese personnel that were employed directly by the company.

All of the U.S. personnel were from the combat arms and for the most part, well-decorated from World War II. The two Parachute Packing Platoon leaders, 1st Lt. Rip Riley and 1st Lt. Ed Thomas, both of whom had been awarded the Silver Star Medal were former members of the 101st Airborne Div. The company commander 1st Lt. John Concordia had started out as a U.S. Army Air Corps parachute rigger at Chanute Field prior to World War II before joining the parachute infantry. All of the NCOs were outstanding and included four members of the original U.S. Army Test Platoon.

I received a bonus shortly after taking over in the form of "Red" King from the original Test Platoon. "Red" apparently had become a little too hot to handle up in Sapparo, and when the Division G-1 asked if we could use "Red," the answer was a quick "Yes." "Red" had been my instructor when I went through the Parachute School in the early days of Airborne, and I had always wanted to get "Red" in my outfit—it finally came true.

When Maj. General Swing left the 11th Airborne he turned it over to Maj. General "Bud" Miley, former CG of the 17th Airborne Div. Both Gen. Swing and Gen. Miley had been concerned about the airborne training status of divisional units because of their widespread locations in Northern Japan and because of the restricted maneuver space adjoining these locations. Also, the 11th was getting a lot of new replacements who had very little unit airborne training due to being in transit and also because of the maneuver space restrictions already mentioned. Jump aircraft were also critical so parachute jumps had to be carefully scheduled and, at times, not under the most ideal conditions.

Some time after Gen. Miley took over, he determined that the 11th was suffering too many jump injuries due to jumping in wind conditions that caused a high degree of oscillation with the T-7 parachute. Wind conditions in that part of Japan tended to be a bit tricky and varied with seasonal changes. To make a long story short, Gen. Miley gave me a directive to determine proper parachute landing techniques to overcome the oscillation problem due to these wind conditions. The directive requested that we do this in winds up to 30 mph. He either thought we were the "hotshots" we thought we were or he wanted to clear out the entire outfit real quick in Roman-Legion-punishment style.

We got a team together of about 10 to 15 experienced jumpers that included most of the officers and senior NCOs and, in particular, "Robbie" Robinson and "Red" King from the original Test Platoon. Our solution was quite simple for we determined that the only way you could do this was to hold a strong parachute slip until just above the ground and then let up very quickly to make what some call a "standing landing." Sounds simple, but try executing it in high wings without some assistance from someone on the ground helping you to judge distance so you can let go of your slip in time to make a decent landing (which we defined as one not resulting in concussion or broken bones). That was only part of the problem, for after landing, a high wind could catch your canopy and send you head-over-heels down field.

We finally placed 1st Lt. Dave Peters on the DZ with a portable sound system that could be heard as the jumper came in for his landing. Lt. Peters managed to become an expert in issuing the proper instructions for letting up on the slip. We finally called a halt at around 25 mph and told Div. Hq. that it was not practical in even relatively low winds for many of the new jumpers and that some of the old ones would have trouble in determining when to slip and when to let up for the landing.

My purpose in recounting the above story is because it was at this point in time that I felt that the parachute rigger was someone special who deserved more than ordinary recognition for the responsibility that came with his turf. Also, I still smarted and was more then a little P.O.'ed at the manner in which the Army was treating its airborne troops—I am referring directly to the removal of our parachute devices from our overseas caps, shirts and jackets. I decided then to come up with some sort of cloth insignia depicting a parachute with a wing reading "Rigger" somewhere on the device. At about this time we received orders for the 11th Airborne Division to return to the USA for station at Fort Campbell, Kentucky, so I placed this project on the back burner until we got back to the USA.

The 11th Airborne Parachute Maintenance Co. entered into a new era after arrival in the Fort Campbell area in 1949. We became acutely aware that we were back in the airborne business and that we had to shake off some of the policies and habits acquired during our Army of Occupation days in Japan. Where we had been the only airborne division in Japan, we were now back in the USA which had been dominated, in an airborne sense, by the 82nd Airborne Division at Fort Bragg. Our biggest problem was the 11th's lack of large-scale unit airborne training and exercises as well as being behind the power curb on the new parachute heavy drop techniques that were being developed by the Army Field Forces Airborne Test Division in conjunction with the 82nd Airborne Division at Fort Bragg.

Fortunately, the 11th Airborne Parachute Maintenance Co. was in a good position to take this new challenge. We had the support of the division commander and his staff of which we took full advantage. The Division G-1 permitted us to set a standard of a 110 AGCT Score, similar to that required for OCS Candidates, as a requirement to be an 11th Airborne Div. Parachute Rigger Candidate for our own Parachute Rigger School. The only exception that was made to this policy was in the event that a candidate was also an exceptional football player and would accept special training to be able to play on the company football team. We, after all, had to maintain our standard as the 11th Airborne Division Football Champions. We were very proud of our football team which was a combination of the 11th Airborne Parachute Maintenance Co. and the 127th Airborne Engineer Bn. We had won the Division Football Championship for two years in a row after beating out the 187th, 188th and 511th Parachute Regiments as well as the 11th Airborne Division Artillery Teams for this title. It was a good recruiting vehicle for the company thanks to a friendly and understanding Div. G-1.

We kept the Parachute Rigger School going at Fort Campbell until such time as the Parachute School at Fort Benning could gear up to handle our needs. This changeover permitted the 11th Airborne Division Parachute Maintenance Co. to turn its priorities to that of getting the necessary equipment and knowledge so as to be able to train the 11th Airborne Div. in the new heavy drop parachute techniques.

We got off to a ragged start on the new heavy drop parachute techniques for we had no training nor did we have the necessary 100-foot diameter parachutes and Allied rigging equipment to even get started. Here is where the "Ol' Boy" Airborne Brotherhood paid off for we were able to make arrangements with good friends at the Army Field Forces Airborne Test Division at Fort Bragg to train our 11th Airborne Div. Personnel. I formed a Parachute Heavy Equipment Drop Section within the 11th PM Co. under the leadership of 1st Lt. Cecil Hospelhorn and Sgt. Richard "Dick" Dunn and placed them on TDY to the AFF Bd No. 1 at Fort Bragg to learn these new techniques. They were accompanied by a small group from the 11th Abn. Div. Arty. The support given by Major Graham Turbeville and CWO John Ward, along with Sgts. Carlton and Swann from the AFF Bd No. 1 Abn. Test Div., was outstanding. We could not have pulled it off without their 100 percent support. When our group finally returned to Fort Campbell they came back with several salvage 100-foot diameter parachutes, heavy drop platforms and rigging material. This salvaged equipment was welcomed with open arms and our 11th Airborne Div. heavy drop training began in earnest.

At about this time the 11th Abn. Division received notice that it would participate, along with the 82nd Airborne Div., in an airborne exercise with the code name of "Exercise Swarmer" at Fort Bragg in early 1950.

It seemed like everything was coming to a head at one time. We were just being introduced to the new C-119 aircraft with its tail loading and heavy drop capability. Additionally, most of the 11th Airborne Div. Staff was new to Airborne as were some of the newly assigned field grade officers that went into the parachute regiments and Division Artillery. Fortunately there was enough remaining old time airborne experience left in the G-3 Section as typified by Capt. Frank Naughton and in units like the 187th as typified by Major Dick Allen. The old World War II airborne experience and "can do" attitude were beginning to pay off.

Our airborne training was stepped up in order to get ship shape for Exercise Swarmer and with this came an expected increase in parachute accidents mostly due to poor jump procedures and techniques. We had not had a serious parachute accident or death for several years and felt that this was no time to start, so the division placed more emphasis on individual jump training as well as on jumpmaster training. This was in addition to the 11th PM Co. Heavy Drop School that was now in full swing. The 11th PM Co. was also assisting the units in their jumpmaster training as well as training on the smaller cargo parachutes and the equipment containers that related to these parachutes. The cargo chutes that were involved were the new 100 and 50 foot diameter parachutes as well as the old smaller World War II G-1 Type Cargo Parachutes.

An increased parachute injury rate is almost certain to start some kind of rumor about the cause of the injury whether it be excessive winds on the DZ, excessive aircraft speed or the parachute itself. Seldom, if ever, will the individual jumper admit to a poor body position or other self induced problem when explaining the cause of his injury and accident. To dispel rumors of poor parachute packing and in an effort to increase the proficiency of the individual parachutist, I formed what I called a Parachute Malfunction/Accident Team that would be present on each DZ, night or day, to immediately investigate any parachute malfunction and accident. The team was equipped with artillery field glasses, the necessary reporting forms and had a 1/4 ton truck that was painted a bright yellow. To start this program off to a good start I assigned Master Sergeant Albert "Robbie" Robinson and Sergeant

First Class William "Red" King, both original members of Army's First Parachute Test Platoon, to take charge of this function. These two were experts in every sense of the word and everyone in the 11th Airborne Division knew it. This program was a great success and in addition to improving the individual proficiency of the average jumper, by observing the type of exits etc. from the jump aircraft, we were also, in the 11th PM Co., able to check out the quality of our parachute packing procedures. Every parachute involved in an accident or injury was minutely inspected on the spot by our team and the name of the parachute rigger and the NCO that signed off on the Parachute Log Book was noted and then referred to the Packing Officer for appropriate action. We started the procedure of pulling out for a detailed inspection all of the parachutes packed by the rigger named in the Incident Report. If the same Rigger's name came up more than once, he was "boarded" by the Parachute Packing Officer and the Senior Parachute Packing NCO. If warranted, the rigger concerned was retrained in parachute packing skills. Additionally, when the reserve parachutes were due for repacking because of the expiration packing date, they were assembled together and each one was pulled, using the rip cord, in front of the Senior Packing NCO. We felt that we could never take a chance on anything going wrong—ever. The last part of the time-honored Riggers Pledge of "I will be sure—always" was respected and rigorously followed.

The airborne training tempo was increased as the deadline for Exercise Swarmer drew closer. We were still concerned about safety, and with the Division Commander's okay, the 11th PM Co. Parachute Riggers were required to personally inspect each parachutist before he entered the jump aircraft. This was in addition to the normal Jumpmasters Check that was performed at plane side. The 11th was not going to take any chances at this critical stage of training for Swarmer. As a result of this the 11th PM Co. Parachute Riggers were in great demand and constantly moving around to assist the jumpmasters to perform their functions.

It was at this time that I borrowed an idea from the crews on the U.S. Navy Aircraft Carriers and decided that the 11th PM Co. personnel would use different colored baseball caps to denote their specific area of airborne expertise. I selected red for the parachute riggers, green for the Heavy Drop Section, yellow for the Parachute Maintenance Section and blue for the Co. Staff. As time went on we finally used the red baseball cap as the only means of identification. These caps did have the Parachute Riggers' Wings centered directly over the bill of the cap. They were operationally effective and a great boost to the riggers' morale. What was really heartwarming was the 11th Airborne Div. liked this idea too. Initially, we were restricted to wearing these caps only in our own area of garrison operations and on the DZs and Aircraft Loading Areas. Later on, in Exercise Swarmer, the restriction was lifted and we were permitted a more liberal policy.

I honestly cannot remember ever asking for permission to adopt the red caps and our new, self-designed Parachute Riggers' Wings—we just did it. We bought the caps and the Riggers Wings out of the Company Fund, and they were supplied to us by Charley Crowe, owner of Crowe's Jewelry Store in Clarksville, TN. This would never have come about without the support and indulgence of the 11th Airborne Div. commanders, Maj. Gen. William Miley and later on in Exercise Swarmer, Maj. Gen. Lemuel Mathewson. Some of the Division Staff thought the 11th PM Co. was spoiled rotten by the CG—which was true, but then we thought that we deserved to be spoiled, apparently the CG and the division thought likewise. It was not until 1986 that the chief of staff of the U.S. Army approved the Parachute Riggers' Wings officially.

Prior to my departure from the 11th Airborne Division in 1950, the U.S. Army Quartermaster Corps influenced by Major Hal Dorsett QMC, had shown a great deal of interest in acquiring the Parachute Maintenance and Parachute Heavy Drop Role for the U.S. Army. The Quartermaster Corps finally took over this mission and provided an excellent career field for the many outstanding officers and NCOs that transferred to this new opportunity. There could not have been a better ending.

I have been asked many times as to why I chose the specific design now adopted as the official Parachute Riggers' Wings. There are two main reasons. The first, is that I felt that the Parachute Rigger, because of the important responsibility entrusted to him and him alone, was truly deserving of a special recognition. The second reason is that I always thought that the regular parachute wings were too small and that they should be more like the British Airborne Wings that are worn on their red berets.

Knowing I could never influence a design change for the regular Parachute Wings, I knew I could do something about our own Parachute Riggers' Wings. The initial design was overly conservative and ended up with a half-wing with a parachute emblazoned with the word "Rigger." This device looked too much like the already-approved Parachute Pathfinder Wings, so we scrapped it and went to a full-wing approach. Within the 11th Airborne Parachute Maintenance Co. we authorized the wearing of the Parachute Riggers Wings on the fatigue jacket and shirt and on the red cap. Only those persons awarded the Parachute Riggers MOS were permitted to wear these wings.

The Rigger's Art: Parachute pack and static line (82nd Airborne Division Association)

AERIAL DELIVERY
THE STORY OF THE 8081ST

By E.G. Miller

In July 1950, shortly after the outbreak of communist aggression in Korea, an order was effected through Headquarters Second Army to activate an aerial delivery company. The primary mission of this organization would be to airdrop supplies by parachute to combat units on the front lines where other means of delivery were not feasible.

This company was to be composed of men hand-picked from the various technical and airborne branches of the Army—men who were well versed in all phases of their assigned duties.

One month following its activation at Fort Campbell, Kentucky the 2348th Quartermaster Airborne Air Supply and Packaging Company (later designated as the 8081st Army Unit) arrived at an airbase in southern Japan ready to assume its role in the Korean War.

Sukchon-Sunchon

An only too vivid picture remains in the memories of the para-quartermasters of their introduction to the functions of an aerial delivery company, for the first big operation they were to encounter was the airborne invasion of Sukchon and Sunchon, North Korea in October 1950. Hard work and sleepless nights were involved in outloading the 187th Regimental Combat Team, then part of the 11th Airborne Division, from Japan to Kimpo Airdrome in Korea which was to be the starting point and resupply hub of the attack.

Not soon to be forgotten was the proud moment when the first supply-laden planes lifted their wheels from the runways of Kimpo signifying that the 2348th's first major airdrop mission was under way.

While the main body of the unit prepared and air-dropped over 650 tons of supplies for a period of three days to the embattled RCT, 17 of the 2348th men parachuted into Sukchon with elements of the 187th in order to recover parachutes and heavy equipment for use again in future operations.

Chosin Reservoir

It began to look as if fate had big things in store for the 2348th in those months, for soon after successfully completing the Sukchon-Sunchon drop Chosin Reservoir became a hellhole for elements of the 7th Infantry and 1st Marine Divisions which were cut off from regular supply lines by surrounding communist forces. As a result, the operational headquarters of the 2348th became a scene of hurried activity as ammunition, food and medical supplies were rigged in back-breaking round-the-clock hours for aerial delivery to the trapped troops.

In the meanwhile, at an advanced airbase in North Korea, to which they had previously been dispatched, a small detachment of 2348th men lost no time in preparing, loading and accomplishing airdrops totaling over 380 tons. Many of the para-quartermasters flew five and six missions a day over the mountainous terrain of the reservoir area.

Included in these drops, in addition to the usual combat supplies, was an M-2 treadway bridge. This was the first time in history that a bridge had ever been airdropped to troops in combat. Here was where the men of the 2348th proved themselves masters of aerial delivery techniques. The mission consisted of parachuting eight bridge sections onto a 240-foot drop zone which left no margin for error in timing for the men who accomplished this modern miracle.

As advancing communist hordes pressed closer to the airbase at Yonpo where resupply activities were in progress, necessity caused a reversal of events and thousands of dollars worth of valuable material had to be loaded on airplanes to be lifted to Japan. Only when the last plane was loaded did the para-quartermasters leave the doomed and burning field.

Munsan-ni

The designation of the unit changed in the months following those record breaking drops at Chosin Reservoir, but, whether the outfit went by 2348th or 8081st, the constant flow of commitments continued without letup. Scores of drop zones like Hongchon, Changto, Yangsang-ni and Singi-ri felt the jarring thud of parachute supply bundles before the last cold days of winter came to an end.

Spring was two days old when the skies over Munsan-ni, Korea became alive with the sight of thousands of 187th parachutes as the second airborne assault of the war was born. For three days the 8081st dropped supplies that literally kept the troopers pushing in spite of heavy enemy resistance. The perseverance of the 8081st served to complete another mission successfully.

Caterpillars and Casualties

However great the airdrop achievements, the task of supplying units in combat is one which cannot be accomplished without endangering the lives of the men who fly the loads and eject them from the aircraft. At any time the aerial delivery technician may find himself drifting to earth with the supplies he is delivering and thus qualify for membership in the "Caterpillar Club." Such was the case of two members of the organization who were knocked from their aircraft by released loads during the Munsan operation.

Altogether, seven 8081st men have had to make emergency jumps while delivering airdrop loads. Out of these events a certain amount of notoriety fell to Cpl. Thurmon Paiva who had packed the parachutes of six of these men.

Heavy ground fire is also a constant threat to a para-quartermaster. Two technicians, Sgt. Floyd N. Alexander and Cpl. Jack A. Beck, were killed when the Flying Boxcars which they were flying were hit by artillery fire.

In addition, three men have received Purple Hearts for wounds received in action. One of these, Sgt. John D. Hodo, narrowly escaped death when his parachute slowed down an armor piercing round and prevented it from inflicting a fatal wound.

While each man of the 8081st understands that his plane may be the next "target for tonight" he must ever bear in mind his creed: "A load of supplies may save lives—if need be, risk yours to deliver it!"

Training Between Missions

In the months succeeding the Munsan operation, as the peace conferences began to be called, the demands for aerial resupply lessened and the para-quartermasters began to find out what a "between-commitment" period meant. Description of aforementioned: training! Extensive, intensive training.

Aerial delivery is an exacting task. If it is to be accomplished successfully it must be practiced repeatedly. With this thought in mind, the 8081st set up schools to refresh the minds of the veteran airdroppers and to teach new members of the company aerial resupply procedures.

Surprisingly enough the para-quartermasters were not the only ones interested in learning airdrop techniques, for common was the sight of members of different branches of the service and other armies of the United Nations (the 1st Commonwealth Division, the Marines, the 187th RCT, the Air Force) being shown how by the 8081st.

Frequent jumps were another important item on the training agenda. Some Airstrip became a gathering place where people came for miles around to watch the awe-inspiring sight of parachutes carrying men and equipment to earth.

Our Secret: Teamwork!

An organization such as the 8081st Army Unit to be capable of handling large operations like those already mentioned must function much the same as the proverbial well-oiled machine. Each section within the company must work in close coordination with the others. The individual himself should know every phase of his assigned duty so that his section may operate smoothly, whether he is rigging heavy drop material with an aerial delivery platoon, packaging small supplies and computing load weights with the packaging and manifesting team or repairing and maintaining cargo and personnel parachutes with the maintenance section.

In short, teamwork is the mainstay of an aerial delivery company and teamwork is what has made it possible for the 8081st to break records in the performance of one of the most singular missions undertaken by Quartermaster troops.

Curtain Call

The history of the 8081st is thus written. Yet, it is not complete, for countless other battle-scarred ridges and valleys that lay as mute testimonies to the aerial delivery miracles performed by the 8081st are not mentioned on these pages. Such places as the Punch Bowl and Heartbreak Ridge are names that will be recalled by posterity whenever mention of the Korean War is made.

UH-1D helicopter of the 2nd Bn., 502nd Regt., 101st Abn. Div. airlifts a M-151 jeep to Deo Mang Pass, 23 August 1965. (Photo by PFC Stanton R. Pratt)

CHRONOLOGY
U. S. ARMY AIRBORNE FORCES

By Geoffrey T. Barker

The concepts of airborne warfare date back far beyond 1940, designated as the official birth year of the United States Airborne Forces. As we celebrate the Golden Anniversary of military parachuting in the United States, let us review the activities that occurred during the first half century of parachuting, commencing with some thoughts of those events and significant activities that led to the evolution of todays airborne forces. Some units listed in this chronology were not airborne units initially, like the six Ranger battalions of World War II, who provided the lineage and demonstrated purpose of the airborne rangers that later served with distinction in Korea, Vietnam, Grenada and Panama. Other units, although not airborne, were assigned or attached to, and supported or transported the Airborne Divisions or Brigades.

The format of this Chronology follows the following sequence by year (when applicable):

1. Pre 1940 events.
2. 1940 through 1990, data sequence by year, as applicable, is as follows:
    Combat Arms (Infantry, Armor, Artillery units)
    Special Operations (Special Forces, Ranger, Psychological Operations, and Combat Support and Combat Service Support areas by year group)
    Aviation (Conventional, airborne and Special Operations Aviation units)
    Medical (Conventional, airborne and Special Operations Medical units)
    Combat Support (Engineers, Military Intelligence, Military Police and Signal Corps units)
    Combat Service (Conventional, airborne and Special Operations Adjutant General, Chemical, Maintenance, Ordnance and Transportation units)
3. It should be noted that as a unit may be converted (ie. change structure from a combat service support unit to become a combat arms headquarters), that unit which is indexed by year group, may be located within either the combat arms, combat support, or sometimes in both areas.

——— Pre-1940 ———

Some of the earliest recorded concepts surrounding the principles of parachuting and vertical envelopment, date back to the 15th Century and the works of Leonardo Di Vinci. He is attributed with stating in 1495, "... if a man had a tented roof of calked linen, twelve braccia broad and twelve braccia high, he will be able to let himself fall from any height without danger to himself." His earlier sketches depict a four sided pyramid design for his proposed parachute. Of course this period was long prior to the evolution of aircraft, requiring Di Vinci to also ponder the conceptual fore-runners of modern aviation.

A magnificent statesman and forefather of our great country, Benjamin Franklin, considered airborne forces when he wrote in 1784, "... where is the Prince who can afford so to cover his country with troops for it's defense, as that 10,000 men descending from the clouds, might not, in many places, do an infinite amount of mischief before a force could be brought together to repel them." Early uses of Psychological Operations, (currently an important element of today's Special Operations inventory), are also attributed to Benjamin Franklin. During the wars between England and France, the emperor of France, Napoleon Bonaparte, dreamed of invading England using hot air balloons, during the infancy of aviation.

As the utilization of balloons increased, parachutes were developed for use as safety devices in the event of emergencies. A successful demonstration of parachuting from a hot air balloon was made by Andre J. Garnerin, in England in 1797. A leading American balloonist of the 19th Century, John Wise, made several successful parachute descents in Pennsylvania in 1838. Balloons became important military observation platforms during the Civil War, and later as transportation for the evacuation of dignitaries from Paris during the Franco-Prussian war of 1870-71.

The advent of powered flight circa 1903, realized the significance of the parachute - as a safety device for aviators, although many US Army aviators did not carry parachutes aboard their aircraft during World War I. The English Air Board rejected the issue of parachutes to pilots, stating that such devices "... invited cowardice in action." German pilots however, discovered parachutes to be critical life savers when facing the onslaught from the Royal Air Force. A container device attached beneath the aircraft wing, was developed by a carnival parachutist - Leo Stevens. A line from the pilot, attached to a parachute within the container, extracted the parachute during emergencies. This greatly influenced the static line of today. The device was successfully demonstrated at Jefferson Barracks, Missouri in 1912 by Albert Berry, but despite the enthusiasm of the assembled military, airborne force concepts were still far from being fully accepted. Two years later in 1914, just prior to the commencement of WWI, Charles Broadwick demonstrated his recently invented 'back-pack' parachute to the US Army. Again, regardless of tremendous enthusiasm by the observers of this demonstration, the US Army elected to support 'tried and trusted' conventional war fighting methodology. Development of the static-line parachute is attributed to Charles Broadwick.

An early visionary of parachute forces was Brigadier General William (Billy) Mitchell. Towards the end of WW I, BG Mitchell proposed a terminal offensive, culminating in a parachute descent by the US 1st Infantry Division, into the city of Mentz, Germany. General Jack Pershing accepted BG (then Colonel) Mitchell's proposal to seriously consider the introduction of airborne forces into the US Army inventory. In 1918, he nominated Major Lewis H. Brereton, to perform the initial feasibility studies on the use and practicality of airborne forces. Preliminary training and logistical procurement had commenced, and was underway when the War-to-end-all-Wars terminated a month later. At this juncture, airborne forces were still a paper concept.

A field test site was established at McCook Field, Ohio in 1919, with an airborne board headed by Major E. L. Huffman. Another established carnival parachutist, Leslie "Sky-High" Irvin, developed a parachute harness, and also experimented in free-fall parachuting. His impressive demonstrations resulted in a contract to supply the US Army (and later the British Army), with military parachutes. Brief experiments were conducted in 1928, when a three man machine gun crew were dropped from four aircraft at Brooks Field, Texas in 1928. Within minutes following their parachute descents, the men had the gun assembled and operating. Limited enthusiasm was developed during the anti-war syndrome of the 1920s and the looming economic depression of the 1930s.

The theories of airborne utilization were strongly supported in Russia in the early 1920s, when parachuting and gliding were introduced as national sports pastimes. A 1,500 person mass tactical airborne demonstration was held in 1936, but shortly thereafter the novelty quickly diminished.

The first country to develop military parachuting techniques was Italy, when in 1927, the theories of Giulio Douhet were employed, using modified versions of the parachutes designed as safety measures for airmen and balloonists. The first unit level airborne operation was conducted in 1927, with battalion-sized airborne elements undergoing training by 1930. Although they did not participate in combat airborne operations during the Second World War, these units were developed into the Folgore and Nembo Divisions of the Italian Army. Italian parachute schools were established in 1938 at Tarquinia, Italy and Tripoli, Libya.

The German government realized the implications of fast deploying airborne forces, and their principal airline - Lufthansa - commenced their acquisition of passenger aircraft that could be quickly modified as troop transporters in 1928. A German parachute training center, was established in Stendal, Bavaria in 1936. A similar school was almost simultaneously opened in Poland, based upon the Russian methodologies. The German institution was allowed significant importance, with Major General Oswalt Bassenge appointed as the Commandant. General Bassenge deliberately arranged for airborne forces participation in the West Prussian major exercises of 1937. A very interested observer of the exercises was Adolph Hitler, and shortly afterwards, the DFS 230 Cargo Glider was developed. On 4 July 1937, Major General Kurt Student was directed by Herman Goring to build an airborne division in preparation of attacking Czechoslovakia. The 7th Flieger Division was constituted, comprised of two parachute battalions and seven airlanded battalions. In late 1938, most of the 7th Flieger Division assets were recalled back to their parent units. General Student commenced the task of rebuilding his airborne division, arranging for himself to be appointed as the Inspector of Parachute and Airborne Forces. He convinced the German Army to redesignate the 22d Infantry Division as an airlanding division. In 1939 he had formed the 1st Parachute Regiment, and already had two battalions assigned to the 2d Parachute Regiment, and presented a 1,000 man mass tactical airborne demonstration onto a single drop zone. In October of that year, Adolph Hitler personally ordered General Student to train for airborne operations in Western Europe.

The French commenced parachute training at Avignon in 1935, organizing and activating the 601st and 602d Air Infantry Groups in 1937. Simultaneously, the German Air Force commenced the development of airborne forces at Stendal, on the outskirts of Berlin.

Not withstanding his failure to introduce airborne forces into World War I, BG Mitchell had not forgotten his dream. Continuing to pursue his efforts, he influenced the War Department to recommend that Major General George A. Lynch, Chief of Infantry, formulate the establishment of the Air Infantry in 1939.

Although the United States remained lethargic, while parachute development was intense in Russia, Germany, Italy, and later in England. Following four years of study, development and modernization, the Airland Corps was developed by the Russian Army, engaging in serious exercises during the next two years. Concurrently, both the French and German armies established parachute training courses. Adolph Hitler continued to encourage and develop the 'fallschirmtruppen' and glider aircraft prior to the onset of World War II. In November 1939, Finland was attacked by the Soviet Army, employing thirty Divisions, and small detachments of airborne forces.

—— 1940 ——

On 9 April, vital military targets in Denmark and Norway, were seized by German troops of the 1st Parachute Battalion. The four objectives of the 1st Parachute Battalion, were the Vordingborg Bridge and the Alborg Airfield in Denmark; and the Sola and Fornebu airfields in Norway. A 30 man platoon parachuted into Alborg, seized the airfield, and within two hours the Luftwaffe was landing aircraft. The Vordingborg bridge was captured in a matter of minutes by 60 paratroopers. Poor weather conditions required the Junkers aircraft to fly in at low level, and the pre-airborne strafing by the Messerschmitts only served to alert machine gun installations. Casualties at Sola were heavy, but the airfield was captured in two hours. The main force inbound to Fornebu was forced to turn back due to the severe weather conditions. Meanwhile one of the strafing Messerschmitts was hit and forced to land. Upon landing, the pilot so effectively employed his machine guns in a ground supporting role, the superior Norwegian force retreated.

In May of 1940, Germany next attacked Belgium and Holland, again employing both airborne and glider borne forces of the 7th Flieger Division, to complement their conventional forces. The German landed their glider forces initially, quickly followed by their parachute troops. In the Maastricht Appendix of Belgium was prime targets; Fort Eben-Emael, and three bridges cross the Albert Canal. Two of the three bridges were secured intact, and Fort Eben-Emael was captured by a force of less than 80 paratroopers. The targets in Holland were manned and waiting, and German airborne forces were repelled as they attempted their invasions adjacent to the Hague and Rotterdam. General Student was himself wounded south of Rotterdam. Despite fierce opposition from the Dutch, the German airborne forces held their targets until relieved by the advancing 9th Panzer Division.

The Prime Minister of England,

Winston Churchill, after observing the success of German Fallschirmjaeger, called for the activation of parachute and seaborne commando units. Anticipating the Prime Minister's directive, had established the Central Landing School at Ringway Airport, Manchester, the previous day. Lieutenant Colonel Dudley Clarke, military assistant to Sir John Dill, Chief of the Imperial General Staff, was appointed Chief of M09, and tasked to raise a parachute force. Airborne and commando personnel were trained at Special Training Schools (STS), the organization responsible for training members of Special Operations Executive (SOE), the British counterparts to the US Office of Strategic Services (OSS). In July, Number 2 Commando and personnel from Number 1 Commando, commenced parachute training at RAF Ringway, and designated as the 11th Special Air Service Regiment. Two British Parachute Brigades were initially envisioned - one in England, and a second one in the Middle East Theater. The Royal Air Force begrudgingly provided six Whitley bomber aircraft - sorely needed for bombing runs - for parachute training techniques and developments. The initial exit methods employed by the British consisted of aperture (jumping through the hole), and the 'pull-off' method from the tail gunners position at the rear of the Whitley. The first official descent, using the pull-off method was recorded on 13 July 1940.

The development of airborne forces and their supporting equipment was not addressed seriously by the War Department until 1940, and the likelihood of US involvement in the ongoing World War II became imminent. Chief of Staff of the US Army, General George C. Marshall, directed that the concepts for the formulation of airborne forces be revisited, and Major William C. Lee was assigned to implement the results of the voluminous parachute feasibility studies. Major Bill Lee had no idea that he would not only become regarded as the 'Father of the US Airborne', but would also command the mighty 101st Airborne Division (the Screaming Eagles).

The 29th Infantry Regiment, supporting the Infantry Board at Fort Benning, Georgia, provided more than 200 volunteers for assignment to the newly activated Parachute Test Platoon. Only 48 volunteers were selected, and US Army airborne training was officially under way.

Lieutenant William T. Ryder, assigned to the Infantry Board, was the first commander of the Parachute Test Platoon. He was selected from 16 of the initial 17 officer volunteers. (The 17th volunteer was Lieutenant William P. Yarborough, who since volunteering, had been reassigned from Fort Benning, Georgia to Fort Jackson, South Carolina). Second Lieutenant James A. Basset was selected as the assistant platoon leader of the Test Platoon. Lieutenant Ryder assumed all responsibilities for the basic research, development and testing of parachute equipment, techniques and doctrine. Warrant Officer Harry Wilson was assigned from Kelly Field, Texas to assist in the development and training of parachute troops. The T-4 parachute was developed at the Air Corps Test Center. Lieutenant Ryder is also credited with the design of the circular cloth parachute insignia worn on the US Army overseas cap. This insignia became the forerunner of the 'glider patch', worn by army personnel on 'jump status' (ie in valid authorized parachute positions within their assigned organization). The overseas cap and glider patch was eventually replaced by the airborne forces maroon beret. Airborne units now wear shield-shaped beret 'flashes', with different designs identifying specific organizations. Officers wear their rank insignia centered on the beret flash, while enlisted personnel wear the distinctive unit insignia (DUI) referred to as the unit crest, centered on the flash.

The Test Platoon, Parachute Troops and Air Infantry, consisted of the following assigned personnel:

First Lieutenant William T. Ryder, Second Lieutenant James A. Bassett, Sergeants John M. Haley, Benedict F. Jacquay, Lloyd McCullough, Lemuel T. Pitts, Grady A. Roberts and Hobart B. Wade. Privates First Class Tyerus F. Adams, Willie F. Brown, Floyd Burkhalter, Donald L. Colee, Louie D. Davis, Edgar F. Dodd, Johnnie A. Ellis, Mitchel Guilbeau, George W. Ivey, John M. Kitchens, Edward Martin, Lester C. McLaney, Tullis Nolin, Joseph L. Peters, John F. Pursley Jr, Benjamin C. Reese, Alsie L. Rutland, Louie O. Skipper, Raymond G. Smith, John A. Ward, Thurman L. Weeks and Obie C. Wilson. Privates John E. Borom, Leo C. Brown, Jules Corbin, Ernest L. Dilburn, Joseph E. Doucet, Aubrey Eberhardt, Frank Kassell Jr, Richard J. Kelly, Sidney C. Kerksis, William N. King, John O. Modisett, Robert H. Poudert, Albert P. Robinson, Thad P. Setman, Robert E. Sheperd, Arthur W. Swilley, Hugh A. Tracey and Steve Voils Jr.

US Army airborne history was officially recorded on 14 August 1940, when the Parachute Test Platoon conducted it's first airborne exercise from a Douglas B-18 aircraft at 1,500 feet. Lt Ryder led the way as the first man in the door. The scheduled first enlisted man to jump froze in the door, allowing the scheduled number two enlisted man, Private William N. (Red) King to become the first enlisted parachutist in the United States Army. Regretfully, Red King passed on in September 1988, missing the 50th Anniversary celebrations, already scheduled to be held in Washington, DC in July 1990. Credit for the adoption of the famous airborne exit war cry of "Geronimo!", is attributed to Private Aubrey Eberhardt, who had been watching a western movie the previous evening.

The month following their epic qualifying exercise, the Parachute Test Platoon became the embryo of the 1st Parachute Battalion. Later that same month (September 1940), the 1st Parachute Battalion was designated to become the 501st Parachute Battalion. Airborne history had commenced - the US Army's apparent notoriety for reorganizing, redesignating and changing the component status of units - had already infiltrated the fledgling airborne force. (Actually, the redesignating from the 1st to 501st Parachute Battalion was to eliminate confusion with the the 1st Parachute Battalion that was activated by the US Marine Corps. The first commander of the 501st Parachute Battalion was Major William M. Miley, assigned in November 1940.

The 501st Parachute Infantry Battalion, in addition to being the embryo for the airborne forces of the United States Army, produced thirteen General Officers from within its ranks. Rising to the General Officer level from the 501st Parachute Infantry Battalion, were Brigadier Generals James W. Coutts, George P. Howell, George M. Jones and William T. Ryder; Major Generals Gerald J. Higgins, Roy E. Lindquist and William M. Miley; Lieutenant-Generals Patrick Cassidy, Julian J. Ewell, Robert F. Sink and William P. Yarborough; and Generals John H. Michaelis and Melvin Zais.

In late 1940, the erecting of the landmark 250-foot jump towers commenced at Fort Benning, Georgia. The towers had originally been designed by the Safe Parachute Company as highlights of the 1939 World Fair, held in New York.

—— 1941 ——

Major William M. Miley, commanding the 501st Parachute Battalion instituted some 'radical' initiatives in the uniform arena. He authorized the wearing of parachute jump boots (the design later to be perfected by Captain William P. Yarborough), with the Army Class 'A' uniform. He also authorized the wearing of the blue cloth circular disk, with a white parachute centered, on the overseas cap, previously designed and recommended by Lieutenant William T. Ryder.

Another young stalwart of the 501st Parachute Battalion was Captain William P. Yarborough, who had 'escaped' from Fort Jackson, South Carolina. Becoming a significant contributor to what have become traditions of today's airborne, Captain Bill Yarborough was tasked to design the parachute badge for the US Army. The limiting restrictions in the design of the badge, included an overall width of no more than two inches. The width limitation was to ensure that the parachutist badge was not mistaken for the revered wings of the intrepid aviator. (Even these restrictions did not compare with those imposed upon their British counterparts, who were required to wear cloth wings sewn on the upper right arm of the uniform. This totally eliminated any comparison with the Royal Air Force.) Undaunted, Captain Yarborough hand carried the approved designs to a jewelry manufacturer in Philadelphia, literally camping on the doorstep of Bailey, Banks and Biddle, until completion of the initial order of 350 sets of airborne wings, of the same design as worn today. It should be noted that also in 1941, Captain Yarborough designed the Model 1941 US Paratrooper Jump Uniform; recommended the adoption of the high-topped paratrooper boots; and designed the original DUI of the 501st Parachute Battalion. The US Army parachute badge was issued to individuals as they became qualified, however in March 1941, the first unit award of the parachute badge was made by the Commanding General of the Infantry School, (then) Brigadier General Omar N. Bradley.

Captain Yarborough developed another initiative. He designed the oval wing background to be worn behind the parachutist wings. Initially this provided unit morale (as each unit background was designed differently), and gave the wings a larger appearance. Today, the wearing of the unit wing background is an indication that the wearer is currently assigned to a parachute duty position. The 501st Parachute Battalion wing background was blue and red, adopted from the unit colors.

Other assignees to the 501st Parachute Battalion included (of course), Lieutenant William Ryder, and a former West Point instructor, Captain James M. Gavin, whose original application to volunteer for parachute duty had not been favorably considered.

The Provisional Parachute Group was activated under the command of (then) Lieutenant Colonel Bill Lee, (who had originally been tasked by General Marshall to implement the results of the parachute feasibility studies.) The initiative to form this umbrella unit is given to Major William M. Miley. The Provisional Parachute Group functioned as the overall command of the parachute battalions and the parachute school, was activated at Fort Benning, Georgia.

The XI Flieger Corps was constituted in the German Army, comprising the seasoned 7th Flieger Division, the 22d Infantry Division (Air-land), the XI Air Command (consisting of the airborne air transportation units), the Corps Troops (medical, signal and supply elements), and the Parachute Assault Regiment (airborne engineers). During the German 'blitzkrieg' of Greece, the German army used airborne forces to successfully seize the bridge at Corinth, holding the bridge to allow their armored and mechanized forces to pass through. The German casualties totalled 240, but resulted in the capture of 2,000 British and Commonwealth troops and 8,000 Greek and Yugoslavian troops.

The German airborne, now fully established, next conducted an airborne invasion against the British defenders on the island of Crete. This costly operation accounted for almost fifty percent (11,000) killed or wounded, and one third of their aircraft lost. The German paratroopers were stopped and rendered ineffective at Retimo, Heraklion and Canae. What appeared to be ideal glider landing zones, were in actuality terraced hills.

The British 11th Special Air Service Battalion (SAS), conducted it's first airborne commando raid in February on Trajino Aqueduct in Italy, which was totally effective, with the raiders escaping by submarine. In September, the 11th SAS became the nucleus of the 1st, 2d and 3d Parachute Battalions, 1st Parachute Brigade. The 2d Parachute Brigade was formed by British Commonwealth troops from India. The 1st Air-Landing Brigade was formed from a brigade recently completing mountain orientation, thus trained in survival with light equipment. With an anticipated shortage of gliders, the British opted for 'Air-Landing' instead of 'Glider', as they envisioned the possibility of air-landing rather than glider landing. With three airborne brigades (two parachute and one air landing), the British 1st Airborne Division was activated on 29 October under the command of Major General F. A. M. (Boy) Browning.

Although not a military organization, the office of the Coordinator of Information, was headed by Colonel William (Wild Bill) Donovan, a Congressional Medal of Honor recipient during World War I. He was appointed by President Theodore Roosevelt to oversea espionage, sabotage and intelligence collection. Many assignees came from the uniformed services, performing airborne infiltrations as part of their assignments.

Combat Arms
Infantry

Another breed of airborne soldiers began to swell the ranks of the Provisional Parachute Group - the glider riders. The 88th and 550th Infantry Airborne Battalions were breaking ground for the 'fearless men in their flying machines' who were to follow in their footsteps. The 88th Infantry Airborne Battalion (originally commanded by Lieutenant Colonel Eldridge G. Chapman), was later reorganized and redesignated as a Glider Infantry Regiment. Let the records show, that whereas the paratrooper has always been required to volunteer, glider soldiers were conscripted from both volunteers and non-volunteers. The 550th Airborne Infantry Battalion was activated from a nucleus of the 501st Parachute Battalion, at Howard Field, Panama under the command of Lieutenant Colonel Harris M. Melansky.

Companies E and F, 51st Infantry (Armored), was activated at Pine Camp, New York, as an element of the 4th Armored Division. Company E, 51st Infantry, had been constituted in the Regular Army in June 1917, and activated at Chickamauga Park, Georgia. The 51st Infantry was assigned to the 6th Division in November 1917, and inactivated at Camp Grant, Illinois in 1921. Relieved from assignment to the 6th Division in 1927, the 51st Infantry was later assigned to the 9th Division in 1933, and back to the 6th Division from 1933 until 1940.

Lieutenant Colonel George M. Jones commanded the 501st Parachute Infantry Battalion. Company C, 501st Parachute Battalion departed Fort Benning, Georgia for Fort Kobbe, Panama Canal Zone.

The 502d, 503d and 504th Parachute Infantry Battalions were activated at Fort Benning, stemming from the original cadre of the 501st Parachute Infantry

Battalion. The first commanders of the 502d and 503d Parachute Infantry Battalions were Major George P. Howell Jr, (502d), (former executive officer of the 501st Parachute Battalion), and Major Robert F. Sink.

Field Artillery

The 4th Battalion, 18th Field Artillery, constituted in the Regular Army, was activated at Fort Sill, Oklahoma.

Medical

Constituted in the Regular Army in 1929, the 41st Hospital Company was redesignated in 1936 as Company H, 14th Medical Regiment, and in 1941 as Company H, 64th Medical Regiment, when it was activated at Camp Bowie, Texas.

—— 1942 ——

Lieutenant General James McNair, commander of US Army Ground Forces, activated the Airborne Command at Fort Benning, Georgia, to which all parachute units were assigned. Lieutenant Colonel Bill Lee, named as the first commander, was promoted to Colonel, and then to Brigadier General. On 9 April, the Airborne Command was relocated from Fort Benning, Georgia to Fort Bragg, North Carolina. As Brigadier General Eldridge G. Chapman was appointed to head the Airborne Command. The missions of the Airborne Command included the organization and training of airborne ground units, such as parachute, air-landing infantry and artillery; continuing airborne training for other units as may be designated; controlling the allocation and training of Air Corps transport or glider units made available from the Army Air Force; coordinate training with Army Air Force combat units; determine operating procedures for airborne operations and supply of large forces; and in coordination with the Navy, determine procedures for joint airborne-seaborne operations.

Two of the World War I divisions, activated in the National Army in 1917 and demobilized in 1919, were allotted to the Organized Reserves in 1921. The 82d and 101st Divisions were both reconstituted in the Army of the United States (AUS) on 15 August 1942, and ordered to Active Military Service at Camp Claiborne, Louisiana. The 82d Infantry Division was commanded by Brigadier General Omar N. Bradley. General Bradley's deputy commander was Brigadier General Matthew B. Ridgeway, who assumed command as General Bradley was assigned to command the 28th Infantry Division. The 82d Infantry Division (the All American), was reorganized and redesignated as The 82d Airborne Division, moving to Fort Bragg, North Carolina on 3 October, and providing cadre as the nucleus of the reorganized and redesignated 101st Airborne Division (the Screaming Eagles). Deputy to Brigadier General Ridgeway was Brigadier General William M. Miley. Assigned to command the fledgling 101st Airborne Division was the 'father of the airborne', Major General Bill Lee, with Brigadier General Don Pratt, former Chief of Staff of the 82d Infantry Division, as his deputy commander.

The 101st Airborne Division stood to on 15 August 1942 at Camp Claiborne, Louisiana, moving to Fort Bragg, North Carolina on 29 September.

This same year, the 11th, 13th, 15th and 17th Airborne Divisions were constituted in the AUS, but not activated.

The Army Air Corps counterpart to the Airborne Command was the Air Transport Command, headed by Colonel Fred Borum established at Stout Field, Indiana. Redesignated twice, the Air Transport Command first was redesignated as I Troop Carrier Command, and then as the Troop Carrier Command. The Troop Carrier Command was tasked with organizing and training glider, medical evacuation and troop carrier crews.

Operation TORCH was planned to be executed in North Africa on 8 November 1942. Major General Mark Clark strongly supported an airborne phase to this operation, and employed Major Bill Yarborough as his airborne staff officer. The 509th Parachute Infantry Battalion, under the command of Lieutenant Colonel Edson Raff would comprise the airborne element.

The first use of airborne troops was by the Japanese in their airborne attacks upon Celebes Islands, Java and Sumatra. On 11 January 1942, the eastern end of the airstrip on the Celebres Islands was initially prepared by fighters and bombers, to be immediately followed by the airborne forces. The Japanese casualties during this five hour operation are reported as being high, but the offensive was successful.

The 1st Parachute Infantry Brigade was activated at Fort Benning, Georgia on 20 July, with the mission of monitoring airborne training under the Airborne Command.

Combat Arms

The Parachute Test Battery was activated at Fort Benning, Georgia, commanded by Second Lieutenant Joseph D. Harris. The Parachute Test Battery researched and tested methods whereby artillery batteries could be inserted in support of the parachute infantry. The successes of airborne artillery employment doctrine, as established by the Parachute Test Battery, resulted in the activation of the first airborne artillery unit - the 456th Parachute Field Artillery Battalion, assigned to the Airborne Command.

Infantry

In February, the 88th Infantry Airborne Battalion was relieved from the Infantry School, being assigned to the newly activated Airborne Command, (organized from the Provisional Parachute Group), and moving on 4 May 1942 to Fort Bragg, North Carolina. On 15 June 1942, the battalion was reorganized and redesignated as the 88th Infantry Regiment (less two battalions). In September, following another reorganization, the 88th became the first of its kind, the 88th Glider Infantry. Colonel Eldridge. G. Chapman, who as a Lieutenant Colonel was the original commander of the 88th Infantry Airborne Battalion, attained the grade of Brigadier General, and command of the Airborne Command in May 1942. Command of the 88th Glider Infantry was assumed by Colonel Robert C. Aloe.

The 325th, 326th and 327th Infantry were constituted and activated in the National Army in 1917, demobilized in 1919 following WW I, and reconstituted in the Organized Reserves in 1921. In 1942, each were called to Active Military Service, assigned to the 82d Division and attached to IV Corps, then redesignated as the 325th, 326th and 327th Glider Infantry, and assigned to the 82d Airborne Division. The 327th Glider Infantry was further assigned to the 101st Airborne Division, and commanded by Colonel George E. Wear.

The 401st Infantry, constituted and disbanded in 1918 in the National Army while assigned to the 101st Division, was reconstituted in the Organized Reserves in 1921. Withdrawn from the Organized Reserves and allotted to the AUS in 1942, the 401st Infantry was reorganized and redesignated as the 401st Glider Infantry, assigned to the 101st Airborne Division, commanded by Colonel Joseph H. Harper, was activated at Camp Claiborne.

The 501st Parachute Battalion, activated on 24 February, was redesignated as the 1st Battalion, 501st Parachute Infantry Regiment on 15 November. The 1st Battalion was inactivated in Australia, and its assets used to replace the 2d Battalion, 503d Parachute Infantry. A new 1st Battalion, 501st Parachute Infantry was activated at Camp Toccoa, Georgia on 15 November. The 501st Parachute Infantry was assigned to the Airborne Command on 15 December.

The 502d and 503d Parachute Infantry Battalions were reorganized and redesignated as the 1st Battalions, 502d and 503d Parachute Infantry Regiments, assigned to the 101st, and 11th Airborne Divisions respectively. The remainder of the 502d Parachute Infantry Regiment was activated at Fort Benning, Georgia on 2 March, and assigned to the 101st Airborne Division on 15 August, moving to Fort Bragg, North Carolina on 24 September. Lieutenant Colonel George P. Howell Jr became the first Regimental Commander of the 502d Parachute Infantry Regiment.

In February, the 504th Parachute Infantry Battalion, constituted and activated in 1941), was consolidated with the 2d Battalion, 503d Parachute Infantry, with the consolidated unit designated as the 2d Battalion, 503d Parachute Infantry. On 6 June 1942, the 2d Battalion, 503d Parachute Infantry, under the command of Lieutenant Colonel Edson D. Raff, sailed for Europe. Later, on 2 November, the 2d Battalion, 503d Parachute Infantry was redesignated as the 2d Battalion, 509th Parachute Infantry. A replacement 2d Battalion, 503d Parachute Infantry was formed from the inactivated 1st Battalion, 501st Parachute Infantry.

Lieutenant Colonel William M. Miley, recalled from Fort Kobbe, Panama, was assigned to command the 503d Parachute Infantry Regiment. The 503d Parachute Infantry Regiment, minus the 2d Battalion, sailed to Cairns, Australia, where they were joined by the new 2d Battalion, raised from Company A, 501st Parachute Infantry. The 2d Battalion, commanded by Lieutenant Colonel Edson D. Raff embarked from New York aboard the luxury liner, Qneen Mary, for Scotland. In July, Colonel Miley was reassigned from the 503d Parachute Infantry to the 96th Infantry Division, and promotion to Brigadier General. His absence from the airborne community was short-lived, as he was recalled in August as the Deputy Commanding General of the recently activated 82d Airborne Division.

A Task Force was established for the airborne invasion of North Africa. The 2d Battalion, 503d Parachute Infantry, and the 1st, 2d and 3d Parachute Battalions, of the British 1st Parachute Brigade comprised the combat arms elements of the force. Only two battalions were flown to North Africa, due to the shortage of available aircraft, and insufficient crews trained in airborne operations. The two battalions selected were the 2d Battalion, 503d Parachute Infantry, commanded by Lieutenant Colonel Edson Raff, and the 3d Parachute Battalion, 1st Parachute Brigade, commanded by Lieutenant Colonel Geoffrey Pine-Coffin. The mission was flown by the US Army Air Corps, and is recorded as the first of many future allied airborne (or combined) operations. The 2d Battalion, 503d Parachute Infantry secured the airfield at Tebessa and Youks les Bains, and awaited First Army to catch up and reinforce them. This became the normal course of action, with the airborne forces leapfrogging ahead of the conventional forces.

The 504th, 505th and 507th Parachute Infantry Regiments were, activated at Fort Benning, Georgia, and assigned to the Airborne Command. The 504th Parachute Infantry was assigned to the 82d Airborne Division on 15 August. The 508th Parachute Infantry Regiment was activated at Camp Blanding, Florida, and assigned to the 2d Airborne Infantry Brigade. Assigned to command the 504th and 505th Parachute Infantry Regiments were two newly promoted Lieutenant Colonels - Reuben H. Tucker and James M. Gavin. The 507th Parachute Infantry Regiment was commanded by Lieutenant Colonel George V. Millett Jr, (another former member of the 501st Parachute Battalion).

The 506th Parachute Infantry was constituted in the AUS and activated at Camp Toombs, Georgia on 20 July, relocating to Fort Benning, Georgia in December, for attachment to the Airborne Command.

The 2d Battalion, 509th Parachute Infantry was reorganized and redesignated from the 2d Battalion, 503d Parachute Infantry.

The 544th Parachute Infantry (redesignated as the 545th Parachute Infantry), was constituted and assigned to the 15th Airborne Division, but never activated.

The 1st Battalion, 551st Parachute Infantry was activated at Fort Kobbe, Panama,

Companies E and F, 51st Infantry (Armored), were redesignated as Companies E and F, 51st Armored Infantry.

The British 3d Parachute Brigade was activated, consisting of the 4th and 5th (Scottish) and 6th (Welsh) Parachute Battalions. A successful raid by the British 2d Parachute Battalion was conducted against the Bruneval Radar Site in Germany in February, and the Parachute Regiment (The Red Devils), were officially constituted and activated in August.

Division Artillery

HHBs, Division Artillery, 11, 13th and 17th Airborne Divisions, were constituted in the AUS.

The 155th, 157th and 15th Field Artillery Brigades were constituted and activated in the National Army in 1917, and respectively assigned to the 80th, 82d and 84th Divisions. The 175th Field Artillery Brigade, constituted and activated in the National Army in 1918, was assigned to the 100th Division. The 15th, 155th, 157th and 175th Field Artillery Brigades were demobilized in 1919, and reconstituted in the Organized Reserves in 1921, (with the 15th being redesignated as the 159th Field Artillery Brigade). Ordered to Active Military Service in 1942, the 155th was organized at Camp Forrest, Tennessee; the 157th at Camp Claiborne, Louisiana; the 159th at Camp Howze, Texas; and the 175th at Camp Jackson, South Carolina. The 155th, 157th, 159th and 175th were respectively designated as the Headquarters and Headquarters Batteries of the 80th Division Artillery, 82d Airborne Division Artillery, 84th Division Artillery and the 100th Division Artillery.

Brigadier General Joseph M. Swing was appointed to command the 82d Airborne Division Artillery.

Field Artillery

Constituted in the National Army in 1917, the 313th, 314th and 315th Field Artillery were assigned to the 80th Division; the 319th, 320th and 321st Field Artillery were assigned to the 82d Division, and activated at Camp Forrest, and the 325th, 326th and 327th Field Artillery were assigned to the 84th Division. Demobilized in 1919, they were reconstituted in 1921 in the Organized Reserves, with the exception of the 315th and 327th Field Artillery, reconstituted in the Organized Reserves in 1929 and the 321st Field Artillery, reconstituted in the Organized Reserves in 1923. Now redesignated, the 313th, 314th and 315th Field Artillery Battalions, and the 325th, 326th and 327th Field Artillery Battalions, they were assigned to the 80th and 84th Infantry Divisions, and activated at Camp Forrest, Tennessee and Camp Howze, Texas, respectively. The 319th, 320th and 321st

Field Artillery were reorganized and redesignated as the 319th, 320th and 321st Glider Field Artillery Battalions, activated at Camp Claiborne, Louisiana, and assigned to the 82d Airborne Division.

The 373d Field Artillery was constituted in the Organized Reserves in 1929, and the 374th and 375th Field Artillery, were constituted in the Organized Reserves in 1921 at Huntington, Clarksburg and Charleston, West Virginia. They were redesignated respectively as the 373d, 374th and 375th Field Artillery Battalions, activated at Camp Jackson, South Carolina, and assigned to the 100th Infantry Division.

The 376th, 456th and 457th Parachute Field Artillery Battalions were constituted in the AUS, and activated at Camp Claiborne, Louisiana and Fort Bragg, North Carolina. The 376th and 457th Parachute Field Artillery Battalions were assigned respectively to the 82d and 11th Airborne Divisions. The 456th Parachute Field Artillery Battalion was assigned to the Airborne Command, performing as the test unit for airborne field artillery deployment and employment doctrine, and assisting with the training of other airborne field artillery organizations.

The 377th Field Artillery, constituted in the Organized Reserves at Green Bay Wisconsin in 1921. Redesignated as the 377th Field Artillery Battalion, and then as the 377th Parachute Field Artillery Battalion, the 377th was activated at Camp Claiborne, Louisiana, allotted to the AUS, and assigned to the 101st Airborne Division.

The 1st Battalion, 472d Field Artillery, constituted in the Organized Reserves in 1921, organized at Buffalo, New York. Initially assigned to Second Corps, the 1st Battalion, 472d Field Artillery was assigned to, and released from the 98th Division in 1929. Called to Active Military Service at Camp Gordon, the 1st Battalion, 472d Field Artillery was reorganized and redesignated as the 472d Field Artillery Battalion.

The 680th and 681st Glider Field Artillery Battalions were constituted in the AUS and assigned to the 17th Airborne Division.

The 305th, 307th and 309th Ammunition Trains were constituted in 1917 in the National Army, and assigned to the 80th, 82d and 84th Divisions. They were demobilized in 1919, reconstituted in the Organized Reserves, and activated respectively in 1921 at Washington, DC (305th); 1922 at Newberry, South Carolina (307th); and in 1921 at Fort Wayne, Indiana (309th). In 1942 they were redesignated as the 905th, 907th and 909th Field Artillery Battalions. Recalled to Active Military Service, the 905th was assigned to the 80th Infantry Division at Camp Forrest, Tennessee; the 907th was assigned to the 101st Airborne Division, (and redesignated as the 907th Glider Field Artillery Battalion); and the 909th was organized at Camp Howze, Texas with the 84th Infantry Division.

The 325th Ammunition Train was constituted in 1921 in the Organized Reserves, assigned to the 100th Division, and activated in West Virginia. The 325th Ammunition Train was redesignated at the 925th Field Artillery Battalion on 30 January, ordered to Active Military Service at Fort Jackson, South Carolina, and assigned to the 100th Infantry Division.

Air Defense Artillery

The 80th and 81st Airborne Antiaircraft Artillery Battalions, (assigned respectively to the 82d and 101st Airborne Divisions), were constituted in the AUS and activated at Camp Claiborne.

Special Operations

The 1st Special Service Force, unique as a joint United States/Canadian commando force, was activated at Fort William Henry Harrison, Montana on 9 July. Trained in parachute, ski, commando, demolitions and cold weather warfare techniques, the 1st Special Service Force, commanded by Brigadier General Robert T. Frederick, was intended to operate behind enemy lines in Norway, Rumania and Northern Italy. Regardless of national origin, the Canadian and US members of this unit wore the same uniforms and insignia; the only time that countrymen of both nations have served together as a single organization.

Although, (like it's five sister Ranger battalions), not designated as an airborne unit, it should be mentioned that the 1st Ranger Battalion was activated in the AUS at Garrickfengus, Northern Ireland. The success of this battalion led to the activations of five more ranger battalions during WW II; forefathers of the airborne rangers of Korea, Vietnam and the 75th Infantry (Ranger), which became the 75th Ranger Regiment following airborne assaults into Grenada in 1983, and later into Panama in 1989.

Aviation

The 317th and 440th Troop Carrier Wings of the 1st Troop Carrier Command, participated in every major operation involving US Airborne Forces and many infiltration flights for the Office of Strategic Services (OSS) personnel, redesignated from the former office of the Coordinator of Information.

The 617th Aviation Detachment was constituted and activated.

Medical

The 305th and 307th Sanitation Trains were constituted in the National Army in 1917, assigned respectively to the 80th and 82d Divisions, and demobilized in 1919. Both were reconstituted in the Organized Reserves in 1921. The 325th and 326th Sanitation Trains, assigned to the 100th and 101st Divisions, were constituted in the National Army and also demobilized in 1918, and also reconstituted in the Organized Reserves in 1921.

The 305th Sanitation Train was organized in 1921, with headquarters in Richmond, Virginia, and redesignated as the 305th Medical Battalion and ordered into Active Military Service in 1942 at Camp Forrest, Tennessee.

The 307th Sanitation Train was organized at Columbia, South Carolina in 1922, and reorganized and redesignated as the 307th Medical Battalion in 1942, remaining assigned to the 82d Division, and ordered to Active Military Service at Camp Claiborne, Louisiana. The 307th Medical Battalion was reorganized and redesignated as the 307th Airborne Medical Company, an element of the 82d Airborne Division.

The 325th Sanitation Train, an element of the 100th Division, was reconstituted in the Organized Reserve in 1921 as the 325th Medical Regiment. The 325th Medical Regiment was reorganized and redesignated as 325th Medical Battalion in 1942, ordered into Active Military Service and activated at Fort Jackson, South Carolina.

The 326th Sanitation Train was organized in 1921 at Milwaukee, Wisconsin, and reorganized and redesignated in 1942 as the 326th Medical Battalion, and also in 1942 further reorganized and redesignated as the 326th Airborne Medical Company. The 326th Airborne Medical Company was concurrently withdrawn from the Organized Reserves and allotted to the AUS, and reorganized at Camp Claiborne, Louisiana as an element of the 101st Airborne Division.

Combat Support
Engineers

The 925th Engineer Airborne Regiment (Provisional) was activated at

Westover Field, Massachusetts on 1 November.

The 37th Engineers were constituted in the National Army in 1917, organized at Fort Meyer, Virginia in 1918, and demobilized at Camp Upton, New York in 1919. Reconstituted in the Regular Army in 1933 as the 37th Engineers (General Service), it was 1941 before the 37th Engineers were activated at Camp Bowie, Texas; and redesignated as the 37th Engineer Combat Regiment in 1942.

The 161st Engineer Battalion was formed from the 161st Engineer Squadron at Fort Bliss, Texas

The 127th and 129th Airborne Engineer Battalions were constituted in the AUS, and assigned to the 11th and 13th Airborne Divisions respectively.

Company C, 139th Airborne Engineer Squadron, constituted in the AUS, and activated at Fort Bliss, Texas.

The 307th and 326th Engineers were constituted respectively in the National Army in 1917 and 1918, and assigned to the 82d and 101st Divisions. They were demobilized in 1919 (307th) and 1918 (326th), and both reconstituted in the Organized Reserves in 1921. The 307th and 326th Engineers were redesignated as the 307th Engineer Battalion and the 326th Engineer Battalion (Combat), and ordered to Active Military Service in 1942. Activated at Camp Claiborne, Louisiana, they were again redesignated, now as the 307th and 326th Airborne Engineer Battalions, and respectively assigned to the (redesignated) 82d and 101st Airborne Divisions.

The 2d Battalions, 309th and 310th Engineers, were constituted in 1917 in the National Army, assigned respectively to the 84th and 85th Divisions, and demobilized in 1919. They were both reconstituted in the Organized Reserves in 1921, and ordered to Active Military Service in 1943. They were reorganized and redesignated as the 876th and 878th Airborne Engineer Battalions, Aviation. The 875th and 876th Airborne Engineer Battalions, Aviation, were activated on 15 November at Camp Claiborne, Louisiana. They were assigned to the IX Engineer Command, a component of the 9th Air Force, with the mission to rapidly assault, capture, and repair essential airfield facilities within supporting distance of the forward lines.

The 871st Airborne Engineer Aviation Battalion was activated on 1 September at Westover Field, Massachusetts, and assigned to the 925th Engineer Airborne Regiment. Also assigned to the 925th Engineer Airborne Regiment was the 874th Airborne Engineer Aviation Battalion, activated at Camp Claiborne on 15 November.

Military Intelligence

The 215th Signal Depot Company was constituted in the AUS, affiliated with the Western Electric Company, and activated at Camp Livingstone.

Military Police

The Military Police Platoons, 11th and 101st Airborne Divisions were constituted in the AUS and activated respectively at Camp Mackall and Fort Bragg, North Carolina.

In 1942, HQ and Military Police Company, 82 Infantry Division, was ordered to Active Military Service at Camp Claiborne, Louisiana. HQ Troop, 82d Division and the 307th Train HQ and Military Police, 82d Division had been concurrently constituted and activated in the National Army, at Fort Gordon Georgia, in 1917. Both demobilized in 1919 at Camp Mills, New York, they were reconstituted in the Organized Reserves in 1921, as HQ Company, 82d Division, and as the 82d Military Police Company, 82d Division. In 1922, HQ Company, 82d Division was organized at Columbia, South Carolina. In 1942, both HQ Company, 82d Division and the 82d Military Police Company were ordered into Active Military Service at Camp Claiborne, Louisiana. HQ Company, 82d Division was reorganized and redesignated as HQ and MP Company (less the MP Platoon), 82d Division. The 82d Military Police Company was redesignated as the MP Platoon, HQ and MP Company, 82d Division. Three months later, the MP Platoon was disbanded at Camp Claiborne, and HQ and MP Company, 82d Division was reorganized and redesignated as HQ Company, 82d Airborne Division. The MP Platoon was constituted in the Organized Reserves and activated at Camp Claiborne.

Signal

The 13th Reserve Field Signal Battalion and the 305th Field Signal Battalion were constituted in the National Army in 1917, and assigned respectively to the 82d and 80th Divisions. That same year, the 13th Reserve Field Signal Battalion was redesignated as the 307th Field Signal Battalion. The 305th and 307th Field Signal Battalions were demobilized in 1917 respectively at Camp Lee, Virginia and Camp Morrison, both being reconstituted in the Organized Reserves in 1921. The 307th Field Signal Battalion was redesignated as the 82d Signal Company and organized at Macon, Georgia in 1922. The 305th Field Signal Battalion and the 82d Signal Company were respectively activated and assigned to the the 80th Division at Camp Forrest, Tennessee, and to the 82d Airborne Division at Camp Claiborne, Louisiana. The 82d Signal Company was reorganized and redesignated as the 82d Airborne Signal Company.

The 246th Signal Operations Company and the 299th Signal Installations Company were constituted in the AUS. The 246th Signal Operations Company was affiliated with the Pennsylvania Telephone Corporation Company of Erie, Pennsylvania; the 299th Signal Operations Company was affiliated with the Western Electric Company, Inc. The 246th and 299th were activated respectively at the Central Signal Corps Training Center, Camp Crowder, Missouri, with both affiliation agreements concurrently terminated.

The 626th Field Signal Battalion, constituted in the National Army in 1918, and assigned to the 101st Division, was demobilized in 1918, and reconstituted in the Organized Reserves in 1921. The 626th Field Signal Battalion was allotted to the AUS in 1942, activated at Camp Claiborne, Louisiana, and assigned to the 101st Airborne Division, as the 101st Airborne Signal Company.

The 511th and 513th Airborne Signal Companies were constituted in the AUS, activated at Camp Mackall, North Carolina, and assigned respectively to the 11th and 13th Airborne Divisions.

Combat Service Support
Adjutant General

Headquarters, Special Troops, 82d Division, constituted in the Organized Reserves in 1921, and organized in 1922 at Columbia, South Carolina, was disbanded.

Bands, 11th and 13th Airborne Divisions Artillery, were constituted in the AUS, and assigned to the 11th and 13th Airborne Divisions.

Ordnance

The 711th, 717th and 782d Airborne Ordnance Maintenance Companies were constituted in the AUS. The 711th and 782d Airborne Ordnance Maintenance Companies were assigned to the 11th Airborne Division, and activated at Miller Field, and the 82d Airborne Division, and activated at Fort Bragg, North Carolina. The 717th Airborne Ordnance Maintenance Company was assigned to the 17th Airborne Division.

In August, the Ordnance Maintenance Platoon Headquarters Company, 409th Quartermaster Battalion, (see 1942, Combat Service Support, Quartermaster), was redesignated as the 784th Ordnance Light Maintenance Company, ordered into Active Military Service, and organized at Camp Howze, Texas.

The Ordnance Maintenance Platoon (Quartermaster), Headquarters Company, 425th Quartermaster Battalion, (see 1942, Combat Service Support, Quartermaster), was redesignated as the 800th Ordnance Light Maintenance Company, and ordered into Active Military Service at Fort Jackson, South Carolina.

Quartermaster

The 82d Division Train was constituted in the Organized Reserves in 1921, and organized in Georgia. The 82d Division Train was reorganized as the 82d Division Quartermaster Train in 1925, and in 1936 reorganized and redesignated as the 407th Quartermaster Regiment, remaining assigned to the 82d Division.

The 307th and 326th Supply Trains were constituted in the National Army and respectively assigned to the 82d Division in 1917, and the 101st Division in 1918. The 307th Supply Train was organized at Camp Gordon, and demobilized in 1919 at Camp Dix, New Jersey. The 326th Supply Train was reconstituted in the Organized Reserves in 1921 as the 101st Division Train, Quartermaster Corps, and organized in 1922 with Headquarters in Milwaukee, Wisconsin. The 307th Supply Train was then consolidated with the 407th Quartermaster Regiment, with the consolidated unit designated as the 407th Quartermaster Regiment in 1936. The 101st Division Train was redesignated in 1925 as the 101st Division Quartermaster Train, and in 1936 as the 426th Quartermaster Regiment. The 407th and 426th Quartermaster Regiments were ordered into Active Military Service in 1942, reorganized and redesignated as the 407th and 426th Quartermaster Battalions. The 407th Quartermaster Battalion was activated at Camp Claiborne, reorganized and redesignated as the 407th Airborne Quartermaster Company, and assigned to the 82d Airborne Division. The 426th Quartermaster Battalion was reorganized and redesignated as the 426th Airborne Quartermaster Company, assigned to the 101st Airborne Division, and allotted to the Regular Army.

The 305th Supply Train was constituted in the National Army in 1917, assigned to the 80th Division, and organized at Camp Lee, Virginia. Demobilized in 1919, the 305th Supply Train was reconstituted in 1921 in the Organized Reserves, and redesignated in 1936 as the 405th Quartermaster Regiment on 1 July 1936. Headquarters, 305th Supply Train was reconstituted and consolidated with HQ Company, 405th Quartermaster Regiment on 21 October 1936. Activated at Camp Forrest, Tennessee, in 1942, the 405th Quartermaster Regiment was initially redesignated as the 405th Quartermaster Battalion, then as the 80th Quartermaster Company, assigned to the 80th Infantry Division. Reorganized and redesignated as the 780th Ordnance Light Maintenance Company on 12 November 1942.

The 309th and 325th Supply Trains were constituted in the National Army in 1917, assigned to the 84th and 100th Divisions. The 309th Supply Train was organized at Camp Dix, New Jersey, however the 325th Supply Train was not organized.

The 309th Supply Train was demobilized in 1919 at Camp Sherman, Ohio, reconstituted in 1936, and consolidated with the 409th Quartermaster Regiment, an element of the 84th Division, stationed in Indiana. The 409th Quartermaster Regiment was reorganized and redesignated as the 409th Quartermaster Battalion in January 1942. Maintenance Platoon, Headquarters Company, 409th Quartermaster Battalion, was redesignated as the 784th Ordnance Light Maintenance Company (see 1942, Combat Service Support, Ordnance).

On 24 June 1921, the 325th Supply Train was reconstituted as an element of Headquarters, 100th Division, allotted to the Organized Reserves, and organized in West Virginia. The 325th Supply Train was redesignated in 1936 as the Division Quartermaster Platoon, Headquarters Company, 425th Quartermaster Regiment. In January 1942, The Division Quartermaster Platoon was redesignated as the Maintenance Platoon, and redesignated again in August as the Ordnance Maintenance Platoon (Quartermaster), Headquarters Company, 425th Quartermaster Battalion. (see 1942, Combat Service Support, Ordnance).

The 408th Airborne Quartermaster Company was constituted in the AUS, and activated at Camp Gruber, Oklahoma. The 409th and 411th Airborne Quartermaster Companies were constituted respectively in the AUS and on the Inactive List, and assigned to the 13th and 17th Airborne Divisions.

The 528th Quartermaster Service Battalion was constituted in the AUS and activated at Camp McCain, Mississippi,

Constituted in the AUS, the 711th Airborne Ordnance Maintenance Company was activated at Miller Field, New York, and assigned as an element of the 11th Airborne Division.

Transportation

The 29th Quartermaster Regiment was constituted in the Regular Army in 1936, activated at Fort Ord in 1942, and concurrently redesignated as the 29th Quartermaster Truck Company.

——— 1943 ———

HHC, XVIII Corps was constituted from the II Armored Corps, activated at the Presidio of Monterey, California, and relocated to Camp Bowie, Texas.

The Airborne Command, under Brigadier General Chapman, moved from Fort Bragg to Camp Huffman. On 1 May, Camp Huffman was renamed as Camp Mackall, named after Private John T. Mackall, who had died of wounds following the invasion of French Northwest Africa, while assigned to Company E, 509th Parachute Infantry Battalion.

Brigadier General Chapman was promoted to Major General in May 1943, assuming command of the 13th Airborne Division, activated at Ft Benning, Georgia. Brigadier General Leo Donovan assumed command of the Airborne Command on 16 November.

The 11th and 17th Airborne Divisions were activated at Camp Mackall, North Carolina, and assigned to the US Airborne Command. The 11th Airborne Division, commanded by Major General Joseph M. Swing, became "paragliders" - more than eighty percent of the glider personnel became parachute qualified. The 17th Airborne Division, commanded by Major General William M. (Bud) Miley was destined for the European Theater of Operations.

The 15th Airborne Division was constituted but never activated.

The 82d Airborne Division departed Fort Bragg, North Carolina via Camp Edwards, Massachusetts, to Europe by sea, arriving at Casablanca on May 10th under the command of Brigadier General Matthew Ridgeway, and air-assaulted into Sicily on 9 July, returning to North Africa 10 days later. The 82d Airborne Division moved back to Sicily on 4 September, and on to Italy, landing on 13

September. In the early morning of 9 September, the Fifth Army had landed by sea at Salerno. Leaving Italy on 19 November, the 82d Airborne Division returned to North Africa 22-30 November, and then to Northern Ireland on 9 December.

The 101st Airborne Division departed from New York on 5 September, arriving in England 10 days later.

The British 6th Airborne Division was activated on 23 April, under the command of Major General Richard N (Windy) Gale. Assigned elements included the 3d and 5th Parachute Brigades and the 6th Air-Landing Brigade.

Combat Arms
Infantry

HQ, 1st and 2d Infantry Brigades were constituted in the National Army in 1917, assigned to the 1st Expeditionary Force, (later to become the 1st Division in 1917 and the 1st Infantry Division in 1960), and organized in New York City, New York. They were reorganized and redesignated in 1921 as HHCs, 1st and 2d Infantry Brigades, until redesignated as HHCs, 1st and 2d Brigades from 1925 until 1936, when they resumed their designations as the HHC, 1st and 2d Infantry Brigades. In 1939, the 1st Infantry Brigade was disbanded at Camp Wadsworth, New York, and the 2d Infantry Brigade was relieved from assignment to the 1st Division, and inactivated at Fort Ontario, New York in 1940.

In 1943, the HHCs, 1st and 2d Infantry Brigades were respectively reconstituted and redesignated as the HHC, 1st Airborne Brigade and HHC, 2d Airborne Infantry Brigade. The 1st Airborne Brigade was activated at Camp Meade, South Dakota, from personnel formerly assigned to the 1st Parachute Infantry Brigade, which had moved from Fort Benning, Georgia to Camp Meade on 6 April. The 1st Airborne Brigade was assigned to the US Airborne Command, and relocated to Camp Mackall, North Carolina. The 2d Airborne Infantry Brigade was activated at Camp Mackall, en route to the European Theater of Operations. Upon arrival in England, the 501st and 508th Parachute Infantry Regiments of the 2d Airborne Infantry Brigade, were respectively attached to the 101st and 82d Airborne Divisions.

The 88th Glider Infantry Regiment was assigned to the 13th Airborne Division. The 187th, 188th, 193d and 194th Glider Infantry Regiments were activated at Camp Mackall, North Carolina, where the 187th and 188th were assigned to the 11th Airborne Division, and the 193d and 194th were assigned to the 17th Airborne Division. The 189th and 190th Glider Infantry Regiments were activated at Fort Bragg, North Carolina and both assigned to the 13th Airborne Division. Although constituted and assigned to the 15th Airborne Division, neither the 191st or 192d Glider Infantry Regiments were activated.

The 325th Glider Infantry Departed New York on 28 April, arriving in North Africa on 10 May, where they remained until assaulting by sea into Salerno, Sicily on 9 July as part of Operation HUSKEY, then back to North Africa on 19 August. From North Africa, the 325th Glider Infantry returned to Sicily on 4 September, and 9 days later into Italy, until returning for refurbishing in Northern Ireland.

The 326th Glider Infantry was relieved from assignment to the 82d Airborne Division, to be assigned to the 13th Airborne Division.

The 327th Glider Infantry was assigned to the 101st Airborne Division, staging from New York on 5 September, to arrive in England ten days later.

The 501st Parachute Infantry moved to Fort Benning, Georgia 23 March, and then to Camp Mackall, North Carolina on 13 April, for assignment to the 2d Airborne Brigade on 3 September.

The 502d Parachute Infantry Regiment staged from Camp Shanks, New York, departing for England in September.

The 503d Parachute Infantry, commanded by Colonel Kenneth H. Kinsler, prepared for their airborne assault into New Guinea, and conducting detailed rehearsals in Australia. They arrived at Port Moseby, New Guinea on 20 August, and conducted an unopposed airborne assault onto Nadzab, on the Markham River on 5 September. Lieutenant Colonels George M. Jones and John L. Tolson commanded the 2d and 3d Battalions. Transported by the 54th Troop Carrier Wing, the 503d Parachute Infantry conducted what may be considered the most accurate infiltration of airborne troops to date.

The 504th Parachute Infantry Regiment assigned from the Airborne Command to the 82d Airborne Division, moving to Fort Bragg, North Carolina in February. The 504th Parachute Infantry staged at Camp Edwards, Massachusetts, departing to Tunisia, North Africa from New York on 28 April, under the command of Colonel Reuben Tucker, arriving in North Africa on 10 May. The 504th Parachute Infantry rehearsed prior to their parachute assault into Gela, Sicily on 9 July, and on to Operation HUSKEY, at Salerno, Italy on 9 September. The 3d Battalion, 504th Parachute Infantry, parachuted onto the high ground near Ponte Olivo, northeast of Gela airstrip in Sicily on 9 July.

The 504th and 505th Parachute Infantry Regiments were initially assigned to the Airborne Command, followed by assignments to the 82d Airborne Division. Both regiments staged out from Camp Edwards, Massachusetts, embarking from New York on 28 April, and arriving at Tunisia, North Africa on 10 May. Both regiments rehearsed their roles for the 9 July assault onto Gela, Sicily. The 504th Parachute Infantry, commanded by Colonel Reuben Tucker, continued on to assault into Salerno, Italy on 9 September, with their 3d Battalion arriving on 18 October.

The 505th Parachute Infantry was initially assigned to the Airborne Command, prior to assignment to the 82d Airborne Division on 10 February 1943, relocating to Fort Bragg, North Carolina on 12 February 1943. Staging from Camp Edwards, Massachusetts, the 505th Parachute Infantry departed New York, arriving in North Africa on 10 May. On 9 July, the 505th Parachute Infantry conducted a parachute assault onto Ponte Olivo airfield north of Gela City, being widely dispersed. Captain Edward Sayre landed closer to their original objective, and after organizing approximately 100 paratroopers, seized a fortified position on the Gela Ridge.

The 506th Parachute Infantry, commanded by Colonel Robert F. Sink, was attached to the 101st Airborne Division from 1 June. Relocating to Sturgis Army Airfield, Kentucky on 6 June, the 506th Parachute Infantry finally reached Fort Bragg, North Carolina on 23 July. They staged out of Camp Shanks, New York, arriving in England on 15 September.

The 507th Parachute Infantry departed Fort Benning, Georgia to Barksdale Field, Louisiana on 7 March, and then to Alliance Field, Nebraska on 23 March, for assignment to the 1st Airborne Brigade on 14 April. The 507th Parachute Infantry staged from Camp Shanks New York, arriving in England on 16 December.

The 2d Battalion, 509th Parachute Infantry Regiment, commanded by Major Doyle Yardley, was attached to the 82d Airborne Division prior to the parachute assault by the 82d Airborne Division into Salerno, Italy on 9 September.

A small force from the 2d Battalion, 509th Parachute Infantry, commanded by Captain Charles W. Howland was amphibiously inserted onto the island of Ventotene. The remainder of the 503d Parachute Infantry carried by the 64th Troop Carrier Group, participated in an airborne assault into the town of Avellino, approximately twenty miles from Salerno. The 2d Battalion, 509th Parachute Infantry Regiment was severely mauled, necessitating reorganization and redesignation at Venafro on 10 December, as the 509th Parachute Infantry Battalion. Lieutenant General Mark W. Clark, commanding the Fifth Army, wrote a personal commendation to the 509th Parachute Infantry following their actions at Avellino. Shortly after Avellino, Lieutenant Colonel William P. Yarborough was assigned to command the 509th Parachute Infantry Battalion. He designed their well known 'gingerbread man' pocket patch. He also instituted the practice of wearing 'combat stars' to signify participation in combat jumps on the parachute wings. (It was not until after the airborne assaults into Grenada in 1983 that this practice was finally and legally approved.) The 509th Infantry Battalion was temporarily pulled back from the front lines following Avellino, to guard the Italian based Fifth Army Headquarters, assigned to the 6615th Ranger Force, Provisional.

The 511th Parachute Infantry Regiment was activated at Camp Taccoa, Georgia on 5 January, and assigned to the Airborne Command. Relocating to Camp Mackall, North Carolina on 21 February, the 511th Parachute Infantry was assigned four days later to the 11th Airborne Division. The 511th Parachute Infantry was reassigned to Fort Benning, Georgia on 14 May, returning to Camp Mackall on 14 June.

The 513th, 515th, 541st and 542d Parachute Infantry Regiments were activated respectively on 11 January, 31 May, 12 August and 1 September at Fort Benning, Georgia. The 513th Parachute Infantry was assigned to the 13th Airborne Division, and moved to Fort Bragg, North Carolina on 1 November.

The 514th and 517th Airborne Infantry Regiments were activated at Camp Mackall, North Carolina and Camp Taccoa, Georgia, and assigned to the 17th Airborne Division.

The 515th, 541st and 542d Parachute Infantry were assigned to the Airborne Command.

Eighteen enlisted volunteers from the 92d Infantry Division, with First Sergeant Walter Morris, and prior volunteer Clarence H. Beavers, were transferred in December to the Alabama Training Area, as the nucleus of the 'Black Test Platoon'. This unit was activated at Fort Benning Georgia, and designated as the 555th Parachute Infantry Company, which became Company A, 555th Parachute Infantry Battalion. The stick order for the first all-black parachute qualification jump included Calvin R. Beal, Clarence H. Beavers, Ned B. Bess, Hubert Bridges, Lonnie M. Duke, McKinley Godfrey Jr, Robert P. Greene, James E. Kornsay, Alvin L. Moon, Walter J. Morris, Leo D. Reed, Samuel W. Robinson, Carstell O. Stewart, Jack D. Tillis, Roger S. Walden and Daniel C. Weil. Brigadier General H. J. Jablonsky, Assistant Commandant for Training, Airborne Command, commended the Black Test Platoon on their outstanding achievement, whereby out of 20 selectees, only one volunteer had failed to meet the course standards, (with two others injured, and one on emergency leave).

Armor

On 15 August 1943, five officers and 70 enlisted volunteers, arrived at Fort Knox, Kentucky from the 20th Armored Division, to be assigned to the first unit of it's kind - the 151st Airborne Tank Company. The unit trained under the guidance of Captain Arthur Cromillion, the assigned Company Commander, and First Sergeant George C. Norton. Additional personnel were assigned from the 20th Armored Division, the Armor School and the Demonstration Regiment (of the Armor School). 1st Lieutenant Richard McCabe assumed command of the 151st Airborne Tank Company, as Captain Cromillion was reassigned to the newly formed 28th Airborne Tank Battalion.

Companies E and F, 51st Armored Infantry, were reorganized and redesignated respectively as Companies C and B, 10th Armored Infantry Battalion, 4th Armored Division.

Corps and Division Artillery

HHB, XVIII Corps Artillery was activated at Camp Cooke, California. Brigadier General Maxwell D. Taylor was assigned as the Commander, 82d Airborne Division Artillery. The 11th and 17th Divisions' Artillery were activated at Camp Mackall, and the 13th Airborne Division Artillery was activated at Fort Bragg.

Field Artillery

The 350th Field Artillery was reorganized and redesignated as the 350th Field Artillery Battalion. The 456th Parachute Field Artillery Battalion was assigned from the Airborne Command to the 82d Airborne Division. The 457th Parachute Field Artillery Battalion was relieved from assignment to the 11th Airborne Division in January, assigned to the Airborne Command at Fort Bragg, North Carolina, and released back to the 11th Airborne Division on the 23d February.

The 4th Battalion, 18th Field Artillery was reorganized and redesignated as the 693d Field Artillery Battalion.

The 458th, 460th, 462d, 464th, 465th and 466th Parachute Field Artillery Battalions were constituted in the AUS, (the 458th and 460th in 1942, and the 462d, 464th, 465th and 466th in 1943). The 458th and 465th Parachute Field Artillery Battalions were activated at Fort Bragg, and the 460th, 462d and 464th Parachute Field Artillery Battalions were activated at Camp Mackall, North Carolina. The 466th Parachute Field Artillery Battalion was assigned to the 17th Airborne Division.

Constituted in the AUS in 1942, the 674th and 675th Glider Field Artillery Battalions were activated at Camp Mackall, and the 676th and 677th Glider Field Artillery Battalions were activated at Fort Bragg, North Carolina. The 680th and 681st Glider Field Artillery Battalions were also activated at Camp Mackall.

Air Defense Artillery

On 10 February, the HHD, 3d Airborne Anti-Aircraft Artillery Battalion constituted in the AUS in 1942, was activated at Camp Stewart, Georgia, and HHD, 9th and 12th Airborne Anti-Aircraft Artillery Battalions were activated at Fort Bliss, Texas on 10 February. The 3d and 9th Airborne Anti-Aircraft Artillery Battalions embarked from San Francisco California for Australia on 26 July, with the 12th Anti-Aircraft Artillery Battalion following on 28 August. They arrived in Australia respectively on 14 and 15 July and 29 August. The 3d, 9th and 12th Anti-Aircraft Artillery Battalions moved to New Guinea, arriving on 11 November.

On 3 September, HHD, 13th Airborne Anti-Aircraft Artillery Battalion was activated at Fort Bliss, and the HHDs, 14th and 15th Airborne Anti-Aircraft Artillery Battalions were activated at Camp Stewart, Georgia. The 13th, 14th and 15th Airborne Anti-Aircraft Artillery Battalions were assigned to the Anti-Aircraft Artillery Command.

The 152d, 153d and 155th Airborne Anti-Aircraft Artillery Battalions constituted in the AUS in 1942 and 1943, were activated respectively at Camp Mackall, Fort Bragg and Camp Mackall, North Carolina, on 25 February, 13 August and 15 April, and assigned to the 11th, 13th and 17th Airborne Divisions.

The 503d Coast Artillery Battalion, constituted in the AUS in 1942, was activated at Camp Stewart, Georgia, to be reorganized and redesignated as the 503d Anti-Aircraft Artillery Gun Battalion. The 154th Airborne Anti-Aircraft Artillery Battalion, was constituted inactive, for assignment to the 15th Airborne Division.

Special Operations

The 1st Special Service Force, was originally intended for employment behind the enemy lines in the cold climates of Norway. They prepared for this contingency, training at Camp Bradford, Vermont in April and Camp Ethan Allen in May. Instead of Scandinavia, the 1st Special Service Force deployed to the west. They departed through San Francisco to the Aleutian Islands on 25 July as part of the Kiska Task Force. The First Regiment arrived at Kiska Island by rubber boat to discover that the Japanese had fled. The same was found by the Third Regiment on Segula Island. The planned airborne assault by the Second Regiment from Amchitka was cancelled. The Force returned to cold weather training at Camp Ethan Allen, Vermont on 9 September, from where they were again misused, deploying to Cassablanca, North Africa on their way to arriving at Naples, Italy on 19 November. Assigned to Fifth Army, under the control of the 36th Division, the 1st Special Service Force clearly demonstrated their capabilities. The attacked on 3-6 December, and seized the seemingly impenetrable and well defended Monte La Difensa, scaling cliffs and accomplishing in a matter of hours, what would normally have been expected of two Divisions in a full week. Without a pause, the Force continued on to capture Monte la Remetanea on 6-9 December, and Mount Sammacro (Hill 720) on 25 December.

Rangers

The 6615th Ranger Force, Provisional was formed in North Africa, with the 1st, 3d and 4th Ranger Infantry Battalions (Provisional), an infantry battalion and a chemical mortar battalion. The mission of the force was to secure the left flank of the Fifth Army.

From within the 1st Ranger Battalion, Companies A and B were redesignated as the 3d Ranger Infantry Battalion, and concurrently consolidated with the 3d Ranger Battalion (Provisional), (organized in North Africa), to be designated the 3d Ranger Battalion.

Companies C and D, 1st Ranger Battalion, consolidated with the 4th Ranger Battalion (Provision), (formed in North Africa), to establish the 4th Ranger Infantry Battalion. and activated at Garrickfengus, Northern Ireland.

Companies E and F remaining as the cadre of the newly redesignated 1st Ranger Infantry Battalion.

The 5th Ranger Battalion was constituted in the AUS, activated at Garrickfengus, and redesignated as the 5th Ranger Infantry Battalion.

On 10 October, the 5307th Composite Unit, Provisional, was organized at Deogarth, India.

Aviation

The XII Troop Carrier Command, with 331 assigned C-47 Douglas Dakota aircraft, arrived in North Africa with the 82d Airborne Division.

The 52d Troop Carrier Wing carried the 504th and 505th Parachute Infantry on Operation HUSKY into Sicily.

The British Air Landing Brigade was carried by the 51st Troop Carrier Wing.

From Sicily, the 61st, 313th and 314th Troop Carrier Groups carried the 504th Parachute Infantry to Salerno in Southern Italy.

Medical

The 221st, 222d and 224th Airborne Medical Companies were constituted in the AUS in 1942, and assigned respectively to the 11th, 13th and 17th Airborne Divisions, The 221st and 224th Airborne Medical Companies were activated at Camp Mackall, North Carolina, and the 222d Airborne Medical Company at Fort Bragg, North Carolina in 1943.

Company H, 64th Medical Regiment was reorganized and redesignated as the 610th Clearing Company.

Combat Support
Engineers

The 925th Engineer Airborne Regiment (Provisional) was disbanded on 1 April, with all personnel and equipment assigned to the concurrently activated 1st Airborne Engineer Aviation Unit Training Center.

Reorganized and redesignated from the 1st Battalion, 37th Engineer Combat Regiment (broken up in 1942), as the 37th Engineer Combat Battalion.

Company C, 161st Engineer Battalion was designated as the Parachute Company, on 5 May 1943. The 161st Engineer Battalion had been relieved from assignment to the 1st Cavalry Division on 15 March. The 161st Engineer Battalion was redesignated as the 161st Airborne Engineer Battalion, and assigned in October to the Airborne Command. Company C, 161st Airborne Engineer Battalion was assigned to the 503d Parachute Infantry Regimental Combat Team, in New Guinea.

The 127th and 129th Airborne Engineer Battalions, assigned to the 11th and 13th Airborne Divisions, were activated respectively at Camp Mackall and Fort Bragg, North Carolina.

Company C, 139th Airborne Engineer Battalion, assigned to the 17th Airborne Division, was activated at Camp Mackall, North Carolina.

The 597th Airborne Engineer Company, constituted in the AUS in 1942, was activated at Camp Mackall, North Carolina.

The 871st Airborne Engineer Aviation Battalion deployed to Australia, en route to New Guinea, on 1 May. The 874th Airborne Engineer Aviation Battalion embarked to New Guinea in January.

At Westover Field, Massachusetts, five Airborne Engineer Aviation Battalions were activated. The 877th Airborne Engineer Aviation Battalion on 1 January; the 878th Airborne Engineer Aviation Battalion on 1 February; and the 879th, 880th and 881st Airborne Engineer Aviation Battalions on 1 March. At Bradley Field, Connecticut, the 882d and 883d Airborne Engineer Aviation Battalions were activated on 1 May, the 884th and 885th Airborne Engineer Aviation Battalion on 1 June, and the 886th Airborne Engineer Aviation Battalion on 1 August. The 886th Airborne Engineer Aviation Battalion was inactivated at Bradley Field on 25 December.

On 25 July, the 876th Airborne Engineer Aviation Battalion embarked from New York to England, and the 875th Airborne Engineer Aviation Battalion embarked on 6 November from San Francisco to Australia. The 879th Airborne Engineer Aviation Battalion departed through Hampton Roads on 14 December for India.

Military Police

The 303d Military Police Battalion, constituted in the Organized Reserves in 1922 and inactivated in 1938, was redesignated the 503d Military Police Battalion in 1940, and activated at Camp Maxey, Texas in 1943.

Signal

The 931st Signal Battalion (Air Support Command), was constituted in the AUS and activated at Esler, Louisiana.

Combat Service Support
Adjutant General

Company A, 22d Replacement Battalion, constituted in the AUS in 1942 and activated at Camp Sutton, North Carolina, was redesignated in November 1943 as the 290th Replacement Company.

The Bands, 11th and 13th Airborne Divisions Artillery, were activated respectively at Camp Mackall and Fort Bragg, North Carolina. The Band, 13th Airborne Division Artillery, was assigned to the Second Army.

Chemical

The 83d Chemical Mortar Battalion was assigned to the 6615th Ranger Force, Provisional, which had been activated during the summer.

Ordnance

The 713th Airborne Ordnance Maintenance Company was constituted in the AUS, activated at Camp Mackall, North Carolina, assigned to the 11th Airborne Division, and relocated to Fort Bragg, North Carolina, and reassigned to the 13th Airborne Division. The 717th Airborne Ordnance Maintenance Company was activated at Beltsville, Maryland.

Quartermaster

The 373d Graves Registration Unit, Quartermaster Corps, was constituted in the Organized Reserves in 1928, redesignated in 1936 as Company D, 714th Quartermaster Battalion, and redesignated again in 1940 as the 612th Quartermaster Company. Redesignated once more in 1942, the 612th Quartermaster Graves Registration Company was ordered into Active Military Service in 1943 and organized at Fort Francis E. Warren in Wyoming.

The 408th Airborne Quartermaster Company was assigned to the 11th Airborne Division, and the 409th Airborne Quartermaster Company was activated at Camp Mackall, North Carolina.

Company D, 687th Quartermaster Laundry Battalion, was constituted in the AUS, and activated at Camp Ellis, Illinois. Company D, 687th Quartermaster Laundry Battalion was reorganized and redesignated as the 600th Quartermaster Laundry Company.

Transportation

HHD, 5th Quartermaster Troop, Transportation Battalion was constituted in the AUS, activated at Camp Swift, Texas, and redesignated as HHD, 5th Quartermaster Battalion, Mobile.

The 29th Quartermaster Truck Company was redesignated as HHD, 29th Quartermaster Battalion, Mobile.

—— 1944 ——

HQ, Combined Airborne Forces, was formed from elements of the Allied Forces, to be redesignated the First Allied Airborne Army, at Ascot, England, with a forward command post near Paris, France. Commanding the First Allied Airborne Army was Lieutenant General Lewis H. Brereton, the major who had been General Pershing's airborne action-officer in 1918. This headquarters was responsible for the movement of airborne elements into France and directing Operation MARKET GARDEN, the allied invasion of Holland. Operation MARKET GARDEN commenced on 17 September 1944. The XVIII Corps (Airborne) contributed significantly to the First Allied Airborne Army by preparing and writing the MARKET (airborne phase) of the Operation, also the airborne insertions of the 82d and 101st Airborne Divisions and the British 6th Airborne Division during the Rhine Crossings

HHC, XVIII Corps relocated from Camp Bowie, Texas to Fort Dupont, Delaware, en route to deployment to England. On 25 August at Osbourne St George, England, HHC, XVIII Corps was reorganized and redesignated as HHC, XVIII Corps (Airborne), and assigned to the First Allied Airborne Army. HHC, XVIII Corps (Airborne). The blue and white airborne Tab was added to the shoulder sleeve insignia (SSI) of the XVIII Corps (Airborne). Initially the traditional triangular armor SSI had been worn, following the redesignation of the II Armored Corps to become the XVIII Corps, but this had been replaced by the square blue and white SSI, with the dragon's head facing directly to the left. On February 14, the square SSI was rotated 45 degrees to the left, allowing the airborne dragon to look down from the skies. In September, Major General Matthew B. Ridgeway assumed command, after leading the 82d Airborne Division during Operation OVERLORD. In addition to commanding the 82d and 101st Airborne Divisions, the XVIII Corps (Airborne) provided logistical and technical airborne planning expertise in preparation for, and during Operation MARKET GARDEN. The Corps forward headquarters moved to Spernay, France, while the main body relocated to Werbonnet, Belgium.

The Airborne Command was relocated from Fort Bragg, North Carolina to Fort Benning, Georgia, with Brigadier General Josiah T. Dalbey assuming command on 22 January. The Airborne Command was redesignated as the Airborne Center on 1 March.

The 1st Airborne Task Force (the provisional Seventh Army Airborne Division), was organized near Rome, Italy on 15 July, to control airborne assets intended to operate in Southern France. An airborne assault was conducted by the 1st Airborne Task Force (as Task Force RUGBY), into the Le Muy-Le Lac area of Southern France on 15 August. This action limited the access of the German Army to the invasion beaches. The 1st Airborne Task Force linked up with the 36th Infantry Division on 17 August. The unit was discontinued in France on 23 November. In addition to US Army airborne elements, the following British units were attached: the 2d Independent Parachute Brigade, the 64th Light Artillery Battalion, the Allied Air Supply Base, and the 1st Independent Parachute Platoon.

In January, the 11th Airborne Division was relocated from Camp Mackall, North Carolina to Camp Polk, Louisiana, on their way to the Pacific Theater. Continuing their paraglider reputation, a parachute was established at Camp Polk for more glidermen to become parachute qualified. Loading out on 20 April for Camp Stoneman, California, the 11th Airborne Division embarked for New Guinea on 8 May. The 11th Airborne Division remained in New Guinea, conducting more airborne training and also jungle and amphibious exercises, from 25 May until 11 November, when they departed for Leyte, Philippine Islands. Landing at Bito Beach, Leyte, the 11th Airborne Division commenced fighting their way across the Philippine Islands. They relieved the 7th Infantry Division in the Barauen-La Pez-Bugho sector, later moving to Mahonag, and pushing west along the Talisayan River to Ormoc Bay. The Japanese forces were decimated at Anonang, and the 11th Airborne Division prepared for Luzon.

The 13th Airborne Division relocated from Fort Benning, Georgia to Camp Mackall, North Carolina, and their assignment to the Airborne Command.

The 17th Airborne Division deployed to the European Theater of Operations, initially to Camp Chisledon, England, then to Mourmelon in the Rhiems area of France, 23-25 December, as part of a spectacular night transport operation. The 17th Airborne Division moved along the Meuse River from Givet to Verdun.

Operation OVERLORD, the allied invasion of Normandy, commenced on D-Day, 6 June 1944. The amphibious assault, code named Operation NEPTUNE included five beach heads. Beach UTAH on the West, was located approximately ten miles from the city of Ste Mere Eglise. Beach SWORD on the east flank, was approximately fifteen miles from the inland city of Caen. Between UTAH and SWORD, were beaches OMAHA, GOLD and JUNO. The 4th Division landed at Utah Beach, while the 1st and 29th Divisions landed at Omaha Beach. Simultaneously, the British 50th Division, Canadian 3d Division and the British 3d Division landed at Gold, Juno and Sword Beaches respectively. The 82d Airborne Division, commanded by Major General Matthew B. Ridgeway parachuted into the town of Ste Mere Eglise, and the 101st Airborne Division, under the command of Major General Maxwell D. Taylor, parachuted into the area of Carentan. The 82d Airborne Division had arrived in England from Ireland on 14 February. The British 6th Airborne Division, commanded by Major General Richard N. Gale conducted a parachute assault into the Caen area.

Prior to the Normandy landings, the largest deception plan ever attempted, was successfully executed by the Allied forces. Misinformation concerning the 'activation' of fifty non-existent Divisions, was carefully leaked to the Axis Forces as part of Operation QUICKSILVER. Included in this 'paper' army were five US Airborne Divisions - the 6th, 9th, 18th, 21st and 135th Airborne Divisions. These units were reportedly involved in the plans to assault the Pas de Calais coastal region of France. So successful was this ploy, the German army diverted many units from the defense of Normandy to the Pas de Calais area.

Following his heart attack in February, it was expected that Major General Bill Lee would be replaced by Brigadier General Don Pratt, the assistant commander of the 101st Airborne Division. It was Brigadier General Maxwell D. Taylor who was selected to replace Major General Lee, with Brigadier Generals Don Pratt and Anthony C. McAuliffe remaining as deputy commander and division artillery commander respectively. Brigadier General Don Pratt was killed during the D-Day landings when the glider in which he was riding, was demolished.

Colonel Edson Raff, former commander of the 509th Parachute Infantry Battalion, headed a Task Force comprised of approximately 100 personnel from the 325th Glider Infantry, with armor and artillery support, joining the 82d Airborne Division after landing at UTAH Beach.

The 82d Airborne Division returned to England on 13 July, made preparations and rehearsed for their airborne assault into Nijmegen-Arnhem, Holland on 17 September. The 82d Airborne Division captured the Maas Bridge at Grave, the Maas-Waal Canal Bridge at Heumen, and the Nijmegen-Groesbeek Ridge. Following their successful airborne assault into Holland, the 82d Airborne Division moved to France on 14 November, and into Belgium on 18 December.

The 101st Airborne Division participated in the airborne assaults during the 6 June Normandy Invasion . The 101st Airborne Division's sector during Operation OVERLORD was in the vicinity of Carentan, where they secured the beach exits in the St Martin-de-Varreville-Pouppeville area. The Division engaged fierce resistance during the battle for Carentan, commencing 8 through 12 June. On 30 June, the 101st Airborne Division moved to relieve the 4th Infantry Division at Cherbourg.

Newly promoted Major General James M. Gavin assumed command of the 82d Airborne Division on 28 August.

The 101st Airborne Division returned to England on 13 July, to commence preparations for Operation MARKET GARDEN, the 17 September airborne invasion into the Nijmegen-Arnhem area of Holland. The 101st Airborne Division seized nine of their eleven objectives across the Rhine River in the vicinity of Eindhoven, to include the bridges at Veghel, (captured intact), and Zon (demolished by the retreating Germans). The 101st Airborne Division continued to march towards Eindhoven, which fell during the first day, and was occupied by the Screaming Eagles. The 101st Airborne Division secured the Eindhoven-Zon-Veghel-St Oedenrode area, and then moved to Schijndel, continuing to push the Germans back. The 101st Airborne Division returned to France for refurbishing on 28 November. The Screaming Eagles enjoyed but a brief respite, being employed in response to the German Ardennes Counteroffensive. The 101st Airborne Division was encircled at Bastogne on 22 December, where Brigadier Anthony McAuliffe succinctly responded "Nuts!" to the German demands for the 101st Airborne Division to surrender. The encirclement was broken at the cost of grat losses, by the advancing 4th Armored Division, led by battalion commander Lieutenant Colonel Creighton W. Abrams.

After Operation MARKET GARDEN, the 101st Airborne Division moved to France in late November, crossing into Belgium on 18 December.

The British 6th Airborne Division, commanded by Major General Robert Urquhart landed at the 'bridge too far' at Arnhem.

The Troop Carrier Command was commanded by Major General Paul L. Williams.

The 1st Airborne Brigade was disbanded at Camp Mackall, North Carolina. The 2d Airborne Infantry Brigade arrived in Northern Ireland on 8 January. The two allocated infantry regiments, (the 501st and 508th Parachute Infantry), were assigned respectively to the 101st and 82d Airborne Divisions, and the 2d Airborne Brigade HHC attached to the 82d Airborne Division.

Combat Arms
Infantry

The First Airborne Task Force, a provisional division-sized airborne organization, was activated to spear-head the allied invasion (Operation DRAGOON) in August 1944. Under command of Major General Robert T. Frederick, whose First Special Service Force constituted a major asset, were the 509th and 517th Combat Teams. The 509th Combat Team comprised the 509th Parachute Infantry Battalion, reinforced by the 463d Parachute Field Artillery Battalion; and the 517th Combat Team consisted of the 517th Parachute Infantry Regiment, and it's attached 460th Parachute Field Artillery Battalion and the 596th Airborne Engineer Company. The 1st Battalion, 551st Parachute Infantry and the 550th Airborne Infantry Battalion were also assigned.

In November, the commander of the 187th Glider Infantry Regiment, Colonel Harry D. Hildebrand was ordered by Major General Joseph Swing, commanding the 11th Airborne Division, to secure the Corps rear area in the vicinity of Bito Beach in the Philippines. The 187th Glider Infantry and 188th Parachute Infantry carried out an amphibious assault. This was a fortunate decision, resulting in the 187th Glider Infantry successfully repelling a full scale attack by the Japanese army. The 188th Glider Infantry (commanded by Colonel Robert H. Soule) was pointed westerly across Leyte. The 2d Battalion, 187th Glider Infantry, under the command of Lieutenant Colonel Arthur H. Wilson Jr, was sent to reinforce the 188th Glider Infantry and 511th Parachute Infantry during their mop-up operations which lasted until the end of the year.

The 193d and 194th Glider Infantry Regiments departed Camp Mackall, North Carolina, for Camp Forrest, Tennessee, on their way to stage from Camp Myles Standish, Massachusetts and then to England. Arriving on 21 August in England, they prepared for their movement to France on 31 December, en route to Belgium.

Continuing the momentum, the 325th Glider Infantry moved from Northern Ireland to England on 14 February, and like the 327th Glider Infantry, prepared for the Normandy invasion (Operation OVERLORD), on 6 June. Landing between Ste Mere Eglise and Carentan, the 325th Glider Infantry crossed the Merderit River and secured a bridgehead at La Fiere on 9 June. Reaching St Sauveur-le-Vicomte, the 325th Glider Infantry and the 505th Parachute Infantry, established a bridgehead at Pont l'Abbe on 19 June. PFC Charles N. DeGlopper, Company C, 325th Infantry, was posthumously awarded the Medal of Honor for his actions on 9 June during the Normandy Invasion.

The 325th and 327th Glider Infantry Regiments participated with the 82d and 101st Airborne Divisions in the airborne invasion of Operation MARKET GARDEN, on 17 September at Nijmegan, Holland, then on to France in November. The 325th Glider Infantry continued on to Belgium on 18 December, the 327th Glider Infantry embarked for Germany on 27 April, and then to France on 30 November. The 325th and 327th Glider Infantry Regiments were credited with participation in the Normandy, Rhineland, Ardennes-Alsace and Central Europe Campaigns. The 325th Glider Infantry was additionally credited with the Sicily and Naples-Foggia Campaigns.

The 327th and the 1st Battalion, 401st Glider Infantry were to link up with the 101st Airborne Division following their amphibious landing at UTAH Beach, during Operation OVERLORD.

The 401st Glider Infantry (minus the 2d Battalion) participated in the airborne assault onto Nijmegan-Arnhem, in Holland, returning to France on 28 November. The 2d Battalion, 401st Glider Infantry, had participated with the 101st Airborne Division in the airborne invasion during Operation OVERLORD.

The 501st Parachute Infantry, commanded by Colonel Howard R. Johnson, staged at Camp Myles Standish, Massachusetts on 2 January, embarking on 18 January from Boston to arrive in England on 31 January, and following a brief initial assignment to the 2d Airborne Infantry Brigade, was attached to the 101st Airborne Division. The 501st Parachute Infantry participated in the Normandy Invasion, capturing the Douve Locks at La Barquette. The 3d Battalion, commanded by Lieutenant Colonel Julian J. Ewell was dispatched by General Maxwell Taylor to capture Pouppeville, which he accomplished by mid-day.

After returning to England, the 501st Parachute Infantry conducted the airborne assault into Nijmegan-Arnhem, in Holland, with the 101st Airborne Division, to capture Beghel. Lieutenant Colonel Patrick Cassidy, commanding the 1st Battalion, was wounded during this operation. Lieutenant Colonel Cassidy had assumed command of the 1st Battalion when Lieutenant Colonel Harry W. O. Kinnard was moved to become the Operations Officer of the 101st Airborne Division. On 8 October, Colonel Howard R. Johnson was killed by enemy artillery fire, at which time Lieutenent Colonel Ewell was assigned to command the Regiment. was Returning to France from 28 November until 18 December, when the 501st Parachute Infantry departed for Belgium.

The 502d Parachute Infantry Regiment, commanded by Colonel George van Horne Mosely, participated with the 101st Airborne Division in the Normandy Invasion, returning to England from France on 13 July. Colonel Moseby broke his leg during the airborne operation, requiring Lieutenant Colonel John H. Michaelis, the Regimental Executive Officer to assume command. On 17 September, again with the 101st Airborne Division, the 502d Parachute Infantry Regiment conducted an airborne assault into Nijmegen-Arnhem, Holland, under the command of newly promoted Colonel Michaelis. During this action, Colonel Michaelis suffered multiple schrapnel wounds. For actions at Best, Holland, Private First Class Joe E. Mann was posthumously awarded the Medal of Honor on 17 September. The 502d Parachute Infantry was attached to the 82d Airborne Division from 4-5 October. For his actions at Carentan, France on 4 October, Lieutenant Colonel Robert Cole was awarded the Medal of Honor.

The 503d Parachute Infantry returned to Australia from New Guinea on 14 April, completed their preparations, and HHC with the 1st Battalion conducted an airborne assault onto Kamiri Airfield on Noemfoor Island, New Guinea on 3 July. For his actions on Noemfoor Island, Sergeant Ray E. Eubanks, Company D, 503d Parachute Infantry, was awarded the Medal of Honor. The 3d Battalion arrived by parachute the following day, and they were joined by the 2d Battalion, which arrived by sea on the 11th July. Following defeat of the Japanese, the 503d Parachute Infantry departed on 9 November for Leyte, Philippine Islands, where they invaded Mindoro at San Agustin and secured San Jose in December.

The battles continued in Italy, where the 504th Parachute Combat Team, still commanded by Colonel Reuben Tucker, fought on the right flank of the VI Corps. In March the 504th Parachute Combat Team withdrew in anticipation of returning to England, however they remained in Italy in support of the Fifth Army. The 504th Parachute Infantry departed Italy for England in April., remaining there until participating with the 82d Airborne Division in the airborne assault into Nijmegen-Arnhem, Holland on 17 September, and returning to France on 15 November.

The 505th Parachute Infantry, commanded by Colonel William E. Eckman, which had arrived in England from Northern Ireland on 14 February, participated with the 82d Airborne Division during Operation OVERLORD. The 505th Parachute Infantry returned victoriously to England from Normandy, France, on 13 July, and after refurbishing, was assigned under the Operational Control of the 82d Airborne Division to participate in Operation MARKET GARDEN and the parachute assault at Nijmegen-Arnhem, Holland on 17 September. For his actions on 21 September, Private John R. Towle was posthumously awarded the Medal of Honor. The 505th Parachute

Infantry again returned briefly to France on 14 November, leaving for Belgium on 18 December. German assaults along the Salm reached the 505th Parachute Infantry in the Trois Ponts area in the latter part of December.

The 506th Parachute Infantry, commanded by Colonel Robert F. Sink, conducted the airborne assaults with the 101st Airborne Division, during Operation OVERLORD, into Normandy and Operation MARKET GARDEN into Holland. Following the airborne assault into Nijmegen-Arnhem, Holland, the 506th Parachute Infantry reopened the Veghel-Uden Highway.

In January the 507th Parachute Infantry Regiment, commanded by Colonel George V. Millett Jr, and the 508th Parachute Infantry Regiment, commanded by Colonel Roy E. Lindquist, were both attached to the 82d Airborne Division on 14 January. The 508th Parachute Infantry had been initially allocated and assigned to the 2d Airborne Infantry Brigade. The 507th and 508th Parachute Infantry Regiments accompanied the 82d Airborne Division during the parachute assaults of Operations OVERLORD and MARKET GARDEN, returning to England on 13 July. The 507th Parachute Infantry was next placed under the Operational Control of the 17th Airborne Division on 27 August., returning through France into Belgium in December. The 508th Parachute Infantry returned to France on 20 November.

The 509th Infantry Battalion waded ashore to occupy the town of Nettuno, south of Anzio, working closely with the 1st Ranger Battalion. It was during a reconnaissance patrol near Carano, Italy on 8 February, that Corporal Paul B. Huff of Company A, 509th Parachute Infantry Battalion, became the first US Army paratrooper to earn the Medal of Honor. The 509th Parachute Infantry Battalion was to be withdrawn, readying themselves for the invasion of Southern France. Assigned to the 1st Airborne Task Force, the 509th Parachute Infantry led the way to occupy Cannes.

The 511th Parachute Infantry relocated to Camp Polk, Louisiana on 5 January, staging from Camp Stoneman, California and departing from San Francisco on 8 Mayu. Arriving in New Guinea on 29 May, the 511th Parachute Infantry, commanded by Colonel Orin D. Haugen, crossed to Leyte, Philippine Islands on 18 November, and commenced a strenuous march from Bito Beach, Leyte, across the mountains from Barauen to Mahonag. On 8 December, Private Elmer Fryer was awarded the Medal of Honor for his actions above and beyond the call of duty.

The 513th Parachute Infantry relocated to Camp Mackall on 15 January, relieved from assignment to the 13th Airborne Division in March, and assigned to the 17th Airborne Division, relocating to Camp Forrest, Tennessee. Staging through Camp Myles Standish and Boston, the 513th Parachute Infantry arrived in England on 28 August.

The 515th and 541st Parachute Infantry moved to Camp Mackall, North Carolina on 19 January and 14 October respectively. The 515th Parachute Infantry was initially assigned under the XIII Corps, and to the 13th Airborne Division on 10 March. The 541st Parachute Infantry was assigned to XIII Corps on 1 March, and then returned to Fort Benning, Georgia on 29 July for assignment to the Replacement School and Command, and attachment to the Airborne Center on 16 November. The 541st Parachute Infantry returned to Camp Mackall, North Carolina on 23 November, under the command of Colonel Ducat M. McEntee, and relocated to France on 24 December, entering Belgium on 25 December.

The 517th Parachute Infantry was released from the 17th Airborne Division on 10 March, and attached to Second Army. Embarking through Hampton Roads, the 517th Parachute Infantry arrived in Italy on 28 May, commencing combat operations under IV Corps. The 517th Parachute Infantry seized St Vallier ten days after their 15 August assault into Southern France. The 517th Parachute Infantry was attached to the XVIII Corps (Airborne), from 22 November through 16 December; then to the 30th Infantry Division from 17 through 27 December; and finally to the 7th Armored Division from 28-29 December. For his actions on 24 December, Private First Class Melvin E. Biddle was awarded the Medal of Honor.

The 542d Parachute Infantry was redesignated as the 542d Parachute Infantry Battalion, and relocated from Fort Benning, Georgia with the reorganized and redesignated 555th Parachute Infantry Battalion, to Camp Mackall, North Carolina. Commanding the 542d Parachute Infantry was Colonel William T. Ryder. The 543d Parachute Infantry was inactivated at Fort Benning.

The 1st Battalion, 551st Parachute Infantry was moved to Italy in June, and participating with the 1st Airborne Task Force airborne assault into southern France on 15 August. The 1st Battalion followed the 509th Parachute Infantry into Cannes, linking up with the 1st Special Service Force. The 551st Parachute Infantry was attached to the 82d Airborne Division from 26 December until 13 January 1945.

Following a visit by negro Brigadier General Benjamin O. Davis, it was decided to expand the all-black 555th Parachute Infantry Company into a negro parachute infantry battalion.

Armor

The 151st Airborne Tank Company, now having six officers and 129 enlisted men assigned, was relocated to Camp Mackall, North Carolina to commence glider training in July 1944. Despite successful airborne field training and maneuvers at Camp Mackall, and passing all tests with an almost perfect rating, the 151st Airborne Tank Company was inactivated in December. The officers were reassigned to the Armored Replacement Pool at Fort Knox, and the enlisted personnel distributed to various armored units across the country.

The 645th Tank Destroyer Battalion was assigned to the 1st Airborne Task Force.

Division Artillery

The 13th Airborne Division Artillery was relocated from Fort Bragg to Camp Mackall, North Carolina and inactivated.

Field Artillery

Assigned to the 504th Parachute Combat Team, the 376th Parachute Field Artillery Battalion remained in Italy in support of the Fifth Army.

The 377th Parachute Field Artillery accompanied the 82d Airborne Division onto Drop Zone ABLE during Operation OVERLORD.

Covering the 11th Airborne Division infantry units in the Philippines, were the 457th Parachute Field Artillery Battalion and the 674th Glider Field Artillery Battalion. Lieutenant Colonel Nicholas G. Stadtheer commanded the 457th Parachute Field Artillery Battalion.

The 460th Parachute Field Artillery Battalion was relieved from assignment to the 17th Airborne Division. The 463d Parachute Field Artillery Battalion was activated at Naples, Italy, and attached to the 101st Airborne Division. The 460th and 463d Parachute Field Artillery Battalions were later respectively assigned to the 517th Parachute Infantry Regiment and the 509th Parachute Infantry Battalion Combat Teams, as part of the First Airborne Task Force.

On 11 March, the 462d Parachute Field Artillery departed San Francisco, California for Brisbane, Australia, and subsequent assignment to the 503d Parachute Infantry Regiment on 29 March.

Activated at Fort Bragg in 1943, the 465th Glider Field Artillery Battalion was reorganized and redesignated as the 465th Field Artillery Battalion. The 467th Parachute Field Artillery Battalion was constituted in the AUS, and activated at Camp Mackall, North Carolina.

The 602d Field Artillery Battalion was attached to the 1st Airborne Task Force.

Battery C, 4th Coast Artillery Battalion, was constituted in the AUS in November, from a consolidation of Battery C, 4th Coast Artillery, (with lineage dating back to the 2d Regiment of Artillery, constituted in 1812), and Battery C, 4th Coast Artillery Battalion, (constituted in October 1944), and activated in the Galapagos Islands.

Air Defense Artillery

The 13th, 14th and 15th Airborne Anti-Aircraft Artillery Battalions were inactivated at Camp Chaffee, Arkansas. The 13th Battalion was inactivated on 30 April and the 14th and 15 Battalions on 5 May.

On 22 June, the 16th Airborne Anti-Aircraft Artillery Battalion was activated in Burma, and HHDs, 17th and 18th Airborne Anti-Aircraft Artillery Battalions, constituted in the AUS, were activated at Moran, India, and redesignated HHBs, 17th and 18th Anti-Artillery Battalions . The 16th, 17th and 18th Airborne Ant-Aircraft Artillery Battalions all participated in the New Guinea campaign, with the 16th Battalion additionally participating in the India-Burma campaign.

The 151st Airborne Anti-Aircraft Artillery Battery, constituted in 1943, was activated in 1944, disbanded, and redesignated as the 151st Anti-Aircraft Artillery Machine-Gun Battery. The 503d Anti-Aircraft Artillery Gun Battalion, deployed from Camp Stoneman, California to Hawaii.

Special Operations

The 1st Special Service Force continued fighting in Italy, overcoming the mountain defenses at Mount Vischiataro on 8 January, followed by Maio, strongly demonstrating the efficiency of their cold weather training during these harsh winter months. In late December, 1943, the battle lines were drawn and prepared for the attack of Monte Cassino.

On January 17, Lieutenant General Mark Clark committed his Fifth Army to a traditional frontal assault, followed by landings beyond the Gustav Line at the fishing village of Anzio. The British X Corps established a bridgehead and crossed the Garigliano River. The US 36th Division attempted to cross an unmarked mined area, with devastating results. The 1st Special Service Force was rushed to secure the right quadrant, (a length of 50 kilometers), under the command of Major General Robert T. Frederick. Replacements for the under strength Force came from the decimated Rangers. With their baggy ski-pants and blackened faces, the Germans named the Force as the Black Devils, (which in later years, following true Hollywood tradition, was to become interpreted as The Devils Brigade). They deployed along the Mussolini Canal until 10 May, overcoming Mounts Arrestino and Rocca Massima by the end of that month. On 4 June, the Black Devils of the Second Regiment, under the command of Colonel J. F. R. (Jake) Akehurst, led the way into Rome. Two months later, on 14 August, the Force invaded the Iles d'Hyeres, off southern France. The force captured Grasse on 27 August, then Vence on 1 September and Mentone on 7 September.

The Force was assigned to the right flank of the 1st Airborne Task Force, replacing the British 2d Independent Parachute Brigade, and linking up with the 509th and 551st Parachute Infantry in Cannes. Crossing the Var River, the 1st Special Service Force drove through Nice and Beaulieu to La Turbia, establishing a stronghold on 2 September. Continuing through Menton, France, the 1st Special Service Force established defensive positions along the Franco-Italian border. The 1st Special Service Force was inactivated at Menton, France on 5 December.

Fighting east along the Franco-Italian border. Major General Frederick assumed command of the 1st Airborne Task Force until the end of November, before moving to Menton, France where the 1st Special Service Force was disbanded on 5 December.

The 2677th OSS Regiment was organized at Algiers, North Africa on 15 July under Fifth Army, and relocated to Caserta, Italy to become part of the OSS Operational Group Command on 27 November.

Rangers

The 1st Ranger Battalion, commanded by LTC William O. Darby, participated in the assault at Anzio, suffering severe casualties. Survivors of the 1st Ranger Battalion were absorbed into the 1st Special Service Force. LTC James E. Rudder led the 2d Ranger Battalion in their heroic assault of Pointe du Hoe during the Normandy invasion. Also during the Normandy invasion, at OMAHA Beach, Major General Norman Cota, commanding the 29th Infantry Division, gave the historic command to Lieutenant Colonel Max Schneider, commanding the 5th Ranger Battalion, "...Rangers, lead the way!"

The 5307th Composite Unit, Provisional, was activated on 1 January, assigned to US Army Forces of the China-Burma-India Theater of Operations. Known as the GALAHAD Task Force or 'Merrill's Marauders', (after the commander, Colonel Frank Merrill), the 5307th Composite Unit comprised three long-range penetration battalions. The 5307th Composite Unit became operational on 12 February in the Hukawing Valley of Burma, and assigned to the Northern Area Combat Command on 8 May. Operating in the rear of the Japanese forces, the 5307th Composite Unit seized the Myitkyina Airfield on 17 May. In retaliation, on 24 May, the 3d Battalion was badly mauled at Charpate. The 2d Battalion was driven from Namkwi on 26 May. Regrouping, the 5307th Composite Unit captured the city of Myitkyina on 3 August, where it was disbanded on 10 August, and personnel were primarily reassigned to the 475th Infantry Regiment. The consolidated unit was designated as the 475th Infantry (Long Range Penetration, Special). The 475th Infantry was assigned to the 5332d Brigade (Provisional), known as the MARS Task Force.

The 98th Field Artillery Battalion, activated at Fort Lewis, Washington in 1941, and assigned to the China-Burma-India Theater of Operations, was converted and redesignated to become the 6th Ranger Infantry Battalion.

Medical

The 676th Medical Collecting Company was assigned to the 1st Airborne Task Force.

Combat Support
Engineers

The US IX Engineer Command HQ was organized in England under the command of Brigadier General James J. Newman, as a component of the 9th Air Force. In addition to commanding four aviation regiments, this headquarters also commanded one camouflage aviation engineer battalion and three airborne aviation engineer battalions. The 876th, 877th and 878th Aviation Engineer Battalions, assigned to the IX Engineer Command, were designed to rapidly assault, capture and repair essential airfields within supporting distance of the forward lines. Parachute status as airborne units, was provided to certain Engineer Battalions, Aviation, in a War Department letter dated 3 February 1943.

The 127th Parachute Engineer Battalion supported the 11th Airborne Division in the Philippines.

Company C, 139th Airborne Engineer Battalion, was relieved from assignment to the 139th Airborne Engineer Battalion, 17th Airborne Division, to be reorganized and redesignated as the 596th Airborne Engineer Company. The 596th Airborne Engineer Company was attached to the 509th Parachute Infantry Battalion Combat Team as part of the First Airborne Task Force.

The 307th Parachute Engineers remained in Italy in support of the Fifth Army, as part of the 504th Parachute Combat Team.

The 887th Airborne Engineer Aviation Company was assigned to the 1st Airborne Task Force.

Military Intelligence

The 77th, 82d and 101st Counter Intelligence Corps Detachments were constituted in the AUS and respectively activated in Hawaii (77th), and England (82d and 101st)

The 356th Interpreter Team was constituted in the Regular Army, assigned to the 315th HQ Intelligence Detachment, and activated at Fort Shafter, Hawaii.

Military Police

The 503d Military Police Battalion deployed to England in January, and then to France in August.

The Provisional Airborne Military Police Platoon was assigned to the 1st Airborne Task Force.

Signal

The 3191st Signal Service Company was constituted in the AUS and activated at Camp Crowder, Missouri.

The 511th Airborne Signal Company deployed to New Guinea, and the 512th Airborne Signal Company was constituted in the AUS, activated in North Africa, and then assigned to the 1st Airborne Task Force.

Combat Service Support
Adjutant General

The Band, 11th Airborne Division Artillery, was redesignated as the 11th Airborne Division Band. Band, 13th Airborne Division Artillery, was relocated to Camp Mackall, North Carolina, assigned to XIII Corps, and disbanded at Camp Mackall.

Chemical

Companies A of the 2d and 83d Chemical Mortar Battalions were attached to the 1st Airborne Task Force.

Quartermaster

The 11th Parachute Maintenance Company, constituted in the AUS in 1944 and assigned to the 11th Airborne Division, was inactivated in New Guinea. The 13th Parachute Maintenance Company was activated in Europe, and assigned to the 13th Airborne Division.

The 334th Quartermaster Depot Supply Company was assigned to the 1st Airborne Task Force.

The HHD, 528th Quartermaster Service Battalion was reorganized and redesignated as HHD, 528th Quartermaster Battalion. (Companies A, B, C and D were redesignated as the 4098th, 4099th, 4100th and 4128th Quartermaster Service Companies, hereafter following separate lineages).

Relocated to Fort Bragg from Camp Mackall, North Carolina, the 713th Airborne Quartermaster Company was relieved from assignment to the 11th Airborne Division and assigned to the 13th Airborne Division.

—— 1945 ——

Hq, First Allied Airborne Army was relocated forward to vicinity of Paris. The First Allied Airborne Army, after successfully planning and executing Operation VARSITY, was disbanded by direction of the Supreme Allied Headquarters. The US elements were assigned to the concurrently designated and activated, First US Airborne Army. The First US Airborne Army was relocated to Berlin, Germany as HQ, US Sector of Berlin, where it was later inactivated this same year.

HHC, XVIII Corps (Airborne), was next assigned to the US First Army, with operational command over the 8th, 78th, 86th and 97th Infantry Divisions, and the 13th Armored Division, and later attached to the British Second Army, with command of the British 6th Airborne Division, and the US 8th Infantry Division, and 82d Airborne Division. This same year, HHC, XVIII Corps (Airborne) redeployed to the United States.

Relocating from Werbonnet, Belgium, to the vicinity of the forward headquarters at Spernay, France, the XVIII Corps (Airborne) moved temporarily to the Huertgen Forest, Germany, returning to Spernay in preparation for the coordination of the Rhine Crossings. The Rhine Crossings involved the coordinated movement and insertion of the (British) 6th Airborne Division and the US 17th Airborne Division. Following the Rhine Crossings, the XVIII Corps (Airborne), was assigned to First Army, with operational command over the 8th, 78th, 86th and 97th Infantry Divisions, and the 13th Armored Division. Later, attached to the British Second Army, the XVIII Corps Airborne commanded the 8th Infantry and 82d Airborne Divisions, prior to the HHC, XVIII Corps (Airborne) redeploying to the United States.

Major General Anthony C. McAuliffe assumed command of the Airborne Center.

The 11th Airborne Division moved to southern Luzon on 24 February, opening Manila Harbor, and later the Balayan and Batangas Bays in March. From there the Division attacked Japanese positions from Lake Taal to Laguna de Bay, with fierce battle in the vicinities of Hill 660, Lipa Hill, Mount Bijang, Bukel Hill, and Mounts Malepunyo, Macoled and Matassna Bundoc. In July, the 11th Airborne Division disembarked for Okinawa, arriving one day before the end of the war. This placed the 11th Airborne Division as available to commence occupation duties in Japan.

The 13th Airborne Division deployed to the European Theater of Operations, and was initially assigned to the XVIII Corps (Airborne). Initially included for participation in Operation VARSITY, the 13th Airborne Division was withdrawn on 10 March, and assigned with the 17th Airborne Division to the First Allied Airborne Army. Two days later, the 13th was written into Operation CHOKER II, but this was cancelled on 24 March, and Operation EFFECTIVE was cancelled in April. The 13th Airborne Division was reassigned back to the XVIII Corps (Airborne), then returned later this same year to the United States, in preparation for deployment to the Pacific Theater. The 17th Airborne Division also redeployed, and was inactivated at Camp Myles Standish, Massachusetts. The 101st Airborne

Division was inactivated at Auxerre, France.

The 17th Airborne Division relieved the 28th Infantry Division in the Neufchateau area, Belgium. The Division moved into line along Dead Man's Ridge between the 87th Infantry and 101st Airborne Divisions, enduring appalling and severe cold conditions and an experienced enemy. Ownership of the town of Flamierge changed several times, but was finally secured by the 17th Airborne Division, who moved on to take Salle. Relieving the 11th Armored Division at Houffalize, the 17th Airborne Division continued on, and by 26 January had seized Compogne, Hautbellain, Watermal, Steinbeck, Limerle and Espeler. On 10 February, the 17th Airborne Division was relieved by the 6th Armored Division.

The next opportunity for the 17th Airborne Division to excel was Operation VARSITY, 24 March 1945. The 17th Airborne Division missions were to seize, clear and secure Dieresfordt and the bridges crossing the Issel River; establish communications with the British 6th Airborne and 15th Infantry Divisions, and the British First Commando Brigade; and protect the southern flank of the XVIII Corps (Airborne). Operation VARSITY was a success, with almost 3,000 enemy prisoners taken by the end of the first day. Brigadier General John L. Whitelaw, Assistant Division Commander, brought in the Division Land Echelon on the 3d day of the operation. The Division moved west of the Rhine River, to enter into serious combat, resulting in the capture of Munster. Relieving the 79th Infantry Division, the 17th Airborne Division was assigned to XVI Corps, establishing a bridgehead across the Rhine-Herne Canal, and continuing south to capture Essen, Mulheim and Driesberg. The 17th Airborne Division received the formal surrender of Duisburg, commencing relief of the 79th Infantry Division on 12 April. Prior to returning to the United States at the close of World War II, the 17th Airborne Division conducted occupational duties, and was inactivated at Camp Myles Standish, Massachusetts on 16 September.

On 30 January, the 82d Airborne Division moved from Belgium into Germany where Comte and Herresback were seized by the All American Division. In early February the 82d Airborne Division reached the Roer River near Bergstein, which was attacked from 7-10 February. Crossing the Roer River, the Division moved to Rheims, France from 19 February through 2 April, preparing for their next offensive. Returning to Germany, the 82d Airborne Division relieved the 86th Infantry Division on the Rhine on 4 April, and performed occupational security duties in Cologne for three weeks. The 82d Airborne Division attacked past the Elbe River in May, and the Bleckede area.

The 101st Airborne Division moved from Bastogne to capture Recogne, the Bois Jacques, Rachamps, Bourcy and Hardigny by 17 January. On 22 January, the 101st Airborne Division had seized Drulingen-Sarraltroff, and Hochfelden by the end of the month. On 25 February, the Screaming Eagles relieved the 36th Infantry Division and moved to Mourmelon, France to re-group. On 31 March, the 101st Airborne Division continued on to the Ruhr Valley to relieve the 97th Infantry Division. The Division moved to Miesbach on 27 April, and three days later assumed the area responsibility for Kaufbeuren-Saulgruber. The Screaming Eagles had continued their advance to Berchtesgaden when the armistice was declared on 7 May.

The 2d Airborne Infantry Brigade was disbanded in England on 15 January.

Combat Arms
Infantry

In the Philippines, the 187th Glider Infantry fought the battle for Hill 660 in March, moving on to Mount Macoled. The 511th Parachute Infantry overcame determined opposition at Mount Bijang, and the 188th Glider Infantry moved to Batangas, and commenced attacking towards Lipa. As the 188th Glider Infantry secured Lipa, the 187th Glider Infantry overcame Bukel Hill and Talisay. From Mount Bijang, the 188th Parachute Infantry seized Mount Malepunyo followed by Atimonan. The most sustained fighting occurred at Mount Matassna Bundoc, which was overcame by the 188th Glider Infantry, the 511th Parachute Infantry, and the attached 8th Cavalry.

The 193d Glider Infantry was attached to the 101st Airborne Division from 3 through 7 January and from 14 through 18th January, and disbandment on 1 March. Campaign participation for the 193d Glider Infantry included the Rhineland, Ardennes-Alsace and Central Europe.

The 194th Glider Infantry returned to France on 11 February, and conducted a parachute assault into Wesel, Germany on 24 March, attacking across the Issel Canal, to eventually occupy positions near the Erle-Schermbeck Road. For his actions on 28 March, Technical Sergeant Clinton M. Hedrick was awarded the Medal of Honor. In early April, the 194th Glider Infantry was attached to the 95th Infantry Division, then back to the 17th Airborne Division. The 194th Glider Infantry disembarked from Europe, arriving in Boston, Massachusetts on 14 September, to be concurrently inactivated at Camp Myles Standish, Massachusetts.

The 325th Glider Infantry transferred from Belgium to Germany on 30 January, then to France in February and returned to Germany on 2 April.

The 327th Glider Infantry was with the 101st Airborne Division at Bastogne, later capturing Bourcy.

The 503d Regimental Combat Team, strengthened by the 3d Battalion, 34th Infantry, prepared for an amphibious/airborne assault into Corregidor Island, Philippine Islands. For his actions at Corregidor, Private Lloyd G. McCarter was awarded the Medal of Honor. This time the 317th Troop Carrier Group was closer to their target. Following Corregidor, the pocket patch of the 503d Parachute Infantry was changed from the snarling wildcat descending by parachute, to a blood red outline of Corregidor Island mounted by a white eagle, and the nickname "the Rock" was attached to the Regiment. With Corregidor subdued by early March, the 503d Regimental Combat Team departed for Palupandan Negros Island on 7 April, to be attached to the 40th Infantry Division until July, when they assumed occupational responsibilities, and finished the clearing of the Island. The 503d Parachute Infantry disembarked for Los Angeles on 23 December, to be inactivated at Camp Anza, California on 25 December.

The 504th Parachute Infantry was attached to the 75th Infantry Division 1-3 January 1945, and moved into Belgium on 26 January, until returning to France in February, en route to Germany, which they reached on 8 April.

The 505th Parachute Infantry moved from Belgium, arriving in Germany on 30 January, and after a short return to France in mid-February, returned to Germany on 2 April.

The 506th Parachute Infantry was officially assigned to the 101st Airborne Division, having been in an attached status since 1 June 1943. Following the encirclement at Bastogne, the 506th Parachute Infantry attacked and seized Recogne, the Bois Jacques, Foy and Neville. The 506th Parachute Infantry moved into Germany on 4 April, and was attached to the 4th Infantry Division during 2-3 May, and was inactivated on 30 November in France.

The 507th and 513th Parachute Infantry Regimental Combat Teams led the way followed by the 194th Glider Infantry, as the vanguard of the 17th Airborne Division into Operation VARSITY, at Wesel, Germany on 24 March, attacking across the Issel Canal. Colonel James W. Coutts commanded the 513th Parachute Infantry Regimental Combat Team. The 507th Parachute Infantry Regimental Combat Team, led by Colonel Edson D. Raff, commenced the first airborne assault at 0945 hours. The last pass of the 194th Glider Infantry was completed within three hours. With the exception of the 1st Battalion, 507th Parachute Infantry, all airborne elements landed on their designated drop zones. The 1st Battalion was approximately two miles from their target area. The 507th Parachute Infantry secured the towns of Wulfen and Bertogne, and the 194th Glider Infantry captured Givroulle. Private George J. Peters was posthumously awarded the Medal of Honor for his actions on 24 March. The 507th Parachute Infantry was attached from 31 March to 2 April to the XIX Corps. Returning to the control of the 17th Airborne Division, the 507th Parachute Infantry attacked across the Rhine-Herne Canal on 8 April, and cleared the Berne Canal the next day, seizing Essen on 10 April. The 507th Parachute Infantry disembarked from the European Theater of Operations, arriving in Boston, Massachusetts on 15 September, and inactivation at Camp Myles Standish, Massachusetts the following day.

The 508th Parachute Infantry was attached to the 7th Armored Division 21-23 January, and returned to Belgium on 26 January with the 82d Airborne Division, (being attached back to the 82d Airborne Division on 25 January). For his actions on 29 January, First Sergeant Leonard Funk was awarded the Medal of Honor. In August, the 508th Parachute Infantry relocated to Sissone, France.

Colonel Orin D. Haugen and the 511th Parachute Infantry, transported by Colonel James Lackey's 317th Troop Carrier Group, made an airborne assault onto Tagaytay Ridge in the Philippine Islands on 3 February. Unfortunately this could not be construed as an accurate insertion, landing in the vicinity of the Manila Hotel Annex. Private First Class Manuel Perez Jr was awarded the Medal of Honor for his actions on 13 February. Later, while assaulting the Genko Line with his 511th Parachute Infantry, Colonel Haugen was mortally wounded and died. Company A, 1st Battalion, 511th Parachute Infantry, conducted the extremely successful airborne assault on the Los Banos Prison. The remainder of the 511th Parachute Infantry and the 188th Glider Infantry, attacked on the ground after an insertion by fifty-nine amphibious tracked vehicles of the 672d Amphibious Tractor Battalion. The insertion of Company A was made by the 65th Troop Carrier Squadron. The 511th Parachute Infantry, while assisting the 37th Infantry Division, conducted the last airborne operation in the Pacific Theater during World War II. The 511th Parachute Infantry assaulted onto Camalaniugan Airfield, south of Aparri in Northern Luzon. On 12 August, the 511th Parachute Infantry arrived in Okinawa on their way to perform occupation duties with the 11th Airborne Division, arriving in Japan on 30 August.

The 513th Parachute Infantry was introduced into the Ardennes Campaign, where in January, in the vicinity of Famierge, for his actions, Sergeant Isadore S. Jackman was awarded the Medal of Honor. returned to France on 11 February, and back to Belgium on 21 March before assaulting with the 17th Airborne Division into Wesel, Germany on 24 March. For his actions on that same day, 24 March, Private First Class Stuart S. Stryker was posthumously awarded the Medal of Honor. The 513th Parachute Infantry returned to France, and later disembarked from Europe to Boston, arriving on 14 September, to be inactivated at Camp Myles Standish that same day.

The 515th Parachute Infantry staged from Camp Shanks, New York on 19 January, arriving in France in February, assigned to the 13th Airborne Division, to return to Fort Bragg, North Carolina on 26 August.

The 517th Parachute Infantry was attached to the 82d Airborne Division from 23-26 January and 3-4 February, moving to Schmidt-Nideggen on 6-7 February (while attached to the 78th Infantry Division). Attached back to the 82d Airborne Division 9-10 February, the 517th Parachute Infantry prepared themselves, in the vicinity of the Huertgen Forest. They were attached to the 13th Airborne Division from 11 February through 1 March, when they became assigned to the 13th Airborne Division. On 25 August the 517th Parachute Infantry returned with the 13th Airborne Division to Fort Bragg, North Carolina on 23 August.

The 541st Parachute Infantry staged at Camp Stoneman, California on 23 May, arriving in Manila, the Philippine Islands, on 10 July. On 10 August the 541st Parachute Infantry was inactivated at Lipa, Luzon, with assigned personnel being used to replenish the 187th and 188th Parachute Infantry Regiments.

Operation FIREFLY was activated, whereby the all-negro 555th Parachute Battalion was dispatched to Pendleton, Oregon and trained as airborne fire fighters, specifically in response to the 10,000 Japanese balloon bombs launched to land along the Western coast of the United States. Although the actual number of balloon bombs located, amounted to about 300, the US Forest Service would have been in dire straights without the Triple Nickels to assist them during the fire season.

HQs, 163d and 164th Infantry Brigades were constituted in the National Army in 1917, organized at Camp Gordon, Georgia, and assigned to the 82d Division. Demobilized in 1919, they were reconstituted as the HHCs, 163d and 164th Infantry Brigades in 1921, assigned to the 82d Division, and organized respectively in 1922 at Macon, Georgia and Jacksonville, Florida. In 1942, the HHCs, 163d and 164th Infantry Brigades were consolidated, converted and redesignated as the 82d Reconnaissance Troop, ordered into Active Military Service, and reorganized at Camp Claiborne, Louisiana. The 82d Reconnaissance Troop was redesignated as the 82d Cavalry Reconnaissance Troop, and disbanded in 1942 at Camp Claiborne. The 82d Cavalry Reconnaissance Troop was reconstituted in the AUS in 1945 as the Reconnaissance Platoon, 82d Airborne Division, and activated in France.

In order to provide early warning of the serious armor threats in the European and Pacific Theaters of Operations, Reconnaissance Platoons were constituted in the AUS, and assigned to the 11th, 13th, 17th and 101st Airborne Divisions. In 1945, the Reconnaissance Platoon, 11th Airborne Division was activated in Luzon, Philippine Islands, and the Reconnaissance Platoons of the 13th and 17th Airborne Divisions were activated in France. HHC, 401st Glider Infantry (constituted in the AUS and activated at Camp Claiborne in 1942), was disbanded on 25 January 1945, to be reconstituted and consolidated in Europe on 1 March, with the Reconnaissance Platoon, 101st Airborne Division.

The US Army airborne forces had not only discharged every vital mission with total courage, they had demonstrated a practical and extremely successful third dimension to the art of war. The establishments of airheads and vertical envelopment became normal strategy considerations in future planning. However, as World War II drew to a close, the might and mobility of the US Army airborne was drawn down drastically. The forerunners of our modern Airborne Rangers, and the Office of Strategic Services - the forefather of US Army Special Forces, were totally disbanded, only to be raised through necessity to perform their unique missions in wars to come.

The demise of the Glider units commenced. The 88th and 193d Glider Infantry were disbanded. The 188th Glider Infantry was redesignated as the 188th Parachute Infantry, and the 194th Glider Infantry was inactivated at Camp Myles Standish, Massachusetts. The 327th and 401st Glider Infantry were disbanded in Auxerre France, with the 1st Battalion, 401st Glider Infantry consolidated with the 3d Battalion, 327th Glider Infantry, and the 2d Battalion, 401st Glider Infantry consolidated with the 3d Battalion, 325th Glider Infantry.

The 501st Parachute Infantry was disbanded in Germany on 20 August. The 502d and 506th Parachute Infantry were inactivated in France. The 503d Parachute Infantry was inactivated at Camp Anza, California, and the 509th Parachute Infantry Battalion was disbanded in France on 1 March. The 507th, 513th and 514th Parachute Infantry, assigned to the 17th Airborne Division, were inactivated at Camp Miles Standish, Massachusetts. The 515th Parachute Infantry was alerted for deployment to Denmark, but did not deploy. The 517th Parachute Infantry was assigned to the 13th Airborne Division. The 541st Parachute Infantry was disbanded in Luzon, and the 542d Parachute Infantry Battalion redesignated as the Airborne Center Training Detachment, prior to inactivation at Fort Benning. The 543d Parachute Infantry, inactivated at Fort Benning in 1944, was officially disbanded.

The 551st Parachute Infantry was relieved from attachment to the 82d Airborne Division on 15 January, but re-attached from 21-27 January, to be inactivated and disbanded in Europe on 10 February.

The 550th Airborne Infantry Battalion was disbanded in the European Theater of Operations.

Corps and Division Artillery

HHB, XVIII Corps Artillery was inactivated at Camp Campbell, Kentucky.

The 17th and 101st Airborne Divisions' Artillery were inactivated at Camp Miles Standish, Massachusetts, and Auxerre, France respectively.

Field Artillery

Parachuting with the 503d Regimental Combat Team was Battery A, 462d Parachute Field Artillery. The 462d Parachute Field Artillery would return later this year to be inactivated at Camp Anza, California.

The 321st Glider Field Artillery Battalion was inactivated in Germany, and the 377th Parachute Field Artillery Battalion was inactivated in France.

The 460th Parachute Field Artillery Battalion was assigned to the 13th Airborne Division. The 463d Parachute Field Artillery Battalion was reorganized, assigned to the 101st Airborne Division, and inactivated in Germany. The 464th Parachute Field Artillery Battalion was assigned to the 11th Airborne Division, then inactivated with the 467th Parachute Field Artillery Battalion at Camp Mackall, North Carolina. The 466th Parachute Field Artillery Battalion was inactivated at Camp Myles Standish, Massachusetts.

Activated at Camp Gordon in 1942, the 472d Field Artillery Battalion was reorganized and redesignated as the 472d Glider Field Artillery Battalion, and assigned to the 11th Airborne Division.

The 674th Glider Field Artillery Battalion was reorganized and redesignated as the 674th Parachute Field Artillery Battalion. The 680th, 681st and 907th Glider Field Artillery Battalions were inactivated at Camp Myles Standish, Massachusetts, and in Germany.

Air Defense Artillery

Battery C, 4th Coast Artillery Battalion, was disbanded at Fort Amador, Panama Canal Zone.

The 3d and 9th Airborne Anti-Aircraft Artillery Battalions moved from New Guinea to the Philippine Islands, arriving on 27 and 16 August respectively. They were both inactivated on 31 August at San Marcellino, Luzon, Philippine Islands.

The 12th Airborne Anti-Aircraft Artillery Battalion was inactivated on 20 June at Nadzab, New Guinea.

The 17th Airborne Anti-Aircraft Artillery Battalion, with it's subordinate 666th, 667th, 668th, 669th and 704th Airborne Antiaircraft Machine Gun Batteries, was relocated to Burma on 12 January and returning to India on 13 May, was disbanded on 10 July at Dinjan, India.

The 18th Airborne Anti-Aircraft Artillery Battalion deployed to Burma on 16 January, returning to India on 29 June. On 9 July, the 16th and 18th Airborne Anti-Aircraft Artillery Battalions were inactivated in Burma and India respectively.

The 152d Airborne Anti-Aircraft Artillery Battalion participated in the Leyte, Luzon and New Guinea Campaigns with the 11th Airborne Division. The 153d Airborne Anti-Aircraft Artillery Battalion was inactivated at Fort Bragg, North Carolina on 25 February, and the 155th Airborne Anti-Aircraft Artillery Battalion, which had participated in the Ardennes Alsace, Central Europe and Rhineland Campaigns with the 17th Airborne Division, was inactivated at Camp Myles Standish, Massachusetts on 14 September.

The 503d Anti-Aircraft Artillery Gun Battalion, deployed from Hawaii to Okinawa.

Special Operations

The 1st Special Service Force was inactivated at Menton France on 5 December, when for the first time, the assigned Canadians stepped out of the assembled ranks. Personnel from the 1st Special Service Force, and Ranger survivors from Anzio were assigned to the 474th Infantry, (constituted in 1944), and activated in 1945. Former cold weather and winter warfare trained members of the 1st Special Service Force were assigned to the 99th Infantry Battalion, 474th Infantry. Later that same year, the 474th Infantry was inactivated and designated as the 74th Infantry, with the 99th Infantry Battalion deploying to - of all places - Norway, to become absorbed into the Office of Strategic Services (OSS), Norway. The 122d Infantry Battalion, 474th Infantry, was moved to Cairo, and operations with the OSS Operational Group, Egypt.

Rangers

The 2d Ranger Infantry Battalion was disbanded at Camp Patrick Henry, Virginia, the 5th Ranger Infantry Battalion was disbanded in Austria, and inactivated at Camp Myles Standish, Massachusetts, and the 6th Ranger Infantry Battalion was inactivated at Kyoto, Japan.

The 475th Infantry, 5332d Brigade (Provisional), was inactivated in China.

Civil Affairs

Commencing in 1945 and continuing through 1946, sixty-three Military

Government Groups, the forerunners of today's Civil Affairs organizations, were constituted in the AUS. Activated at the Civil Affairs Holding Area at the Presidio of Monterey, California, they were the nation rebuilders during the periods of occupation following WW II. More than ninety percent of Military Government and Civil Affairs units are not included as they were not designated as airborne units. Included within this Chronology are those units that did transition from Military Government to airborne Civil Affairs Groups and Battalions. Comprehensive Civil Affairs lineages and insignia may be found in A Concise History of US Army Special Operations Forces by Geoffrey T. Barker, (Anglo-American Publishing Company, Suite 196, 813 East Bloomingdale Avenue, Brandon, Florida 33511). Included in the sixty-three units organized at the Civil Affairs Holding Area were the 95th, 96th and 97th Military Government Groups. The 28th Military Government HHC was constituted in the AUS, activated on Okinawa, and relocated to Korea.

Medical

The 325th Medical Battalion was inactivated at Camp Kilmer, New Jersey. HHD, 325th Medical Battalion was redesignated as the 325th Airborne Medical Company, an element of the 100th Airborne Division (concurrently assigned to Second Army); concurrently the remainder of the 325th Medical Battalion was disbanded. The 325th Airborne Medical Company was activated at Parkersburg, West Virginia

The 224th Airborne Medical Company was inactivated (less 1st Platoon), at Camp Myles Standish, Massachusetts. The 1st Platoon was inactivated at Camp Mackall, North Carolina).

The 326th Airborne Medical Company was inactivated at Auxerre, France.

Combat Support
Engineers

The IX Engineer Command was now assigned (with the XVIII Corps (Airborne) and the British Ist Airborne Corps) to the First Allied Airborne Army, which had now re-located to Paris, where the forward headquarters had been operating since 1944.

The 37th Engineer Combat Battalion was inactivated at the New York Port of Embarkation.

Company B, 129th Airborne Engineer Battalion, consolidated on 6 April with the 596th Airborne Engineer Company (disbanded in France on 1 March 1945), with the consolidated unit designated as Company B, 129th Engineer Battalion.

The 161st Airborne Engineer Battalion was inactivated at Fort Knox, Kentucky on 22 February.

The 326th Airborne Engineer Battalion was inactivated in Germany.

The 874th Airborne Engineer Aviation Battalion was inactivated in the Philippines after participating in the Luzon and New Guinea Campaigns. The 876th, 877th and 878th Airborne Engineer Aviation Battalions were inactivated on 21 December at Camp Myles Standish, Massachusetts. The 876th Airborne Engineer Aviation Battalion was credited with Central Europe, Normandy, Northern France and Rhineland campaign participation. The 877th Airborne Engineer Aviation Battalion participated in the Normandy and Northern France campaigns, and the 878th Airborne Engineer Aviation Battalion in the Rhineland campaign. The 882d Airborne Engineer Aviation Battalion was inactivated on New Guinea, following participation in the Leyte and New Guinea Campaigns.

Military Intelligence

The 101st Counter Intelligence Corps Detachment was inactivated in France; the 215th Signal Depot Company was inactivated at Camp Kilmer, New Jersey; and the 356th Interpreter Team was inactivated at Fort Shafter, Hawaii.

The 162d Language Detachment was constituted in the AUS and activated on Luzon, Philippine Islands.

The 319th HQ Intelligence Detachment was activated in Germany.

The 389th Translator Team was activated at Fort Shafter, Hawaii, en route to Korea.

Military Police

The Military Police Platoon, 101st Airborne Division, was inactivated at Auxerre, France.

Signal

Constituted in the Regular Army in 1899, Company H, Signal Corps, was activated in the Philippine Islands, to be redesignated in 1913 as Telegraph Company H, Signal Corps. They were further redesignated as the 51st Telegraph Battalion in 1917. The 51st Telegraph Battalion was inactivated at Houston in 1921, activated at Fort Sheridan, Illinois in 1940, and inactivated in Germany this year.

The 101st Airborne Signal Company was inactivated at Auxerre, France.

The 346th Signal Operations Company was inactivated at Fort Monmouth, New Jersey. The 299th Signal Installations Company was redesignated as the 299th Signal Service Company.

The 512th Airborne Signal Company, inactivated in North Africa in 1944, was disbanded in France, to be reconstituted in the AUS and consolidated with the 112th Airborne Army Signal Battalion, (with the consolidated unit designated as the 112th Airborne Army Signal Battalion). Assigned to the First US Airborne Army, the 112th Airborne Army Signal Battalion was later this year inactivated at Camp Patrick Henry, Virginia.

The 931st Signal Battalion (Air Support Command), was inactivated in India, and the 3191st Signal Service Company was inactivated in the Philippine Islands.

Combat Service Support
Adjutant General

Headquarters, Special Troops, 82d Division, was reconstituted in the Organized Reserves as HQ Special Troops, 82d Airborne Division, and activated in France.

Ordnance

The 713th Airborne Ordnance Maintenance Company was redeployed from France, relocating at Camp Mackall, and moving to Fort Bragg, North Carolina. The 801st Airborne Ordnance Maintenance Company was inactivated at Auxerre, France. The 717th Airborne Ordnance Maintenance Company was inactivated at Camp Myles Standish, Massachusetts.

Quartermaster

The 82d and 101st Quartermaster Parachute Supply and Maintenance Company organized in the AUS.

The 426th Airborne Quartermaster Company was inactivated at Auxerre, France.

Deploying briefly to France, the 713th Airborne Quartermaster Company redeployed to Fort Bragg, North Carolina.

—— 1946 ——

The Airborne Center, (formerly the Airborne Command), was inactivated and discontinued at Fort Benning, Georgia.

The 13th Airborne Division did not embark to the Pacific Theater, and was inactivated.

HQ, 80th and 84th Divisions, were

constituted in the National Army in 1917, and organized respectively at Camp Lee, Virginia and Camp Zachary Taylor, Kentucky. The 80th and 84th Divisions were demobilized in 1919 following World War I, at Camp Lee, Virginia and Camp Zachary Taylor, Kentucky. HQ, 80th and 84th Divisions were reconstituted in the Organized Reserves at Richmond, Virginia in 1921. In 1942, HQ 80th and 84th Divisions were redesignated as Division HQs, 80th and 84th Divisions, and ordered into Active Military Service, to be reorganized at Camp Forrest, Tennessee as HQ, 80th Infantry Division, and at Camp Howze, Texas as HQ, 84th Infantry Division.

Inactivated in 1946 at Camp Kilmer, New Jersey, HQs, 80th and 84th Infantry Divisions returned to their Organized Reserves status, to be reorganized and redesignated as HQs, 80th and 84th Airborne Divisions.

HQ, 80th Airborne Division was activated at Richmond, Virginia and assigned to Second Army. HQ, 84th Airborne Division was assigned to Fifth Army, but not activated.

HQ, 100th Division was constituted in the National Army in 1918 and organized at Camp Bowie, Texas. Demobilized in 1919, HQ, 100th Division was reconstituted in the Organized Reserves at Charleston, West Virginia, relocating to Huntington, West Virginia in 1924, and back to Charleston in 1937. Redesignated in 1942, the Division HQ, 100th Division was ordered into Active Military Service, to be reorganized and redesignated at Fort Jackson, South Carolina as HQ, 100th Infantry Division. Inactivated at Camp Patrick Henry, Virginia, HQ, 100th Infantry Division was reorganized and redesignated as HQ, 100th Airborne Division.

HQ, 100th Airborne Division was assigned to Second Army, and activated at Louisville, Kentucky.

One of the mythical allied divisions, listed under Operation QUICKSILVER, (the US portion of the overall Operation FORTITUDE), was the 108th Infantry Division (see 1944). HQ, 108th Airborne Division was constituted in 1946 in the Organized Reserves, and assigned to Seventh Army. Activated at Atlanta, Georgia, HQ, 108th Airborne Division was relieved from assignment to Seventh Army, and assigned to Third Army.

Combat Arms
Infantry

The 317th, 318th and 319th Infantry, were constituted in the National Army in 1917, and assigned to the 80th Division. The 317th and 318th were demobilized at Camp Lee, Virginia in 1919, and the 319th demobilized the same year at Camp Upton, New York. All three were allotted to, and reconstituted in the Organized Reserves in 1921, remaining assigned to the 80th Division. They were ordered into Active Military Service for the duration of WW I, being inactivated in 1946 at Camp Kilmer, New Jersey. They were reorganized and redesignated in the Organized Reserves as the 317th Parachute Infantry, and the 318th and 319th Glider Infantry, assigned to the also newly reorganized and redesignated 80th Airborne Division. The 317th and 319th were activated respectively in Washington, DC and Baltimore, Maryland.

The 325th and 326th Glider Infantry Regiments disembarked to Fort Bragg, North Carolina, where the 326th Glider Infantry was inactivated.

The 333d, 334th and 335th Infantry, were constituted in the National Army in 1917, and assigned to the 84th Division. All three were demobilized at Camp Zachary Taylor, Kentucky in 1919. They were allotted to, and reconstituted in the Organized Reserves in 1921, remaining assigned to the 84th Division. They were ordered into Active Military Service for the duration of WW I, being inactivated in 1946 at Camp Kilmer, New Jersey. They were reorganized and redesignated in the Organized Reserves as the 333d Glider Infantry, and the 334th and 335th Parachute Infantry, assigned to the also newly reorganized and redesignated 84th Airborne Division. The 333d was activated at Grand Rapids, Michigan.

The 397th, 398th and 399th Infantry, were constituted in the National Army in 1917, and assigned to the 100th Division. They were demobilized 1918. All three were allotted to, and reconstituted in the Organized Reserves in 1921, remaining assigned to the 100th Division. They were ordered into Active Military Service for the duration of WW I, with the 397th inactivated in 1946 at Camp Patrick Henry, Virginia; and the 398th and 399th being inactivated at Camp Kilmer, New Jersey. They were reorganized and redesignated in the Organized Reserves as the 397th and 398th Parachute Infantry, and the 399th Glider Infantry, assigned to the also newly reorganized and redesignated 100th Airborne Division. The 397th and 398th were activated respectively in Lexington, Kentucky and Charleston, West Virginia.

The 485th Glider Infantry and the 518th and 519th Parachute Infantry were constituted in the Organized Reserve in 1946, assigned to the 108th Airborne Division. The 108th Division had originally been listed as one of the 50 fictitious allied divisions under Operation QUICKSILVER in 1944. The 108th Airborne Division was constituted in the Organized Reserve and activated at Atlanta, Georgia, initially assigned to the Seventh Army, but later assigned to Third Army. The 485th Glider Infantry was activated in Jacksonville, Florida. The 518th and 519th Parachute Infantry were activated in Charlotte, North Carolina and Atlanta, Georgia, respectively.

The 501st Parachute Infantry Battalion was reconstituted in the AUS, and activated at Fort Benning, Georgia. The 504th Parachute Infantry disembarked from the European Theater of Operations for New York on 3 January, returning to Fort Bragg, North Carolina by 16 January.

The 505th Parachute Infantry returned through New York to Fort Bragg, North Carolina, arriving home on 16 January, after participating in every airborne operation conducted by the 82d Airborne Division during World War II.

The 508th Parachute Infantry returned from the European Theater of Operations through New York on 24 November, and was inactivated at Camp Kilmer, New Jersey on 25 November. The 515th and 517th Parachute Infantry were inactivated at Fort Bragg, North Carolina on 25 February.

The Reconnaissance Platoon, 13th Airborne Division, was inactivated at Fort Bragg, North Carolina.

Armor

Companies B and C, 10th Armored Infantry Battalion, were converted and redesignated respectively as Troop B and Company C, 10th Constabulary Squadron, assigned to the 14th Constabulary Regiment.

Division Artillery

HHB, 80th and 84th Division Artillery, were inactivated at Camp Kilmer, New Jersey, and redesignated as HHB, 80th and 84th Airborne Division Artillery. HHB, 100th Division Artillery was inactivated at Camp Patrick Henry, redesignated as HHB, 100th Airborne Division Artillery, and activated at Louisville, Kentucky. The 108th Airborne Division Artillery was constituted in the Organized Reserves.

Field Artillery

Battery B, 26th Field Artillery, constituted in the National Army in 1918, was demobilized in 1919, to be reconstituted in the Regular Army in 1921. Activated at Fort Bragg, North Carolina in 1940, Battery B was reorganized and redesignated as Battery B, 26th Field Artillery Battalion, and inactivated in Germany this year.

The 23d Cavalry, constituted in the Regular Army in 1916, was converted and redesignated as the 81st Field Artillery Battalion in 1940. Activated at Fort Lewis, Washington, and assigned to the 8th Division, the 81st Field Artillery Battalion was inactivated at Camp Kilmer in 1945.

Constituted in the Regular Army in 1933, the 89th Field Artillery was activated in Hawaii, assigned to the 25th Infantry Division in 1941, and reorganized and redesignated as the 89th Field Artillery Battalion. The 89th Field Artillery Battalion was inactivated in Japan.

The 313th, 314th and 315th Field Artillery Battalions, assigned to the 80th Division, were inactivated at Camp Kilmer, New Jersey, assigned to the concurrently redesignated 80th Airborne Division, as the 313th and 314th Parachute Field Artillery Battalions, and the 315th Glider Field Artillery Battalion.

The 325th, 326th and 327th Field Artillery Battalions were inactivated at Camp Kilmer, New Jersey, redesignated as the 325th and 327th Glider Field Field Artillery Battalions and the 326th Parachute Field Artillery Battalion, and all assigned to the concurrently redesignated 84th Airborne Division.

The 373d and 375th Field Artillery Battalions were inactivated at Camp Patrick Henry, Virginia. The 374th Field Artillery Battalion was inactivated at Camp Kilmer, New Jersey. The 373d Field Artillery Battalion was redesignated as the 373d Glider Field Artillery Battalion, activated at Louisville, Kentucky, and assigned to the concurrently redesignated 100th Airborne Division. The 374th and 375th Field Artillery Battalions, also assigned to the 100th Airborne Division, were redesignated as the 374th and 375th Parachute Field Artillery Battalions and activated respectively at Charleston, West Virginia, and Louisville, Kentucky.

The 458th and 460th Parachute Field Artillery Battalions were inactivated at Fort Bragg, North Carolina. The 462d Parachute Field Artillery Battalion, inactivated at Camp Anza last year, was redesignated as the 462d Field Artillery Battalion, allotted to the Organized Reserves, assigned to Second Army and activated at Toledo, Ohio. The 465th Field Artillery Battalion was inactivated in the Philippines.

The 506th and 507th Parachute Field Artillery Battalions were constituted in the Organized Reserves, assigned to the 108th Airborne Division, and activated respectively at Charlotte, North Carolina and Atlanta, Georgia.

The 693d Field Artillery Battalion was inactivated at Camp Kilmer, New Jersey. The 676th and 677th Glider Field Artillery Battalions were inactivated at Fort Bragg, North Carolina.

Inactivated at Camp Kilmer, New Jersey, the 905th and 909th Field Artillery Battalions were redesignated as the 905th Glider Field Artillery Battalion, assigned to the 80th Airborne Division, and the 909th Parachute Field Artillery Battalion, assigned to the 84th Airborne Division. The 925th Field Artillery Battalion was inactivated at Camp Patrick Henry, Virginia, redesignated as the 925th Glider Field Artillery Battalion, activated at Lexington, Kentucky, and assigned to the 100th Airborne Division.

Air Defense Artillery

The 3d Anti-Aircraft Artillery Battalion, disbanded in 1945 in the Philippines, was reorganized and redesignated as the 84th Airborne Anti-Aircraft Artillery Battalion, assigned to the 84th Airborne Division. The 503d Anti-Aircraft Artillery Gun Battalion was inactivated at Camp Anza, California.

Special Operations
Civil Affairs

The 95th Military Government Group was inactivated at Kurume, Japan. The 28th Military Government HHC was inactivated in Korea.

Medical

The 305th Medical Battalion was inactivated at Camp Kilmer, New Jersey, and redesignated as the 305th Airborne Medical Company.

The 222d Airborne Medical Company was inactivated at Fort Bragg, North Carolina.

Combat Support
Engineers

Company B, 129th Airborne Engineer Battalion inactivated at Fort Bragg, North Carolina.

The 871st Airborne Engineer Aviation Battalion was inactivated on 15 May in Japan after serving in New Guinea, the Biak Islands and the Philippines, receiving credit for the Luzon and New Guinea Campaigns. The 879th Airborne Engineer Aviation Battalion disembarked for New York from the Pacific Theater on 2 January, after being credited with campaign participation in Central Burma, the China Offensive and the India-Burma Campaign.

Military Intelligence

The 525th Military Intelligence Group was constituted in the AUS, organized at Heidelberg, Germany, and assigned to US Forces, European Theater.

The 162d Language Detachment was inactivated at Kyoto, Japan.

The 77th Counter Intelligence Corps Detachment was inactivated in Hawaii.

The 319th HQ Intelligence Detachment was inactivated in Germany, redesignated as the 319th Military Intelligence Company, and activated in Japan.

The 356th Interpreter Team was activated in Hawaii, attached to IX Corps, and deployed to Sendai, Japan.

The 389th Translator Team was relocated from Seoul to Pusan, Korea.

Military Police

The Military Police Platoon, 84th Division constituted in the National Army in 1917, and assigned to the 84th Division, was demobilized in 1919, and reconstituted in the Organized Reserves in 1921. Reorganized and redesignated, the 84th Airborne Military Police Company was assigned to the concurrently reorganized and redesignated 84th Airborne Division.

The 503d Military Police Battalion was inactivated at Camp Kilmer, New Jersey, and reactivated at Florence, Italy.

Signal

The 299th Signal Service Company was inactivated in France.

The 305th Field Signal Battalion was inactivated in New York, and redesignated as the 80th Airborne Signal Company.

Combat Service Support
Adjutant General

The 290th Replacement Company was inactivated in Italy.

Ordnance

The 713th Airborne Ordnance Maintenance Company was inactivated at Fort Bragg, North Carolina (see 1967, Combat Service Support, Maintenance).

The 780th Ordnance Light Maintenance Company was inactivated at Camp Kilmer, New Jersey, and redesignated as the 780th Airborne Ordnance Maintenance Company. The 784th and 800th Ordnance Light Maintenance Companies were respectively inactivated at Camp Kilmer, New Jersey, and Camp Patrick Henry, Virginia, They were redesignated as the 784th and 800th Airborne Ordnance Maintenance Companies. The 784th was assigned as an organic element of the 84th Airborne Division. The 800th was assigned to Second Army and activated at Buckhannon, West Virginia as an organic element of the 100th Airborne Division.

Quartermaster

The 80th Quartermaster Company was inactivated at Camp Kilmer, New Jersey, and redesignated as the 80th Airborne Quartermaster Company. The 84th Maintenance Company was redesignated as the 84th Parachute Maintenance Company.

The 409th Airborne Quartermaster Company was inactivated at Fort Bragg, North Carolina.

HHD, 528th Quartermaster Battalion was redesignated on 24 May, as HHD, 528th Quartermaster Battalion, Mobile, and later converted and redesignated on 1 August as HHD, 528th Transportation Corps Truck Battalion.

The 600th Quartermaster Laundry Company was inactivated in France.

Transportation

The 5th and 29th Quartermaster Battalions, Mobile were respectively inactivated in Germany and the Philippine Islands, and redesignated as the 5th and 29th Transportation Corps Truck Battalions.

HHD, 528th Transportation Corps Truck Battalion was activated following the conversion and redesignation of HHD, 528th Quartermaster Battalion.

—— 1947 ——

HQ, 84th Airborne Division was activated at Madison, relocating to Milwaukee, Wisconsin.

Combat Arms
Infantry

The 318th Glider Infantry was activated in the Organized Reserves at Richmond, Virginia; assigned to the 80th Airborne Division. The 334th and 335th Parachute Infantry were activated at Milwaukee, Wisconsin and Chicago, Illinois, and assigned in the Organized Reserves to the 84th Airborne Division. The 399th Glider Infantry, assigned in the Organized Reserves to the 100th Airborne Division, was activated at Louisville, Kentucky.

The 325th Glider Infantry, with Companies A, B, C, D and E, remaining assigned to the 82d Airborne Division, were redesignated as the 325th Infantry. The 504th and 505th Parachute Infantry were redesignated as the 504th and 505th Airborne Infantry.

The 509th Parachute Infantry Battalion and Companies A, B, and C, were disbanded in France on 1 March 1945, and reconstituted in the Regular Army on 12 May 1947 as the 509th Parachute Infantry Battalion. The 1st Battalion, 551st Parachute Infantry, was also reconstituted in the Regular Army.

The 519th Parachute Infantry, (108th Airborne Division), was assigned from Seventh Army to Third Army.

The 555th Parachute Infantry Battalion was inactivated at Fort Benning, Georgia.

The Reconnaissance Platoon, 82d Airborne Division, was reorganized and redesignated as the 82d Reconnaissance Company, and relieved from assignment to the 82d Airborne Division.

Division Artillery

The 80th Airborne Division Artillery was allotted to the Organized Reserves, and activated at Richmond, Virginia, and the 108th Airborne Division Artillery was activated at Atlanta, Georgia.

Field Artillery

Battery B, 26th Field Artillery Battalion, was activated at Fort Dix, New Jersey, and the 81st Field Artillery Battalion was activated at Fort Sill, Oklahoma.

The 89th Field Artillery Battalion was relieved from assignment to the 25th Infantry Division, and the 350th Field Artillery Battalion was redesignated as the 98th Field Artillery Battalion, and activated at Fort Bragg, North Carolina.

The 313th and 314th Parachute Field Artillery Battalions and the 315th Glider Field Artillery Battalion were activated respectively at Richmond, Roanoke and Norfolk, Virginia. The 319th Glider Field Artillery Battalion was reorganized and redesignated as the 319th Field Artillery Battalion.

The 325th and 327th Glider Field Artillery Battalions were activated at Battle Creek, Michigan and Chicago, Illinois; the 326th Parachute Field Artillery Battalion was activated at Beloit, Wisconsin.

The 376th and 456th Parachute Field Artillery Battalions were redesignated as the 376th and 456th Airborne Field Artillery Battalions

The 462d Field Artillery Battalion was relocated from Toledo to Canton, Ohio. The 465th Field Artillery Battalion was allotted to the Organized Reserves, assigned to Fifth Army, and activated at Minot, North Dakota. The 467th Parachute Field Artillery Battalion was redesignated as the 467th Field Artillery Battalion, allotted to the Organized Reserves, assigned to Fifth Army and activated at Grand Rapids, Michigan.

The 693d Field Artillery Battalion was redesignated as the 544th Field Artillery Battalion. The 905th Glider Field Artillery Battalion was activated at Bristol, Virginia, and the 909th Parachute Field Artillery Battalion was activated at Sheboygan, Wisconsin.

Air Defense Artillery

The 84th Airborne Antiaircraft Artillery Battalion was activated at Milwaukee, Wisconsin. The 152d Airborne Antiaircraft Artillery Battalion was inactivated in Japan. The 503d Antiaircraft Artillery Gun Battalion, inactivated at Camp Anza in 1947, was redesignated the 503d Airborne Antiaircraft Artillery Battalion (Negro), allotted to the Regular Army, assigned to the 82d Airborne Division, and activated at Fort Bragg, North Carolina.

Medical

The 305th Airborne Medical Company was activated in the Organized Reserves at Richmond, Virginia, assigned to the 80th Airborne Division. The 307th Airborne Medical Company was reorganized and redesignated as the 307th Medical Battalion.

The 610th Clearing Company was inactivated in Germany.

Combat Support
Engineers

The 209th Engineer Combat Battalion, an element of the 37th Engineer Combat Regiment (broken up in 1943), was inactivated at Camp Kilmer, New Jersey in 1945; and is redesignated as the 27th Engineer Combat Battalion.

Military Intelligence

The 77th Counter Intelligence Corps Detachment was withdrawn from the AUS and concurrently allotted to the Organized Reserves, to be assigned to First Army and activated at New York City, New York.

The 356th Interpreter Team was assigned to I Corps and attached to the 24th Infantry Division, relocated to Japan, and redesignated as the 356th Intelligence Detachment.

The 389th Translator Team was inactivated at Pusan, Korea.

Military Police

The 503d Military Police Battalion was inactivated at Florence, Italy.

The Military Police Platoon, 82d Airborne Division, activated at Camp Claiborne in 1942, was redesignated as the 82d Airborne Military Police Company.

Signal

The 80th Airborne Signal Company was assigned to the 80th Airborne Division.

Combat Service Support
Adjutant General

The 290th Replacement Company was redesignated as the 82d Replacement Company, and assigned to the 82d Airborne Division.

HQ, Special Troops, 82d Airborne Division, was consolidated with the 82d Replacement Company, with the consolidated unit designated as the 82d Replacement Company, an element of the 82d Airborne Division.

Ordnance

The 780th Airborne Ordnance Maintenance Company and the 784th Airborne Ordnance Light Maintenance Company were activated respectively at Richmond, Virginia and Milwaukee, Wisconsin.

Quartermaster

The 80th Airborne Quartermaster Company was activated in Richmond, Virginia and assigned to the 80th Airborne Division. The 84th Parachute Maintenance Company was activated in the Organized Reserves, and assigned to the 84th Airborne Division.

HHD, 528th Transportation Corps Truck Battalion was converted back on 1 February, and redesignated as HHD, 528th Quartermaster Battalion, to be inactivated in France on 20 November.

1948

The 11th Airborne Division was withdrawn from the AUS and allotted to the Regular Army. The 17th Airborne Division was also allotted to the Regular Army from the AUS, and activated as a Training Division at Camp Pickett, Virginia.

The 82d and 101st Airborne Divisions were withdrawn from the Organized Reserve Corps and allotted to the Regular Army. On 27 March, Major General Clovis E. Byers assumed command of the 82d Airborne Division.

HQ, 84th Airborne Division relocated to Madison, Wisconsin.

The 101st Airborne Division was activated at Camp Breckinridge, Kentucky, as a Training Division.

Combat Arms
Infantry

The 187th Glider Infantry and the 188th Parachute Infantry were allotted as Regular Army. The 194th Glider Infantry was redesignated as the 514th Airborne Infantry. The 325th Glider Infantry was allotted as Regular Army, and redesignated as the 325th Airborne Infantry. The 327th Glider Infantry was also allotted to the Regular Army, but redesignated as the 516th Airborne Infantry, and assigned to the 101st Airborne Division at Camp Breckinridge, Kentucky.

Companies A and B, 501st Parachute Infantry Battalion were inactivated at Fort Benning, Georgia. The 502d Parachute Infantry, (Companies A through F), and the 506th Parachute Infantry (Companies A, B and C), were allotted to the Regular Army, assigned to the 101st Airborne Division, and activated at Camp Breckinridge, Kentucky as the 502d and 506th Airborne Infantry.

The 504th and 505th Airborne Infantry were allotted to the Regular Army. The 507th, 513th and 514th Parachute Infantry were redesignated as the 507th, 513th and 514th Airborne Infantry, allotted to the Regular Army, assigned to the 17th Airborne Division, and activated at Camp Pickett, Virginia.

The 550th Infantry Airborne Battalion, disbanded in the European Theater of Operations in 1945, was reconstituted in the regular Army as the 550th Airborne Infantry Battalion. Activated at Camp Pickett, Virginia, the 550th was assigned to the 17th Airborne Division.

The Reconnaissance Platoons, 11th, 17th and 101st Airborne Divisions, were allotted to the Regular Army, with the Reconnaissance Platoons, 17th and 101st Airborne Divisions being redesignated as the Anti-Tank Platoons.

Armor

Troop B and Company C, 10th Constabulary Squadron, inactivated in Germany, and concurrently converted and redesignated as Companies B and C, 10th Armored Infantry Battalion, assigned to the 4th Armored Division.

Division Artillery

The 11th, 17th, 82d and 101st Airborne Divisions' Artillery were allotted to the Regular Army. The 17th Airborne Division Artillery was activated at Camp Pickett, Virginia, and the 101st Airborne Division Artillery was activated at Camp Breckinridge, Kentucky. The 84th Airborne Division Artillery was activated at LaCrosse, Wisconsin.

Field Artillery

The 81st Field Artillery Battalion was inactivated at Fort Sill, Oklahoma.

The 319th, 320th and 321st Glider Field Artillery Battalions were allotted to the Regular Army. The 319th was reorganized and redesignated as the 319th Airborne Field Artillery Battalion, and the 320th was inactivated at Fort Bragg, North Carolina. The 321st was redesignated as the 518th Airborne Field Artillery Battalion, and activated at Camp Breckinridge, Kentucky. The 375th Parachute Field Artillery Battalion was relocated from Louisville to Fort Thomas, Kentucky.

The 376th and 456th Airborne Field Artillery Battalions were allotted to the Regular Army. The 377th, 457th, 463d, 464th and 674th Parachute Field Artillery Battalions were allotted to the Regular Army; and also the 675th and 680th Glider Field Artillery Battalions were allotted to the Regular Army. The 377th, 463d, 464th and 466th Parachute Field Artillery Battalions were redesignated as the 549th, 516th, 515th and 550th Airborne Field Artillery Battalions respectively. The 680th and 681st Glider Field Artillery Battalions were respectively redesignated as the 551st Airborne Field Artillery Battalion (680th), and consolidated with the 513th Airborne Infantry Regiment (681st).

The 515th, 516th and 518th Airborne Field Artillery Battalions were activated at Camp Breckinridge, Kentucky. The 549th and 551st Airborne Field Artillery

Battalions were activated at Camp Pickett, Virginia. The 550th Airborne Field Artillery Battalion was assigned to the 17th Airborne Division, and activated at Camp Pickett, Virginia.

The 907th Glider Field Artillery Battalion was relieved from assignment to the 101st Airborne Division, redesignated at the 907th Field Artillery Battalion, activated at Jacksonville, Florida, and assigned to Third Army.

Air Defense Artillery

The 80th Airborne Antiaircraft Artillery Battalion was allotted to the Regular Army, and inactivated at Fort Bragg, North Carolina. Batteries C, D, E and F, 81st Airborne Antiaircraft Artillery Battalion, (inactivated in 1945), were converted and redesignated as Support Company and Companies K, L and M, 502d Airborne Infantry, 101st Airborne Division, and activated at Camp Breckinridge, Kentucky. The 152d Airborne Antiaircraft Artillery Battalion, inactivated in Japan in 1947, was allotted to the Regular Army, released from assignment to the 11th Airborne Division, redesignated as the 88th Airborne Antiaircraft Artillery Battalion, and activated at Fort Bliss, Texas.

Special Operations
Rangers

The 1st Ranger Infantry Battalion, inactivated in 1944, was reconstituted as the 1st Infantry Battalion, and activated in the Panama Canal Zone.

Civil Affairs

The 95th Military Government Group was activated at Fort Bragg, North Carolina. The 28th Military Government HHC, inactivated in Korea in 1946, was redesignated as the 28th Military Government Company, and activated at Fort Bragg.

Medical

The 307th Medical Battalion was reorganized and redesignated as the 307th Airborne Medical Battalion, withdrawn from the Organized Reserve Corps, and allotted to the Regular Army.

The 326th Airborne Medical Company was reorganized and redesignated as HHC, 501st Airborne Medical Battalion; (Medical Detachment, 907th Glider Field Artillery Battalion was reorganized and redesignated as Clearing Company; and the 595th Motor Ambulance Company was redesignated as Ambulance Company). The 501st Airborne Medical Battalion was allotted to the Regular Army, and activated at Camp Breckinridge, Kentucky.

The 221st Airborne Medical Company was withdrawn from the AUS and allotted to the Regular Army.

The 224th Airborne Medical Company was redesignated on 18 June as HHC, 17th Airborne Medical Battalion, also allotted to the Regular Army, and activated on 6 July at Camp Pickett, Virginia. (Medical Detachment, 681st Glider Field Artillery Battalion, and the 634th Clearing Company (activated in 1942 as Company D, 60th Medical Battalion), redesignated as Ambulance Company and Clearing Company, 17th Airborne Medical Battalion.)

Combat Support
Engineers

The 127th and 307th Airborne Engineer Battalions were allotted to the Regular Army.

Company C, 139th Airborne Engineer Battalion, was redesignated as the 517th Engineer Battalion.

The 326th Airborne Engineer Battalion was consolidated with the 49th Engineer Combat Battalion, with the consolidated unit designated as the 49th Airborne Engineer Battalion; allotted to the Regular Army, activated at Camp Breckinridge, Kentucky; and assigned to the 101st Airborne Division.

Military Intelligence

The 356th Intelligence Detachment was relieved from attachment to the 24th Infantry Division, and attached to the 25th Infantry Division and relocated to Maizuro, Japan.

Military Police

Military Police Platoons, 11th, and 101st Airborne Divisions, were allotted to the Regular Army. The Military Police Platoon, 101st Airborne Division, was reorganized as the 101st Airborne Military Police Company. The 82d Airborne Military Police Company (designated in 1947), was withdrawn from the Organized Reserves and allotted to the Regular Army.

Signal

The 82d, 101st and 511th Airborne Signal Companies were allotted to the Regular Army. The 101st Airborne Signal Company was activated at Camp Breckinridge, Kentucky.

Combat Service Support
Adjutant General

The 11th Airborne Division Band was allotted to the Regular Army.

The 82d Replacement Company was withdrawn from the Organized Reserve Corps and allotted to the Regular Army.

Ordnance

The 711th and 801st Airborne Ordnance Maintenance Companies were allotted to the Regular Army. The 801st was activated at Camp Breckinridge, Kentucky.

The 782d Airborne Maintenance Company was withdrawn from the AUS, and allotted to the Regular Army. The 800th Airborne Ordnance Light Maintenance Company was relocated from Buckhannon to Elkins, West Virginia.

Quartermaster

The 11th and 13th Parachute Maintenance Companies were withdrawn from the AUS and allotted to the Regular Army.

The 407th, 408th and 426th Airborne Quartermaster Companies were withdrawn from the Organized Reserves and allotted to the Regular Army. The 426th Airborne Quartermaster Company was redesignated as the 101st Airborne Quartermaster Company and activated at Camp Breckinridge, Kentucky.

The 411th Airborne Quartermaster Company, constituted on the inactive list in 1942 and assigned to the 17th Airborne Division, was redesignated as the 17th Airborne Quartermaster Company.

On 12 October, HHD, 528th Quartermaster Battalion was redesignated as HHD, 426th Quartermaster Battalion, allotted to the Organized Reserve Corps, and assigned to Second Army. Ten days later, on 22 October, HHD, 426th Quartermaster Battalion was activated at Clarksburg, West Virginia.

The 612th Quartermaster Graves Registration Company was inactivated at Karlsruhe, Germany.

Transportation

HHD, 29th Transportation Corps Truck Battalion was redesignated as HHD, 29th Truck Battalion.

—— 1949 ——

The 11th Airborne Division returned to the United States, from occupation duties in Japan, to Fort Campbell, Kentucky.

The 17th and 101st Airborne Divisions were inactivated respectively at Camp Pickett, Virginia, and Camp Breckinridge, Louisiana.

The command of the 82d Airborne Division was assumed by Brigadier General Ridgely Gaither from 19 July until 31 October, when Major General Williston B. Palmer was appointed commander on 1 November.

Combat Arms
Infantry

The 187th and 188th Glider Infantry Regiments and the 511th Parachute Infantry Regiment, disembarked with the 11th Airborne Division, from occupation assignments in Japan to Fort Campbell, Kentucky.

The 187th Glider Infantry and the 188th, 317th and 511th Parachute Infantry, were redesignated as the 187th, 188th, 317th and 511th Airborne Infantry, with the 188th Airborne Infantry being disbanded at Fort Campbell, Kentucky.

The 327th, 502d, 506th and 516th Airborne Infantry were inactivated at Camp Breckinridge, Kentucky. The 507th, 513th and 514th Airborne Infantry, and the 550th Airborne Infantry Battalion were inactivated at Camp Pickett, Virginia.

The 82d Reconnaissance Company was allotted to the Regular Army.

The Reconnaissance Platoon, 11th Airborne Division, was reorganized and redesignated as the Anti-Tank Platoon, 11th Airborne Division. The Anti-Tank Platoons, 17th and 101st Airborne Divisions, were inactivated at Camps Pickett, Virginia and Breckinridge, Kentucky.

Division Artillery

The 17th Airborne Division Artillery was inactivated at Camp Pickett, Virginia, and the 101st Airborne Division Artillery inactivated at Camp Breckinridge, Kentucky.

Field Artillery

The 472d Glider Field Artillery Battalion and the 475th Airborne Field Artillery Battalion were inactivated at Fort Campbell, Kentucky.

The 515th, 516th and 518th Airborne Field Artillery Battalions were inactivated at Camp Breckinridge, Kentucky. The 544th Field Artillery Battalion was activated at Fort Campbell, Kentucky. The 549th, 550th and 551st Airborne Field Artillery Battalions were inactivated at Fort Pickett, Virginia.

The 674th and 675th Glider Field Artillery Battalions were redesignated as the 674th and 675th Airborne Field Artillery Battalions.

Air Defense Artillery

The 80th Airborne Antiaircraft Artillery Battalion was activated at Fort Bragg, North Carolina with assigned Negro personnel, and assigned to the 82d Airborne Division. The 81st Airborne Antiaircraft Artillery Battalion was inactivated at Camp Breckinridge, Kentucky. The 88th Airborne Antiaircraft Artillery Battalion was assigned to the 11th Airborne Division, and the 503d Airborne Antiaircraft Artillery Battalion was relieved from the 82d Airborne Division, and inactivated at Fort Bragg, North Carolina. The 17th and 81st Airborne Antiaircraft Artillery Battalions were respectively inactivated at Camps Pickett, Virginia and Breckinridge, Kentucky.

Special Operations
Rangers

The 2d Ranger Infantry Battalion, inactivated in 1945, was redesignated as the 2d Infantry Battalion, (with Companies E and F disbanded), and activated in the Panama Canal Zone.

Civil Affairs

The 96th and 97th Military Government Groups were inactivated in Korea.

Medical

The 501st Airborne Medical Battalion was inactivated at Camp Breckinridge, Kentucky.

The 221st Airborne Medical Company was reorganized and redesignated as HHC, 11th Airborne Medical Battalion. Concurrently, organic elements formed by the redesignating and activation of existing companies included the 489th Motor Ambulance Company and the 610th Clearing Company being redesignated as the Ambulance Company and Clearing Company, 11th Airborne Medical Battalion.

HHC, 17th Airborne Medical Battalion was inactivated at Camp Pickett, Virginia.

Combat Support
Engineers

The 49th Airborne Engineer Battalion was inactivated at Camp Breckinridge, Kentucky, and the 517th Engineer Battalion was inactivated in Germany.

Military Police

The Military Police Company, 11th Airborne Division was redesignated as the 11th Airborne Military Police Company. The 17th and 101st Airborne Military Police Companies were inactivated at Camps Pickett, Virginia and Breckinridge, Kentucky.

The 503d Military Police Battalion was activated at Fort Bragg and assigned to the XVIII Corps (Airborne).

Signal

The 101st and 517th Airborne Signal Companies were inactivated respectively at Camp Breckinridge, Kentucky, and Camp Pickett, Virginia.

The 246th Signal Operations Company was redesignated as the 516th Signal Company, and activated at Salzburg, Austria.

Combat Service Support
Adjutant General

The 101st Replacement Company and the 101st Airborne Division Band were inactivated at Camp Breckinridge, Kentucky. The 17th Replacement Company and the 17th Airborne Division Band were inactivated at Camp Pickett, Virginia.

Maintenance

The 17th and 101st Airborne Maintenance Companies were inactivated prior to June, at Camps Pickett, Virginia and Breckinridge, Kentucky.

Ordnance

The 717th Airborne Ordnance Maintenance Company was inactivated at Camp Pickett, Virginia. The 800th Airborne Ordnance Light Maintenance Company was relocated from Elkins to Morgantown, West Virginia. The 801st Airborne Ordnance Maintenance Company was inactivated at Camp Breckinridge, Kentucky.

Quartermaster

The 11th Parachute Maintenance Company was redesignated as the 11th Airborne Parachute Maintenance Company, and the 17th Airborne Quartermaster Company was inactivated at Camp Pickett, Virginia. The 101st Airborne Quartermaster Company was inactivated at Camp Breckinridge, Kentucky.

The 612th Quartermaster Graves

Registration Company was redesignated as the 612th Quartermaster Service Company, assigned to the Fifth Army, and activated in Fargo, North Dakota.

1950

Lieutenant General John W. Leonard assumed command of XVIII Corps (Airborne).

Major General Thomas P. Hickey was appointed to command the 82d Airborne Division. The 101st Airborne Division was inactivated at Camp Breckinridge, Louisiana.

Combat Arms
Infantry

The 187th Airborne Infantry was reorganized and became the first airborne unit to be designated as a Regimental Combat Team, deploying to Korea. Participating in the Inchon invasion, the 187th Regimental Combat Team made an airborne assault into Sukchon-Sunchon. Private First Class Richard G. Wilson was posthumously awarded the Medal of Honor for actions on 21 October at Opari, Korea.

The 188th Airborne Infantry was activated at Fort Campbell, and assigned to the 11th Airborne Division.

The 318th Parachute Infantry and the 319th Glider Infantry, (assigned to the 80th Airborne Division), were redesignated as the 318th and 319th Airborne Infantry.

The 397th and 398th Parachute Infantry, and the 399th Glider Infantry (assigned to the 100th Airborne Division), were redesignated as the 397th, 398th and 399th Airborne Infantry.

The 475th Infantry was designated as the 75th Infantry, allotted to the Regular Army, and activated on Okinawa.

The 516th Airborne Infantry (formerly the 327th Glider Infantry), the 502d and 506th Airborne Infantry, were again activated with the 101st Airborne Division at Camp Breckinridge, Kentucky.

The 555th Parachute Infantry Battalion, inactivated at Fort Benning in 1947, was disbanded.

The 82d Reconnaissance Company was reorganized and redesignated as the 82d Airborne Reconnaissance Company, and assigned to the 82d Airborne Division.

Armor

The 42d Armored Regiment was constituted in the AUS in 1942, and activated at Camp Polk, Louisiana, assigned to the 11th Armored Division. Reorganized and redesignated as the 42d Tank Battalion in 1943, it was disbanded in Germany in 1945. Reconstituted in the Regular Army in 1950, the 42d Tank Battalion was assigned to the 101st Airborne Division.

The 44th Armored Regiment was constituted and activated at Camp Campbell, Kentucky, in the AUS in 1942, and assigned to the 12th Armored Division. The 44th Armored Regiment was broken up in November 1943, with the Regiment reorganized and redesignated as the 44th Tank Battalion. The 3d Battalion was redesignated as the 714th Tank Battalion, to be reassigned to the 12th Armored Division on 7 March 1944, until inactivated at Camp Kilmer in 1945, and relieved from that assignment. The 44th Tank Battalion was inactivated in Japan in 1946, until redesignated and activated at Fort Bragg, North Carolina, as the 44th Heavy Tank Battalion. The 714th Tank Battalion was allotted to the Regular Army in 1950, and activated at Fort Bragg. The 44th and 714th Tank Battalions were assigned to the 82d Airborne Division.

Assigned to the 11th Airborne Division in 1950, was the 76th Tank Battalion. Activated in 1944 at Fort Ord, California, as the 727th Ammunition Train Battalion, this unit was later inactivated in 1946 at Okinawa. Redesignated as the 76th Heavy Tank Battalion, and allotted to the Regular Army in 1948, the 76th Heavy Tank Battalion was activated at Fort Campbell, Kentucky in 1949, to be redesignated as the 76th Tank Battalion in 1950.

Constituted in the Regular Army in 1941, the 77th Tank Battalion was activated at Fort Ord, California to be reorganized and redesignated on 8 May 1941 as the 757th Tank Battalion until inactivated in Europe in 1945. The 77th Tank Battalion was redesignated on 10 October 1950 as the 65th Tank Battalion, and assigned to the 101st Airborne Division, the 65th Tank Battalion was activated on 25 August 1950 at Camp Breckinridge, Kentucky.

The 710th Tank Battalion, activated in the National Army in 1918 and assigned to the 60th Infantry Regiment, 15th Division, was demobilized in 1919. Activated at Pine Camp, New York, and assigned to the 4th Armored Division, the 710th was twice redesignated in 1941, as the 8th Armored Regiment, and then as the 80th Armored Regiment. It was elements of the 4th Armored Division that broke through the cordon and rescued the entrapped 101st Airborne Division at Bastogne. The 80th Armored Regiment was inactivated at Pine Camp in 1942, and activated at Fort Knox, newly assigned to the 8th Armored Division, to be redesignated in 1943 as the 3d Battalion, 80th Armored Regiment. That same year (1943), the 3d Battalion, 80th Armored Regiment was redesignated to again become the 710th Tank Battalion, and inactivated at Camp Anza, California in 1945. Activated in 1950 at Camp Campbell, the 710th Tank Battalion was assigned to the 11th Airborne Division, until inactivated at Fort Stewart, Georgia in 1958.

Cavalry

The 57th Cavalry Reconnaissance Troop, Mechanized, was constituted in the AUS in 1944 and activated at Camp McIntosh, Texas, and in 1945 was inactivated at Camp Bowie, Texas. Activated in the Philippines from 1946 until inactivation in 1947, the 57th Cavalry Reconnaissance Troop, Mechanized, was redesignated in 1948 as the 57th Reconnaissance Company, when it was activated and inactivated that same year at Fort Knox, Kentucky. In 1950, the 57th Reconnaissance Company was redesignated as the 101st Airborne Reconnaissance Company, activated and assigned to the 101st Airborne Division at Camp Breckinridge, Kentucky.

Division Artillery

The 101st Airborne Division Artillery was activated at Camp Breckinridge, Kentucky.

Field Artillery

The 81st Field Artillery Battalion was assigned to the 101st Airborne Division, and activated at Camp Breckinridge, Kentucky, and the 98th Field Artillery Battalion was assigned to the 82d Airborne Division.

The 313th and 314th Parachute Field Artillery Battalions, and the 315th Glider Artillery Battalion, were redesignated as the 313th, 314th and 315th Airborne Field Artillery Battalions, remaining assigned to the 80th Airborne Division.

The 320th Glider Field Artillery Battalion was relieved from assignment to the 82d Airborne Division, and the 321st Airborne Field Artillery Battalion was inactivated at Camp Breckinridge.

The 373d Glider Field Artillery Battalion, together with the 374th and 375th Parachute Field Artillery Battalions, assigned to the 100th Airborne Division, were redesignated as the 373d, 374th and 375th Airborne Field Artillery Battalions.

The 465th Field Artillery Battalion was ordered to Active Military Service, and the 467th Field Artillery Battalion was inactivated at Grand Rapids, Michigan.

The 515th, 516th and 518th Airborne Field Artillery Battalions were activated at Camp Breckinridge, Kentucky. The 544th Field Artillery Battalion was assigned to the 11th Airborne Division, and activated at Fort Campbell, Kentucky.

The 674th Airborne Field Artillery Battalion was attached to the newly formed 187th Airborne Regimental Combat Team, and deployed to Korea.

The 905th Glider Field Artillery Battalion was relieved from assignment to the 80th Airborne Division, redesignated as the 905th Field Artillery Battalion, and assigned to the concurrently redesignated 80th Infantry Division. The 925th Glider Field Artillery Battalion, while remaining assigned to the 100th Airborne Division, was redesignated as the 905th Airborne Field Artillery Battalion.

Air Defense Artillery

Battery C, 4th Coast Artillery Battalion, consolidated with Battery C, 4th Antiaircraft Artillery Automatic Weapons Battalion, with the consolidated unit designated as Battery C, 4th Antiaircraft Artillery Automatic Weapons Battalion. Redesignated as Battery C, 4th Antiaircraft Artillery Battalion.

The 81st Airborne Antiaircraft Artillery Battalion was activated as an element of the 101st Airborne Division at Camp Breckinridge, Kentucky.

Special Operations

The UNPIK (United Nations Partisan Infantry, Korea) were formed. As part of UNPIK, the 8157th and 8240th Army Units organized Korean partisans along similar lines that had been demonstrated by the Office of Strategic Services in Europe and the Far East, during the Second World War.

Rangers

The 1st and 2d Infantry Battalions were inactivated in the Panama Canal Zone. Company A, 1st Infantry Battalion was activated at Fort Benning, Georgia; redesignated as the 1st Ranger Infantry Company; deployed and attached to the 2d Infantry Division in Korea, commanded by Captain John Striegel. Company B became the 5th Ranger Company, commanded by Captain John C. Scagnelli, and also activated at Fort Benning.

The Ranger Training Center (Airborne), 3340th Army Service Unit, was organized and established under Third Army, at the Infantry School, Fort Benning, Georgia, on 29 September.

Companies A, B, E and F, 2d Infantry Battalion were activated at Fort Benning, with Company A reorganized and redesignated as the 2d Ranger Infantry Company, commanded by First Lieutenant Warren E. Allen, (deploying and attached to the 7th Infantry Division in Korea).

Company B was reorganized and redesignated as the 6th Ranger Infantry Company, under the command of Captain James S. (Sugar) Cain, relocated and attached to the VII Corps, in the Republic of Germany. Captain Cain had earned his commission while assigned to the 1st Special Service Force in Europe during WW II.

Company E was reorganized and redesignated as 9th Ranger Infantry Company, assigned to the US Army Infantry Center, Fort Benning, Georgia.

Company F was reorganized and redesignated as the 10th Ranger Infantry Company, commanded by Captain Charles E. Spragins, and relocated to Japan. Volunteers for the 10th Ranger Infantry Company had been selected from more than 600 volunteers from the 45th Infantry Division.

The Eighth Long Range Patrol was organized under the authority of the US Army Far Eastern Command, to be redesignated as the Eighth Army Ranger Company, 8213th Army Unit, and activated at Camp Drake, Japan. Deployed to Korea, the Eighth Army Ranger Company was assigned to IX Corps, Eighth Army, then reassigned to the 25th Infantry Division. Captain Ralph S. (Ranger) Puckett, commanding the Eighth Army Ranger Company, was awarded the Distinguished Service Cross for actions in Korea; and in 1967-68, commanded the 2d Battalion, 502d Infantry, 1st Brigade, 101st Airborne Division in Vietnam. On 5 December, Captain John P. Vann assumed command of the Eighth Army Ranger Company. Also assigned was Captain Bob Sigholtz, who had served with the 5307th Composite Unit (Merrill's Marauders) in the Pacific during World War II.

Companies A through E, 3d Ranger Infantry Battalion, were reconstituted in the Regular Army and activated at Fort Benning, Georgia.

Company A was redesignated as the 3d Ranger Infantry Company, commanded by Captain Jesse M. Tidwell, was assigned to the 3d Infantry Division in Korea.

Company B was redesignated as the 7th Ranger Infantry Company, to be commanded by Captain Robert W. Eikenberry, and assigned to the US Army Infantry Center, Fort Benning, Georgia.

Company C was redesignated as the 11th Ranger Infantry Company, and relocated to Japan.

Company D became the 12th Ranger Infantry Company, and relocated to Camp Atterbury, Indiana.

Company E was redesignated as the 13th Ranger Infantry Company, and relocated to Camp Pickett, Virginia.

Companies A and B, 4th Ranger Infantry Battalion were reconstituted in the Regular Army and activated at Fort Benning, Georgia. Companies A and B were redesignated as 4th and 8th Ranger Infantry Companies, deployed to Korea, and assigned respectively to the 1st Cavalry and the 24th Infantry Divisions. The 4th Ranger Company, commanded by Captain Dorsey B. Anderson, who later was awarded the Silver Star Medal in Korea, was the only black Ranger company in the US Army. The 8th Ranger Company was commanded by Captain James A. Herbert.

Psychological Operations

The US Army Psychological Warfare Division and School was constituted as the Psychological Division (Propaganda) (UW), and activated at Fort Bragg, North Carolina. The 1st Radio Broadcasting and Leaflet Company was organized and activated at Fort Riley, Kansas.

Civil Affairs

The 28th Military Government Company was relocated from Fort Bragg, North Carolina to Yokohama, Japan.

Medical

The 305th Airborne Medical Company was reorganized and redesignated as the 305th Airborne Medical Battalion.

Companies A and B, 325th Medical Battalion were reconstituted and redesignated respectively as Ambulance Company and Clearing Company, 325th Airborne Medical Battalion. The 325th Airborne Medical Company was reorganized and redesignated as HHC, 325th

Airborne Medical Battalion, an element of the 100th Airborne Division.

The 501st Airborne Medical Battalion was activated at Camp Breckinridge, Kentucky.

Combat Support
Engineers

The Battalion of Engineer Troops was constituted in the Regular Army in 1861, redesignated as the Battalion of Engineers in 1866, to become the 2d Battalion of Engineers in 1901. The 2d Battalion of Engineers was expanded in 1916 to form the 2d Regiment of Engineers, and again in 1917, when Company D, (organized at Ojo Federico, Mexico in 1916 with part of the Headquarters expanded in 1916), was redesignated to form the 5th Regiment of Engineers. The 5th Regiment of Engineers was redesignated in 1917 as the 5th Engineers and assigned to the 7th Division. Inactivated in 1921 at Camp Humphries, Virginia, the 5th Engineers were relieved from assignment to the 7th Division in 1936, and activated at Fort Belvoir. In 1942, the 5th Engineers were redesignated as the 5th Engineer Combat Regiment, which was reorganized and redesignated, with HHSC, 5th Engineer Combat Regiment redesignated as HHC, 1128th Engineer Combat Group. In 1946, HHC, 1128th Engineer Combat Group was inactivated in Germany, to be redesignated as HHC, 20th Engineer Brigade, and activated at Fort Leonard Wood, Missouri in 1950.

The 27th Engineer Combat Battalion was activated for one month at Fort Lewis, Washington, prior to inactivation.

The 49th Airborne Engineer Battalion was activated at Camp Breckinridge, Kentucky.

Military Intelligence

The 77th Counter Intelligence Corps Detachment was inactivated at New York City, New York.

The 297th Interpreter Detachment was constituted in the Organized Reserve Corps, assigned to Sixth Army, and activated at Pasadena, California.

The 82d Military Intelligence Detachment was reorganized and redesignated as the 82d Military Intelligence Company.

The 356th Intelligence Detachment was inactivated in Japan.

The 389th Translator Team was withdrawn from the Regular Army and allotted to the Organized Reserves; activated and inactivated at Fort Worth, Texas; reorganized and redesignated as the 389th Translator Detachment, assigned to Second Army, and activated at Cincinnati, Ohio.

Military Police

The Military Police Platoon, 82d Infantry Division (disbanded in 1942 at Camp Claiborne, Louisiana), was reconstituted and consolidated with the 82d Airborne Military Police Company, with the consolidated unit designated as the 82d Airborne Military Police Company.

The 101st Airborne Military Police Company was activated at Camp Breckinridge, Kentucky.

Signal

The 101st Airborne Signal Company was activated at Camp Breckinridge, Kentucky.

Combat Service Support

The 1st Logistical Command was constituted in the Regular Army, and activated at Camp MacPherson, Georgia.

Chemical

The 808th Chemical Technical Services Intelligence Detachment, constituted in the Organized Reserve Corps in 1950, and assigned to First Army, was activated at Niagara Falls, New York.

Ordnance

The 801st Airborne Ordnance Maintenance Company was activated at Camp Breckinridge, Kentucky.

Quartermaster

The 11th Airborne Parachute Maintenance Company was reorganized and redesignated as the 11th Airborne Quartermaster Parachute Maintenance Company.

The 84th Parachute Maintenance Company was redesignated as the 84th Airborne Parachute Maintenance Company.

The 101st Airborne Quartermaster Company was activated at Camp Breckinridge, Kentucky.

The 426th Quartermaster Battalion was inactivated on 8 November at Clarksburg, West Virginia.

The 612th Quartermaster Service Company was inactivated at Fargo, North Dakota.

Transportation

HHD, 5th Transportation Corps Truck Battalion was redesignated as HHC, 5th Transportation Corps Truck Battalion, allotted to the Regular Army, and activated at Fort Eustis, Virginia.

—— 1951 ——

HQ, XVIII Corps (Airborne) was assigned to the US Strike Command, and activated at Fort Bragg, North Carolina.

Combat Arms
Infantry

The 187th Regimental Combat Team was relieved from assignment to the 11th Airborne Division on 1 February, and participated in an airborne assault at Munsan-Ni. For his actions near Wontong-ni on 31 May, Corporal Rodolpho P. Hernandez, assigned to Co G, 187th Airborne Regimental Combat Team, was awarded the Medal of Honor.

The 191st and 192d Glider Infantry Regiments, and the 545th Parachute Infantry (all assigned to the 15th Airborne Division, but never activated) were disbanded.

The 333d Glider Infantry and the 334th and 335th Parachute Infantry (assigned to the 84th Airborne Division), were redesignated as the 333d, 334th and 335th Airborne Infantry Regiments. The 398th Airborne Infantry, 100th Airborne Division, relocated to South Charleston, West Virginia.

The 485th Glider Infantry, 108th Airborne Division, was redesignated as the 485th Airborne Infantry.

The 501st Parachute Infantry Battalion was redesignated as the 501st Airborne Infantry Regiment, with Companies A and B becoming the 1st and 2d Battalions, 501st Airborne Infantry.

The 503d Parachute Infantry was redesignated as the 503d Airborne Infantry, allotted to the Regular Army and assigned to the 11th Airborne Division, and activated at Fort Campbell, Kentucky.

The 508th Parachute Infantry was redesignated as the 508th Airborne Infantry, allotted to the Regular Army, was activated at Fort Bragg, North Carolina, and reorganized and redesignated as the 508th Airborne Regimental Combat Team.

The 518th and 519th Parachute Infantry, 108th Airborne Division, were redesignated as the 518th and 519th Airborne Infantry.

Corps Artillery

XVIII Corps (Airborne) Artillery was

allotted to the Regular Army, and activated at Fort Bragg, North Carolina.

Field Artillery

Inactivated in Japan in 1946, and relieved from assignment from the 25th Infantry Division in 1947, the 89th Field Artillery Battalion was redesignated as the 89th Airborne Field Artillery Battalion, activated at Camp Breckinridge, Kentucky, and assigned to the 101st Airborne Division.

The 320th Glider Field Artillery Battalion was redesignated as the 320th Airborne Field Artillery Battalion, and activated at Fort Benning, Georgia.

The 325th and 327th Glider Field Artillery Battalions and the 326th Parachute Field Artillery Battalion, assigned to the 84th Airborne Division, were redesignated as the 325th, 326th and 327th Airborne Field Artillery Battalions. The 326th Airborne Field Artillery Battalion relocated from Berloit to Racine, Wisconsin.

The 506th and 507th Parachute Field Artillery Battalions were redesignated as the 506th and 507th Airborne Field Artillery Battalions.

The 674th Airborne Field Artillery Battalion was relieved from assignment to the 11th Airborne Division.

Remaining assigned to the 84th Airborne Division, the 909th Parachute Field Artillery Battalion was redesignated as the 909th Airborne Field Artillery Battalion.

Special Operations
Rangers

On 5 April, the Ranger Training Command was redesignated as the Ranger Training Center, Fort Benning, Georgia, where Rangers were surreptitiously wearing black berets and black boots with their khaki uniforms. This was a double no-no as the US Army was still wearing brown boots.

Captain Charles L. Carrier assumed command of the 1st Ranger Infantry Company.

The 3d, 5th and 8th Ranger Infantry Companies embarked for Korea on 5 March. The 5th Ranger Infantry Company; deployed and consolidated with the (inactivated) Eighth Army Ranger Company in Korea, and attached to the 25th Infantry Division. This same year, the 1st and 5th Ranger Infantry Companies were later inactivated on 1 August in Korea.

Captain Theodore C. Thomas assumed command of the 9th Ranger Infantry Company.

The 12th Ranger Infantry Company, selected from volunteers from the 28th Infantry Division (Pennsylvania ARNG), was activated and commanded by Captain Harold V. Kays.

The 13th Ranger Infantry Company (activated from 43d Infantry Division volunteers), was activated and commanded by Captain Victor K. Harwood.

Companies C and D, 2d Infantry Battalion were activated at Fort Benning, Georgia. Company C was redesignated as the 14th Ranger Infantry Company, commanded by Captain Sam L. Amato. The 14th Ranger Infantry Company had been selected from 4th Infantry Division volunteers, and was relocated to Camp Carson, Colorado. Company D was redesignated as 15th Ranger Infantry Company, assigned to the US Army Infantry Center, Fort Benning. Commanded by Captain Paul W. Kopitzke, the 15th Ranger Infantry Company comprised volunteers from the 47th Infantry Division (Minnesota and North Dakota ARNG).

Companies A through F, 2d Infantry Battalion, also the 2d, 3d, 4th, 7th and 8th, Ranger Infantry Companies were inactivated in Korea on 1 August.

The 11th Ranger Infantry Company, attached to the 40th Infantry Division, was inactivated in Japan on 21 September.

The 12th, 13th, and 14th Ranger Infantry Companies were inactivated respectively in October at Camp Atterbury, Indiana; Camp Pickett, Virginia and Camp Carson, Colorado.

The 15th Ranger Infantry Company was inactivated on 5 November at Fort Benning, Georgia.

The 2d and 4th Ranger Companies participated with the 187th Regimental Combat Team in their airborne combat assault at Munsan-Ni.

Captain John P. Vann was reassigned to the Ranger Training Command, and the Eighth Army Ranger Company gained Captain Charles G. Ross (the former Executive Officer), as the new commander.

Psychological Operations

The 1st Radio Broadcasting and Leaflet Company was reorganized as the 1st Radio Broadcasting and Leaflet Group (8239th Army Unit).

Civil Affairs

The 98th Military Government Group and the 28th Military Government Company were inactivated in Korea.

Combat Support
Engineers

The 27th Engineer Combat Battalion was activated at Fort Campbell, Kentucky.

Military Intelligence

The 356th Intelligence Detachment was allotted to the Regular Army, assigned to the Connecticut Military District, and activated at Hartford, Connecticut.

The 389th Translator Detachment relocated from Cincinnati, Ohio to Fort Thomas, Kentucky.

Signal

The 51st Telegraph Battalion was redesignated as the 50th Signal Battalion, Corps, and activated in Japan.

The 3191st Signal Service Company was redesignated as the 358th Communications Reconnaissance Company, allotted to the Regular Army, and activated at Fort Devens, Massachusetts.

The 516th Signal Company was allotted to the Regular Army.

Combat Service Support
Quartermaster

The 17th Airborne Quartermaster Company was inactivated.

—— 1952 ——

Lieutenant General Thomas P. Hickey turned command of the 82d Airborne Division to Major General Charles D. W. Canham on 1 February. General Hickey assumed command of XVIII Corps (Airborne).

Combat Arms

The short-term fix for the retention of a sizeable airborne force within the United States Army quickly ceased as the 80th, 84th, 100th and 108th Airborne Divisions, were reorganized and redesignated as the 80th, 84th, 100th and 108th Infantry Divisions. HQ, 108th Infantry Division relocated from Atlanta, Georgia to Charlotte, North Carolina.

Infantry

The 317th, 318th and 319th Airborne Infantry, (80th Airborne Division); 333d, 334th and 335th Airborne Infantry, (84th Airborne Division); 397th, 398th and 399th

Airborne Infantry, (100th Airborne Division); and the 485th, 518th and 519th Airborne Infantry, (108th Airborne Division), were reorganized as non-airborne units, and redesignated as the 317th, 318th, 319th, 333d, 334, 335th, 397th, 398th, 399th, 485th, 518th and 519th Infantry Regiments. The 519th Infantry was assigned to the the 81st Infantry Division.

Division Artillery

The 80th, 84th, 100th and 108th Airborne Divisions, Artillery, were reorganized and redesignated as the 80th, 84th, 100th and 108th Infantry Divisions' Artillery.

Field Artillery

The 313th, 314th and 315th Airborne Field Artillery Battalions assigned to the 80th Airborne Division; the 325th, 326th, 327th and 909th Airborne Field Artillery Battalions assigned to the 84th Airborne Division; the 373d, 374th, 375th and 925th Airborne Field Artillery Battalions assigned to the 100th Airborne Division; and the 506th and 507th Airborne Field Artillery Battalions assigned to the 108th Airborne Division; were redesignated as the 313th, 314th, 315th, 325th, 326th, 327th, 373d, 374th, 375th, 506th, 507th, 909th and 925th Field Artillery Battalions, and assigned to the concurrently redesignated 80th, 84th, 100th and 108th Infantry Divisions.

Designated for assignment to the 15th Airborne Division, but never activated, the 459th Parachute Artillery Battalion was disbanded.

The 467th Field Artillery Battalion, inactivated in 1950, was also disbanded.

Air Defense Artillery

The 84th Airborne Antiaircraft Artillery Battalion was relieved from the 84th Airborne Division, and inactivated at Madison, Wisconsin.

The 154th Airborne Antiaircraft Artillery Battalion, designated for assignment to the 15th Airborne Division, was never activated, and finally disbanded.

Special Operations
Special Forces

The 10th Special Forces Group was activated under the Command of Colonel Aaron Bank, at Fort Bragg, North Carolina, assigned to the Psychological Warfare Division and School.

Special Forces Operational Detachments (USAR) 17 through 25, were activated at Fayetteville, North Carolina.

Rangers

The 5th Ranger Infantry Company (inactivated in Korea in 1951), was designated, (but not activated), as the 1st Ranger Infantry Battalion.

The 2d Infantry Battalion was redesignated as the 2d Ranger Infantry Battalion. The 2d, 3d and 4th Ranger Infantry Battalions were reconstituted in the Regular Army, (but not activated), with all former elements restored.

Psychological Operations

The Psychological Division (Propaganda) (UW) was redesignated as the US Army Psychological Warfare School at Fort Riley Kansas. The Psychological Warfare Division was relocated from Fort Riley to Fort Bragg, North Carolina, at which time Colonel Charles Karlstad was named as the Commander.

The first special warfare unit assigned, was the 10th Special Forces Group. The 9th Loudspeaker and Leaflet Company (Army), was constituted in the Regular Army and activated at Fort Riley, Kansas.

Medical

The 305th Airborne Medical Battalion was reorganized and redesignated as the 305th Medical Battalion, with Companies A and D reconstituted on the inactive list, and redesignated respectively as Ambulance Company and Clearing Company, 305th Medical Battalion.

The 325th Airborne Medical Battalion was redesignated as the 325th Medical Battalion. Company D, 325th Medical Battalion, was reconstituted and redesignated as Medical Detachment, Division Headquarters, 100th Infantry Division.

Combat Support
Military Intelligence

The 319th Military Intelligence Company was reorganized and redesignated as the 319th Military Intelligence Service Company, and allotted to the Regular Army.

The 389th Translator Detachment was inactivated at Fort Thomas, Kentucky.

Military Police

The 84th Airborne Military Police Company was redesignated as the 84th Military Police Company, assigned to the concurrently reorganized and redesignated 84th Infantry Division.

Signal

The 80th Airborne Signal Company was inactivated in Virginia.

Constituted in the Regular Army, the 336th Communications Reconnaissance Company was activated at Fort Devens, Massachusetts.

Combat Service Support

HQ Company, 1st Logistical Command (HQ, 1st Logistical Command was constituted and activated in 1950), was constituted in the Regular Army, and activated at Fort Bragg, North Carolina.

Ordnance

The 780th, 782d, 784th and 800th Airborne Ordnance Maintenance Companies were reorganized as HHD, 780th, 782d, 784th and 800th Ordnance Maintenance Companies. The 780th Ordnance Maintenance Company was then redesignated as HHC, 780th Ordnance Battalion.

Quartermaster

Redesignated, the 80th Airborne Quartermaster Company became the 80th Quartermaster Company, assigned to the concurrently reorganized and redesignated 80th Infantry Division. The 84th Airborne Parachute Maintenance Company was disbanded.

The 600th Quartermaster Laundry Company was withdrawn from the AUS, allotted to the Regular Army, and activated at Fort Devens, Massachusetts.

The 612th Quartermaster Service Company was redesignated as the 612th Quartermaster Aerial Supply Company; concurrently withdrawn from the Organized Reserve Corps and allotted to the Regular Army, and activated at Fort Bragg, North Carolina.

HHD, 426th Quartermaster Battalion was withdrawn from from the Organized Reserve Corps on 15 January, allotted to the Regular Army, and redesignated as HHD, 528th Quartermaster Battalion. Activated at Camp Atterbury, Indiana on 1 February.

Transportation

HHC, 5th Transportation Corps Truck Battalion was redesignated as HHC, 5th Amphibious Truck Battalion, then redesignated as HHC, 5th Transportation Corps Battalion (Amphibious Truck).

1953

Lieutenant General Joseph P. Cleland assumed command of XVIII Corps (Airborne).

Major General Gerald J. Higgins replaced General Canham as the interim Commander, 82d Airborne Division on 14 September, with Major General Francis W. Farrell assuming command on 16 October.

The 101st Airborne Division was inactivated at Camp Breckinridge, Kentucky.

Combat Arms
Infantry

For his actions on 14 August near Kumwha, Korea, Corporal Lester Hammond Jr, assigned to Company A, 187th Airborne Regimental Combat Team, was awarded the Medal of Honor.

The 501st, 502d 506th and 516th Airborne Infantry were inactivated at Camp Breckinridge, Kentucky. The 333d Infantry was relocated to Detroit, Michigan, and assigned to the 70th Infantry Division.

Colonel Charles W. Davis, recipient of the Medal of Honor while assigned to the 25th Infantry Division on Guadalcanal, commenced his three year command of the 503d Airborne Infantry Regiment.

Armor

The 42d and 65th Tank Battalions, and the Anti-Tank Platoon, 101st Airborne Division, was inactivated at Camp Breckinridge, Kentucky.

Companies B and C, 10th Armored Infantry Battalion were redesignated as Companies B and C, 510th Armored Infantry Battalion, 4th Armored Division.

Cavalry

The 101st Airborne Reconnaissance Company was inactivated at Camp Breckinridge, Kentucky.

Division Artillery

The 101st Airborne Division Artillery was inactivated at Camp Breckinridge, Kentucky.

Field Artillery

Remaining assigned to the 101st Airborne Division, the 81st Field Artillery Battalion was inactivated at Camp Breckinridge, Kentucky.

The 98th Field Artillery Battalion, assigned to the 82d Airborne Division, was redesignated as the 98th Airborne Field Artillery Battalion.

The 515th, 516th and 518th Airborne Field Artillery Battalions were inactivated at Camp Breckinridge, Kentucky.

Assigned to the 11th Airborne Division, the 544th Field Artillery Battalion was redesignated as the 544th Airborne Artillery Battalion.

Air Defense Artillery

The 80th Airborne Antiaircraft Artillery Battalion was inactivated with the 101st Airborne Division at Camp Breckinridge, Kentucky.

Special Operations
Special Forces

The 77th Special Forces Group was activated at Fort Bragg, North Carolina, and the 10th Special Forces Group relocated from Fort Bragg to Flint Kaserne, Bad Toelz, Germany. Lieutenant Colonel Jack T. Shannon, Executive Officer of the 10th Special Forces Group under Colonel Aaron Bank, was assigned as the first commander of the 77th Special Forces Group.

USAR Special Forces Operational Detachments 17 through 25 were inactivated, and Operational Detachment 26 was activated at Fayetteville North Carolina.

Psychological Operations

Colonel Gordon Singles replaced Colonel Charles Karlstad as the Commander, US Army Psychological Warfare Division and School at Fort Bragg, North Carolina.

The 9th Loudspeaker and Leaflet Company (Army), was reorganized and redesignated as 9th Loudspeaker and Leaflet Company.

Medical

The 501st Airborne Medical Battalion was inactivated at Camp Breckinridge.

Combat Support
Engineers

The 27th Engineer Combat Battalion was redesignated as the 27th Engineer Battalion.

The 49th Airborne Engineer Battalion was inactivated at Camp Breckinridge, Kentucky.

Military Intelligence

The 297th Interpreter and the 356th Intelligence Detachments were inactivated respectively at Pasadena, California and Hartford, Connecticut.

Military Police

The 101st Airborne Military Police Company was inactivated at Camp Breckinridge.

Signal

The 101st Airborne Signal Company was inactivated at Camp Breckinridge, Kentucky.

The 299th Signal Service Company was redesignated as the 299th Signal Company, allotted to the Regular Army, and activated at Camp Gordon Georgia.

Combat Service Support
Adjutant General

The 101st Replacement Company was inactivated at Camp Breckinridge, Kentucky.

Ordnance

Reorganized, the 711th Airborne Ordnance Maintenance Company was redesignated as the 711th Airborne Ordnance Battalion, with organic elements concurrently constituted and activated at Fort Campbell, Kentucky.

The 782d Airborne Ordnance Maintenance Company was reorganized and redesignated as HHD, 782d Airborne Ordnance Battalion.

The 784th Ordnance Maintenance Company was reorganized and redesignated as the 784th Ordnance Battalion.

The 801st Airborne Ordnance Maintenance Company was inactivated at Camp Breckinridge.

Quartermaster

The 11th Airborne Quartermaster Parachute Maintenance Company was redesignated as the 11th Airborne Quartermaster Company. The 101st Airborne Quartermaster Company was inactivated at Camp Breckinridge, Kentucky.

The 600th Quartermaster Laundry Company was reorganized and redesignated as the 600th Quartermaster Company, at Fort Devens, Massachusetts.

The 612th Quartermaster Aerial Supply Company was reorganized and redesignated to become the 612th Quartermaster Company. The 600th and 612th Quartermaster Companies were assigned to the 1st Logistical Command.

Transportation

HHC, 29th Truck Company was reorganized and redesignated as HHC, 29th Transportation Battalion.

1954

The 101st Airborne Division was activated in May at Fort Jackson, South Carolina.

Combat Arms
Infantry

The 333d and 334th Infantry relocated from Detroit, Michigan to Flint, Michigan, and from Milwaukee to Madison, Wisconsin respectively.

The 485th Infantry was relocated to Birmingham, Alabama.

The 501st, 502d and 506th Airborne Infantry were assigned to the 101st Airborne Division, and activated at Fort Jackson, South Carolina.

The 516th Airborne Infantry was relieved from the 101st Airborne Division in April, and activated at Camp Jackson, South Carolina in May.

The 101st Airborne Reconnaissance Company was activated at Camp Jackson, South Carolina, (see 1956, Combat Arms, Cavalry).

Armor

The 42d and 65th Tank Battalions, and the Anti-Tank Platoon, 101st Airborne Division, were activated at Camp Jackson, South Carolina.

On 11 October, the 714th Tank Battalion was relieved from assignment to the 82d Airborne Division, and assigned on 2 December, to the 23d Infantry Division.

Companies B and C, 510th Armored Infantry Battalion were activated at Fort Hood, Texas.

Division Artillery

The 84th Infantry Division Artillery was relocated to Milwaukee, Wisconsin. The 101st Airborne Division Artillery was activated at Fort Jackson, South Carolina.

Field Artillery

Activated at Fort Jackson, South Carolina, the 81st Field Artillery Battalion, assigned to the 101st Airborne Division, was redesignated as the 81st Airborne Field Artillery Battalion.

The 325th Field Artillery Battalion was inactivated at Fort Custer, Michigan.

The 507th Field Artillery Battalion was inactivated at Atlanta, Georgia.

The 515th, 516th and 518th Airborne Field Artillery Battalions were activated at Fort Jackson, South Carolina.

Air Defense Artillery

The 81st Airborne Antiaircraft Artillery Battalion was activated at Fort Jackson, South Carolina, and assigned to the 101st Airborne Division.

Special Operations
Rangers

Elements of the 5307th Composite Unit (Provisional), (organized in the AUS in 1943), were consolidated into Companies C and F, 475th Infantry in 1944. The consolidated units, designated as Companies C and F, 475th Infantry, were inactivated in China in 1945, to be designated as Companies C and F, 75th Infantry, allotted to the Regular Army and activated on Okinawa this year.

Psychological Operations

From July through December, Colonel Thomas A. McAnsh commanded the US Army Psychological Warfare Division and School, at Fort Bragg, North Carolina. In December he was replaced by Colonel Edson D. Raff.

Civil Affairs

The 95th Military Government Group was allotted to the Regular Army from the AUS.

Medical

HHC, 501st Airborne Medical Battalion was activated at Fort Jackson, South Carolina, and reorganized and redesignated as the 326th Airborne Medical Company (organic companies (Clearing and Ambulance) remained active, but unfilled; subsequently inactivated and redesignated as the 226th and 595th Medical Companies respectively).

Combat Support
Engineers

The 37th Engineer Combat Battalion was redesignated at the 37th Engineer Battalion, and activated in Germany.

The 49th Airborne Engineer Battalion was activated at Fort Jackson, South Carolina.

Military Intelligence

The 319th Military Intelligence Service Company was inactivated in Japan.

Military Police

The 101st Airborne Military Police Company was activated at Fort Jackson, South Carolina.

Signal

The 50th Signal Battalion, Corps was reorganized and redesignated as the 50th Signal Battalion.

Combat Service Support
Adjutant General

The 101st Replacement Company was activated at Fort Jackson, South Carolina.

Chemical

The 808th Chemical Technical Services Intelligence Detachment was redesignated as the 808th Chemical Detachment.

Ordnance

The 801st Airborne Ordnance Maintenance Company was redesignated as the 801st Airborne Ordnance Battalion (with organic elements concurrently constituted), and activated at Fort Jackson, South Carolina (see 1957, Combat Service Support, Maintenance).

Quartermaster

The 101st Airborne Quartermaster Company was activated at Fort Jackson, South Carolina.

1955

Lieutenant General Ridgley Gaither, former commander of the 82d Airborne Division, assumed command of XVIII Corps (Airborne), followed by LTG Paul D. Adams.

Major General T. J. H. Trapnell was named as commander of the 82d Airborne Division.

Combat Arms
Field Artillery

The 325th Field Artillery Battalion was activated at Fort Custer, Michigan, and the 327th Field Artillery Battalion relocated to Champaign, Illinois.

The 465th Field Artillery Battalion was released from Active Military Service, and inactivated at Minot, North Dakota.

Special Operations

HQ, Support Operations, Task Force Europe (SOTFE), was constituted and activated at Patch Barracks, Stuttgart, Germany, as the joint special operations coordinating headquarters for the US European Command (EUCOM).

Special Forces

USAR Special Forces Operational

Detachment 300 was activated at Fayetteville, North Carolina.

Rangers

The 2d Ranger Infantry Battalion was redesignated as the 2d Infantry Battalion, and activated in Iceland.

Psychological Operations

The 1st Radio Broadcasting and Leaflet Group was discontinued and consolidated with HHC, 1st Radio Broadcasting and Leaflet Battalion, with the consolidated units designated as HHC, 1st Radio Broadcasting and Leaflet Battalion, and activated at Fort Bragg, North Carolina.

Civil Affairs

The 95th Military Government Group was activated at Camp Gordon, Georgia.

Combat Support
Military Intelligence

The 215th Signal Corps Co was reorganized and redesignated as HHC, 313th Communications Reconnaissance Battalion; allotted to the Regular Army; and activated at Fort Bragg, North Carolina.

The 319th Military Intelligence Service Company was reorganized and redesignated as the 319th Military Intelligence Battalion, and activated at Fort Meade, Maryland.

The 358th Communications Reconnaissance Company was reorganized and redesignated as Company A, 313th Army Security Agency Battalion.

The 162d Language Detachment was redesignated as the 162d Military Intelligence Platoon, allotted to the Regular Army, and activated at Fort George G. Meade, Maryland.

Signal

The 299th Signal Company was inactivated at Camp Gordon, Georgia.

The 336th Communications Reconnaissance Company was reorganized and redesignated as Company A, 311th Communications Reconnaissance Battalion.

The 516th Signal Company was inactivated at Camp Roeder, Austria.

Combat Service Support
Transportation

HHC, 5th Transportation Corps Battalion (Amphibious Truck), was inactivated at Fort Eustis, Virginia.

——1956——

Major General John W. Bowen replaced General Trapnell as commander of the 82d Airborne Division.

On 21 September, the 101st Airborne Division was reorganized as the first Pentomic Division in the United States Army.

Combat Arms
Infantry

187th Airborne Infantry redeployed from the Pacific Theater, to be assigned to the 101st Airborne Division.

The 516th Airborne Infantry was redesignated as the 327th Airborne Infantry, and also assigned to the 101st Airborne Division.

The 485th Infantry was inactivated at Birmingham, Alabama.

The 508th Airborne Regimental Combat Team was assigned to the Pacific Theater.

Cavalry

The 101st Airborne Reconnaissance Company was reorganized and redesignated as the 101st Airborne Reconnaissance Troop.

Division Artillery

HHB, 101st Airborne Division Artillery, was reorganized and redesignated as HHSB, 101st Airborne Division Artillery.

Field Artillery

The 376th Airborne Field Artillery Battalion was inactivated at Fort Bragg, North Carolina.

The 515th, 516th and 518th Airborne Field Artillery Battalions were redesignated respectively as the 377th, 463d and 321st Airborne Artillery Battalion.

Activated at Fort Campbell, Kentucky, the 907th Field Artillery Battalion, was redesignated as the 907th Airborne Field Artillery Battalion, allotted to the Regular Army, and also assigned to the 101st Airborne Division. The 674th Airborne Field Artillery Battalion was also assigned to the 101st Airborne Division.

Special Operations
Special Forces

Special Forces Operational Detachment 14 (Area), was activated in Hawaii, using the cover designation as the 8251st Army Unit. Concurrently the 8231st Army Unit with Detachments A, B and C, (12, 13 and 16), were activated at Camp Drake, Japan.

SF Operational Detachment A-301 (SF ODA) was activated in Boise, Idaho.

Rangers

The 75th Infantry was activated on Okinawa.

Psychological Operations

The US Army Psychological Warfare Division and School was redesignated as the US Army Special Warfare School at Fort Bragg, North Carolina. In April, Colonel William J. Mullen assumed command, following Colonel Edson D. Raff.

Combat Support
Engineers

The 49th Airborne Engineer Battalion was redesignated as the 326th Airborne Engineer Battalion.

Intelligence

Companies A, 311th and 313th Communications Reconnaissance Battalions were redesignated as Companies A, 311th and 313th Army Security Agency Battalions.

The 101st Counter Intelligence Corps Detachment was allotted to the Regular Army and activated at Fort Campbell, Kentucky.

Signal

The 101st Airborne Signal Company reorganized and redesignated as HHSD, 501st Airborne Signal Battalion, an element of the 101st Airborne Division. (The 516th Signal Company and the 299th Signal Company, were concurrently redesignated as the Operations Company and the Installation Company, 501st Airborne Signal Battalion, and respectively activated at Fort Campbell, Kentucky).

Combat Service Support

HHSC, 101st Airborne Division Support Group was constituted in the Regular Army, assigned to the 101st Airborne Division, and activated at Fort Campbell, Kentucky.

Adjutant General

The 101st Replacement Company was reorganized and redesignated as the 101st Airborne Administrative Service Company.

Ordnance

The 801st Airborne Ordnance Battalion was reorganized and redesignated as the 801st Airborne Maintenance Battalion., (see 1957, Combat Service Support, Maintenance).

Quartermaster

The 101st Airborne Quartermaster Company was reorganized and redesignated as the 426th Airborne Quartermaster Company.

The 600th Quartermaster Company was inactivated at Fort Devens, Massachusetts.

—— 1957 ——

Lieutenant General Robert F. Sink assumed command of XVIII Corps (Airborne).

HQ, 11th, 82d and 101st Airborne Divisions were reorganized and redesignated as HHC, Command and Control Battalions, 11th, 82d and 101st Airborne Divisions.

HQ, 84th Infantry Division relocated from Madison to Milwaukee, Wisconsin.

The 101st Airborne Division was deployed to Little Rock, Arkansas to assist in civil disorder activities.

Combat Arms

This was the year to reorganized the Airborne Infantry Regiments into Airborne Battle Groups.

Infantry

The 187th Infantry was reorganized as a parent regiment under the Combat Arms Regimental System (CARS). Companies A, B, C, D and E, 187th Infantry were redesignated as the 1st, 2d, 3d, 4th and 5th Airborne Battle Groups, 187th Infantry, with the 3d, 4th and 5th Airborne Battle Groups, 187th Infantry, concurrently inactivated. The 1st Airborne Battle Group was assigned to the 11th Airborne Division and the 2d Airborne Battle Group remaining assigned to the 101st Airborne Division.

The 188th Airborne Infantry was released from assignment to the 11th Airborne Division and inactivated in Germany, with the exception of Company D, redesignated as the 4th Battalion, 188th Airborne Infantry.

The 325th Airborne Infantry, (minus Company A), was relieved from assignment to the 82d Airborne Division, and reorganized as a parent regiment under the CARS. Companies A, B, C, D and E were redesignated as the 1st, 2d, 3d, 4th and 5th Airborne Battle Groups, 325th Infantry. The 2d, 3d, 4th and 5th Airborne Battle Groups were inactivated, while the 1st Airborne Battle Group, 325th Infantry remained assigned to the 82d Airborne Division.

The 327th Airborne Infantry, (minus Company A), was relieved from assignment to the 101st Airborne Division, and reorganized as a parent regiment under the CARS. Companies A, B, C, D and E were redesignated as the 1st, 2d, 3d, 4th and 5th Airborne Battle Groups, 327th Infantry. The 2d, 3d, 4th and 5th Airborne Battle Groups were inactivated, while the 1st Airborne Battle Group, 327th Infantry remained assigned to the 101st Airborne Division.

The 501st and 502d Airborne Infantry were relieved from assignment to the 101st Airborne Division, and reorganized as parent regiments under the CARS. Companies A and B, 501st Airborne Infantry were redesignated as the 1st and 2d Airborne Battle Groups, 501st Infantry. The 1st Airborne Battle Group, 501st Infantry remained assigned to the 101st Airborne Division, and the 2d Airborne Battle Group was assigned to the 82d Airborne Division. Captain Donald E. Rosenblum, formerly assigned to the 1st Battalion, 505th Airborne Infantry, commenced his two year command of Company C, 2d Airborne Battle Group, 501st Infantry. Companies A, B, C, D, E and F, 502d Airborne Infantry were reorganized and redesignated as the 1st, 2d, 3d, 4th, 5th and 6th Airborne Battle Groups, 502d Infantry. The 1st Airborne Battle Group, 502d Infantry remained assigned to the 101st Airborne Division, the 2d Airborne Battle Group, 502d Infantry was assigned to the 11th Airborne Division.

The 503d Airborne Infantry was relieved from assignment to the 11th Airborne Division, and reorganized as a parent regiment under the CARS. Companies A, B, C and D were reorganized and redesignated as the 1st, 2d, 3d and 4th Airborne Battle Groups, 503d Infantry. The 1st Airborne Battle Group, 503d Infantry remained assigned to the 11th Airborne Division, while the 2d Airborne Battle Group was assigned to the 82d Airborne Division. The 3d and 4th Airborne Battle Groups, 503d Infantry were inactivated in Germany.

The 504th and 505th Airborne Infantry were relieved from assignments to the 82d Airborne Division, and reorganized as parent regiments under the CARS. Companies A, B, C, D and E of the 504th Airborne Infantry were reorganized and redesignated as the 1st, 2d, 3d and 5th Airborne Battle Groups, 504th Infantry. Companies A, B, C and D, 505th Airborne Infantry were reorganized and redesignated as the 1st, 2d, 3d and 4th Airborne Battle Groups, 505th Infantry. The 1st Airborne Battle Groups, 504th and 505th Infantry remained assigned to the 82d Airborne Division, the 2d Airborne Battle Groups, 504th and 505th Infantry were assigned to the 11th Airborne Division. The 3d, 4th and 5th Airborne Battle Groups, 504th Infantry, and the 3d and 4th Airborne Battle Groups, 505th Infantry were inactivated.

The 506th Airborne Infantry was relieved from assignment to the 101st Airborne Division, and reorganized as a parent regiment under the CARS. Companies A, B and C were reorganized and redesignated as the 1st, 2d and 3d Airborne Battle Groups, 506th Infantry. The 1st Airborne Battle Group, 506th Infantry was assigned to the 101st Airborne Division, the 2d and 3d Airborne Battle Groups, 506th Infantry were inactivated.

The 508th Airborne Infantry was inactivated at Fort Campbell, Kentucky, and the 511th Airborne Infantry was relieved from the 11th Airborne Division and inactivated in Germany.

Armor

The 42d and 65th Tank Battalions, assigned to the 101st Airborne Division, and the Anti-Tank Platoons, 11th and 101st Airborne Divisions, were inactivated at Fort Campbell, Kentucky. The 44th Tank Battalion was relieved from assignment to the 82d Airborne Division.

Companies B and C, 510th Armored Infantry Battalion, were inactivated at Fort Hood, Texas. The 510th Armored Infantry Battalion was concurrently relieved from assignment to the 4th Armored Division.

Cavalry

Troops A, B and C, 17th Cavalry, were constituted in the Regular Army in 1916, and organized at Fort Bliss, Texas. Inactivated in 1921 at the Presidio of Monterey, California, Troops A and B, 17th Cavalry were disbanded on 9 March 1951. In 1957, Troops A, B and C, 17th Cavalry were reconstituted in the Regular Army, and consolidated respectively with the 82d Airborne Reconnaissance Company, the 101st Airborne Reconnaissance Troop, and the 11th Airborne Reconnaissance Company. The consolidated units were designated as Troops A, B and C, 17th Cavalry, and assigned respectively to the 82d, 101st and 11th Airborne Divisions. Troops A, B and C were concurrently activated at Fort Bragg, North Carolina, Fort Campbell Kentucky and Germany.

Division Artillery

HHSB, 101st Airborne Division Artillery was redesignated as HSB, 101st Airborne Division Artillery.

Field Artillery

Inactivated at Fort Carson, Colorado, Battery B, 26th Field Artillery Battalion, was relieved from assignment to the 9th Infantry Division, to be redesignated as HHB, 1st Observation Battalion, 26th Artillery. Battery A, 39th Field Artillery Battalion, was inactivated at Fort Benning, Georgia, relieved from assignment to the 3d Infantry Division, and redesignated HHB, 1st Battalion, 39th Artillery. The 81st Field Artillery Battalion was inactivated at Fort Campbell, Kentucky and relieved from assignment to the 101st Airborne Division. The 89th and 98th Airborne Field Artillery Battalions, assigned to the 11th and 82d Airborne Divisions, were inactivated in Germany and at Fort Bragg, North Carolina.

The 319th and 320th Airborne Field Artillery Battalion was reorganized as a parent regiment under the Combat Arms Regimental System, and redesignated as the 319th and 320th Artillery Regiments.

The 321st Airborne Field Artillery Battalion was relieved from assignment to the 101st Airborne Division.

The 374th Field Artillery Battalion relocated to Owensboro, Kentucky.

The 377th Airborne Field Artillery Battalion was reorganized as a parent regiment under the Combat Arms Regimental System, and redesignated as the 377th Artillery Regiment.

The 456th, 457th and 463d Airborne Field Artillery Battalions were inactivated respectively at Fort Bragg, North Carolina, in Germany, and at Fort Campbell, Kentucky.

The 550th Airborne Artillery Battalion was redesignated as the 466th Airborne Field Artillery Battalion.

The 507th Field Artillery Battalion, relocated from Gainesville, Georgia in 1955, was moved to Atlanta, Georgia.

The 544th and 675th Airborne Field Artillery Battalions were relieved from assignment to the 11th Airborne Division, and inactivated in Germany.

Assigned to the 101st Airborne Division, the 674th and 907th Airborne Field Artillery Battalions were inactivated at Fort Campbell, Kentucky.

HHS Battery, 905th Field Artillery Battalion was reorganized and redesignated as HHD, 2d Brigade, 80th Division (Training).

Air Defense Artillery

Battery C, 4th Antiaircraft Artillery Battalion, inactivated in England. The 80th and 81st Airborne Antiaircraft Artillery Battalions were relieved from respective assignments to the 82d and 101st Airborne Division. The 81st Airborne Antiaircraft Artillery Battalion was inactivated at Madison, Wisconsin. The 88th Airborne Antiaircraft Artillery Battalion was inactivated in Germany.

Special Operations
Special Forces

The 1st Special Forces Group, activated in Japan, was constituted in the AUS from the 2d Company, 1st Battalion, First Regiment, 1st Special Service Force, and manned from personnel assigned from the 77th Special Forces Group at Fort Bragg, North Carolina, and the 8231st Army Unit, Camp Drake, Japan.

Special Forces Detachment 14 in Hawaii, relocated to Okinawa, and was redesignated Special Forces Detachment 16 (District B). Detachments A, B and C, 8231st Army Unit became Special Forces Detachments (Regiment) 12 and 13. Special Forces Operational Detachment 303 was activated at Kearney, New Jersey.

Special Forces (SF) Operational Detachments A (ODA) 12 and 13 were relocated from Camp Drake, Japan to Okinawa and inactivated. Special Forces Detachment 16 was inactivated.

Aviation

The 11th and 82d Aviation Companies were constituted in the Regular Army, and respectively activated in Germany and and at Fort Bragg, North Carolina., The 11th Aviation Company was assigned to the 11th Airborne Division, and the 82d Aviation Company was assigned to the 82d Airborne Division.

Medical

HHC, 11th Medical Battalion, was reorganized and redesignated as the 111th Medical Company. The 307th Airborne Medical Battalion was reorganized and redesignated as the 82d Medical Company, an element of the 82d Airborne Division.

The 326th Airborne Medical Company redesignated as the 326th Medical Company.

Combat Support
Engineers

The 127th Airborne Engineer Battalion was reorganized and redesignated as the 127th Engineer Battalion (Airborne Division).

The 307th and 326th Airborne Engineer Battalions were redesignated as the 307th and 326th Engineer Battalions.

Military Intelligence

The 525th Military Intelligence Group relocated from Heidelberg, Germany to Fort George G. Meade, Maryland.

HHCs, 311th and 313th Communications Reconnaissance Battalions were redesignated as HHCs, 311th and 313th Army Security Agency Battalions. Companies A, 311th and 313th Army Security Agency Battalions were inactivated respectively at Camp Wolters, Texas and Fort Bragg, North Carolina.

Military Police

The 82d and 101st Airborne Military Police Companies were inactivated at Fort Bragg, North Carolina, and Fort Campbell, Kentucky respectively, and concurrently relieved organic elements of the 82d and 101st Airborne Divisions.

Signal

The 50th Signal Battalion was reorganized and redesignated as the 50th Airborne Signal Battalion.

The 501st Airborne Signal Battalion was reorganized and redesignated as the 501st Signal Battalion, with Operations and Installation Companies concurrently reorganized and redesignated as Companies A and B respectively).

The 82d and 511th Airborne Signal Companies were reorganized and redesignated as the 82d and 511th Signal Battalions. The 82d Signal Battalion was activated at Fort Bragg, North Carolina, and the 511th Signal Battalion relocated to Germany.

Combat Service Support
Adjutant General

The 11th and 82d Replacement Companies, assigned to the 11th and 82d Airborne Divisions, were reorganized and redesignated as the 11th and 82d Administration Companies.

The 101st Airborne Administrative Service Company was reorganized and redesignated as the 101st Administration Company.

Maintenance

The 711th Airborne Ordnance Battalion was reorganized and redesignated as the 711th Maintenance Battalion.

The 782d Airborne Ordnance Battalion was reorganized and redesignated as the 782d Maintenance Battalion.

The 801st Airborne Maintenance Battalion, (see 1956, Combat Service Support, Ordnance), was reorganized and redesignated as the 801st Maintenance Battalion. HHD, and activated at Fort Jackson, South Carolina.

Quartermaster

HHC, Support Groups, 11th and 82d Airborne Divisions were constituted in the Regular Army, and respectively activated in Germany and assigned to the 11th Airborne Division, and activated at Fort Bragg, North Carolina, and assigned to the 82d Airborne Division. HHSC, 101st Airborne Division Support Group, was reorganized and redesignated as HHC, Support Group, 101st Airborne Division.

The 11th Airborne Quartermaster Company was reorganized and redesignated as the 11th Quartermaster Parachute Supply and Maintenance Company.

The 407th, 408th and 426th Airborne Quartermaster Companies were reorganized and redesignated as the 407th, 408th and 426th Supply and Transportation Companies.

—— 1958 ——

The 11th Airborne Division was inactivated in Germany.

The airborne elements assigned to the 24th Infantry Division in Germany, deployed to participate in combat operations in the Lebanon.

Major General Hamilton H. Howze assumed command of the 82d Airborne Division on 2 January.

Combat Arms
Infantry

HHC, 1st Airborne Brigade and HHC, 2d Airborne Infantry Brigade were reconstituted in the Regular Army as HHCs, 1st and 2d Infantry Brigades, and activated respectively at Fort Benning, Georgia and Fort Devens, Massachusetts.

The 1st Airborne Battle Group, 187th Infantry, the 2d Airborne Battle Group, 502d Infantry, and the 1st Airborne Battle Group, 503d Infantry, were relieved from assignment to the 11th Airborne Division. The 1st Airborne Battle Group, 187th Infantry and the 1st Airborne Battle Group, 503d Infantry were assigned to the 24th Infantry Division, later deploying to the Lebanon as part of the peacekeeping force. The 2d Airborne Battle Group, 502d Infantry, was inactivated in Germany.

The 1st Airborne Battle Group, 504th Infantry was relieved from assignment to the 82d Airborne Division, and assigned to the 8th Infantry Division in Germany. The 2d Airborne Battle Groups, 504th and 505th Infantry were relieved from assignment to the 11th Airborne Division, and inactivated in Germany.

Armor

The 44th Tank Battalion was inactivated at Fort Bragg, North Carolina.

Cavalry

Troop C, 17th Cavalry was relieved from assignment to the 11th Airborne Division, and inactivated in Germany.

Division Artillery

HHB, 11th Airborne Division Artillery was inactivated in Germany.

Field Artillery

HHB, 1st Observation Battalion, 26th Artillery, was activated at Fort Bragg, North Carolina. HHB, 1st Battalion, 39th Artillery, was redesignated as HHB, 1st Missile Battalion, 39th Artillery.

Batteries A, B and C, 320th Artillery were inactivated in Germany, and relieved from assignments to the 11th Airborne Division.

Air Defense Artillery

Battery C, 4th Antiaircraft Artillery Battalion consolidated with Battery C, 4th Field Artillery Battalion, (organized in 1901), with the consolidated unit designated as HHB, 3d Howitzer Battalion, 4th Artillery, activated in June 1958, and inactivated in September at Fort Sill, Oklahoma.

The last of it's kind, the 80th Airborne Anti-Aircraft Artillery Battalion was inactivated at Fort Bragg, North Carolina.

Special Operations

The command of the US Army Special Warfare School was passed in January to Colonel George H. Jones.

Special Forces

Special Forces Operational Detachment 302 was activated in Chicago, Illinois, and Operational Detachment 303 was relocated from Kearney, New Jersey to Camp Kilmer, New Jersey.

Aviation

The 11th Aviation Company was relieved from assignment to the 11th Airborne Division, and inactivated in Germany.

Medical

The 111th Medical Company was inactivated in Germany.

Combat Support
Engineers

HHC, 20th Engineer Brigade was inactivated at Fort Bragg, North Carolina.

The 37th Engineer Battalion, and the 127th Engineer Battalion (Airborne Division), were inactivated in Germany.

Military Intelligence

HHC, 313th Communications Reconnaissance Battalion, reorganized and redesignated as the 313th Army Security Agency Battalion, (with the 358th and 337th Communications Reconnaissance Companies inactivated and concurrently reorganized, redesignated, and activated as Companies A and B).

The 162d Military Intelligence Platoon was reorganized and redesignated as the 162d Military Intelligence Company.

The 82d Counter Intelligence Corps Detachment was reorganized and redesignated as the 82d Military Intelligence Detachment.

Signal

The 50th Airborne Signal Battalion was reorganized and redesignated as the 50th Signal Battalion.

The 511th Signal Battalion was inactivated in Germany.

Combat Service Support

HHCs, Support Groups, 82d and 101st Airborne Divisions, were reorganized and redesignated as HHDs, Support Groups, 82d and 101st Airborne Divisions.

Maintenance

The 711th Maintenance Battalion was inactivated in Germany.

Quartermaster

The 11th Quartermaster Parachute Supply and Maintenance Company was relieved from assignment to the 11th Airborne Division, and assigned to the 24th Infantry Division.

The 408th Supply and Transport Company was inactivated in Germany.

1959

Major General Dwight F. Beach was appointed as Commander, 82d Airborne Division.

The 80th, 84th, 100th and 108th Infantry Divisions were reorganized and redesignated as the 80th, 84th, 100th and 108th Divisions (Training).

Combat Arms
Infantry

Companies B and C, 510th Armored Infantry Battalion, were redesignated respectively as HHCs, 5th and 6th Battle Groups, 51st Infantry.

The 1st Airborne Battle Group, 187th Infantry, and the 1st Airborne Battle Group, 503d Infantry, were relieved from assignments to the 24th Infantry Division, and assigned to the 82d Airborne Division.

The 1st Airborne Battle Group, 505th Infantry, was relieved from assignment to the 82d Airborne Division, and assigned to the 8th Infantry Division.

The 317th, 318th, and 319th, Infantry were reorganized and redesignated as the 317th, 318th and 319th Regiments (Basic Combat Training), assigned to the 80th Division (Training).

The 333d Infantry was reorganized and redesignated as the 333d Regiment (Basic Combat Training), and assigned to the 70th Division (Training).

The 334th and 335th Infantry were reorganized and redesignated as the 334th and 335th Regiments (Basic Combat Training), and assigned to the 84th Division (Training).

The 397th, 398th and 399th Infantry were reorganized and redesignated as the 337th, 338th and 339th Regiments (Basic Combat Training), and assigned to the 100th Division (Training).

The 485th and 518th Infantry were reorganized and redesignated as the 485th and 518th Regiments (Basic Combat Training), and assigned to the 108th Division (Training).

Cavalry

Troop E, 17th Cavalry was constituted in the Regular Army in 1916, and organized at Fort Bliss, Texas. Inactivated in 1921 at the Presidio of Monterey, California, Troop E, 17th Cavalry was disbanded in 1951, but reconstituted in the Regular Army in 1959.

Division Artillery

HHB, 80th, 84th, 100th and 108th Infantry Divisions' Artillery, were converted, reorganized and redesignated as HHC, 80th, 84th, 100th and 108th Regiment (Common Specialist Training), respectively assigned to the 80th, 84th, 100th and 108th Divisions (Training).

Field Artillery

The 81st Artillery Battalion was inactivated, reorganized as a parent regiment under the Combat Arms Regimental System, and designated as the 81st Artillery Regiment.

The 313th, 314th and 315th Field Artillery Battalions were disbanded respectively at Richmond, Roanoke and Norfolk, Virginia.

The 321st Airborne Field Artillery Battalion was inactivated and reorganized as a parent regiment under the Combat Arms Regimental System, and designated the 321st Artillery Regiment.

The 325th, 326th, 327th, 373d, 374th, 375th, and 506th Field Artillery Battalions were disbanded respectively at Fort Custer; Racine, Wisconsin; Champaign, Illinois; Louisville, Owensboro and Fort Thomas, Kentucky; and Durham, North Carolina.

The 507th Field Artillery Battalion was inactivated at Atlanta, Georgia.

The 544th Airborne Field Artillery Battalion was consolidated with the 18th Artillery.

The 909th and 925th Field Artillery Battalions were disbanded in Sheboygan, Wisconsin, and Lexington, Kentucky.

Special Operations
Special Forces

Special Forces (SF) Operational Detachment 301 was relocated from Boise, Idaho to Seattle, Washington, and inactivated.

SF Operational Detachments 302 and 303 were inactivated at Chicago, Illinois and Camp Kilmer, New Jersey. SF Operational Detachments 304 and 305 were activated and inactivated at Fayetteville, North Carolina and Lima, Ohio respectively.

SF Operational Detachments 306, 307, 308 and 309 were activated respectively at Stillwater and Tulsa, Oklahoma; Little Rock, Arkansas; and Albuquerque, New Mexico.

SF Operational Detachments 310, 311, 312, 313, 314, 317, 318, 319, 324, 325, 326, 328 and 330 were activated and inactivated respectively at Columbus, Ohio; Fort Thomas, Kentucky; Columbus, Ohio; San Francisco, California; Fort MacArthur, California; Boston, Massachusetts; Manchester, New Hampshire (318th and 319th); Rutland, Vermont; South Bend, Indiana; Kewaunee, Wisconsin; Nankato, Minnesota; Chicago, Illinois; and Boise, Idaho.

SF Operational Detachment 316 was activated at Meadeville, Pennsylvania, relocated to Washington, DC, and moved again to Columbus, Ohio.

SF Operational Detachment 320 was activated at Rutland Vermont, and relocated to Boston, Massachusetts.

SF Operational Detachments 321, 322, 323 and 329 were activated in Decatur, Alabama; Jacksonville, Florida; Miami, Florida and Fort DeRussey, Hawaii.

Civil Affairs

The 95th Military Government Group was reorganized and redesignated as the 95th Civil Affairs Group, assigned to the US Army Special Warfare School, and activated at Fort Bragg, North Carolina.

Medical

The 325th Medical Battalion, assigned to the 100th Infantry Division, was disbanded.

Combat Support
Military Intelligence

The 319th Military Intelligence Battalion was activated in Hawaii.

The 101st Counter Intelligence Corps Detachment was reorganized and redesignated as the 101st Military Intelligence Detachment.

Combat Service Support
Chemical

The 808th Chemical Detachment was inactivated at Niagara Falls, New York.

Quartermaster

The 11th Quartermaster Parachute Supply and Maintenance Company was relieved from assignment to the 24th Infantry Division, and assigned to the 8th Infantry Division.

The 80th Quartermaster Company was disbanded at Richmond, Virginia.

Transportation

HHC, 29th Transportation Battalion was reorganized and redesignated as HHD, 29th Transportation Battalion.

1960

Lieutenant General Thomas J. H. Trapnell, former commander of the 82d Airborne Division, assumed command of XVIII Corps (Airborne).

Combat Arms
Infantry

The 317th Regiment (Basic Combat Training), was relocated to Washington, DC.

The 2d Airborne Battle Group, 503d Infantry, was relieved from the 82d Airborne Division and assigned to the 25th Infantry Division.

The 2d Airborne Battle Group, 504th Infantry, was relieved from assignment to the 11th Airborne Division, and assigned to the 82d Airborne Division, and activated at Fort Bragg, North Carolina.

Field Artillery

HHB, 3d Howitzer Battalion, 4th Artillery, assigned to the 2d Infantry Brigade, and activated at Fort Devens, Massachusetts.

Battery C, 319th Field Artillery, (formally Battery C, 319th Airborne Field Artillery in 1948), relieved from assignment to the 82d Airborne Division, and assigned to the 25th Infantry Division.

Battery C, 320th Field Artillery, (redesignated as such in 1951), was relieved from assignment from the 11th Airborne Division, and assigned to the 82d Airborne Division.

Special Operations
Special Forces

The 7th Special Forces Group was activated at Fort Bragg, North Carolina.

The 1st and 2d Companies, 1st Battalion, 3d Regiment, 1st Special Service Force, were consolidated with Companies A and B, 5th Ranger Battalion, with the consolidated units designated as the 19th and 20th Special Forces Groups. The 19th and 20th Special Forces Groups were withdrawn from the Regular Army and allotted to the Army National Guard.

Special Forces Operational Detachment 323 was inactivated in Miami, Florida.

Rangers

In order to preserve the lineage and honors of the six Ranger Infantry Battalions, who performed so magnificently during World War II, the 1st, 2d, 3d, 4th, 5th and 6th Ranger Infantry Battalions were consolidated with the 1st Special Service Force, and the eighteen Special Forces Groups of the active army (1st, 3d, 5th, 6th, 7th, 8th and 10th; the ARNG (16th, 19th, 20th and 21st); and the USAR, (2d, 9th, 11th, 12th, 13th, 17th and 24th), to form the 1st Special Forces, a parent regiment under the Combat Arms Regimental System. The eighteen Special Forces Groups were constituted from elements of the 1st Special Service Force.

The 2d Infantry Battalion was inactivated at Fort Hamilton, New York.

Psychological Operations

HHC, 1st Radio Broadcasting and Leaflet Battalion reorganized and redesignated as the 1st Psychological Warfare Battalion.

Combat Support
Military Intelligence

Detachments 1, 2, 3 and 4, 80th US Army Security Agency Special Operations Unit, were organized in the Regular Army and activated at Vint Hill Farms, Virginia. Detachment 1 relocated to Okinawa, and assignment to the 1st Special Forces Group. Detachment 2 was relocated to Langgries, Federal Republic of German, for assignment to the 10th Special Forces Group. Detachment 3 was redesignated as Operational Detachment 3, and assigned to the 7th Special Forces Group at Fort Gulick, Panama. Detachment 4 relocated to Fort Bragg, North Carolina, and was assigned to the 5th Special Forces Group.

Combat Service Support
Transportation

HHC, 5th Transportation Battalion (Amphibious Truck) was redesignated as the 5th Transportation Battalion, and activated on Okinawa.

1961

Lieutenant General Hamilton H. Howze, former Commander, 82d Airborne Division, assumed command of XVIII Corps (Airborne). Major General Theodore J. Conway became the commander of the 82d Airborne Division.

Lieutenant General (USA Retired) James M. Gavin was appointed by President John F. Kennedy to the US Ambassador to France.

Combat Arms
Infantry

The 397th, 398th and 399th Regiments (Basic Combat Training), were ordered to Active Military Service.

The 2d Airborne Battle Group, 503d Infantry, was relieved from assignment to the 25th Infantry Division.

Field Artillery

HHB, 1st Observation Battalion, 26th Artillery, redesignated as the 2d Target Acquisition Battalion, 26th Infantry.

Battery C, 319th Artillery relieved from assignment to the 25th Infantry Division.

Special Operations

Command of the US Army Special Warfare Center was provided to Brigadier General William P. Yarborough.

President John F. Kennedy toured Fort Bragg, North Carolina, and inspected the XVIII Airborne Corps, the 82d Airborne Division, and the US Army Special Warfare Center. He had indicated his approval of a green beret for wear by the US Army Special Forces, however the coup-de-theater was accomplished when Brigadier General Yarborough, marched his Special Forces soldiers past the reviewing stand wearing their long fought for headgear.

Special Forces

The 5th Special Forces Group was activated at Fort Bragg, North Carolina, and the 2d, 9th, 11th, 12th 13th, 17th, 21st and 24th Special Forces Groups, were activated in the US Army Reserve at Columbus, Ohio; Little Rock Arkansas; Boston, Massachusetts; Oak Park, Illinois; Jacksonville, Florida; Boise, Idaho; and Fort DeRussy, Hawaii.

The 16th Special Forces Group was organized and activated from personnel assigned to the 170th Special Forces Detachment and the 403d Ordnance Detachment.

The 19th and 20th Special Forces Groups were activated in the ARNG at Salt Lake City, Utah and Homewood, Alabama.

Special Forces Operational Detachments 306, 307, 308, 309 and 316 were inactivated respectively at Stillwater and Tulsa, Oklahoma; Little Rock, Arkansas; Albuquerque, New Mexico; and Columbus, Ohio. Special Forces Operational Detachments 320, 321, 322 and 329 were respectively inactivated at Boston, Massachusetts; Decatur, Alabama; Jacksonville, Florida; and Fort DeRussey, Hawaii.

Civil Affairs

The 97th Military Government Group was redesignated as the 97th Civil Affairs

Group; concurrently allotted to the Regular Army from the AUS; activated on Okinawa and assigned to the US Army Pacific and attached to the 1st Special Forces Group.

Combat Service Support

Headquarters Company, 1st Logistical Command disbanded at Fort Bragg, North Carolina. Headquarters, 1st Logistical Command reorganized and redesignated as HHC, 1st Logistical Command.

—— 1962 ——

Command of the 82d Airborne Division was given to Major General John L, Throckmorton. His son Edward (Russ) Throckmorton would later serve as a Lieutenant with the 101st Airborne Division in Vietnam, and retire as a Lieutenant Colonel at Fort Bragg in 1988.

The 82d and 101st Airborne Divisions were deployed to the University of Mississippi to assist in civil disorder activities.

The Howze Board, named after the Commanding General, XVIII Airborne Corps commenced to examine in earnest the feasibility of airmobile operations, conducting a series of fast deploying, rapid maneuvering airmobile-concept exercises.

Major General Charles J. Timmes assumed command of the Military Assistance Advisory Group, Vietnam (MAAG-Vietnam) in July. He was the last MAAG-Vietnam commander, as the Military Assistance Command, Vietnam (MACV), had arrived in Saigon on 8 February, commanded by General Paul Harkins. MAAG-Vietnam was the redesignated MAAG-Indochina, being the proponent for providing advisor support for the South Vietnamese Army. MAAG-Indochina had been active from 17 September 1950 until 31 October 1955, and redesignated MAAG-Vietnam on 1 November 1955.

Combat Arms
Infantry

HHCs, 1st and 2d Infantry Brigades were inactivated respectively at Fort Benning, Georgia and Fort Devens, Massachusetts.

The 397th, 398th and 399th (Regiments), Basic Combat Training, were released from Active Military Service.

The 508th Airborne Infantry Regiment was reorganized and redesignated as the 508th Infantry Regiment, a parent regiment under the Combat Arms Regimental System. Companies A, B and C, 508th Airborne Infantry Regiment, were reorganized and redesignated as the 1st, 2d and 3d Battalions, 508th Infantry. The 3d Battalion, 508th Airborne Infantry was activated at Fort Kobbe, Panama Canal Zone on 8 August.

Cavalry

Troop C, 17th Cavalry, was activated at Fort Knox, Kentucky.

Field Artillery

HHB, 3d Howitzer Battalion, 4th Artillery, inactivated at Fort Devens, Massachusetts, and relieved from assignment to the 2d Infantry Brigade.

The 2d Target Acquisition Battalion, 26th Artillery was redesignated as the 2d Battalion, 26th Artillery.

Batteries A and B, 320th Artillery were redesignated as HHB, 1st and 2d Battalions, 320th Artillery, activated respectively at Fort Bragg, North Carolina and Fort Campbell, Kentucky, and assigned to the 82d Airborne Division and the 101st Airborne Division. Battery C, 320th Artillery was relieved from assignment to the 82d Airborne Division and inactivated at Fort Bragg.

Special Operations
Special Forces

US Army Special Forces, Vietnam (Provisional), commanded by Colonel George Morton, was established to provide the interim headquarters for US Army Special Forces units and personnel assigned temporarily to the Republic of Vietnam.

The Special Forces Training Group was constituted, activated at Fort Bragg, North Carolina, and assigned to the US Army Special Warfare School.

Civil Affairs

The 28th Military Government Company was redesignated as the 28th Civil Affairs Company and activated in the Federal Republic of Germany.

Aviation

The 82d Aviation Company was redesignated as the 82d Aviation Battalion. Companies A and B, 82d Aviation Battalion were activated at Fort Bragg, North Carolina, and assigned to the 82d Airborne Division.

The 23d Aviation Detachment was constituted, activated at Fort Bragg, North Carolina, and assigned to the US Army Special Warfare School.

Combat Support
Military Intelligence

Company A, 313th Army Security Agency Battalion was activated at Fort Bragg, North Carolina.

The 77th Counter Intelligence Corps Detachment was reorganized and redesignated as the 77th Intelligence Corps Detachment.

The 297th Interpreter Detachment was reorganized and redesignated as the 297th Military Intelligence Detachment, assigned to Sixth Army, and activated at Santa Monica, California.

The 389th Translator Detachment was reorganized and redesignated as the 389th Military Intelligence Detachment.

Detachment 1, 80th US Army Security Agency Special Operations Unit (80th USASASOU), was redesignated as the 10th Radio Research Unit. Detachment 2, USASASOU was redesignated as the 12th Radio Research Unit. Detachment 3, USASASOU was redesignated as the 11th Radio Research Unit, and was deployed to Fort Gulick, Panama. Detachment 4, USASASOU was redesignated as the 13th Radio Research Unit.

The 5th Counterintelligence Corps Detachment, activated in August 1944, was redesignated as the 5th Military Intelligence Detachment, (5th MID). The 5th MID was activated at Fort Carson, Colorado, and assigned to the 5th Infantry Division.

Military Police

The 503d Military Police Battalion were the first Federal Troops deployed to Oxford, Mississippi and the campus of the University of Mississippi, as part of the peace-keeping task force during civil disturbances.

Combat Service Support

HHC, 1st Logistical Command was reorganized and redesignated as HHD, 1st Logistical Command.

—— 1963 ——

Lieutenant General William C. Westmoreland assumed command of XVIII Corps (Airborne).

The 11th Airborne Division was reorganized and redesignated as the 11th Air Assault Division, and activated at Fort Benning, Georgia. This opened a new chapter in the annals of airborne warfare. The concepts developed by the 11th Air Assault Division would prove to become the initial doctrine for airmobile operations, executed magnificently by the 1st Cavalry Division (Airmobile) when they deployed in 1965 to the Republic of Vietnam. HHC, Command and Control Battalion, 11th Airborne Division was reorganized and redesignated as HHC, 1st Brigade, 11th Air Assault Division. HHCs, 2d and 3d Brigades, 11th Air Assault Division were constituted in the Regular Army and activated at Fort Benning.

HQ, 173d Brigade, was constituted in the National Army in 1917, assigned to the 87th Division, and Organized at Camp Pike, Arkansas. Demobilized in 1919, following WW I, HQ, 173d Infantry Brigade was reconstituted in the Organized Reserves in 1921 as HHC, 173d Infantry Brigade, remaining assigned to the 87th Division, and organized at Mobile, Alabama. From 1925 until 1936, HHC, 173d Infantry Brigade was redesignated as HHC, 173d Brigade, until reverting back to being an infantry brigade. In 1942, HHC, 173d Infantry Brigade was converted and redesignated as the 87th Reconnaissance Troop (less the 3d Platoon), and ordered into Active Military Service at Camp McCain, Mississippi. The 87th Reconnaissance Troop was redesignated one more time during 1942 - as the 87th Cavalry Reconnaissance Troop, and in 1943 was reorganized and redesignated as the 87th Reconnaissance Troop, Mechanized until inactivated in 1945. Redesignated again in 1947, the 87th Mechanized Cavalry Reconnaissance Troop was activated at Birmingham, Alabama, redesignated as the 87th Reconnaissance Company in 1949, until inactivated in 1951. In 1963, the 87th Reconnaissance Company was converted and redesignated (less the 3d Platoon), as HHC, 173d Airborne Brigade, relieved from assignment to the 87th Division and concurrently withdrawn from the Army Reserve and allotted to the Regular Army, and activated on Okinawa.

Combat Arms
Infantry

HQ Troop, 8th Division, was constituted in the National Army in 1917, and organized at Camp Fremont, California in 1918. Demobilized after WW I at Camp Dix, New Jersey, HQ Troop was reconstituted in the Regular Army in 1923 as HQ and MP Company (less the MP Platoon), 8th Division, but was not activated until 1940 at Camp Jackson, South Carolina. In 1942, HQ and MP Company was assigned to the reorganized and redesignated 8th Motorized Division, until becoming HQ Company, 8th Infantry Division in 1943. HQ Company, 8th Infantry Division was inactivated at Fort Leonard Wood, Missouri in 1919, following WWII, and activated at Fort Jackson, South Carolina from 1950 until disbanded in 1960. HQ Company was reconstituted in the Regular Army in 1963 as HHC, 1st Brigade, 8th Infantry Division, activated in Germany, and redesignated HHC, 1st Brigade (Airborne), 8th Infantry Division.

HHCs, 1st and 2d Infantry Brigades, (inactivated in 1962), were reorganized and redesignated as HHCs, 1st and 2d Brigades, 1st Infantry Division.

The 1st Airborne Battle Group, 187th Infantry was inactivated at Fort Bragg, North Carolina. The 3d Airborne Battle Group, 187th Infantry was redesignated as the 3d Battalion, 187th Infantry, assigned to the 11th Air Assault Division, and activated at Fort Benning, Georgia.

The 188th Airborne Infantry Regiment was reorganized and redesignated as the 188th Infantry Regiment, a parent regiment under the Combat Arms Regimental System (CARS). Companies A, B and C were redesignated as the 1st, 2d and 3d Battalions, 188th Infantry, activated and assigned to the 11th Air Assault Division at Fort Benning, Georgia. The 4th Battalion, 188th Airborne Infantry, was redesignated as the 4th Battalion, 188th Infantry, and also assigned to the 11th Air Assault Division.

The 2d Airborne Battle Group, 502d Infantry, was relieved from the 11th Airborne Division, and assigned to the 11th Air Assault Division at Fort Benning, Georgia.

The 1st Airborne Battle Group, 503d Infantry, was relieved from assignment to the 82d Airborne Division, redesignated as the 1st Battalion, 503d Infantry, and assigned to the 173d Airborne Brigade.

The 1st Airborne Battle Groups, 504th and 505th Infantry, were released from the 8th Infantry Division, and assigned to the 82d Airborne Division.

The 509th Parachute Infantry Battalion was reorganized and redesignated as the 509th Infantry Regiment, a parent regiment under the Combat Arms Regimental System. Companies A, B, and C were reorganized and redesignated as the 1st, 2d and 3d Battalions, 509th Infantry. The 1st and 2d Battalions, 509th Infantry were assigned to the 1st Brigade, 8th Infantry Division, and activated in Germany on 1 April. The 1st Brigade, 8th Infantry Division functioned essentially as an airborne brigade within a mechanized infantry division, assuming the airborne functions and responsibilities that the 24th Infantry Division had adopted from the 11th Airborne Division in 1958.

The 511th Airborne Infantry was reorganized and redesignated as the 511th Infantry Regiment, a parent regiment under the CARS. Company A was reorganized and redesignated as the 1st Battalion, 511th Infantry, assigned to the 11th Air Assault Division, and activated at Fort Benning, Georgia.

Cavalry

Companies A and B, 8th Cavalry were constituted in the Regular Army on 28 July 1866, and organized at the Presidio of San Francisco, California. (Cavalry companies became designated as Troops in 1883). The 8th Cavalry was assigned in 1917 to the 15th Cavalry Division; relieved from that assignment in 1918 and assigned to the 1st Cavalry Division in 1921. The 8th Cavalry dismounted in 1943 and was reorganized under both Cavalry and Infantry Tables of Organization and Equipment (TOE), and reorganized as Infantry on 20 July 1945, while retaining Cavalry designations. In 1949, Troops A and B, 8th Cavalry were redesignated as Companies A and B, 8th Cavalry. Company B, 8th Cavalry was consolidated in 1957 with the 4th Reconnaissance Company, with the consolidated unit designated as HHT, 2d Reconnaissance Squadron, 8th Cavalry, an element of the 4th Infantry Division. The 8th Cavalry was relieved from assignment to the 1st Cavalry Division in 1957. Company A was reorganized and redesignated as HHC, 1st Battle Group, 8th Cavalry and assigned to the 1st Cavalry Division. The 1st Battle Group, 8th Cavalry was redesignated on 1 September 1963 as the 1st Battalion, 8th Cavalry. HHT, 2d Reconnaissance Squadron, 8th Cavalry, was reorganized and redesignated, (also on 1 September), as the 2d Battalion, 8th Cavalry, and relieved from assignment from the 4th Infantry Division. Both the 1st and 2d Battalions, 8th Cavalry, were assigned to the 1st Brigade, 1st Cavalry Division.

Troop A, 12th Cavalry was constituted in the Regular Army in 1901 and organized at Fort Hood, Texas. The 12th Cavalry was assigned to the 2d Cavalry Division from 1923 until 1933, and was reassigned to the 1st Cavalry Division, and dismounted in 1943, being reorganized under a combination of Cavalry and Infantry Tables of Organization and Equipment (TOE), and reorganized wholly as Infantry in 1945, while retaining Cavalry designations. Troop A, 12th

Cavalry was inactivated in 1949 in Japan, and relieved from assignment to the 1st Cavalry Division. In 1957, Troop A, 12th Cavalry was consolidated with HHSC, 81st Reconnaissance Battalion (active), with the consolidated unit concurrently redesignated as HHT, 1st Reconnaissance Squadron, 12th Cavalry. HHT, 1st Reconnaissance Squadron was assigned to the 1st Armored Division in 1957, and reorganized at Fort Polk, Louisiana (Companies A, B, C and D, 81st Reconnaissance Battalion were concurrently redesignated as Troops A, B, C and D, 1st Reconnaissance Squadron, 12th Cavalry, with Troops C and D being inactivated later that same year). HHT, 1st Reconnaissance Squadron was reorganized and redesignated in 1959 as Troop A, 12th Cavalry, (with Troops A and B, 1st Reconnaissance Squadron, 12th Cavalry being inactivated). Troop A, 12th Cavalry was inactivated in 1962 at Fort Hood, Texas and relieved from assignment to the 1st Armored Division. In 1963, Troop A, 12th Cavalry was reorganized and redesignated as HHC, 1st Battalion, 12th Cavalry, and assigned to the 1st Cavalry Division, (Troops A, B and C, 1st Reconnaissance Squadron were redesignated as Companies A, B and C, 1st Battalion, 12th Cavalry). The 1st Battalion, 12th Cavalry was activated in Korea.

Troop B, 3d Squadron, 17th Cavalry was activated at Fort Rucker, Alabama. Troop C, 17th Cavalry was inactivated at Fort Knox, Kentucky, and redesignated as HHT, 3d Squadron, 17th Cavalry, and assigned to the 11th Air Assault Division, with organic elements concurrently constituted. Troop E, 17th Cavalry was assigned to the 173d Airborne Brigade, and activated on Okinawa.

Division Artillery

HHB, 11th Airborne Division Artillery reorganized and redesignated as HHB, 11th Air Assault Division Artillery, and activated at Fort Benning, Georgia.

Field Artillery

Battery C, 319th Artillery, was reorganized and redesignated as HHB, 3d Battalion, 319th Artillery, and assigned to the 173d Airborne Brigade.

Special Operations

Colonel Theodore Leonard assumed command of the HQ, Special Forces Vietnam (Provisional) in November, from Colonel George C. Morton.

The 3d, 6th and 8th Special Forces Groups were activated at Fort Bragg, North Carolina, and Fort Gulick, Panama Canal Zone.

Headquarters, 11th Special Forces Group was relocated from Boston, Massachusetts to Staten Island, New York.

The 13th Special Forces Group was inactivated at Jacksonville, Florida.

The 21st Special Forces Group was inactivated at New Orleans, Louisiana.

Rangers

The US Army's VII Corps Long Range Reconnaissance Patrol (LRRP) Company was constituted and activated in the Federal Republic of Germany.

Psychological Operations

Activated at Fort Gulick, Panama Canal Zone, the 9th Loudspeaker and Leaflet Company was redesignated as the 9th Psychological Warfare Company.

Civil Affairs

The 28th Civil Affairs Company was inactivated in the Federal Republic of Germany.

Aviation

HHC, 10th Air Transportation Brigade; the HHC, 11th Air Assault Aviation Group; the 226th Surveillance and Attack Battalion; and the 11th Aviation Company were constituted in the Regular Army, all assigned to the 11th Air Assault Division, and activated at Fort Benning, Georgia.

Medical

The 111th Medical Company redesignated as HHC, 11th Medical Battalion, and assigned to the 11th Air Assault Division. The 11th Medical Battalion, less Companies B and C, were activated at Fort Benning, Georgia.

Combat Support
Engineers

The 127th Engineer Battalion (Airborne Division), inactivated in Germany in 1958, was activated at Fort Benning, Georgia, and assigned to the 11th Air Assault Division.

The 173d Engineer Company was constituted in the Regular Army, activated on Okinawa, and assigned to the 173d Airborne Brigade.

Military Intelligence

The 77th Intelligence Corps Detachment was assigned to Fifth Army.

The 297th Military Intelligence Detachment was reorganized and redesignated as the 297th Military Intelligence Detachment (Special Forces Group), and attached to the 19th Special Forces Group.

The 389th Military Intelligence Detachment was assigned to the 11th Special Forces Group, and activated at Louisville, Kentucky.

The 10th Radio Research Unit was redesignated as the 400th Army Security Agency Detachment. The 11th Radio Research Unit was reassigned from the 7th Special Forces Group to the 8th Special Forces Group.

Military Police

As the Combat Arms elements of the US Army continued to reorganize and redesignate, the 503d Military Police Battalion maintained it's vigil in Mississippi, being ordered to Birmingham in May.

The 11th Airborne Military Police Company was assigned to the 11th Air Assault Division.

Signal

HHD, 511th Signal Battalion was redesignated as HHC, 511th Signal Battalion, and assigned to the 11th Air Assault Division.

Combat Service Support
Chemical

The 445th Chemical Detachment, constituted in the USAR in 1962 and assigned to the Sixth Army, was activated in Seattle, Washington. The 808th Chemical Detachment was assigned to First Army and activated in the USAR at New York City, New York.

Maintenance

The 711th Maintenance Company was activated at Fort Bragg, North Carolina, and assigned to the 11th Air Assault Division.

Quartermaster

The 408th Supply and Transport Company, (inactivated in 1958), was redesignated as the 408th Supply and Service Company, and activated at Fort Bragg, North Carolina.

Transportation

The 5th Transportation Corps Battalion was inactivated on Okinawa.

The 170th Transportation Battalion was constituted in the Regular Army, activated at Fort Benning, Georgia, and assigned to the 11th Air Assault Division.

1964

General William C. Westmoreland assumed command of the Military Assistance Command, Vietnam, from General Paul Harkins in June, having served as Deputy Commander since January. In August Lieutenant John L. Throckmorton assumed the duties as Deputy Commander.

Lieutenant General John C. Bowen assumed command of XVIII Corps (Airborne). The reins of the 82d Airborne Division were passed to Major General Robert H. York.

Both the 82d and 101st Airborne Divisions were converted from the Pentomic Organizational Concept of Battle Groups, to the Re-Organization Army Division (ROAD), concept of Battalions. HHC, Command and Control Battalions, 82d and 101st Airborne Divisions were redesignated as HHC, 1st Brigades, 82d and 101st Airborne Divisions.

HQ, 155th and 159th Infantry Brigades, were constituted in the National Army in 1917, respectively assigned and organized to the 78th and 80th Divisions at Camp Dix, New Jersey and Camp Lee, Virginia. They were both demobilized following WWI, in 1919, again at Camps Dix and Lee, and reconstituted in the Organized Reserves as HHCs, 155th and 159th Infantry Brigades. They were redesignated as HHC, 155th and 159th Brigades in 1925, returning to the HHC, 155th and 159th Infantry Brigades designation in 1936. In 1942, they were converted to be the 78th and 80th Reconnaissance Troops (less their respective 3d Platoons), ordered to Active Military Service, at Camp Butner, North Carolina and Camp Forrest, Tennessee. They were reorganized and redesignated as the 78th and 80th Cavalry Reconnaissance Troops, and assigned respectively to the 78th and 80th Divisions. In 1943 they became the 78th and 80th Reconnaissance Troops, Mechanized until 1945 when 78th Reconnaissance Troop, Mechanized was redesignated as the 78th Mechanized Cavalry Reconnaissance Troop, until inactivation in Germany in 1956. The 80th Reconnaissance Troop was reorganized and redesignated as the Reconnaissance Platoon, of the recently reorganized and redesignated 80th Airborne Division. In 1947, the Reconnaissance Platoon was redesignated as the 80th Airborne Reconnaissance Platoon, and the 78th Mechanized Cavalry Reconnaissance Troop was allotted to the Organized Reserves, assigned to the 79th Infantry Division, activated at Plainfield, and relocated to Newark, New Jersey. The 80th Airborne Reconnaissance Platoon was redesignated as the Reconnaissance Platoon, 80th Airborne Division in 1948, and in 1949, the 78th Mechanized Cavalry Reconnaissance Troop was redesignated as the 78th Reconnaissance Company. In 1950, the Reconnaissance Platoon, 80th Airborne Division was reorganized and redesignated as the 80th Airborne Reconnaissance Company, becoming the 80th Reconnaissance Company in 1952 as the 80th Airborne Division was reorganized and redesignated to be the 80th Infantry Division, until disbanded in 1959. In 1951, the 78th Reconnaissance Company had relocated to Irvington, New Jersey, where it was also disbanded in 1959. Following disbandment, the 78th and 80th Reconnaissance Companies were reconstituted as HHC, 155th and 159th Infantry Brigades (less 3d Platoons), and in 1964, were respectively reorganized, redesignated, and activated at Fort Bragg, North Carolina and Fort Campbell, Kentucky, as HHC, 2d Brigades, 82d and 101st Airborne Divisions.

HQ, 156th and 160th Infantry Brigades were constituted in the National Army in 1917, assigned respectively to the 78th and 80th Divisions, and organized at Camp Dix, New Jersey and Camp Lee, Virginia. Following WW I, they were demobilized respectively at Camps Dix and Lee. In 1921 they were reconstituted in the Organized Reserves as HHCs, 156th and 160th Infantry Brigades, assigned to the 78th and 80th Divisions, and organized at Newark, New Jersey and Baltimore, Maryland. In 1925 they were redesignated as HHC, 156th and 160th Brigades until redesignated as HHC, 156th and 160th Infantry Brigades in 1936. In 1942 they were reorganized and redesignated as the 3d Platoons, 78th and 80th Reconnaissance Troops, and ordered into Active Military Service at Camp Butner, North Carolina and Camp Forrest, Tennessee, where they were reorganized into the 78th and 80th Reconnaissance Troops. They were reorganized and redesignated as the 78th and 80th Reconnaissance Troops, Mechanized in 1943, and prior to both being inactivated in 1946, the 78th Reconnaissance Troop, Mechanized was redesignated as the 78th Mechanized Cavalry Reconnaissance Troop in 1945. In 1947 the 78th Mechanized Cavalry Reconnaissance Troop was allotted to the Organized Reserves, assigned to the 78th Infantry Division, and activated in Plainfield New Jersey, moving to Newark, New Jersey. Reorganized as the 78th Reconnaissance Company in 1949, they relocated to Irvington, New Jersey, and inactivation in 1953. The 80th Reconnaissance Troop, Mechanized had been inactivated at Camp Kilmer, New Jersey in 1946 and reorganized and redesignated as the Reconnaissance Platoon, 80th Airborne Division. Activated in 1947, the Reconnaissance Platoon was redesignated as the 80th Airborne Reconnaissance Platoon, and assigned to the 80th Airborne Division in 1948. In 1950, the 80th Airborne Reconnaissance Platoon was redesignated back to being the Reconnaissance Platoon, 80th Airborne Division, but eventually returning to the designation of the 80th Airborne Reconnaissance Platoon in 1952, concurrent with the reorganization and redesignating of the 80th Infantry Division from the 80th Airborne Division. In 1959, both the 78th and the 80th Reconnaissance Companies were disbanded. In 1963, the 3d Platoons, 78th and 80th Reconnaissance Companies, were reconstituted in the Regular Army, being redesignated and activated in 1964, as the HHCs, 3d Brigade, 82d and 101st Airborne Divisions.

During airborne exercises with the Army of the Republic of Nationalist China, the 173d Airborne Brigade was given the local nickname of 'Tien Bing', which literally translates to 'Sky Soldier'.

Combat Arms
Infantry

HHC, 1st Brigade, 1st Infantry Division, was activated at Fort Riley, Kansas.

The 1st Airborne Battle Group, 187th Infantry, was inactivated at Fort Bragg, North Carolina on 25 May, and consolidated with the 1st Battalion, 187th Infantry (activated at Fort Benning, Georgia in February), with the consolidated unit designated as the 1st Battalion, 187th Infantry, and activated at Fort Benning on 1 March 1964. Assigned to the 11th Air Assault Division, Fort Benning, Georgia, the 2d Airborne Battle Group, 187th Infantry was relieved from assignment to the 101st Airborne Division, and inactivated in February at Fort Campbell, Kentucky. The 3d Battalion, 187th Infantry, was relieved from assignment to the 11th Air Assault Division, and assigned to the 101st Airborne Division.

The 1st, 2d and 3d Airborne Battle Groups, 325th Infantry were reorganized and redesignated as the 1st, 2d and 3d Battalions, 325th Infantry. The 2d and 3d Battalions, 325th Infantry were activated at Fort Bragg, North Carolina, and assigned to the 82d Airborne Division. The 1st Battalion continued to be assigned to the 82d Airborne Division.

The 1st and 2d Airborne Battle Groups, 327th Infantry were reorganized and redesignated as HHC, 1st and 2d Battalions, 327th Infantry, with the 2d Battalion being reassigned to the 101st Airborne Division.

The 1st and 2d Airborne Battle Groups, 501st, 502d, 504th and 505th Infantry were reorganized and redesignated as the 1st and 2d Battalions, 501st, 502d, 504th and 505th Infantry.

The 2d Battalion, 501st Infantry was relieved from assignment to the 82d Airborne Division, and assigned to the 101st Airborne Division.

The 2d Battalion, 502d Infantry was relieved from assignment to the 11th Airborne Division, activated at Fort Campbell, Kentucky, and assigned to the 101st Airborne Division.

The 1st and 2d Battalions, 504th Infantry and the 1st, Battalion, 505th Infantry remained assigned to the 82d Airborne Division.

The 2d Battalion, 505th Infantry was relieved from the 11th Air Assault Division and assigned to the 82d Airborne Division.

The 2d Airborne Battle Group, 506th Infantry was reorganized and redesignated as the 2d Battalion, 506th Infantry, assigned to the 101st Airborne Division, and activated at Fort Campbell, Kentucky. The 1st and 2d Battalions, 508th Infantry were activated at Fort Bragg and assigned to the 82d Airborne Division.

Cavalry

Troops A and B, 17th Cavalry were reorganized and redesignated as HHTs, 1st and 2d Squadrons, 17th Cavalry (with organic elements constituted and activated). The 3d Squadron, 17th Cavalry (less Troop B), was activated at Fort Benning, Georgia.

Field Artillery

HHB, 1st Missile Battalion, 39th Artillery, was inactivated in Germany.

Batteries A and B, 319th Artillery, were reorganized and redesignated as HHSBs, 1st and 2d Battalions, 319th Artillery. The 2d Battalion, 319th Artillery was relieved from assignment to the 82d Airborne Division, and assigned to the 101st Airborne Division.

Batteries A and B, 321st Artillery were reorganized and redesignated as HHSBs, 1st and 2d Battalions, 321st Artillery. The 2d Battalion, 321st Artillery was relieved from assignment to the 101st Airborne Division, and assigned to the 82d Airborne Division. Battery C, 321st Artillery was relieved from the 101st Airborne Division, and inactivated at Fort Campbell, Kentucky.

Special Operations

The Military Assistance Command Vietnam, Studies and Observations Group (MACV-SOG), was constituted and manned with personnel assigned to the Special Operations community in South East Asia. This classified, joint command was originally activated to assist the Central Intelligence Agency to advise the Republic of Vietnam Special Exploitation Service. Missions included highly classified Special Operations, psychological operations, sabotage and many other diversified classified missions. MACV-SOG was activated in Cholon, Vietnam.

The US Army Special Warfare School was redesignated as the US Army Center for Special Warfare at Fort Bragg, North Carolina.

Special Forces

The 5th Special Forces Group deployed from Fort Bragg, North Carolina to Nha-Trang, Republic of Vietnam, under the command of Colonel John H. Spears, replacing the US Army Special Forces, Vietnam (Provisional). The HQ, 12th Special Forces Group was relocated from Chicago to Oak Park, Illinois. The 36th, 37th and 38th Special Forces Detachments were constituted and activated in the Alaska ARNG, and activated at Anchorage, Alaska.

Operational Detachment A-503, 5th Special Forces Group organized the 5th Mobile Strike Force (Mike Force), in Saigon, comprised of ethnic minorities trained in airborne operations. The 5th Mike Force functioned as a reaction force to endangered Special Forces camps and Civilian Irregular Defense Group (CIDG) camps throughout Vietnam. Operational Detachment B-55 was assigned to provide command and control of the 5th Mike Force.

Captain Roger H. C. Donlon, commanding Special Forces Detachment A-726, was wounded four times during the five hour long battle at Nam Dong, where his Special Forces camp was taken under siege. He was awarded the Medal of Honor for his actions on 6 July.

Operational Detachment B-56, 5th Special Forces Group organized Project LEAPING LENA, training CIDG personnel in Long Range Reconnaissance Patrol techniques. ODB-56 was commanded by Captain William J. Richardson, assisted by Sergeant Major Paul Payne.

Operational Detachment B-52, 5th Special Forces Group, activated the Military Assistance Command Vietnam RECONDO School.

Aviation

HHC and Companies A and B, 229th Assault Helicopter Battalion were constituted in the Regular Army, assigned to the 11th Air Assault Division, and activated at Fort Benning, Georgia.

Medical

The 82d Medical Company was reorganized and redesignated as Headquarters and Company A, 307th Medical Battalion, (the remaining organic elements concurrently constituted and activated).

The 326th Medical Company was reorganized and redesignated as Headquarters and Company A, 326th Medical Battalion, (Companies B, C and D concurrently constituted and activated).

Combat Support
Military Intelligence

The 525th Military Intelligence Group relocated from Fort George G. Meade, Maryland to Fort Bragg, North Carolina.

The 11th, 12th and 13th Radio Research Units were inactivated respectively at Fort Gulick, Panama; Bad Toelz, German; and Fort Bragg, North Carolina. Reorganized and redesignated as the 401st, 402d and 403d Army Security Agency Detachments. The 401st and 402d ASA Detachments were activated and reassigned back to the 1st and 10th and 5th Special Forces Groups at Fort Gulick, Panama and Bad Toelz, Germany. The 403d ASA Detachment was reassigned from the 5th to the 7th Special Forces Group at Fort Bragg, North Carolina.

Military Police

The 82d and 101st Airborne Military Police Companies were redesignated as the 82d and 101st Military Police Companies, assigned organic elements of the 82d and 101st Airborne Division, and activated at Fort Bragg, North Carolina, and Fort Campbell, Kentucky.

Signal

HHC, 511th Signal Battalion was activated at Fort Benning, Georgia.

Combat Service Support
Quartermaster

HHDs, Support Groups, 82d and 101st Airborne Divisions, consolidated respectively with the 82d and 101st Airborne Division Bands, then reorganized and redesignated as HH and Band, 82d Airborne Division Support Command, and HHC and Band, 101st Airborne Division Support Command.

The 170th Transportation Battalion was reorganized and redesignated as the 170th Aircraft Maintenance Battalion.

The 407th and 426th Supply and Transportation Companies, were reorganized and redesignated as the 407th and 426th Supply and Transportation Battalions. The 408th Supply and Service Company was reorganized and redesignated as HHC, 408th Supply and Service Battalion (with organic elements concurrently activated).

The 82d Quartermaster Parachute Supply and Maintenance Company was reorganized and redesignated as Company B, 407th Supply and Transportation Battalion.

Companies A and B, 426th Supply and Transportation Battalion were organized and activated at Fort Campbell, Kentucky.

The 101st Quartermaster Parachute Supply and Maintenance Company was reorganized and redesignated as Company B, 426th Supply and Transportation Battalion.

1965

HHC, XVIII Corps (Airborne) was redesignated as HQ, XVIII Airborne Corps. Lieutenant General Bruce Palmer assumed command of XVIII Airborne Corps.

General William C. Westmoreland assumed command of the US Army Vietnam (USARV), while concurrently remaining assigned as Commander, Military Assistance Command, Vietnam (MACV). General Creighton W. Abrams was assigned as his deputy.

The 11th Air Assault Division was inactivated at Fort Benning, Georgia.

Major General Joe S. Lawrie was appointed to command the 82d Airborne Division on 2 August.

The 1st Brigade, 101st Airborne Division, commanded by Colonel Joseph D. Mitchell, and the 173d Airborne Brigade (Separate), commanded by Brigadier General Ellis W. Williamson, were amongst the early contingent of the US Army to deploy to the Republic of Vietnam in South East Asia. Brigadier General James S. Timothy replaced Colonel Mitchell as the 1st Brigade, 101st Airborne Division commander during August and September, until Brigadier General Willard Pearson assumed command.

Also deployed to Vietnam in September in an airborne status was the 1st Brigade, 1st Cavalry Division (Airmobile). As the 11th Air Assault Division was inactivated, it's units were reassigned to the 1st Cavalry Division (the First Team), which was configured to assume the airmobile role, developed by the 11th Air Assault Division. In addition to the airmobile role of the Division, the 1st Brigade was designed to enhance that role with an airborne capability. The 1st Cavalry Division (Airmobile) deployed under the command of Major General Harry W. B. Kinnard. (See 1965, Combat Arms, Cavalry).

The 3d Brigade, 82d Airborne Division led the initial deployment of the 82d Airborne Division to the Dominican Republic, as Operation POWER PACK.

The 1st Battalion, Royal Australian Regiment, with supporting artillery, armored personnel carriers, light aircraft and engineers, deployed from Australia, and was attached to the 173d Airborne Brigade in the Republic of Vietnam.

Three Medals of Honor were awarded for actions above and beyond the call of duty this year by three members of the 173d Airborne Brigade: Sergeant Larry Pierce for his actions on 20 September; Private First Class Milton L. Olive (posthumously), for his actions of 22 October; and Specialist Fifth Class Lawrence Joel, for his deeds of 8 November.

Combat Arms

The 11th Air Assault Division was inactivated at Fort Benning, Georgia. The concepts developed by the 11th Air Assault Division were noted and taken with the 1st Cavalry Division, (Airmobile), for practical employment in the Republic of Vietnam. The colors of the 1st Cavalry Division had been transferred to Fort Benning for the activation ceremony, and personnel formerly assigned to the 11th Air Assault Division were assigned to the concurrently activated 1st Cavalry Division (Airmobile). The 1st Brigade, 1st Cavalry Division (Airmobile) was initially configured as the airborne brigade.

Infantry

1st Battalion, 187th Infantry and the 1st, 2d, 3d and 4th Battalions, 188th Infantry, and the 1st Battalion, 511th Infantry, were all relieved from assignment to the 11th Air Assault Division, and were inactivated at Fort Benning, Georgia.

The 1st, 2d, 3d and 4th Battalions, 188th Infantry were redesignated as Companies A, B, C and E, 188th Infantry.

The 1st and 2d Battalions, 327th Infantry and the 2d Battalion, 502d Infantry, deployed with the 1st Brigade, 101st Airborne Division, to the Republic of Vietnam in July.

Personnel assigned to the Long Range Reconnaissance Patrol unit of the 1st Brigade, 101st Airborne Division, attended the MACV-RECONDO School provided by B-52, 5th Special Forces Group, in Nha-Trang.

Deploying to the Dominican Republic on Operation POWER PACK, from the 82d Airborne Division, were the 1st, 2d and 3d Battalions, 325th Infantry; the 1st and 2d Battalions, 504th Infantry; the 1st and 2d Battalions, 505th Infantry; and the 1st and 2d Battalions, 508th Infantry. The ground forces of the 82d Airborne Division were spearheaded by the 1st Battalion, 508th Infantry.

The 1st and 2d Battalions, 503d Infantry, were assigned to and deployed with the 173d Airborne Brigade as two of the first US ground elements, arriving in the Republic of Vietnam on 31 May.

Armor

Company D, 16th Armor deployed with the 173d Airborne Brigade from Okinawa to the Republic of Vietnam. Company D fulfilled an airborne anti-tank reconnaissance role with the 173d Airborne Brigade.

Cavalry

HQ, 1st Cavalry Brigade was constituted in the Regular Army in 1917, organized at Fort Sam Houston in 1918, where it was assigned to, and relieved from assignment to, the 15th Cavalry Division. In 1919, the 1st Cavalry Brigade was demobilized at Brownsville, Texas. Reconstituted in the Regular Army in 1921, HHT, 1st Cavalry Brigade was assigned to the 1st Cavalry Division, and organized at Camp Harry J. Jones, Arizona. Following WW II, in 1949, HHT, 1st Cavalry Brigade was relieved from the 1st Cavalry Division Special, inactivated in Japan, converted and redesignated as HQ, 1st Constabulary Brigade, US Constabulary, and activated in Germany until relieved the following year, 1950, and inactivation in 1951. In 1963, HHT, 1st Cavalry Brigade was reconstituted in the Regular Army as HHC, 1st Brigade, 1st Cavalry Division, and activated in Korea. Deploying with the 1st Cavalry Division to Vietnam in 1965, HHC, 1st Brigade, 1st Cavalry Division, was was redesignated as HHC, 1st Brigade (Airborne), 1st Cavalry Division.

The 1st and 2d Battalions, 8th Cavalry, and the 1st Battalion, 12th Cavalry deployed with the 1st Brigade, 1st Cavalry Division to An Khe, Republic of Vietnam.

The 1st Reconnaissance Squadron, 17th Cavalry, deployed in support of the 82d Airborne Division to the Dominican Republic, on Operation POWER PACK. The 1st Squadron provided reconnaissance and mobile fire power support during this crisis deployment. The 3d Squadron, 17th Cavalry was relieved from assignment to the 11th Air Assault Division, and inactivated at Fort Benning, Georgia.

Division Artillery

HHB, 11th Air Assault Division Artillery, was inactivated at Fort Benning, Georgia.

The 82d Airborne Division Artillery supported the 82d Airborne Division, deploying to the Dominican Republic as part of Operation POWER PACK.

Field Artillery

Deployed to the Dominican Republic were the 1st Battalion, 319th Artillery; 1st Battalion, 320th Artillery; and the 2d Battalion, 321st Artillery.

The 3d Battalion, 319th Artillery and the 2d Battalion, 320th Artillery deployed to the Republic of Vietnam, assigned respectively to 173d Airborne Brigade, and the 1st Brigade, 101st Airborne Division.

Special Operations

Following Brigadier General William P. Yarborough, command of the US Army Center for Special Warfare, was passed to Brigadier General Joseph W. Stillwell.

Special Forces

Colonel William A. McKean assumed command of the 5th Special Forces Group.

Second Lieutenant Charles Q. Williams was posthumously awarded the Medal of Honor for actions on 10 June with the 5th Special Forces Group.

Elements of the 7th Special Forces Group participated in Operation POWER PACK, deploying to the Dominican Republic.

The 1st, 2d, 3d and 4th Mobile Strike Forces (Mike Forces), were activated respectively in the I, II, III and IV Corps areas, providing reaction forces and conducting ambushes, raids and intelligence collection activities.

Rangers

Company D (Long Range Patrol), 17th Infantry was constituted in the Regular Army, and activated at Fort Benning, Georgia prior to deployment to the Federal Republic of Germany as the Long Range Patrol Company of the US Army V Corps.

The VII Corps LRRP Company was inactivated in Germany, redesignated as Company C (Long Range Patrol), 58th Infantry, to be activated back in Germany.

Psychological Operations

The 1st Psychological Warfare Battalion was reorganized and redesignated as the 1st Psychological Operations Battalion. HHC, 6th Psychological Operations Battalion was constituted in the Regular Army and activated at Fort Bragg, North Carolina.

Aviation

The 226th Aerial Surveillance and Escort Battalion was inactivated at Fort Benning, Georgia.

HHC and Companies B and D, 229th Assault Helicopter Company, were reorganized and redesignated as HHC and Companies B and D, 229th Aviation Battalion; relieved from assignment to the 11th Air Assault Division, assigned to the 1st Cavalry Division, and deployed to the Republic of Vietnam.

The 173d Aviation Company was activated at Fort Benning and assigned to the 173d Airborne Brigade.

Medical

The 11th Medical Battalion was inactivated at Fort Benning, Georgia.

The 307th Medical Battalion deployed and supported the 82d Airborne Division in the Dominican Republic.

Combat Support
Engineers

The 307th Engineer Battalion deployed and supported the 82d Airborne Division in the Dominican Republic during Operation POWER PACK.

The 173d Engineer Company (Combat) (Airborne), deployed from Okinawa to assignment with the 173d Airborne Brigade, at An Khe, Republic of South Vietnam.

Military Intelligence

The 525th Military Intelligence Group deployed to the Republic of Vietnam.

The 172d Military Intelligence Detachment deployed to the Republic of Vietnam with the 173d Airborne Brigade.

The 403d Army Security Agency Detachment was reassigned from the 7th Special Forces Group, to the 3d Special Forces Group at Fort Bragg, North Carolina.

Military Police

In March, the 503d Military Police Battalion was charged with the security of the 'Freedom March' from Selma to Montgomery, Alabama. After successfully prosecuting their responsibilities in Mississippi, the 503d Military Police Battalion redeployed to Fort Bragg, North Carolina in time to deploy in May with the XVIII Airborne Corps and elements of the 82d Airborne Division, to the Dominican Republic, returning in November (minus Company C).

The 11th Airborne Military Police Company was inactivated at Fort Benning, Georgia.

Signal

Deploying to the Dominican Republic, the 82d Signal Battalion, supported the 82d Airborne Division during Operation POWER PACK.

HHC, 511th Signal Battalion was inactivated at Fort Benning, Georgia.

The 173d Signal Company (Airborne), deployed with the 173d Airborne Brigade to the Republic of Vietnam.

Combat Service Support

The 101st Support Battalion arrived in the Republic of Vietnam in a Provisional status, in July, in support of the 1st Brigade, 101st Airborne Division. These functions included supply, maintenance, medical services and miscellaneous field service support.

The 173d Support Battalion deployed from Okinawa on 6 May, to provide brigade level service and support to the 173d Airborne Brigade in the Republic of Vietnam. The brigade level services provided, are as described for the 101st Support Battalion, above.

Maintenance

The 711th Maintenance Battalion was inactivated.

Quartermaster

The 170th Aircraft Maintenance Battalion was inactivated at Fort Benning, Georgia concurrent with the inactivation of the 11th Air Assault Division.

The 407th and 426th Supply and Transportation Battalions were inactivated respectively at Forts Bragg and Jackson. Company B, 426th Supply and Transportation Battalion was redesignated as the 101st Quartermaster Company.

1966

Final elements of HQ, XVIII Airborne Corps redeployed from the Dominican Republic to Fort Bragg, North Carolina.

Brigadier General Paul F. Smith commanded the 173d Airborne Brigade from February until December, when Brigadier General John R. Deane Jr assumed command.

I and II Field Forces, Vietnam (I and II FFV), were activated on 15 March. I FFV was formed from Task Force ALPHA (Provisional), which had been organized on 1 August 1965, under Lieutenant General Stanley R. (Swede) Larsen. I FFV exercised operational control (OPCON) over US and allied forces in the II Corps Tactical Zone. On 29 March, the 1st Brigade, 101st Airborne Division, and Company B, 5th Special Forces Group, were assigned OPCON to I FFV.

II FFV was to become the biggest combat command of the US Army in Southeast Asia, and was activated under the command of Major General Jonathan O. Seaman. Company A, 5th Special Forces Group was assigned OPCON to II FFV on 15 March, and the 173d Airborne Brigade assigned OPCON on 22 March.

For his actions on 7 February with the 1st Brigade, 101st Airborne Division, First Lieutenant James A. Gardner was awarded the Medal of Honor. Sergeant Charles B. Morris, assigned to the 173d Airborne Brigade, was awarded the Medal of Honor for his actions on 29 June.

Combat Arms
Infantry

The 4th Airborne Battle Group, 503d Infantry, was activated at Fort Campbell, as the 4th Battalion, 503d Infantry, assigned to the 173d Airborne Brigade, and deployed to the Republic of Vietnam.

Company E (Long Range Patrol), 30th Infantry, was constituted in the AUS, activated at Fort Rucker, Alabama, and assigned to the US Army Aviation Center and School.

Cavalry

In November, HHC, 1st Brigade (Airborne), 1st Cavalry Division, was redesignated as HHC, 1st Brigade, 1st Cavalry Division.

The 1st and 2d Battalions (Airborne), 8th Cavalry and the 1st Battalion (Airborne), 12th Cavalry reverted to non-airborne airmobile status concurrent with the redesignating of the 1st Brigade, 1st Cavalry Division.

The 3d Squadron, 17th Cavalry was activated at Fort Knox, Kentucky.

Special Operations

Brigadier General Joseph W. Stillwell and his aircraft disappeared, never to be accounted for. Brigadier General Albert E. Malloy assumed command of the US Army Center for Special Warfare.

Military Assistance Command, Studies and Observations Group (MACV-SOG) relocated from Cholon to Saigon, Republic of Vietnam.

The Provincial Operations Units (PRU), were formed in the Republic of Vietnam from the former Counter Terrorist Teams (previously known as the Armed Propaganda Teams). The PRU were advised in the North (I Corps) by US Marine Corps Force Reconnaissance personnel, in II and III Corps by US Army Special Forces personnel, and in the Delta region of IV Corps by US Navy SEALs. In addition to conducting intelligence gathering reconnaissance patrols, ambushes and raids, the PRU acted as the primary action arm of the PHOENIX Program.

Special Forces

The 2d, 9th, 13th, 16th, 17th and 24th Special Forces Groups, were inactivated at Columbus, Ohio; Little Rock, Arkansas; Charleston, West Virginia; Seattle, Washington; and Fort DeRussy, Hawaii.

In June, Colonel Francis J. (Splash) Kelly assumed command of the 5th Special Forces Group.

Operational Detachment B-50, (ODB-50) 5th Special Forces Group organized Project OMEGA to exploit the North Vietnamese Army and Viet Cong trail networks through specialized Long Range Reconnaissance Patrol activities.

The ODB-52, 5th Special Forces Group (Project DELTA), operated MACV-RECONDO School, and commenced providing instruction to US personnel assigned to Long Range Reconnaissance Patrol units.

Project SIGMA was activated by Operational Detachment B-56, providing special Long Range Reconnaissance Patrol activities for II Field Force, Vietnam.

Company D (Augmented), 1st Special Forces Group, was constituted in the Regular Army, and activated at Fort Bragg, North Carolina.

The 38th Special Forces Detachment was reorganized and consolidated with the 36th and 37th Special Forces Detachments, Alaska ARNG.

Aviation

HHD, 269th Aviation Battalion was constituted in the Regular Army and activated at Fort Bragg, North Carolina.

The 173d Aviation Company was deployed to the 173d Airborne Brigade in the Republic of Vietnam.

Medical

The Field Epidemiological Survey Team (FEST), was organized in the Regular Army and activated at Walter Reed Army Hospital, Washington, DC. FEST relocated to Fort Bragg, North Carolina, en route to assignment to the 5th Special Forces Group in the Republic of Vietnam.

Combat Support
Military Intelligence

Company A, 311th Army Security Agency Battalion was disbanded.

Companies A, B and C, 313th Army Security Agency Battalion reorganized and redesignated respectively as the 358th, 337th and 71st Army Security Agency Companies.

The 77th Intelligence Corps Detachment was assigned to the 12th Special Forces Group, reorganized and redesignated as the 77th Military Intelligence Detachment.

The 403d Army Security Agency Detachment was reassigned from the 3d to the 5th Special Forces Group, deploying from Fort Bragg, North Carolina to Nha Trang, Republic of South Vietnam, and using the 'cover' designation of the 403d Radio Research Unit.

Military Police

HHD, 16th Military Police Group was constituted in the Regular Army and activated at Fort George G. Meade, Maryland, deploying to the Republic of Vietnam.

In September, Company C, 503d Military Police Battalion closed in to Fort Bragg, North Carolina from the Dominican Republic.

Signal

The 543d Signal Company deployed from Fort Benning, Georgia, to provide airborne forward area communications capabilities to the 173d Airborne Brigade in the Republic of Vietnam.

Combat Service Support
Quartermaster

HHD, 528th Quartermaster Battalion was inactivated on 25 November.

Lieutenant General John I. Throckmorton assumed command of XVIII Airborne Corps, followed by Lieutenant General Robert H. York. Major General Richard J. Seitz was appointed to command the 82d Airborne Division.

Lieutenant General Bruce Palmer Jr was appointed as Commander, II Field Force, Vietnam (II FFV) in March, and later appointed as the Deputy Commander, US Army Vietnam (USARV) in July. The 3d Brigade, 101st Airborne Division was placed under OPCON of II FFV on 20 December.

Task Force OREGON was activated in Vietnam on 12 April, under the command of Major General William B. Rosson, until command was passed to Major General Richard T. Knowles in June. Task Force OREGON, operating from Quang Ngai, allowed US Marine elements to fight further north, and the 1st Cavalry Division to extend their operations from Binh-Dinh to Da-Nang. Task Force OREGON was the forerunner of the 23d Infantry (Americal) Division, and comprised 3d Brigade, 25th Infantry Division; the 1st Brigade, 101st Airborne Division; and the 196th Infantry Brigade (Light). The brigades of the 25th Infantry and 101st Airborne Divisions were released back to their respective Divisions upon assignment of the 11th and 198th Infantry Brigades.

In February, Brigadier General Salve H. Matheson (the Iron Duke), assumed command of the 1st Brigade, 101st Airborne Division. The remainder of the 101st Airborne Division, commanded by Major General Olinto M. Barsanti, and including the 2d and 3d Brigades deployed from Fort Campbell in November to the Republic of Vietnam. For his actions on 18 May, Specialist Fourth Class Dale E. Wayrynen, assigned to the 101st Airborne Division, was posthumously awarded the Medal of Honor. Similarly, for his actions on October 15 with the 101st Airborne Division, Staff Sergeant Webster Anderson was awarded the Medal of Honor.

Brigadier General Leo H. Schweiter assumed command of the 173d Airborne Brigade. Operation JUNCTION CITY was implemented on 22 May, involving the airborne insertion of the 2d Battalion, 503d Airborne Infantry, supported by artillery, engineers and military police of the 173d Airborne Brigade. The airborne task force provided a blocking force for a multi-divisional sweep conducted by the 1st, 4th and 25th Infantry Divisions and the 11th Armored Cavalry Regiment. The 173d Airborne Brigade was placed under the operational control of I Field Force, Vietnam on 23 May. For his actions on 8 April, while assigned to the 173d Airborne Brigade, Specialist Fourth Class Don L. Michael was posthumously awarded the Medal of Honor.

HHDs, 1st Brigades, 80th, 84th and 100th Divisions (Training), were reconstituted in the Army Reserve from the former HQ Companies, 80th, 84th and 100th Infantry Divisions. HHD, 1st Brigade, 108th Division (Training), was constituted in the Army Reserve and activated in 1967. (Units constituting the 2d, 3d and 4th Brigades, 80th, 84th, 100th and 108th Divisions (Training), are listed under 1967, Combat Arms: Artillery; Combat Support: Engineers and Combat Service Support: Ordnance). It should be noted that the with one exception, included in the lineage of each of the HHDs of the four brigades of the 80th, 84th, 100th and 108th Divisions (Training), are former airborne organizations. The exception is the 235th Engineer Battalion, (originally constituted as the 2d Battalion, 48th Engineers at Camp Gruber, Oklahoma in 1942), was reorganized and redesignated to become the HHD, 3d Brigade, 108th Division (Training), and never served in an airborne capacity.

Privates First Class John A. Barnes III and Carlos J. Lozada and Major Charles J. Watters were posthumously awarded the Medals of Honor for their respective actions on 12, 20 and 19 November, while assigned to the 173d Airborne Brigade.

Combat Arms
Infantry

HHCs, 5th and 6th Battle Groups, 51st Infantry, redesignated as Companies E and F, 51st Infantry, were both activated in the Republic of Vietnam. Company E was assigned to the 23d Infantry Division, and Company F to the 199th Infantry Brigade (Light).

Deploying to the Republic of Vietnam with the remainder of the 101st Airborne Division, the 1st and 2d Battalions, 501st Infantry and the 1st Battalion, 502d Infantry, were assigned to the 2d Brigade.

The 3d Battalion, 187th Infantry and the 1st and 2d Battalions, 506th Infantry, were assigned to the 3d Brigade.

Other infantry units assigned to the 101st Airborne Division in Vietnam, included Company F, 58th Infantry (Long Range Patrol).

The 3d Battalion, 506th Infantry was activated at Fort Campbell, Kentucky, and deployed to the Republic of Vietnam for assignment to the 1st Brigade, 101st Airborne Division.

Cavalry

The 2d Squadron, 17th Armored Cavalry (later Air Cavalry), deployed with the 101st Airborne Division to the Republic of Vietnam.

Field Artillery

The 2d Battalion, 11th Artillery (155mm); 1st Battalion, 39th Artillery (155mm); 4th Battalion, 77th Artillery (Aerial Rocket); 2d Battalion, 319th Artillery; 2d Battalion, 320th Artillery; 1st Battalion, 321st Artillery; and Battery A, 377th Artillery (Aviation), all deployed with the 101st Airborne Division to the Republic of Vietnam.

HHBs, 506th, 905th 909th and 925th Field Artillery Battalions were reconstituted respectively in the Army Reserve as the HHDs, 2d Brigades, 108th, 80th, 84th and 100th Divisions (Training).

Special Operations

For his actions on 7 February, while assigned to Military Assistance Command Vietnam, Studies and Observations Group (MACV-SOG), First Lieutenant George K. Sisler was posthumously awarded the Medal of Honor.

Command and Control North, Central and South (CCN, CCC and CCS) were respectively constituted and activated at Da-Nang, Kontum and Ban-Me-Thuot, Republic of Vietnam. CCN, CCC and CCS were assigned as forward deployed headquarters of Military Assistance Command Vietnam, Studies and Observations Group (MACV-SOG) and provided command and control of US/ARVN Special Forces, organized as Hatchet Forces and Search-Locate-Annihilation-Mission (SLAM) Teams. CCN was organized from personnel assigned to MACV-SOG Forward Operations Bases (FOB) 1, 3 and 4, conducting missions into Laos and North Vietnam, leading Chinese Nung Tribes in Vietnam and Meo Tribes in Laos. CCC was organized from personnel assigned to MACV-SOG FOB 2, and conducted missions along the Laos-Cambodian-Vietnamese borders. CCS personnel were assigned from the 5th Special Forces Group, conducting missions and cross-border operations against the North Vietnamese and Viet-Cong in Cambodian sanctuaries.

Special Forces

Colonel Jonathan F. Ladd commanded the 5th Special Forces Group from June 1967 through June 1968.

Master Sergeant Charles E. (Snake) Hosking Jr was posthumously awarded the Medal of Honor while assigfned to the 5th Special Forces Group.

Company D (Augmented), 1st Special Forces Group, was discontinued at Fort Bragg, North Carolina, redesignated as the 46th Special Forces Company, activated and deployed to Lop Buri, Thailand.

Project GAMMA was activated under Operational Detachment B-57, 5th Special Forces Group, providing intelligence through Long Range Reconnaissance Patrols into the North Vietnamese Army bases in Cambodia.

Rangers

Companies C, F and H, 75th Infantry (inactivated in Okinawa in 1951 and 1956), were designated as Companies E, F and H (Long Range Patrol), 20th, 50th and 52d Infantry, assigned respectively to the 4th and 9th Infantry Divisions and the 1st Cavalry Division, were activated in the Republic of Vietnam.

Company D (Long Range Patrol), 151st Infantry, was constituted from Companies B and C, 1st Battalion (Airborne), 151st Infantry. This Indiana ARNG unit deployed after training at Fort Benning, Georgia, and was the only National Guard unit to serve in the Republic of Vietnam, (continuing to wear the shoulder sleeve insignia of it's parent Indiana National Guard unit - the 38th Infantry Division). Company D (Long Range Patrol) was attached to the 199th Infantry Brigade (Separate) (Light).

Activated in Vietnam, Company E (Long Range Patrol), 51st Infantry, was constituted and assigned to the 23d Infantry (American) Division.

Company F (Long Range Patrol), 52d Infantry, was constituted in the AUS, activated and assigned to the 1st Infantry Division in Vietnam.

The 71st Infantry Detachment (Long Range Patrol), was constituted in the AUS, activated in Vietnam, reorganized as Company F (Long Range Patrol), 51st Infantry, and assigned to the 199th Infantry Brigade (Separate)(Light).

Psychological Operations

The 4th Psychological Operations Group and the 8th Psychological Operations Battalion were constituted in the Regular Army and activated in the Republic of Vietnam. The 9th Psychological Warfare Company was expanded, reorganized and redesignated as the 9th Psychological Warfare Battalion.

Civil Affairs

The 96th Military Government Group, inactivated in Korea in 1949, was redesignated as the 96th Civil Affairs Company, and activated at Fort Lee, Virginia. The 28th Civil Affairs Company was reorganized and redesignated as the 28th Civil Affairs Battalion, and activated at Fort Campbell, Kentucky.

Aviation

Assigned to the 101st Aviation Group, (formerly the 160th Aviation Group), were the 101st, 158th and 159th Aviation Battalions, the 163d and 478th Aviation Companies, the 5th Transportation Battalion (Aviation), all deployed to the Republic of Vietnam with the 101st Airborne Division.

HHD, 269th Aviation Battalion deployed to Vietnam, and assignment with the 12th Aviation Group, 25th Infantry Division.

Medical

The Field Epidemiological Survey Team (FEST), which had been jointly sponsored by Special Forces and the Walter Reed Army Hospital, completed their study of diseases with a special interest to the Army Medical Department. These interests included the entomological aspects of tropical sprue, schistosomiasis, febrile illness, dengue, malaria and filarisis. (The critical data obtained was published in a Vietnam Studies Report by Major General Spurgeon Neel, and was included in a Department of the Army publication in 1973, entitled Medical Support of the Army in Vietnam 1965-1970). FEST was inactivated in the Republic of Vietnam.

The 326th Medical Battalion, deployed with the 101st Airborne Division to the Republic of Vietnam.

Combat Support
Engineers

HHC, 20th Engineer Brigade was activated at Fort Bragg, North Carolina, and deployed to the Republic of Vietnam.

The 326th Engineer Battalion deployed to the Republic of Vietnam with the 101st Airborne Division.

The HHSCs, 305th, 309th, 325th and 235th Engineer Battalions were reorganized, redesignated and activated respectively as the 3d Brigades, 80th, 84th, 100th and 108th Divisions (Training).

Military Intelligence

The 101st Military Intelligence Company, and the 265th ASA Company deployed with the 101st Airborne Division to the Republic of Vietnam.

Military Police

In October, the 503d Military Police Battalion deployed from Fort Bragg, North Carolina to the Pentagon, Washington DC, to conduct security operations.

Signal

The 501st Signal Battalion deployed to the Republic of Vietnam with the 101st Airborne Division.

The 931st Signal Corps Battalion (Air Support Command), (inactivated in India in 1945), was redesignated as HHD, 35th Signal Group. Allotted to the Regular Army, the 35th Signal Group was activated at Fort Bragg, North Carolina, and assigned to the XVIII Airborne Corps.

The 173d Signal Company (Airborne), deployed from Anh Khe to Bien Hoa, Republic of Vietnam, with the 173d Airborne Brigade.

Combat Service Support
Adjutant General

The 101st Admin Company, the 22d Military History Detachment, and the 25th, 34th and 45th Public Information Detachments deployed to the Republic of Vietnam with the 101st Airborne Division.

Chemical

The 10th Chemical Platoon; 20th and 36th Chemical Detachments deployed with the 101st Airborne Division to the Republic of Vietnam.

Maintenance

The 713th Airborne Quartermaster Company (see 1945, Combat Service Support, Quartermaster), was redesignated as the 713th Maintenance Company, allotted to the Regular Army and assigned to the Sixth Army, and activated at Fort Lewis, Washington.

Ordnance

The HHDs, 780th, 784th, 800th and 808th Ordnance Battalions, were reorganized, redesignated and activated as the HHD, 4th Brigades, 80th, 84th, 100th and 108th Divisions (Training).

Quartermaster

The 101st Airborne Division combat service support elements deploying to the Republic of Vietnam, included the 426th Supply and Transport Battalion and the 801st Maintenance Battalion.

—— 1968 ——

Lieutenant General John L. Tolson, former commander of the 1st Cavalry Division in the Republic of Vietnam), assumed command of the XVIII Airborne Corps.

The XXIV Corps was consolidated with the Provisional Corps, and activated in Vietnam on 15 August, under the command of Lieutenant General Richard G. Stilwell. The Provisional Corps was created from assets of the Military Assistance Command, Vietnam (Forward), to stop the influx of North Vietnamese, following the TET Offensive. On 9 March, the Provisional Corps had been organized under the command of Lieutenant General William B. Rosson, with operational command of several major US units, including the 101st Airborne Division and the 3d Brigade, 82d Airborne Division.

The command of the 82d Airborne Division was passed on 14 October, to Major General John R. Deane Jr.

The 101st Airborne Division was reorganized and redesignated as the 101st Airborne Division (Airmobile), and command was passed to Major General Melvin Zais. Colonel John (Rip) Collins assumed command of the 1st Brigade, 101st Airborne Division from Brigadier General Matheson.

Brigadier General Salve H. Matheson (the Iron Duke), was reassigned from commanding the 1st Brigade, 101st Airborne Division, to become a Senior US Army Advisor to the I Corps Tactical Zone, Republic of Vietnam.

Eight Medals of Honor, the highest award for valor of the United States, were awarded to men of the 101st Airborne Division in 1968. For their actions on 21 February, the Medal of Honor was awarded to Staff Sergeant Clifford C. Simms (posthumously), and to Sergeant Joe R. Hooper; for his actions on 19 March, Captain Paul W. Bucha; for his actions on 26 April, posthumously to Private First Class Milton A. Lee; for his actions on 6 May, to Sergeant Robert M. Patterson; for his actions on 18 May, posthumously to Specialist Fourth Class Peter M. Guenette; for his actions on 29 June, to Private First Class Frank A. Herda; and for his actions on 11 July, to Specialist Fourth Class Gordon R. Roberts.

The 3d Brigade, 82d Airborne Division deployed to the Republic of Vietnam, (and to combat for the first time by C-141 aircraft), commanded by Colonel Thomas R. Bolling Jr, from February through December, when Brigadier General George W. Dickerson took command. The brigade was initially attached to the 101st Airborne Division in the vicinity of Hue-Phu Bai, in the northern I Corps area, under the operational command of the XXIV Corps. Replacing the now deployed 3d Brigade, the 4th Brigade, 82d Airborne Division was activated at Fort Bragg.

Command of the 173d Airborne Brigade was given to Brigadier General Richard J. Allen from April through December, when Brigadier General John W. Barnes assumed command.

HHC, 1st Battalion, 36th Infantry Brigade (Texas ARNG), was consolidated with HHC and Company B, 7th Battalion, 112th Armor Regiment (Texas ARNG), with the consolidated unit designated as HQ, 71st Infantry Brigade (Texas ARNG), and activated at Houston, Texas.

Combat Arms
Infantry

Assigned and deploying to Vietnam with the 3d Brigade, 82d Airborne Division, were the 1st and 2d Battalions, 505th Infantry, and the 1st Battalion, 508th Infantry.

As the 4th Brigade, 82d Airborne Division was activated at Fort Bragg. Elements activated and assigned to the 4th Brigade included the 4th Battalion, 325th Infantry; 3d Battalion, 504th Infantry and the 3d Battalion, 505th Infantry.

The 3d Battalion, 506th Infantry was relieved from the 101st Airborne Division, and from May through June, was assigned to I Field Force, Vietnam (I FFV), until being assigned in July to Task Force SOUTH, I FFV.

The 3d Battalion, 508th Infantry was inactivated at Fort Kobbe, Panama.

Company F, 51st Infantry, was inactivated in the Republic of Vietnam.

Armor

Back at Fort Bragg, newly assigned to the 82d Airborne Division, was the 4th Battalion, 68th Armor. This battalion was organized in 1918 in the American Expeditionary Force as Company C, 327th Battalion, Tank Corps, in France; to be redesignated that same year as Company C, 345th Battalion, Tank Corps. In 1921, it was allotted to the Regular Army and assigned to the 2d Division as the 2d Tank Company, until relieved from the 2d Armored Division in 1939, and redesignated on 1 January 1940 as Company D, 68th Infantry (Light Tanks), an element of the 2d Armored Division. Inactivated at Fort Benning in 1942, Company D was concurrently relieved from assignment to the 2d Armored Division, and assigned with the 68th Infantry (Light Tanks), to the 6th Armored Division - activated at Fort Knox, Kentucky. In 1943, Company D was reorganized and redesignated as Company A, 68th Tank Battalion. The 68th Tank Battalion was relieved from assignment to the 6th Armored Division in 1945, and inactivated at Camp Patrick Henry, Virginia. Redesignated as Company A, 68th Medium Tank Battalion, this unit was activated at Fort Leonard Wood, Missouri, remaining there until inactivation in 1956. In 1957, Company A, 68th Medium Tank Battalion, was concurrently redesignated as HHC, 4th Medium Tank Battalion, 68th Armor, and relieved from assignment to the 6th Armored Division. The 4th Medium Tank Battalion was activated in 1958 at Fort Bragg, North Carolina, to be reorganized and redesignated as the 4th Battalion, 68th Armor, and assigned to the 82d Airborne Division on 22 March.

Cavalry

Troop K, 16th Cavalry was constituted in the Regular Army in 1916 at Fort Bliss, Texas and demobilized in 1921 at The Presidio of Monterey, California. Troop K, 16th Cavalry was reconstituted in the Regular Army in 1959 and redesignated as Troop K, 17th Cavalry, but not activated until 1968 when assigned to the 82d Airborne Division at Fort Bragg, North Carolina.

Assigned to the 3d Brigade, 82d Airborne Division, deploying to the Republic of Vietnam, was Troop B, 1st Squadron, 17th Cavalry (Armored).

Field Artillery

The 2d Battalion, 26th Artillery was redesignated as the 2d Battalion, 26th Field Artillery, and inactivated at Fort Bragg, North Carolina. The 1st Battalion, 39th Artillery was redesignated as the 1st Battalion, 39th Field Artillery.

Battery C, 320th Artillery, was redesignated as HHB, 3d Battalion, 320th Artillery, assigned to the 4th Brigade, 82d Airborne Division, and activated at Fort Bragg, North Carolina.

The 2d Battalion, 321st Artillery deployed to Phu Bai, Republic of Vietnam with the 3d Brigade, 82d Airborne Division. Some personnel, assigned to the 2d Battalion, 321st Artillery had only rotated beck to Fort Bragg from assignment with the 1st Battalion, 320th Artillery, 101st Airborne Division, just weeks prior.

Battery A, 377th Artillery (Aviation), deployed to Gia Le, Republic of Vietnam, and assignment to the 101st Airborne Division.

Special Operations

Brigadier General Edward M. Flanagan was named to command the US Army Center for Special Warfare at Fort Bragg, North Carolina.

Medals of Honor were awarded to three soldiers assigned to the Military Assistance Command, Studies and Observations Group (MACV-SOG). The recipients were Staff Sergeant Fred W. Zabitosky, for his actions on 19 February; Specialist Fifth Class John J. Kedenberg (posthumously), for actions on 13 June; and First Lieutenant Robert L. Howard, for his actions on 30 December.

Special Forces

Colonel Harold R. Aaron assumed command of the 5th Special Forces Group in June.

The 10th Special Forces Group relocated from Bad Toelz, Germany to Fort Devens, Massachusetts, with the 1st Battalion, 10th Special Forces Group remaining at Flint Kaserne.

The 11th and 12th Special Forces Groups were respectively assigned to the 97th and 86th Army Reserve Commands.

Sergeant First Class Eugene Ashle Jr, was posthumously awarded the Medal of Honor for his actions on 7 February; and Staff Sergeant Roy Benavidez was awarded the Medal of Honor for his actions on 2 May. Both of these Non-Commissioned Officers were assigned to the 5th Special Forces Group. Sergeant Gordon D. Yntema was posthumously awarded the Medal of Honor while assigned to the 5th Special Forces Group, for his deeds of 18 January.

Operational Detachment B-57 (Project GAMMA), 5th Special Forces Group, relocated from Saigon to Nha-Trang, Republic of Vietnam.

Rangers

Airborne Ranger Companies of the 75th Infantry were activated from Long Range Patrol (LRP) Companies.

Companies A and B were formerly Company D (LRP), 17th Infantry, and Company C (LRP), 58th Infantry, were both relocated from Germany. Company A (Ranger), was assigned to the 197th Infantry Brigade, at Fort Benning, Georgia. Company B (Ranger), was assigned to Fifth Army at Fort Carson, Colorado.

Other LRP companies were assigned to the Republic of Vietnam, and eventually to the 75th Infantry.

The Medal of Honor was posthumously awarded to Staff Sergeant Laszlo Rabel, for his actions on 13 November, while assigned to the 74th Infantry Detachment (LRP).

The 78th Infantry Detachment (LRP), was constituted, activated and assigned to HQ, US Army, Vietnam (USARV).

Company D (LRP), 151st Infantry was inactivated in Vietnam.

Company E (Ranger), 75th Infantry was formed from Company E (LRP), 50th Infantry at Dong Tam, IV Corps, Republic of Vietnam, and assigned to the 9th Infantry Division.

The inactivated Company F (LRP), 52d Infantry, was designated as Company I (Ranger), remaining assigned to the 1st Infantry Division.

Company F (LRP), 58th Infantry, was constituted, activated in Vietnam, and assigned to the 101st Airborne Division.

The 79th Infantry Detachment (LRP) was constituted in the AUS, activated in Vietnam, and assigned to the US Army, Vietnam (USARV).

Company E (Ranger), 425th Infantry, was constituted and activated in the Michigan ARNG, from Company B, 1st Battalion (Airborne), 225th Infantry.

Civil Affairs

The 3d Civil Affairs Group, constituted in the AUS in 1967, was assigned to US Army South (USARSO), and activated at Fort Clayton, Panama.

The 28th Civil Affairs Battalion was reorganized and redesignated as the 28th Civil Affairs Company, relocated to Fort Bragg, North Carolina, and assigned to the JFK Center for Special Warfare.

Aviation

HHC 160th Aviation Group and HHC, 159th Aviation Battalion were constituted in the Regular Army, assigned to the 101st Airborne Division (Airmobile), and activated in the Republic of Vietnam.

The 101st Aviation Battalion was assigned initially to the 160th Aviation Group (later redesignated as the 101st Aviation Group).

Company A, 82d Aviation Battalion, provided aviation support to the 3d Brigade, 82d Airborne Division in the Republic of Vietnam.

Combat Support
Engineers

Company C, 307th Engineer Battalion deployed to provide engineering support to the 3d Brigade, 82d Airborne Division in the Republic of Vietnam.

Military Intelligence

The 319th Military Intelligence Battalion was inactivated in Hawaii.

The 77th Military Intelligence Detachment was relocated with the 12th Special Forces Group to Oak Park Illinois.

The 389th Military Intelligence Detachment was assigned to the 83d Army Reserve Command, First US Army.

The 402d Army Security Agency Detachment relocated with the Hqs, 10th Special Forces Group from Bad Toelz, Germany to Fort Devens, Massachusetts.

The 405th Army Security Agency (ASA) Detachment was activated in the Republic of Vietnam on 1 June. During the assignment of the 405th ASA Detachment in Vietnam, it provided support to the 3d Brigade, 82d Airborne Division and the 3d Brigade, 1st Cavalry Division.

The 408th Army Security Agency Detachment and the 518th Military Intelligence Detachment deployed in support of the 3d Brigade, 82d Airborne Division, to the Republic of Vietnam.

Military Police

On 7 April, the 503d Military Police Battalion again deployed to Washington, DC for civil disturbance control operations following the assassination of Dr Martin Luther King, participating in extensive saturation patrolling. The 503d Military Police Battalion remained at Fort Belvoir, Virginia for two months during the period of 'Resurrection City' and the funeral of Senator Robert Kennedy.

Signal

The 58th Signal Company deployed with the 3d Brigade, 82d Airborne Division to the Republic of Vietnam.

Combat Service Support

Headquarters, Headquarters and Band, 82d Airborne Division Support Command was redesignated as HHC and Band, and deployed with the 3d Brigade, 82d Airborne Division to the Republic of Vietnam.

Chemical

The 808th Chemical Detachment was assigned to the 77th Army Reserve Command, First US Army.

The 52d Chemical Detachment deployed with the 3d Brigade, 82d Airborne Division to the Republic of Vietnam.

Quartermaster

The 426th Supply and Transportation Battalion, was reorganized and redesignated as the 426th Supply and Service Battalion, and activated in the Republic of Vietnam, (concurrently, the 101st Quartermaster Company was reorganized and redesignated as Company B, 426th Supply and Service Battalion).

Transportation

The 5th Transportation Truck Company was reorganized and redesignated as the 5th Transportation Corps Battalion, assigned to the 101st Airborne Division (Airmobile), and activated in the Republic of Vietnam.

—— 1969 ——

General William B. Rosson was assigned as the Deputy Commander, Military Assistance Command, Vietnam, and Lieutenant General Melvin Zais assumed command of the XXIV Corps.

Major General John M. Wright assumed command of the 101st Airborne Division (Airmobile), and Brigadier General Hubert S. Cunningham assumed command of the 173d Airborne Brigade.

The 101st Airborne Division was reorganized and redesignated as the 101st Airborne Division (Airmobile) on 1 July. The first soldier to be awarded the Medal of Honor within the redesignated 101st Airborne Division (Airmobile), was Staff Sergeant John G. Gertsch who was posthumously awarded the Medal of Honor for his actions on 19 July. The Medal of Honor was also awarded posthumously to Specialist Fourth Class Joseph G. LaPointe Jr for actions on 2 June, while assigned to the 101st Airborne Division.

As the 3d Brigade, 82d Airborne Division redeployed from the Republic of Vietnam, the 4th Brigade was inactivated at Fort Bragg.

While assigned to the 173d Airborne Brigade, Specialist Fourth Class Michael R. Blanchfield was posthumously awarded the Medal of Honor, for his actions on 3 July. Also assigned to the 173d Airborne Brigade, when posthumously awarded the Medal of Honor for his actions on 20 March, was Corporal Terry T. Kawamura.

Combat Arms
Infantry

Inactivations concurrent with the inactivation of the 4th Brigade, 82d Airborne Division, included the 4th Battalion, 325th Infantry, 3d Battalion, 504th Infantry, and the 3d Battalion, 505th Infantry.

The 3d Battalion, 506th Infantry was assigned from Task Force SOUTH to the 173d Airborne Brigade.

Company E (Long Range Patrol), 30th Infantry, assigned to the US Army Aviation Center and School, was removed from airborne status, and redesignated as Company E, 30th Infantry.

On 1 February, Company E, 51st Infantry was inactivated in the Republic of Vietnam, to be activated and redesignated as Company G (Ranger), 75th Infantry.

Cavalry

Troop K, 17th Cavalry was relieved from assignment to the 82d Airborne Division.

Field Artillery

The 1st Missile Battalion, 39th Artillery, inactivated in Germany in 1964, was redesignated as the 1st Battalion, 39th Artillery, and activated in the Republic of Vietnam.

The 3d Battalion, 320th Artillery was inactivated concurrent with the inactivation of the 4th Brigade, 82d Airborne Division.

Special Operations

The US Army Center for Special Warfare was reorganized and redesignated as (both) the US Army John F. Kennedy Center for Military Assistance, (providing Command and Control of US Army Special Forces), and as the US Army Institute for Military Assistance, (as the proponent for Special Forces, Psychological Operations and Civil Affairs schools, and security assistance).

Special Forces

The 3d Special Forces Group was inactivated at Fort Bragg, North Carolina.

Colonel Robert B. Rheault commanded the 5th Special Forces Group from May through July, when he was relieved by General Creighton W. Abrams, over an extremely controversial issue concerning a Vietnamese double agent. Colonel Alexander Lemeres became the interim commander until Colonel Michael D. Healy was assigned as the commander in August.

Sergeant First Class William M. Bryant was posthumously awarded the Medal of Honor for his actions on 24 March, while assigned to the 5th Special Forces Group.

Rangers

Company E, 51st Infantry was inactivated in the Republic of Vietnam, to be activated and redesignated as Company G (Ranger), 75th Infantry, on 1 February.

Formed from personnel assigned to the concurrently inactivated Company E (LRP), 52d Infantry, the newly activated Company H (Ranger) was assigned to the 1st Cavalry Division.

Companies E and F (LRP), 58th Infantry, were inactivated to form Companies K and L (Ranger), continuing their assignments to the 4th Infantry Division and the 101st Airborne Division respectively.

The 75th Infantry was reorganized as a parent regiment under the Combat Arms Regimental; System (CARS), with additional Long Range Patrol (LRP) companies assigned.

Company C (Ranger), 75th Infantry, assigned to the 4th Infantry Division, was formerly Company E (LRP), 20th Infantry, I Field Force Vietnam.

Company D (Ranger), 75th Infantry was previously Company D (LRP), 151st Infantry, II Field Force Vietnam.

Company F (Ranger), 75th Infantry was the inactivated Company F (LRP), 50th Infantry, and assigned back to the 25th Infantry Division.

While assigned to Company G (Ranger), 75th Infantry, attached to the 23d Infantry Division, Staff Sergeant Robert J. Pruden was posthumously awarded the Medal of Honor for his actions on 29 November.

Specialist Fourth Class Robert D. Law was posthumously awarded the Medal of Honor for his actions on 22 February, while assigned to Company I (Ranger), 75th Infantry, attached to the 1st Infantry Division.

Company M (Ranger), 75th Infantry was activated from Company F (LRP), 51st Infantry, and assigned to the 199th Infantry Brigade (Separate)(Light).

The inactivated 74th Infantry Detachment (LRP) was concurrently redesignated as Company N (Ranger), remaining assigned to the 173d Airborne Brigade.

Company O (Ranger), came from the personnel of the inactivated 78th Infantry

Detachment (LRP), and attached to the 3d Brigade, 82d Airborne Division prior to inactivation later this year.

The 79th Infantry Detachment (LRP) was inactivated in Vietnam, and redesignated as the concurrently activated Company P (Ranger), attached to the 1st Brigade, 5th Infantry Division (Mechanized).

Company E (Ranger), 200th Infantry was constituted in the Alabama ARNG, and formed from personnel assigned to the 788th Light Maintenance Company (DS).

Company F (Ranger), 425th Infantry, was constituted, organized from Company C, 1st Battalion (Airborne), 225th Infantry, and activated in the Michigan ARNG.

Civil Affairs

The 97th Civil Affairs Group was inactivated on Okinawa.

Aviation

HHC, 160th Aviation Group was redesignated as the HHC, 101st Aviation Group.

The 158th Aviation Battalion was constituted in the Regular Army, activated at Fort Carson Colorado, assigned to the 160th (and the redesignated) 101st Aviation Groups, 101st Airborne Division (Airmobile), and deployed to the Republic of Vietnam.

The 23d Aviation Detachment was redesignated as the JFK Center for Military Assistance/JFK Institute for Military Assistance Aviation Detachment.

Combat Support
Military Police

In January, a reinforced company of the 503d Military Police Battalion deployed to Washington DC, in support of the Armed Forces Police during the inauguration of President Richard Nixon. The Battalion committed a 50 person contingent to aid in the disaster relief in Mississippi following hurricane Camille. The 503d Military Police Battalion again returned to Washington in November, remaining on 'stand-by' during the Peace Moratorium marches.

Combat Service Support
Chemical

The 445th Chemical Detachment was relocated from Seattle to Fort Lawton, Washington.

Quartermaster

HHD, 528th Quartermaster Battalion was redesignated as HHC, 528th Quartermaster Battalion, and activated in the Republic of Vietnam.

Transportation

HHC, 5th Transportation Corps Battalion was redesignated as HHC, 5th Transportation Corps Battalion (Aircraft Maintenance).

——1970——

Major General George S. Blanchard assumed command of the 82d Airborne Division.

Medals of Honor were awarded to three valorous soldiers of the 101st Airborne Division during 1970. For his actions on 7 May, the Medal of Honor was awarded to Private First Class Kenneth M. Kays; Lieutenant Colonel Andre C. Lucas was posthumously awarded the Medal of Honor for his actions on 23 July; and Corporal Frank R. Fratellinico received the Medal of Honor posthumously for actions on 19 August.

Staff Sergeant Glenn H. English Jr was posthumously awarded the Medal of Honor for his actions on 7 September, while assigned to the 173d Airborne Brigade.

Combat Arms
Infantry

From April through July, the 3d Battalion, 506th Infantry, continuing its exodus across Vietnam, was reassigned from the 173d Airborne Brigade to the 1st Brigade, 4th Infantry Division. During the following month (August), the 3d Battalion was assigned to I Field Force Vietnam, prior to returning to the 1st Brigade, 101st Airborne Division.

Cavalry

Troop K, 17th Cavalry was activated in the Republic of Vietnam on 1 October, and briefly assigned to the 17th Aviation Group prior to inactivation on 15 December.

Special Operations

The Forces Armee Nationales Khmeres (FANK) Training Command was organized in the AUS under the cover designation, US Field Training Command (FTC). Formed from members of Operational Detachment B-36, 5th Special Forces Group. The FTC was commanded by Major (soon to become Lieutenant Colonel) Ola Lee Mize, a recipient of the Medal of Honor from the Korean war. The mission of the FTC was to train Cambodians loyal to the new Premier of Cambodia, Marshall Lon Nol.

For his actions on 5 January, Staff Sergeant Franklin D. Miller was awarded the Medal of Honor, while assigned to the Military Assistance Command Vietnam, Studies and Observations Group (MACV-SOG).

Major Paul Ogg and Captain Geoffrey T. Barker were respectively assigned as the Senior advisors to the Provincial Reconnaissance Units (PRU) within the III and IV Corps Tactical Zones. The PRU were target against the Viet Cong infrastructure, and were habitually employed as the action force for the PHOENIX Program.

Special Forces

The Medal of Honor was awarded to Sergeants Garry B. Beikirch and Brian L. Buker (posthumously) their respective actions on April 1st and 5th, while assigned to the 5th Special Forces Group.

HQs, 11th and 12th Special Forces Groups were relocated respectively from Staten Island to Tappan, New York, and from Oak Park to Arlkington Heights, Illinois.

ODB-52 (MACV-RECONDO School); ODB-55 (Mike Force); ODB-56 (Project DELTA); and ODB-57 (Project GAMMA), all assigned to the 5th Special Forces Group, were inactivated. ODB-56 was inactivated in June; ODB-57 in March; and ODB-52 and 55 in December.

Rangers

Company D (Ranger), 75th Infantry, redeployed to home assignment with the Indiana ARNG, redesignated as Company D, 151st Infantry.

Companies E, K and L (Ranger), 75th Infantry were inactivated in Vietnam.

Company H (Ranger), 75th Infantry was reassigned within the 1st Cavalry Division to the 9th Air Cavalry Brigade (Provisional) (Air Cavalry Combat).

Company I (Ranger), 75th Infantry was inactivated at Fort Riley, Kansas following redeployment with the 1st Infantry Division.

Company N (Ranger), 75th Infantry redeployed with the 199th Infantry Brigade (Separate) (Light) to Fort Benning, Georgia, and inactivation.

Company O (Ranger), 75th Infantry was activated at Fort Richardson, Alaska, and assigned to the 172d Infantry Brigade.

Combat Support
Intelligence
The 77th Military Intelligence Detachment was relocated with the 12th Special Forces Group from Oak Park to Arlington Hieghts, Illinois.

The 403d Army Security Agency Detachment was inactivated concurrent with the redeployment of the 5th Special Forces Group from the Republic of Vietnam to Fort Bragg, North Carolina.

Military Police
HHC, 16th Military Police Group redeployed from the Republic of Vietnam, where it had commanded the 93d, 97th and 504th Military Police Battalions, to Fort Bragg, North Carolina.

The 503d Military Police Battalion once again returned to Washington, DC as the 1970 Peace Moratorium march commenced. The battalion returned to Fort Bragg, North Carolina, two days later.In May, the 503d Military Police Battalion was alerted for participation in Operation FONDA's FOLLIES, when radical peace activist Jane Fonda and approximately 250 anti-war protesters formed in different locations at Fort Bragg. As she had desired, Miss Fonda was duly apprehended and later released.

In November, Companies A, B and C, 503d Military Police Battalion, provided cadre for the activation of the 21st, 65th and 108th Military Police Companies, activated at Fort Bragg in support of the XVIII Airborne Corps.

Combat Service Support
HHD, 1st Logistics Command redeployed from Vietnam to Fort Lee, Virginia, to be reorganized and redesignated as HHC, Special Troops, 1st Field Army Support Command.

—— 1971 ——
Lieutenant General John H. Hays Jr assumed command of the XVIII Airborne Corps.

Specialist Fourth Class Michael J. Fitzmaurice was awarded the Medal of Honor for his actions on 23 March, while assigned to the 101st Airborne Division (Airmobile).

The 3d Brigade, 101st Airborne Division (Airmobile), redeployed from Vietnam to Fort Campbell, Kentucky on 21 December.

The 1st Brigade, 1st Cavalry Division redeployed from Vietnam to Fort Hood, Texas.

In September, the 173d Airborne Brigade redeployed to Fort Campbell, Kentucky, to be integrated into the 101st Airborne Division (Airmobile).

Combat Arms
Infantry
The 1st Battalion, 327th Infantry was assigned in November from the 1st Brigade, 101st Airborne Division (Airmobile), to the 196th Infantry Brigade at Hue/Phu-Bai, Republic of South Vietnam.

The 3d and 4th Battalions, 503d Infantry redeployed back to the United States in August, from assignment with the 173d Airborne Brigade in the Republic of South Vietnam, to be inactivated at Fort Lewis, Washington.

In May, the 3d Battalion, 506th Infantry had departed from assignment to the 1st Brigade, 101st Airborne Division (Airmobile), and inactivation at Fort Lewis Washington.

Cavalry
The 1st Battalion, 8th Cavalry redeployed to Fort Hood, Texas with the 3d Brigade, 1st Cavalry Division.

Field Artillery
The 319th, 320th and 321st Artillery Regiments were redesignated as the 319th, 320th and 321st Field Artillery Regiments.

The 2d Battalion, 26th Artillery was redesignated as the 2d Battalion, 26th Field Artillery, and inactivated at Fort Bragg, North Carolina.

The 1st Battalion, 39th Artillery was redesignated as the 1st Battalion, 39th Field Artillery.

Air Defense Artillery
The 3d Howitzer Battalion, 4th Artillery, inactivated at Fort Devens, Massachusetts in 1962, was redesignated as the 3d Battalion, 4th Air Defense Artillery.

Special Operations
In January, Brigadier General Henry E. (Hank) Emmerson, was assigned to command the US Army John F. Kennedy Center for Military Assistance, and concurrently as Commandant,US Army Institute for Military Assistance, both at Fort Bragg, North Carolina.

Military Assistance Command Vietnam, Studies and Observations Group (MACV-SOG) was inactivated in Saigon, Republic of South Vietnam. MACV-SOG was redesignated as Team 158, Strategic Technical Directorate, MACV, and activated in Saigon.

First Lieutenant Loren D. Hagen was posthumously awarded the Medal of Honor for his actions on 7 August, and Staff Sergeant Jon R. Cavaiani was awarded the Medal of Honor for his actions on 5 June. Both Special Forces soldiers were assigned to the Vietnam Training Advisory Group.

Special Forces
The 5th Special Forces Group, commanded by Colonel Michael D. Healy, redeployed to Fort Bragg, North Carolina on 3 March.

The 6th Special Forces Group was inactivated at Fort Bragg on 6 March.

Colonel Selby F. (Buster) Little assumed command of the 10th Special Forces Group at Fort Devens, Massachusetts from Colonel Reginald Woollard.

Operational Detachment B-56 (Project SIGMA), 5th Special Forces Group, was inactivated on 2 May.

The 38th Special Forces Detachment, Alaska ARNG, was reorganized and consolidated with the 36th and 37th Special Forces Detachments, to be redesignated as the 38th Special Forces Company, remaining assigned to the Alaska ARNG.

The 46th Special Forces Company, commanded by Lieutenant Colonel George Marecek, was inactivated at Lop Buri, Thailand, to be absorbed into the 3d Battalion, 1st Special Forces Group, Torri Station, Okinawa.

Rangers
Company E (Ranger), 65th Infantry, Puerto Rico ARNG, was formed from personnel assigned to the 755th Transportation Company (Medium Truck, Cargo), and activated at Fort Buchanan, Puerto Rico.

Companies C, G, N and P (Ranger), 75th Infantry were inactivated in the Republic of Vietnam.

Company H (Ranger), 75th Infantry redeployed with the 1st Cavalry Division to Fort Hood, Texas.

Company A (Ranger), 259th Infantry, Delaware ARNG, was formed from personnel assigned to the 1049th Transportation Company (Aircraft Maintenance) (GS), and activated at the Greater Wilmington Airport, Delaware.

Psychological Operations
The 4th Psychological Operations Group redeployed from the Republic of South Vietnam, and was inactivated at Fort Lewis, Washington.

The 6th and 8th Psychological Operations Battalions were inactivated respectively in the Republic of Vietnam and at Fort Lewis, Washington.

Aviation

HHC, 11th Aviation Group was relieved from assignment to the 1st Cavalry Division, and assigned to the 1st Aviation Brigade.

The 158th and 269th Aviation Battalions redeployed respectively from the Republic of South Vietnam to Fort Campbell, Kentucky and Fort Bragg, North Carolina.

The 11th Aviation Company redeployed to Fort Bliss, Texas and was inactivated.

The 617th Aviation Detachment was inactivated.

Combat Support
Engineers

HHC, 20th Engineer Brigade was inactivated in the Republic of Vietnam.

Military Intelligence

The Forces Command Intelligence Group was reorganized and redesignated as the 18th Combat Intelligence Group (Provisional.

Combat Service Support
Quartermaster

HHD, 528th Quartermaster Battalion was inactivated in the Republic of Vietnam.

---- 1972 ----

HHC, 11th Air Assault Division and the HHCs, 1st, 2d and 3d Brigades, 11th Air Assault Division, were redesignated, but not activated, as HHC, 11th Airborne Division, and HHCs, 1st, 2d and 3d Brigades, 11th Airborne Division.

Major General Frederick J. Kroesen, Senior US Army Advisor to the I Corps Tactical Zone, Republic of Vietnam, returned to Fort Bragg, North Carolina, and assumed command of the 82d Airborne Division on 17 July.

Major Generals Michael D. Healy and Thomas M. Tarpley were assigned respectively as Senior US Army Advisors to the II Corps and IV Corps Tactical Zones, Republic of South Vietnam.

The 1st Brigade, 101st Airborne Division (Airmobile), redeployed from the Republic of South Vietnam to Fort Campbell, Kentucky on 17-19 January. The 2d Brigade, 101st Airborne Division followed in February, with HQ, 101st Airborne Division closing out in March for Fort Campbell.

HHC, 173d Airborne Brigade was inactivated at Fort Campbell, Kentucky.

Combat Arms
Infantry

The 2d Battalion, 327th Infantry was assigned from the 1st Brigade, 101st Airborne Division (Airmobile) to the US Army Support Command, Cam Ranh Bay, Republic of South Vietnam, from January through March, and then assigned to US Army Forces, Military Region II.

The 1st and 2d Battalions, 327th Infantry redeployed respectively in January and April to Fort Campbell, Kentucky, and assignments with the 101st Airborne Division (Airmobile).

The 1st and 2d Battalions of the 501st and 502d Infantry redeployed from the Republic of Vietnam to Fort Campbell, Kentucky, and assignments with the 101st Airborne Division (Airmobile). The 2d Battalion, 501st Infantry was inactivated.

The 1st and 2d Battalion, 503d Infantry, relieved from assignments with the 173d Airborne Brigade, were assigned to the 101st Airborne Division (Airmobile).

The 2d Battalion, 506th Infantry, redeployed from assignment with the 1st Brigade, 101st Airborne Division (Airmobile) in the republic of Vietnam, to be inactivated at Fort Campbell, Kentucky.

Cavalry

The 2d Battalion, 8th Cavalry, and the 1st Battalion, 12th Cavalry, redeployed from the Republic of South Vietnam to Fort Hood, Texas with the 3d Brigade, 1st Cavalry Division, commanded by Brigadier General James Hamlet. Other combat arms assets of the 3d Brigade were the 1st Battalion, 7th Cavalry and the 2d Battalion, 5th Cavalry. This completed the redeployment of the 1st Cavalry Division.

Troop E, 17th Cavalry was relieved from assignment with the 173d Airborne Brigade on 14 January, and assigned to the 101st Airborne Division (Airmobile). Troop E was inactivated at Fort Campbell, Kentucky on 1 July, and relieved from assignment to the 101st Airborne Division (Airmobile).

Division Artillery

The HHB, 101st Airborne Division (Airmobile), Division Artillery redeployed from South Vietnam to Fort Campbell, Kentucky on 17-19 January.

Field Artillery

The 2d Battalion, 319th Field Artillery was relieved from assignment to the 101st Airborne Division (Airmobile), and inactivated at Fort Campbell, Kentucky. The 3d Battalion, 319th Field Artillery was relieved from assignment to the 173d Airborne Brigade, and assigned to the 101st Airborne Division (Airmobile).

Special Operations

Both Command and Control North and Command and Control Central (CCN and CCC), were respectively inactivated at Da-Nang and Kontum, Republic of South Vietnam, to be redesignated as Task Force 1 and Task Force 2, Advisory Element, MACV-SOG. They were both inactivated later this same year.

Special Forces

The 3d Battalion, 7th Special Forces Group relocated from Fort Bragg, North Carolina to the Panama Canal Zone.

The 8th Special Forces Group was inactivated in the Panama Canal Zone, with personnel being assigned to the recently arrived 3d Battalion, 7th Special Forces Group.

The Special Forces Training Group, Fort Bragg, North Carolina, was redesignated as the Military Assistance Training Battalion.

Project OMEGA was discontinued in the Republic of South Vietnam.

Major James Horak and personnel assigned to the inactivated Company E (Ranger), 200th Infantry, were assigned to Company A, 1st Battalion, 20th Special Forces Group, Alabama ARNG.

Company E (Ranger), 425th Infantry, Michigan ARNG, was consolidated with Company F (Ranger), 425th Infantry, and inactivated.

Psychological Operations

The 4th Psychological Operations Group was assigned to the John F. Kennedy Center for Military Assistance, and activated at Fort Bragg, North Carolina.

The 6th and 8th Psychological Operations Battalions were activated at Fort Bragg, North Carolina.

The 1st, 6th and 8th Psychological Operations Battalions were assigned to the 4th Psychological Operations Group.

Aviation

HHC and Companies B and D, 229th Aviation Battalion redeployed from the Republic of South Vietnam to Fort Hood, Texas, where they were relieved from assignment to the 1st Cavalry Division and inactivated.

The 173d Aviation Company redeployed to Fort Campbell, Kentucky from the Republic of South Vietnam and was inactivated.

The US Army Center for Military Assistance/John F. Kennedy Institute for Military Assistance Aviation Detachment was inactivated at Fort Bragg, North Carolina.

Medical

The Joint Casualty Resolution Center (JCRC), was organized in Saigon, Republic of South Vietnam, as a humanitarian Special Operations organization, to account for missing and/or captured US military personnel. JCRC was relocated to Nakhom Phanom Royal Thai Air Force Base, Thailand.

Combat Support
Engineers

The 173d Engineer Company was relieved from assignment to the 173d Airborne Brigade, and inactivated.

Military Intelligence

The 297th Military Intelligence Detachment (Special Forces Group), USAR, was assigned to the 63d Army Reserve Command, while remaining attached to the 19th Special Forces Group, Utah ARNG.

The 401st Army Security Agency Detachment was relieved from assignment to the 8th Special Forces Group, and inactivated at Fort Gulick, Panama.

Military Police

On 8 July and 11 August, the 503d Military Police Battalion provided the 108th Military Police Company, and one platoon of the 65th Military Police Company, from Fort Bragg, North Carolina, during the Democratic and Republican Party conventions. These units were deployed to assist in quelling any potential civil disturbances.

From February through July, the 503d Military Police Battalion was instrumental in providing security for returning Prisoners of War and their families, during Operation EGRESS.

Combat Service Support

HHC, Special Troops, 1st Field Army Support Command was reorganized and redesignated as HHC, 1st Corps Support Command (1st COSCOM). Although not designated as an airborne command, several airborne positions were designated within the staff elements. 1st COSCOM provided command and control guidance to the (assigned) two largest airborne rigger companies in the US Army, in support of the XVIII Airborne Corps.

HQ, 101st Airborne Division (Airmobile) Support Command redeployed from the Republic of South Vietnam to Fort Campbell, Kentucky on 17-19 January.

Chemical

The 445th Chemical Detachment was inactivated at Fort Lawton, Washington

Transportation

HHC, 5th Transportation Battalion (Aircraft Maintenance), redeployed to Fort Campbell, Kentucky with the 101st Airborne Division (Airmobile).

—— 1973 ——

Lieutenant General Richard J. Seitz assumed command of the XVIII Airborne Corps.

The 71st Airborne Brigade, Texas ARNG, was redesignated as the 36th Airborne Brigade.

Combat Arms
Infantry

HHC, 1st Brigade (Airborne), 8th Infantry Division was redesignated as HHC, 1st Brigade, 8th Infantry, remaining in Germany assigned to US Army Europe.

The 1st and 2d Battalions (Airborne), 143d Infantry, Texas ARNG, were reassigned from the 71st Airborne Brigade to the 36th Airborne Brigade. The 3d Battalion (Airborne), 143d Infantry was inactivated.

The 1st Battalion, 509th Infantry was relieved from assignment to the 1st Brigade (Airborne), 8th Infantry Division, and relocated to Vincenza, Italy to be assigned to the Southern European Task Force (SETAF). The 2d Battalion, 509th Infantry was inactivated in Germany. The 3d Battalion, 509th Infantry was activated in Germany but later inactivated in Vincenza, Italy.

Cavalry

The 3d Squadron, 17th Cavalry was inactivated at Fort Lewis, Washington.

Special Operations

Brigadier General Michael D. Healy was concurrently assigned as the Commanding General of the US Army Center for Military Assistance and Commandant, the John F. Kennedy Institute for Military Assistance, at Fort Bragg, North Carolina.

Command and Control South (CCS), was inactivated at Ban-Me-Thout, Republic of Vietnam, and redesignated as Task Force 3, assigned to the Advisor Element, Military Assistance Command Vietnam, Studies and Observations Group, but was inactivated shortly after activation.

Special Forces

HQ, 11th Special Forces Group, USAR, was relocated from Tappan, New York to Fort Meade, Maryland.

Aviation

HHC, 11th Aviation Group redeployed from the Republic of South Vietnam to Fort Benning, Georgia.

Combat Support
Military Intelligence

The 525th Military Intelligence Group redeployed from the Republic of South Vietnam to Oakland, California, and inactivation.

Combat Service Support
Adjutant General

The 101st Administration Company was reorganized and redesignated as the 101st Adjutant General Company.

HHC and Band, 101st Airborne Division (Airmobile) Support Command (less Band), was reorganized and redesignated as HHC, 101st Airborne Division (Airmobile) Support Command (101st DISCOM). The Band was withdrawn from HHC and Band, and assigned to the 101st Adjutant General Company.

—— 1974 ——

Major General Thomas H. Tackaberry was named to command the 82d Airborne Division.

The 101st Airborne Division (Airmobile) was redesignated as the 101st Airborne Division (Air Assault).

Special Operations
Special Forces

The 1st Special Forces Group was inactivated at Fort Bragg, North Carolina. Special Forces Detachment-Korea (Det K), was constituted in the AUS, from personnel assigned to the inactivated 1st Special Forces Group. Det-K was activated at Seong-Nam, Republic of Korea.

Rangers

Company B (Ranger), 75th Infantry was inactivated.

Company C (Ranger), 75th Infantry, (inactivated in Vietnam in 1971), was consolidated with Company A, 1st Ranger Battalion (inactivated in Korea in 1951), with the consolidated unit designated as HHC, 1st Battalion (Ranger), 75th Infantry. HHC, 1st Battalion (Ranger), was activated at Fort Stewart, Georgia.

Company H (Ranger), 75th Infantry, (inactivated at Fort Hood, Texas in 1972), was consolidated with Company A, 2d Infantry Battalion, (inactivated at Fort Hamilton, New York in 1960), with the consolidated unit designated as HHC, 2d Battalion (Ranger), 75th Infantry. HHC, 2d Battalion (Ranger), was activated at Fort Lewis, Washington.

Company A (Ranger), 259th Infantry, Delaware ARNG, was inactivated at the Greater Wilmington Airport, Delaware.

Psychological Operations

The 9th Psychological Operations Battalion was inactivated at Fort Gulick, Panama.

Civil Affairs

The 3d and 5th Civil Affairs Groups were inactivated respectively at Fort Clayton, Panama and Fort Bragg, North Carolina.

The 96th Civil Affairs Company was inactivated at Fort Lee, Virginia, with personnel assets used to form the 96th Civil Affairs Battalion. The 96th Civil Affairs Battalion was activated at Fort Bragg, North Carolina, and assigned to the John F. Kennedy Institute for Military Assistance.

The 28th Civil Affairs Company was inactivated at Fort Bragg, North Carolina.

Combat Support
Engineers

HHC, 20th Engineer Brigade was activated at Fort Bragg, North Carolina, and assigned to the XVIII Airborne Corps.

Military Intelligence

The 400th Army Security Agency Detachment was inactivated on Okinawa, concurrent with the inactivation of the 1st Special Forces Group. The 400th Army Security Agency Detachment was activated at Fort Bragg, North Carolina and assigned to the 5th Special Forces Group.

Combat Service Support

HHC and Band, 82d Airborne Division Support Command was reorganized and redesignated (less Band), as HHC, 82d Airborne Division Support Command (82d DISCOM).

Quartermaster

The 407th Supply and Transportation Company was reorganized and redesignated as the 407th Supply and Service Battalion, and activated at Fort Bragg, North Carolina, assigned to the 82d Airborne Division.

—— 1975 ——

Lieutenant General Henry E. (Hank) Emmerson was appointed to command the XVIII Airborne Corps.

On 11 October, Major General Roscoe Robinson Jr became the first black Commanding General of the 82d Airborne Division.

Combat Arms
Infantry

The 3d Battalion, 509th Infantry was redesignated as Company C, 509th Infantry, and assigned as the Pathfinder Company for the US Army Aviation Center and School, Fort Rucker, Alabama.

Special Operations

Brigadier General Robert C. Kingston was named to assume the concurrent responsibilities as Commander, US Army Center for Special Warfare and as Commandant, John F. Kennedy Institute for Military Assistance, at Fort Bragg, North Carolina.

Combat Support
Military Intelligence

The 77th Military Intelligence Detachment was reorganized and redesignated as the 77th Military Intelligence Company (Special Action Force).

The 297th and 389th Military Intelligence Detachments (Special Forces Group), were reorganized and redesignated as the 297th and 389th Military Intelligence Detachments (Special Action Force).

—— 1976 ——

Combat Arms
Artillery

The 2d Battalion, 26th Field Artillery was reorganized and redesignated as Battery B, 26th Field Artillery, assigned to the 82d Airborne Division, and activated at Fort Bragg, North Carolina.

Combat Support
Military Intelligence

The 400th Army Security Agency Detachment was inactivated at Fort Bragg, North Carolina.

Medical

The Joint Casualty Resolution Center was relocated from Nakhom Phanom Royal Thai Air Force Base, Thailand, to Barbera Point Naval Air Station, Hawaii. A liaison was retained in Bangkok, Thailand.

—— 1977 ——

Lieutenant General Volney E. Warner assumed command of the XVIII Airborne Corps.

Combat Arms
Infantry

The 5th Airborne Battle Groups, 325th and 504th Infantry, (both inactivated in 1957), were redesignated as Companies E, 325th and 504th Infantry. They were assigned as TOW guided missile anti-tank companies, of their respective regiments, assigned to the 82d Airborne Division, and activated at Fort Bragg, North Carolina.

Company E, 151st Infantry, Indiana ARNG, was inactivated in Indiana.

Special Operations

Brigadier General Jack V. Mackmull was appointed commander and commandant, of the US Army John F. Kennedy Center for Military Assistance and the US Army Institute for Military Assistance, Fort Bragg, North Carolina.

Augmentation, First Special Forces Operational Detachment, Delta, was constituted in the Regular Army on 19 December 1977, at Fort Bragg, North Carolina, and assigned to US Army Forces Command.

Combat Service Support
Chemical

The 445th Chemical Detachment (NBC Reconnaissance) (Special Forces) was activated at Columbus, Georgia, and assigned to the 81st Army Reserve Command, First US Army.

The 801st Chemical Detachment (NBC Staff), was constituted in the USAR, assigned to the 121st Army Reserve Command, First US Army, and activated at Anniston, Alabama.

The 445th and 801st Chemical Detachments were both affiliated with the 5th Special Forces Group.

1978

Major General G.S. Meloy assumed command of the 82d Airborne Division.

Combat Arms
Corps Artillery

XVIII Airborne Corps Artillery was reorganized and redesignated as the 18th Field Artillery Brigade.

HHC, 36th Airborne Brigade, Texas ARNG, was inactivated at Houston, Texas.

Special Operations

Augmentation, Special Forces Operational Detachment, Delta was redesignated as the First Special Forces Operational Detachment, Delta (Airborne) on 20 July, under the command of Colonel Charlie Beckworth. Colonel Richard Potter was appointed as the Deputy Commander

Special Forces

The 6th Battalion, 20th Special Forces Group, Rhode Island ARNG, was constituted as Company D, 16th Special Forces Group, Rhode Island ARNG in 1962, and redesignated as Company D, 19th Special Forces Group in 1966. The 6th Battalion, 20th Special Forces Group (Rhode Island ARNG) in 1972 was activated in 1972, and inactivated in 1978 at Camp Fogarty, Rhode Island. Personnel were assigned to Command and Control, Rhode Island Army National Guard, Providence, Rhode Island.

Aviation

HHC and Companies B and D, 229th Aviation Battalion were assigned to the 101st Airborne Division (Air Assault), and activated at Fort Campbell, Kentucky.

Combat Support
Military Intelligence

The 5th Military Intelligence Detachment was reorganized and redesignated as the 15th Military Intelligence Company.

The 82d Military Intelligence Company was assigned to the 82d Airborne Division.

Company A, 311th Army Security Agency Battalion, disbanded in 1966, was reconstituted in the Regular Army as the 335th Army Security Agency Company.

1979

Lieutenant General Thomas H. Tackaberry assumed command of the XVIII Airborne Corps. The XVIII Airborne Corps was designated as the Army component of the US Rapid Deployment Joint Task Force.

Combat Arms
Infantry

The German Long Range Reconnaissance Patrol School was formally inaugurated in the Danude Valley by Major General Heyde, General of Combat Troops, Army of the Federal Republic of Germany.

Aviation

Companies C and D, 82d Aviation Battalion were constituted in the Regular Army, activated at Fort Bragg, North Carolina, and assigned to the 82d Airborne Division.

Company D, 159th Aviation Battalion was constituted and activated at Fort Campbell, Kentucky, and assigned to the 101st Airborne Division (Air Assault).

Combat Support
Military Intelligence

The 18th Combat Intelligence Group (Provisional) was reorganized and redesignated as the 525th Military Intelligence Group (Corps) (Airborne), relocated to Fort Bragg, North Carolina, and assigned to the XVIII Airborne Corps.

The 405th Army Security Agency Detachment was reorganized and redesignated as the 405th Army Security Agency Company, activated at Fort Polk, Louisiana, and assigned to the 5th Infantry Division (Mechanized).

HHC, 313th Army Security Agency Battalion was reorganized and redesignated as HHC, 313th Military Intelligence Battalion, and assigned to the 82d Airborne Division.

The 358th Army Security Agency Company and the 82d Military Intelligence Company were reorganized and redesignated as Companies A and B, 313th Military Intelligence Battalion.

The 336th Army Security Agency Company was activated at Fort Bragg, North Carolina.

Signal

HHD, 35th Signal Group was reorganized and redesignated as the 35th Signal Brigade (Corps) (Airborne), remaining assigned to the XVIII Airborne Corps.

Combat Service Support

The 808th Chemical Detachment was inactivated at New York City, New York.

1980

Major General Donald E. Rosenblum, the first commander of the 24th Infantry Division upon it's reactivation in 1975, was appointed as Deputy Commanding General, XVIII Airborne Corps.

HHC, 36th Airborne Brigade (Texas ARNG), (inactivated at Houston, Texas in 1978), was converted and redesignated as HHC, 386th Engineer Battalion (Texas ARNG).

Combat Arms
Infantry

The German Long Range Reconnaissance Patrol School was redesignated as the International Long Range Reconnaissance Patrol (ILRRP) School, and relocated in January from Neuhausen Ob Eck to the Welfenkaserne, Weingarten, Germany. A Memorandum of Agreement was signed in April, whereby cadre from Belgium, Germany and the United Kingdom were jointly assigned to the ILRRP School at Neuhausen Ob Eck.

Special Operations

The Joint Special Operations Command (JSOC) was constituted in the Regular Army and activated at Fort Bragg, North Carolina, designed to study the special operations requirements of all military services, to ensure standardization. Brigadier General Richard A. Scholtes, former Assistant Division Commander, 82d Airborne Division, was assigned as the first JSOC Commander, and promoted to Major General.

Brigadier General Joseph C. Lutz assumed command of the US Army John F. Kennedy Center for Military Assistance, and concurrently as Commandant as the US Army Institute for Military Assistance at Fort Bragg, North Carolina. Colonel Charles Norton, past commander of the 7th Special Forces Group, and former alumni of the 5th and 10th Special Forces Groups, was appointed as the Deputy Commander.

The First Special Operations Detachment - Delta, under the command of Colonel Charlie Beckworth, participated in the unsuccessful attempt to release US hostages from Iran, code named Operation EAGLE CLAW. The action was aborted on 25 April, following aircraft collisions in the Iranian desert, at operational phase site 'Desert One'.

The Military Assistance Student Battalion, assigned to the JFK Institute for Military Assistance, Fort Bragg, North Carolina, was redesignated as the Special Forces School Battalion.

Rangers

Company G (Ranger), 143d Infantry, was constituted in the Texas ARNG, and activated at Houston, Texas, from elements of Company A, 2d Battalion, 143d Infantry, as that battalion was inactivated. The lineage and heritage of Company G (Ranger), may be traced back to the Houston Light Guard, Texas Militia, participants in the Mexican Wars.

Combat Support
Signal

The Joint Communication Unit was concurrently constituted, activated, and assigned to the Joint Special Operations Command.

Combat Service Support

HHC, 1st Corps Support Command (1st COSCOM), was redesignated as the 1st Support Command (1st SUPCOM). Colonel William J. Richardson assumed command of the 1st COSCOM.

Chemical

The 306th Chemical Detachment (NBC Staff) (Special Forces), and the 382d Chemical Detachment (NBC Reconnaissance) (Special Forces), were constituted in the USAR, activated at Arlington Heights, Illinois, and assigned to the 86th Army Reserve Command.

―― 1981 ――

Lieutenant General Jack V. Macmull, (former Commander of the US Army John F. Kennedy Center for Military Assistance and the US Army Institute for Military Assistance); and commander of the 101st Airborne Division (Air Assault), assumed command of XVIII Airborne Corps. HQ, XVIII Airborne Corps was assigned to Third US Army in support of US Central Command. Major General James J. Lindsay was named to Command the 82d Airborne Division.

Combat Arms
Infantry

Greece became the fourth active cadre provider to the International Long Range Reconnaissance Patrol School, with Greek instructors assigned at Welfenkaserne, Weingarten, Germany.

Aviation

Company D, 159th Aviation Battalion, assigned to the 101st Airborne Division (Air Assault), reorganized and redesignated as the 160th Aviation Battalion.

Medical

The Special Forces Branch was organized at the Academy of Health Sciences, Fort Sam Houston, Texas.

Combat Support
Military Police

HHD, 16th Military Police Group was reorganized and redesignated as HHC, 16th Military Police Brigade, assigned to XVIII Airborne Corps at Fort Bragg, North Carolina.

Combat Service Support
Chemical

The 426th Chemical Detachment (NBC Reconnaissance), was constituted in the USAR, activated at Fort Meade, Maryland, and assigned to the 11th Special Forces Group under the 97th Army Reserve Command, First US Army.

The 808th Chemical Detachment was redesignated as the 808th Chemical Detachment (NBC Staff) (Special Forces), was also assigned to the 11th Special Forces Group, and activated at Fort Meade.

The 900th and 901st Chemical Detachments were constituted in the USAR, assigned to the 63d Army Reserve Command, attached to the 19th Special Forces Group, and activated at Tucson, Arizona.

―― 1982 ――

Lieutenant General James A. Lindsay was relieved from assignment as Commander, 82d Airborne Division, to become Commanding General, XVIII Airborne Corps.

Major General Donald E. Rosenblum, Deputy Commanding General, XVIII Airborne Corps, was promoted to Lieutenant General, to command the First US Army, Fort Meade, Maryland.

Major General Edward L. Trobaugh assumed command of the 82d Airborne Division.

Combat Arms
Infantry

As the US Army Light Infantry Divisions became more robust, An added capability designed to enhance the Light Infantry Divisions - the Long Range Surveillance Units (LRSU) - gradually commenced to be adopted and expanded within the US Army. The normal LRSU organization comprises either four or six airborne reconnaissance teams (tailored for Light or Heavy Divisions), with six personnel assigned per team.

The International Long Range Surveillance Patrol (ILRRP) School hosted the first LRRP/Special Forces Symposium, with 11 NATO nations participating. the ILRRP School provides Long Range Reconnaissance Patrolling, Survival, Leadership and Recognition course for NATO members. Lieutenant Colonel Wendeborn was appointed as the commander on 26 September.

Combat Arms
Infantry

The Scout Company, 9th Infantry Division was established to perform Long Range Reconnaissance Patrol missions for the 9th Infantry Division.

Special Operations

Command and Control, Rhode Island ARNG, was redesignated as Troop Command, Rhode Island ARNG.

Aviation

The 45th Aviation Battalion was constituted and activated in the Oklahoma ARNG, at Sperry, Oklahoma.

Combat Support

HHC, 1st Support Command (1st SUPCOM), was redesignates as HHC, 1st Corps Support Command (1st COSCOM).

Military Intelligence

The 319th Military Intelligence Battalion was activated at Fort Bragg, North Carolina. Concurrently, the 336th Army Security Agency Company (activated at Fort Bragg in 1979), was redesignated as Company B, 319th Military Intelligence Battalion.

The 15th Military Intelligence Company and the 405th Army Security Agency Company were concurrently inactivated and consolidated. The consolidated unit was designated as the 105th Military Intelligence Battalion, assigned to the 5th Infantry Division (Mechanized).

The 82d Airborne Division employed the Reconnaissance Platoon, Company B, 313th Military Intelligence Battalion to perform Long Range Reconnaissance Patrol duties.

1983

HQ, XVIII Airborne Corps commanded US Army elements, including the 82d Airborne Division, deployed to Grenada in support of Operation URGENT FURY. Special Operations Forces deployed to Grenada under the command of the 1st Special Operations Command.

Combat Arms
Infantry

The 1st Battalion, 187th Infantry, was relieved from assignment to 11th Airborne Division, and assigned to the 193d Infantry Brigade in Panama. The 2d Airborne Battle Group, 187th Infantry (inactivated in 1964), was redesignated as the 2d Battalion, 187th Infantry, activated in Panama and assigned to the 193d Infantry Brigade, replacing the 1st Battalion.

The 325th Infantry was withdrawn from the Combat Arms Regimental System (CARS), and reorganized as a parent regiment under the US Army Regimental System (USARS).

The 4th Battalion, 325th Infantry was activated at Vincenza, Italy and assigned to SETAF.

Companies C, D, E and F, 327th Infantry (inactivated at Fort Campbell in 1957), were reorganized and redesignated as the 3d, 4th, 5th and 6th Battalions, 327th Infantry. The 3d Battalion, 327th Infantry was activated at Fort Campbell, and assigned to the 101st Air Assault Division. The 4th, 5th and 6th Battalions were assigned to the 172d Infantry Brigade, Alaska, with the 4th and 5th Battalions located at Fort Richardson, and the 6th Battalion at Fort Wainwright.

The 2d Battalion, 503d Infantry was relieved from assignment to the 101st Air Assault Division, and inactivated at Fort Campbell, Kentucky. The 1st Battalion, 509th Infantry was inactivated at Vincenza, Italy.

Cavalry

The 1st Battalion, 12th Cavalry was inactivated at Fort Hood, Texas, and relieved from assignment to the 1st Cavalry Division.

Special Operations

The 1st Special Operations Command (Airborne) (1st SOCOM), was activated at Fort Bragg, North Carolina from the former US Army John F. Kennedy Center for Military Assistance (USAJFKCENMA). Brigadier General Joseph C. Lutz, former commander of the USAJFKCENMA and concurrently Commandant of the US Army Institute for Military Assistance, became the first Commander of 1st SOCOM. 1st SOCOM at Fort Bragg, was assigned under the US Army Forces Command.

The US Army Institute for Military Assistance, also at Fort Bragg, was redesignated as the John F. Kennedy Special Warfare Center, and assigned to the US Army Training and Doctrine Command The interim commander was named as Colonel David L. Pemberton, (former Deputy Commandant under Brigadier General Lutz), until the assignment of Brigadier General Robert Wiegand was accomplished.

The Special Forces School Battalion was redesignated as the Special Forces Training Battalion.

Brigadier General Joseph C. Lutz, Commander, 1st Special Operations Command, coordinated and commanded all Special Operations Forces involved in Operation URGENT FURY in Grenada.

Special Forces

The 5th, 7th and 10th Special Forces Groups were assigned to the 1st Special Operations Command.

Rangers

The 1st and 2d Battalions, 75th Infantry (Ranger), assigned to the 1st Special Operations Command (1st SOCOM), conducted airborne assaults into Grenada.

The 3d Reconnaissance Company (Provisional), was established as the Long Range Reconnaissance Patrol unit of the 3d Infantry Division.

Psychological Operations

Elements of the 4th Psychological Operations Group, assigned to the 1st SOCOM, deployed during Operation URGENT FURY to Grenada.

Civil Affairs

Elements of the 358th Civil Affairs Brigade (USAR), volunteered and deployed to Grenada from Pennsylvania, to work with the 96th Civil Affairs Battalion, (1st SOCOM), deployed from Fort Bragg, during Operation URGENT FURY.

Combat Service Support

The 82d Airborne Division Support Command deployed with the 82d Airborne Division to Grenada, on Operation Urgent Fury.

Chemical

The 306th Chemical Detachment (NBC Staff) (Special Forces) and the 382d Chemical Detachment (NBC Reconnaissance) (Special Forces), were assigned to the 12th Special Forces Group, 86th Army Reserve Command, Fifth US Army.

The 426th Chemical Detachment (Reconnaissance), was redesignated as the 426th Chemical Detachment (Reconnaissance) (Special Forces). The 445th Chemical Detachment (Reconnaissance) (Special Forces), was reassigned with the 81st Army Reserve Command from First US Army to Second US Army.

The 801st Chemical Detachment (NBC Staff) was redesignated as the 801st Chemical Detachment (NBC Staff) (Special Forces).

The 900th and 901st Chemical Detachments were redesignated respectively as the 900th Chemical Detachment (NBC Reconnaissance) (Special Forces) and the 901st Chemical Detachment (NBC Staff) (Special Forces).

1984

Lieutenant General Donald E. Rosenblum retired, following command of the First US Army.

Major General John Foss was designated as Commander, 82d Airborne Division.

Combat Arms
Infantry

The 5th Airborne Battle Group, 187th Infantry, (inactivated in 1957), was reorganized and redesignated as the 5th Battalion, 187th Infantry, activated at Fort Campbell, Kentucky, and assigned to the 101st Air Assault Division.

Company E, 325th Infantry, was inactivated at Fort Bragg, North Carolina.

The 3d, 4th, 5th and 6th Airborne Battle Groups, 502d Infantry, (relieved from the 101st Airborne Division and inactivated at Fort Campbell, Kentucky in 1957), were activated, reorganized and redesignated as the 3d, 4th, 5th and 6th Battalions, 502d Infantry.

The 3d Battalion, 502d Infantry was assigned to the 101st Air Assault Division at Fort Campbell, Kentucky.

The 4th, 5th and 6th Battalions, 502d Infantry were assigned to the Berlin Brigade and activated at Berlin, Germany.

The 1st Battalions, 503d and 506th Infantry were relieved from assignment to the 101st Air Assault Division, and inactivated at Fort Campbell, Kentucky.

Armor

The 4th Battalion, 68th Armor was relieved from assignment to the 82d Airborne Division on 15 February, and assigned to the 4th Infantry Division. The 4th Battalion, 68th Armor was replaced by another Sheridan tank battalion, with an airborne capability - the 3d Battalion, 73d Armor.

The 3d Battalion, 73d Armor was constituted in the AUS in 1941 as Company C, 76th Tank Battalion, redesignated as Co C, 756th Tank Battalion, and activated at Fort Lewis, Washington. Inactivated at Camp Kilmer, and activated at Fort Benning in 1946, Company C was redesignated in 1948 as Company C, 756th Heavy Tank Battalion. Redesignated as Company C, 73d Heavy Tank Battalion in 1949, they were again redesignated, to become Company C, 73d Tank Battalion in 1959. Company C was assigned to the 7th Infantry Division from 1951 until inactivated in Korea in 1957. In 1984, Company C, 73d Tank Battalion was reorganized and redesignated as HHC, 3d Battalion, 73d Armor, assigned to the 82d Airborne Division, and activated at Fort Bragg, North Carolina.

Special Operations

Major General Carl W. Stiner was assigned as the second commander of the Joint Special Operations Command, Fort Bragg, North Carolina. The outgoing commander, Major General Richard A. Scholtes was appointed as the commander of the 2d Armored Division at Fort Hood, Texas.

The Joint Special Operations Command was awarded the Joint Meritorious Unit Award by the General John Vessey, Chairman, Joint Chiefs of Staff, dated 17 January 1984, for actions in Grenada.

Brigadier General Joseph C. Lutz was promoted to Major General, and assigned as the Chief, Military Advisory Assistance Group (MAAG), to Greece. His replacement at the 1st Special Operations Command was Major General Leroy N. Suddath Jr, former Provost Marshall for Berlin, Germany, and former Company, Battalion and Brigade Commander with the 82d Airborne Division. Retaining institutional knowledge, Colonels Paul Fisher, as the Deputy Commander, Colonel Donald J. Soland as the Chief of Staff, and Command Sergeant Major James Hargraves, were key figures during this transition.

Special Forces

The 1st Battalion, 1st Special Forces Group was activated at Fort Bragg under the command of Lieutenant Colonel James Estep, and assigned to the 1st Special Operations Command. The 1st Battalion relocated to Torri Station, Okinawa, and the 1st Special Forces Group, commanded by Colonel David Barrato, was activated at Fort Lewis, Washington.

Rangers

The 75th Infantry, assigned to the 1st Special Operations Command, was reorganized, redesignated and activated at Fort Benning, Georgia, as the Regimental Headquarters, 75th Infantry, under the command of Colonel Wayne P. Downing.

Concurrently activated at Fort Benning, Georgia, was HHC, 3d Battalion, 75th Infantry. The 3d Battalion's lineage is defined from a consolidation of the former Company F, 75th Infantry (inactivated in March 1971 in Vietnam), and Company A, 3d Ranger Battalion (inactivated in Korea in 1961).

Medical

The Special Forces Branch, Academy of Health Sciences, Fort Sam Houston, Texas, commanded by Major Warner 'Rocky' Farr, was redesignated as Company F, 3d Battalion, Academy of Health Sciences.

Combat Support
Military Intelligence

The 77th, 297th and 389th Military Intelligence Companies (Special Action Force), were reorganized and redesignated as the 77th, 297th and 389th Military Intelligence Companies (Combat Electronic Warfare Intelligence) (Airborne). The 356th and 389th Military Intelligence Detachments were reorganized and redesignated as the 356th and 389th Military Intelligence Detachments (Combat Electronic Warfare Intelligence) (Special Forces Group). The 389th Military Intelligence Detachment was assigned to the 125th Army Reserve Command.

Combat Service Support
Chemical

The 306th Chemical Detachment (NBC Staff) (Special Forces), and the 382d Chemical Detachment (NBC Reconnaissance) (Special Forces), were reassigned with the 12th Special Forces Group and the 86th Army Reserve Command, from the Fifth US Army to the Fourth US Army.

The 801st Chemical Detachment (NBC Staff) (Special Forces), was reassigned with the 121st Army Reserve Command, from First US Army to Second US Army.

—— 1985 ——
Combat Arms
Infantry

The 507th Infantry was relieved from assignment to the 17th Airborne Division, and reorganized as a parent regiment under the US Army Regimental System (USARS). At last the 507th Infantry was placed on line with the other airborne infantry regiments. Company A, 507th Airborne Infantry Regiment, (inactivated at Camp Pickett in 1949), was reorganized and redesignated as the 1st Battalion, 507th Infantry. Assigned to the US Army Training and Doctrine Command (TRADOC), the 1st Battalion, 507th Infantry was assigned to the Airborne Department, US Army Infantry Center and School, and activated at Fort Benning, Georgia.

Cavalry

The LRSDs, 2d Squadron, 10th Cavalry and the 1st Squadron, 167th Cavalry, was assigned as the Long Range Surveillance Detachments of the 7th Infantry Division (Light) at Fort Ord, California, and the 35th Infantry Division (Nebraska ARNG), at Omaha, Nebraska. The LRSDs provided the primary Human Intelligence assets for Deep Battle operations. The LRSD, 2d Squadron, 10th Cavalry provides the only airborne capability within the 7th Infantry Division (Light).

Troop E, 17th Cavalry was redesignated as HHT, 5th Squadron, 17th Cavalry and activated at Fort Hood, Texas, with organic elements constituted and activated.

Special Operations

HQ, Support Operations, Task Force Europe, activated at Stuttgart, Germany in 1955, was redesignated as Special Operations Command, Europe (SOCEUR).

Constituted in the USAR, (as conceived by Lieutenant Colonel Geoffrey T. Barker, 1st Special Operations Command, Fort Bragg), the 1st Special Operations Command Augmentation, was activated at Fort Bragg, North Carolina, assigned to the 120th US Army Reserve Command.

The first commander was Colonel Joseph C. Hurteau. This organization was designed to replace those elements of 1st Special Operations Command required to deploy in support of global contingencies.

Another element coordinated to augment 1st SOCOM was the Troop Command, Rhode Island ARNG, which was redesignated as Detachment 1, (Special Forces), Troop Command, RI ARNG.

Discussions to establish Detachments 2 and 3, (Special Forces), from the Troop Commands, Maryland and West Virginia ARNG, were addressed, but not finalized. The Troop Commands of Rhode Island, Maryland and West Virginia consist of many Special Operations qualified personnel, formerly assigned to the ARNG Special Forces elements within their respective states. (The deployment capability of 1st SOCOM was no longer required with the advent of the unified combatant US Special Operations Command, activated in 1987.

The Special Forces Training Battalion, JFK Special Warfare Center, Fort Bragg, was redesignated as the 1st Special Warfare Training Battalion (Airborne).

Rangers

Company F (Ranger), 425th Infantry, (Michigan ARNG), was reorganized and redesignated as Company F (Long Range Reconnaissance Patrol), 425th Infantry.

Psychological Operations

The 9th Psychological Operations Battalion was activated at Fort Bragg, North Carolina, and assigned to the 4th Psychological Operations Group.

Aviation

The 160th Aviation Battalion, Fort Campbell, Kentucky, was assigned to the 1st Special Operations Command.

Medical

Company F, 3d Battalion, Academy of Health Sciences was redesignated as Company F (Airborne), 3d Battalion, Academy of Health Sciences Brigade, Fort Sam Houston, Texas.

Combat Support
Military Intelligence

The 356th and 389th Military Intelligence Detachments (Combat Electronic Warfare Intelligence) (Special Forces Group) were reorganized and redesignated as the 356th and 389th Military Intelligence Companies (Combat Electronic Warfare Intelligence) (Special Forces Group) (Airborne).

1986

Lieutenant General John W. Foss, former commander of the 82d Airborne Division, assumed command of the XVIII Airborne Corps.

Major General Carl W. Stiner was assigned to command the 82d Airborne Division.

Combat Arms
Infantry

The 4th, 5th and 6th Battalions, 327th Infantry, were relieved from assignment to the 172d Infantry Brigade, remaining in Alaska, and assigned to the 6th Infantry Division (Light).

The 503d, 504th and 509th Infantry were withdrawn from the Combat Arms Regimental System (CARS), and reorganized as parent regiments under the US Army Regimental System (USARS), with the 503d Infantry Headquarters in Korea, the 504th Infantry Headquarters at Fort Bragg, North Carolina, and the 509th Infantry headquarters at Fort Rucker, Alabama.

The 2d Battalion, 503d Infantry was assigned to the 2d Infantry Division, and activated in Korea.

The 3d Battalions, 504th and 505th Infantry were activated at Fort Bragg; and Companies E, 504th and 505th Infantry (TOW), were inactivated at Fort Bragg.

The 1st and 2d Battalions, 508th Infantry were relieved from assignment to the 82d Airborne Division and inactivated at Fort Bragg, North Carolina.

The United States entered into a Memorandum of Understanding with the International Long Range Reconnaissance Patrol (ILRRP) School, Weingarten, Germany, whereby the US would provide instructor cadre.

Company F (LRSC), 51st Infantry, was activated as the Long Range Surveillance Company assigned to VII Corps in Germany. Company F (LRSC), 51st Infantry was commanded by Captain Robert Kennedy, assisted by First Sergeant Kenneth Floyd.

Company G (Ranger), 75th Infantry, inactivated in the Republic of Vietnam in 1971, was redesignated as Company E (LRSC), 51st Infantry, and attached to the 165th Military Intelligence Battalion, in support of V Corps in Germany.

Company D (Ranger), 151st Infantry, (inactivated in 1977), was redesignated as Troop F (LRSU), 1st Squadron, 238th Cavalry and activated at Muncie, Indiana. The first commander was Captain Joel M. Wierenga. The Detachment Sergeant was Sergeant First Class William E. Butler, who had previously served with Company D (Ranger), 151st Infantry in the Republic of Vietnam, and prior to that, with the 5th Ranger Company in Korea.

Cavalry

Second Lieutenant Mark Renfrow and SFC Ronald Deitch headed the newly organized LRSD, 1st Squadron, 18th Cavalry, 40th Infantry Division. The LRSD was formed from the Reconnaissance Platoon, Troop D, 1st Squadron, 18th Cavalry.

The LRSD, Troop E, 5th Squadron, 117th Cavalry, assigned to the 50th Armored Division (New Jersey ARNG), was organized at Westfield, New Jersey on 2 October. Primarily personnel were assigned from the Reconnaissance Platoon, Troop D (Air), 5th Squadron, 117th Cavalry. The first commander was Captain Richard J. Murphy, formerly of Troop B, and the Detachment Sergeant was named as Sergeant First Class Steven A. Halasz, assigned through the Active Guard Reserve (AGR) program.

Company D (Ranger), 151st Infantry, (inactivated in 1977), was redesignated as Troop F (LRSU), 1st Squadron, 238th Cavalry and activated at Muncie, Indiana. The first commander was Captain Joel M. Wierenga. The Detachment Sergeant was Sergeant First Class William E. Butler, who had previously served with Company D (Ranger), 151st Infantry in the Republic of Vietnam, and prior to that, with the 5th Ranger Company in Korea.

Field Artillery

The 320th Field Artillery Regiment was withdrawn from the Combat Arms Regimental System, and reorganized as a parent regiment under the US Army Regimental System, with headquarters at Fort Campbell, Kentucky.

Special Operations

Command of the Joint Special Operations Command was passed to Major General Gary A. Luck, as Major General Stiner assumed command of the 82d Airborne Division.

The Joint Special Operations Command was awarded the Joint Meritorious Unit Award by Admiral William Crowe, Chairman of the Joint Chiefs of Staff, dated 17 September 1986.

Special Operations Command, Central (SOCCENT), was constituted as the joint special operations headquarters for the Unified US Central Command (CENTCOM), commanded by Colonel Jerry King. Activated at MacDill AFB, Florida, SOCCENT is responsible for special operations forces assigned to support the CENTCOM contingency plans for South West Asia.

Special Operations Command, South (SOCSOUTH) was activated in the Panama Canal Zone, assigned as the Special Operations command element for US Southern Command (USSOUTHCOM). Colonel Gary Weber, former Division Chief, G-3, 1st Special Operations Command, and Battalion Commander, 7th Special Forces Group, was appointed as the first commander of SOCSOUTH.

Special Operations Detachment, Korea (Det-K), Seong-Nam, Republic of Korea, was assigned to the 1st Special Forces Group.

The 1st Special Operations Command Augmentation was redesignated as the 1st Special Operations Command Augmentation Detachment, (1st SOCOM Aug Det).

While remaining at Fort Bragg, the 1st SOCOM Aug Det was reassigned from the 120th to 97th US Army Reserve Command at Fort Meade, MD.

The JFK Special Warfare Center, Fort Bragg, North Carolina, was redesignated as the John F. Kennedy Special Warfare Center and School.

Company F (Airborne), 3d Battalion, Academy of Health Sciences Brigade, was assigned with the 3d Battalion, Academy Brigade to the Special Operations Forces Division, US Army Institute of Special Operations Medical Training, Fort Sam Houston, Texas.

Rangers

The 75th Infantry (Ranger), was redesignated as the 75th Ranger Regiment under the command of Colonel Wayne P. Downing in 1984, and withdrawn in 1986 from the 1st Special Forces under the Combat Arms Regimental System, to be reorganized as a parent regiment under the US Army Regimental System, and assigned to the 1st Special Operations Command.

The 75th Ranger Regiment, as a Major Subordinate Command of the 1st Special Operations Command, actively commands the 1st, 2d and 3d Battalions, 75th Infantry, concurrently redesignated as the 1st, 2d and 3d Ranger Battalions.

As Secretary of the Army, the Honorable John O. Marsh Jr directed during the activation ceremony at Fort Bragg, North Carolina, the battle streamers and honors of the six Ranger battalions and the Ranger companies of Korea and Vietnam, were transferred back to the 75th Ranger Regiment from the colors of the 7th and 10th Special Forces Groups, his remarks included "...that one elite organization (1st SOCOM and Special Forces), has safeguarded the traditions of the other..." . The Battle Streamers for Grenada had already been presented directly to the 1st and 2d Ranger Battalions.

The John F. Kennedy Special Warfare Center and School at Fort Bragg, North Carolina ran the initial pilot program for the Long Range Surveillance Leader's Course from 31 March through 23 May. This was proven so successful that the course was exported to the Ranger Department, Fort Benning, Georgia, and implemented by the Long Range Surveillance Unit Division of the Ranger Department.

Psychological Operations

The 4th Psychological Operations Group and the 1st, 6th, 8th and 9th Psychological Operations Battalions were redesignated as the 4th Psychological Operations Group (Airborne), and the 1st, 6th, 8th and 9th Psychological Operations Battalions (Airborne).

Elements of the 4th Psychological Operations Group participated in Operation URGENT FURY in Grenada.

Aviation

The Aviation Brigade, 82d Airborne Division was constituted in the Regular Army, activated at Fort Bragg, and assigned to the 82d Airborne Division.

The 101st Aviation Group was redesignated as the Aviation Brigade, 101st Airborne Division (Air Assault).

The 160th Aviation Battalion was reorganized and redesignated as the 160th Aviation Group.

The 129th Aviation Company (Special Operations), was constituted in the Regular Army, assigned to the 1st Special Operations Command, and activated at Hunter Army Airfield, Georgia.

Combat Support
Military Intelligence

The Long Range Surveillance Detachment, 125th Military Intelligence Battalion, was activated in support of the 25th Infantry Division (Light), at Schofield Barracks, Hawaii. This is the first airborne unit ever assigned to the 25th Infantry Division, in the past, airborne Ranger companies had been attached.

Signal

The 112th Airborne Army Signal Battalion was allotted to the Regular Army and concurrently reorganized and redesignated as the 112th Signal Battalion. Assigned to the 1st Special Operations Command, the 112th Signal Battalion was activated at Fort Bragg, North Carolina, under the command of Lieutenant Colonel David Bryant.

Combat Service Support
Quartermaster

The 528th Quartermaster Battalion was redesignated as the 13th Support Battalion (Airborne), activated at Fort Bragg, North Carolina, and assigned to the 1st Special Operations Command (Airborne). Lieutenant Colonel Louis Mason was assigned as the battalion commander.

—— 1987 ——
Combat Arms
Infantry

The 1st Battalion, 187th Infantry, was relieved from assignment to the 193d Infantry Brigade, and assigned to the 101st Air Assault Division. The 5th Battalion, 187th Infantry was relieved from assignment to the 101st Air Assault Division, and inactivated at Fort Campbell, Kentucky.

The 6th Battalion, 327th Infantry was relieved from assignment to the 6th Infantry Division (Light), and inactivated at Fort Wainwright, Alaska.

The 506th Infantry was withdrawn from the Combat Arms Regimental System (CARS), and reorganized as a parent regiment under the US Army Regimental System (USARS), with Headquarters in Korea. The 1st Battalion, 506th Infantry was assigned to the 2d Infantry Division and activated in Korea.

The 1st Battalion, 508th Infantry was assigned to the 193d Infantry Brigade and activated in Panama.

The US was formally recognized in ceremonies at Weingarten, Germany, as the fifth participating member of the International Long Range Reconnaissance Patrol (ILRRP) School, providing two Special Forces officers and six LRRP qualified non-commissioned officer instructors.

Attached to HHC, 10th Mountain Division, the LRSD was organized in December, under the command of Captain Marshall E. McKinney and SFC Mark Diering.

Captain J. Rex Hastey and First Sergeant Gerald I. Parks succeeded Captain Robert Kennedy and First Sergeant Kenneth Floyd at the helm of Company F (LRSC), 51st Infantry.

Company G (Ranger), 143d Infantry, (Texas ARNG), was reorganized and redesignated as Company G (Long Range Reconnaissance Patrol), 143d Infantry.

Company F (Long Range Reconnaissance Patrol), 425th Infantry, (Michigan ARNG), redesignated as Company F (Long Range Surveillance Unit), 425th Infantry.

Cavalry

On 1 July, the LRSD, 2d Squadron (Reconnaissance), 10th Cavalry, 7th Infantry Division (Light), was reassigned as the LRSD, 107th Military Intelligence Battalion, 7th Infantry Division (Light).

HHT, 5th Squadron, 17th Cavalry was inactivated at Fort Hood, Texas.

In January, the LRSD, 1st Squadron, 101st Cavalry was activated at Staten Island, New York, with selected personnel assigned from within the 42d (Rainbow) Infantry Division, New York Army National Guard. The unit was organized initially with 12 personnel (known locally as the Twelve Angry Men), commanded by Captain Gilbert Metsler, assisted by Detachment Sergeant, Sergeant First Class Richard Riehle.

LRSD, Troop E, 5th Squadron, 117th Cavalry, was redesignated as the LRSD, 5th Squadron, 117th Cavalry.

The Reconnaissance Platoon, 194th Cavalry (Air), was reorganized and redesignated on 1 October, as the Long Range Surveillance Detachment (Airborne), 1st Squadron, 194th Cavalry (Air). A month later, on 1 November, Captain John D. McGowan II was assigned as the first commander of the LRSD, 1st Squadron, 194th Cavalry (Air), assisted by Staff Sergeant Donald F. Shroyer, the first Detachment Sergeant and Training NCO.

Troop F (LRSU), 1st Squadron, 238th Cavalry was redesignated as the Long Range Surveillance Detachment, 1st Squadron, 238th Cavalry.

Corps Artillery

HHB, XVIII Airborne Corps Artillery was constituted in the Regular Army, and activated at Fort Bragg, North Carolina.

Field Artillery

The 319th and 321st Field Artillery Regiments were withdrawn from the Combat Arms Regimental System, and reorganized as parent regiments under the US Army Regimental System. The 319th Field Artillery Regiment Headquarters were transferred to Fort Bragg, North Carolina; the 321st Field Artillery Regiment Headquarters were transferred to the United States Training and Doctrine Command, and organized at Fort Sill, Oklahoma.

Battery C, 321st Field Artillery, was redesignated as HHB, 3d Battalion, 321st Artillery.

Special Operations

The Unified Combatant US Special Operations Command was activated at MacDill AFB, Florida, with General James J. Lindsay appointed as the Commander in Chief. His Chief of Staff was named as Major General Joseph C. Lutz. A Washington office for the US Special Operations Command was opened at the Pentagon, with Brigadier General Wayne P. Downing assigned as the USSOCOM liaison with the Pentagon and the Washington community. Subordinate service components of the USSOCOM included the Joint Special Operations Command, the 1st Special Operations Command and the JFK Special Warfare Center and School (JFKSWCS), all at Fort Bragg, North Carolina; the 23d Air Force, located at Hurlburt Field, Florida; and the Navy Special Warfare Command, Coronado, California.

General Robert Sennewald, commanding Forces Command, made the decision, following a Special Operations Forces (SOF) briefing by the 1st Special Operations Command (1st SOCOM), to positively improve the command lines between 1st SOCOM and the USAR SOF. He reduced the USAR intermediate headquarters in the peacetime chains of command, from nineteen down to five. Each of the five Continental US Armies (First, Second, Fourth, Fifth and Sixth Armies), were directed to designate a single SOF Major US Army Reserve Command in each of their geographic areas of responsibility. The result was the nomination and assignment of the 97th, 120th, 86th and 90th Army Reserve Commands respectively for the First, Second, Fourth and Fifth US Armies; and the 351st Civil Affairs Command for the Sixth US Army. Colonel Donald J. Soland was appointed as the Deputy Commanding Officer, 1st SOCOM. He was the former Chief of Staff and ACofS G-3 (Assistant Chief of Staff for Operations), 1st SOCOM, and prior to that he was the Director, Security Assistance and Training Management Office, JFKSWCS.

Command Sergeant Major Stephen Holmstock replaced Command Sergeant Major James Hargraves as the senior noncommissioned officer at the !st Special Operations Command, Fort Bragg, North Carolina.

Colonel James Estep was named as the first Commander of the provisionally organized Special Operations Command-Korea (SOC-K), activated at Yeong-San, Republic of Korea.

Special Forces

The 1st Battalion, 5th Special Forces Group, commanded by Lieutenant Colonel John C. Woolshlager, relocated to Fort Campbell, Kentucky.

Rangers

The 4th Ranger Training Battalion (4th RTB) was activated at Fort Benning, Georgia from the former Ranger Department of the Infantry School. The mission of the 4th RTB is to develop the technical leadership skills of selected male officer and enlisted personnel through the Ranger Course, the Long Range Surveillance Leader's Course (LRSLC) and the Light Leaders Course. Hands on training and written examinations are culminated in a realistic tactical field environment, providing potential leaders the opportunity to demonstrate their leadership abilities under mental and physical stress. Company D, 4th RTB was activated in November, under the command of Lieutenant Colonel Hitchcock and First Sergeant Burdenshaw, and provides course instructors for the Long Range Surveillance Unit (LRSU) volunteers. Company D was previously identified as the LRSU Division of the Ranger Department. Companies A, B and C, 4th RTB provide instructors for the Ranger School at Fort Benning, and Company E instructs the Light Leader Course for the Light Infantry Divisions and Brigades.

The 5th Ranger Training Battalion was activated at the Ranger mountain training camp at Camp Frank D. Merrill; and the 6th Ranger Training Battalion was activated at Camp Rudder in Florida.

Psychological Operations

Company D, 1st Psychological Operations Battalion (Airborne), was forward deployed to Quarry Heights, Panama.

Aviation

HHC, 18th Aviation Brigade was formed from the reorganized and redesignated HHC, 269th Aviation Battalion, and assigned to XVIII Airborne Corps, and HHC, Aviation Brigade, 82d Airborne Division, was activated at Fort Bragg, North Carolina.

The 160th Aviation Group was redesignated as the 160th Aviation Group (Airborne).

The 45th Aviation Battalion (Oklahoma ARNG), was first redesignated as the 45th Aviation Battalion (Special Operations) (Airborne), then later redesignated as the 1st Battalion, 245th Aviation (Special Operations) (Airborne).

The 82d Aviation Battalion was relieved from assignment to the 82d Airborne Division, and the 158th and 159th Aviation Battalions were relieved from assignments to the 101st Airborne Division (Air Assault).

HHCs, 82d, 158th and 159th Aviation Battalions, were reorganized as parent regiments under the US Army Regimental System, and redesignated as the 82d, 158th and 159th Aviation Regiments, with headquarters at Fort Bragg (82d and 159th), and Fort Hood (158th).

Company A, 82d Aviation Battalion was reorganized and redesignated as HHC, 1st Battalion, 82d Aviation, assigned to the 82d Airborne Division. Companies B, C and D, 82d Aviation Battalion, were reorganized and redesignated as Companies B, C and D (Armed Helicopter Company), 82d Aviation, and concurrently remained assigned to the 82d Airborne Division.

The 226th Aerial Surveillance and Escort Battalion was redesignated as the 226th Aviation Battalion.

Companies A and D, 229th Aviation Battalion, were relieved from assignment to the 101st Airborne Division (Air Assault), and inactivated at Fort Campbell, Kentucky.

The 617th Aviation Detachment, inactivated in 1971, was redesignated as the 617th Aviation Detachment (Special Operations) (Airborne), activated at Fort Bragg, North Carolina, and assigned to the 1st Special Operations Command, then relocated to USSOUTHCOM, Panama Canal Zone.

Medical

Company F (Airborne), was reassigned with the 3d Battalion, Academy Brigade, to the Special Operations Forces Division, US Army Institute of Special Operations Medical Training, at Fort Sam Houston Texas.

Combat Support
Engineers

The 37th Engineer Battalion was activated at Fort Bragg, North Carolina, and assigned to the 20th Engineer Brigade.

Military Intelligence

The LRSD, 107th Military Intelligence Battalion was activated internally within the 7th Infantry Division (Light), from the former LRSD, 2d Squadron (Reconnaissance), 10th Cavalry.

Company D, 124th Military Intelligence Battalion, 24th Infantry Division, was reorganized, redesignated, and activated as LRSD, 124th Military Intelligence Battalion, Fort Stewart, Georgia. The first Commander of the LRSD was Captain Daniel J. Keefe, and the Detachment Sergeant was Sergeant First Class Charles Thomas (both former members of the 1st Battalion, 75th Ranger Regiment.

The 279th Military Intelligence Company (Combat Electronic Warfare Intelligence) (Special Forces Group) (Airborne) was reassigned from the 63d Army Reserve Command to the 351st Civil Affairs Command, (remaining attached to the 19th Special Forces Group).

The 389th Military Intelligence Detachment (Combat Electronic Warfare Intelligence) (Special Forces Group) (Airborne), was assigned to the 11th Special Forces Group, under the 97th Army Reserve Command, First US Army.

Signal

The 112th Signal Battalion was redesignated as the 112th Signal Battalion (Airborne).

Combat Service Support
Chemical

The 382d Chemical Detachment (NBC Staff) (Special Forces), was inactivated at Arlington Heights, Illinois.

The 445th and 801st Chemical Detachments, (NBC Reconnaissance) (Special Forces) and (NBC Staff) (Special Forces), were both assigned to the 97th Army Reserve Command, with the 801st Chemical Detachment later being inactivated at Anniston, Alabama.

The 808th Chemical Detachment (NBC Staff) (Special Forces) was inactivated at Fort Meade, Maryland.

The 900th and 901st Chemical Detachments (NBC Reconnaissance) (Special Forces) and (NBC Staff) (Special Forces), were both reassigned from the 63d Army Reserve Command to the 351st Civil Affairs Command, while remaining attached to the 19th Special Forces Group.

Quartermaster

The 407th Supply and Service Battalion, assigned to the 82d Airborne Division at Fort Bragg, North Carolina, was reorganized and redesignated as the 407th Supply and Transport Battalion.

The 13th Support Battalion (Airborne), was redesignated as the 528th Support Battalion (Special Operations) (Airborne), remaining assigned to the 1st Special Operations Command.

—— 1988 ——

Lieutenant General Carl W. Stiner, former commander of the 82d Airborne Division, and the Joint Special Operations Command, assumed command from Lieutenant General John Foss, of the XVIII Airborne Corps. HQ XVIII Airborne Corps participated in Operation GOLDEN PHEASANT in Honduras. General Foss was assigned to the Pentagon as the US Army Deputy Chief of Staff for Operations, replacing General Norman Schwarzkopf, who was reassigned to become the Commander in Chief, US Central Command.

Recently promoted Major General James H. Johnson Jr, former Assistant Division Commander for Operations, under (then) Major General Stiner, assumed command of the 82d Airborne Division.

Major General Michael F. Spigelmire was reassigned from command of the 24th Infantry Division, to become the Commanding General of Fort Benning, Georgia, replacing retiring Major General Kenneth Leuer.

Combat Arms
Infantry

Company E, 51st Infantry, commanded by Captain Donald J. Lash Jr, was redesignated as Company C, 165th Military Intelligence Battalion.

The 1st and 2d Battalions, 187th Infantry were relieved from assignment to the 193d Infantry Brigade in Panama, inactivated and assigned to the 101st Airborne Division (Air Assault), Fort Campbell, Kentucky. The 4th and 5th Battalions, 187th Infantry were inactivated at Fort Campbell, Kentucky.

Lieutenant Colonel Mark Lewis assumed command of the 1st Battalion, 507th Parachute Infantry (more commonly referred to as the Airborne School, Fort Benning, Georgia), from Lieutenant Colonel Jimmie Holt. Lieutenant Colonel Lewis served with the 82d Airborne Division during Operation URGENT FURY as the

Operations Officer, 3d Battalion, 325th Airborne Infantry. He later served as an operations officer at HQ, 82d Airborne Division.

The Headquarters of the 509th Infantry Regiment were transferred under the US Army Regimental System from Fort Rucker, Alabama to Headquarters, Training and Doctrine Command (TRADOC), specifically to Little Rock Air Force Base, Alabama.

The LRSD, 10th Mountain Division commenced qualification training, and was assigned in October to the newly activated 110th Military Intelligence Battalion, 10th Mountain Division.

Company G (Long Range Reconnaissance Patrol), 143d Infantry, (Texas ARNG), was redesignated as Company G (Long Range Surveillance Unit), 143d Infantry.

Cavalry

The Cavalry has traditionally maintained the role of providing the eyes and ears capabilities in the forward battle areas, locating targets for their larger forces to engage. Continuing this tradition, a number of the current divisions configured/reorganized their cavalry elements to support a Long Range Surveillance capability as follow:

The 2d Squadron, 10th Cavalry was relieved from Long Range Surveillance duties with the 7th Infantry Division (Light), at Fort Ord, California.

Troop E (LRSD), 1st Squadron, 17th Cavalry and the LRSD, 2d Squadron, 17th Cavalry were respectively assigned as the Long Range Surveillance Detachments for the 82d Airborne Division and 101st Airborne Division (Air Assault).

The 3d and 5th Squadrons, 17th Cavalry were respectively assigned to the 10th Mountain Division and activated at Fort Drum, New York, and to the 2d Infantry Division, and activated in Korea.

Captain William Brabazon, assisted by Platoon Sergeant James V. Walker assumed command of the LRSD, 1st Squadron, 18th Cavalry, in support of the 40th Infantry Division (California ARNG).

Captain Gerard Noone replaced Captain Gilbert Metsler as commander of the LRSD, 1st Squadron, 101st Cavalry, 42d Infantry Division (New York ARNG), in October, with Sergeant First Class Robert Jenks assuming the duties as Detachment Sergeant in December.

The LRSD, 5th Squadron, 117th Cavalry, 50th Armored Division (New Jersey ARNG), was placed on jump status, with the first airborne operation being conducted at Lakehurst Naval Air Station, New Jersey, on 23 April.

The following Long Range Surveillance Detachments were assigned to support their respective Divisions: LRSD, 1st Squadron, 104th Cavalry, Chambersburg, Pennsylvania supporting the 28th Infantry Division (Pennsylvania ARNG); LRSD, 1st Squadron, 158th Cavalry, Annapolis, Maryland, supporting the 29th Infantry Division (Maryland ARNG); and Company H (LRSU), 122d Infantry, Cartersville, Georgia.

Special Operations

The US Special Operations Command, by direction of General Lindsay, organized the Joint Mission Analysis under the Joint Studies Directorate, headed by Colonel James L. Zachary.

Brigadier General James A. Guest, (former commander of the JFK Special Warfare Center and School, and prior commander of the 5th Special Forces Group), assumed command of the 1st Special Operations Command as Major General Leroy N. Suddath retired.

Brigadier General Sidney Shacknow, former Chief of Staff, 1st Special Operations Command, (and prior alumni of the 502d Airborne Infantry, 101st Airborne Division in Vietnam), was appointed to replace Brigadier General Wayne P. Downing to head the Washington Liaison Office for the US Special Operations Command.

Brigadier General David J. Barrato (former commander, 1st Special Forces Group), replaced Brigadier General James A. Guest as the commander of the JFK Special Warfare Center and School. Brigadier General Joseph C. Hurteau, former Commander, 1st Special Operations Command Augmentation Detachment (USAR), was appointed as the first USAR Individual Mobilization Augmentee (IMA), Deputy Commanding General, 1st Special Operations Command.

Special Operations Commands, Central, Europe and the Pacific (SOCCENT, SOCEUR and SOCPAC), were designated as a General Officer Commands, (with a USAR Individual Mobilization Augmentee (IMA) commander for SOCCENT) and active army commanders for SOCEUR and SOCPAC.

The assigned IMA commander of SOCCENT was Brigadier General William C. Cockerham, former Deputy Commanding General (DCG IMA) to 1st Special Operations Command (1st SOCOM), and former Commander, 12th Special Forces Group. The Regular Army SOCCENT Deputy Commander was Colonel Larry Dugan, (formerly Commanding the 5th Special Forces Group).

Assigned to command SOCEUR was Brigadier General James T. (Terry) Scott, and Brigadier General Barry J. Sottak was assigned to SOCPAC.

Special Operations Command - Korea (SOC-K), provisionally organized at Yeong-San, Republic of Korea in 1987, assigned to the US Pacific Command (USPACOM), providing close coordination to Combined Forces Command, Korea, for the employment of Special Operations Forces. Colonel Robert Howard, recipient of the Medal of Honor when assigned to the 5th Special Forces Group in Vietnam, became the second commander.

Special Forces

The HQ, 2d and 3d Battalions, 5th Special Forces Group relocated to join the 1st Battalion at Fort Campbell, Kentucky.

Rangers

Command of Company D (LRSU), 4th Ranger Training Battalion, assigned to the Infantry School, Fort Benning, Georgia, was assumed by Major Harvey H. Latson III, and later by Major Bill Ball. The assigned First Sergeant became First Sergeant McVey.

Aviation

The 160th Aviations Group (Airborne), was redesignated as the 160th Aviation Group (Special Operations) (Airborne).

Companies B and D, 229th Aviation Battalion, were reorganized and redesignated as the 2d and 4th Battalions, 229th Aviation, and activated at Fort Hood, Texas.

Combat Support
Military Intelligence

Providing long range reconnaissance intelligence gathering capabilities, Long Range Surveillance Units were organized within Military Intelligence Battalions supporting the Light and Heavy Infantry Divisions and Armored Divisions:

Company E, 51st Infantry, commanded by Captain Donald J. Lash Jr, (attached to the 165th Military Intelligence Battalion), was redesignated on 16 October, as Company C, 165th Military Intelligence Battalion.

Company D (LRSU), 101st Military

Intelligence Battalion, supports the 1st Infantry Division, Fort Riley, Kansas.

Company D (LRSU), 102d Military Intelligence Battalion, 2d Infantry Division, Korea, was activated in February 1988, commanded by Captain Kevin P. McGrath, and assisted by Staff Sergeant (P) Robert C. Carlson. In June, Captain Thomas A. Perkins assumed command, and the First Sergeant became Sergeant First Class Michael A. Lingo in July. Company D was officially recognized as an airborne unit on 1 October, making its first jump (with the Republic of Korea parachutes) on 4 October.

LRSD, 103d Military Intelligence Battalion, Wurzburg, Germany; Company D (LRSU), 104th Military Intelligence Battalion, supporting the 4th Infantry Division, Fort Carson, Colorado; Company D (LRSU), 105th Military Intelligence Battalion, supporting the 5th Infantry Division, Fort Polk, Louisiana; and Company C (LRSD), 106th Military Intelligence Battalion, supporting the 6th Infantry Division (Light), Fort Richardson, Alaska.

Company D (LRSU), 108th Military Intelligence Battalion, supporting the 8th Infantry Division, Germany; Company E (LRSU), 109th Military Intelligence Battalion, supporting the 9th Infantry Division, Fort Lewis, Washington; and the LRSD, HHC, 10th Mountain Division, which was reassigned to become the LRSD, 110th Military Intelligence Battalion, newly activated and assigned to support the 10th Mountain Division, Fort Drum, New York.

Company D, 312th Military Intelligence Battalion was reorganized and redesignated as the LRSD, 312th Military Intelligence Battalion on 1 November. Personnel were predominantly assigned from the 1st and 2d Battalions, 5th Cavalry, 1st Cavalry Division. Captain Jeffrey P. Holt, assisted by Sergeant First Class Ronnie L. Rawls, assumed command, in support of the 1st Cavalry Division, Fort Hood, Texas. At the time of writing, and although beret flash and oval wing background designs have been submitted to the Institute of Heraldry, the LRSD, 312th MI Battalion is still pending placement on airborne status.

The LRSD, 501st Military Intelligence Battalion, supporting the 1st Armored Division, Germany; Company D (LRSU). The LRSD, 501st Military Intelligence Battalion, at the time of writing, had not been designated as an airborne unit; the 522d Military Intelligence Battalion, supporting the 2d Armored Division, Fort Hood, Texas; and Company D (LRSU), 533d Military Intelligence Battalion, 3d Armored Division, Germany.

Combat Service Support
Chemical

The 901st Chemical Detachment (NBC Staff) (Special Forces), the last of this type of organization, was inactivated at Tucson, Arizona.

—— 1989 ——

The Unites States invaded Panama, under the overall ground command of Lieutenant General Carl W. Stiner, Commanding General, XVIII Airborne Corps.

Major XVIII Airborne Corps units included the 82d Airborne Division (Commanded by Major General James H. Johnson Jr), the 16th Military Police Brigade, the 18th Aviation Brigade, the 35th Signal Brigade and the 525th Military Intelligence Brigade.

Major General J. H. Binford Peay assumed command of the 101st Airborne Division (Air Assault) in August, from Major General Teddy G. Allen. Major General Allen was assigned as the Deputy Inspector General (Inspections), Washington, DC. Major General Peay served as the Assistant Division Commander, 101st Airborne Division (Air Assault), from July 1987 through July 1988, prior to assignment as the Deputy Command, US Army Command and General Staff College, Fort Leavenworth, Kansas.

Combat Arms
Infantry

The 1st Battalion, 508th Infantry (Airborne), assigned to the 193d Infantry Brigade, stationed in Panama, participated in Operation JUST CAUSE. Conducting an airborne assault with the 1st Brigade Headquarters element of the 82d Airborne Division, from Fort Bragg, North Carolina, came the 1st and 2d Battalion, 504th Parachute Infantry; the 4th Battalion, 325th Airborne Infantry; and Company A, 3d Battalion, 505th Parachute Infantry. The 3d Battalion, 504th Parachute Infantry deployed from their training site with the US Army Jungle Warfare School.

Company C, 165th Military Intelligence Battalion (TE), was redesignated as Company E (LRS), 51st Infantry, and assigned to the 165th Military Intelligence Battalion (TE). Command of Company F (LRS), 51st Infantry, was passed to Captain Chris L. Mueller and First Sergeant George Tillman Jr.

Artillery

The 3d Battalion, 319th Field Artillery, parachuted into Panama with the 82d Airborne Division, during Operation JUST CAUSE.

Armor

Twelve M-551 Sheridan Tanks were air dropped with the element of the 3d Battalion, 73d Armor, (commanded by Lieutenant Colonel James J. Grazioplene), 82d Airborne Division, during the early air assaults into Operation JUST CAUSE.

Cavalry

Captain Guy Sands assumed command of the LRSD, 1st Squadron, 101st Cavalry, 42d Infantry Division.

The Long Range Surveillance Detachment, 1st Squadron, 238th Cavalry was redesignated as the 151st Infantry Detachment (Long Range Surveillance).

Special Operations

Participating airborne elements during Operation JUST CAUSE of the US Special Operations Command, included the Joint Special Operations Command, with Special Operations Forces from the Army, Navy and Air Force.

The US Army Special Operations Command (USASOC) was activated on 1 December at Fort Bragg, North Carolina, replacing the 1st Special Operations Command (1st SOCOM), as the Army Component headquarters for the US Special Operations Command. Lieutenant General Gary E. Luck (former commander of the Joint Special Operations Command), was appointed as the commander The Command Sergeant Major was CSM Ronnie Strahan, former CSM of 1st SOCOM. The deputy commander became Brigadier General William F. Garrison, former Deputy Commander/Chief of Staff, 1st SOCOM, and Commander, First Special Forces Operational Detachment, Delta. The Deputy Commanding General, ARNG, was named as Major General Joseph C. Boyersmith.

Command Sergeant Major Joe Dennison, former CSM of the 5th Special Forces Group, replaced CSM Ronnie Strahan as the CSM, 1st Special Operations Command.

Brigadier General Joseph C. Hurteau, incumbent Deputy Commander, USAR, 1st Special Operations Command, was appointed as the acting commander, US Army Reserve Special Operations Command.

Major General Wayne A., Downing,

assumed command of the Joint Special Operations Command, and coordinated activities of the Special Operations Forces of all services, participating in Operation JUST CAUSE in Panama.

Brigadier General Richard Potter was reassigned from Deputy Commander, JFK Special Warfare Center and School, to command Special Operations Command, Europe, at Patch Barracks, Stuttgart, Germany. General Potter had previously commanded the 10th Special Forces Group, and had been Deputy Commander of the 1st Special Forces Detachment - Delta. Brigadier General Sidney Shacknow was reassigned from the US Special Operations liaison in Washington, to command the Berlin Brigade.

Colonel Robert Jacobelly was assigned to replace Colonel Gary Weber as the Commander, Special Operations Command, South (SOCSOUTH).

In order to clearly delineate the responsibilities for coordination and provision of in-theater support to the deployed special operations units, Special Operations Support Commands (Theater Army), were identified for activation.

Numbered 3d through 7th Special Operations Support Commands, the 3d SOSC was assigned to support US Army South (USARSO), and provisionally activated in Panama. The 4th SOSC was activated from the Special Operations Division of G-3, US Army Western Command (WESTCOM). WESTCOM was inactivated this year to become US Army Pacific (USARPAC).

At Bad Toelz, Germany, the Army Special Operations Support Element (ARSOSE), became the 7th SOSC, assigned to US Army Europe (USAREUR), under the US European Command (USEUCOM).

Special Forces

The 3d Battalion, 7th Special Forces Group, stationed in Panama, assisted by Company A, 2d Battalion, 7th Special Forces Group, from Fort Bragg, participated as early deploying, pre-positioned units in Operation JUST CAUSE.

Rangers

The 75th Ranger Regiment HQ, commanded by Colonel William F. Kernan, and the 1st, 2d, and 3d Ranger Battalions participated in airborne assaults into Rio Hato and the Torrijos International Airport during Operation JUST CAUSE in Panama.

Captain Scott A. Francis assumed command of Company D (LRSU), 4th Ranger Training Battalion, assisted by First Sergeant Powell.

Aviation

The 160th Aviation Group (Special Operations) (Airborne), from Fort Campbell, Kentucky, and the 617th Aviation Detachment (Special Operations)(Airborne), based in Panama, provided tremendous support to both conventional and Special Operations Forces during Operation JUST CAUSE.

The 82d Aviation Brigade provided support to the 82d Airborne Division during Operation JUST CAUSE, with elements parachuting into Panama.

The 129th Aviation Company (Special Operations), at Hunter Army Airfield, Georgia, was reorganized and designated as the 3d Battalion, 160th Aviation Regiment.

Medical

Company B, 307th Medical Battalion deployed to Panama with the 1st Brigade, 82d Airborne Division, during Operation JUST CAUSE.

Combat Support
Engineers

The 307th Engineer Battalion accompanied the 1st Brigade, 82d Airborne Division in the parachute assault during Operation JUST CAUSE in Panama.

Military Intelligence

On 1 April, Sergeant First Class Melvin L. Kinkade became the First Sergeant of Company D (LRSU), 102d Military Intelligence Battalion, and Captain William J. Medley assumed command on 5 May.

The Long Range Surveillance Detachment, 107th Military Intelligence Battalion, deployed as part of the 7th Infantry Division (Light) contingency in Operation JUST CAUSE into Panama.

Captain Mark S. Crowson assumed command of the LRSD, 110th Military Intelligence Battalion, 10th Mountain Division, in August.

In March, SFC Ronald Callahan was assigned as the Detachment Sergeant, and on 5 May, Captain Jay W. Peterson was selected as the new commander of Company D (LRSD), 124th Military Intelligence Battalion, 24th Infantry Division. Company D (LRSU), 124th Military Intelligence Battalion was redesignated as the LRSD, 2d Squadron, 4th Cavalry Regiment. At the time of publication, although designated by TOE as an airborne unit, the LRSD, 2d Squadron, 4th Cavalry has not been placed upon airborne status by the 24th Infantry Division.

Company C, 165th Military Intelligence Battalion (TE), was redesignated as Company E (LRS), 51st Infantry, and assigned to the 165th Military Intelligence Battalion (TE).

The 313th Military Intelligence Battalion participated in the airborne assault into Panama with the 82d Airborne Division during Operation JUST CAUSE.

Military Police

The 503d Military Police Battalion was flown in from Fort Bragg, North Carolina, to reinforce the Panama-based 92d Military Police Battalion, during Operation JUST CAUSE. In addition, the 519th Military Police Battalion had been flown in from Fort Meade, together with five of it's assigned Military Police Companies.

Signal

The Joint Communications Unit of the Joint Special Operations Command, and the 112th Signal Battalion, 1st Special Operations Command, provided communications support for Special Operations Forces during Operation JUST CAUSE in Panama.

The 82d Airborne Division communications support were provided by Company B, 82d Signal Battalion, participants in the airborne invasion of Panama.

Combat Service Support

The 82d Airborne Division Support Command participated in the airborne operation into Panama.

Quartermaster

Supporting Special Operation Forces deployed to Panama during Operation JUST CAUSE, was the Fort Bragg, North Carolina based 528th Support Battalion of the 1st Special Operations Command.

The 82d Airborne Division was supported during the airborne operation by Companies A, 407th Supply and Transportation Battalion and Company A, 782d Maintenance Battalion.

—— 1990 ——

Elements of the US Special Operations Command and the XVIII Airborne Corps redeployed from Panama to their home bases. HQ XVIII Airborne Corps and HQ, 82d Airborne Division, redeployed to Fort Bragg, North Carolina on 12 January. On 19 January, HQ US Army

Personnel Command published the approval for the wearing of organizational shoulder sleeve insignia (SSI), on the right shoulder of the Army uniform, by personnel who participated in Operation JUST CAUSE. The organizations authorized to wear their unit SSI as a 'combat patch', included the US Army Special Operations Command; XVIII Airborne Corps; 82d Airborne Division; 1st Special Operations Command; 1st Corps Support Command; 16th Military Police Brigade; 18th Aviation Brigade; 35th Signal Brigade; 7th Special Forces Group; HQ and 1st, 2d and 3d Battalions, 75th Ranger Regiment; and the 525th Military Intelligence Brigade. The period required for authorization to wear the SSI combat patch and receipt of the Armed Forces Expeditionary Medal was approved for service member participants during the period 20 December 1989 and 13 January 1990.

General Theodore J. Conway, former Commanding General, 82d Airborne Division; Chief, Joint US Military Advisory Group, Thailand; Commanding General, Seventh US Army; and Commander in Chief, US Strike Command, passed away on 11 September at St Petersburg, Florida.

Lieutenant General James Maurice (Slim Jim) Gavin passed away on 23 February, and was buried at West Point on 28 February. A memorial service was held at Fort Myer, Arlington, Virginia, on 6 March 1990.

As elements were still returning from Operation JUST CAUSE, Iraq invaded its neighbor, Kuwait in August, proceeding towards Saudi Arabia. Operation DESERT SHIELD was implemented, with the 82d Airborne Division spearheading the United States force build up in Saudi Arabia, to protect our allies in Saudi Arabia, help to re-establish the legal government of Kuwait, and protect US citizens in the Middle East. The 101st Airborne Division (Air Assault) and the 24th Infantry Division (as elements of the XVIII Airborne Corps contingency force for the US Central Command), quickly followed, to be later reinforced by the 2d Armored Division, the 1st Cavalry Division, and the 1st Corps Support Command. At the time of writing, US troop strengths (of all services), topped more than 200,000, with an expectation of at least 100,000 to be added.

Combat Arms
Infantry
The 1st and 2d Battalions, 504th Infantry redeployed to Fort Bragg on 12 January. The 3d Battalion had redeployed on 3 January.

The two primary objectives of the 1st Battalion, 508th Infantry, assigned to the 193d Infantry Brigade at Fort Kobbe, Panama, included the successful capture of Fort Amador (with more than one hundred Panama Defense Force prisoners and more than 1,500 weapons). A second objective, where the two soldiers were killed and six wounded, was the seizure of La Comandancia.

The 4th Battalion, 325th Infantry and Company A, 1st Battalion, 505th Infantry redeployed to Fort Bragg, North Carolina on 12 January.

The 325th Airborne Infantry Regiment with the 504th and 505th Parachute Infantry Regiments deployed with the 82d Airborne Division to Saudi Arabia in August. The 187th, 327th and 502d Infantry Regiments deployed with the 101st Airborne Division (Air Assault), on Operation DESERT SHIELD in September.

Norway became the sixth participating nation in the International Long Range Reconnaissance Patrol (ILRRP) School, Weingarten, Germany, on 1 January.

The 142d Infantry Company (Long Range Surveillance)(Airborne) was the new designation of the former LRSD, 1st Squadron, 101st Cavalry, 42d Infantry Division. Sergeant First Class James M. Brown, 151st Infantry Detachment (Long Range Surveillance), replaced Sergeant First Class William E. Butler as the Detachment Sergeant.

The 150th and 194th Infantry Detachments (Long Range Surveillance), were formed and redesignated respectively from the LRSDs, 5th Squadron, 117th Cavalry, 50th Armored Division (New Jersey ARNG), and the 1st Squadron, 194th Cavalry (Air), (Iowa ARNG).

Armor
The 3d Battalion, 73d Armor, 82d Airborne Division, commanded by Lieutenant Colonel Thomas A. Bruno, deployed to Saudi Arabia. Company D, 3d Battalion, 73d Armor was activated to remain at Fort Bragg, North Carolina, while the battalion was deployed.

Artillery
Lieutenant Colonel W. H. Groening assumed command of the 1st Battalion, 39th Artillery (Airborne), which deployed to Saudi Arabia with the XVIII Airborne Corps.

The 319th Airborne Artillery Regiment deployed with the 82d Airborne Division, and the 320th Artillery Regiment deployed with the 101st Airborne Division (Air Assault), on Operation DESERT SHIELD.

Cavalry
The LRSDs, 1st Squadron, 101st Cavalry, the 5th Squadron, 177th Cavalry, and the 1st Squadron, 194th Cavalry (Air), were reorganized and redesignated as the 142d Company (LRS) (Airborne), the 150th Infantry Detachment (LRS), and the 194th Infantry Detachment (LRS). They were respectively assigned to the 42d Infantry, 50th Armored, and 47th Infantry Divisions.

The 1st and 2d Squadrons, 17th Cavalry, with their assigned Long Range Surveillance Units, deployed respectively with the 82d Airborne Division and the 101st Airborne Division (Air Assault).

Special Operations
General Carl W. Stiner, who led Operation JUST CAUSE into Panama, was promoted and reassigned from the XVIII Airborne Corps to become the senior airborne officer in the United States Army. He replaced General James J. Lindsay (who retired on 30 June), as the Commander in Chief of the US Special Operations Command (USSOCOM), MacDill AFB, Florida. Other great losses from the Special Operations community were Major General Joseph C. Lutz, who retired from his position as Chief of Staff, USSOCOM in September, and Colonel Jim Zachary who retired in July.

Lieutenant General Gary E. Luck, was assigned from the US Army Special Operations Command (USASOC), to command the XVIII Airborne Corps.

Replacing General Luck at USASOC is Lieutenant General Michael F Spigelmire. General Spigelmire had served with the 10th Special Forces Group from April 1963 until April 1965. He served with the 2d Battalion (Airborne), 8th Cavalry, assigned to the 1st Brigade, 1st Cavalry Division in Vietnam from May 1966 until June 1967. His later assignments included assignment with the Infantry School, Fort Benning, Georgia; Commander, 5th Battalion, 7th Cavalry and 1st Battalion, 5th Cavalry; Assistant Chief of Staff (G-3), VII Corps; Assistant Deputy Chief of Staff for Operations, FORSCOM; Assistant Division Commander and Commanding General, 24th Infantry Division (Mechanized).

Colonel Jesse Johnson finished command of the 10th Special Forces Group, and concurrently assumed command of

Special Operations Command-Central, MacDill AFB, Florida. Special Operations Command-Central (SOCCENT) deployed to Saudi Arabia on Operation DESERT SHIELD, augmented by personnel from the USSOCOM and the 5th and 10th Special Forces Groups. Colonel Johnson elected not to request activation of the Individual Mobilization Augmentees, under the IMA Wartime Commander, assigned to SOCCENT.

The 5th and 6th Special Operations Support Commands (SOSC) were activated in 1990, assigned respectively to US Army Central (ARCENT) of US Central Command (USCENTCOM), and US Army Atlantic (ARLANT) of US Atlantic Command (USARLANT).

The responsibilities for ARLANT fall upon the Commander in Chief (CINC) US Forces Command (USFORSCOM), who has dual responsibilities as CINCFOR and CINCLANT. Thus the Special Operations Division of USFORSCOM, headed by Colonel James Shepherd, was inactivated, to provide the nucleus of the 6th SOSC, with Colonel Shepherd appointed as the first commander.

Lieutenant Colonel James F. Johnson assumed command of the Support Battalion, 1st Special Warfare Training Group (Airborne).

Elements of the 112th Signal Battalion (Airborne) and the 528th Support Battalion (Special Operations) (Airborne) deployed to Saudi Arabia.

Special Forces

Colonel Richard A. Todd, former Director of Development, John F. Kennedy Special Warfare Center and School, replaced Colonel Eddie J. White on 25 July, as Commander, 1st Special Forces Group. Colonel Todd's assignments with Special Forces have included the 5th Special Forces in Vietnam; the 7th Special Forces at Fort Bragg; the 10th Special Forces in Germany; and Deputy Commander, 1st Special Forces Group. Colonel White returned to Fort Bragg.

The 3d Special Forces Group was reactivated at Fort Bragg, North Carolina on 29 June, under the command of Colonel Peter Stankovich, from his assignment with J-3, US Special Operations Command. The 3d Special Forces Group was assigned to the 1st Special Forces Command (Provisional).

The 5th Special Forces Group, and elements of the 10th Special Forces Group deployed to Saudi Arabia on Operation DESERT SHIELD.

Change of command of the 10th Special Forces Group was effected from Colonel Jesse Johnson to Colonel William Tangney.

Rangers

HQ, 75th Ranger Regiment, and the 3d Battalion, 75th Ranger Regiment, redeployed to Fort Benning, Georgia on 7 January 1990. The 1st and 2d Battalions, 75th Ranger Regiment, redeployed respectively to Hunter Army Airfield, Georgia and Fort Lewis, Washington on the 3 and 11-12th January.

The 75th Ranger Regiment was relieved from assignment to the 1st Special Operations Command, which became the 1st Special Forces Command, and assigned directly to the US Army Special Operations Command.

Command of the 4th Ranger Training Battalion was provided to Lieutenant Colonel Michael J. Pearce and Command Sergeant Major Donald Purdy. Lieutenant Colonel Pearce previously served as the Executive Officer and Deputy Commander of the 75th Ranger Regiment. Sergeant Major Purdy served with Company F (Ranger), 75th Infantry in Vietnam, and was one of the original cadre of Company C, 1st Battalion, 75th Infantry, when activated in 1974. He later served as First Sergeant, HHC, 3d Battalion, 75th Ranger Regiment.

Lieutenant Colonel Harvey H. Latson III was reassigned from S-2, 75th Ranger Regiment to J-2, US Special Operations Command.

Psychological Operations

Headquarters elements of the 4th Psychological Operations Group, and the 9th Psychological Operations Battalion deployed to Saudi Arabia.

Civil Affairs

Company D, 96th Civil Affairs Battalion, deployed to Saudi Arabia.

Aviation

The 160th Aviation Group (Airborne), was redesignated as the 160th Special Operations Aviation Regiment (Airborne), (160th SOAR), on 28 June, and assigned from the 1st Special Operations Command to the newly activated US Army Special Operations Command.

Companies C and D, 160th Aviation Group were consolidated, reorganized and redesignated as the 1st Battalion, 160th SOAR. Company E was reorganized and redesignated to become the 2d Battalion, 160th SOAR. The former Company F, 160th Aviation Group, was reorganized to become Companies D (Maintenance) for the 1st, 2d and 3d Battalions, 160th SOAR.

The 3d Battalion, 160th SOAR deployed to Saudi Arabia on Operation DESERT SHIELD.

The 18th Aviation Brigade deployed with the XVIII Airborne Corps, and the 82d and 101st Aviation Battalions deployed to Saudi Arabia with their respective divisions.

Medical

The 44th Medical Brigade deployed with the 1st Corps Support Command to Saudi Arabia, taking the 28th Combat Support Hospital, the 32d Medical Supply - Optical, and the 56th Medical Battalion.

Combat Support
Engineers

The 20th Engineer Brigade (Airborne) with 27th and 37th Engineer Battalions deployed to Saudi Arabia.

The 307th Engineer Battalion deployed with the 82d Airborne Division, and the 326th Engineer Battalion deployed with the 101st Airborne Division (Air Assault) on Operation DESERT SHIELD.

Military Intelligence

The 525th Military Intelligence Brigade returned to Fort Bragg on 14 January 1990, deploying in September to Saudi Arabia.

Sergeants First Class Gary D. Fowler and Frank D. Williamson, were appointed as First Sergeants, respectively, of Company D (LRSU), 102d Military Intelligence Battalion, 2d Infantry Division, in the Republic of Korea, and the LRSD, 110th Military Intelligence Battalion, 10th Mountain Division, Fort Drum, New York.

Company D (Long Range Surveillance Unit), 124th Military Intelligence Battalion; the Long Range Surveillance Detachment, 312th Military Intelligence Battalion; and Company D, (Long Range Surveillance Detachment), 522 Military Intelligence Battalion, respectively deployed to Saudi Arabia with the 24th Infantry Division, 1st Cavalry Division, and the 2d Armored Division.

Military Police

The 16th Military Police Brigade and the 82d Military Police Company returned from Panama, and further deployed to Saudi Arabia.

Signal

The 35th Signal Brigade (Airborne) and the 82d Signal Battalion (Airborne), deployed to Saudi Arabia on Operation DESERT SHIELD. The 25th Signal Battalion provided the primary communications link between Saudi Arabia and Fort Bragg.

Elements of the 112th Signal Battalion (Airborne) deployed on Operation DESERT SHIELD.

The 501st Signal Battalion deployed with the 101st Airborne Division (Air Assault) to Saudi Arabia.

Combat Service Support

Elements of the 1st Corps Support Command commenced returning from Panama to Fort Bragg in small increments, and later deployed to Saudi Arabia to support Operation DESERT SHIELD.

Maintenance

Lieutenant Colonel Thomas D. Bortner assumed command of the 782d Maintenance Battalion from Lieutenant Colonel Laney M. Pankey on 31 July. Lieutenant Colonel Pankey's previous assignments with the 82d Airborne Division included the 320th Field Artillery; S-4 and Commander, Company D, 782d Airborne Division Maintenance Battalion; and Center Commander and Division Materiel Management Officer, 182d Division Materiel Management Center. The 782d Maintenance Battalion deployed to Saudi Arabia.

The 189th Maintenance Battalion, 1st Corps Support Command deployed to Saudi Arabia.

The 801st Maintenance Battalion deployd with the 101st Airborne Division (Air Assault) on Operation DESERT SHIELD.

Quartermaster

The 46th Support Group, 1st Corps Support Command deployed from Fort Bragg, North Carolina to Saudi Arabia, with elements of the 530th Supply and Service Battalion and the 259th Field Service Company.

The 407th Supply and Transportation Battalion, 82d Airborne Division, and the 782d Maintenance Battalion, 101st Airborne Division (Air Assault), deployed on Operation DESERT SHIELD.

The 528th Support Battalion (Special Operations) (Airborne) deployed elements to Saudi Arabia in support of Operation DESERT SHIELD.

Copyright 1990 by the Anglo-American Publishing Company. All rights reserved. No part of this chronology may be reproduced or transmitted in any form or by any means, electronic or mechanical, including photocopying, recording, or by any information storage and retrieval system, without the express written permission of Anglo-American Publishing Company, Brandon, Florida, except where permitted by law. Used by permission.

A rifle squad of the 3rd Battalion, 187th Infantry, 101st Airborne Division, Ft. Campbell, Kentucky leave their UH-1H (Huey) helicopter during "Quick Eagle I", a field training exercise of the division's 3rd Brigade near Ft. Campbell. (U.S. Army photo)

Campaign Streamers awarded to airborne units since the establishment of United States Airborne Forces.

WORLD WAR II
ASIATIC-PACIFIC THEATER

India-Burma	2 April 1942—28 January 1945
Papua	23 July 1942—23 January 1943
Guadalcanal	7 August 1942—21 February 1943
New Guinea	24 January 1943—31 December 1944
Leyte	17 October 1944—1 July 1945
Luzon	15 December 1944—4 July 1945
Central Burma	29 January—15 July 1945
Southern Philippines	27 February—4 July 1945

WORLD WAR II
EUROPEAN-AFRICAN-MIDDLE EASTERN THEATER

Algeria-French Morocco	8-11 November 1942
Tunisia	
Air	12 November 1942—13 May 1943
Ground	17 November 1942—13 May 1943
Sicily	
Air	14 May—17 August 1943
Ground	9 July—17 August 1943
Naples-Foggia	
Air	18 August 1943—21 January 1944
Ground	9 September 1943—21 January 1944
Anzio	22 January—9 September 1944
Rome-Arno	22 January—9 September 1944
Normandy	6 June—24 July 1944
Southern France	15 August—14 September 1944
Rhineland	15 September 1944—21 March 1945
Ardennes-Alsace	16 December 1944—25 January 1945
Central Europe	22 March—11 May 1945

KOREAN WAR

UN Defensive	27 June—15 September 1950
UN Offensive	16 September—2 November 1950
CCF Intervention	3 November 1950—24 January 1951
First UN Counteroffensive	25 January—21 April 1951
CCF Spring Offensive	22 April—8 July 1951
UN Sumer-Fall Offensive	9 July—27 November 1951
Korea, Summer-Fall 1952	1 May—30 November 1952
Korea, Summer 1953	1 May—27 July 1953

VIETNAM MILITARY OPERATIONS

Vietnam Advisory Campaign	15 March 1962—7 March 1965
Vietnam Defense Campaign	8 March 1965—24 December 1965
Vietnam Counteroffensive	25 December 1965—30 June 1966
Vietnam Counteroffensive Phase II	1 July 1966—31 May 1967
Vietnam Counteroffensive Phase III	1 June 1967—29 January 1968
Tet Counteroffensive	30 January 1968—1 April 1968
Vietnam Counteroffensive Phase IV	2 April 1968—30 June 1968
Vietnam Counteroffensive Phase V	1 July 1968—1 November 1968
Vietnam Counteroffensive Phase VI	2 November 1968—22 February 1969
Tet 1969 Counteroffensive	23 February 1969—8 June 1969
Vietnam Summer-Fall 1969	9 June 1969—31 October 1969
Vietnam Winter-Spring 1970	1 November 1969—30 April 1970
Sanctuary Counteroffensive	1 May 1970—30 June 1970
Vietnam Counteroffensive Phase VII	1 July 1970—30 June 1971
Consolidation I	1 July 1971—30 November 1971
Consolidation II	1 December 1971—29 March 1972
Vietnam Cease Fire	30 March 1972—28 January 1973

GRENADA CAMPAIGN

Grenada	23 October—21 November 1983

PANAMA CAMPAIGN

Panama	20 December 1989—31 January 1990

THE USA AIRBORNE FIFTIETH ANNIVERSARY

By William E. Weber, Col., USA, Ret.

"Where is the prince who can afford to so cover his country with troops for its defense, as that ten thousand men descending from the clouds might not, in many places, do an infinite deal of mischief before a force could be brought together to repel them" (Benjamin Franklin, 1784).

PROLOGUE

That hypothesis came to traumatic reality for America's enemies in the fifty years (1940-1990) of American Airborne history. Throughout the world: from the sand of Africa; the hedgerows and fields of Europe; the islands and atolls of the Pacific; the mountains and rice paddies of Korea; the jungles of Vietnam; the internecine strife of Lebanon; the isolation of Grenada; the dichotomy of Panama; and, today the deserts of the Middle East, those who would disrupt the good order of the world have felt the might of America's Airborne power!

In honor of that record of valorous service and dedication to our Nation and the ideals of the American way of life, over 37,000 veterans and active duty, Reserve and Guard wearers of the "silver wings of courage", gathered together in Washington, D.C., 4-8 July 1990, to celebrate the Fiftieth anniversary of the founding of America's Airborne forces. Joined by their families and thousands of interested spectators, airborne made a vertical assault on the nation's capital for a five day celebration to honor their history and to pay homage to their heritage!

Never before has such an assemblage taken place! A muster of every airborne unit and every fraternal Airborne Association, gathered together to demonstrate their unity and the camaraderie and comradeship of arms that distinguishes them from the ordinary. Thier message was simple "We Are Airborne, All The Way, Always And In All Ways!"

For those who read this book and are not of the airborne fraternity it seems appropriate to define "airborne" in more detail. It is more, much more, than those who use parachutes to reach the ground. In brief, it is any form or mode of vertical insertion that gets a combat trooper from the transporting vehicle to the ground. It can be by glider, helicopter, troop carrier aircraft, air landing aircraft, hover craft or lighter than air transport such as balloons, air ships and the like.

And, it is not only the troopers whose mission is combat when they reach the ground that are airborne! Those who fly the vehicles that transport the combat trooper are also airborne, as are those who set up and mark the drop and landing zones. An airborne operation can be the insertion of one man or the mass of an entire field army. there is no prescribed limitation or formula save that of the availability of the air lift and the availability of trained and properly equipped troopers.

As stated in the Program of the fiftieth Anniversary by Col. Weber, "We are a brotherhood of those who had to give more of themselves to perform their duties, and who are recognized by the many varieties of the silver wings of courage that each wear on the breast. Airborne wings mark a comradeship that transcends the normal expected in a military organization. You know that those who wear the silver wings of courage are those on whom you can depend, no matter what the circumstances. The units in which you served added pages of valor and accomplishments to our nation's history and heritage that are unsurpassed by others".

In a letter to honor the Fiftieth Anniversary of USA Airborne, President George Bush stated: "I have always had the greatest respect for the unique talent, courage, and magnificent camaraderie of our American Airborne Forces. From Normandy to Panama, whether by glider, parachute, air assault, air landing or special operations, Airborne has always risen to the occasion. Your special fraternity has set—and continues to set—the standard for elite forces of the world."

BACKGROUND

The idea of a fifty year muster originated in 1983 and was manifested by Col. William E. Weber, USA-Ret, who presented his idea to his contemporary leaders of the various fraternal airborne associations. These men, intrigued by the idea, agreed to take the leadership within their respective associations and unite them in support of making the dream come to reality.

To plan, organize and conduct the affair an organization was formed, The USA Airborne Fiftieth Anniversary Foundation, and was incorporated in Washington, D.C. This organization was tasked by the affiliated associations to produce the needed planning documents to achieve the goal of a meaningful celebration and to provide the leadership and funds to underwrite the events.

Each affiliated association provided seed funds to enable the foundation to begin work and appointed a representative to the Advisory Board, in which was vested the authority to execute certain planning and programming.

From the Advisory Board of almost thirty members, an Executive Committee of thirteen members volunteered to give substance to the decisions of the Board and to provide a forum to address issues that required new direction or determination. this Committee elected four of its members to function as an Executive Group of four officers: two Co-Chairmen, Dr. Charles E. Pugh, (517th PRCT) and Col. William E. Weber, USA-Ret (187th ARCT & 11th Abn Div); a Secretary, Mr. Joe Quade (17th Abn Div); and, a Treasurer, Dr. D. A. Parks, (187th ARCT).

In them was vested the necessary authority to carry out the myriad of tasks agreed to by the Board, and to take action as appropriate to conduct the day to day duties of the Foundation. As well, several separate working sub-Committees and Chairmen and members were designated by the Executive Group to plan and execute the several foundation sponsored events that would transpire.

Though agreeing to cooperate, as possible, with the plans of the Foundation, each affiliated association remained autonomous and assumed full responsibility for the conduct of their independent activities conducted as part of their annual reunion programs. It was through the scheduling of their annual reunions for the same period in Washington that the muster became a possibility as no other means would have achieved a similar attendance of the airborne fraternity.

No event such as this could have been possible without the full cooperation of the Department of Defense, agencies of the federal government and the District of Columbia and, beyond question, the active duty, Reserve and National Guard units of the Armed Forces whose roles and mission are airborne oriented. Add to that the dedication and support of the fraternal associations and their individual memberships and it was a given that success would follow.

There are too many individuals from the airborne fraternity who gave of their time and money in support of this Anniversary celebration to name them, but it is also a given that without them it could not have been accomplished! Suffice to say that all have a rightful share in whatever credit is due! To them our fraternity is indebted!

And a much greater debt is owed to Charlie Pugh for his diplomacy, reasoned leadership and, willingness to take the static; to Joe Quade for his dedication above and beyond the call of duty; and, to Don Parks for his common-sense determination to keep us on an even keel in fiscal matters. This applies to the members of the Advisory Board, Executive Committee and the sub-committee Chairmen and members, for without their loyalty to the mission there would have been no Fiftieth Anniversary Celebrations! The dedication of these men is a credit to the airborne fraternity. For three years they met on a quarterly basis, coming from all over the country at their own expense to serve their fraternity! We could have asked no more of them. They gave their all! The sum total of the efforts of all is equal to the degree of dedication airborne has demonstrated to America for fifty years!

THE PROGRAM

Though simple in scope, the program was complex in coordination. But, it all came together! The joint activities consisted of a ceremony at Arlington National Cemetery at the Tomb of the Unknown Soldier on 4 July; memorial services by appropriate associations at the Vietnam Memorial and site of the Korean War Memorial on 5 and 6 July; a mass parade on 7 July; and, the capstone event, a muster and Memorial Program at Robert F. Kennedy Stadium on 8 July.

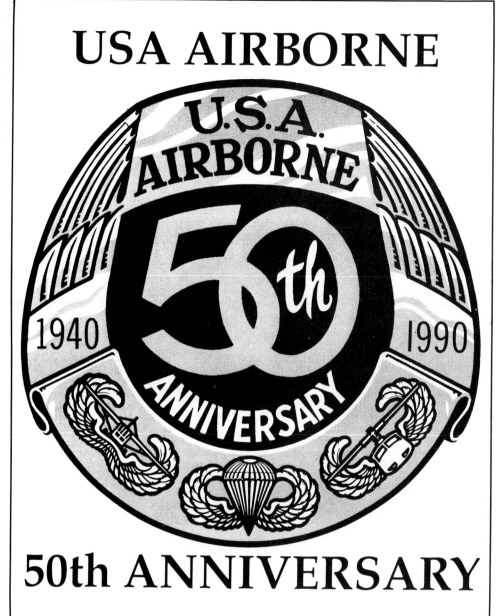

50th Anniversary U.S. Airborne Commemorative Program.

In retrospect the events programmed seem to be not so difficult to arrange. But the truth is that it took the better part of three years to arrange the coming together of the many elements necessary of the Advisory Board, the special committees and the individual affiliate associations all labored with dedication to make it happen. There were no guidelines because this had never been done before! We were breaking new ground.

Though not a part of the scheduled activities by the Foundation, also included in the Fiftieth Anniversary was the opening, on 2 July 1990, of a special exhibit at the Smithsonian Institution's Museum of American History. The exhibit will remain open for a five-year period and is exclusively oriented to honor American Airborne. Included are artifacts from the fifty years of airborne and a pictoral montage of major airborne operations during the period 1940 to 1990. Though instituted at the request of the Foundation, funding for the exhibit was provided by Coors Brewing Company as their gesture to honor the fifty years of American Airborne.

Interspersed between the foregoing joint Foundation events, the individual associations filled the time with their independent reunion programs. As might be expected the activities of the associations were as varied as are they themselves. While all share the common bond of airborne, their history and service vary from war to war, theater to theater and the uniqueness of their membership. But a common thread runs through all in that each established some form of hospitality area where "war stories" abound and old comrades see each other for the first time or recapture the camaraderie last shared at a previous reunion.

Those associations whose historical unit is still in the active structure make possible the attendance of active personnel from those units. These young troopers bring the vitality and "gung ho" aspects of youth to the reunions. In turn, they are afforded an opportunity to associate with the veterans of their unit who made possible the history they cherish and the heritage they guard. A casual observer would find it difficult to ascertain which of these groups, the veterans or the active troopers, gain the most from the relationship of bonding that binds the past with the present!

FOUNDATION ACTIVITIES
2 July 1990: "GO!"

The Foundation opened a Command Post south of the Potomic River in Virginia at the Holiday Inn Crowne Plaza in Crystal City at 0800 hours. From there all programmed activities were coordinated and the Press Center provided releases and set up interviews as requested by the media. Active duty personnel from US Special Operations Command, XVIIIth Airborne Corps and Military District of Washington established control and communication facilities to coordinate the activities of the active, reserve and guard units and assist the foundation events sub-committees.

The affiliate association leaders set up their control elements at their respective reunion hotels. As is typical, the "early birds" began arriving as early as 1 July, ready to make airborne's fiftieth birthday the biggest bash ever! From all over the country—every state and territory and overseas—planes, buses and private vehicles began converging on Washington disgorging troopers and their families and friends by the thousands.

At Forts Bragg, Campbell, Benning, Rucker, Chaffee and Lewis, and from Reserve and Guard armories, units finalized their movement plans. The 82nd and 101st made last minute preparations for their respective battalion-sized parachute and air assault operations into Andrews Air Force Base, Maryland, and Camp A.P. Hill, Virginia. Special Forces, Rangers and contingents from the separate regiments did likewise. At Forts Belvoir, Meade, Myers and McNair, MDW ensured that the means to house, feed and transport the arriving contingents were on track.

All was in a readiness state. Airborne was awaiting the command to "go"!

3 JULY 1990: FINAL CHECKS

More of the same! Last minute checking and rechecking! The stamps honoring airborne that were issued by the island nation of Grenada were distributed to affiliates and sample covers of the Fiftieth Anniversary Volume were sent to each reunion hotel so that attendees could see the quality and extent of this book.

Bill Story, of the First Special Service Force, put the press center into high hear and ensured that the local and national media could cover all the events. The Command Post took on an atmosphere of high intensity, similar in every respect to the excitement and sense of dedication of a wartime operation. And, in reality, it was just that! We were breaking new ground and everything had to be done from scratch. There were no guidelines to follow. Airborne was setting the standard that others would have to emulate!

4 JULY 1990: ARLINGTON CEMETERY WREATH LAYING

A day for honoring and remembering! The wreath-laying ceremony at Arlington National Cemetery (The Tomb of the Unknown) was attended by thousands of troopers and their families. The full Honor Guard company from the 3rd Infantry Regiment, "The Old Guard," opened the ceremony with their tradi-

Commemorative stamp issued by Grenada honoring fifty years of U.S. Airborne.

555th PIR Association Wreath. (Photo by Chuck Christianson)

Gen. Roscoe Robinson and Big Jim Queen. (Photo by Chuck Christianson)

Honor Guard. (Photo by Chuck Christianson)

Color Guard, Tomb of the Unknown. (Photo by Chuck Christianson)

Crowd at wreath laying, Arlington National Cemetery, 4 July 1990. (Photo by Chuck Christianson)

Spectators at the wreath laying. (Photo by Chuck Christianson)

Dr. Charles Pugh with Col. William E. Weber and Col. Vito S. Pedone. (Photo by Chuck Christianson)

Dr. Charlie Pugh, Col. William E. Weber, Col. Vito S. Pedone and spectators. (Photo by Chuck Christianson)

tional parade up the steps below the Tomb. Each affiliate association had designated three representatives to be a part of the official airborne contingent participating in the ceremonial activities. They formed a cordon of honor in an arc facing the Tomb and each of the affiliate groups were attired in their associations distinctive "uniform". It was clear to see that the individualism of each affiliate blended perfectly with the commonality of airborne pride to send a message of pride and dignity!

The Co-Chairmen of the Foundation, and Col. (Ret) Vito Pedone representing, respectively, the ground and air elements of airborne accompanied the wreath as it was placed on the Tomb. the wreath was a mass of roses surrounding the official logo of the USA Airborne Fiftieth Anniversary Foundation. It was both beautiful and appropriate and thanks must be given to Carmella La Spada from "No Greater Love" for her efforts in obtaining the wreath.

Upon placement of the wreath, three vollies were fired and a bugler from the Army Band ("Pershing's Own") played Taps. The ceremony never fails to move those who are present and today was no exception. Every trooper present had thoughts of his own comrades that fell in battle or who had made their last formation. Collectively, their thoughts encompassed all of our airborne comrades whose presence, silent, unseen, but in every way, pervasive, is always with each of us! We cannot forget them! We will not forget them!

Following the ceremony the representatives of the affiliates and the foundation representatives joined Ray Costanzo, the Superintendent of Arlington Cemetery (and a 508th veteran) in the Rotunda display room for the presentation and placing of a plaque honoring the occasion and airborne's tribute to the Unknown.

Late in the day we began to converge on the Washington Monument grounds to witness the traditional Independence Day fireworks and entertainment. A real extravaganza and, as always, the firing battery from the 3rd Inf. Regt. wowed the crowd during the playing of the 1812 Overture by the United States Marine Corps Band.

Elsewhere during the day, affiliates did their own thing. some had separate ceremonies at Arlington dedicating plaques and trees. Others followed routine reunion schedules.

Throughout the day, troopers continued to arrive and register for their reunions. One thing became evident at the

Vietnamese Airborne Association with jumpers of the 3/325th Abn. Inf. (Photo by Chuck Christianson)

Command Post! It was apparent early on during the day that we were gong to achieve an overall attendance that would set a record for this kind of gathering in Washington! We were not assembling to petition the government. We were not assembling to object to something. We were not assembling to complain. We were not assembling to advance a cause. We were not assembling to harass.

Our assemblage was to honor our brotherhood and to acknowledge our fiftieth birthday! We were doing something that had never been done before in this city and we were showing them how it should be done! Throughout the city and within the federal government agencies that we dealt with, you could hear the repeated statements of amazement that so many of us would go to such lengths to satisfy only our personal desire to honor our heritage.

Throughout the day, troopers and their families visited the many memorials in the city and the Smithsonian. The highlight of their touring the Smithsonian Institution buildings was the display honoring the Fifty Years of Airborne in the Museum of American History. The display was funded by a donation from Coors Brewery and is the first time any separate entity of the Armed forces has been so honored.

The bulk of the items displayed were loaned by individual troopers and the centerpiece bronze of the trooper at the moment of spilling his chute upon landing was sculptured and loaned by Trooper William Porteus (11th Abn Div).

The display will be there for a period of five years. In that time frame hundreds of thousands of American and foreign

Part of the Airborne display at the Smithsonian Institution.

visitors will view the display and they will come away aware that such fraternity as airborne does exist and created a history and heritage in a way that has never been done before!

5-6 JULY 1990: MARK TIME

At the Vietnam War Memorial and the site of the soon-to-be-erected Korean War Memorial, sculpted by a former member of the 513 P.I.R., various units assembled their membership to conduct memorial and dedication events. The 101st Airborne Division, 173rd Airborne Brigade, 1/8th Calvalry Regiment and Special Forces oriented on the Vietnam memorial and the RAKKASANS (187th

A Member of the 503rd Parachute Infantry Regiment Association with Gen. William C. Westmoreland. (Photo by Chuck Christianson)

Colors and lead element, 509th Parachute Infantry. (Photo by Chuck Christianson)

Airborne Regimental Combat Team) and Airborne Infantry Rangers Companies at the site of the Korean War Memorial. While there, the members of these two associations pledged over $30,000 in contributions to this memorial fund.

Throughout the various reunion hotels each associationn oriented on independent activities and individual troopers took advantage of the time to sightsee and enjoy the capital and hospitality suites. War stories, the staple of reunions, abounded, and it was the rare trooper who didn't have one of his own to relate to strain the believability of some! Yet, even those on the cutting edge of belief have, at heart, an incident that did occur. Put in context with the totality of all, the sage of airborne is unequalled in the annals of war and peace!

The highlight of these two days (aside from the banquet events by some association), was the drop of a battalion from the 82nd Airborne Division at Andrews AFB. Fully combat loaded, a battalion of the 325th Airborne Regiment replicated their jump into Panama in January 1990. Thousands of troopers were on hand to witness the jump and you could feel the pride these veterans had in the young troopers who were "hitting the silk." Upon assembly, the battalion doubled-timed to the spectator area and one could see the same kind of pride reciprocated by the young troopers to the veterans. In all, the bonding was complete. Let no one question the mutual commitment that each generation has for the other!

Sadly, the air assault insertion of a battalion of the 101st Airborne Division was too far from Washington to enable the bulk of veteran troopers to view the operation. Those who did and have memories of Vietnam shared the same kind of thoughts as their comrades who witnessed the jump.

The "bottom line" lesson we learn as we observe this present generation of active duty airborne troopers, is that they are superb soldiers! Strong, tough, proud, professional and ready! They are what we remember we were. There may be differences in equipment and practices but the hard core of the airborne spirit and dedication is still there! And, for those who take the trouble to compare, the same is true of the pilots and crew who fly and man the air craft and "choppers" that airborne uses today, just as it was with the glider pilots and transport pilots of our day. Anybody contrasting the C-5 with the C-47, or the Huey with the CG-4A has got to realize that you've got to be good and you've got to be motivated to get your troopers on target without foul-ups.

7 JULY 1990! THE PARADE

"Oh what a beautiful morning! Oh what a wonderful day!" How true was this phrase from the popular show tune from "Oklahoma", as dawn broke on 7 July. Once again God demonstrated his partiality for airborne troopers and their mission. In the nation's capital in July you can anticipate high temperatures and humidity, but not this day! Airborne was blessed with a temperature range from the mid-70s to low 80s and humidity levels at 40% with a gentle breeze to make wearing a jacket tolerable and the myriad of colors to flow from their staffs without straining the bearers.

The crowd of spectators estimated at almost 40,000 started lining up along Constitution Avenue early in the morning. they stretched from the start point at 7th Street up through 14th Street and then along 14th to Pennsylvania Avenue with a heavy concentration at the Reviewing Stand between 13th and 14th Streets on Pennsylvania Avenue.

While the spectators were early those of the airborne fraternity who would parade, all 7,000 of them, were there ahead of them. But this time it was not solely because of the old Army adage of, "Hurry up and wait"! Rather, it was because no one wanted to be late for this once-in-a-lifetime chance to be a part of something that airborne had never done before.

Earlier than any was John Costello (509th P.I.B.) and his Parade sub-committee, and marshals from each marching contingent. John and his committee had the full responsibility to make sure it all went off without a hitch. In company with the MDW staff, John hit a "ten" on the target.

By 0900, association after association had arrived at their assembly points along 7th Street between Independence and Constitution Avenues. The active duty marching contingents in battle dress uniforms with bayonets fixed to the weapons and wearing helmets presented an imposing sight. No less imposing were the members of the various associations. Most had adopted a distinctive form of dress for the parade, ranging from blazers and slacks to WWII jump suits. Colors and guidons as well as banners marked each Association's parent unit.

The RAKKASANS added a unique touch with their own bagpipe band complete with kilts, bearskin headgear and a Pipe Major almost seven feet tall! The famed "Ambassadors in Blue", Canada's Niagara Regional Police Pipe and Drum Band paraded the length of 7th Street to honor the various marching units and serenade them with "Blood on the Risers" played on the bagpipes! Surely a first, if there ever was one!

Yet, even an uninformed observer could see the thread of commonality present amongst the assembled Associations. You could not mistake it! Without exception, the unique symbols of airborne were present in every unit. Association flags, special banners, placards and associationn and chapter patches abounded. No two exactly alike except as they personified the esprit of the airborne trooper!

Wings of every airborne mode, many with as many as five combat jump stars, Combat Infantry Badges with three stars, campaign ribbons from every war of the past fifty years and every theater of those wars. Decorations for valor and service were profuse, from the Medal of Honor to the Good Conduct Medal. (Good Conduct Medal? Wonder how the kind of troopers we were ever earned those!) I kid. What was important and impressive was that every one in the line-up was ready to "strut his stuff" this day!

From one end of 7th Street to the other you could hear the same admonition being passed by association leaders, "Remember! When we pass the Reviewing Stand we'll be dong an eyes left, not right! Stay in step—Listen to the cadence—Dress your ranks and files! Suck in those guts—March tall and proud!

Men, some who had not marched a step in over forty years, were beginning to remember. You could see the memories of past parades coming back to them in so

Reviewing Stand with Col. Lewis Millett, Col. Vito Pedone, S/Sgt. Longerbeam, Gen. Westmoreland, Gen. Stiner, and MG Joseph C. Lutz. (Photo by Chuck Christianson)

Guidons of the Airborne Regiment Combat Team Association. (Photo by Chuck Christianson)

many subtle ways. The way they checked their lines; the guidon bearers practicing signaling the preparatory command and command of execution; each man double checking to make sure his headgear was squared off; one, telling another how something should be done; eyeing the other outfits and commenting on that "route step" outfit; the silent, and at times, vocal pride they felt; and, the growing anticipation that "cross the IP time" was close at hand. Yes, they were going to march again: Only, this time it might be the last time and they were determined to make it the best!

One could sense the young troopers from the active units taking all this in and they looked on in wonderment. But nowhere did you see a knowing look or hear a sarcastic remark. Instead you could almost hear them thinking that these old troopers still had it! There was a real pride for these men in their eyes. The feeling of oneness between the active troopers and the veteran troopers. More than just mutual respect; more than just pride! It was a wave of comradeship and kinship that rolled down 7th Street this day uniting all that it touched with the spirit of airborne, brothers-in-arms sharing a once-in-a-lifetime experience. To see it was infectious. To be a part of it was inspiring. To know that it was finally going to happen was incredible!

At the head of the assembly area were the troops of the Joint Services Honor Guard. A platoon each from the Army, Navy, Marine Corps, Air Force and Coast Guard. At their head was the USMC Band and the Joint Service Color Party with the national colors those of the services and, directly behind them the colors of the

General William C. Westmoreland. (Photo by Chuck Christianson)

Niagara Ontario Police Pipe and Drum Corps (Canada). 187th ARCT Association Honorary Band. (Photo by Chuck Christianson)

Colors bearers, 1st Special Service Force Association. (Photo by Chuck Christianson)

Alumni of the 501st Parachute Infantry Battalion Association. (Photo by Chuck Christianson)

Old Guard with massed airborne colors. (Photo by Chuck Christianson)

509th Parachute Infantry marchers. (Photo by Chuck Christianson)

Parade Marshal General Westmoreland and General Ryder in Airborne Parade.

Colors, 542nd Parachute Infantry Association with John Grady (Association founder) center. (Photo by Chuck Christianson)

517th Parachute Infantry Regiment Combat Team Association members. (Photo by Classic Photos)

USA Airborne Fiftieth Anniversary Foundation!

Behind them in the parade line up was contingent of color bearers from the 3rd Inf Regt who were carrying the massed colors of the more than sixty famous units of airborne's history whose colors had not been unfurled since they were cased following the inactivation of these units after WWII. From these colors flew streamers and citations of every campaign in which airborne played a role during WWII. A litany of valor and service unequaled among the airborne forces of any nation anywhere in the world!

When viewed in company with the massed colors, down to battalion level, of the 82nd and 101st Airborne Divisions and the colors of the separate active airborne Regiments and the Special Forces, Rangers and Reserve and Guard units, there were almost 200 unit colors present!

Waiting on the corner of 7th & Constitution was a "gussied up" jeep which was the vehicle in which the Parade Grand Marshal and Senior Reviewing Officer, Gen. (Ret) William C. Westmoreland, would lead the parade. Joining him in that honor was BG (Ret) William Ryder, the first man of the Test Platoon out of the aircraft on their virgin jump in the summer of 1940.

"Westy" and Bill Ryder arrived promptly at 1045 and, after a quick "troop the line", they mounted the jeep ready to move out.

Right on the dot, John Costello gave the "word", the bands started playing, color bearers cupped their colors and the Honor Guard gave the command "Forward March". We were off and running, writing another chapter to airborne's history. What a day!

For those who marched in the parade or viewed it as spectators no description is needed. Those reading this who were not there and who have not seen the video tape of the parade, an adequate description may not be possible!

The emotion was overpowering. The cheers from the crowd started as soon as the parade stepped off and they didn't stop until the last unit passed the reviewing stand almost 50 minutes later" Charlie Pugh, whose duty station was the reviewing stand area, stated that he knew instantly when the parade actually started. Almost a mile away he could hear the sound rising over the government buildings on Constitution Avenue and bouncing down the length of Pennsylvania Avenue.

General Officers marched at the head of their commands. LTG Luck headed up

13th Airborne Division alumni. (Photo by Chuck Christianson)

Korean Rangers with guidons. (Photos by Chuck Christianson)

173rd Airborne Brigade alumni. (Photo by Chuck Christianson)

Airborne Troop Carrier Command alumni. (Photo by Chuck Christianson)

550th Airborne Infantry Battalion Association. (Photo by Chuck Christianson)

20th Special Forces Group, Maryland Army National Guard. (Photo by Chuck Christianson)

Navy marchers. (Photo by Chuck Christianson)

3rd Special Forces Group contingent from Ft. Bragg, NC (Photo by Chuck Christianson)

104th Infantry Det. (Long Range Surveillance). (Photo by Chuck Christianson)

Ranger Infantry Companies, Airborne-Korea, contingent. (Photo by Chuck Christianson)

Army Band (28th Inf. Div.-Pennsylvania National Guard). (Photo by Chuck Christianson)

Robert F. Kennedy Stadium. (Photo by Chuck Christianson)

82nd Airborne Division Chorus. (Photo by Chuck Christianson)

Special Forces Association led by Charlie Norton, Pres. Chapter XI, Washington, DC. (Photo by Chuck Christianson)

Spectators at the 50th Airborne Anniversary Parade. (Photo by Chuck Christianson)

Lt. Gen. Vaught, USA, with US Advisors and members of the Vietnam Airborne Association. (Photo by Chuck Christianson)

the XVIIIth Airborne Corps contingent followed by MG Johnson leading the 82nd Abn Div and, in turn, MG Peay leading the 101st Airborne Division (Air Assault). The same pattern prevailed amongst the associations. They were led by their presidents or, in many cases, the war-time commanders of the units the associations represented.

Troopers in wheel chairs, on crutches and other disabled riding in the restored tactical vehicles of the WWII era, were as much a part of the parade as those who marched on two good legs. They ranged in age from men pushing 80 years to the young troopers, some of whom were barely 18. Today, age and conditioning were unimportant; motivation had taken over.

When Gen. Westmoreland reached the reviewing stand the crowd of well-wishers was so dense and so intent on shaking his hand that he had difficulty getting to the stand to accept the salute of the passing units. He made it however and was joined by the five men who were chosen to represent the fraternity of airborne for this Fiftieth Anniversary, and Gen Carl W. Stiner, CINC, US SOCOM, who was to deliver the keynote address at RFK Stadium on Sunday. The men who symbolized the airborne fraternity were:

Col. Lewis A. Millett, Medal of Honor, USA Ret: representing all veteran paratroopers.

Former S/Sgt. Herman L. Longerbeam, Distinguished Service Cross: representing all veteran glider troopers.

Col. Vito S. Pedone, USAF Ret: representing all veteran airborne troop carrier pilots and crewmen and glider pilots and all active duty airlift personnel.

PV2 James Held, the youngest graduate of Air Assault School at Fort Campbell just prior to 4 July 1990: representing all active duty air assault troopers.

PVT Roberto Ruiz, the youngest graduate of Jump School at Fort Benning just prior to 4 July 1990: representing all active duty parachute troopers.

As each unit neared the reviewing stand the parade announcer identified them and gave a brief rendition of their history. The crowd roared as each unit passed an it truly was a sight to behold! They were in step, marching proud. Their pride was like an explosion of emotion. You could not avoid the impact. The active units were a testament to America's power. The veterans, a testament to America's past glories on the battlefield. If there was a difference between them it was only that of the years that separated

17th Airborne Division Association colors. (Photo by Chuck Christianson)

542nd Parachute Infantry Association colors. (Photo by Chuck Christianson)

555th Parachute Infantry Battalion colors and marchers. (Photo by Chuck Christianson)

11th Airborne Division alumni. (Photo by Chuck Christianson)

The Riderless horse symbolizing a fallen warrior.

The RAKKASAN Color Party.

them, for otherwise on this day, they were equals in every other respect!

No unit was lacking in pride this day and none were spared the cheers of the crowd. When the members of the Vietnamese Airborne Division led by the US personnel from the Advisor Group, passed the reviewing stand a forest of South Vietnamese flags sprang into view. Their countrymen, now all US citizens, journeyed from all over this land to pay homage to these brave former allies. Of great significance was the fact that their presence in the parade was their means of honoring USA Airborne. As with our veterans, these men came from all over the US, not to celebrate an anniversary unique to them, but rather to honor us on ours! That message was not lost on anybody.

At the rear of the parade and part of the 1/8th Cavalry Association contingent, a riderless coal-black horse with boots reversed in the stirrups, was led by a trooper from the 3rd Infantry. A fitting touch, as the symbolism represents those who have fallen in battle. And yet, if one looked very closely, it seemed as if the ranks of the passing units were filled with these silent warriors marching in company with their comrades. That they were there is undeniable! That we could sense their presence is unremarkable, for they live in our memories everlastingly, as close to us today as when we served together!

The parade ended, but the memory is never ending. Airborne had its day on the streets of the Nation's Capital and nothing can change that!

8 JULY 1990:
THE MEMORIAL SERVICE

Another unbelievably perfect day for airborne! The sun rose early this Sunday, as did most troopers. None wanted to be late at RFK. The sky was laced with fluffy white clouds that marched to the cadence of a breeze to temper what otherwise could have been a typical "dog day" in Washington's summertime.

In the early morning hours, as they did all Saturday afternoon, the bands, color parties and production crew ran through yet another rehearsal. Bob Sommers (13th Arirborne Division), Chairman of the Memorial Program committee and his staff, in company with Bob Johnson and his staff from Johnson Associates, the firm hired to stage manage our extravaganza, wanted to be certain that all was in readiness. From MDW, Mark Murray, ceremonial events expert, and MDW officers and NCOs were put-

50th Logo on the field at Robert F. Kennedy Stadium.

RAKKASAN Reserves on parade.

Mort Walker, of the comic strip "Beetle Bailey" fame, created this special cartoon honoring airborne on its fiftieth birthday.

ting the final touches on the premiere event of the day, the Colors Presentation ceremony.

Like nervous mother hens, Pugh, Weber, Quade and Parks, hovered on the periphery of the action, trying not to get in the way but unable to stay away!

In the center of the stadium field was a sixty-foot-in-diameter logo of the Fiftieth Anniversary. Ringing the logo were fifty conical receptacles, each capable of holding fifty red roses to be placed by fifty small children, each a grandchild of an airborne trooper. This "Tribute to the Fallen" portion of the ceremony was conceived by Carmella La Spada of "No Greater Love". Her dedication to our Anniversary was further demonstrated by her persuading the cartoonist Mort Walker, of the comic strip "Beetle Bailey" fame, to create a special cartoon honoring airborne on its fiftieth birthday.

Troopers, their families, citizens of the city and tourists continued to pour into RFK until the shady portion of the lower deck was filled. Scattered throughout the uncovered areas were those who wanted to be close to the action and those who missed out on the shaded seats. Down in front, the battalion of troopers from the 101st were massed in four seating sections. Sadly, the battalion from the 82nd had to miss the show as their airlift back to Fort Bragg could not lay over an additional day. But, the 82nd was still well represented by their massed colors party and the 82nd band and they were joined by the massed colors party and band from the 101st and the color contingents from the other active, Guard and Reserve units.

The VIP section was directly behind homeplate centered on line with the stage set up behind second base in short centerfield. The VIPs were seated by 1115, though sadly, President Bush was absent because of his priority need to attend the World Economic Conference. The crowd seemed to understand the need for his absence and riveted their attention to the field, curious as to what the day would bring. Even though the Commemorative Program was passed out to all attendees, the description of programmed events was kept brief so as not to detract from the element of surprise.

Decorated with bunting and a huge logo of the USA Airborne Fiftieth Anniversary Foundation, the stage was a beehive of last minute activity as the MC for the day (Col. Bill Weber) and Bob Johnson went over the cues to signal the various events.

At exactly 1140 Bob Sommers was satisfied that all was a "go" and he radioed to all to get cracking. We were about to be a part of the last chapter in airborne's first fifty years!

The USMC Band, "The President's Own," resplendent in their red dress uniforms marched on and, positioning themselves by second base, began a medley of songs to warm up the crowd and get them tapping their feet in time to the music. At the end of the medley the off-stage announcer acknowledged the attending VIPs with particular emphasis on the members of the original Test Platoon and the original black Test Platoon. Also acknowledged, though unknown, were the first men to ride a military glider and the first to test air assault techniques. As well, the first glider pilot and the first airborne troop carrier pilot were acknowledged. Though unknown, these airborne men are joined with the known members of the Test Platoon as the honored men of airborne, the true pioneers who first did that which those who followed had to learn to emulate.

In his opening remarks Weber stated, "We are gathered here in this, the first mass muster of our Corps, to celebrate fifty years of service to our nation and our people. I welcome each of you and I share with you the honor of representing the more than 1.75 million of our comrades whose presence is with us today as is that of the thousands of our fallen comrades who join us in spirit. They fill the seats next to you today just as they filled our ranks and files as we marched yesterday! Unseen, but not unknown, they are with us. What we have accomplished these past three days and what we shall do here today as the capstone of our Fiftieth Anniversary will be the standard against which others will be measured when they attempt to emulate what we have done. That is how it should be: that is how it has always been! Airborne leads the way!".

With music from the Marine Band, the show opened with the posting of the Joint Armed Forces Colors, the 53 State and territorial colors and lastly, the colors of the USA Airborne Fiftieth Anniversary Foundation. During the march on and recognizing of the colors Weber asked all to meet those sitting next to them. He emphasized the fact that a microcosm of America was in the stadium and emphasized that the diversity of our backgrounds was matched by the commonality of our dedication to our country and to airborne.

As he closed the opening remarks the audience rose to join in the singing of our National Anthem, followed by Chaplain Wood, a four combat jump veteran of the 82nd who gave the invocation. As the

Reviewing Stand: Col. Millett, S/Sgt. Longerbeam, Col. Pedone, PFC Held, Pvt. Ruiz, Col. Weber and Gen. Stiner. (Photo by Chuck Christianson)

Children's Tribute by "No Greater Love". (Photo by Chuck Christianson)

Uncasing newly reactivated unit colors. (Photo by Chuck Christianson)

Colors presented. (Photo by Chuck Christianson)

Britt Small and Festival. (Photo by Chuck Christianson)

General Carl W. Stiner, CINC US Special Operations Command. (Photo by Chuck Christianson)

Britt Small and Festival followed. Britt formed his group from Vietnam veterans, having served himself with the 173rd, shortly after the US came home from Nam. His program is aimed at honoring the veteran and stimulating patriotic pride. His rendition of the "Ballad of the Green Berets" brought the crowd to their feet and generated a terrific roar from the Special Forces seating section.

As Britt left the stage, Foundation Secretary Joe Quade came forward to present a copy of the USA Airborne Fiftieth Anniversary Foundation colors to Commissioner Anthony Capizzi of the city of Dayton, Ohio. They were enshrined in the Dayton Aviation Hall of Fame on 27 July 1990.

As an adjunct to fund raising for the Foundation to defray the costs of the anniversary celebration, the 82nd Abn Div Association volunteered to conduct a raffle, the prize being a "shorty" (airborne) version of the M-16 rifle. No ordinary weapon, this is gold plated and is serial numbered 0001 of a collectors edition. Shirley Gossett, a past president of the 82nd coordinated the exercise. Almost $100,000 was raised by the raffle—a real life saver to the Foundation budget. Clark Archer (517th) won the prize and one could say it was well deserved for, early on in the planning for the anniversary, Clark developed a concept for fund raising that was of inestimable value. The current president of the 82nd Association, Larry Law, did the honors for the raffle drawing.

Returning to a serious vein, Weber read letters from General Matthew B. Ridgway (82nd Airborne Division), and Maj. Gen. William "Bud" Miley (17th Airborne Division), the only surviving members of the fraternity to lead airborne divisions in WWII. Ridgway's letter said, "As I think back when this fraternity had its birth, I clearly recall my feelings as I watched the airborne pass in review. I felt then, as I do now, that deep understanding, respect and admiration which a veteran experiences for all those whose dangers and deprivations he has been privileged to share, that feeling which Rudyard Kipling so well expressed:

I've eaten your bread and salt,
I've drunk your water and wine.
The deaths you died I lived beside,
And the lives you led were mine.
My heart goes out to you all."

And from Maj. Gen. Miley, 17th Air-

state and territorial colors and the Marine Band departed from the field one could hear the beginning stains of "Airborne" the introductory song of the famed 82nd Abn Div Chorus. Doing their famous airborne shuffle and attired in WWII jump suits and brown jump boots they moved to center stage and began a medley of famous airborne songs. Moving the crowd as only they can, the stamping of feet and clapping of hands soon filled the stadium with the sound of the crowd "getting it on"!

The chorus was followed by the "Ambassadors in Blue", the famous bagpipe band of the Canadian Niagara Regional Police whose kilts swayed in time to their intricate marching pattern. Led by their Pipe Major, Jim Caddis, they soon had the crowd on their feet cheering every tune.

borne Division, "Best wishes to all my airborne comrades assembled in Washington, D.C. on the 50th Anniversary of Airborne. I salute you and wish you a most successful reunion. God bless you all."

The next event presented the 82nd and 101st Abn Div Bands in order. Each played a medley of three songs whose words and music honor the various units of airborne's past and present. From the song of the 82nd, the "All Americans", to that of the 101st, "Rendezvous With Destiny", the songs of airborne units are known by all who are airborne. Go to any gathering of more than two airborne types and you'll hear one or more of the songs. As an evening of festivity wears on, the songs are sung louder and more often, some will say the level of the volume rises disproportionately with the decreasing level of the liquid refreshments that are a mainstay of any airborne gathering. The origin of that hypothesis is lost in history. You are on your own in any attempt to prove or disprove the validity!

As each of the division bands left the field most of the crowd missed a true example of comradeship and mutual respect. The band members of the alternating bands were standing at the sally port and as their counterparts marched off the field they applauded them in recognition of their soldierly and musical professionalism. As Weber commented to the crowd, "As you listen to these bands remember that these airborne troopers are combat troopers first and bandsmen second."

The centerpiece event of the program was about to begin. As a preliminary to the event the men who symbolized the airborne fraternity were gain presented to the audience. Millett, Longerbeam, Pedone, Held and Ruiz were each acknowledged as Weber said of them, "Every event of man can be translated with symbols that will forever recall that event to mind. For our Fiftieth Anniversary, we have chosen to symbolize our brotherhood by living symbols. It is appropriate that we do so as our family is one of people—people who became airborne soldiers of their country. These five men represent the spirit of airborne. Selected not only for their individual accomplishments, which are significant, but also because they are representative of each component of our airborne fraternity. By recognizing them, we honor the past, present and future of airborne. Together these men are our airborne fraternity. Those they symbolize and the spirit they represent constitute the Corps of Honor that wear the silver wings of courage. In each of them is each of you and collectively they are we! In honoring them, we honor ourselves!"

As the crowd cheered their living symbols the men were greeted by Gen. Stiner who was scheduled to give the main address of the day. Before Gen. Stiner rose to speak he joined Weber at the podium to take the salute of the massed colors of airborne when presented. At the command of the color sergeant, almost 200 unit colors, 60 of them still cases, marched on the field joined by the state and territorial colors. In three huge arcs extending from left field to right field the colors centered on the state with the cased colors in the first arc, the active unit colors in the second and the state and territory flags in the rear. The US and USA Airborne Fiftieth Anniversary Foundation colors took a center position directly in front of the stage.

Joe Quade, Commissioner Capizzi and Mr. Reiling. (Photo by Chuck Christianson)

The children with the roses during the Children's Tribute. (Photo by Chuck Christianson)

Massing of Airborne colors reactivated for the Memorial Ceremony.

82nd Airborne Division Band. (Phot by Chuck Christianson)

325th Glider Infantry Regiment alumni. (Photo by Chuck Christianson)

On arriving in position, a ceremonial activation ceremony took place.

A truly historical moment was about to occur. For the first time in Armed Forces history a mass ceremonial activation, designation and/or constitution of inactive units was about to take place. In a few moments the colors of every airborne unit ever activated or contemplated were about to fly together; something that had never been seen before. Through the auspices of Department of Army, orders were published on 14 June 1990, the 215th anniversary of the United States Army and the fiftieth anniversary of USA Airborne! The orders authorized the temporary designation, constitution and/or activation of every airborne unit not now on the active, Guard or Reserve rolls, from 0001 hours 7 July 1990, to 2400 hours 8 July 1990, at which time the units are again inactivated. The orders were read and, on command of the color sergeant the appropriate colors were uncased and unfurled! If you were there or saw the tape of the ceremony you know the high emotion of the moment. If you missed the ceremony, no words can describe how the veterans of these units felt at seeing the colors they fought under again flying proudly under American skies!

As the last color unfurled, the color sergeant brought the massed colors to Present Arms and reported the colors Present to Gen. Stiner. He took the salute and ordered the colors to be posted. On command, each arc of colors split in the middle and extended left and right of the stage. In unison, each arc faced about, saluted the audience and grounded colors. The cheers were deafening and it truly looked as if the whole of the field of RFK Stadium was a sea of proud military colors. God again smiled on airborne, as a mild breeze brought all colors to a gentle fly! A stirring sight—one that will never be forgotten and may never again be duplicated!

As the crowd quieted, Gen. Stiner began his address. He opened by stating. "As I look at these colors, I cannot imagine any greater honor, or privilege, than to speak here today, to this assembled family of the Airborne as we celebrate and commemorate the first half century of dedication, sacrifice and service to our nation. And, I want to thank each one of you for being here, and for the commitment that you have made to the preservation of our freedom." He extended the apologies of the Chief of Staff, Gen. Carl Vuono, who accompanied the President, for being unable to join us on this glorious day. Briefly covering the history of the development of airborne and the growing role played by airborne forces in WWII, Gen. Stiner emphasized that it was the elan and valor of airborne men and units that made so great a contribution to victory. He noted that three factors stand out in the annals of airborne that were paramount in making airborne what it was and is. These are the rigorous training that airborne undergoes, the kind of leadership from the squad up to field army level and, the outstanding quality of the airborne soldier. In closing Gen. Stiner opined, "The best way we can honor these and all other great men who shed their blood for freedom, is to make sure our country's defense remains strong—and our forces are ready to meet any challenge to our free and democratic way of life. In so doing, we can help to guarantee that those whom we praise and honor today shall not have served in vain."

Gen. Stiner was followed to the lectern by Chaplain Wood who conducted the Memorial Service eulogy and a separate segment honoring the fallen men of airborne by era. The eras covered were WWII, Korea, Vietnam, Cold War, Grenada and Panama. As he tolled each era those unit colors with service in those eras were dipped in salute to the men who died serving those colors. As Chaplain Wood ended his remarks he initiated the beginning of the Children's Tribute to the fallen of airborne. Under the leadership of Carmella La Spada, assisted by Sea Cadets of the Northern Virginia Division of the US Naval Sea Cadet Corps, the children began their march on to the field and encircling the 50th Anniversary logo. While they were marching on Ms Bebe Gribble sang the beautiful song, "Let There Be Peace On Earth." Each of the 50 children was a grandchild of a airborne trooper and each carried fifty roses to be placed as their tribute to the fallen.

Colonel William E. Weber. (Photo by Chuck Christianson)

Dr. Charles Pugh

As the children completed placing the roses, Chaplain Wood offered the benediction. Upon ending the benediction two buglers, one each from the 82nd and 101st Bands began an "echo Taps". The lead bugler was positioned at the center of the 50th logo on the field and the echo bugler was in the upper stadium deck. Their rendition of echo Taps was perfect! Not a sound or movement could be heard or seen in the entire stadium for a full minute after Taps! No one who was there will ever forget that moment!

The colors then reformed their closed arcs facing the stage and the combined bands of the 82nd and 101st conducted a full Retreat ceremony, after which the colors were again presented for the furling and casing of the colors of units that were again to be inactivated. It was a sad and yet proud moment to see all those magnificent colors, representing so much courage and honor, again being cased and returned to inactive status. Perhaps never again will such a sight occur.

While the colors were being retired from the field the 82nd and 101st bands positioned themselves to the left and right of the stage respectively. Weber closed by saying, "We have come to the end of the last page of the chapter on our 50 years of service to this nation. It is highly possible that never gain will so much history be written by so few men. Airborne was, is and will remain, a breed apart! As good as the best and better than the rest. You have a right to be proud! Be Proud! Be Airborne, all the way, always and in all ways!.

The ceremonies and the last day of the USA Airborne's Fiftieth Anniversary ended with the massed bands playing and the audience joining in the song, "God Bless America"! Nobody wanted to leave. They didn't want it to end! But, as with all things, the end was inevitable!

EPILOGUE

In addressing the young soldiers in the audience Weber stated, on behalf of the Foundation, its officers, Committee and Advisory Board, and all airborne veterans, "There are in this audience today men who shall live to see the year 2040. To you, those of us who will be gone, charge you to muster this brotherhood in the year 2040 to celebrate the centennial year of USA Airborne! And, in the interim, maintain the standards that set us apart. Let the name "airborne" continue to be representative of the values that make our nation, our armed forces and our airborne units the best in the world! Every American fighting man who has earned and worn the silver wings of courage is with us today and joins us in giving you this mission."

That those who are the inheritors of our history, and the heritage we have passed to them, will carry on our airborne traditions is indisputable! One need only observe the present generation to realize that they are imbued with the spirit we possess and, that they are extending that spirit to ever greater heights.

No one can predict with certainty what the future portends for airborne. However, of this we can be certain, airborne will remain a way of soldiering that will attract only the best. Those who will give their all to keep our colors in the forefront of our nation's Armed Forces. Today, "Operation Desert Shield" in the Persian Gulf is a testament to what the future has in store for airborne. That here will be other such operations is as certain as is the history of our past.

As our numbers diminish it is imperative that we ensure the filling of our ranks with those who are the veterans of tomorrow! There must never be a gap in the generations. Our fraternity must be regenerated with new blood in a continuous flow from us to others. We are but the first fifty years in a line of American airborne that must extend into the distant future without any perceptible break in that line. It is we, the past of airborne, who must take the lead in the bonding process that ensures continuity. We cannot assume we were the exception to the rule! We must assure perpetuation of our corps by bringing the present of Airborne into our fraternal brotherhood as equals! To do less would be to deny that airborne is immortal!

Our Fiftieth Anniversary has ended—but it will never really be over! It is seared into our memories and we shall relive every moment a thousand times over. What you accomplished is almost unbelievable—that is, until you remember what "airborne" is all about.

'TIL WE FORM AGAIN—
AIRBORNE!

DEPARTMENT OF THE ARMY
OFFICE OF THE SECRETARY OF THE ARMY
WASHINGTON, D.C. 20310-0101

CEREMONIAL ORDERS   90-1   14 June 1990

First Allied Airborne Army
- 11th   Airborne Division - all inactive organic units
- 13th   Airborne Division - all inactive organic units
- 17th   Airborne Division - all inactive organic units
- 173d   Airborne Brigade (Separate) - all inactive organic units
- 1st   Special Forces Regiment (Airborne) - all inactive organic units
- 9th   Special Forces Group (Airborne) - all inactive organic units
- 15th   Special Forces Group (Airborne) - all inactive organic units
- 1st   Special Service Forces (American-Canadian) - all inactive organic units
- 88th   Glider Infantry Regiment - all inactive organic units
- 188th Glider Infantry Regiment - all inactive organic units
- 189th Glider Infantry Regiment - all inactive organic units
- 190th Glider Infantry Regiment - all inactive organic units
- 191st Glider Infantry Regiment - all inactive organic units
- 192d   Glider Infantry Regiment - all inactive organic units
- 193rd Glider Infantry Regiment - all inactive organic units
- 194th Glider Infantry Regiment - all inactive organic units
- 326th Glider Infantry Regiment - all inactive organic units
- 401st Glider Infantry Regiment - all inactive organic units
- 511th Parachute Infantry Regiment - all inactive organic units
- 513th Parachute Infantry Regiment - all inactive organic units
- 514th Parachute Infantry Regiment - all inactive organic units
- 515th Parachute Infantry Regiment - all inactive organic units
- 516th Parachute Infantry Regiment - all inactive organic units
- 517th Parachute Infantry Regiment - all inactive organic units
- 518th Parachute Infantry Regiment - all inactive organic units
- 519th Parachute Infantry Regiment - all inactive organic units
- 541st Parachute Infantry Regiment - all inactive organic units
- 542d   Parachute Infantry Regiment - all inactive organic units
- 543d   Parachute Infantry Regiment - all inactive organic units
- 544th Parachute Infantry Regiment - all inactive organic units
- 545th Parachute Infantry Regiment - all inactive organic units
- 550th Airborne Infantry Battalion - all inactive organic units
- 551st Parachute Infantry Regiment - all inactive organic units
- 555th Parachute Infantry Battalion - all inactive organic units
- 89th   Airborne Field Artillery Battalion - all inactive organic units
- 98th   Airborne Field Artillery Battalion - all inactive organic units
- 215th Glider Field Artillery Battalion - all inactive organic units
- 313th Parachute Field Artillery Battalion - all inactive organic units
- 314th Parachute Field Artillery Battalion - all inactive organic units
- 315th Glider Field Artillery Battalion - all inactive organic units
- 321st Glider Field Artillery Battalion - all inactive organic units
- 325th Glider Field Artillery Battalion - all inactive organic units
- 326th Parachute Field Artillery Battalion - all inactive organic units
- 327th Glider Field Artillery Battalion - all inactive organic units
- 373d   Glider Field Artillery Battalion - all inactive organic units
- 374th Parachute Field Artillery Battalion - all inactive organic units
- 375th Parachute Field Artillery Battalion - all inactive organic units

CEREMONIAL ORDERS   90-1   14 June 1990

376th Airborne Field Artillery Battalion - all inactive organic units
377th Parachute Field Artillery Battalion - all inactive organic units
407th Parachute Field Artillery Battalion - all inactive organic units
456th Parachute Field Artillery Battalion - all inactive organic units
457th Parachute Field Artillery Battalion - all inactive organic units
458th Parachute Field Artillery Battalion - all inactive organic units
459th Parachute Field Artillery Battalion - all inactive organic units
460th Parachute Field Artillery Battalion - all inactive organic units
462d   Parachute Field Artillery Battalion - all inactive organic units
463rd Parachute Field Artillery Battalion - all inactive organic units
464th Parachute Field Artillery Battalion - all inactive organic units
465th Glider Field Artillery Battalion - all inactive organic units
466th Parachute Field Artillery Battalion - all inactive organic units
467th Parachute Field Artillery Battalion - all inactive organic units
472d   Glider Field Artillery Battalion - all inactive organic units
507th Parachute Field Artillery Battalion - all inactive organic units
515th Airborne Field Artillery Battalion - all inactive organic units
516th Airborne Field Artillery Battalion - all inactive organic units
518th Airborne Field Artillery Battalion - all inactive organic units
549th Airborne Field Artillery Battalion - all inactive organic units
550th Airborne Field Artillery Battalion - all inactive organic units
551st Airborne Field Artillery Battalion - all inactive organic units
674th Glider Field Artillery Battalion - all inactive organic units
675th Glider Field Artillery Battalion - all inactive organic units
676th Glider Field Artillery Battalion - all inactive organic units
677th Glider Field Artillery Battalion - all inactive organic units
680th Glider Field Artillery Battalion - all inactive organic units
681st Glider Field Artillery Battalion - all inactive organic units
907th Glider Field Artillery Battalion - all inactive organic units
3d   Airborne Anti-Aircraft Artillery Battalion - all inactive organic units
9th   Airborne Anti-Aircraft Battalion - all inactive organic units
12th   Airborne Anti-Aircraft Battalion - all inactive organic units
13th   Airborne Anti-Aircraft Battalion - all inactive organic units
14th   Airborne Anti-Aircraft Battalion - all inactive organic units
15th   Airborne Anti-Aircraft Battalion - all inactive organic units
16th   Airborne Anti-Aircraft Battalion - all inactive organic units
17th   Airborne Anti-Aircraft Battalion - all inactive organic units
18th   Airborne Anti-Aircraft Battalion - all inactive organic units
80th   Airborne Anti-Aircraft Battalion - all inactive organic units
81st   Airborne Anti-Aircraft Battalion - all inactive organic units
84th   Airborne Anti-Aircraft Battalion - all inactive organic units
88th   Airborne Anti-Aircraft Battalion - all inactive organic units
89th   Airborne Anti-Aircraft Battalion - all inactive organic units
150th Airborne Anti-Aircraft Battalion - all inactive organic units
151st Airborne Anti-Aircraft Battalion - all inactive organic units
152d   Airborne Anti-Aircraft Battalion - all inactive organic units
153rd Airborne Anti-Aircraft Battalion - all inactive organic units
154th Airborne Anti-Aircraft Battalion - all inactive organic units
155th Airborne Anti-Aircraft Battalion - all inactive organic units
444th Airborne Anti-Aircraft Artillery Battalion - all inactive organic units
492d   Airborne Anti-Aircraft Artillery Battalion - all inactive organic units
503d   Airborne Anti-Aircraft Artillery Battalion - all inactive organic units
651st Airborne Anti-Aircraft Artillery Battalion - all inactive organic units
125th Parachute Engineer Battalion - all inactive organic units
127th Parachute Engineer Battalion - all inactive organic units
128th Parachute Engineer Battalion - all inactive organic units

CEREMONIAL ORDERS   90-1   14 June 1990

Units Ceremonially Activated in the Army Reserve

80th Airborne Division - all inactive organic units
84th Airborne Division - all inactive organic units
100th Airborne Division - all inactive organic units
108th Airborne Division - all inactive organic units
135th Airborne Division - all inactive organic units

Units Ceremonially Constituted

15th Airborne Division - all inactive organic units

Units Ceremonially Designated

6th Airborne Division - all inactive organic units
9th Airborne Division - all inactive organic units
18th Airborne Division - all inactive organic units
21st Airborne Division - all inactive organic units

The above organizations are directed to participate in the 50th Anniversary of the American Airborne in Washington, DC for the following purpose:

Action: Rededication to the Airborne.

Mission: Ceremonial

Effective Date: 0001 07 July 1990

Authorized Strength: 0

Authority: By confirmation of the Secretary of the Army, the Honorable Michael P.W. Stone, and General Carl E. Vuono, United States Army Chief of Staff.

Additional Instructions: Status terminates at 2400 08 July 1990.

Robert Arter

Robert Arter
Lieutenant General, USA (Ret)
Special Consultant to the
Secretary of the Army for the
50th Anniversary of World War II

DISTRIBUTION:
Unit
HQDA (DAMO-FD)
HQDA (DAMH-HSO)
HQDA (TAPC-PDH)
Cdr, USAMDW 9ANC & SE)

Airborne Units Affiliated with the U.S.A. Airborne Fiftieth Anniversary Foundation

UNIT	REPRESENTATIVE	UNIT	REPRESENTATIVE
11th Airborne Div.	Frederick Thompson George Hurchalla	**517th PRCT**	Charlie Pugh
13th Airborne Div.	Robert Sommers	**542nd Parachute Rgt. and Airborne Command**	John Grady Ed Segassie
17th Airborne Div.	Joseph Quade	**551st Parachute Rgt.**	Doug Dillard
82nd Airborne Div.	Dan Campbell	**555th Parachute Bn.**	Cliff Allen
101st Airborne Div.	Robert E. Jones	**Special Forces Assoc.**	Clyde Sincere
173rd Airborne Bde.	Gene Weeks	**First Special Service Force**	Bill Story
187th ARCT	Don Parks Bill Weber	**Assn. of Ranger Inf. Cos. - Korea**	Joseph C. Watts
503rd PRCT	Charles Rambo Gerald Hasbargen	**Society of Vietnamese Airborne**	Paul DeVries
507th Parachute Rgt.	John Marr	**1/8, 1st Cav. Div.**	Gerrell Plummer
509th Parachute Bn.	John Costello	**Airborne Troop Carrier**	Vito Pedone

The San Leandro, California memorial, built by the 82nd's "Golden Gate" Chapter, commemorates all Airborne Divisions. Courtesy of the Golden Gate Chapter

UNIT ACTIVITIES DURING THE CELEBRATION

It was a milestone in American history; a once-in-a-lifetime event. And as always, the camaraderie and brotherly love was there in full force as the U.S.A. Airborne celebrated its 50th Anniversary. Thousands of men from the Airborne fraternity gathered at the nation's capital July 4-8, 1990 to participate in this spectacular celebration. This was more than a veteran's memorial service—it was a phenomenon.

Events included a parachute drop by the 82nd Airborne Division members at Andrews AFB, an air assault insertion at Camp A. P. Hill by a battalion of the 101st Airborne Division (Air Assault), a Smithsonian Airborne exhibit in the Museum of American History, wreath presentations at the Tomb of the Unkowns, the Vietnam Veterans Memorial and the future site of the Korean War Memorial. Individual units celebrated with banquets, conventions, tours and participated in the grand parade on Saturday. A Memorial Muster at Robert F. Kennedy stadium was held Sunday along with a children's tribute by No Greater Love, placing flowers around the giant 50th anniversary Airborne logo. The mass colors of Airborne were presented for the first time in history. In closing the ceremonies, the colors were retired, until called again during our great country's time of need.

Airborne members will remember these events, as they are now a part of Airborne history. But it is the men and women of U.S.A. Airborne who will be remembered the most. They share a common bond of courage. They will do whatever it takes to go "all the way."

The following pages contain photographs of the units that participated in the 50th Anniversary Celebration.

499

AIRBORNE MEMORIALS

United States Airborne Forces earned their reputation as hardened fighters on battlefields around the globe during the past half century. To commemorate the units, the battles, and the men who made Airborne history, scores of monuments have been dedicated throughout the United States and the world. The following is a photographic essay that illustrates only a few of the many monuments that exist.

PUBLISHER'S MESSAGE

Dave Turner, President - Turner Publishing Company

It started with a vision. Col. Bill Weber saw the need for a history book to commemorate the 50th Anniversary of all U.S. Airborne. With this idea he enlisted Lt. Col. Bart Hagerman, chief editor, who contacted Turner Publishing Company, previous publisher of the 17th Airborne and other Airborne histories.

This began the creation and realization of a historic book on USA Airborne units. Two years of hard work, careful planning, and special touches from many creative and insightful people carried out Col. Weber's vision.

Special thanks goes to Lt. Col. Hagerman for writing several sections of the book and making numerous trips to the publisher's office to review copy, write captions, place photos, and follow through to help create this project and see it through to completion.

If it had not been for the efforts of the Airborne Executive Committee–men dedicated to their country and to the cause and actions of Airborne troops–this book would not have achieved the historic, high quality that it has. Special thanks to those sharing in the success of USA Airborne: Col. Bill Weber, Co-Chairman (187th ARCT & 11th Abn. Div.)–whose meticulous attention to detail created a comprehensive and accurate history; Dr. Charles Pugh, Co-Chairman, (517th PRCT)–for his patience, cordiality and professionalism in assisting with the completion of the book; Mr. Joe Quade, (17th Abn. Div.)–for his untiring assistance and careful analysis in all areas to ensure the quality of the book; and Dr. Donald Parks, Treasurer, (187th ARCT)–who coordinated fiscal matters between the association and Turner Publishing Company.

The USA 50th Anniversary Commemorative History Book will forever preserve the achievements and sacrifices of all Airborne troops for the past half century.

USA Airborne is a comprehensive history of US Airborne forces from its inception in 1940 to "Operation Desert Shield" in 1990. It also includes the 50th Anniversary Commemorative Celebration of U.S.A. Airborne in Washington, D.C.

As a publisher and a veteran, I have a personal interest in all of Turner Publishing Company's military history books, that of Airborne history being especially close to my heart. My brother, John H. Turner, once a member of the 101st Airborne Division, was KIA in Vietnam.

Airborne is a precious reminder and treasured keepsake for those, like me, who lost family and friends–a book to pass down to children and grandchildren. Airborne is not only a preservation of the past, but also a legacy for future Airborne men and women to read and realize that they have a great history, and that thousands of brave men before them made the supreme sacrifice that created the proud Airborne Units of the United States Armed Forces–units that continue to ensure the freedom we now enjoy.

It is your effort and dedication, along with ours, that has preserved the story of Airborne for today's America, as well as future generations. Turner Publishing Company salutes each and every one of you.

Having published over 100 military association histories, Turner Publishing Company continues to lead in documenting the legacy of veterans. For a free full-color catalog on military titles and information on publishing your military association history at no cost please write or call:

THE FRONT LINE
Turner Publishing Company
P. O. Box 3101
Paducah, Kentucky, 42002-3101
(502) 443-0121

INDEX

1st Air Cav. Div., 226, 227, 247, 443, 444
1st Air Commando Force, 78
1st Abn. Battle Gp., 247, 437, 439, 440, 443, 445, 446
1st Abn. Bde., 439
1st Abn. Div., 203
1st Abn. Inf. Bde., 227, 442
1st Abn. Task Force, 51, 293, 296, 306, 416, 417
1st Allied Abn., 53, 62, 203, 289, 394
1st Armored Div., 444
1st ARVN Inf. Div., 210, 221
1st Aviation Bde., 457
1st Bn. Combat Team, 257, 259
1st British Abn., 291, 293
1st Cav. Div., 65, 93, 138, 146, 227, 228, 229, 232, 233, 236,240, 316, 447, 449, 454, 456, 457, 458
1st Corps Support Cmd., 473
1st Fld. Army Support Cmd., 456, 458
1st Inf. Bn., 430, 432
1st Inf. Div., 89, 220, 312, 401,445, 453
1st Logistical Cmd., 442, 456
1st Marine Div., 362
1st Military Intelligence Co., 332
1st Missile Bn., 454
1st Mobile Strike Forces, 448
1st Observation Bn., 439, 441
1st Parachute Bn., 34, 35, 361, 362, 364, 403, 404
1st PIB, 405
1st Parachute Regt., 30, 35, 363, 364, 402
1st Psychological Op. Bn., 457, 465
1st Psychological Warfare Bn., 448
1st Raider Bn., 362
1st Ranger Co., 65, 316, 342
1st Ranger Inf. Bn., 427, 441
1st Ranger Inf. Co., 432
1st Recon. Squadron 17th Cav., 448
1st Spec. Forces Cmd., 472
1st Spec. Forces Gp., 438, 450, 463
1st Spec. Op. Cmd., 328, 464, 465, 470, 472
1st Spec. Service Force, 111, 119, 144, 318, 322, 324, 325, 326, 327, 330, 331, 332, 333, 337, 339, 340, 341, 412, 458, 470
1st SS Panzer Division, 262
1st U.S. Army, 204
2d Abn. Bn., 356, 430, 441
2d Abn. Battle Gp., 437, 441, 443, 445, 446
2d Armored Div., 89
2d Army, 18
2d Cav., 236
2d FA Signal Corps, 304
2d Inf. Bn., 433, 459
2d Inf. Bde., 410, 441
2d Inf. Div., 65, 106, 227, 250, 288, 316
2d Mobile Strike Force, 448
2d Parachute Bn., 35, 203, 361, 362, 363, 364, 404
2d Parachute Bde., 297
2d Parachute Regt., 402
2d Ranger Co., 65, 316
2d Ranger Inf. Bn., 428, 432, 436, 441
2d Spec. Forces Gps., 449
3d Armored Div., 293, 298
3d Bn. Combat Team, 259
3d Howitzer Bn., 441, 442, 456
3d Inf. Div., 89, 106, 218, 240, 316, 438
3d Marine Div., 362, 363
3d Parachute Bn., 35, 362, 363, 364, 404
3d Parachute Bde., 410
3d Ranger Inf. Bn., 441
3d Ranger Co., 65, 432
3d Spec. Force Gp., 324, 327, 333, 334, 444, 454, 472
3d Spec. Op. Support Cmd., 470
4th Abn. Battle Gp., 437, 449
4th Abn. Ranger Co., 65
4th Antiaircraft Art. Bn., 438, 439
4th Armored Div., 212, 218, 429, 434
4th Cav., 470
4th Inf. Div., 46, 105, 109, 216, 280, 317, 414, 420, 450, 454, 455
4th Med. Tank Bn., 452
4th Mobile Strike Force Cmd., 335, 448
4th Psychological Op. Gp., 451, 456, 457, 464, 465, 472
4th Ranger Bn., 412
4th Ranger Co., 65, 316, 430, 432
4th Ranger Inf. Bn., 441
4th Ranger Training Bn., 466, 468, 472
4th Sp. Op. Support Cmd., 470
5th Abn. Battle Gp., 437, 459
5th Army, 41, 320
5th Army Abn. Training Ctr., 291, 321
5th Cav. Regt., 2339
5th Div., 236
5th Inf. Div., 268, 304, 455, 469
5th Marine Div., 35, 364
5th Military Intelligence Det., 460
5th Mobile Strike Force, 446
5th Mountain Div., 34
5th Parachute Bde., 410
5th Parachute Div., 272
5th Quartermaster Bn., 425
5th Ranger Co., 65, 316, 331, 334, 430, 464
5th Ranger Inf. Bn., 412, 421, 441
5th Sp. Forces Gp., 75, 76, 324, 326, 327, 331, 334, 335, 336, 337, 338, 339, 341, 441, 446, 450, 453, 456, 466, 468, 472
5th Sp. Op. Support Cmd., 470, 472
5th Transportation Bn., 431, 436, 441, 444, 454, 455, 458
6th Abn. Div., 38, 46, 203, 414, 418
6th Armored Div., 272
6th Inf. Div., 464
6th Parachute Div., 273
6th Parachute Regt., 216
6th Psychological Op. Bn., 448, 457, 465
6th Ranger Co., 65, 316, 317, 329,339
6th Ranger Inf., 430, 441
6th Spec. Forces Gp., 324, 326, 327, 339, 340, 444, 456
6th Spec. Op. Support Com., 472
6th Spec. Op. Support Gp., 470
7th Armored Div., 206, 208, 288, 293, 298
7th Army, 296
7th Cav., 94, 138, 233
7th Inf. Div., 55, 65, 90, 92, 153, 155, 157, 316, 399, 413, 430, 466, 468
7th Ranger Co., 65, 316, 432
7th Spec. Forces Gp., 324, 326, 327, 335, 339, 341, 342, 441, 457, 470, 471, 472
7th Spec. Op. Support Cmd., 470
8th Abn. Div., 226, 227
8th Armored Regt., 429
8th Army, 65
8th Army Ranger Co., 316
8th Cav., 443, 449, 456, 457, 471
8th Cav. RCT, 94, 226, 228, 229, 230, 232, 233, 236, 237, 238
8th Inf., 65, 267, 298, 418, 443
8th Inf. Div., 458
8th Parachute Div., 273
8th Psychological Op. Bn., 451, 457, 465
8th Ranger Co., 65, 316, 430, 432
8th Spec. Forces Gp., 324, 327, 339, 341, 342, 444, 457
8th U.S. Army, 316
9th Air Cav. Bde., 455
9th Air Force, 43, 53
9th Abn. Div., 38, 46, 414
9th Inf. Div., 450
9th Marines, 363
9th Panzer Div., 402
9th Psychological Op. Bn., 459, 464, 465
9th Ranger Co., 65
9th Ranger Inf., 430
9th Spec. Forces Grp., 449
9th Trp. Carrrier Cmd., 203
9th Trp. Carrier Pathfinder Gp., 371
10th Air Trans. Bde., 444
10th Armored Div., 217
10th Armored Inf. Bn., 423, 426, 434
10 Cav., 463, 466, 468
10th Chem. Platoon, 451
10th Mt. Div., 465, 468, 469, 470
10th Ranger Co., 65
10th Spec. Forces Gp., 324, 325, 326, 327, 342, 343, 344, 345, 434, 456, 471, 472
11th Air Assault Div., 226, 227, 228, 233, 236, 247, 444, 446, 447, 448, 457
11th Abn. Div., 21, 38, 42, 55, 56, 59, 63, 65, 90, 92, 162, 176, 177, 178, 180, 181, 183, 227, 242, 246, 247, 300, 391, 392, 396, 397, 398, 399, 405, 406, 407, 408, 412, 414, 415, 416, 418, 426, 427, 431, 432, 437, 439, 457
11th Abn. Div. Artillery, 439, 444
11th Abn. Quartermaster Co., 439
11th Abn. Med. Bn., 428, 448
11th Abn. Parachute Maintenance Co., 396, 397, 427, 431
11th Artillery, 450
11th Abn. Quartermaster Parachute Maintenance Co., 434, 439, 440
11th Aviation Gp., 457, 458
11th Aviation Co., 438, 439, 444, 457
11th Med. Co., 439
11th Parachute Maintenance Co., 398
11th Ranger Co., 65, 316, 430
11th Replacement Co., 438
11th Rifle Div., 28
11th Spec. Forces Gp., 324, 327, 444, 458
12th Air Force, 365
12th Abn. Div., 226
12th Cav. Div., 227, 228, 443, 444
12th Marines, 363
12th Ranger Co., 65, 430
12th Spec. Forces Gp., 324, 327, 446, 463
13th Abn. Div., 21, 38, 42, 63, 184, 190, 242, 249, 250, 293, 299, 309, 391, 392, 405, 408, 410, 412, 413, 414, 418, 420, 422
13th Abn. Div. Art., 416
13th Armored Div., 418
13th Parachute Maint. Co., 418, 427
13th Ranger Co., 65, 430, 432
13th Spec. Forces Gp., 444, 449
13th Support Bn., 467
14th Inf., 304
14th Med. Detachment, 405
14th Ranger Co., 65, 317, 432
14th Spec. Forces Op. Det., 326, 331, 340
15th Abn. Div., 38, 405, 406, 412, 431, 433
15th Div., 273
15th FA Bde., 406
15th Ranger Co., 65, 432
16th Armor, 239
16th Art., 237
16th Cav., 452
16th Inf. Bde., 79
16th MP Bde., 471, 472
16th MP Gp., 449, 456
16th Spec. Forces Gp., 441, 449, 460
17th Abn. Div., 20, 21, 38, 42, 59, 62, 63, 75, 76, 90, 129, 131, 133, 134, 137, 162, 165, 167, 175, 184, 186, 190, 203, 209, 242, 250, 251, 265, 271, 294, 295, 297, 300, 304, 369, 391, 392, 405, 406, 407, 411, 412, 414, 416, 419, 421, 426, 463
17th Abn. Maint. Co., 428
17th Abn. Med. Bn., 427, 428
17th Art. Battery, 237
17th Aviation Gp., 455
17th Cav., 210, 222, 239, 437, 439, 440, 442, 444, 446, 452, 454, 455, 457, 466, 468
17th Div., 275
17th Spec. Forces Gp., 449
18th Abn. Corps, 203, 209
18th Abn. Div., 38, 414
18th Art., 238
18th Aviation Bde., 466, 469, 471, 472
18th Combat Intelligence Gp., 457, 460
18th VG Div., 307
19th Art., 236, 237, 238
19th Inf. RCT, 255
19th Sp. Forces Gp., 20, 324, 327, 346, 456
20th Chem. Det., 451
20th Eng. Bde., 66, 431, 439, 451, 457, 459, 467, 472
20th Spec. Forces Gp., 324, 327, 346, 347, 460
21st Airborne Division, 38, 414
21st Army Gp., 62, 394
21st Marines, 363
21st Military Police Co., 456
21st Spec. Forces Gp., 444
22d Inf. Div., 402, 404
23d Inf., 288, 312
24th Inf. Div., 65, 247, 316, 427, 430, 439, 443, 470, 471, 472
24th Spec. Forces Gp., 449
25th Inf. Div., 65, 210, 316, 424, 427, 432, 434, 441, 450
26th Artillery, 456, 459
26th FA Bn., 424, 425, 438, 452
26th Inf. Div., 279
27th Eng. Bn., 434, 472
28th Inf. Div., 405, 419
29th Inf. Regt., 19, 32, 215, 403
29th Quartermaster Bn., 425
29th Quartermaster Truck Co., 413
29th Trans. Bn., 427, 435, 440
30th Div. Art., 298
30th Inf. Div., 207, 275, 298, 307
31st "Dixie" Div., 316
32d Inf. Regt., 92
33d Inf., 304
34th Inf., 256
35th Signal Bde., 471, 473
35th Signal Gp., 460
36th Abn. Bde., 393, 460
36th Chem. Det., 451
36th Div. Art., 296
36th Inf. Bde., 452
36th Inf. Div., 295, 413
36th Inf. Regt., 65
37th Eng., 408
37th Eng. Bn., 439, 467, 472
37th Eng. Combat Bn., 422, 435
37th Eng. Combat Regt., 412, 425
37th Inf. Det., 210
37th Inf. Div., 59
38th Inf. Div., 59, 393, 450
38th Spec. Forces Det., 449, 456
39th Inf. Art., 439, 446, 450, 456, 471
39th FA Bn., 438
39th RCT, 89
40th Div. Artillery, 258
40th Inf. Div., 258, 316, 464
42d Armored Regt., 429
42d Inf. Div., 469
42d Tank Bn., 434, 435, 437
42d (Rainbow) Inf. Div., 466, 471
43d Inf. Div., 317, 432
44th Armored Regt., 429
44th Med. Bde., 472
44th Tank Bn., 429, 439
45th Aviation Bn., 467
45th Div., 83, 85
45th Inf. Div., 316
46th Inf. Div., 393
46th Spec. Forces Co., 324, 327, 331, 333, 450, 456
46th Support Gp., 472
47th Inf. Div., 290, 471
49th Abn. Eng. Bn., 427, 428, 431, 434, 436
50th Armored Div., 471
50th Inf. Det., 239, 312
50th Signal Bn., 438, 439
50th Trp. Carrier Wings, 203
51st Armored Inf., 411
51st Inf., 404, 406, 454, 464, 466, 467, 468
51st Inf. Det., 312
51st Telegraph Bn., 422
52d Chem. Det., 210, 454
52d Inf. Det., 312
52d Trp. Carrier Wing, 18, 203, 261
53d Wing, 306
55th Spec. Forces Grp., 341
56th Div., 321
56th Med. Bn., 472
57th Cav. Recon. Trp., 429
58th Inf. Det., 312
58th Signal Co., 210
60th Trp. Carrier Gp., 80
62d VG Div., 307
64th Light Art. Bn., 413
64th Medical Regiment, 412
64th Troop Carrier Group, 411
65th Infantry Ranger Company, 393, 456
65th Military Police Company, 456, 458
65th Tank Bn., 429, 434, 435, 437
65th Trp. Carrier Squadron, 420
66th NVA Regt., 230
68th Armor, 452, 461
68th Inf., 452
68th Med. Tank Bn., 452
70th Inf. Div., 434
71st Abn. Bde., 458
71st Inf. Det., 451
73d Armor, 461
73d Heavy Tank Bn., 463
74th Inf. Det., 312, 453, 454
75th Inf., 436, 453, 454, 455, 456, 459, 463, 464, 465
75th Inf. Div., 210, 288, 298
75th Rangers, 66, 67, 75, 76, 312, 314
75th Rangers Regt., 76, 465, 470, 472
76th Tank Bn., 463
77th Army Reserve Cmd, 454
77th Art., 450
77th Intelligence Corps Det., 444
77th Counter-Intelligence Corps Det., 418, 424, 426, 431, 442
77th Intelligence Corps Det., 449, 453, 456, 459, 463
77th Spec. Forces Gp., 324, 325, 331, 340, 341, 342, 343, 346, 434, 438
77th Squadron, 385
77th Tank Bn., 429
78th Cav. Recon. Trps., 445
78th Inf. Det., 312
78th Inf. Div., 298, 299, 445
79th Inf. Det., 312, 455
79th Inf. Div., 419
80th Abn. Anti-Aircraft Art. Bn., 434, 439
80th Abn. Anti-Aircraft Bn., 281, 316, 407, 426
80th Abn. Div., 38, 425, 432, 433
80th Abn. Signal Co., 424, 426
80th Cav. Recon. Trps., 445
80th Div., 450
80th Div. Art., 406, 423
80th Inf. Div., 407, 409, 423, 433, 440, 445
80th Quartermaster Co., 425, 426, 440
81st Abn. Anti-Aircraft Bn., 211, 407, 426, 435
81st Art., 236, 424, 426, 429, 434, 440
81st Recon. Bn., 444
81st Trp. Carrier Squadron, 369
82d Abn. Div., 18, 21, 38, 40, 41, 42, 45, 54, 55, 59, 65, 66, 68, 72, 74, 75, 76, 83, 89, 97, 98, 99, 100, 101, 106, 107, 108, 128, 148, 149, 151, 152, 153, 157 158, 160, 162, 164, 167, 180, 184, 190, 193, 194, 195, 196, 197, 198, 200, 203, 204, 207, 208, 209, 215, 216, 217, 222, 227, 246, 247, 260, 272, 277, 278, 283, 286, 287, 288, 289, 290, 291, 293, 298, 302, 307, 309, 316, 340, 366, 368, 371, 390, 391, 392, 394, 396, 397, 407, 410, 412, 413, 414, 416, 419, 420, 426, 428, 429, 432, 434, 438, 439, 441, 452, 454, 455, 461, 464, 465, 467, 468, 469, 470, 471, 473
82d Abn. Div. Support Cmd., 459
82d Abn. MP Co., 438
82d Abn. Recon. Co., 437
82d Airborne Signal Co., 427
82d Aviation Bn (Abn., 467
82d Aviation Bde., 470
82d Aviation Co., 438, 442
82d Counter Intelligence Corps Det., 418, 439
82d Div., 18, 38, 268, 282, 287, 405, 408, 447
82d Med. Co., 446
82d Military Intelligence Co., 460
82d Military Intelligence Det., 431
82d MP Co., 210, 408, 427, 431
82d Paratrooper, 204
82d Quartermaster Parachute Sply. and Maint. Co., 422, 447
82d Signal Bn., 470, 473
82d Support Bn. (Abn.), 210
83d Chemical Mortar Bn., 413, 418
83d Inf. Div., 394
84th Abn. Anti Aircraft Art. Bn., 425, 433
84th Abn. Div., 38, 406, 423, 425, 431, 431, 433
84th Div., 450
84th Div. Art., 406, 423
84th Inf. Div., 423, 433, 435, 437, 440
84th Parachute Maint. Co., 426, 431
86th Army Reserve Command, 466
86th Inf. Div., 418, 419
87th Inf., 150, 154
88th Abn. Anti-Aircraft Art. Bn., 438
88th Abn. Inf. Bn., 32, 304, 390, 405
88th Glider Inf., 249, 250, 405, 410, 421
88th Inf. Abn. Bn., 249, 404
88th Inf. Regt., 250, 405
89th Field Artillery, 424, 432
89th Abn. FA Bn., 438
90th Army Reserve Command, 466
90th Div., 270, 282
91st German Div., 266
92nd Infantry Div., 42
95th Military Government Gp., 424, 427, 435, 436, 440
95th Regt., 238
96th Civil Affairs Co., 459
96th Military Government Gps., 428, 450
96th Ranger Inf. Bn., 339
97th Army Reserve Command, 466
97th Civil Affairs Gp., 455
97th Inf. Div., 418
97th Military Government Gp., 428, 441
97th MP Bn, 456
98th FA Bn., 339, 429, 434, 438
99th Inf. Bn., 421
100th Abn. Div., 38, 406, 423, 424, 425, 429, 433
100th Div., 450
100th Div. Art., 406, 423
100th Inf., 306, 407
100th Inf. Div., 433, 440
101st Air Cavalry Div., 65, 214, 220
101st Abn. Administrative Service Co., 458
101st Abn. Div., 19, 21, 38, 45, 54, 55, 59, 68, 72, 74, 76, 105, 106, 107, 109, 120, 162, 163, 184, 186, 203, 210, 211, 212, 214, 215, 216, 217, 219, 220, 242, 246, 247, 248, 277, 280, 281, 283, 289, 293, 297, 366, 368, 369, 371, 373, 374, 384, 385, 390, 391, 392, 394, 396, 403, 405, 407, 408, 410, 414, 415, 419, 420, 421, 422, 426, 427, 429, 430, 434, 435, 437, 438, 439. 441, 444, 448, 449, 450, 452, 453, 454, 455, 457, 458, 467, 468, 471, 473
101st Abn. Div. Art., 428, 434, 438
101st Abn. Div. Bands, 447
101st Abn. Div. Support Command, 458
101st Abn. Maint. Co., 428
101st Abn. MP, 431, 435, 438
101st Abn. Quartermaster Co., 431, 437
101st Abn. Recon. Co., 429, 434, 435
101st Aviation Gp., 451, 455, 465
101st Counter Intelligence Corps Detachment, 418, 422, 436, 440
101st Inf. Div., 38, 211, 212, 214, 221, 222, 405, 409
101st Military Intelligence Co., 451
101st Parachute Maint. Bn., 211
101st Pathfinders, 215
101st Quartermaster Co., 449
101st Signal Co., 211, 427, 431
101st Quartermaster Parachute Sply. and Maint. Co., 447
102d Military Intelligence Bn., 469
103d Military Intelligence Bn., 469
104th Cav., 468
106th Div., 104
106th Inf. Div., 298

107th Military Intelligence Bn., 467
108th Abn. Div., 38, 423, 431, 450
108th Inf. Div., 423, 433, 440
108th MP Co., 456, 458
11th Med. Co., 444
112th Signal Bn., 465, 467, 470, 472, 473
115th Eng. Bn., 258
115th Med. Bn., 258
116th Spec. Forces Op. Det., 346
117th Cav., 464, 466, 471
120th Army Reserve Cmd., 466
121st Army Reserve Cmd., 460
123d Quartermaster Co., 347
124th Military Intelligence Bn., 467, 470, 472
125th Military Intelligence Bn., 465
127th Abn. Eng., 91, 242, 243, 244, 245, 397, 408, 427,438, 444
127th Parachute Eng. Bn., 418
129th Abn. Eng. Bn., 408, 424
129 Aviation Co., 465, 470
132d Inf. Regt., 363
135th Abn. Div., 38, 414
139th Abn. Eng. Bn., 272, 294, 295, 418, 427
142d Inf. Co., 471
143d Inf., 458, 466, 468
143d Inf. Ranger Co., 393
147th FA Bn., 255
150th Inf. Det., 471
151st Inf., 393, 459, 464
151st Inf. Long Range, 312
152d Abn. Anti-Aircraft Art. Bn., 412
153d Airborne Anti-Aircraft Art. Bn., 412
154th Airborne Anti-Aircraft Art. Bn., 433
155th Anti-Aircraft Bn., 273, 412
155th FA Bde., 406
155th Inf. Bde., 455
155th RCT, 259
156th Inf. Bde., 445
157th FA Bde., 406
158th Aviation Bn., 455, 457, 467
158th RCT, 94, 254
159th Aviation Bn., 460, 467
159th Inf. Bde., 455
160th Aviation Bn., 464, 465
160th Aviation Gp., 453, 455, 467, 468, 470, 472
160th Inf. Bde., 445
160th RCT, 258, 259
161st Abn. Eng. Bn., 253, 255, 256, 258
161st Parachute Eng. Co., 253, 256
162d Military Intelligence Platoon, 439
164th RCT, 258
165th Military Intelligence Bn., 469, 470
170th Spec. Forces Det., 441
170th Transportation Bn., 447
172d Military Intelligence Det., 448
173d Abn. Bde., 21, 65, 239, 277, 449, 454
173d Aviation Co., 448, 449, 458
173d Brigade, 443, 447, 455, 457
173d Eng. Co., 444, 458
173d Signal Company, 451
175th FA Bde., 406
177th Cav., 471
180th Inf., 85
185th RCT, 258, 259
187th Abn. Inf., 138, 140, 143, 242, 244, 245, 246, 247, 429, 436, 437, 447, 465
187th Abn. RCT, 21, 63, 64, 72, 75, 76, 139, 141, 143, 144, 166, 244, 247, 317, 400, 430, 431, 434
187th Glider Inf. Regt., 56, 90, 92, 178, 242, 410, 415, 426, 428
187th Inf., 219, 220, 222, 227, 228, 242, 243, 248, 467
187th Parachute Glider Inf. Regt., 244, 397
187th Rakkasan Regt., 248
188th Glider Inf. Regt., 56, 242, 410, 419, 420, 428
188th Inf., 94, 228, 229, 243, 429, 437
188th Parachute Regt., 397, 415
188th RCT, 92
189th Maint. Bn., 473
190th Regt., 307
193rd Glider Regt., 272, 410, 420
193rd Inf. Bde., 247, 465
194th Cav., 471
194th Glider Inf. Regt., 64, 276, 304, 410, 420, 421, 426
194th Inf. Det., 471
196th Inf. Bde., 456
199th Inf. Bde., 455
200th Inf. Ranger Co., 393
209th Eng. Combat Bn., 425
213th FA Bn., 258
215th Signal Corps, 436
221st Abn. Med. Co., 412, 427, 428
222d Abn. Med. Co., 412, 424
222d FA Bn., 258
224th Abn. Med. Co., 412, 422, 427
225th Inf., 455
226th Med. Co., 435
229th Attack Helicopter Bn., 222, 446
229th Aviation Bn., 448 460, 467, 468
235th Eng. Bn., 451
238th Cav., 464, 266
239th Regt., 64, 245
245th Aviation, 467
245th FA Bn., 258
259th Inf. Ranger Co., 393, 456, 459
297th Military Intelligence Det., 444, 458, 459, 463
299th Signal Op. Co., 408, 436
299th Signal Svc. Co., 424, 434
303rd MP Bn., 413
305th Abn. Med. Co., 430
305th Field Signal Bn., 408, 424
305th Med. Bn., 407, 424, 425, 433
306th Chem. Det., 463
307th Abn. Eng. Bn., 203, 207, 210, 281, 282, 408, 427, 438, 448, 453, 470, 472
307th Med. Bn., 407, 427, 438, 448, 470
307th Parachute Eng., 418
309th Eng. Bn., 451
312th Military Intelligence Bn., 469, 472
313th Abn. FA Bn., 433
313 FA, 406, 424, 440
313th Parachute FA Bn., 425, 429
313th Troop Carrier Gp., 369, 385
314th Abn. FA Bn, 433, 440
314th FA, 406, 424
314 Parachute FA Bn, 425, 429
315th Air Div., 245
315th Abn. FA Bn., 433, 440
315th FA, 406, 424
315th Glider FA Bn., 425
316th Troop Carrier Gp., 384
317th Abn. Inf., 432, 440
317th Inf. Regt., 433, 441
317th Trp. Carrier Wing, 256, 407
318th Abn. Inf., 429, 432, 440
318th Glider Inf., 425
318th Inf. Regt., 433
319th Art., 441, 444, 448, 450
319th Abn. Art. Bn., 240
319th Art. Regt., 454, 471
319th FA, 441, 457, 469
319th Abn. FA Bn., 466, 438
319th Abn. Inf., 429, 432, 440
319th Glider FA Bn., 280, 287, 407, 425, 426
319th Inf. Regt., 433
320th Art., 215, 218, 290, 406, 438, 440, 441, 442, 446, 450, 452, 453, 456, 464, 466, 473
322d Trp. Carrier Wing, 258
325th Abn. Inf., 437, 445, 447, 468, 469
325 Eng. Bn., 451
325th FA, 406, 424, 435
325th Glider Inf., 107, 109, 123, 128, 153, 204, 206, 211, 268, 270, 280, 282, 291, 298, 393, 405, 410, 414, 415, 423, 426
325th Inf., 196, 269, 288, 452, 459, 471
325th RCT 208
326th Abn. Med. Co., 211, 215, 422, 427, 438, 446
326th Eng. Bn., 451
326th FA, 406, 424, 432
326th Glider Inf., 250, 405, 423
326th Inf. Regt, 196
326th Med. Bn., 407
327th Abn. Inf., 428, 437, 446, 457, 464
327th FA, 406, 424, 425, 435, 432
327th Glider Infantry, 107, 211, 214, 219, 405, 415, 426, 429, 456, 465
333d Inf., 423, 431, 433, 435, 440
334th Inf., 423, 431, 433, 435, 440
335th Inf., 423, 440
335th Inf. Regt., 433
335th Parachute Inf., 423, 431
356th Intelligence Det., 427, 431, 434
356th Military Intelligence Det. 464
373d Abn. FA Bn., 433
373d FA, 407, 424, 429
373d Glider FA Bn., 429
374th Abn. FA Bn., 407, 424, 429, 433, 438
376th Abn. FA Bn., 426, 436
376th Parachute FA Bn., 203, 407, 416, 425
377th Art., 450
377th FA, 215, 407, 416, 426, 438, 453
382d Chem. Det., 467
389th Military Intelligence Det., 444, 453, 459, 463, 464, 467
397th Inf. Regt., 433, 440
397th Parachute Inf., 429
398th Inf., 440
398th Inf. Regt., 433
398th Parachute Inf., 429
399th Glider Inf., 429
399th Inf., 440
399th Inf. Regt., 433
401st Glider Inf. Regt., 211, 215, 415, 421
401st Inf. Div., 405
402d Parachute Inf., 215
403d Ordnance Det., 441
407th Abn. Quartermaster Co., 427, 439
407th Quartermaster Regt., 409
408th Abn. Quartermaster Co., 396, 409, 413, 427
409th Abn. Quartermaster Co., 413
425th Inf., 393, 455, 457, 464, 466
425th Quartermaster Bn., 409
426th Quartermaster Co., 211, 215, 422, 427, 431,437, 439
427th Abn. Anti-Aircraft Battery, 290
437th Gp., 297, 306
438th Gp., 297
439th Trp. Carrier Group, 116, 296
440th Troop Carrier Wings, 407
445th Chem. Det., 444, 455, 458, 459, 460, 467
456th Abn. FA Bn., 426, 438
456th Parachute Art., 86, 405, 407, 425
457th Abn. FA Bn., 438
457th Parachute FA Bn., 91, 93, 407, 416, 426
458th Parachute FA Bn., 424
460th Parachute Art., 111, 112, 294, 295, 297, 424
462d FA , 95, 253, 255, 256, 258
463d Abn. FA Bn., 438
463d Parachute FA Bn., 414, 426
464th Parachute FA Bn., 273, 274, 426
465th FA Bn., 430, 435
467th FA Bn., 430
467th Parachute FA, 301, 417
472d FA, 94, 428
474th Inf., 421
475th Abn. FA Bn., 428
475th Inf. Regt., 429
485th Abn. Inf., 431, 433, 440
485th Glider Inf., 423, 431
485th Inf., 435, 436
485th Inf. Regt., 433
501st Abn. Inf., 435, 437, 446, 457
501st Abn. Med. Bn., 428, 431, 434, 435
501st Abn. Signal Bn., 438, 473
501st Glider Inf., 107
501st Inf., 222, 434
501st PIB, 31, 32, 251, 252, 305, 381, 383, 403, 404, 405, 414, 423, 426, 431
501st PIR, 107, 172, 175, 211, 215, 217, 218, 406
502d Inf., 216, 219, 222, 426, 428, 429, 434, 435, 437, 446, 457
502d Parachute Bn., 252, 383, 404, 405, 406
502d PIR, 107, 108, 211, 217, 218, 264, 278, 421
503d Abn. Inf., 437
503d Abn. Inf. Regt., 179, 255, 434
503d Inf., 439, 449, 456, 457, 463, 464
503d MP Bn., 413, 424, 426, 442, 448, 455, 456, 458, 470
503d Parachute Bn., 252, 253, 383, 404, 405
503d PIR, 41, 74, 75, 76, 80, 86, 95, 145, 160, 161, 239, 252, 253, 254, 258, 264, 291, 305, 406, 411, 417, 431
503d Parachute RCT, 21, 55, 56, 253, 255, 256, 258, 259
504th Combat Team, 199, 200
504th Inf., 206, 207, 208, 288, 437, 446, 447, 452, 454, 459, 464, 471
504th MP Bn., 456
504th Parachute Bn., 252, 253, 291, 404, 406
504th Parachute Combat Team, 416, 418
504th Parachute Infantry, 74, 89, 99, 100, 120, 124, 128, 152, 153, 156, 160, 196, 197, 198, 204, 206, 264, 285, 286, 292, 302, 307, 321, 371, 394, 406, 410, 412, 425,469,471
505th Abn. Inf. Bde., 166
505th Inf., 42, 206, 207, 210, 262, 267, 304, 309, 437, 446, 447, 452, 454, 464, 471
505th Parachute Combat Team, 196, 197, 208
505th Parachute Inf. Regt., 40, 74, 83, 89, 99, 100, 101, 107, 152, 160, 163, 196, 199, 200, 204, 260, 282, 289, 299, 302, 371, 396, 406, 416, 423,425, 469, 471
505th Regt., 270
506th Abn. FA Bn., 433
506th Inf., 219, 426, 428, 429, 434, 435, 437, 454, 455, 457, 461, 465
506th Parachute Inf. Regt., 107, 108, 211, 215, 218, 416, 421
506th Parachute FA Bn., 424
507th Airborne FA Bn., 433, 435, 438
507th Combat Team, 275
507th Inf., 463
507th Parachute Inf..., 21, 63, 107, 131, 133, 134, 161, 163, 188, 190, 264, 265, 266, 269, 270, 271, 272, 273, 274, 277,282, 406, 416, 420, 424,
508th Abn. RCT, 247, 290, 436
508th Inf. (Abn.) 150, 163, 207, 210, 267, 287, 288, 289, 290, 414, 437, 447, 464, 465, 468, 471
508th Parachute Inf. Regt., 107, 108, 123, 204, 206, 207, 209, 266, 278, 282, 289, 290, 307, 423, 431
509th Abn. Div., 21
509th Inf., 458, 459, 464
509th Inf. Regt., 468
509th PIB, 40, 42, 74, 76, 80, 82, 98, 100, 107, 112, 113, 115, 117, 119, 160, 253, 291, 292, 293, 296, 297, 306, 371, 405, 406, 409, 411, 414, 416, 425, 443
509th Parachute Inf. Bn. Combat Team, 418
509th PIR, 19, 21
510th Armored Inf. Bn., 434, 435, 437., 440
511th Abn. Inf., 443
511th Signal Bn., 444, 446
511th PIR, 59, 75, 76, 90, 92, 93, 94, 177, 178, 180 ,182, 242, 243, 244, 397, 411, 420, 428
513th Abn. Inf., 426
513th PIR, 63, 411
513th Parachute Inf. RCT, 56, 131, 275, 420
514th Abn. Inf., 426
515th Abn. FA Bn., 428, 434, 435, 436
515th PIR, 184, 249, 411, 423
516th Abn. FA Bn., 428, 434, 436
516th Abn. Inf., 426, 428, 429, 434, 435, 436
516th Signal Co., 436
517th Combat Team, 414
517th Engineer Bn., 428
517th PIR, 76, 112, 115, 160, 294, 295, 297, 299, 423
517th Parachute Regimental Combat Team, 21, 110, 111, 113, 115, 116, 117, 118, 119, 294, 295, 296, 297, 298, 299, 306, 307
518th Abn. FA Bn., 428, 434, 436
518th Abn. Inf., 431, 433, 440
518th Military Intelligence Det., 210
518th Parachute Inf., 423
519th Abn. Inf., 431, 433
519th Airborne Quartermaster Co., 290
519th Inf. Regt., 433
519th MP Bn., 470
519th Parachute Inf., 423, 425
522d Military Intelligence Bn., 472
525th Military Intelligence Bde., 472
525th Military Intelligence Gp., 424, 438, 446, 448, 458
528th Quartermaster Bn., 409, 418, 425, 427, 450, 455, 457, 465
528th Support Bn., 470, 473
528th Transportation Corps Truck Bn., 425, 426
530th Sply. and Service Bn., 473
533d Military Intelligence Bn., 469
534th Eng. Det., 333
539th Eng., 331
541st PIR, 300, 411, 421
542d PIB, 301, 302
542d PIR, 21, 301, 302, 303, 392, 411
543d Signal Co., 449
544th FA Bn., 430, 438
544th Parachute Inf., 406
545th Parachute Inf., 406
550th Abn. Art. Bn., 438
550th Abn. Inf Bn., 296, 304, 390
550th Glider Inf. Bn., 111, 113, 116, 119, 206
550th Inf. Abn. Bn., 404, 426, 428
551st PIR, 21, 76, 119, 405, 416
551st PIB, 110, 113, 116, 117, 160, 295, 296, 298, 301, 305, 307
555th PIB, 42, 309, 411, 425, 429
555th Parachute Inf. Co., 309, 311
55th PIR, 21
55th Parachute Inf. Test Platoon, 309
595th Med. Co., 435
596th Abn. Eng. Co., 414, 418, 422
598th Abn. Eng. Co., 290
600th Quartermaster Laundry Company, 413, 425, 434, 437
601st Abn. Inf. Gp., 402
602d Abn. Inf. Gp., 402
610th Clearing Co.y, 412, 425
612th Quartermaster Company, 413, 427, 429, 431, 434
617th Aviation Detachment, 457, 467, 470
626th Field Signal Battalion, 408
634th Clearing Co., 427
667th Abn. Anti-Aircraft Machine Gun Battery, 421
668th Abn. Anti-Aircraft Machine Gun Battery, 421
669th Abn. Anti-Aircraft Machine Gun Battery, 421
674th Abn. FA Bn., 242, 243, 245, 430, 432, 438
674th Glider FA Bn., 428
674th Parachute FA Art. Bn., 426
675th FA, 94, 428, 438,
676th Glider FA, 250, 411
677th Glider FA Bn., 411
680th Glider FA Bn., 407
681st Glider FA Bn., 407, 427
687th Quartermaster Laundry Bn., 413
693d FA Bn., 425
704th Abn. Anti-Aircraft Machine Gun Battery, 421
705th Intelligence Group Det., 333
705th Tank Destroyer Bn., 217
711th Abn. Ordnance Bn., 439
711th Abn. Ordnance Maint. Co., 396, 427, 434
711th Inf. Div., 46
711th Maint. Bn., 439
711th Maint. Co., 444
713th Abn. Ordnance Maint. Co., 413, 422, 424
713th Abn. Quartermaster Co., 418, 422, 451
714th Quartermaster Bn., 413
714th Tank Bn., 429, 435
716th Tank Bn., 258, 259
717th Abn. Ordnance Maint. Co., 409, 413, 428
740th Tank Bn., 208
757th Tank Bn., 429
778th Maint. Co., 393
780th Ordinance Bn., 433
780th Ordinance Light Maint. Co., 409, 425, 426
782d Maint. Bn., 473
782d Abn. Maint. Co., 427
782d Abn. Ordnance Bn., 439
782d Abn. Ordnance Maint. Co., 434
784th Ordnance Light Maint. Co., 409, 426
784th Ordnance Maint. Co., 434
788th Light Maint. Co., 455
800th Ordnance Light Maint. Co., 425, 428
801st Abn. Maint. Bn., 439, 452, 473
801st Abn. Ordinance Maint. Co., 422, 427, 431, 434, 435
801st Chem. Det., 459, 460, 463, 467
801st Ordnance Co., 211
808 Chem. Det., 431, 435, 440, 444, 454, 460
871st Abn. Eng. Aviation Bn., 424
874th Abn. Eng. Aviation Bn., 412
875th Abn. Eng. Bn., 408
876th Abn. Eng. Bn., 408, 422
877th Airborne Engineer, 418, 422
877th Aviation Eng. Bn., 418
878th Abn. Eng., 422
878th Aviation Eng. Bn., 418
886th Abn. Eng. Aviation Bn., 412
887th Abn. Eng. Aviation Co., 418
900th Chem. Det., 467
901st Chem. Det., 467, 469
905th FA Bn., 430, 432, 438
905th Glider FA Bn., 425, 430
907th Abn. FA Bn., 438
907th Glider FA, 215, 427
909th Parachute FA Bn., 432
925th Eng. Abn. Regt., 407, 408, 412
925th FA Bn, 407, 424, 433
925th Glider FA Bn., 430
931st Signal Corps., 413, 451
969th FA Bn., 217
1049th Transportation Co., 393, 456
1128th Eng. Combat Gp., 431
2348th Quartermaster Abn. Air Sply. and Packaging Co., 399
2671st Special Recon. Bn., 354
3340th Army Service Unit, 430
4128th Quartermaster Svc. Co., 418
5307th Composite Provisional Unit, 312, 430, 435
6615th Ranger Force, 411, 412, 413
8081st Air Resupply, 399, 400
8213th Army Unit, 430
8231st Army Unit, 438
8240th Army Unit, 430
8321st Army Spec.l Op. Det., 325
8521st Army Service Unit, 331

— A —

A Shau Valley, 214, 220, 221, 232
Aaron, Harold R., 332, 338, 453
Abood, Edmond P., 223
Abrams, Creighton W., 414, 446, 454
Adair, Allan, 123
Adams, Paul D., 322, 435
Adams, Tyerus F., 403
Adams, William H., 353
Adderley, Tyrone, 330
Adkins, Bernard G., 338
Aerial Doughboys, 381
Africa, 17, 38, 40, 43, 80, 83, 89, 97, 98, 103, 198, 254, 291, 302, 304, 319, 327, 330, 333, 334, 338, 341, 344, 350, 351, 353, 365, 367, 371, 387, 392, 405, 409, 410, 412, 417, 418, 422
Agnew, Spiro T., 221
Air Assault Division, 242, 373
Airborne Casualties, 159
Airborne Firefighters, 311
Airborne Museum, 388
Airborne Pathfinders, 371
Airborne ROAD Division, 219
Airborne Troop Carrier Command, 21, 365
Air Force Air Commando Squadron, 333
Akehurst, J.F.R., 322, 417
Al-badr, Mohammed, 344
Alamo, Gabriel R., 342
Alarcon, Ricardo, 156
Albert Canal, 31, 402
Alden, Carlos C., 291
Aleutian Islands, 319, 322, 334, 339, 342, 412
Alexander, Floyd N., 399
Alexander, Foster, 341
Alexander, J.H., 139
Alexander, Mark, 83, 101, 197
Algeria, 19, 40, 74, 76, 80, 291, 302, 334, 339, 367
Algiers, 417
Alicki, John, 294
All American Division, 209
Allen, Clifford, 21
Allen, Dick, 397
Allen, Everett T., 353
Allen, George C.D., 338
Allen, James 235
Allen, Lewis D., 339
Allen, Paul, 345
Allen, Richard J., 452
Allen, T.G., 224, 469
Allen, Warren E., 430
Allison, John 336
Almond, Edward M., 138
Aloe, R.C., 249, 405
Alpha Battery, 238
Alpha Company, 230
Alsace, 212
Alsop, Stewart D., 353
Amalfi, 100
Ambleve River, 204, 207
American Rangers, 50
American Rapid Deployment Force, 209
Amerman, Walter G., 223
Ames, Charles, 266
An Khe, 214
Anderson, Douglas F., 346
Anderson, Robert J., 353
Anderson, Webster, 219, 223, 450
Anderson, Wilford C., 294
Andross, Elgin, 385
Anstett, Robert M., 353
Anti-tank Platoon, 242
ANVIL, 51, 109, 296
Anzio, 101, 195, 209, 334, 339, 342, 378
APACHE SNOW, 214, 220
Archer, Clark, 20
Archipelago, 228, 233
Ardennes, 161, 165, 186, 193, 195, 207, 217, 218, 265, 271, 272, 277, 281, 286, 288, 291, 293, 307, 339, 369, 414, 420
Ardennes-Alsace, 211, 264, 290, 294, 297, 304, 305, 415, 419, 421
Ardziejewski, 269
Argentina, 333
Armijo, Roberto, 155, 157

Army Air Corps, 306
Army Medical Department, 451
Army of Occupation, 289
Army Rangers, 326
Army Special Forces, 324, 325
Arnhem, 55, 203, 286
Arnhem Bridge, 127, 284
Arnold, Henry H. "Hap", 18, 311, 352, 386
Arntz, John P., 334
Arvin, Carl R., 359
Ashley, Eugene, Jr., 328, 338, 453
Asia, 18, 325, 326, 331, 447, 465
Asiatic-Pacific Operations, 55
ATLANTIC CITY, 241
ATTLEBORO, 241
Attu, 330
AURORA I, 241
AURORA II, 241
AUSTIN II, 214
AUSTIN VI, 214
Austin, Hudson, 148
Austin, B. McDonald, 353
Australia, 240, 253, 303, 380, 384, 410, 411, 417
Austria, 218, 370, 421, 436
Austrian 5th Hochgebirgesjager Division, 306
AVALANCHE, 41, 97, 98, 198

— B —

Bacik, Albert V., 353
Badoglio, Pietro, 99
Baer, John, 235
Bagnal, Charles W., 224
Bagnoli, 295
Bailey, George E., 301
Bailey, Hal, 21
Bailey, Henry, 332, 350
Baird, Robert R., 353
Baker, Eldon L., 223
Baldwin, Bob, 274
Baldwin, Norman E., 338
Baldwin, Samuel, 26
Baldwin, Thomas, 26
Ballard, Ray, 266
Bandoglio, Marshall, 198
Bank, Aaron, 32, 325, 328, 342, 345, 353, 434
Baranek, Mike, 291
Baratto, David J., 332
Barbados, 66
Barker, Geoffrey T., 148, 150, 401, 421, 455, 463
Barker, Jack L., 223
Barkley, Harold E., 274
Barnes, John A., III, 240, 450
Barnes, John W., 452
Barrato, David, 463, 468
Barrett, Charles R., 301
Barsanti, Olinto M., 224, 450
Bartelt, Robert H., 333
Bartholomees, James B., 333, 334
Barton, Raymond F., 223
Bassenge, Oswalt, 402
Bassett, James, 172, 174, 251, 390, 403
Baster, Bruce R., 338
Bastogne, 62, 167, 188, 204, 211, 212, 217, 218, 222, 271, 293, 304, 365, 369, 378, 385, 387, 392, 419, 429
Bataan, 56
Bataan Death March, 339
Battipaglia, 100
Battle of Bloody Ridge, 362
Battle of Bonneville-Renouf, 264
Battle of Burp Gun Corner, 385
Battle of Dead Man's Ridge, 264
Battle of Dorsten, 264
Battle of Hamburger Hill, 245
Battle of Iron Triangle, 239
Battle of Kursk, 343
Battle of La Poterie Ridge, 264
Battle of Leyte, 243
Battle of Ta Quan II, 238
Battle of Tam Quan, 232
Battle of the Bulge, 59, 129, 162, 164, 165, 167, 186, 207, 262, 271, 288, 304, 305, 307, 309, 369
Battle of the Merderet River, 264
Battle of the New York to Paris Line, 264
Battle of the Our River, 264
Battle of the Ourthe River Junction, 264
Battle of the Ruhr Pocket, 264
Battle of Trang Bang, 245
Battle of Vindefontaine, 264
Battle on Cake Hill, 271, 272
Baumgold, Theodore, 353
Baupte Peninsula, 270
Bavaria, 218, 344, 402
Bazaar, Andrew, 277
Bazata, Douglas D., 353
Beach, Dwight E., 440
Beal, Calvin R., 441
Beauregard, 214
BEAVER, 215
Beavers, Clarence H., 411
Beck, Ernest, 277
Beck, Jack A., 399
Beckerman, Arnold S., 338
Becket, Wilson, 322
Beckworth, Charlie, 461
Bedell-Smith, Walter, 103
Beikirch, Gary B., 328, 338, 455
Belgium, 19, 74, 76, 137, 216, 218, 222, 262, 271, 272, 286, 288, 289, 293, 294, 298, 299, 304, 305, 306, 307, 345, 351, 367, 386, 402, 413, 418, 419, 420
Bell, Christopher, 223
Bell, Lory, 332
Benavidez, Roy P., 328, 338, 453
BENTON, 214
Berardi, Pat, 176
Berkey, Russell S., 255
Berlin, 17, 29, 36, 272, 326, 392, 402, 418
Berlin, James R., 353
Berne Canal, 420
Berry, Albert, 26, 401
Berry, Sidney B., 221, 224
Bertogne, 272
Bess, Ned B., 411
Bethlehem, 248
Beynon, Willard, 353
Biak Island, 95
Biazza Ridge, 88, 89, 100, 197
Biddle, Melvin E., 32, 294, 299, 416
Bien Hoa, 240, 241
Bien Hoa Airfield, 239
Bier, Leo J., 274
Bieringer, M.G., 306
BIG OAKIE, 333
BIG RED, 241
Bigger, Warner T., 363, 364
Biggs, Bradley, 309
BIG SPRING, 241
Billingsley, James R., 353
Biscari Airfield, 89
Bishop, Maurice, 148
Bismarck, 228, 233
Black Devils, 417
Blackburn, Donald D., 329, 330, 340, 341
Blackburn, Donald V., 336
BLACKJACK 21, 336
Blackjack Forces, 335
Blackman, Milt, 271
Blair, John D., IV, 332, 338
Blakely, William, 223
Blanchard, George S., 455
Blanchard, Jean Pierre, 26
Blanchfield, Michael R., 454
Blanchford, Michael R., 240
Block, Harlon H., 35
Bloody Nose Ridge, 317
BLUE SPOON, 151
Boe, O'Niel, 271
Boggs, William C., 353
Boland, Gerhard, 264
BOLLING, 241
Bolling, Alexander R., 195, 209
Bolling, Thomas R., Jr., 452
Bols, E., 62
Bonaparte, Napoleon, 401
Bonner, William E., 359
Bonsall, John H., 353
Boos, Mike, 344
Borja, Domingo R.S., 338
Borkowski, Janusz, 338
Borom, John E., 403
Bortner, Thomas D., 473
Borum, Fred, 405
Bott, Russell P., 338
Bourgoin, Lucien, 353
Bourne, John, 322
Bowell, William D., 266
Bowen, Frank S., Jr., 138, 224, 244
Bowen, John C., 445
Bowen, John W., 436
Bowlby, Herbert, 118
Bowser, Myron J., 332, 338
Boyersmith, Joseph C., 469
Boykin, Arnold, 238
Boyle, William J., 111, 118, 294, 295, 296, 297
Brabazon, William, 468
Bradley, Omar N., 103, 105, 167, 193, 196, 251, 282, 404, 405
Bradner, John L., 353
Brakonecke, Morgan, 269, 276
Brandenburg, John N., 224
Bravo Company, 235
Bray, Linda, 155
Brazelton, Rusell W., 353
Brereton, Lewis H., 43, 53, 62, 203, 283, 402, 413
Brett, G.H. 386
Breuer, William B., 379
Bridges, Hubert, 411
Brightman, Paul F., 353
BRIGHTSTAR, 222
British 1st Airborne Division, 18, 54, 55, 283
British 2nd Army, 54
British 6th Airborne Division, 46, 48, 50, 62
British 51st Highland Division, 272
British Commandos, 18, 50
British Guards Armored Division, 262
British Isles, 43
Britten, John W., 255
Broadwick, Charles, 401
Brooks, Elton E., 223
Brown, Charles E., III, 353
Brown, Herman Lee, 223
Brown, James F., 270, 394
Brown, James M., 471
Brown, Leo C., 174, 403
Brown, Lewis, 251
Brown, Walter R., 223
Brown, Willie F., 403
Brown, Wyburn D., 177
Browning, F.A.M., 41, 42, 283, 404
Brucker, Leslie L., Jr., 339
Brummitt, Leroy D., 269, 271
Bruno, Thomas A., 471
Brux, Cary H., 359
Bryant, David, 465
Bryant, William M., 328, 338, 454
Bucha, Paul W., 223, 245, 452
Buchanan, Michael D., 339
Buckley, George "Pete", 384
Bucsek, Charles J., 269
Bukel Hill, 418, 419
Buker, Brian L., 328, 338, 455
Bulgaria, 74, 76
Burauen Airstrip, 243
Burgess, Henry A., 176, 178, 182, 183
Buri Airstrip, 243
Burkhalter, Floyd, 403
Burma, 74, 76, 324, 350, 351, 353, 365, 387, 417, 421, 424
Burnett, Charles, 340
Burnham, William P., 193
Bush, Elbert W., 358, 359
Bush, Francis, 236, 237
Bush, George, 67, 150, 151, 157
Bustamante, Manuel C., 338
Butler, John 384
Butler, William E., 464, 471
Butts, Lonnie R., 223
Byer, James, 255
Byers, Clovis E., 426
Bynam, Holland E., 338

— C —

Cain, Jerry A., 223, 430
Calabria, 97
Calderon, Ricardo Arias, 150
Caldwell, James J., 353
Callahan, Joe, 344
Callahan, Ronald, 470
Cam Ranh Bay, 214
Camalaniugan Airfield, 59
Cambodia, 220, 227, 336, 337, 451
Camp Lucky Strike, 250
Camp Mackall, 81, 112, 176, 177, 180, 183, 186, 242, 246, 247, 253, 278, 291, 294, 295, 300, 301, 302, 304, 306, 309, 389, 391, 392, 408, 409, 410, 411, 412, 413, 414, 415, 416, 417, 418, 422
Campbell, Dan, 21
Canada, 345, 386
CANARY/DUCK, 241
Caniff, Milton, 78
Cape Torokina, 364
Carboni, G., 99
CARENTAN I, 210, 214, 222
CARENTAN II, 210, 214
Caribbean, 306
Carlson, Robert C., 469
Carman, Charles M., Jr., 353
Carn, Robert M., Jr., 359
Carpenter, James J., 353
Carpenter, Michael F., 338
Carpenter, William S., Jr., 223
Carter, Norman, 268
Carter, Tennis, 338
Casablanca, 291
Casey, Maurice A., 338
Cassidy, Patrick F., 223, 403, 415
Castillero, Narcisa, 157
Castro, Fidel, 148
Castro, Mario, 157
Catherman, Robert T., 223
Cato, Raymond L., 294, 297
Cavaini, Jon R., 328, 338, 456
Cavanaugh, Stephen, 336, 346
Cavezza, Carmen J., 153
Charbourne, Philip H., Jr., 353
Chamberlain, Craig R., 338
Chambers, Wallace L., 92
Champagne Campaign, 367
Chanute Field, 172, 174
Chapman, Eldridge G., 184, 249, 390, 391, 404, 405
Chapman, M.G., 306
Chappuis, Steven A., 223
Charlie Battery, 238
Chase, Richard, 251
Chattahoochee, 265
Chavis, Kenneth, 341
CHECKERBOARD, 214
CHECKMATE II, 214
Chef-Du-Pont, 267, 268, 279, 280
Cheneux, 204, 206
Cherbourg, 261, 266, 267, 282, 414
Cherbourg Peninsula, 282
CHEROKEE TRAIL, 333
China, 74, 75, 76, 227, 326, 340, 350, 351, 365, 421, 445
China-Burma-India, 78, 365
Chisholm, Robert, 285
Chitwood, Auten, 271, 272
Choiseul Island Diversion, 364
CHOKER, 184
CHOKER II, 418
Chu Lai Airbase, 209
Church, John, 357, 359
Churchill, Winston, 31, 40, 43, 63, 78, 110, 199, 318, 350, 352, 403
CIA, 75, 76
CINCINNATI, 241
Civilian Irregular Defense Group (CIDG), 326, 331
Clark, David, 341
Clark, Hal C., 18
Clark, Harlow G., 227, 234
Clark, Mark, 98, 99, 100, 101, 198, 199, 291, 405, 411, 417
Clark, Richard A., 339
Clarke, Dudley, 403
Clegg, Carl R., 291
Cleland, Joseph P., 290, 434
Cline, Paul H., 223
Clough, Timothy W., 338
Coard, Bernard, 148
Coard, Phyllis, 148
Cobb, Henry, 346
COBRA, 50
COCHISE, 241
Cochran, Phillip, 78, 291, 384
Cockerham, William C., 468
Coehlo, Antonio J., 339
Cofforth, Alfred P., 223
Colby, William E., 353
Cole, Francis J., 353
Cole, John M., 332
Cole, Robert G., 216, 223, 415
Colee, Donald L., 403
Coleman, J.D., 236
Collins, Boggs C., 291
Collins, J. Lawton, 143, 144, 261, 316
Collins, John (Rip), 452
Collins, Kenneth G., 223
Collins, Lawton, 65
Columbia, 150
Combat Arms Regimental System (CARS), 437, 454
Combat Jumps, 72, 74, 75, 77
COMBINE, 209, 309, 311
COMBINE II, 309, 311
Combs, Alfred H., 359
Comein, Lucien E., 353
Comerford, Steven, 332
Conable, John, 180
Concordia, John, 396
Condamire Valley, 304
Conner, DeForest S., 223
Connors, John, 154
Consolvo, Bill, 274
Contreras, Albert, Jr., 223
Conway, James B., 338
Conway, Theodore J., 441, 471
COOK, 214
Cook, Julian, 124, 340
Coombs, Paul H., Jr., 333
Corbin, Jules, 403
Cordes, Herman J., 223
Corinth Canal, 32
Corregidor, 19, 56, 59, 95, 139, 162, 164, 167, 252, 253, 257, 365, 378, 392, 419
CORREGIDOR, 256
Costello, John, 21
Cota, Norman, 417
Cote, Roger E., 353
Cotentin Peninsula, 266, 270, 278, 366, 371
COURAGEOUS, 141
Coutts, James W., 186, 403, 420
Cowan Creek, 272
Coyle, Jimmy, 262
Crabtree, Daniel F., 338
Crafton, James, 235
Craig, Hillary, 232
Cramer, Harry G., 332
Crawford, Leo, 183
Creek, Roy E., 264, 268, 269, 272, 276
Crete, 19, 32, 34, 35, 36
CRIMP, 241
Critz, Harry W., 224
Cross, Thomas R., 110, 396
Crerar, John A., 345
Cromillion, Arthur, 411
Crouch, Cleo, 275
Crouch, Joe, 273
Crouch, Joel, 89, 371
Crowe, Charley, 398
Crowe, William J., 150
Crowson, Mark S., 470
Crum, Robert, 232
Cuba, 66, 148, 150, 156, 233
Cummings, Jackie, 236
Cunningham, Hubert S., 454
Cushman, John W., 224
Cutler, Stuart, 224
Cutulo, Edward V., 345
Cyr, Paul, 353
Czechoslovakia, 74, 76, 343, 402

— D —

D Day, 19, 45, 48, 50, 261, 262, 273, 274, 278, 283, 292, 296, 305, 306, 307, 342, 351, 352, 365, 366, 379, 414
Dahlia, Joseph A., 267
Dalbey, Josiah T., 391, 413
Dalrymple, Robert W., 111, 294, 297
Danforth, Virgil E., 223
Dangerfield, Steven J., 346
DANIEL BOONE, 336, 337
Daniel, Robert G., 339
Danielsen, Ted, 227
Danielson, Theodore, 229, 230, 231
DARBY CREST, 241
DARBY MARCH, 241
DARBY PUNCH III, 241
DARBY TRAIL, 241
Darby, William O., 417
Darby's Rangers, 321
Darcy, Paul, 338
Darlac Province 334
Daugitan River, 91
Davan, Benedict M., 339
David, Kenneth J., 223
David, Leslie B., 291
Davies, Dennis M., 371
Davis, Arthur H., 157
Davis, Benjamin O., 416
Davis, Charles W., 434
Davis, Douglas C., 176
Davis, Harley, 338
Davis, John, 254, 264, 270, 271
Davis, Louis D., 350, 403
Davis, Paris, 345
DAVY CROCKET, 235
Dawson, Marion L., 361
DAYTON, 241
De Glopper, Charles N., 109, 193, 200, 415
De Ploeg, 284
De Witt, Paul, 224
Dead Man's Ridge, 419
Deane, Bill, 358
Deane, John R., Jr., 145, 449, 452
Deane, William L., 359
Decarria, Joe T., 346
DECKHOUSE II, 214
DeGavre, Chester B., 392
Del Cid, Luis, 150, 156
Delameter, Benjamin F., 340
DELAWARE, 214
DeLeo, Dan A., 291
Delta Company, 230
Delta Force, 312
Dempsey, Miles, 103, 106
Denmark, 19, 30, 63, 345, 402
Dennard, Danny, 223
Denneau, Anthony J., 353
Dennison, Joe, 469
Dennison, Joseph L., 338
Dent, William L., 223
Denton, Edward, 340, 341
DENVER, 241
Depue, Robert, 341
Desert One, 158, 314
DESERT SHIELD, 68, 471, 472, 473
DESERT STRIKE, 214
Deshayes, Albert P., 294
Deuel, William T., 359
Devils Brigade, 319, 324, 325, 330, 417
Devil Dogs, 242
Devil's Hill, 285, 286
Devlin, Gerard, 384
DeVries, Paul, 21
Dews, Hampton, 339
DEXTER, 241
Dexter, George E., 342
Dexter, Herbert J., 223
Dickerson, George W., 195, 452
Die, Danny, 236
Diering, Mark, 465
Diersfordt, 174, 273
Dietrich, Frank L., 223
Dilburn, Ernest L., 403
Dill, Sir John, 403
Dillard, Douglas C., 21, 305
Dillon, Conrad C., 353
Dinkins, Clifford, 223
Division Reconnaissance Platoon, 206
Dix, Drew D., 328, 338
Dnieper River, 29
Dod, Halsey H., Jr., 267
Dodd, Edgar F., 403
Dolby, David C., 227, 231, 240
Dominican Republic, 209
Donlon, Roger H.C., 326, 328, 341, 342, 446
Donne, John, 17
Donovan, Leo, 409
Donovan, Philip W., 353
Donovan, William J., "Wild Bill", 350, 404
Donovan, William O., B, 324, 342
Doolittle, James, 384
Dorrity, Edward, 137
Dorsett, Hal, 398
Dorsey, Harry A., 353
Doucet, Joseph E., 403
Douglas, Donald, 387
Douve River, 48, 266, 278, 279, 282
Downing, Wayne A., 153, 469
Downing, Wayne P., 463, 465, 466, 468
Doyle, Hubert M., 345
DRAGOON, 51, 53, 110, 113, 115, 119, 296, 305, 367, 414
DRAGOON/ANVIL, 306
Dragseth, Raymond, 157
Drennan, Fred O., 223
Dreux, William B., 353
Duarte Bridge, 209
Duff, Rolland J., 266
Dugan, Larry, 468
Duggan, Lawrence W., 338
Duke, Lonnie M., 411
Dunaway, George W., 332, 338
Duncan, George B., 193
Dunlap, Robert H., 35
Dunn, Richard "Dick", 397
Dunn, Theodore L., 196
DuPont, Richard, 386
Durst, Jay B., 338, 340
Dussaq, Rene, 353

— E —

EAGLE, 215
EAGLE CLAW, 461
Eagle Creek, 363
EAGLE THURST, 214, 219
EAGLEWING, 212, 219
Eben Emael, 19, 30, 31, 36
Eberhardt, Aubrey, 403
Eberhardt, Charles, 381
Eberhardt, Guy, 229
Eckman, William E., 107, 342
Eddy, William A., 350
Edson, Merritt A., 362
Edwards, Irwin A., 340, 341
EFFECTIVE, 184, 418
Egypt, 28, 32, 222, 242, 247, 248, 421
Egypt-Israeli PeaceTreaty, 222
Ehly, Erle, 118
Eichelberger, Robert L., 92
Eifler, Carl F., 353
Eikenberry, Robert W., 430
Eindhoven, 368, 378, 414
Eisenhower, Dwight, 41, 42, 43, 45, 53, 54, 63, 79, 80, 98, 110, 167, 180, 209, 212, 218, 242, 247, 260, 283, 286, 289, 322, 340, 351, 353, 366, 368
Ekman, William E., 122, 309, 325, 345
El Salvador, 327, 341
Elbe River, 29, 419
Ellis, Johnnie A., 403
Emerson, Henry E., 328
Emmerson, Henry E., (Hank), 456, 459
Endara, Guillermo, 150, 151, 157
Endhoven, 222
England, 18, 34, 40, 80, 148, 203, 212, 216, 261, 262, 265, 271, 278, 283, 286, 291, 293, 296, 297, 303, 343, 350, 351, 367, 369, 371, 384, 385, 401, 402, 403, 410, 413, 414, 415, 417, 418
English, Glenn H., Jr., 240, 455
Ennis, Riley F., 224
Erickson, John, 254, 256
Erwin, James, 193
Escaut Canal, 283
Esch, Elmer B., 353
Escobar, Gasper, 267
Estep, James, 332, 463, 466
Ethiopia, 333, 340
Ettman, Paul, 343
Eubanks, Ray, 95, 255, 415
EUREKA, 43, 110
Europe,17, 18, 36, 38, 39, 40, 43, 53, 55, 59, 62, 63, 79, 143, 211, 212, 214, 236, 260, 264, 265, 277, 284, 290, 294, 301, 303, 318, 324, 325, 326, 332, 339, 342, 344, 350, 365, 366, 370, 371, 387, 393, 409, 415, 419, 420, 421, 422, 430, 463, 468, 470
Evans, Billy D., 338
Evans-Smith, William, 341
Evule, George, 223
Ewell, Julian, 108, 223, 403, 415

— F —

Fagan, Richard, 364
Faistenhammer, Ludwig, Jr., 344
Farr, Warner 'Rocky', 463
FARRAGUT, 214
Farrell, Francis W., 176, 177, 340, 434

Farren, James, 176, 177
Faust, Joseph, 266, 272, 274
Felder, Louis F., 342
Fenlon, James A., 336
Ferguson, Charles R., 333, 340, 345
Ferguson, William G., 338
Fergusson, Rogert C.L. 223
Fessler, Jack, 277
Field Epidemiological Survey Team (FEST), 449, 451
Fiji Islands, 248, 362
Fike, Emmett E., 316
FILLMORE, 214
Finland, 29, 340, 343, 381, 402
FIREFLY, 309, 311, 420
Fisher, Paul, 463
Fitzgerald, B.R., 154
Fitzmaurice, Michael J., 223, 456
Flanagan, Edward M., 138, 145, 176, 328, 453
Fletcher, Larry A., 223
Floyd, Kenneth, 464, 466
Floyd, Richard C., 353
Flynn, Robert 253
Foggia, 291
Foley, John P., 285
FONDA'S FOLLIES, 456
Fontaine, Sully, 343
Foote, Oliver C., 271
Forbes, Ian, 353
Ford, Edward E., 223
Ford, Guillermo, 150
Foreman, Forest K., 338
FORTITUDE, 105, 423
Fort Wayne, 241
Foss, John, 462, 464, 467
Foster, Ray H., 353
Fowler, Gary D., 472
Fowler, James, 231
FRAG ORDER, 240, 241
FRAG ORDER VOGG, 241
Fraiture Ridge, 206
France, 29, 31, 40, 45, 46, 48, 51, 53, 72, 74, 76, 131, 133, 137, 184, 204, 209, 212, 215, 217, 218, 250, 261, 265, 266, 270, 271, 272, 276, 277, 279, 280, 282, 284, 286, 289, 290, 291, 293, 294, 297, 299, 304, 305, 306, 307, 318, 322, 324, 330, 333, 334, 339, 340, 342, 343, 345, 350, 353, 354, 365, 366, 367, 369, 371, 380, 385, 387, 392, 394, 401, 413, 414, 415, 416, 417, 418, 419, 420, 421, 422, 424, 426, 441
Francis, Clyde W., 338
FRANCIS MARION, 241
Francis, Scott A., 470
Franklin, Benjamin, 401
Franklin, Norman R., 353
Fraser, Donald, 118
Fratellenico, Frank R., 223, 455
Freddie's Freighters, 320
Frederick, Robert T., 111, 118, 297, 318, 322, 324, 407, 417
Freed, William, 277
FREEDOM DEAL, 337
Freels, Tom, 231
Friedrich, Robert L., 223
Friele, Berent E., 353
French Underground, 266
Frost, John, 127, 203
Fry, Jerry R., 223
Fryar, Elmer E., 92, 176, 181
Frye, John W., 325
Fryer, Elmer, 416
Fucik, Julius, 17
Fuller, Horace W., 353
Funk, Leonard, 193, 203, 207, 288, 420
Furman, Robert E., 341
Furst, Henry, 340

— G —

Gabriel Demonstration Area, 333
Gabrys, Stephen, 339
Gaeta, Gulf of, 97
Gairy, Eric, 148
Gaither, Ridgely, 179, 428, 435
Gale, Richard N., 105, 106, 410, 414
GALLANT SHIELD, 221
Gandy, Michael L., 223
Gannon, Timothy G., 341
Garcia, Arthur F., 332
GARDEN, 368
Gardner, James A., 219, 223
Garigliano River, 417
Garnerin, Andre J., 401
Garrett, Chester, 338
Garrett, Robert W., 332
Garrett, Scott, 25
Garrison, William F., 469
Garwood, Charles E., 341
Gaspard, George W., 339
Gates, Joseph, 311
Gavin, James M., 40, 43, 83, 97, 108, 120, 193, 197, 199, 200, 203, 204, 209, 228, 260, 262, 267, 269, 270, 280, 283, 302, 307, 309, 369, 387, 390, 404, 406, 414, 441, 471
Gaydosik, Edward, 341
Gaytan, Eliezer, 156
Gaze, Roy E., 291
Genko Line, 93
Gennerich, Charles J., 353
Geraci, John C., 332
German Long Range Reconnaissance Patrol School, 460
Germany, 18, 19, 29,34, 36, 39, 43, 75, 76, 80, 131, 137, 159, 184, 188, 195, 203, 209, 218, 221, 222, 242, 247, 272, 274, 284, 289, 293, 299, 317, 325, 327, 339, 340, 343, 344, 345, 366, 368, 369, 393, 394, 402, 418, 419, 420, 422, 424, 425, 428, 434, 435, 438, 439, 444, 446, 448, 453, 458, 460, 461, 463, 469, 470, 471
Gertsch, John G., 223, 454
Getzinger, John M., 345
GI Olympics, 276
GIANT I, 98
GIANT II, 98
GIBRALTAR, 214
Gilbert, Robert F., 312
Gilday, T., 322
Gildee John J., 353
Gillespie, Vernon, 238
Gillmore, Willia M.M., 224
Givens, James C., 301
Glemser, James P., 223
Glider Pilots, 384, 386
Goddard, Lewis F., 353
Godfrey, McKinley, Jr., 411
Godsey, James F., 338
Godwin, Harry M., 223
Gogos, Sporos, 294
Golden Brigade, 210
GOLDEN EAGLE, 222
Gonzalez, Gonzalo, 157, 231
Gonzalez, Rodrigo, 235
Goodwin, Earl, 388
Goranson, Ralph, 109
Gorham, Art, 88, 197
Goring, Herman, 402
Grady, John C., 301
Granlund, Arthur J., 269
Grant, Gerald, 338
Graves, Rupert D., 110, 118, 294, 295, 296, 298, 299, 305
Gray, W.S., 322
Grazioplene, James J., 152, 469
Great Britain, 39, 40, 43, 65, 214, 222, 290, 318, 365, 368
Greece, 34, 214, 228, 343, 345, 350, 353, 354, 463
GREELEY, 241
Green, Gerald D., 223
GREEN LIGHTNING, 241
Green, Vernon, 344, 345
Greene, Robert P., 411
GREENE STORM, 241
GREEN SURE, 241
Grenada, 18, 19, 65, 66, 72, 75, 76, 148, 149, 152, 158, 195, 401, 407, 411, 462, 463, 465
Griffith, Wilbur, 121
Grimsley, Lee G., 359
Griner, George W., Jr., 184
Groening, W.H., 471
Gruen, Arthur, 353
Guadalcanal, 35, 161, 362, 363, 364, 434
Guadalcanal-Tulagi, 364
Guam, 226, 228, 234, 346
Guards Armored Division, 203, 286
Guderian, Heinz, 218
Guenette, Peter M., 223
Guernsey, 279
Guest, James A., 328, 338, 468
Guilbeau, Mitchel, 403
Guillot, Nelson E., 353
Gulf of Agaba, 248
Gulf of Salerno, 41
Gulf of Suez, 248
Guppy Island, 363
Guzzo, Dante "Tony", 266
Gypsy Task Force, 94
GYROSCOPE, 179, 247, 290

— H —

H-Hour, 296, 379
Hacksaw Ridge, 92, 181
Hagan, "Pappy", 262
Hagen, Loren D., 328, 338, 456
Hagen, William, 84
Hagerman, Bart, 20, 21, 78, 129, 186, 300, 506
Hakala, Robert W., 341
Halasz, Steven A., 464
Hale, Bartley E., 275
Haley, John M., 251, 403
Hall, 98
Hall, Billy, 338
Hallford, James L., 341, 345
Hamilton, Gilbert L., 339
Hamlet, James, 357
Hammond, Lester, 245, 434
Hand, Michael J., 338
Hanlon, John, 230, 231
Hanoi, 329, 330
Hanna, Walter C., Jr., 353
Hansen, Marcus, 359
Hansen, Richard B., 346
Hanson, Frank A., 353
Harbough, Francis L., 223
"HARDIHOOD", 241
Hardy, Herbert F., Jr., 332
Hargraves, James, 463, 466
Harkins, Paul D., 336, 441, 445
Harper, Joseph H., 405
Harrell, Ben, 224
Harris, Audley C., 338
Harris, Ernest O., 223
Harris, Hugh P., 179, 184, 249, 390
Harris, Joseph D., 405
Harris, T., 272
Harrison, 214
Harrison, Bailey, 223
Harrison, William E., 345
Hasbarger, Gerald, 23
Hassan, Eduardo Herrera, 155, 157
Hastey, J. Rex, 466
Hatch, Brooke H., 364
Haugen, Orin D., 91, 176, 177, 181, 182, 416
Hawley, Donald R., 359
HAWTHORNE, 214
Hays, John H., Jr., 456
Hayward, Richard W., 364
Healy, Michael D., 456, 457, 458
Healy, Mike, 326, 328, 332, 338, 454
Heaps, George H., 338
Heartbreak Ridge, 400
Hebbe, Frederick, 270
Heckard, George W., 294
Hedrick, Clinton M., 419
Heilman, Gus, 322
Helin, Gertrude, 155
Hell's Highway, 54, 369
Hemmingway, Charles L., 359
Hemphill, John, 235
Henderson, Alvin, 284
Henderson, Arthur D., 301
Henderson, Arthur M., 301, 302
Henely, Michael F., 353
Hennessey, John J., 224, 227
HENRY CLAY, 236
Henry, Patrick, 295
Herakleion, 34
Herbert, James A., 430
Herda, Frank A., 223
Hermann Goering Panzer Division, 40, 260
Hernandez, Rodolfo P., 245, 431
Hetzler, Walter G., 339
Heyns, Robert E., 353
Hickey, Thomas P., 429
Hicks, Edwin C., 291
Higgins, Gerald J., 403, 434
HIGHLAND, 214
Hildebrand, Harry, 176, 177, 181, 415
Hill 75.9, 285
Hill 95, 282
Hill 131, 282
Hill 148, 362
Hill 205, 316, 317
Hill 228, 65
Hill 229, 65
Hill 255, 245
Hill 383, 317
Hill 400, 289, 299
Hill 628, 317
Hill 660, 418
Hill 670, 255
Hill 720, 321
Hill 832, 317
Hill, James, 106
Hilty, James R., 350
Hinchcliff, John J., 265
Hinton, Deane R., 157
Hinton, William 344
Hitler, Adolf, 30, 31, 32, 34, 36, 38, 43, 217, 260, 262, 365, 378, 402
Hodge, Harmon D., 341
Hodo, John d., 399
Hofatrour, William L., 223
Hoffman, Phil, 338, 340
Hogan, John, 223
Hoiser, Robert, 332
Holcombe, Frank S., 34, 92, 176, 361
Holland, 19, 30, 54, 74, 76, 120, 124, 137, 164, 165, 195, 204, 207, 209, 212, 262, 281, 283, 284, 286, 288, 289, 351, 365, 366, 368, 369, 371, 385, 387, 392, 402, 414, 415, 416
HOLLANDIA, 241
Holloway, Milton R., 178
Holm, William N., 305
Holmstock, Steve, 345, 466
Holt, Jeffrey P., 469
Holtan, Irvin N., 270
Honduras, 222
Honeycutt, Weldon F., 223
HOOD RIVER, 214
Hooper, Garnett D., 347
Hooper, Joe R., 223, 452
Hopkins, Harry, 318
Horak, James L., 347, 457
Hoska, Lukas E., Jr., 176, 178, 180, 181
Hosking, Charles E., Jr., 328, 338, 451
Hosking, Charles, 343
Hospelhorn, Cecil, 397
Houffalize, 204
Houten, Van, 316
Houston, Robert J., 223
Howard, Robert L., 453, 468
Howell, George P., 403
Howard, John D., 355
Howard, Marcellus J., 34, 361, 363, 364
Howard, Robert L., 328, 338
Howell, George P., 251, 405, 406
Howland, Charles C.W., 292
Howze, Hamilton H., 439, 441
Huddleston, Louis D., 248
Huempfner, MiloC., 305
Huff, Paul B., 416
Huff, Paul H., 291, 292
Huffman, E.L., 402
Hufft, Ray, 322
Hugart, Clarence, 267
Hughes, Les, 394
HUMP, 241
Hungary, 74, 76
Hurchalla, George, 21
Hurteau, Joseph C., 157, 464, 468, 469
HUSKEY, 410
HUSKEY I, 40, 197, 302, 412
HUSKEY II, 197
Hyatt, Robert A., 345
Hyde, James W., 365
Hynds, Henry G., 176

— I —

Iker, Gilbert H., 346
ILVER CITY, 241
Imjin River, 64, 65
Inch'on, 141
India, 324, 350, 384, 421, 424
Indiana National Guard, 312
Indochina, 65, 75, 341, 355
Indonesia, 75, 76, 331
Infantry Board, 31
Inter Allied, 74, 76
International Long Range Reconnaissance Patrol (ILRRP), 464
International Long Range Surveillance Patrol (ILRRP), 461
Iran, 312, 343
Iraq, 28, 68, 333
Ireland, 265, 278, 342, 407, 410, 412, 414
Iron Curtain, 325
IRON TRIANGLE, 241
Irvin, Leslie "Sky-High", 402
IRVING, 236
Irwin, Dave, 269
Isaac, Jesse A., 223
Isler, Jack, 344
Israel
Issel River, 273, 419
Italy, 27, 29, 36, 40, 51, 74, 76, 89, 100, 164, 184, 195, 198, 200, 209, 214, 260, 292, 294, 295, 296, 299, 304, 319, 320, 330, 343, 345, 353, 365, 366, 367, 371, 402, 410, 412, 413, 415, 416, 417, 424, 426, 458, 462
Ivanov, Ivan, 341
Ivey, Glen S., 359
Ivey, George W., 403
Iwo Jima, 35

— J —

Jablonsky, Harvey J., 247, 411
Jackson, Joseph J., 341
Jackson, William N., 341
Jacobelly, Robert C., 341, 470
Jacquay, Benedict F., 251, 403
Jager, Thomas W., 338
Jamaica, 66
Japan, 36, 43, 209, 218, 228, 242, 243, 245, 246, 247, 300, 303, 311, 316, 325, 331, 350, 370, 392, 396, 397, 399, 424, 425, 426, 430, 432, 436, 438, 444, 447
JEB STUART, 214
Jedburgh Team, 74, 76
JEFFERSON GLEN, 220
Jella, Elbert, 385
Jenkins, Gilbert K. "Joe", 336
Jenks, Robert, 468
Jennings, Delbert, 238
Jersey Islands, 279
Jerusalem, 248
Jeziorski, Edward J., 266
JIM BOWIE, 235
Joel, Lawrence, 240, 447
Joerg, Wood G., 110, 301, 302, 305
Johnson, Charles, 269
Johnson, Harold K., 226, 239
Johnson, Howard R., 107, 415
Johnson, James F., 472
Johnson, James H., Jr., 467, 469
Johnson, Jesse L., 345, 471
Johnson, Lyndon B., 228
Johnson, Peter W., 338
Johnson, Stephen R., 333
Johnson, Donald R., 232
Joint Casualty Resolution Center (JCRC), 458
Joint Special Operations Command (JSOC), 460, 463, 464, 470
Jones, Bob, 21
Jones, George G., 439
Jones, George M., 56, 95, 204, 252, 254, 257, 259, 328, 403, 404, 410
Jones, Harry J., 447
Jones, Kyle D., 223, 255
Jones, Robert E., 345
Jordan, 343
Joseph, John T., 266, 271, 272
Joubert, Donald L., 223
JUMPLIGHT, 212
JUNCTION CITY ALTERNATE, 65, 145, 241, 450
JUNCTION CITY II, 241
JUST CAUSE, 67, 150, 151, 157, 158, 469, 470, 471

— K —

Kall River, 299
Kamiri Airfield, 95, 415
Kane, Francis B., 340
Karlstad, Charles H., 328, 434
Kasler, Charles L., 339, 340
Kassel, Frank, Jr., 403
Kasun, David R., 223
Katz, Morton N., 291
Kawamura, Terry T., 240, 454
Kays, Kenneth M., 223, 455
Kedenberg, John J., 453
Keefe, Daniel J., 467
Keerans, C.L., 249
Keesling, Earl L., 338
Kehoe, Robert R., 353
Kellem, F.C., 108
Kelly Field, 17
Kelly, Francis J., 332, 335, 338, 449
Kelly, Richard J., 403
Kemmer, Thomas, 330
Kendengburg, John J., 328, 338
Kennedy, Herman J., 332
Kennedy, John F., 324, 326, 341, 441, 456, 457, 458, 459, 460, 461, 462, 465, 472
Kennedy, Leslie D., 223
Kennedy, Robert, 453, 454, 466
Kenney, George C., 95
Kent, Alan, 223
KENTUCKY JUMPER, 214
Kerksis, Sidney C., 403
Kernan, William F., 470
Kerrigan, William F., 153
Kesselring, Albert, 218
Key Albert Canal, 386
Kherly, Stephen J., 353
Kimpo Airfield, 139, 140, 141, 244, 399
King, Ernest, 352
King, Felix, 235
King, Jerry, 465
King, Martin Luther, 453
King, William N. "Red", 31, 174, 251, 396, 398, 403
KINGPIN, 329
Kingston, Robert C., 328, 459
Kinkade, Melvin L., 470
Kinnard, Harry W.O., 179, 223, 226, 227, 228, 233, 234, 236, 415, 447
Kinnes, Ralph, 378
Kinney, Howard D., 340
Kinsler, Kenneth H., 252, 254, 410
Kirkpatrick, Ronald I., 359
Kiser, Walter W., 35
Kiska, 319, 320, 322, 330, 412
Kit Carson Scouts, 312
Kitchens, John M., 403
Kittleson, Galen C., 332, 341
KLAMATH FALLS, 219
Knowlen, Chueck, 230
Knowles, Richard T., 226, 450
Knox, Bernard M.W., 353
Knox, Cameron, 254
Kochanic, John, 280
Koje-do, 65
Kolek, Joe, 268
Korea, 17, 63, 64, 72, 75, 76, 138, 141, 143, 144, 146, 148, 159, 161, 166, 167, 222, 227, 229, 233, 236, 242, 244, 245, 247, 311, 312, 316, 317, 325, 326, 331, 332, 346, 392, 399, 401, 407, 422, 424, 426, 429, 430, 432, 434, 444, 447, 458, 459, 463, 464, 465, 466, 469, 472
Korean War, 19
Kornsay, James E., 411
Koun, Charles W., 197
Kozlowski, Charles J., 178
Krause, Edward, 83, 107
Kreckel, John E., 223
Kroesen, Frederick J., 457
Kurlak, Victor H., 363, 364
Kuhn, William A., 246, 276, 277
Kunert, Gerhardt, 341
Kurg, Alfred, 223
Kuwait, 68, 471

— L —

La Bonneville, 270
La Chaussee, Charles, 118
La Difensa, 320
La Fiere, 267, 268, 270, 279, 280
La Fiere Bridge, 277
La Flamme, Ernest H., 90, 176, 177, 178, 181
La Poterie Bridge, 270, 271
La Senia, 40, 80
Laboa, Jose Sebastian, 156, 157
Lackery, John, 256
Ladd, Jonathan F., 332, 338, 451
Laguna de Bay, 365
Lahti, Edward H., 90, 93, 176, 177, 178, 181, 182
Laird, Melvin, 329
Lajeunesse, Lucian E., 353
Lam Son, 220, 221
Lamm, George D., 284, 288
Langen, Robert E., 223
Laos, 75, 76, 246, 326, 331, 336, 337, 340
LaPointe, Joseph G., Jr., 223, 454
Larsen, Stanley R., 449
Lash, Donald J., Jr., 467, 468
Latson, Harvey H. III, 468, 472
Law, Robert D., 312, 454
Lawrence, William, Jr., 223
Lawrie, Joe, 95, 259, 290, 447
Lawton, John P., 223
Lawton, William, 267
Le Havre, 46
Le Motey, 267
Le Port, 280
Leal, Carlos, 333
Lebanon, 242, 247
Ledbetter, William, 339
Lee, Bill, 403, 404, 405, 414
Lee, Milton, A., 223, 452
Lee, William C, 18, 31, 32, 36, 38, 45, 214, 215, 224, 390, 391, 403
Lehew, Donald L., 338
Leigh-Mallory, Trafford, 103
Leino, Loyd J., 223
Leisy, Robert, 232
Leitch, Joseph D., 184
Lek, 53
Lemeres, Alexander, 454
Lemnitzer, Lyman, 179
Leonard, John W., 429
Leonard, Theodore, 338, 444
Leuer, Kenneth, 467
Lev, Orel H., 223
Lewis, Harry, 109
Lewis, Mark, 467
Leyte, 55, 56, 90, 95, 176, 177, 181, 228, 233, 243, 246, 253, 255, 339, 413, 415, 416, 421, 422
Libya, 402
Lilienthal, Otto, 36
Lillge, Karl, 265
Lillyman, Frank L., 105, 215, 223, 371
Lima, Paul E., 345
Lincoln, 235
Lindquist, Roy, 107, 108, 122, 200. 247, 251, 267, 278, 280, 282, 288, 383, 403, 416
Lindsay, James A., 461
Lindsay, James J., 328, 461, 466, 471
Lindsey, J.B., 249, 356
Lingayen, 56
Lingo, Michael A., 469
Link, John F., 338
Lipa Hill, 418
Lippe River, 275
Lisk, Percy, 396
Little, Selby F., 345, 456
Livorno Division, 40
Loeper, Charles P., 176
Long Range Patrol (LRP), 454
Long Range Reconnaissance Patrol, 393, 447, 449, 468
Long Range Surveillance Detachment, 465, 466, 469, 470
Long Range Surveillance Unit, 393
Lopez, Edgardo, 157
Lord, William G., II (2), 278
Los Banos, 19, 59, 94
Loughrin, Richard N., 90
Love, Joseph B., 341
Lowe, Mills T., 176
Lozada, Carlos J., 240, 450
Lucas, Andre C., 223, 455
Lucas, John Porter, 101
Lucas, Robert A., 353

Luck, Gary E., 328, 464, 469, 471
Luftwaffe (German Air Force), 29, 32
Luke, Ronald, 234
Lupyak, Joseph W., 338
Luthy, Henry D., 341
Lutz, Joseph C., 148, 328, 460, 462, 463, 466, 471
Luxembourg, 272, 345
Luzon, 19, 56, 59, 94, 139, 176, 177, 178, 181, 183, 228, 233, 243, 244, 246, 253, 255, 339, 378, 387, 420, 421
Lynch, George A., 31, 402
Lynch, Marshall, 333
Lyons, James L., 338
Lysek, Henry, 268
Lytton, Balfour O., Sr., 339

—M—

Maas River, 53, 203
Maas Waal Canal, 203, 284, 414
MACARTHUR, 241
MacArthur, Douglas, 63, 64, 95, 138, 139, 141, 143, 179, 183, 228, 233, 246, 253, 254, 257, 303, 316, 392
MacAuliffe, Anthony, 108
MacDill Field, 249
Mackall, John "Tommy", 81, 181, 291, 409
Mackmull, Jack V., 222, 224, 328, 459, 461
MacVicar, Norman, 279
MacWilliam, Tom, 322
Madding, Scott, 331, 332
Maddox, R., 338
Madigan, Donald F., 255
Maestas, Alan, 153
Magonyrk, James R., 223
Maharg, D., 271
Mahonag, 91
Mahoney, Jiggs, 322
Maksin, Francis A., 274
Maladowitz, Raymond, 338
Malaya, 74, 76
Marleme Airfield, 34
Malheur I, 214
Malheur II, 214
Malinta Hill, 256
Malby, Albert E., 449
Malmedy, 262
Maloney, Arthur A., 264, 268, 270
Maloney Hill, 92
Maloney, John S., 223
Malvar, 94
Manarawat, 91
Mancil, Edward C., 138
Manierre, Cyrus, 353
Manila, 56, 59, 181, 182, 183, 242, 392
Manila Bay, 56
Mann, Joe E., 217, 223, 415
Mann, Thomas L., 176, 178
Manor, Leroy J., 329
Mansfield, Gordon A., 223
Mapstrike, 333
Maracek, George, 333, 338
Marauder, 241
Mari, Louis A., 338
Marine Parachute Battalions, 361
Mark, Marion L., 223
MARKET, 368, 369, 413
MARKET-GARDEN, 55, 59, 120, 128, 203, 216, 217, 368, 369, 413, 414, 415, 416
Markham River, 410
Markham Valley, 19, 41
Marne River, 272
Marr, John, 21, 267, 268, 273
Marseilles, 53
Marsh, John, 175, 277, 312, 465
Marshall, Cookson, 322
Marshall, George, 18, 31, 41, 43, 95, 105, 149, 234, 242, 318, 352, 367, 403
Marshall Islands, 331
Martin, Edward, 403
Martin, James G., 346
Martin, Linwood D., 339
Martin, Roy, 227, 229, 231
Martin, Thomas E., 223
Martinez, Arnold J., 277
Maseda, Gerald L., 359
Mason, Louis, 465
Mass River, 284
Massachusetts Striker, 214, 220
Massad, Ernest L., 176, 178
MATADOR, 235
Matheson, Salve H., 345, 450, 452
Matthewson, Lemuel, 179, 398
Mattingly, James, 267
Mattox, Robert D., 338, 340
May, Robert, 86
Mayer, Arthur C., 223
Mayer, Edward E., 341
Maynard, Thomas, 235
Mayor, Robert G., 223
Mazak, Stefan, 343, 344
McAnsh, Thomas A., 328, 435
McAuliffe, Anthony C., 217, 218, 223, 224, 228, 369, 414
McCabe, Richard, 411
McCarter, Lloyd G., 419
McCarthy, John E., 338
McCarthy, Thomas W., 359
McClure, Robert, 342
McCollum, Timothy P., 223
McColloch, Joseph, 340
McCullough, Lloyd, 251, 403
McDaniel, Derrill, 179
McDonough, Robert C., 364
McDougall, Edward F., 333, 341
McDurmott, Michael A., 223
McEnery, John W., 224
McEntee, Ducat M., 178, 300
McFadden, Jerry, 322
McGee, Tildon S., 223
McGill, Ralph, 266
McGovern, George W., 338
McGowan, John D. II, 466
McGowan, John E., 353
McGrath, Kevin P., 469
McGuire, Gregory T., 332
McHugh, John J., 339
McIntosh, Henry D., 353
McKean, William A., 338, 448
McKinney, Marshall E, 465
McLallen, Richard, 353
McLaney, Lester C., 403
McMahon, Robert E., 294
McNair, James, 405
McNair, Lesley J., 36, 38, 41, 42, 181, 215, 242, 390
McNamara, Robert S., 247
Meade, George C., 436
Meade, George G., 446
Meadows, Richard J., 329, 330
Medal of Honor, 35, 92, 109, 166,176,181, 182, 183, 186, 193, 194, 195, 200, 204, 207, 227, 231,232, 238, 240, 245, 255, 264, 265, 274, 277, 280, 288, 291, 292, 293, 294, 299, 312, 326, 328, 337, 339, 341, 342, 343, 379, 404, 415, 416, 419, 420, 429, 431, 434, 446, 447, 448, 449, 450,451, 452, 453, 454, 455, 456
Meddaugh, Bill, 262
Medical Detachment, 289
Mediterranean, 32, 40, 41, 51, 260, 350
Medley, William J., 470
Merrman, W., 394
Mekong River, 337
Melansky, Harris M., 404
Meloy, G.S., 460
Menard, Noel A., 332, 340
Mendez, Louis G., Jr., 207, 279, 280, 282, 284, 286, 288
MERCURY, 32
Merderet River, 46, 48, 266, 267, 279, 280
Merkerson, Willie, Jr., 338
Merrill, Frank D., 78, 324, 417, 466
Merrill's Marauders, 312, 324, 331, 353, 417
Mersereau, Charles P., 353
Mertel, Kenneth D., 226, 229, 230, 237
Metheny, Fred R., 223
Metsler, Gilbert, 466, 468
Meuse River, 104
Michael, Don L., 240, 450
Michaelis, John H., 403, 415
Mick, Fred C., 332
Mickle, Gerald St. Clair, 224
Mike Forces, 335, 336, 446
Milender, John E., 359
Miley, William "Bud", 62, 175, 179, 186, 251, 252, 275, 382, 383, 396, 398, 403, 404, 405, 406, 409
Military Assistance Advisory Group (MAAG), 334
Military Police Platoon, 211
Millener, Raymond D., 223
Miller, Charles A., 362, 364
Miller, E.G., 399
Miller, Franklin D., 328, 338, 455
Miller, Ray, 345
Miller, Robert H., 302
Miller, William J., 271, 274
Millett, George V. Jr., 107, 108, 200, 264, 265, 267, 277, 406, 416
Milloy, Albert E., 328
Mills, Francis B., 332
Mindoro, 95, 257, 258
MINDORO (OPERATION), 255
Minnoch, Jack M., 346
Mitchell, Billy, 19, 27, 31, 43, 401
Mitchell, Harris T., 251
Mitchell, Jachin, 223
Mitchell, Joseph D., 447
Mitchell, William C., 159
Mize, Ola Lee, 240, 339, 455
Mobile Trng. Teams, 333
Modisett, John O., 403
Moir, Willian J., 291
Moir, William P., 291
Mojave Desert, 219
Monger, Elmer, 334, 341
Monkey Point, 257
Monroe, James H., 232
Monte Camino, 321
Monte LaDifensa, 319
Montgomery, Bernard, 53, 103, 203, 209, 276
Montgomery, Marshal, 167
Montgomery Rendezvous, 214, 220
Montgomery, Robert K., 353
Montilio, George, 223
Moody, William D., 109
Moon, Alvin L., 411
Moore, Charles J., 359
Moore, Dennis F., 223
Moore, Joseph W., 338
Moore, Raymond E., 353
Moore, Robert, 322
Marakon, Peter, 341
Morocco, 74, 76
Morris, Charles B., 240, 449
Morris, Melvin, 339
Morris, Thompson J., 266
Morris, Walter J., 411
Morrison, Manley, 234, 236
Morton, George C., 326, 334, 338, 441, 444
Mosely, George Van Horne,107, 108, 415
Moskaluk, George, 345
Mountbattan, Lord Louis, 318
Mountel, Robert A., 338
Mousseau, Lloyd F., 339
Mozey, Bill, 229
Mrazek, James E., 184, 249
Mrozek, Steven, 193, 393
Mt. Bataulao, 92
Mt. Carilao, 92
Mt. Macolod, 94
Mt. Malepunyo, 94
Mt. Sungay, 94
Mud Rats, 243
Mueller, Chris L., 469
Mueller, Ed, 322
Muer River, 235
Mulcahy, Robert, 345
Mullen, William J., 328, 340, 436
Mundinger, Robert G., 353
Munsan-ni, 64 65, 141, 143, 144, 166, 244, 317, 399
Munson, 378
Murphy, D., 394
Murphy, Lyle, 254
Murphy, Michael, 384
Murphy, Richard, 464
Murray, Bob, 290
Murray, Kenneth E., 223
Murray, Wayne H., 338
Mussolini Canal, 321
Myitkyina, 353
Mynatt, Cecil F., 353

—N—

Nadzab, 41, 95, 253, 256, 365
Nagel, Gordon, 274
Naples, 260, 291, 292, 293, 295, 304, 322, 324, 334, 339, 342, 367, 412, 415, 416
Nasugbu, 92
NATAHAN HALE, 214, 236
National Guard, 20
Naughton, Francis E., 269
Naughton, Frank, 397
Neel, Spurgeon, 451
NEGROS OPERATION, 258, 259
Nelson, Raider E., 270
Neo Marauders, 312
NEPTUNE, 105, 200, 265, 266, 366, 414
Neptune, Herman R., 271
Netherlands, 211, 222, 262, 290, 345
NEVADA EAGLE, 214, 219, 220
New Bilibid, 182
New Caledonia, 35, 362
New Guinea, 41, 55, 74, 76, 90, 95, 176, 182, 228, 233, 242, 252, 253, 339, 365, 384, 410, 411, 412, 413, 415, 416, 417, 418, 421, 422
NEW HOPE, 241
NEW LIFE, 241
New River, 34
New Zealand, 35, 239, 240, 362
NEWARK, 241
Newfoundland, 222, 248
Newman, James J., 417
Newman, James T., 223
Nichols Field, 93, 171, 178, 182, 242, 256
Nichols, John H., 274
Nichols, J.S., 394
Nieves, Mike, 150
Nijmegen, 54, 262, 284, 285, 286, 368, 378
Nijmegan-Arnhem, 415, 416
Nijmegen Bridge, 122, 127
Nishi Ridge, 35
Nix, James, 233, 234, 235
Nixon, Richard, 246, 329, 357, 358, 455
Nocera, 97
NOEMFOOR, 254
Noemfoor Island, 19, 95, 255, 256
Nolin, Tullis, 403
Nomands of Vietnam 214, 219
Nongtakoo, 333
Noone, Gerard, 468
Noriega, Manuel Antonio, 67, 150, 151, 153, 154, 155, 158, 506, 507
Normandy, 19, 36, 45, 46, 48, 50, 51, 53, 79, 101, 102, 103, 105, 109, 110, 120, 128, 161,162, 163, 164, 166, 167, 193, 195, 203, 207, 209, 211, 212, 216, 260, 261, 262, 264, 265, 266, 277, 278, 279, 281, 283, 284, 290, 339, 353, 366, 367, 369, 371, 372, 378,379, 384, 385, 387, 392, 414, 415, 417, 422
North Atlantic Treaty Organization (NATO), 222
North Sea, 284
Northern Solomons, 364
Northrup, Ralph A., 223
Norton, Charles W., 341
Norton, George C., 411
Norton, Jack, 261
Norton, John, 89, 101
Norway, 19, 30, 74, 76, 343, 345, 402
Novak, Frank J., 279, 280

—O—

Oberg, Karl, 218
O'Buckley, Donald J. "Pat", 271
Odom, George W., 338
Odom, Hubert, 223
Odum, George W., 223
Office of Strategic Services (OSS), 350
Ogg, Paul, 455
Ojala, S.V., 322
Okinawa, 179, 230, 242, 246, 312, 325, 326, 327, 331, 332, 333, 339, 340, 418, 421, 429, 438, 441, 444, 447, 455, 456, 459, 463
Old Abe, 211
Old Airborne Spirit, 20
Olive, Milton L., 240, 447
Olmsted, John M., 353
OLYMPIA, 63
OMEGA, 457
O'Neill, Daniel L., 223
ON GUARD, 241
OPORD, 240, 241
Oran, 40
Orne River, 46
O'Rourke, Donald C., 275, 340
Ortiz, Alberto, 359
Ortiz, Raymond, 231
OSS, 72, 74, 75, 76
Ostberg, Edwin J., 264, 267, 268
Osterberg, Norman A., 223
Otway, Terrence B.H., 106
Our River, 272
Ourthe River, 272
OVERLORD, 43, 50, 105, 106, 215, 352, 413, 414, 415, 416

—P—

Pacific Operations, 35, 38, 90
Pacific Ranger Company, 324
Paget, Bernard, 103
Paiva, Thurmon, 399
Pakistan, 340, 343, 344
Palmer, Bruce, Jr., 450
Palmer, George E., 345
Palmer, Harold T., Jr., 338
Palmer, Howard V., 353
Palmer, William T., 341
Palmer, Williston B., 428
Panama, 18, 19, 67, 75, 76, 150, 151, 155, 156, 157, 158, 222, 247, 253, 265, 304, 306, 339, 342, 401, 406, 407, 441, 458, 459, 462, 465, 469, 470, 471, 472
Panama Canal, 28, 67, 228, 234, 252, 305, 326, 327, 341, 346, 362, 390, 391, 404, 427, 430, 444, 457, 463
Panamanian Defense Force (PDF), 67
Panay Island, 258
Pankey, Laney M., 473
Papal Nuncia, 156, 157
Parachute Maintenance Detachment, 242
Parachute Test Platoon, 31, 175
Paramarines, 361
Paranaque River, 93, 178
Park, William Forest, 350
Parker, Arthur M., III, 359
Parker, George W., 223
Parker, James, 278
Parker, Jesse J., 223
Parker, Otis, 339
Parks, Donald, 21, 72, 475, 488, 497, 506
Patch, Lloyd E., 223
Pathfinder, 74, 76
PATOG HILL 4055, 259
Patrick, Burton, D., 224
Patten, Clarence W., 340
Patterson, Robert, 180, 223, 452
Patton, George, 40, 41, 46, 89, 184, 198, 209, 262, 353, 367, 368
Paulick, "Iron Mike", 343, 345
Pawlik, Joseph E., 272
Paxton, Forest S., 294, 298
Payne, Paul, 446
Paz, Robert, 150
Pearce, Michael J., 472
Pearce, T.E., 322
Pearl Harbor, 36, 260, 265, 318, 350, 386, 391, 392
Pearson, Benjamin F., 264, 270, 272
Pearson, George, 176, 178, 181, 243, 290
Pearson, Willard, 447
Peay, J. H. Binford, 469
Peay, J.H. Binford, III, 224
Peck, Dana F., 346
Peck, Paul N., 274
Pedone, Vito S., 21, 365
Pemberton, David L., 462
Pendola, Tony, 230
Peoples, Leon, 223
Peres, Luis Delfin, 156
Perez, Manuel Jr., 93, 176, 182
Perkins, Thomas A., 469
Perkins, Stuart, 341
Perry, Michael P., 223
Pershing, Jack, 402
Pershing, John J., 27, 195
Peters, Dave, 397
Peters, George J., 264, 273, 277, 382, 420
Peters, Joseph L., 403
Peterson, Jay W., 470
Petry, Charles R. "Slats", 340
Petry, Charles L., 339
Pettingill, Patrick E., 332
Pezzelle, Roger, 343
Phelps, Joseph V., 186
Philippines, 55, 56, 74, 75, 76, 90, 139, 171, 242, 246, 252, 253, 254, 300, 327, 329, 331, 339, 365, 380, 384, 392, 413, 415, 416, 418, 419, 420, 421, 422, 424, 425
Phillips, Clifton, 340
PHOENIX, 241
PHU DUNG, 336, 337
Pick, Robert G., 223
Pickles, James C., 341
Pierce, Bob, 186
Pierce, James R., 186
Pierce, Larry S., 240, 447
Pierce, Willie, 235
Pierce, Roger L., 353
Pierson, Albert, 176, 177
Pietsch, William H., Jr., 353
Pike, Emory J., 193
Pindkerton, B.J., 342
Pioletti, John F., 338, 341, 345
Piombino Valley, 296
Pitts, Lemuel T., 403
Plato, Robert D., 338
Pleiku, 231
Pleiku Campaign, 234, 241
Plummer, Gerrell, 21, 229, 232
Plummer, Jerry, 227
PLUNDER, 272
Poche, Francis M., Jr., 353
Poett, J.H.N., 105
Point Salines Airport, 66
Poland, 29, 340, 343, 370, 402
Polish Airborne Brigade, 54
Polish Parachute Brigade, 18
Pollette, Floyd, 282
Pollette, Harold L., 284
Pond, Donald, 235
Porada, E, 394
Porter, James H., 311
Porter, Ray E., 224
Portugal, 345
Potter, P., 394
Potter, Richard, 345, 460, 470
Poudert, Robert H., 403
Powell, Beverly E., 224
Powell, Colin J, 150
Powell, Nolan L., 294
Powell, Thomas, 330
Power Pack, 209
POWER PACK, 448
Poynter, Ray, 229, 232
PRAIRIE FIRE, 336, 337
Pratt, Don, 48, 108, 216, 224, 379, 384, 405, 414
Pritchard, C.H.V., 110, 116
PROMOTE LIBERTY, 158
Provisional Parachute Group, 251, 390
Pruden, Robert J., 312, 454
Pruitt, James N., 339
Pryor, Robert D., 339
Puckett, Ralph, Jr., 223, 430
PUERTO PINE, 214
Puerto Rico, 214, 242, 393
Pugh, Charles, 21, 475, 488, 497, 506
Punch Bowl Ridge, 400
Purdy, Donald, 472
Purple Heart, 240, 242, 265, 294, 299, 330, 332, 339, 341, 352, 353, 399
Purple Heart Hill, 243
Pursley, John F., Jr., 403
Pusan, 65, 141

—Q—

Quade, Joseph, 21, 475,488, 490, 497, 506
Quandt, Douglass P., 176, 180
Quang Tri, 222
QUICK EAGLE, 221
QUICK KICK, 333
QUICKSILVER, 414, 423
QUICKSTRIKE, 219

—R—

Rabel, Laszlo, 240, 453
Radtke, William P., 333
Rae, Robert D., 264, 269
Raff, Edson D., 40, 80, 82, 98, 186, 200, 264, 270, 273, 274, 276, 277, 291, 325, 328, 340, 405, 406, 414, 420 435
Raker, Jeffrey H., 332
Rakkasans, 227, 242, 243, 244, 245, 246, 247, 248
Rambo, Charles, 21
Ramsey, Bertram, 103
Randall, Michael P., 359
RANDOLPH GLEN, 220
RANDOLPH GREEN, 214
Randstein, Knut H., 223
RANGER BULLDOZER, 214
Raymond, J.E., 394
Reagan, Ronald W., 149, 150, 341
Reault, Robert B., 332
Rebecca, 43
Red Army, 29
Red Devils, 278, 279, 280, 282, 283, 290, 291, 406
Red Hats, 355, 359
Reed, Edward B., 176, 269
Reed, James, 154
Reed, Leo D., 411
Reed, Pip, 270
Reese, Benjamin C., 403
REFORGER 75, 222
REFORGER 76, 221, 222
REFORGER 78, 222
Reichswald Forest, 203
Renfrow, Mark, 464
REPUBLIC SQUARE, 214
Resor, Stanley, 312
Retimo Airfield (Rhethymnon), 34
Rheault, Robert B., 338, 454
Rhine, 54, 55, 62, 63, 129, 130, 131, 133, 137, 184, 208, 272, 274, 275, 289, 385, 387,392, 419
Rhine-Herne Canal, 276, 420
Rhineland, 161, 186, 211, 264, 265, 277, 290, 291, 293, 294, 304, 305, 322, 334, 339, 342, 369, 378, 415, 419, 421, 422
Rhine River, 190, 203, 207, 284, 368, 419
Rhone Valley, 297
Rhur Valley, 190, 301
Rich, Charles W.G., 224
Rich, David F., 223
Richards, Edward T., 341
Richardson, Ray L., 223, 235
Richardson, William J., 446, 461
RICHLAND SQUARE, 214
Ridgway, Matthew, 62, 89, 97, 98, 101, 106, 120, 141, 143, 167, 193, 196, 198, 203, 242, 260, 261, 283, 297, 298, 369, 405, 409, 413, 414
Riehle, Richard, 466
Riley, Rip, 396
Ringler, John, 180, 182
RIPPER, 64
Riveer, Ken, 230, 231
Rivera, R.S., 338, 345
Roberts, Elvey, 226
Roberts, Grady A., 403
ROBIN, 241
Robinson, Albert P., 403
Robinson, Albert "Robbie", 396, 397
Robinson, Clifford L., 359
Robinson, Roscoe, Jr., 459
Robinson, Samuel W., 411
Rocca, Vincent M., 353
ROCK FORCE, 256
Rock, John, 45
Rock Hill, 92
Rocket Ridge, 246
Rodela, Jose, 339
Roer River, 289, 419
Rogers, Robert, 324

Roll, William C., 227
Rome-Arno, 294, 305, 334, 339, 342
Rommel, Erwin, 40, 81, 82, 104, 109
Rondiak, Roman, 345
Roosevelt, Franklin, 40, 78, 318, 350
Roosevelt, Theodore, 404
Roper, Delbert C., 286
Roper, George J., 275
Rose, Gary M., 339
Rose, Harold, 234
Rosemond, St. Julian F., 223
Rosen, David, 302
Rosenblum, Donald E., 460, 461, 462
Rosenstock, Steve, 21
Rosson, William B., 450, 452, 454
Rouse, Henry F., 291
Ray Creek, 272
Royal Air Force, 386
Royal Navy, 32, 34
Roye, Herbert F., 338
Roye, J., 341
Rozelle, Joseph H., 223
Rucker, Lucius O., 350
Rudd, Jack, 223
Rudder, James E., 109, 417
Ruhr Pocket, 208, 212, 275
Ruhr River, 276
Rundstedt, Von, 204
Runge, Kurt, 154
Rupertus, William H., 362
Russell, Clyde, 336, 341
Russia, 18, 39, 43, 148, 227, 402
Rutland, Alsie L., 403
Ryals, Kenneth E., 141
Ryan, Cornelius, 124
Ryder, William T., 31, 171, 251, 301, 302, 390, 403, 404

—— S ——

Sabalauski, Walter J., 224
Sabine River, 265
Sachs, Edward I., 111, 186
Sadler, John "Skip", 336
Sage, Jerry M., 345, 350
Sage, Terry, 357, 359
Saigon, 210, 219, 222, 356, 357, 456
Saint Jacques, 298
SALEM HOUSE, 336, 337
Salerno, 19, 97, 100, 209, 378
Salm River, 206, 262, 287, 298, 307
Salmon, Carl, 296
Salol, Al, 344
Sampson, Francis L., 224
San Isidro Airfield, 209
San Jose, 415
San Pablo Airstrip, 243
Sanchez, Thmas J., 338
Sands, Guy, 469
Santarsiefo, Charles J., 224
Santo Tomas, 94
Sarno, 97
SATURATE, 214
Saudi Arabia, 68, 343, 471, 472, 473
SAYONARA, 214
Sayre, 88
Sayre, Edward, 410
Schartzwalder, Ben, 273
Schimmelpfenning, Irving, 176
Schlenker, Leonard, 237
Schmidt, William R., 224
Schneider, Max, 67
Schoch, Nicholas W., 224
Scholes, Ed, 151
Scholtes, Richard A., 460
Schroeder, Bud, 305
Schungel, Daniel F., 339, 341
Schwarzkapf, Norman, 356, 467
Schwartzwalder, Floyd B., 267, 270, 276
Schweiter, Leo H., 338, 450
Scotland, 105, 265, 278, 291
Scott, Carl A. 353
Scott, James T. (Terry), 468
Scowcroft, Brent, 157
Screaming Eagles, 248, 414, 419
Scuggs, Hugh F., 341
SEAGULL, 214
Seal Detachment, 75, 76
Seaman, Jonathan O., 449
Segassie, Ed, 21
Segula Island, 412
Seitz, Richard, 111, 118, 295, 296, 449, 458
Sele River, 99
Senkewich, Arthur R., 333, 340, 344, 345
Sennewald, Robert, 466
Service Battalian, 320
Setman, Thad P., 403
Severson, Paul R., 338
Seymour, Roger G., 345
Shackleton, Ronald A., 341
Shacknow, Sidney, 468

Shalikashville, Othar J., 345
Shannon, Jack T., 325, 340, 353, 434
Sharm el-Shiekh, 248
Shaw, Harry, 379
Shaw, Robert E., 341
Sheathelm, Glenn H., 228, 232, 236
Shelton, Harmon, 338
Shepard, Charles E., Jr., 361, 364
Sheperd, Robert E., 403
Shepherd, James, 472
Sherburne, Thomas L., 224
SHINGLE, 199
SHINING BRASS, 336
Shipley, Norman, 91, 92, 176, 177, 181
Shoemaker, Robert, 226, 228
Shorthy Ridge, 93
Shroyer, Donald F., 466
Sicily, 19, 40, 41. 42, 83, 89, 97, 100, 102, 110, 128, 163, 164, 165, 180, 193, 197, 198, 199, 209, 254, 260, 291, 292, 302, 304, 306, 333, 339, 365, 366, 378, 384, 387, 392, 409, 410, 412, 415
Siedenburg, Joe R., 176
Siegfried Line, 53, 54, 188, 289
Sietz, Richard J., 294
SIFT STRIKE I, 219
SIFT STRIKE II, 219
Silver Wings of Courage, 16
Simmons, Burnell, 224
Simms, Clifford C., 452
Simons, Arthur D. "Bull", 329, 330, 342
Simpson, Charles M., 326, 332
Simpson, Dewey C., 332
Sims, Clifford C., 223
Sanai Desert, 247, 248
Sincere, Clyde J., 21, 324, 338
Singlaub, John K., 336, 353
Single, James C., 224
Singles, Gordon, 328, 434
Sink, Robert F., 107, 108, 109, 251, 403, 405, 410, 416, 437
SIOUX CITY, 241
Sisler, George K., 328, 338
Skamser, Harold P., 339
Skau, George, 182
Skipper, Louie A., 403
Sky Soldier, 445
Sky Soldiers, 239, 240
Slifer, Richard, 234
SMASH, 241
Smith, Edward, 251, 350, 351
Smith, Eugene "Mike", 340
Smith, Gordon K., 267
Smith, J.J., 261
Smith, Magnus L., 342
Smith, Paul F., 244, 267, 268, 272, 273, 277, 449
Smith, Raymond G., 403
Smith, Wayne C., 179
Smith, Arthur W., 403
Smudin, George, 277
Snell, Robert M., 224
Snowden, Steve, 344
Soland, Donald J., 463, 466
Sollenberger, McCord, 353
Solomon Islands, 35
Somerset Plain, 214, 220
Sommers, Bob, 21
Sommers, Robert, 21
Son Cong River, 329
Son Tay, 329, 330
Sorge, Arthur L., 267
Sosnack, Andrew, 224
Sottak, Barry, J., 468
Soule, Robert H., 176, 177, 178, 182, 415
Soviet Union, 19, 28, 29, 38, 43, 66
Soy-Hotton Mission, 298
Spain, 343, 345, 371
Spears, Don A., 353
Spears, John H., 338, 446
Special Action Force (SAF), 333
Special Allied Reconnaissance Force, 394
Special Forces, 16, 65, 72, 75, 76,326, 328, 329, 436
Special Forces Detachment 14, 438
Special Forces Operational Detachments A 12 & 13, 438
Special Forces Operational Detachment 302, 439
Special Forces Training Group, 457
Special Operations Detachment, 465
Special Recon. Force, 75, 76
Special Service Force, 297, 318, 324
Speer, Robert E., 301
Speers, Max D., 338
Spicer, Marion A., 338
Spigelmire, Michael F., 467
Spitz, James D., 224
Spitzer, Jerzy S., 130
Spradling, Donald, 346
Spragins, Charles E., 430

Sprecher, Kenneth N., 224
St. Saveur Le Vicomte, 270
Stadtherr, Nicholas G., 176
Stahl, Phillip, 338
Stalin, Joseph, 39
Stankovich, Peter, 472
Stanley, Leroy S., 333, 334
Stark, Peter M., 338
Starring, Mason B., 353
Ste. Mere Eglise, 18, 48, 261, 262, 264, 278, 280, 371, 384, 414, 415
Stearns, Clarence L., 338
Steele, John, 176
Steele, Robert C., 176
Stephens, Albert L., 274
Stephens, Howard A., 266, 271
Sternberg, Ben, 224
Stevens, Forestad A., 338
Stewaert, Carstell O., 411
"STILLWELL", 241
Stilwell, Joseph, 78, 328, 353, 448, 449
Stilwell, Richard G., 450
Stimson, Henry L., 41, 42, 392
Stiner, Carl W., 151, 463, 464, 467, 469, 471
STING RAY, 241
Stockton, John, 230
Stone, Harry J., 291
Story, Bill, 21, 318
Stoyka, John L., 353
STRAC Forces, 229
Stradtherr, Nick, 178
Strahan, Ronnie, 469
Straight of Gibraltar, 40
Streicher, Julius, 218
Strong, John, 176
Strubbs, Maurice G., 186
Student, Kurt, 34, 41, 54, 402
Subic Bay, 257
Sucre, Heraclides, 157
Suddath, Leroy N., 468
Suddath, Leroy N., Jr., 328, 332, 463
Sukchon, 64, 139, 141, 378
Sukchon-Sunchon, 244, 245, 399
Sultan, Daniel I., 353
SUMMERALL, 214
Summers, John W., 353
Sunchon, 64, 139, 140, 141, 166
Support Company, 242
Supreme Headquarters Allied Expeditionary Force (SHAEF), 204
Swank, Lawrence E., 353
Swann, Thomas L., 338, 397
SWARMER, 209, 246, 247
Swetish, John, 251, 350, 351
Swift, Eben, 193
SWIFT ACTION II, 214
SWIFT STRIKE I, 214, 333
SWIFT STRIKE III, 214
Swilley, Arthur W., 403
Swing, Joseph, 38, 90, 176, 177, 179, 180, 181, 183, 242, 246, 396, 409, 415
Swingler, Harold H., 87
Swiski, Edward A., 275
Sydnor, Elliot P., 329, 330, 332
Sylvester, Gene R., 270
SWITCHBACK, 334
Szarck, Henry, 343

—— T ——

Table of Contents, 5
Tackaberry, Thomas H., 236, 458, 460
Taegu, 228
Tafaraoui, 40, 80, 81
Tagaytay Ridge, 92, 93, 178, 181, 243, 392
Taiwan, 331
Talmadge, Roger, 229
Tanauan, 94
Tan Son Nhut Air Base, 210
Tangney, William, 472
Tarpley, Thomas M., 224, 457
Task Force ALPHA, 449
Task Force Ivory Coast, 330
Task Force OREGON, 450
Tason, Marcela, 157
Tate, Con, 322
Taylor, Allen W., 264, 267, 271, 272
Taylor, James T., 338
Taylor, Maxwell, 18, 45, 99, 106, 120, 198, 214, 215, 216, 224, 280, 414, 415
Taylor, Robert, 341
Taylor, Royal R., 282
Taylor, Wesley H., 148, 149
Teach, William, 231
Tedder, Arthur, 98, 103
Teevens, Richard P., 338
Tennessee Maneuvers, 215
Terry, Ronald, 332
Test Airborne Landing Detachment, 28, 29
Test Platoon, 18, 171, 251
TEXAS STAR, 220

Thailand, 326, 327, 330, 331, 333, 340, 353, 451, 458, 471
THAYER, 236
Theriault, Samuel S., 338
Thomas, Calvin, 341
Thomas, Charles, 467
Thomas, Ed, 322, 396
Thomas, Ronald F., 302
Thompson, Floyd, 341
Thompson, Frederick, 21
Thompson, Kimberly, 157
Thompson, James, E., 224
Thompson, John S., 120
Thompson, Ulrich, 270
Thompson, William T., 353
Thomson, George, 353
Thorenche, Molen, 286
Thornton, Steve W., Jr., 353
THOT NOT, 337
Throckmorton, John I., 450
Throckmorton, John L., 441, 445
Throne, Larry, 343
Thurman, Maxwell, 150, 156, 157, 332
Tien Bing, 445
Tierney, Tom, 269
TIGER, 215
Tilley, Leonard W., 338
Tillis, Jack D., 411
Tillman, George, Jr., 469
Timmes, Charles J., 224, 264, 267, 268, 269, 270, 273, 441
Timothy, James S., 447
Tippelskirch, Von, 208
Tipton, Norman E., 176, 178
Todd, Harvey A., 353
Todd, Richard A., 472
TOLEDO, 241
Tolson, John J., 146, 254
Tolson, John L., 410, 452
Tomasik, Edmund J., 291, 293
Toneguzzo, Dante, 264, 275
Tonkin Gulf, 329
TORCH, 40
Torgerson, Harry L., 364
Toth, Samuel, 249, 285
TOTOHATCHEE, 333
Tougher, Martin J., 268
Towle, John R., 193, 204
Townsend, John G., 345
Tracey, Hugh A., 403
Tracy, Thomas J., 353
Trapnell, James J.H., 245, 441
Trapnell, T.J.H., 435
Trees, Paul M., 338
Trent, Herman T., 224
TRICAP Division, 238
Trice, Harley N., 300
Triple Nickles, 42, 309, 311, 420
Triplett, Grady T., 359
Trobaugh, Edward L., 149, 195, 461
Trois-Ports, 204
Troop Carrier Command, 43
Troth, Paul, 390, 392
Trotti, Tom M., 364
Truman, Harry S., 64, 138, 207, 288, 316
Trumps, Shirley R., 353
Truscott, Lucien, 118
Tryone, James, 338
Tubbs, Herbert A., 224
Tucker, Reuben, 121, 122, 196, 197, 198, 199, 203, 204, 206, 406, 410, 415
Tulagi, 35
Tullidge, George, 265, 268, 277
Tunisia, 19, 74, 76, 81, 82, 83, 197, 291, 339
Turberville, Graham, 397
Turkey, 214, 247, 343
Turner, William L., 107, 224
Tweedy, Stuart, 230, 231
Tyan, Conelius E., 224
Tyrrenian Sea, 321

—— U ——

Ugalde, Jesse G., 334
Ukraninian Military District, 29
Underwood, Victor, 338
Unit Awards, 193
United Kingdom, 345, 460
United Nations Partisan Infantry Korea (UNPIK), 72, 75, 76
Updegraph, Charles L., Jr., 361
Urban, Del, 319
Urgent fury, 66, 148, 149, 462, 467
Urquhart, Robert F., 120, 127, 414
USAAF Med Evac, 74, 76
USAAF Med. Opns., 76
US Army Light Infantry Divisions, 461
US Army Parachute Riggers, 396
US Army Psychological Warfare Division, 436
US Army Quartermaster Corps, 398

US Army Rangers, 312
US Army Regimental System (USARS), 462
US Army Reserve Special Operations Command, 469
US Army Special Operations Command, 469, 471
US Infantry and Armored Divisions, 18
USMC 1st Recon, 75, 76
US Military Combat Jumps, 72
US Rangers, 18
US Special Operations Command, 464

—— V ——

Van Hart, John, 353
Van Buren, 214
Vance, Robert T., 362, 364
Vandeleur, Giles, 124, 127
Vanderbilt, Cornelius, 184
Vandervoort, Benjamin, 107, 108, 122, 123, 127, 260
Vargennes Valley, 117
VARSITY, 62, 63, 129, 130, 131, 134, 137, 165, 188, 264, 272, 273, 274, 369, 385, 418, 419, 420
Vatican, 156, 157
Vella, Paul, 294
Verhaeghe, George M., 353
Verret, John J., 271
Vessey, John, 463
Viau, Wallace, 332
Vielslam, 204
Vierk, Duane C., 332
Vierk, John, 345
Viet Cong, 236, 239, 246, 326, 335, 336, 449
Vietnam, 17, 18, 19, 65, 72, 75, 76, 145, 159, 161, 166, 167, 195, 209, 210, 211, 214, 219, 220, 221, 222, 226, 227, 228, 229, 232, 234, 236, 237, 238, 239., 240., 242, 244, 245, 246, 247, 311, 312, 325, 326, 327, 328, 329, 331, 334, 335, 336, 337, 339, 240, 343, 344, 355, 356, 358, 359, 370, 378, 392, 401, 407, 442, 443, 445, 446, 447, 448, 450, 451, 452, 454, 455, 456, 457, 458, 459, 463, 464, 468,
Vietnamese Abn. Advisory Detachment, 355
Villarosa, Paul H., 339
Vinson, Wilbur, 237
VIOLET, 394
Vittoria, 260
Voils, Steve, Jr., 403
Volckman, Russ, 342
Volturno River, 292
Von Luttwitz, Heinrich, 218
Voss, Robert L., 332

—— W ——

Waal River, 53, 203, 286
WACO, 241
Wade, C., 338
Wade, Hobart B., 251, 403
Waghelstein, J.P., 341
Walden, Jerry T., 224
Walden, Roger S., 411
Walkabout, Billy B., 224
"WALKER", 241
Walker, Edwin A., 322
Walker, Fred L., 295
Walker, James V., 468
Walker, Walton, H., 138
Wallace, John A., 190
Wallach, Marshall, 332
Wallau, Alex L., 301
Walsh, Louis A., Jr., 294, 295
Walters, Harmon "Tex", 276
Walthall, Charles, 273
Walton, Bill, 319
Wante, Lucien H., 353
Ward, John A., 403
Ward, Johnnie, 267, 397
Ward, William, 234
Wareing, Jerry, 333
Warfield, G., 394
Warner, Volney E., 459
Warran, Thomas E., 224
Warren, Shields, Jr., 203, 279, 281, 284
Warrior River, 363
Wasco, Joseph Jr., 224
WASHINGTON GREEN, 241
WATER MOCCASIN, 333
Waterhouse, Frederick J., 242
Waters, S.W., 322
Watson, Gerald L., 339
Watson, Payton, 235
Watters, Charles J., 240, 450
Watters, Lee J., 353
Watts, Joseph C., 21
Waymire, Jackie L., 338
Wayne, John, 228
Wayrynen, Dale E., 219, 223, 450
Wear, George E., 405

Weber, Eldred E. "Red", 339, 342
Weber, Gary, 465, 470
Weber, James, 237
Weber, William E., 16, 21, 475, 488, 490, 491, 493, 497, 506
Wechsler, Ben L., 83, 84
Weeks, Gene, 21
Weeks, Thurman L., 403
Wehrmacht, 32, 261
Weil, Daniel C., 411
Wentzel, John A., 304
Werbomont, 204
Werner, Carlton G., 224
Wesel, 62, 164, 378, 387
Wessel, Leon Mack, Jr., 224
Westmoreland, Loner, 341
Westmoreland, William C., 209, 214, 221, 224, 228, 245, 384, 441, 445, 447
Weyand, Frederick C., 359
Wheeler, Earle, 329
WHEELER, 214
White, Eddie J., 332, 472
White, Everett C., 340
White, Guy, 277
White, John A., 353
WHITE WING, 235
Whitelaw, John L., 137, 186, 419
Whitfield, George A., 223
Wickham, John A., 224
Wiegand, Robert, 462
Wicrcnga, Joel M., 464
Williams, Charles Q., 326, 328, 338, 448
Williams, D.D., 322
Williams, Jack L., 338
Williams, Michael J., 224
Williams, Paul L., 113, 369, 414
Williams, Robert H., 362, 364
Williamson, Ellis W., 239, 447
Williamson, Frank D., 472
Williamson, Mac, 345
Wills, Lloyd E., 338
Wilmer, Douglas W., 386, 389
Wilson, Arthur H., 90, 415
Wilson, Charles H., 251
Wilson, Donald, 253
Wilson, Harry L., 141, 172, 176, 178, 403
Wilson, Obie C., 403
Wilson, Richard G., 245, 429
Wilson, Robert, 234
Wilson, Samuel V., 340
Wilson, Steve, 233, 235
WINCHESTER, 241
Windle, Paul R., 359
Wingate, Charles, 78, 340
Winters, Richard D., 224
Wise, John, 401
Wishik, Jeffrey, 224
Witcher, Earl, 345
Woerner, Frederick, 150
Wolfe, Martin, 365
Wolford, Grover, 224
Wolverton, Robert M., 107
Woodruff, J.R., 206
Woods, Luther L., 224
Woolard, R.W., 345
Woolshlager, John C., 466
Worley, Morris G., 338
Worrall, P., 394
Worrell, Iran, 21
Wright, Jack, 226
Wright, John M., 224, 454
Wright, Leroy N., 339
Wright, Orville, 36
Wright, Robert L., 224
Wright, Ronald J., 224
Wright, Wilbur, 36

—— Y ——

Yalu River, 64, 316
Yarborough, William P., 32, 80, 81, 110, 171, 251, 291, 292, 326, 328, 339, 381, 403, 404, 405, 411, 448
Yardley, Doyle R., 199, 291, 292
Yntema, Gordon D., 328, 338, 453
York, Alvin C., 193, 195
York, Don J., 356, 359
York, Robert H., 194, 445, 449
Yorktown, Victor, 210, 241
Young, Neil, 277
Yound, Willard "Tex", 269
Yugoslavia, 74, 76

—— Z ——

Zabitosky, Fred W., 328, 338, 453
Zachary, James, 345, 468
Zachary, Jim, 471
Zacher, George A., 341
Zahn, Donald E., 224
Zais, Melvin, 111, 118, 224, 294, 295, 296, 403, 452, 454
Zerbe, Roy, 277
Zhukov, Marshall, 209
Zielski, William L., 353